AWS/TR-95/001

FORECASTERS GUIDE
to
TROPICAL METEOROLOGY
AWS TR 240 Updated

By
Colin S. Ramage

AUGUST 1995

AIR WEATHER SERVICE
102 West Losey Street
Scott Air Force Base, Illinois 62225-5206

REVIEW AND APPROVAL STATEMENT

AWS/TR–95/001, *Forecasters Guide to Tropical Meteorology, AWS TR 240 Updated,* August 1995, has been reviewed and is approved for public release. There is no objection to unlimited distribution of this document to the public at large, or by the Defense Technical Information Center (DTIC) to the National Technical Information Service (NTIS).

CLIFFORD R. MATSUMOTO, Colonel, USAF
AWS Director of Technology, Plans, and Programs

FOR THE COMMANDER

CAROL L.WEAVER, PhD
AWS Scientific and Technical Information
Program Manager
24 August 1995

ii

REPORT DOCUMENTATION PAGE

2. Report Date: August 1995

3. Report Type: Technical Report

4. Title: Forecasters Guide to Tropical Meteorology–AWS TR 240 Updated

6. Author: Colin S. Ramage, DSc

7. Performing Organization:: United States Air Force Environmental Technical Applications Center, Scott AFB, IL 62225-5116

8. Performing Organization Report Number: AWS/TR–95/001

11. Supplementary Notes: Supersedes AWS TR 240, AD-723392

12. Distribution/Availability Statement: Approved for public release; distribution is unlimited.

13. Abstract: AWS TR 240, by Maj Gary D. Atkinson, has served as the reference manual for USAF weather forecasting in the tropics since it was first published in 1971. Although it has endured for the past 20 years, HQ Air Weather Service recognized the need for an update and contracted with tropical forecasting authority Dr Colin S. Ramage to produce one. Although a great deal of new material has been added to reflect new techniques and new technology, it still covers the basic facts of climatology, circulation, and synoptic models, with emphasis on analysis and forecasting techniques for the tropics. Physical factors that control tropical circulations are discussed briefly. The climatologies of pressure, winds, temperature, humidity, clouds, rainfall, and disturbances are presented in a form specially suitable for forecasters. Analysis and forecasting of disturbances, cyclones, severe weather, terminal weather, etc., are treated at length. The uses of climatology and the interpretation and use of weather satellite imagery are emphasized. Numerous figures adapted from the literature or prepared by the author illustrate all the essential facts and principles discussed. A summary of the state of art and future outlook of tropical meteorology is included, along with an extensive bibliography.

14. Subject Terms: WEATHER, WEATHER FORECASTING, METEOROLOGY, TROPICAL METEOROLOGY, ANALYSIS, TROPICAL REGIONS, TROPICAL CYCLONES, SATELLITE METEOROLOGY, CLIMATOLOGY, ATMOSPHERIC PHYSICS

15. Number of Pages: XXX

17. Security Classification of Report: Unclassified

18. Security Classification of this Page: Unclassified

19. Security Classification of Abstract: Unclassified

20. Limitation of Abstract: UL

Standard Form 298

Typhoon Russ. DMSP visible image at 2245Z, 19 December 1990, centered 130 NM (200 km) southeast of Guam. Maximum winds: 125 knots.

The effects of Typhoon Russ as it passed south of Guam, 19 December 1990

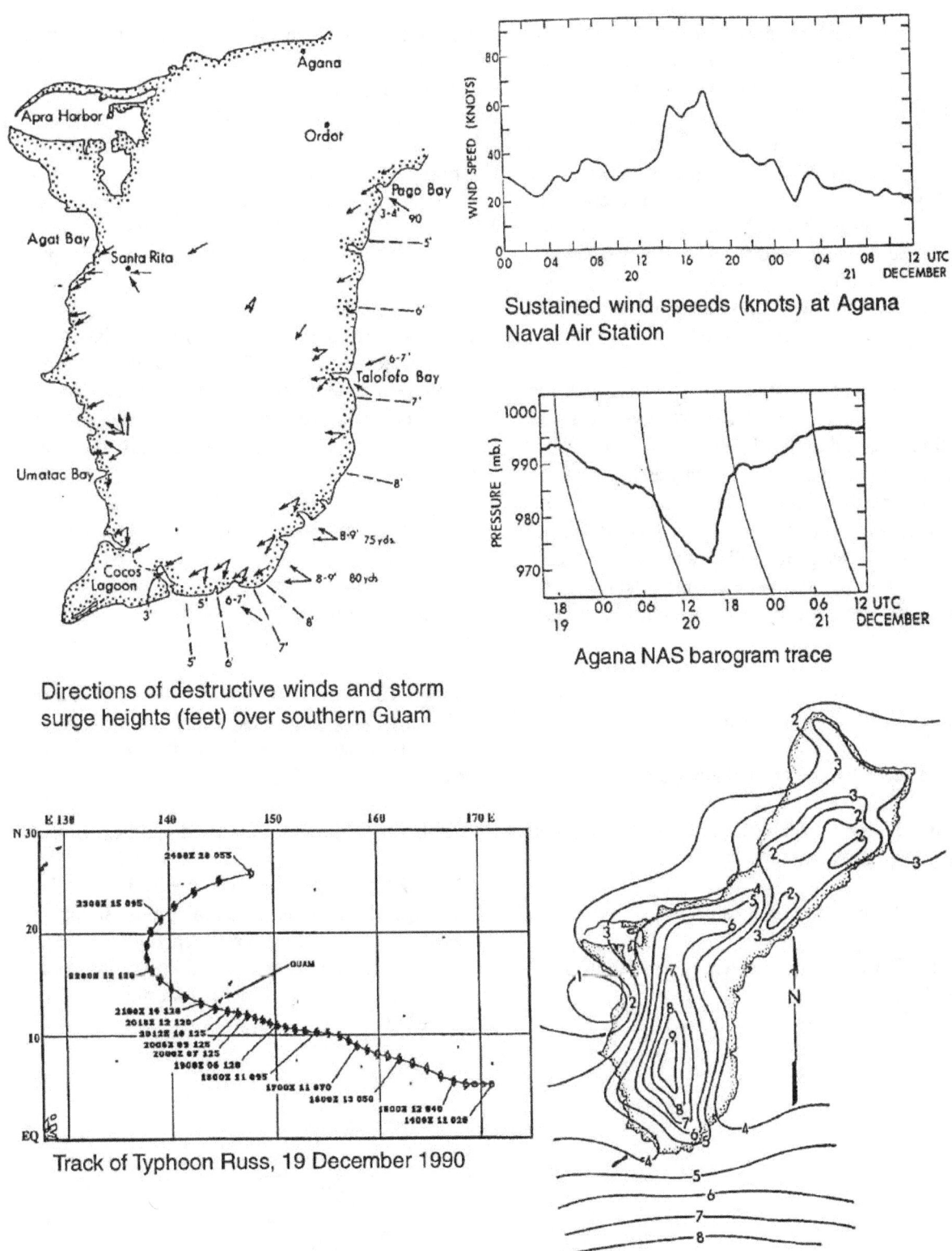

Directions of destructive winds and storm
surge heights (feet) over southern Guam

Sustained wind speeds (knots) at Agana
Naval Air Station

Agana NAS barogram trace

Track of Typhoon Russ, 19 December 1990

Estimated total rainfall (inches) over Guam

PREFACE

As correctly forecast in the Preface to the first (1971) edition, *Forecasters' Guide to Tropical Meteorology,* by Major Gary D. Atkinson, AWS TR 240 was used not only to train Air Weather Service meteorologists, but also as a text for university courses. It was even translated into Chinese. Technological advances in the last 20 years, especially in satellite-based observations, have not seriously affected views expressed in the first edition, but have allowed them to be expanded and refined. Both climatology and synoptic meteorology have benefitted, while many hitherto obscure parts of the tropics have been exposed to the global eye.

This edition was prepared under contract with the Department of Meteorology, University of Hawaii. Satellite data and wide-ranging research now allow the tropics to be treated more evenly and extensively than before. Sections dealing with Africa, the Americas, and south and southwest Asia have been expanded. Satellite pictures illustrate tropical systems and processes. Besides these changes, the following topics have been added; those in italics have been *deleted.*

Chapter 1. Terminology; heat lows.

Chapter 2. Stress-differential along a coast.

Chapter 3. *Constant-level balloons;* surface reports from buoys; Doppler radar; Profiler.

Chapter 4. Low-level jet streams.

Chapter 5. Upwelling; tradewinds; low-latitude westerlies.

Chapter 6. Fog; character of significant rain in the tropics; interannual variation of rainfall and El Niño; seasonal distribution of rainfall and the monsoons; pentad (5-day) distribution of rainfall.

Chapter 7. A new chapter on diurnal variation, combining the scattered discussions of the first edition, and expanding and adding to them. New sections on vertical mixing, stratocumulus and fog, and squall lines.

Chapter 8. (formerly Chapter 7) Equatorial waves; monsoon depressions; squall lines; surges; troughs in the upper-tropospheric westerlies; South Pacific convergence zone; superposition of tropical and extra-tropical disturbances; near-equatorial tradewind convergence.

Chapter 10. (formerly Chapter 9). Duststorms; volcanic ash.

Chapter 11. (formerly Chapter 10). *Isogon analysis;* kinematic frontal analysis.

Chapter 12. (formerly Chapter 11) *Stability indexes* (reduced); 30-60 day cycles; Atlantic tropical storms (seasonal forecasts); preparing to forecast in a new area.

FIGURES. 136 were retained and 173 were added.

REFERENCES. 175 were retained and 390 added. The range of topics and locations cannot simply be dealt with sequentially; to avoid repetition, extensive cross-referencing is used. Detailed subject and geographic indexes, along with a locator map, have been added.

ACKNOWLEDGMENTS

Many present and former members of Air Weather Service contributed to the preparation of this edition, including the 20 people who carefully and constructively reviewed the first and second drafts, and those who provided weather satellite pictures, bibliographies, and publications. Lieutenant Colonels Charles Guard, Charles Holliday, Clifford Matsumoto and Harry White were particularly helpful, and I greatly appreciated Gary Atkinson's comments.

Captain Randall Skov effectively coordinated the information. In Hawaii, Louis Oda drafted and revised the figures and Thomas Schroeder handled administration. Others who helped include Mike Gentry, William Gray, Carole Hahn, Barry Hinton, Greg Holland, Richard Orville, Patrick Sham, Norman Wagner, J.F. White, Scott Woodruff, and X. Ziang. My wife Alice proofread.

CONTENTS

CHAPTER 1   **INTRODUCTION**   Page

1.1 General 1
1.2 Terminology 3
1.3 Units of Measurement 8
1.4 Locator Charts 8

CHAPTER 2   **PRIMARY PHYSICAL CONTROLS OF THE TROPICAL CIRCULATION**

2.1 General 10
2.2 Earth's Heat-Energy Balance 11
2.3 Tropical General Circulation 15
2.4 Primary Physical Factors 20
    2.4.1 General 20
    2.4.2 Land and Water Distribution 20
    2.4.3 Stress-Differential Along Coasts 21
    2.4.4 Sea-Surface Temperature 21
    2.4.5 Interactions with Mid-Latitude Flow 22
2.5 Small-Scale Controls 23
2.6 Summary 24

CHAPTER 3   **THE OBSERVATIONAL BASIS**

3.1 Current Data Sources 25
    3.1.1 Surface Reports from Land Stations 25
    3.1.2 Surface Reports from Ships, Moored Buoys, and Automatic
           Island Stations 27
    3.1.3 Surface Reports from Drifting Buoys 28
    3.1.4 Upper-Air Data 29
    3.1.5 Meteorological Satellite Data 33
    3.1.6 Weather Radar Data 35
    3.1.7 Lightning Detection 35
3.2 Probable Future Data Sources 36
    3.2.1 Meteorological Satellites 36
    3.2.2 Radar 37
3.3 Summary 38

CHAPTER 4   **PRESSURE AND WINDS**

4.1 General 39
4.2 Pressure 40
    4.2.1 Mean Sea-level Pressure 40
    4.2.2 Large-Scale Pressure Changes 45
4.3 Winds 48
    4.3.1 General 48
    4.3.2 Resultant Gradient-Level Winds 48
    4.3.3 Resultant 200-mb Winds 57
    4.3.4 Tropical Jet Streams 62
           4.3.4.1 Subtropical Jet Streams 62
           4.3.4.2 Tropical Easterly Jet (TEJ) 62
           4.3.4.3 Low-Level Jet Streams (LLJs) 62
    4.3.5 Tropical Stratospheric Winds 63

CHAPTER 5 **TEMPERATURE AND WATER VAPOR**
    5.1 General 66
    5.2 Surface Air Temperature and Water Vapor 67
        5.2.1 Means 67
        5.2.2 Upwelling/Downwelling 73
        5.2.3 Elevation Effects 74
    5.3 Upper-Air Temperature and Water Vapor 75
        5.3.1 Means 75
        5.3.2 Mean Precipitable Water 76
        5.3.3 Mean Vertical Soundings 78
        5.3.4 Equivalent Potential Temperature 81
        5.3.5 Tradewinds 84
        5.3.6 Low-Latitude Westerlies 87

CHAPTER 6 **CLOUDINESS AND RAINFALL**
    6.1 General 88
    6.2 Cloudiness 89
        6.2.1 Mean Cloudiness 89
        6.2.2 Annual Cloudiness Variation 89
        6.2.3 Year-to-Year Cloudiness Variations 92
        6.2.4 Zonal Averages 92
        6.2.5 Large-scale Organization of Deep Tropical Cloud Systems 92
        6.2.6 Tropical Cloud Types 98
            6.2.6.1 Cumulus 98
            6.2.6.2 Cumulonimbus (Cb) 101
            6.2.6.3 Stratus and Stratocumulus 101
            6.2.6.4 Fog 101
            6.2.6.5 Altocumulus and Altostratus 105
            6.2.6.6 Cirrus and Cirrostratus 105
        6.2.7 The Relationship of Summer Cloudiness to Other Elements
            Over the Tropical Atlantic 105
    6.3 Rainfall 109
        6.3.1 The Character of Significant Rain in the Tropics 109
            6.3.1.1 Thunderstorm Frequency and Rainfall 109
            6.3.1.2 Continuous Thunderstorms 112
        6.3.2 Significant Tradewind Rainfall 113
        6.3.3 Mean Rainfall 114
        6.3.4 Rainfall Variability 115
        6.3.5 Interannual Variation and El Niño 115
        6.3.6 Seasonal Distribution—The Monsoons 119
            6.3.6.1 Definition and Extent of the Monsoons 119
            6.3.6.2 Role of the Himalayan-Tibetan Massif in the Monsoons 121
            6.3.6.3 Rain During the Monsoons 123
            6.3.6.4 March of the Seasons 124
            6.3.6.5 Summary 126
        6.3.7 Monthly Distribution 127
        6.3.8 Monthly Variability 130
        6.3.9 Pentad Distribution and Beginning and Ending of Rainy Seasons 131
        6.3.10 Daily Distribution 142
        6.3.11 Rainfall Variation with Elevation 144

        6.3.12 Mesoscale Distribution 145

        6.3.13 Rainfall Extremes 149

        6.3.14 Rainfall Associated with Tropical Cyclones 151

CHAPTER 7   **DIURNAL VARIATIONS**

  7.1 General 153

  7.2 Diurnal Temperature Variation 154

  7.3 Diurnal Pressure Variation 156

  7.4 Diurnal Wind Variation 157

      7.4.1 Open Ocean 157

      7.4.2 Land-Sea Breezes 157

          7.4.2.1 Observations 157

          7.4.2.2 Numerical Models 161

      7.4.3 Lake Winds 164

      7.4.4 Mountain-Valley Winds 165

      7.4.5 Vertical Mixing 167

          7.4.5.1 *General* 167

          7.4.5.2 *Low-Level Jets (LLJs)* 168

          7.4.5.3 *Interactions Along Coasts* 169

      7.4.6 Summary 170

  7.5 Diurnal Cloudiness Variation 171

      7.5.1 Open Ocean Tradewind Cloud 171

      7.5.2 Open Ocean Stratocumulus 171

      7.5.3 Open Ocean and Coastal Fog 172

      7.5.4 Overland Stratocumulus and Fog 172

      7.5.5 Popcorn Cumulus 173

  7.6 Diurnal Variation of Rain 174

      7.6.1 Over Open Oceans 174

          7.6.1.1 *Squall Lines* 175

          7.6.1.2 *Summary* 175

      7.6.2 Over Land 176

          7.6.2.1 *General* 176

          7.6.2.2 *Synoptic-Scale Weather Distrubances* 176

          7.6.2.3 *Squall Lines* 176

          7.6.2.4 *Tradewind Regimes* 178

          7.6.2.5 *Dry Regimes* 178

          7.6.2.6 *Lake Regimes* 178

          7.6.2.7 *Mountain Slopes* 180

          7.6.2.8 *Moist Near-Equatorial Regimes* 180

  7.7 Summary 186

CHAPTER 8   **TROPICAL SYNOPTIC MODELS**

  8.1 General 187

  8.2 Waves 188

      8.2.1 Waves in the Easterlies 188

  8.3 Vortexes 189

      8.3.1 General 189

      8.3.2 Monsoon Depressions 189

      8.3.3 West African Cyclones 189

      8.3.4 Mid-Tropospheric Cyclones 193

          8.3.4.1 *Subtropical Cyclones* 193

          8.3.4.2 Arabian Sea Cyclones 196

     8.3.5 Upper-Tropospheric Cyclones 201

     8.3.6 Temporal Storms of Central America 207

  8.4 Linear Disturbances 208

     8.4.1 General 208

     8.4.2 Squall Lines 208

     8.4.3 Surface Cold Fronts and Shear Lines 213

     8.4.4 Surges 214

          8.4.4.1 Winter Surges 214

          8.4.4.2 Other Surges 220

          8.4.4.3 Summary 222

     8.4.5 South Pacific Convergence Zone (SPCZ) 223

          8.4.5.1 Winter 223

          8.4.5.2 Summer 224

     8.4.6 Troughs in the Upper-Tropospheric Westerlies 225

          8.4.6.1 Moist-Dry-Moist "Sandwiches" 229

     8.4.7 Superposition of Tropical and Extratropical Disturbances 230

     8.4.8 The Near-Equatorial Tradewind Convergence (NETWC) 234

          8.4.8.1 Convergence Over the Oceans 234

          8.4.8.2 Convergence Over Northern South America 237

     8.4.9 Mid-summer Dry Spell 239

CHAPTER 9   TROPICAL CYCLONES

  9.1 General 241

  9.2 Structure of Mature Tropical Cyclones 242

     9.2.1 General 242

     9.2.2 Winds 242

     9.2.3 Temperatures 245

     9.2.4 Clouds 246

  9.3 Classification and Definition of Tropical Disturbances 248

  9.4 Global Climatology of Tropical Cyclones 249

  9.5 Tropical Cyclone Formation 260

     9.5.1 Adequate Source of Surface Energy 260

     9.5.2 Wind Shear 260

     9.5.3 Preexisting Disturbance 260

     9.5.4 Earth's Rotation 260

     9.5.5 Upper-Tropospheric Outflow 262

  9.6 Tropical Cyclone Dissipation 264

     9.6.1 Surface Energy Source Removed 264

     9.6.2 Excessive Wind Shear 264

     9.6.3 Upper Tropospheric Convergence 264

  9.7 Tropical Cyclone Movement 265

  9.8 Tropical Cyclone Forecasting 266

     9.8.1 General 266

     9.8.2 Positioning 267

     9.8.3 Forecasting Techniques 267

          9.8.3.1 Movement 267

          9.8.3.2 Intensity 268

     9.8.4 Position Forecast Accuracy 268

  9.8 Summary 272

CHAPTER 10   SEVERE WEATHER IN THE TROPICS

10.1 General 273
10.2 Thunderstorms 274
    10.2.1 Thunderstorm Frequency 274
    10.2.2 Thunderstorm Duration 281
    10.2.3 Thunderstorm Heights 282
10.3 Tornadoes and Waterspouts 283
    10.3.1 Hurricane-generated Tornadoes 287
10.4 Hail 290
10.5 Extreme Winds 293
    10.5.1 General 293
    10.5.2 Thunderstorm Winds 293
    10.5.3 Duststorms 296
        10.5.3.1 General 296
        10.5.3.2 Winter Duststorms 297
        10.5.3.3 Summer Duststorms 299
        10.5.3.4 Dust Devils 301
    10.5.4 Tropical Cyclone Winds 301
    10.5.5 Extreme Wind Analysis 304
10.6 Turbulence 307
    10.6.1 General 307
    10.6.2 Mechanical Turbulence 307
    10.6.3 Convective Turbulence 307
    10.6.4 Clear Air Turbulence 307
10.7 Icing 310
10.8 Ocean Waves 310
    10.8.1 Storm Surges 310
    10.8.2 Tsunamis 312
10.9 Volcanic Ash 315

CHAPTER 11   TROPICAL ANALYSIS

11.1 General 316
11.2 Data Collection and Evaluation 316
    11.2.1 General 317
    11.2.2 Surface Observations 318
        11.2.2.1 Pressure 318
        11.2.2.2 Temperature and Dew Point 318
        11.2.2.3 Wind 319
        11.2.2.4 Cloudiness, Rain and Visibility 321
    11.2.3 Upper-Air Observations 322
        11.2.3.1 Rawinsonde and Pilot-Balloon 322
        11.2.3.2 Aircraft 324
        11.2.3.3 Weather Reconnaissance 325
        11.2.3.4 Satellite-Derived Wind and Temperature 325
11.3 Manual Analysis Techniques 326
    11.3.1 General 326
    11.3.2 Instruction in Kinematic Analysis 326
        11.3.2.1 General 326
        11.3.2.2 Streamline Analysis 328
        11.3.2.3 Isotach Analysis 329

11.3.2.4 Frontal Analysis 330
11.3.3 Use of Aircraft Reports 334
11.3.4 Use of Climatology 336
    11.3.4.1 General 336
    11.3.4.2 Streamlines 336
    11.3.4.3 Isotachs 337
11.3.5 Use of Satellite Data 338
11.3.6 Post-Analysis Programs 339
11.3.7 Examples of Kinematic Analyses 339
11.3.8 Recommended Analysis Levels 343
11.3.9 Frequency and Scale of Analysis 346
11.3.10 Operational Procedures 346
    11.3.10.1 Surface/Gradient-Level Charts 347
    11.3.10.2 Upper-Level Charts 347
11.4 Automated Analysis Techniques 349
11.4.1 General 349
11.4.2 Objective Wind Analysis 349
11.4.3 Numerical Weather Prediction Analysis 350
11.4.4 Computer-Aided Analysis 350
11.5 Auxiliary Aids 351
11.5.1 Time Cross Sections 351
    11.5.1.1 Wind Direction 351
    11.5.1.2 Wind Speed 351
    11.5.1.3 Moisture 351
    11.5.1.4 Temperature/Height 351
11.5.2 Space Cross Sections 353
11.5.3 Thermodynamic Diagrams 353
11.5.4 Checkerboard Diagrams 353
11.5.5 Pressure-Change Charts 353
11.5.6 Wind-Shear Charts 355
11.5.7 Rainfall Analyses 355

CHAPTER 12 TROPICAL FORECASTING
12.1 General 356
12.2 Short-Range Forecasting Techniques 358
12.2.1 General 358
12.2.2 Circulation and Cloud Prognostic Charts 358
    12.2.2.1 Circulation Prognostic Charts 358
    12.2.2.2 Cloud Prognostic Charts 358
12.2.3 Climatological Aids 359
    12.2.3.1 General 359
    12.2.3.2 Conditional Climatology Summaries 359
    12.2.3.3 Diurnal Pressure and Temperature Data 362
    12.2.3.4 Variation of Weather 362
    12.2.3.5 Sources of Climatological Data 363
12.2.4 Local Forecast Studies 363
    12.2.4.1 General 363
    12.2.4.2 Objective Forecast Studies 364
    12.2.4.3 Synoptic Studies 366
12.2.5 Use of Stability Indexes 365

12.2.5.1 *General* 369
12.2.5.2 *Stability-Index Studies* 369
12.2.5.3 *Thunderstorm Wind Gusts* 369
12.2.6 Radar 372
12.2.6.1 *General* 372
12.2.6.2 *Tropical Storm Tracking* 372
12.2.6.3 *Echo Movement* 372
12.2.6.4 *Severe Weather Forecasting* 376
12.2.6.5 *Visibility Forecasting* 376
12.2.6.6 *Rainfall Estimates* 377
12.2.6.7 *Radar Climatology* 377
12.3 Tropical Numerical Weather Prediction 380
12.4 Medium and Long-Range Forecasting 384
12.4.1 General 384
12.4.2 Medium-Range Forecasting (MRF) 384
12.4.3 Long-Range Forecasting (LRF) 386
12.4.3.1 *General* 386
12.4.3.2 *LRF of Summer Rainfall in India* 386
12.4.3.3 *Atlantic Tropical Storms* 388
12.4.3.4 *Other Forecasts* 389
12.5 Preparing to Forecast in a New Area 390
12.6 Summary 391

REFERENCES A-1

ABBREVIATIONS AND ACRONYMS B-1

GEOGRAPHICAL INDEX C-1

SUBJECT INDEX D-1

DISTRIBUTION E-1

FIGURES

Figure 1-1. Schematic representations of monsoon and tradewind troughs and near-equatorial tradewind convergence (top) and near-equatorial buffer zones (bottom) 3

Figure 1-2. Electromagnetic energy categories, frequencies, and wavelengths (based on Elrick and Meade, 1987) 4

Figure 1-3. Model of the low-level summer circulation of the tropical eastern North Pacific and associated meridional rainfall and pressure patterns (after Sadler, 1967) 5

Figure 1-4. Four definitions of the "intertropical convergence zone" (ITCZ) over the tropical North Atlantic (Sadler, 1975a) 6

Figure Moved to End of Book

Figure 1-5a. Locator chart, Eastern Hemisphere 9

Figure Moved to End of Book

Figure 1-5b. Locator chart, Western Hemisphere 10

Figure 2-1. Average annual meridional distribution of the net radiation fluxes of the Earth's surface, atmosphere, and Earth-atmosphere system (from Sellers, 1965) 11

Figure 2-2. Average annual meridional distribution of precipitation, evaporation, and precipitation minus evaporation (from Sellers, 1965) 12

Figure 2-3. Average annual meridional distribution of the components of the poleward heat energy transport (from Sellers, 1965) 13

Figure 2-4. Mean vertical velocity in units of 10^{-4} mb s^{-1} or 1.5 mm s^{-1} at 500 mb in the tropics (from Kyle, 1970). 16

Figure 2-5a. December-May Meridional cross-sections of zonal wind speed (m s^{-1}); westerlies positive; subtropical jets are labeled "J" (from Newell et al., 1972). 17

Figure 2-5b. June -November Meridional cross-sections of zonal wind speed (m s^{-1}); westerlies positive; subtropical jets are labeled "J" (from Newell et al., 1972). 18

Figure 2-6. Meridional cross-sections of the mean meridional circulation in the tropics in the form of mass-flow streamlines. Units: 10^{12} g s^{-1} (from Newell et al., 1979) 19

Figure 2-7. Zonal circulation cells along the equator (from Newell, 1979). 20

Figure 2-8. Schematic of the effects of coastal friction on low-level wind flow (Fett and Bohan, 1986). 21

Figure 2-9. Approximate dimension and time scales of tropical systems (after Orlanski, 1975). 24

Figure 3-1. Boundaries of the six WMO regions (World Meteorological Organization, 1986). 25

Figure 3-2. Average daily number of shipboard surface weather observations received at the National Meteorological Center in June 1988, totaled for 4-degree latitude/longitude rectangles 27

Figure 3-3. Drifting buoy observations in the tropical Pacific during August 1989. Lines indicate weekly displacements (Climate Analysis Center, 1989). 28

Figure 3-4. Tropical stations making radiosonde (·) and rawin/pibal (o) observations at the beginning of 1991 (at least one observation a day on two-thirds of the days). 29

Figure 3-5. Maximum weekly number of international commercial airline flights crossing each 10° rectangle between 30° N and 30° S (based on International Civil Aviation Organization charts). 31

Figure 3-6. Number of aircraft weather observations received at AFGWC for each 10 degree rectangle between 30° N and 30° S from 0000 to 1400Z, 19 December 1990. 32

Figure 3-7. Cloud motion vectors (full, low-level; dashed, cirrus-level) obtained from picture sequences made by GOES at 0°, 135° W, on 20 December 1989 (from NESDIS). 34

Figure 4-1a. January mean sea-level pressure in millibars (From Naval Oceanographic Command, 1981) 41

Figure 4-1b. January standard deviation of sea-level pressure in millibars (From Naval Oceanography Command, 1981) 41

Figure 4-2a. April mean sea-level pressure in millibars (From Naval Oceanography Command, 1981) 42

Figure 4-2b. April standard deviation of sea-level pressure in millibars (From Naval Oceanography Command, 1981) 42

Figure 4-3a. July mean sea-level pressure in millibars (From Naval Oceanography Command, 1981) 43

Figure 4-3b. July standard deviation of sea-level pressure in millibars (From Naval Oceanography Command, 1981) 43

Figure 4-4a. October mean sea-level pressure in millibars (From Naval Oceanography Command, 1981) 44

Figure 4-4b. October standard deviation of sea-level pressure in millibars (From Naval Oceanography Command, 1981) 44

Figure 4-5. Average annual meridional distribution of sea-level pressure for January and July (derived from Crutcher and Davis, 1969) 45

Figure 4-6. Simultaneous hourly surface pressure anomalies for three Pacific stations (from Brier and Simpson, 1969) 46

Figure 4-7. Surface pressure change (mb) between 00Z 21 July and 00Z 23 July 1966 (from Sadler et al., 1968) 47

Figure 4-8. Correlations between daily surface pressure at Tarawa (1° N, 173° E) and other Pacific stations for May 1986 (from Lander, 1991) 46

Figure 4-9a. Resultant gradient winds (knots) for January, 120° W to 60° E (from Atkinson and Sadler, 1970) 50

Figure 4-9b. Resultant gradient winds (knots) for January, 60° E to 120° W (from Atkinson and Sadler, 1970) 50

Figure 4-10a. Resultant gradient winds (knots) for April, 120° W to 60° E (from Atkinson and Sadler, 1970). 52

Figure 4-10b. Resultant gradient winds (knots) for April, 60° E to 120° W (from Atkinson and Sadler, 1970) 52

Figure 4-11a. Resultant gradient winds (knots) for July, 120° W to 60° E (from Atkinson and Sadler, 1970). 54

Figure 4-11b. Resultant gradient winds (knots) for July, 60° E to 120° W (from Atkinson and Sadler, 1970). 54

Figure 4-12a. Resultant gradient winds (knots) for October, 120° W to 60° W (from Atkinson and Sadler, 1970). 56

Figure 4-12b. Resultant gradient winds (knots) for October, 60° E to 120° W (from Atkinson and Sadler, 1970). 56

Figure 4-13. Resultant 200-mb winds (knots) for January (from Sadler and Wann, 1984). 58

Figure 4-14. Resultant 200-mb winds (knots) for April (from Sadler and Wann, 1984). 59

Figure 4-15. Resultant 200-mb winds (knots) for July (from Sadler and Wann, 1984). 60

Figure 4-16. Resultant 200-mb winds (knots) for October (from Sadler and Wann, 1984). 61

Figure 4-17. Time-height section of monthly mean zonal wind components (m s⁻¹) at Canton Island, 4° S, 174° W (January 1953 to August 1967); Gan (1° S, 73° E (September 1967 to December 1975); and Singapore, 1° N, 104° E (January 1976 to April 1985) (Naujokat, 1986) 64

Figure 4-18. Mean monthly zonal wind speed at Darwin (12° S, 131° E) at 70,000 feet (from Hopwood, 1968). 65

Figure 4-19. Average amplitude (knots) of the quasi-biennial and annual cycles of the mean monthly zonal wind speed in the tropical stratosphere (from Reed, 1964). 65

Figure 5-1a. Surface air temperature means (° C) for January (from Naval Oceanography Command, 1981). 68

Figure 5-1b. Surface air temperature standard deviations (° C) for January (from Naval Oceanography Command, 1981). 68

Figure 5-1c. Surface air temperature mean (° C) for April (from Naval Oceanography Command, 1981).. 69

Figure 5-1d. Surface air temperature standard deviations (° C) for April (from Naval Oceanography Command, 1981) 69

Figure 5-1e. Surface air temperature means (° C) for July (from Naval Oceanography Command, 1981). 70

Figure 5-1f. Surface air temperature standard deviations (° C) for July (from Naval Oceanography Command, 1981). 70

Figure 5-1g. Surface air temperature means (° C) for October (from Naval Oceanography Command, 1981). 71

Figure 5-1h. Surface air temperature standard deviations (° C) for October (from Naval Oceanography Command, 1981). 71

Figure 5-2. Annual march of mean monthly temperatures at selected tropical stations along meridians through the continents. 72

Figure 5-3. Schematic showing the effects of surface wind blowing parallel to the west coast of a southern hemisphere continent such as South America or southern Africa. 73

Figure 5-4. Upwelling regions (stippled) (Meigs, 1973). 74

Figure 5-5. Mean 700-mb temperature (° C, solid lines) and relative humidity (percent, dashed lines) for winter (top) and summer (bottom) (after Crutcher and Davis, 1969). 75

Figure 5-6. Annual ranges of mean 700-mb temperature (° C) and relative humidity (percent), NH summer (SH winter) minus NH winter (SH summer) (derived from Crutcher and Davis, 1969). 76

Figure 5-7. Mean precipitable water (inches and mm) for January and July (after Tuller, 1968). 77

Figure 5-8. Annual mean relationship between precipitable water and SST (Stephens, 1990). 77

Figure 5-9. Standard atmosphere adiabatic diagrams for the tropics (ESSA, et al., 1967). 79

Figure 5-10. Mean vertical temperature and dew-point soundings for (A) Bangkok, and (B) Papeete, during the dry and wet season 80

Figure 5-11. Vertical structure of equivalent potential temp (θ_e) for disturbed and undisturbed days on the tropical Atlantic (after Garstang, et al., 1967) and Saigon (after Harris and Ho, 1969). 81

Figure 5-12. Average vertical distribution of equivalent potential temperature (θ_e) during the two warmest and two coldest months (after Gray, 1968). 83

Figure 5-13. Average sounding for the southern California coast in summer (solid line, Neiburger et al., 1961) and a typical summer sounding (dashed line, 19 June 1976) at Lihue, 22° N, 159° W 84

Figure 5-14. GOES visible image of the eastern Pacific, about 1500L, 19 January 1984. 85

Figure 5-15. Tradewind clouds photographed from the Space Shuttle in June 1985. 86

Figure 6-1. Mean cloudiness (oktas) for January (from Sadler et al., 1984) (from Janowiak et al., 1985, and Duvel, 1989) 90

Figure 6-2. Mean cloudiness (oktas) for April (from Sadler et al., 1984) (from Janowiak et al., 1985, and Duvel, 1989) 90

Figure 6-3. Mean cloudiness (oktas) for July (from Sadler et al., 1984) (from Janowiak et al., 1985, and Duvel, 1989) 91

Figure 6-4. Mean cloudiness (oktas) for October (from Sadler et al., 1984) (from Janowiak et al., 1985, and Duvel, 1989) 91

Figure 6-5. Mean zonal cloudiness in oktas for each month (from Sadler et al., 1984) 92

Figure 6-6. GMS infrared (IR) image for 0900Z 19 July 1985, showing cloud clusters over and east of southeast Asia 93

Figure 6-7. GOES IR images for: (A) 1100L 25 October 1988 and (B) 1600L 25 October 1988, showing development of "popcorn" cumulus over northern South America 94

Figure 6-8. Cloud-cluster frequency distributions of width and separation distance for months shown in the tropical North Pacific (adapted from Hayden, 1970) 96

Figure 6-9. Idealized view of the larger-scale tropical cloud systems on a typical summer day in each hemisphere 98

Figure 6-10. Schematic hierarchy of tropical cumulus (from Simpson and Dennis, 1972). 99

Figure 6-11. Cumulus leaning with shear vertically. 99

Figure 6-12. Mean zonal tradewind speed in the lowest 3 km for three tropical Pacific stations (derived from Adams, 1964). 100

Figure 6-13. Percent frequencies of cumulus top heights for the area 2.5° N to 25° N and 155° E to 170° W (after Palmer et al., 1955). 101

Figure 6-14. Measurements made along about 50° E by R.V. Discovery between 16 and 21 August 1964 (Ramage, 1971) 103

Figure 6-15. DMSP visible image for 7 August 1986 showing clear skies off Somalia and a patch of fog or stratus along the southeast coast of Arabia 104

Figure 6-16a. Mean summer (June-August) percent frequency of cloudiness by type over the tropical Atlantic (from Warren et al., 1989) 106

Figure 6-16b. August mean values of various elements used to assist in interpreting the data in Figure 6-16a 107

Figure 6-17. Ratio of the mean number of thunderstorm days to the mean annual rainfall in decimeters. 109

Figure 6-18. Monthly means of thunderstorms, rainfall, lapse-rate, and wind shear in the vertical at four stations in western India. 110

Figure 6-19. Comparison of July mean rainfall and mean sea-level pressure over northern India   (Data from the India Meteorological Department.) 111

Figure 6-20. Satellite view of storm cloud over Singapore (1° N, 104° E) at 1300L, 2 December 1978 112

Figure 6-21. Rainfall record at Mt Waialeale (22° N, 159° W) for 14 and 15 October 1976 113

Figure 6-22. Mean annual outgoing longwave radiation in Wm^{-2} (from Janowiak et al., 1985). 114

Figure 6-23. Mean annual zonal cloudiness in oktas (from Sadler et al., 1984) and outgoing longwave radiation in Wm^{-2} (from Janowiak et al., 1985). 115

Figure 6-24. The top graph shows the Southern Oscillation Index (SOI), calculated by subtracting surface pressure (mb) at Darwin (12° S, 131° E) from surface pressure at Easter Island (27° S, 109° W), then smoothing and plotting departures from the long-term mean difference. The bottom graph shows departures of monthly sea-surface temperatures from long-term means at Puerto Chicama (8° S, 79° W) 116

Figure 6-25. Schematic representation of principal rainfall anomalies found to accompany El Niño and anti-Niño years. 118

Figure 6-26. Tracks of South Pacific tropical cyclones from tropical depression stage onward for December 1982 through May 1983 (Sadler and Kilonsky, 1983). 119

Figure 6-27. Schematic representation of the vertical circulation associated with the summer (top) and winter monsoons. 120

Figure 6-28. The area satisfying monsoon criteria is enclosed by the solid line 121

Figure 6-29. Mean annual cloudiness in oktas, based on 3 years of weather satellite photographs (from Sadler, 1969). 122

Figure 6-30. July monsoon circulation 122

Figure 6-31. January monsoon circulation. 123

Figure 6-32. Circulation components of the monsoon, arranged schematically according to weather, divergence, and vertical motion. 124

Figure 6-33. Annual latitudinal variation of lower-tropospheric pressure troughs over the Indian Ocean ... 124

Figure 6-34. GS IR image of a Mei-Yu front extending from northern Indochina to Southern Taiwan at 1800Z, 28 May 1982 126

Figure 6-35. Mean monthly rainfall for selected stations in Asia/Australia and the Americas. 128

Figure 6-36. Mean monthly rainfall for selected stations in the Pacific and Africa. 129

Figure 6-37. Mean monthly rainfall versus coefficient of variation for stations in India (derived from Mooley and Crutcher, 1968). The dashed line is from a similar study for stations in the Philippines (after Coligardo, 1967) 130

Figure 6-38. Pentad rainfall means (solid lines) for nine stations in southern China and five stations in or south of Japan 131

Figure 6-39. Pentad mean rainfalls and monthly means (bars) for selected Indian stations (Ananthakrishnan and Pathan, 1970) 132

Figure 6-40. Onset dates of southwest monsoon rains over Bombay (19° N, 73° E) from 1879 through 1975 (Rao, 1976). 134

Figure 6-41. Number of years each pentad received more than 25 mm of rain at San Salvador (14°N, 89° W) (40-year POR) (after Griffiths, 1964). 135

Figure 6-42. Periods each year 1918-57 at San Salvador when the soil-water budget exceeded 25 mm (after Griffiths, 1964). 136

Figure 6-43. Mean date of start of the rainy pentads (more than 1 inch/25 mm) over Central America (Gramzow and Henry, 1972). 137

Figure 6-44. Mean date of end of the rainy pentads (last pentad with more than 1 inch/25 mm) over Central America (Gramzow and Henry, 1972). 137

Figure 6-45. Histograms showing the climatological annual progression of fractional low radiance (FLR) based on OLR data for 1974-88 (Horel et al., 1989) 138

Figure 6-46. Time-latitude diagram of the 8-year (1980-87) climatology of pentad OLR across the Americas (Horel et al., 1989) 139

Figure 6-47. The durations of wet seasons over the tropical Americas (top) and Africa (bottom) (Horel et al., 1989) 140

Figure 6-48. Consecutive 3-day means of the total rainfall at seven stations near Ho Chi Minh City (11° N, 107° E) for each year from 1955 to 1968 (less 1963) and the 13-year average (Adapted and expanded from Fukuda, 1968) 141

Figure 6-49. Relationship between cumulative rainfalls and cumulative rainfall frequency (after Martin, 1964) 142

Figure 6-50. Generalized profile of mean annual precipitation (cm) vs. altitude in the tropics 144

Figure 6-51. Composite curve of Contingency Index (CI) as a function of distance between Central American rainfall stations. Mesoscale clouds and circulations drawn atop the distance scale illustrate physical relationships (after Henry, 1974) 146

Figure 6-52. Typical rainfall associated with a single mesoscale convective system (from Henry, 1974) 146

Figure 6-53. Frequency distribution of rainshower amounts and durations (minutes) at Piarco, Trinidad (11° N, 61° W) during dry and wet seasons (after Garstang, 1959) 147

Figure 6-54. Nomogram based on mean annual rainfall to determine mean number of clock-hours that a specified rainfall rate is equaled or exceeded (after Winner, 1968). 147

Figure 6-55. World record observed point rainfalls for various time periods (after Paulus, 1965; Dhar, 1977; and Chaggar, 1984). 149

Figure 6-56. Probable extreme daily rainfall for selected tropical stations (adapted from Atkinson, 1968; Lockwood, 1967; and Griffiths, 1967). 150

Figure 6-57. Rainfall at radial distances from tropical cyclone centers, as computed from moisture convergence (asterisks) Riehl, 1979a; as measured in Hurricane Donna (x's), Riehl, 1979a; as measured in Florida around tropical cyclones (deltas), Miller, 1958; as measured on west Pacific islands around typhoons (diamonds), Frank, 1976 151

Figure 6-58. Average 48-hour rainfall (inches) for 46 Gulf Coast hurricanes (after Goodyear, 1968) 152

Figure 7-1. Instrument sites on Willis Island (Neal, 1973) 154

Figure 7-2. Diurnal variation of surface air temperature at a buoy in the North Atlantic tradewinds averaged between 4 and 13 February 1969 (Prumm, 1974), and at Willis Island (main site and beach site) for 59 undisturbed tradewind days in July-September 1964 (Neal, 1973). 154

Figure 7-3. Mean diurnal August and March temperature and August dew-point curves at Luang Prabang (20° N, 102° E). Annual temperature and dew-point curves at Enewetak (11° N, 162° E) 155

Figure 7-4. Adjustments needed to eliminate the diurnal variation of surface pressure (mb) over the tropical oceans (from Meteorological Office, 1968; Jenkins, 1945). 156

Figure 7-5. Mean diurnal January variation of surface pressure for Hong Kong (22° N, 114° E) 156

Figure 7-6. Hodographs of summer diurnal variation of surface winds at Hilo (20°, 155° W) on windward Hawaii (solid line) and at Waikola Beach (20° N, 156° W) on leeward Hawaii (dashed line) (Ramage, 1978a) 158

Figure 7-7. Artist's conception of east-west land- and mountain-sea breeze circulations during summer over the eastern part of the island of Hawaii (Garrett, 1980) 159

Figure 7-8. Isotachs (m s⁻¹) for the land-sea breeze at Djakarta (6° S, 107° E) (after van Bemmelen, 1922). 159

Figure 7-9. Synthesized empirical model of the land-sea breeze along the Texas Gulf Coast (modified from Hsu, 1970). 160

Figure 7-10. Western Bight of Benin (Atlantic coast of equatorial Africa) during the afternoon (NASA photo) 160

Figure 7-11. Numerical model of the sea breeze (1100 and 1700L) with three prevailing synoptic flows: (A) calm, (B) wind offshore 10 knots, and (C) wind onshore 10 knots (from Estoque, 1962). 161

Figure 7-12. Numerical model of south Florida sea breeze (from Pielke, 1974). 163

Figure 7-13. Meteosat IR image of Lake Victoria (2° S, 32° E) at 1400L 26 July 1990, showing effects of lake winds on clouds 164

Figure 7-14. Surface circulation over Lake Victoria (from Leroux, 1983). 164

Figure 7-15. Lake Volta in the afternoon (NASA) 165

Figure 7-16. Surface winds on the island of Hawaii at 0500 (top) and 1700L (bottom) on 2 August 1990. Land contours are at 1 km intervals. 165

Figure 7-17. Diurnal circulation and cloudiness pattern in the Mt. Kenya area (0°, 37° E) (Hastenrath, 1985). 166

Figure 7-18. Cloud pattern associated with the effect of Mt. Haleakala (21° N, 156° W; 3,111 meters altitude) showing anabatic (upslope) and katabatic (downslope) winds. (Lyons, 1979) 166

Figure 7-19. Resultant surface streamlines over the Red Sea/Gulf of Aden in the morning (left) and afternoon (right) during July and August (modified from Flohn, 1965). 167

Figure 7-20. Hodograph of diurnal variation of surface winds at Honolulu Airport (21° N, 158° W) (Ramage, 1978a). 167

Figure 7-21. Composite 3-hour profiles of wind speeds at Obbia (A) and Burao (B) (from Ardanuy, 1979). 168

Figure 7-22. Schematic diagram of circulations over and near large rivers of Amazonia: (A) night, and (B) day (Garstang et al., 1990). 169

Figure 7-23. Nocturnal ocean-land interface with *onshore* prevailing wind. 169

Figure 7-24. Nocturnal ocean-land interface with *offshore* prevailing wind. 170

Figure 7-25. Average diurnal variation of the inversion height (I) and the cloud base height (B) for a period of undisturbed tradewinds during ATEX (Brill and Albrecht, 1982). 171

Figure 7-26. Mean November 1978 cloudiness (C) and cloud top temperature (T_c) for a 250 x 250 km^2 region at 21.4° S, 86.3° W (Minnis and Harrison, 1984). 172

Figure 7-27. Estimated rates of net radiational warming within a tropical disturbance and in the surrounding clear or mostly clear regions (Gray and Jacobson, 1977)... 174

Figure 7-28. Idealized slopes of pressure surfaces between a disturbance and the surrounding clear air during the day (dashed lines) and at night (solid lines), along with the resultant night vs. day inward-outward radial wind patterns (Gray and Jacobson, 1977). 174

Figure 7-29. Diurnal variation of satellite-observed radiating cloud surfaces associated with tropical cyclones. Areas of "cold" (<-65° C), "intermediate" (-40° C to -65° C), "warm" (-15° C to -40° C), and "total" (<15° C) (from Zehr, 1987) 175

Figure 7-30. Percent of the total thunderstorm activity occurring during any hour at Majuro Atoll (7° N, 171° E) and Clark AB (15° N, 121° E)... 176

Figure 7-31. Diurnal variation of rainfall frequency during the month of maximum rainfall at selected stations in southeast Asia. 177

Figure 7-32. Mean percent of total rainfall for 6-hourly periods at 11 stations on the Korat Plateau, Thailand (Jun-Aug 1967 and 1968) (from Ing, 1971)... 177

Figure 7-33. Diurnal variation in frequencies of rainfall at stations on the island of Hawaii (Schroeder et al., 1978) 179

Figure 7-34. Positions of the rainfall maxima over northeast Brazil by observation times (Kousky, 1980). 180

Figure 7-35. Mean horizontal divergence over West Malaysia for 25 days in August 1957 and 1958 derived from coastal pilot balloon observations 180

Figure 7-36. Areal extent of diurnal rainfall regimes over West Malaysia in August (adapted from Ramage, 1964). 181

Figure 7-37a. 1900L GOES IR image, Central America, 6 July 1985 182

Figure 7-37b. 0700L GOES IR image, Central America, 7 July 1985 182

Figure 7-38. Three-dimensional presentation of the island of Sulawesi, as viewed from the southeast. 183

Figure 7-39. Centers of cumulonimbus activity over and around the island of Sulawesi on 1-15 December 1978 are marked by the "x" symbols. Based on GMS infrared photographs... 184

Figure 7-40. Diurnal variation of ratio (normalized by area) between numbers of cumulonimbus centers over Sulawesi and the neighboring sea, based on 3-hourly GMS IR photos. 184

Figure 7-41. GMS infrared images of eastern Borneo and Sulawesi on 20 July 1985 185

Figure 8-1. Summer. Mean areal frequency of surface monsoon depressions (Ramage, 1968). 190

Figure 8-2. Schematic of the vertical circulation across a monsoon depression (Douglas, 1987). 190

Figure 8-3. Streamline analysis for 850 mb over the Bay of Bengal for 14 and 16 July 1966 (after Sadler, 1967). 191

Figure 8-4. DMSP visible image of a monsoon depression developing over the northern Bay of Bengal on 12 August 1986 191

Figure 8-5. Streamlines at 10,000 and 2,000 feet (3 km and 600 meters) and sea-level pressure analysis over west Africa for 1200Z, 24 August 1967 (adapted from Carlson, 1969a). 192

Figure 8-6. Disturbances over west Africa. (A) Schematic flow pattern at 10,000 feet (3 km) (solid lines) and 2,000 feet (600 meters) (dashed lines). (B) Mean percentage cloudiness with respect to wave axis (after Carlson, 1969b) 193

Figure 8-7. Surface and 700-mb pressure-height charts for 7 January 1949 illustrating a subtropical cyclone in the N. Pacific (after Simpson, 1962). 194

Figure 8-8. Typical tracks of N. Pacific subtropical cyclones (adapted from Simpson, 1952). 194

Figure 8-9. Composite surface winds (knots) around a North Pacific subtropical cyclone, 4-5 April 1960(adapted from Ramage, 1962). 194

Figure 8-10. Schematic radial cross section of clouds, weather, and vertical motion in a symmetrical subtropical cyclone (from Ramage, 1962). 195

Figure 8-11. GOES visible image of a subtropical cyclone centered near 23° N, 139° W, on 1 January 1986 196

Figure 8-12. Mean July positions of the monsoon trough over India at 3,000 feet and 500 mb (adapted from Miller and Keshavamurthy, 1968). 196

Figure 8-13. Meridional cross section along western India of the July mean zonal winds (knots) (adapted from Miller and Keshavamurthy, 1968). 197

Figure 8-14. Composite kinematic analyses (knots) for 1-10 July 1963 198

Figure 8-15a Vertical east-west cross-section of the mid-tropospheric cyclone shown in Figure 8-14 199

Figure 8-15b. Horizontal cloudiness and rainfall distribution of the cyclone shown in Figure 8-14 199

Figure 8-16. DMSP visible image of a mid-tropospheric cyclone over the Arabian Sea on 8 August 1986 200

Figure 8-17. GOES visible image of a "dry" upper-tropospheric (cold) cyclone over the central North Pacific on 8 July 1979 201

Figure 8-18. Composite mean cloudiness (tenths) derived from 13 cases of "wet" upper-tropospheric (cold) cyclones in the N Atlantic, 1961-66 (adapted from Frank, 1970). 202

Figure 8-19. GOES moisture channel image of a cold low centered west of Hawaii on 24 May 1984 202

Figure 8-20. Mean locations of 200-mb ridges and troughs over the Northern Hemisphere in August (after Sadler, 1964). 203

Figure 8-21. Schematic model of a tropical cyclone initiated by an upper-tropospheric low (Sadler, 1976a). 204

Figure 8-22. Time-altitude section and 12-hourly surface observations for Midway Island (28° N, 177 W°), 27-31 July 1970 (Sadler, 1976a). 205

Figure 8-23. Model of the lower-tropospheric contour and weather patterns of a temporal (adapted from Pallman, 1968). 207

Figure 8-24. Cross-section of typical tropical squall line; airflow is relative to the squall line, which is moving from right to left. Circled numbers are typical values of θ_w in ° C (Zipser, 1977). 208

Figure 8-25. Schematic streamlines of airflow relative to convective disturbances in the Line Islands area. (C) represents the N-S section of the dissipating phase (after Zipser, 1969). 210

Figure 8-26. Isochrones of the leading edge of the downdraft air from a disturbance in the Line Islands area on 1 April 1967 (after Zipser, 1969). 211

Figure 8-27. Schematic of cloud development over a ridge west of an approaching west African squall line (from Leroux, 1983). 211

Figure 8-28. DMSP images of west Africa on 18 July (top) and 19 July (bottom) 1987 212

Figure 8-29. Model of the surface kinematic pattern associated with a cold front extending into a tropical ocean shear line (adapted from Palmer, et al., 1955) 213

Figure 8-30. Goes visible image at 0000Z 25 December 1985 showing a cold front crossing Hawaii 214

Figure 8-31. Daily values for January 1967 over the South China Sea 215

Figure 8-32. Schematic meridional cross-section over the South China Sea during a surge of the winter (northeast) monsoon (adapted from Navy Weather Research Facility Staff, 1969) 216

Figure 8-33. GMS IR image for 0900Z 9 December 1985, prior to a cold surge across the South China Sea 217

Figure 8-34. GMS IR image for 0900Z 10 December 1985, following a cold surge across the South China Sea 217

Figure 8-35. Topography of Central America (Sadler and Lander, 1986). 218

Figure 8-36a. January isotachs—ship data climatology (Sadler and Lander, 1986). 219

xxvi

Figure 8-36b. January kinematics—ship data climatology (Sadler and Lander, 1986). 219

Figure 8-36c. January sea-surface temperature—ship data climatology (Sadler and Lander, 1986). 219

Figure 8-36d. January sea-level pressure—ship data climatology (Sadler and Lander, 1986). 219

Figure 8-37. Synoptic sequence over North Africa for a 48-hour period centered on 23 February 1943. 219

Figure 8-38. Schematic vertical profile of winds and clouds associated with weak, moderate, and strong surges of the southwest (summer) monsoon (Guard, 1985). 221

Figure 8-39. Schematic time sequence (according to Leroux, 1983) of a surge (dashed line) progressing from Egypt to west Africa during the warm season 222

Figure 8-40. Three-dimensional synoptic model of an altostratus layer overlying a surface front in the region of the southwest Pacific convergence zone (SPCZ) during winter (Hill, 1964) 223

Figure 8-41. GMS IR image at 1600Z 1 July 1985 of an SPCZ in its normal climatological position 224

Figure 8-42. 10-18 January SPCZ in the southwest Pacific 225

Figure 8-43. Circulation at 300 mb for 0000Z 6 Feb 1968 226

Figure 8-44. Soundings, relative humidities, and winds (one full barb denotes 10 knots) at 0000Z 6 February 1968 226

Figure 8-45. Clouds photographed by ESSA 3 at 0500Z 6 February 1968 227

Figure 8-46. Clouds over southern Africa photographed from ESSA 3, 17 April 1968 (Fox, 1969) 228

Figure 8-47. Sounding at Lihue (22° N, 159° W) at 0000Z 19 April 1974 229

Figure 8-48. Frequency distribution of heights of the dry layer base as a function of 200-mb wind direction, Lihue, 1976. 229

Figure 8-49. Lihue Sounding at 0000Z, 15 October 1976. Solid line: temperature; dashed line, dew point. 229

Figure 8-50. TIROS VI cloud photos taken at 0900Z 4 December 1962 when an upper trough and a lower disturbance were superposed 231

Figure 8-51 Total rainfall (mm) over India from 1 through 6 December 1962 232

Figure 8-52. Time cross-section for Bombay (19° N, 73° E) from 3 to 6 December 1962. 232

Figure 8-53. Role of the Atlantic upper-tropospheric trough in the stimulation or constraint of mass circulation in a tropical cyclone 233

Figure 8-54. Flight tracks and altitudes flown by NOAA P-3 research aircraft, November 1977-January 1978 (Ramage et al., 1981). 234

Figure 8-55. Time latitude cross-section of GOES images along a strip 2-degrees wide centered at 150° W for 29 November 1977 through 5 January 1978 (Ramage et al., 1981) 235

Figure 8-56. Schematic meridional cross-section of a vigorous near-equatorial tradewind convergence. 236

Figure 8-57. GOES visible image (0000Z 2 December 1985) of an upper-tropospheric trough interacting with a near-equatorial threatened convergence 237

Figure 8-58a. Positions of the 700-mb ridge on alternate days from 11 to 21 July 1967 (adapted from Sadler et al., 1968). 239

Figure 8-58b. 700-mb analysis for 11 July 1967 (adapted from Sadler et al., 1968). 239

Figure 8-58c. 700-mb analysis, 15 July 1967 (adapted from Sadler et al., 1968). 239

Figure 8-58d. 700-mb analysis, 19 July 1967 (adapted from Sadler et al., 1968). 239

Figure 8-59. GMS IR image (1600Z 11 July 1985) showing a midsummer dry spell over south China and the South China Sea 240

Figure 9-1. Vertical radial cross section of the mean tangential velocity (knots) in 14 Pacific typhoons Iafter Izawa, 1964) 242

Figure 9-2. Kinematic analyses of (A) the lower-tropospheric and (B) the upper-tropospheric circulation in Hurricane Donna on 10 September 1960. Areas of speed maxima are shaded (after B.I. Miller, 1967) 243

Figure 9-3. Fujita typhoon models (after Fujita et al., 1967). 244

Figure 9-4. Vertical cross section of temperature anomalies (° C) relative to the mean tropical atmosphere, Hurricane Cleo, 18 August 1958 (after LaSeur and Hawkins, 1963). 245

Figure 9-5. Radarscope photo of Typhoon Nelson, centered 157 km southeast of Kadena (RODN) at 1138Z 6 October 1988 246

Figure 9-6a. Enhanced GMS IR image of Super Typhoon Flo, centered near 26° N, 129° E at 0538Z 17 September 1990 (from NOAA/NESDIS, 1983) 247

Figure 9-6b. The enhancement curve used to shade-code the image in Figure 9-6a (from NOAA/NESDIS, 1983) 247

Figure 9-7. Average monthly number of tropical cyclones of tropical storm intensity relative to calendar and solar years (from Gray, 1978). 249

Figure 9-8. Average monthly number of tropical cyclones that reach tropical storm or greater intensity in each development area for 1958 to 1977 (from Gray, 1978). 250

Figure 9-9. The tracks of tropical cyclones for a 3-year period (Gray, 1978). 250

Figure 9-10. Locations of near-equatorial tradewind convergence (solid lines) and the monsoon troughs (dashed lines) at the gradient level over land and at sea level over the oceans during February, May, August, and November (adapted from Atkinson and Sadler, 1970) 252

Figure 9-11. Mean position of the 200-mb ridge in July and August and mean August 26.7° C sea-surface isotherm in the eastern North Pacific (Sadler, 1964). 253

Figure 9-12. Tracks of surface cyclones over the tropical Atlantic during August 1963 (after Aspliden et al., 1965-1967). 254

Figure 9-13. Model of low-level cyclones in the tropical North Atlantic depicting either a chain of cyclones or the life history of one cyclone   (after Sadler, 1967) 254

Figure 9-14. Twenty-two cross-equatorial named tropical cyclone pairs over the Pacific Ocean, September 1971 through January 1980. (from Keen, 1982) 255

Figure 9-15. DMSP visible image of twin cyclone development at 2330Z, 17 December 1990. Observed surface winds are plotted. 256

Figure 9-16. Tracks of pairs of tropical cyclones over the Indian Ocean, 1964 to 1974 (Mukerjee and Padmanabham, 1977). 257

Figure 9-17. (A) Latitude at which initial disturbances that later became tropical storms were first detected in each development area. Average number of years of data are given in parentheses. (B) Combined data for the NH, SH, and Globe (adapted from Gray, 1968) 258

Figure 9-18. Frequency distribution of the annual number of tropical cyclones of tropical storm intensity or greater in various development areas (Gray, 1978). 258

Figure 9-19. Average zonal wind-shear (knots) between 850 mb and 200 mb over the tropics for January, April, July, and October (after Gray, 1968). 261

Figure 9-20. Mean recurvature latitudes and ranges for western North Pacific tropical cyclones, 1965-1982 (from Guard, 1983). 265

Figure 9-21. Mean monthly recurvature positions for tropical storms and hurricanes in the North Atlantic (after Colòn, 1953). 265

Figure 9-22. 1989 error statistics for selected objective tropical cyclone forecasting techniques in the western North Pacific. 266

Figure 9-23. Developmental cloud pattern types used in intensity analysis of satellite images (Dvorak, 1984) 268

Figure 9-24. Yearly averages of tropical cyclone position-forecast errors compared to best tracks, 1970 through 1989 for NHC, Miami (solid lines, and JTWC, Guam (dashed lines) 269

Figure 9-25. How to use average root-mean-square vector errors to establish areal confidence limits for typhoon forecasts. 271

Figure 10-1. Mean annual number of thunderstorm days over the globe (World Meteorological Organization, 1956b). 274

Figure 10-2. One year of midnight lightning locations detected on DMSP images for 365 consecutive days from September 1977-August 1978 (Orville and Henderson, 1986). 275

Figure 10-3. Mean annual percentage of days with thunder by 5° longitude intervals for the equatorial belt between 10° N and 10° S (adapted from Ramage, 1968). 275

Figure 10-4. Mean annual number of thunderstorm days over Southeast Asia (after Atkinson, 1967a). 276

Figure 10-5. Mean annual number of thunderstorm days over India during the period 1931-1960 (after Alvi and Punjabi, 1966). 277

Figure 10-6a. Frequency distributions for selected categories of monthly means of the number of thunderstorm days for stations in Thailand. 278

Figure 10-6b. Standard deviations for selected categories of monthly means and line of best fit in the number of thunderstorm days for stations in Thailand. 278

Figure 10-6c. Cumulative frequency distributions for the monthly number of thunderstorm days according to the mean values for stations in Thailand. 279

Figure 10-6d. Cumulative frequency distributions for the annual number of thunderstorm days according to the mean values. for stations in Thailand. 280

Figure 10-7a. Thunderstorm occurrence, Jun-Aug 1965 at Mactan Air Base, Philippines. 281

Figure 10-7b. Frequency distribution of thunderstorm durations considering all separate occurrences at Mactan Air Base, Philippines. 281

Figure 10-7c. Frequency distribution of thunderstorm duration derived by considering successive periods of thunderstorm activity separated by less than 2 hours as one occurrence at Mactan Air Base, Philippines 281

Figure 10-8. Percentage frequency of occurrence of tornadoes over India, 1951 to 1980; total number, 42 (Singh, 1981). 283

Figure 10-9. Locations of reported funnels aloft (V) and waterspouts/tornadoes (∇) for the Hawaiian Islands, 1961-1974 (Schroeder, 1976) 284

Figure 10-10. Occurrence of funnel clouds, waterspouts, and tornadoes within 140 km of Miami, Florida, 1957 to 1966; (A) monthly, (B) Diurnal variation (hourly values smoothed by a 3-hourly running mean) (adapted from Gerrish, 1967) 285

Figure 10-11. Distribution of tornadoes, waterspouts, and funnel clouds in the south Florida area, 1957 to 1966 (after Gerrish, 1967). 286

Figure 10-12. Distribution of hurricane tornadoes, 1948 to 1972 (Novlan and Gray, 1974). 287

Figure 10-13. Hurricane tornadoes over the US, 1955-1964, located with reference to the center and direction of movement of the hurricanes (adapted from Pearson and Sadowski, 1965) 288

Figure 10-14. A Hurricane-tornado forecasting checklist (Novlan and Gray, 1974) 289

Figure 10-15. Frequency of hail occurrences at the surface in parts of the tropics (adapted from Frisby and Sansom, 1967). 291

Figure 10-16. Hail regimes for three tropical zones. The X's indicate months in which hail at the surface is most likely (adapted from Frisby, 1966). 292

Figure 10-17. Climatology of thunderstorm squalls and related phenomena at Nagpur (21° N, 79° E), India: (A) mean and maximum squall frequency, (B) monthly extreme gust speeds with squalls, and frequency of distribution of (C) maximum squall winds, (D) pressure changes and (E) temperature changes associated with squalls (adapted from Sharma, 1966) 294

Figure 10-18. Vertical cross section of the evolution of the microburst wind field (from Wilson et al., 1984). 296

Figure 10-19. Schematic for 1200Z 31 January 1991 showing duststorms accompanying a cold front crossing Saudi Arabia 297

Figure 10-20. DMSP IR image for 19 October 1984 of a duststorm generated by strong northwest winds (Shamal) over the Persian Gulf and northeast Arabia 298

Figure 10-21. DMSP IR image for 25 June 1985, showing widespread dust covering Arabia and the neighboring seas 300

Figure 10-22. DMSP images for 25 August 1989, showing a duststorm extending from the Nubian Desert over the Red Sea. (Brooks, 1990) 302

Figure 10-23. Curve of best fit between maximum sustained surface wind speeds and minimum sea-level pressures for 76 western Pacific tropical cyclones (from Atkinson and Holliday, 1977) 303

Figure 10-24. Double-exponential distribution (full line) fitted to the annual peak gusts at Kadena Air Base (26° N, 128° E), Okinawa, 1945-1988. 304

Figure 10-25. Expected extreme wind gusts (knots) for 2-year and 100-year return periods for selected tropical stations. 305

Figure 10-26. Nomogram for determining expected extreme wind gusts for various return periods based on the 2-year and 100-year return-period values (based on the Gumbel double-exponential distribution). 306

Figure 10-27. Relative frequency of various types of turbulence versus altitude (after Saxton, 1966)..... 307

Figure 10-28. Regression of maximum storm-surge height on central pressure for Atlantic and Gulf Coast hurricanes (adapted from Hoover, 1957 and Conner et al., 1957)... 311

Figure 10-29. Storm-tide height profiles for four hurricanes that entered the U.S. Gulf coastline west of Tallahassee, Florida, from the south-southeast, south, and south-southwest. (adapted from Hoover, 1957)... 312

Figure 10-30. Storm-surge heights on the U.S. coasts south of 35° N for great hurricanes (central pressures 950 mb or less) 313

Figure 11-1. Hourly temperature, dew point, and rainfall for Canton Island for 12 to 13 April, 1953 (after Palmer et al., 1955). 319

Figure 11-2. Example of cyclonic and anticyclonic cusps in an east-to-west current (after Palmer et al., 1955).. 327

Figure 11-3. Models of pure outdrafts, pure indrafts, and six types of vortexes possible in streamline analysis (after Palmer, et al., 1955). 327

Figure 11-4. Wind data for kinematic analysis shown in Figure 11-5 (wind speeds in knots) (Dept. of Weather Training, Chanute AFB, IL). 329

Figure 11-5. Kinematic analysis based on wind data in Figure 11-4 (Dept. of Weather Training, Chanute AFB, IL). 330

Figure 11-6. Schematic cross section of two types of front. (A) cold and warm air both rising. (B) cold air sinking, warm air rising.. 330

Figure 11-7. Schematic kinematic analysis of surface winds around a moving front (Ramage, 1957) 331

Figure 11-8. Same as Figure 11-7 (V$_s$ does not change direction through the front) 331

Figure 11-9. Surface chart for 1200Z 22 July 1956. (A) Isobaric analysis (mb). (B) Kinematic analysis (isotachs labeled in knots.) 332

Figure 11-10. Chart for 250-mb winds (knots) and temperatures (° C), for 1200Z 26 November 1985....... 333

Figure 11-11. Illustrating how AIREP wind speeds can be adjusted to analysis level using the computed wind-shear in the vertical from surrounding rawinsonde stations... 334

Figure 11-12. Root-mean-square variability (knots) of upper-air wind speeds as a function of wind speed and time or distance (adapted from Lenhard, 1967). 335

Figure 11-13. Illustrating how AIREP positions can be adjusted according to the movement of synoptic-scale features. 335

Figure 11-14. Chart of the mean positions of major circulation features over the western North Pacific and southern Asia for October. 336

Figure 11-15. Relationship between the monthly mean resultant wind speed at the gradient level and the "steadiness."(Atkinson and Sadler, 1970) 337

Figure 11-16. Relationship between the monthly mean resultant wind speed at the gradient level and the percent of time that the wind is within ±45 percent of the resultant direction. (Atkinson and Sadler, 1970) 337

Figure 11-17. 850-mb kinematic analysis and nephanalysis for the western North Pacific and Asia for 0000Z 26 July 1967. (from analyses made at the University of Hawaii) 340

Figure 11-18. 500-mb kinematic analysis and nephanalysis for the western North Pacific and Asia for 0000Z 26 July 1967. (from analyses made at the University of Hawaii) 341

Figure 11-19. 200-mb kinematic analysis and nephanalysis for the western North Pacific and Asia for 0000Z 26 July 1967. (from analyses made at the University of Hawaii) 342

Figure 11-20. Recommended plotting model for composite lower-tropospheric wind charts. 343

Figure 11-21. January long-term mean (A) ship winds and (B) satellite cloud motion vectors; (C) wind shear in the vertical between the two data sets (m s^{-1}) (Sadler and Kilonsky, 1985). 344

Figure 11-22. Recommended plotting models for composite upper-tropospheric wind charts. 345

Figure 11-23. Objective kinematic analyses for 0000Z 2 December 1978. Wind speed in knots (A) at 950 mb; (B) at 200 mb (Davidson and McAvaney, 1981). 349

Figure 11-24. Vertical time cross-section of winds (knots) above Johnston Island (17°N, 170° W) on 2-6 May 1954 (Dept of Weather Training, Chanute AFB, IL) 352

Figure 11-25. Checkerboard plot of hourly surface observations at Don Muang (14°N, 101°E), Thailand, 19 to 31 January 1968 (Dept of Weather Training, Chanute AFB, IL). 354

Figure 12-1. Monthly climatologies of weather variables at Saigon/Ho Chi Minh City (11° N, 107° E)... 360

Figure 12-2. Conditional climatology summaries for Saigon (11° N, 107° E) and Danang (16° N, 108° E). 361

Figure 12-3. Climatological display for Saigon (11° N, 107° E) in August. 362

Figure 12-4. Mean May 1959 300-mb wind vector errors over the tropical Pacific for climatology, persistence, and forecasts, based on 50 percent climatology plus 50 percent persistence (after Lavoie and Wiederanders, 1960) 365

Figure 12-5. Relationship between maximum wind speed in the lowest 10,000 feet (3 km) at 0800L and maximum surface wind gust observed in subsequent 12 hours at Mactan AB (10° N, 124° E), Philippines (First Weather Wing, 1968) 365

Figure 12-6. Hourly radar index percent coverage of radar echoes within 50 NM (90 km) of station for Saigon (11° N, 107° E) for June, July, and August 1967 (after Conover, 1967) 367

Figure 12-7. Kinematic analyses of the gradient-level resultant wind for: (A) days with above average (left) and (B) below average (right) radar-index values for the Saigon area (shown by shading). Isotachs are labeled in knots (after Conover, 1967) 368

Figure 12-8. Percent probability of: (A) ceiling/visibility less than 1,500 feet (460 meters) and/or 3 miles (5 km), and (B) less than 200 feet (60 meters) and/or a half-mile (800 meters) during a 1-hour period at Kadena Air Base (26° N, 128° E), Okinawa, in relation to tropical cyclone-center location (adapted from Atkinson and Penland, 1967) 368

Figure 12-9. Relation between Showalter-Index and Lifted-Index values at 0600L and probability of thunderstorms during the following 24 hours at New Delhi (29° N, 77° E), and within radii of 50 and 100 NM (93 and 185 km) (adapted from Subramaniam and Jaim, 1966) 370

Figure 12-10. Reported maximum thunderstorm-produced surface wind gusts (knots) at stations in the southeastern United States and southeast Asia according to observed values of (A) the Delta-T Index, and (B) the T_1 Index (adapted from R.C. Miller, 1967) 371

Figure 12-11. Radar-echo frequencies measured simultaneously by 3-cm radars located at Udon (17°N, 103°E) and Ubon (15°N, 105°E), for June-August 1967 and 1968 (adapted from Ing, 1971) 372

Figure 12-12. Hourly center positions of Hurricane Carla, 10-12 September 1961, based on radar fixes from Galveston (29 ° N, 95° W) 373

Figure 12-13. Kinematic model of a hurricane spiral rainband (after Hardy et al., 1964) 374

Figure 12-14. Statistics of large radar-echo groups (greater than 50 NM (93 km) wide) near Saigon/Ho Chi Minh City (11° N, 107° E) during the southwest monsoon (adapted from Conover, 1967). 375

Figure 12-15. Surface visibility (V) in showers related to reflectivity (Z) measured by radar (adapted from Wilson, 1968, Cataneo, 1969, and ESSA, 1967) 377

Figure 12-16. Radar climatology for New Delhi, India: (A) monthly distribution of echo tops, (B) time of occurrence of maximum areal coverage by month, (C) frequency distributions of echo sizes, and (D) interspacing of echoes during the hot season (March-May) (adapted from Kulshrestha and Jain, 1968) 378

Figure 12-17. Radar climatology of the southeast Florida area: (A) areas and grid used to compile the climatology, (B) diurnal variation of echo frequency, (C) percent frequency variation by time of day and distance from the coast, and (D) relative frequencies of heights of echo tops for 1200 to 1800L (adapted from Gerrish (1971) 379

Figure 12-18. Wind forecasts for March and April 1991 at 32 tropical stations. Comparison of NMC numerical prediction of 850-mb winds 24 hours in advance to objective—½ (persistence + climatology); persistence; and climatological predictions 381

Figure 12-19. Wind forecasts for March and April 1991 at 32 tropical stations. Comparison of NMC numerical prediction of 200-mb winds 24 hours in advance to objective—½ (persistence + climatology); persistence; and climatological predictions 382

Figure 12-20. Composite 30-60 day anomalies of outgoing long-wave radiation between the equator and 10° N (isopleth interval 5 W m⁻¹) for May to October (left side) and November to April (right side) (from Knutson and Weickmann, 1987) 385

Figure 12-21. Fifteen-year running mean correlation coefficients between: (1) average of April-May surface pressure at three South American stations near 30° S and the subsequent June-September rainfall over Peninsular India (dashed line); (2) the difference between the January mean sea-level pressures at Irkutsk (52° N, 104° E) and Tokyo (36° N, 114° E) (solid line) 387

Figure 12-22. Predictions of North Atlantic hurricanes and tropical storms made each May for the following season by Gray (1990) (solid line), compared to the outcomes (dashed line) and to climatology (dot-dashed line) 388

Figure 12-23. Iterative procedure for evaluating and modifying case studies through forecast performance. 392

TABLES

TABLE 2-1. Poleward transport of heat energy by ocean currents according to Sellers (1965) and Vonder Haar and Oort (1973), with percentages of the total (air plus ocean) fluxes (units of 10^{13} W) 14

TABLE 3-1. Implementation of WMO synoptic observing program 1988 (from WMO, 1989) 26

TABLE 3-2. Implementation of WMO radiosonde observing program, 1987 (from WMO, 1989) 30

TABLE 3-3. Operational weather satellites (AWS/XOOO, 1995) 33

TABLE 3-4. Worldwide weather radar station distribution (from WMO, 1989 35

TABLE 3-5. Proposed satellite missions (AWS/XOO, 1995). 36

TABLE 3-6. Tropical stations beyond the continental United States at which 10-cm Doppler radars (WSR-88D) are either installed or being installed 37

TABLE 5-1. Weather effect on electrooptical devices 66

TABLE 5-2. Normal atmosphere for 30° N (January and July) and 15° N (Annual) 78

TABLE 6-1. Starting years of El Niño and anti-El Niño events (1877-1988) (from Kiladis and Diaz, 1989) 115

TABLE 6-2. Relationship of moderate or strong Niño to same-year Indian summer monsoon rainfall, 1875-1980 (Ramage, 1983) 117

TABLE 6-3. Advance of Indian summer monsoon (normal starting dates for summer monsoon rains) 134

TABLE 6-4. Data for the universal rainfall distribution curve (from Martin, 1964) 143

TABLE 6-5. Examples of applying the universal rainfall distribution curve to determine distribution of daily rainfall 143

TABLE 6-6. The relationship of daily rainfall between station pairs (Henry, 1974) 145

TABLE 6-7. Empirical relationships between the mean annual rainfall given by X and the mean number of hours per year given by Y that various clock-hour rainfalls are equalled or exceeded 148

TABLE 8-1. Average monthly and seasonal frequency of Bay of Bengal monsoon depressions for the period 1891-1970 (Rao, 1976) 190

TABLE 9-1. Areas of occurrence of intense tropical cyclones and regional terminology (obtained from World Meteorological Organization, 1979) 248

TABLE 9-2. Average monthly frequency of tropical cyclones of at least tropical storm intensity for each major basin (storm-development area) 251

TABLE 9-3. Average errors (NM) and trends in tropical cyclone position forecasts made at JTWC, Guam (13° N, 145° E) and NHC, Miami (26° N, 80° W) for 1970 through 1989 269

TABLE 9-4. Average errors in 24-hour persistence/climatology movement forecasts of tropical cyclone centers for tropical cyclone basins (Pike and Neumann, 1987) 270

TABLE 10-1. Thunderstorm duration, shown by the ratio of the mean annual number of hourly observations reporting thunderstorms to the mean annual number of thunderstorm days for stations in southeast Asia 282

TABLE 10-2. Frequency distribution of squall duration at Nagpur 294

TABLE 10-3. Peak gusts seasonal averages and absolutes at Cape Kennedy 295

TABLE 10-4. Frequency of duststorms at Khartoum (16° N, 33°E) based on 8 years of observations (from Sutton, 1925) 299

TABLE 10-5. Saffir/Simpson Damage Potential Scale ranges 304

TABLE 10-6. Percent frequency of various categories of CAT by 5 degree latitude-longitude rectangles in the tropics, for flights above 20,000 feet (6 km) (for the 1964-65 ICAO collection periods) 309

TABLE 11-1. Relationship of Beaufort wind-scale number and wind-speed (knots) as included in synoptic reports (from World Meteorological Organization, 1966b) 320

TABLE 11-2. Ratio of the wind speed at various levels to the wind speed at 66 feet (20 meters) elevation, based on the one-seventh power law 321

TABLE 11-3. Root-mean-square (RMS) pressure-height errors (meters) arising in radiosonde observations with an RMS temperature error of 1.0° C and an RMS pressure error of 2.0 mb 323

TABLE 11-4. Root-mean-square errors (knots) in rawinsonde wind speeds according to altitude and the magnitude of the mean-wind vector (Meteorology Working Group, 1965 324

TABLE 11-5. Percentage frequency of modal wind directions differing by less than one compass point from resultant wind directions (after Wiederanders, 1961) 338

TABLE 12-1. March and April 1991 average tropical wind forecasting errors (in knots) 380

TABLE 12-2. March and April 1991 average tropical wind forecasting errors (in knots), stratified by latitude 380

TABLE 12-3. Predictors for summer rainfall in India (Walker, 1924) 386

Chapter 1

INTRODUCTION

1.1 GENERAL

The urge to understand tropical meteorology has been driven by a variety of forces that include the devastation caused by tropical cyclones, demands by military operations (for example, World War II, the Marshall Islands weapons tests, and the conflict in southeast Asia), and the belief of influential mid-latitude meteorologists that the tropics may hold the key to successful forecasts of global weather patterns and climate change. Efforts by World Weather Watch (WWW) and the Global Atmospheric Research Program (GARP), including the GARP Atlantic Tropical Experiment (GATE) of 1974 and the Winter and Summer Monsoon Experiments (WMONEX and SMONEX) of December 1978 and June-July 1979, gave many meteorologists their first experience of the tropics. Research papers proliferated, concentrating on the small time and space scales described by the special observing systems (aerial reconnaissance, rawinsondes, and radars) set up for the experiments.

But where does all this leave the field forecaster? A case study dependent on data not operationally available and supported by brief, doubtful statistics, doesn't help much in the day-to-day grind of tropical analysis and forecasting.

Predicting tropical cyclones taps almost all the resources available to the tropical forecaster. Riehl (1979a) devotes a chapter to numerical hurricane prediction but nothing to other tropical forecasting. On an average day in the tropics, less than one tenth of one percent of the area is affected by a tropical cyclone. Only about 80 cyclones develop each year, but since 1950 more than 5,000 operational and research reconnaissance flights have been made into cyclone centers. In contrast, only one Arabian Sea mid-tropospheric cyclone, one monsoon depression, and two upper-tropospheric cold lows have been reconnoitered. Fifteen single-aircraft research traverses have been made across the near-equatorial trade-wind convergence, but none into the mysterious weak circulations that cause disastrous floods (see

6.3.1.2). For most generating regions, about 100 years of daily tropical cyclone positions provide the basis for detailed, authoritative climatologies. Apart from monsoon depressions, not even 1 year of tracks has been plotted for other tropical vortices. This disproportion, although slightly reduced by squall-line studies in GATE, Venezuela, and the Amazon, and a 1987 field study of early-summer rain systems off southeast China (Taiwan Area Mesoscale Experiment–TAMEX), means that a meteorologist trying to forecast tropical cyclones is served by a variety of subjective, numerical, and statistical techniques, special services from global prediction centers, and at least hourly satellite updates. But for the other 99.9 percent of the time, forecasters are pretty much on their own. This manual tries to redress the balance. It devotes one chapter to tropical cyclones, but emphasizes other weather Systems, background climatologies, and forecasting hints and pitfalls.

Observational information on the tropics increased significantly through the 1970s, but progress has been patchy over Africa and the Americas. Weather radars operate at many tropical stations, and there is great potential for more aircraft weather reports. Compared to 20 years ago, there are fewer human weather observations, while automated observations, especially those from satellites, are vastly increased and not yet fully exploited.

There is an urgent need to synthesize from these diverse observational systems a coherent basis for the analysis and prediction of tropical circulations and associated weather. Our ability to digest and use this information in tropical numerical models has not kept pace with numerical weather prediction in the middle latitudes. This is partially due to the greater complexity of weather systems and interactions between various scales of motion in the tropics. Thus, although worldwide predictions are being made, tropical weather forecasting (at least in the near future) will continue to depend heavily on subjective judgment, aided by selected numerical products such as objective wind analyses, global long-wave predictions, and computer-processed satellite data.

West Pacific typhoons and Atlantic hurricanes are the same phenomenon; they need not be described separately. Thunderstorms also possess the same characteristics worldwide. What is not so well-known is that some other phenomena occur throughout the tropics. These include surges, cold lows, linear convergence lines and shear lines, squall lines, low-level jets, heat lows, upwelling and dust storms. Once each of these phenomena is explained, all forecasters need to know are distributions and frequencies.

Although this manual discusses various theoretical hypotheses, it uses little mathematical formulation. It assumes that readers are familiar with the pressure/ wind and thermal wind relationships; with divergence, convergence, and vertical motion; and with vorticity and lapse rate. It also assumes that they generally understand Mid-latitude synoptic models. The comprehensive list of references in the bibliography serves to facilitate further study on selected aspects of tropical meteorology in much greater detail than is possible here.

Weather forecasters are viewed as applied researchers who can evaluate hypotheses and come up with new ideas. They should realize that weather mechanisms in the tropics are not often understood. "Authoritative" opinions may differ, for example, on the prevalence of easterly waves, on the cause of the quasi-biennial oscillation, on predicting El Niño, and on the generation, movement, and dissipation of surges. Even when there *is* general agreement, such as on tropical cyclone dissipation, diurnal variations along mountainous coasts, and the role of surface cooling in causing fog, apparent exceptions may occur.

Chapters 2 through 6 focus on annual cycles and monthly averages while making some attempt to interpret them in terms of weather. Chapter 3 discusses data. Chapter 7 covers diurnal effects. The remaining chapters concern weather analysis and forecasting where astronomical control by the annual and diurnal cycles is muted by complex feedbacks among atmosphere, ocean, and terrain.

1.2 TERMINOLOGY

Definitions in the Glossary of Meteorology (Huschke, 1959) are generally used here, but this section expands some of these definitions, adds new terms introduced to the tropics since 1958, and discusses terms that have been made misleading by a variety of definitions. Terms used only once in the text (e.g., "hodograph") are defined where they appear.

Buffer Zone was defined by Conover and Sadler (1960). In Figure 1-1, a buffer zone straddles the equator and separates two oppositely-directed wind streams: The easterly trades and the monsoon westerlies. The sense of rotation in the buffer zone is clockwise in the Northern Hemisphere (NH) summer and anticlockwise in the NH winter. The zone tends to persist; it may contain circulation cells that are generally stationary and it may last 3 to 4 days. If it moves more than a few degrees from the equator it takes on the character of a *trough* if the curvature is cyclonic, or a *ridge* if the curvature is anticyclonic. The term has not been widely accepted, and has been criticized by Leroux (1983) "because there is generally no interruption of the stream."

Cold Low (upper troposphere) In comparison to its environment, this low is most intense and coldest in the upper troposphere. It is therefore weaker at lower levels and can be hard to find below 500 mb. Apart from a few cumulonimbus clouds that may form near the center, it is generally subsident and cloud-free, except for rising motion and cloud on its eastern periphery. In summer over the tropical Atlantic and Pacific Oceans, as well as the South Indian Ocean, a line of cold lows may comprise the tropical upper tropospheric trough (TUTT)–see Figures 8-17 to 8-22.

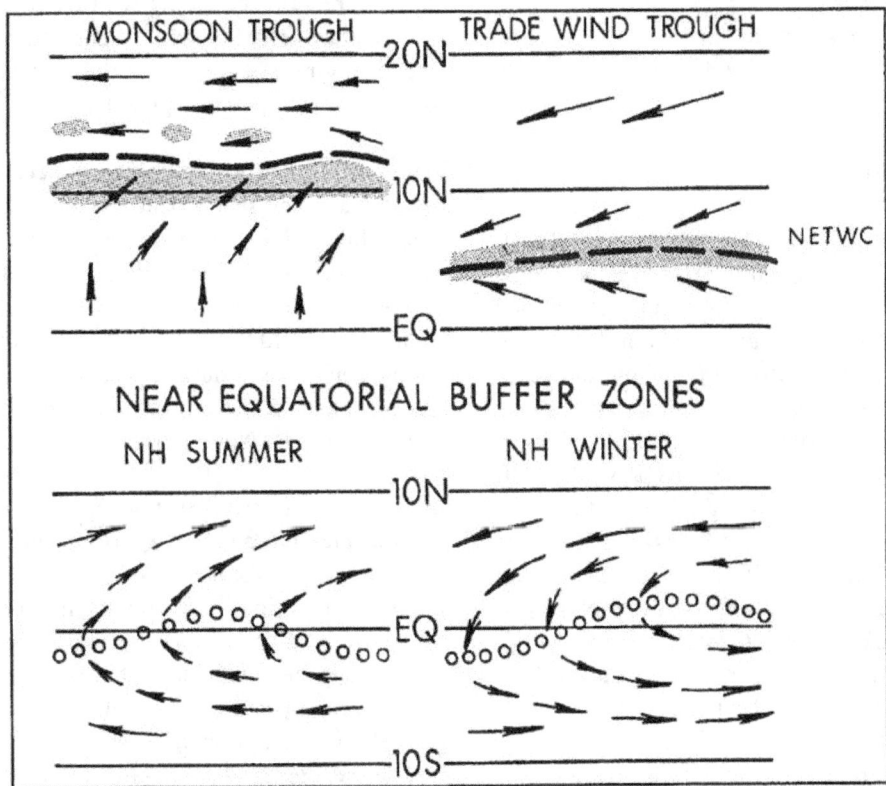

Figure 1-1. Schematic representations of monsoon and tradewind troughs and near-equatorial tradewind convergence (top) and near-equatorial buffer zones (bottom). Stippling denotes areas of unsettled weather.

Electromagnetic Radiation. All electrooptical devices and all the instruments on meteorological satellites are triggered by electromagnetic waves that are emitted from, or reflected by, objects or gases in the atmosphere. These waves are usually specified in terms of *wavelength* (e.g., a 10-cm radar) or *frequency* (e.g., 3 GHz) or, more descriptively, as *visible* or *infrared.* Frequency (in cycles per second)

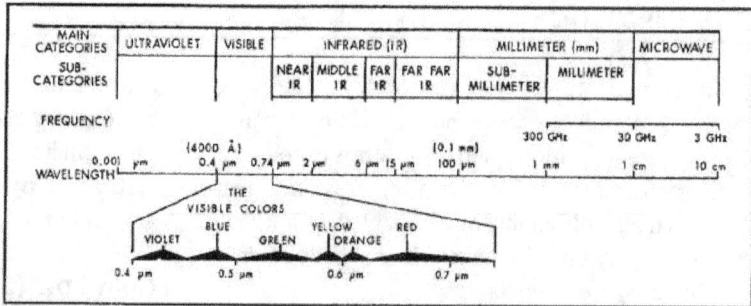

Figure 1-2. Electromagnetic energy categories, frequencies, and wavelengths (based on Elrick and Meade, 1987).

equals (speed of light (2.997 x 10^{10} cm s^{-1}) divided by the wavelength in cm. For example, a 10-cm radar operates at a frequency of 2.977 GHz. Figure 1-2 illustrates the range of electromagnetic waves of interest to the meteorologist.

The following units are used:

Frequency:
  1 Hertz = 1 Hz = 1 cycle sec^{-1}
  1 Gigahertz = 1 GHz = 10^9 Hz

Wavelength:
  1 micrometer = 1 μm = 10^{-6} m

Heat Low/Heat Trough Surface heat lows, occasionally strung along a trough (as over the deserts of North Africa and southwest Asia in summer, and over southern South America, southern Africa and along the Great Rift of Africa throughout the year– see Figures 4-1 to 4-4), form above surfaces made relatively hot by insolation through clear skies. Heating of the overlying air causes its pressure to fall to a relative minimum and a depression to form. Air converging into the heat low rises, but a strong subsidence inversion, reflecting convergence in the upper troposphere, allows only shallow clouds to form; this is quickly evaporated through mixing with very dry air from above the inversion. Axes of heat troughs have been termed "heat equators."

In the family of tropical vortexes (see Figure 6-32), heat lows/troughs are linked to surface anticyclones and upper tropospheric (cold) lows in which mid-

tropospheric air sinks and weather is fine. Over the deserts, as winter follows summer and as surface air cools and becomes denser, surface pressure increases and anticyclones replace heat lows. On the eastern and western edges of heat troughs, moister air and some clouds reduce insolation; the surface pressure minimum, especially over the oceans, is less marked. However, the troughs are still heat troughs, with the lowest pressure coinciding with the highest surface temperature. That, in turn, is ensured by a relative cloud minimum and light variable winds (oceanic doldrums) along the trough axis.

The term "monsoon trough" has been incorrectly applied to all persistent near-equatorial troughs with westerlies on the equatorward side and easterlies on the poleward side–refer to Figure 1-1. Although this usage is too well-established to be readily changed, readers should remember that in this report, "monsoon" troughs may be year-long, (as over South America and southern Africa), and not just summer features (as over North Africa).

Figure 1-3 (opposite) is a schematic model of the rainfall, pressure, and low-level circulation associated with the eastern North Pacific "monsoon" trough. The meridional section, based on satellite observations, reveals that maximum cloudiness and rainfall do not occur along the trough line, but in the westerlies south of the trough (see Figure 6-16). A slight secondary rainfall maximum is evident just north of the trough. Along the trough line, clouds are scattered or even absent. This confirms that the trough is really a heat trough where light, variable winds and fewer clouds allow the sun to maintain a relatively warm sea surface and relatively low air density. Similar distributions have been observed over central India (see Figure 6-19), in the south Indian Ocean (Ramage, 1971, 1974) and across West Malaysia (Lim, 1979).

Intertropical Convergence Zone (ITCZ)
Although this term is still widely used, its meaning has been hopelessly corrupted. Besides "ITCZ," the extensive low-latitude surface pressure trough has been called "equatorial trough," "intertropical front," "heat trough," and "monsoon trough." As shown in Figure 1-4 (next page), the ITCZ has been variously defined as coinciding with (a) the axis of the surface trough; (b) a zone determined by "wind observations" (Stechnovsky and Krouskova,1970); (c) "the fluence asymptote associated with the equatorial trough, without regard to cloudiness" (Simpson, 1967); and (d) the line of maximum cloudiness (Sadler, 1975a). Finally, the semantic impossibility of two ITCZs along the same meridian has also been perpetrated. This manual, then, replaces "ITCZ" with more specific terms: "heat trough," "monsoon trough" (already somewhat contaminated), and "near-equatorial tradewind convergence" (NETWC).

Maritime Continent This region includes Indonesia, Malaysia, Mindanao, the western Caroline Islands, Papua New Guinea, and the Solomon Islands. Along with equatorial Africa and equatorial South America, it is the source of most tropical thunderstorms (see Figures 10-1, 10-2, and 10-3). The large, high islands generate continental-like convection, while the warm surrounding ocean is an inexhaustible moisture source. The Maritime Continent is outlined in the locator chart (Figure 1-5a, Page 1-10).

Mid-Tropospheric (Subtropical) Cyclone This system originates in baroclinic environments and is most intense in the middle troposphere where horizontal temperature gradients are least. It weakens both upward and downward from this level because, relative to the environment, it is warmer in the upper troposphere and cooler in the lower troposphere. Although a vertical eye does not develop, mid-tropospheric cyclones may be mistaken for tropical cyclones in a weather satellite picture. Extensive nimbostratus with embedded cumulonimbus sometimes gives heavy rain. These systems are most common over the subtropical North Pacific and North

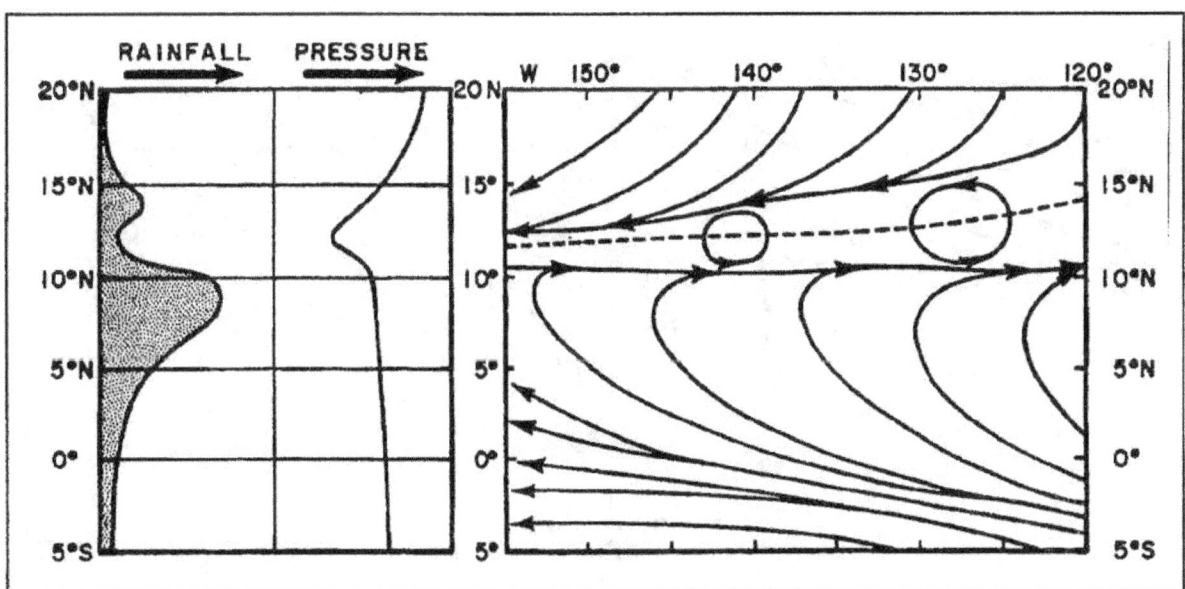

Figure 1-3. Model of the low-level summer circulation of the tropical North Pacific and associated meridional rainfall and pressure patterns. The dashed line marks the monsoon trough (after Sadler, 1967).

Atlantic (where they are known as "subtropical cyclones") and in summer over the northern Arabian Sea where the term "mid-tropospheric cyclone" is generally used. See Figures 8-7 through 8-16.

Monsoon Trough See Heat Low/Heat Trough.

Near-Equatorial Tradewind Convergence (NETWC) Throughout the year, surface pressure gradients are directed from east to west in near-equatorial parts of the central Pacific and eastern Atlantic/South America. Coupled with small equatorward-directed meridional pressure gradients, this ensures that the northeast tradewinds of the NH converge with the southeast tradewinds of the SH in a persistent zone of unsettled weather—the near-equatorial tradewind convergence. Over the oceans, the NETWC shifts between about 5° N in April and 10° N in September. Over northern South America, the annual range is from about 2° S to 10° N. Between February and May, a secondary NETWC sometimes forms near 5° S over the eastern Pacific—see Figures 8-55, 8-56 and 9-10.

Quasi-Biennial Oscillation (QBO) Winds in the tropical stratosphere undergo a downward-propagating oscillation between easterlies and westerlies on an average of once every 26 months (hence the name). The oscillation maximum occurs at 75,000 to 85,000 feet (23-26 km) above the equator and decreases downward and poleward. The cause is still debated. See Figure 4-17.

Rains/Showers These terms, as used in this manual, are given the following meanings:

• *Rains* fall from deep nimbostratus with embedded cumulonimbus when wind shear in the vertical and lower tropospheric convergence are both large. They occur in the circulations of tropical cyclones, mid-tropospheric cyclones, and monsoon depressions, as well as in squall lines and vigorous cloud clusters. Rains vary in intensity, but since skies stay mainly overcast, heaviest falls are most likely during the night and around dawn.

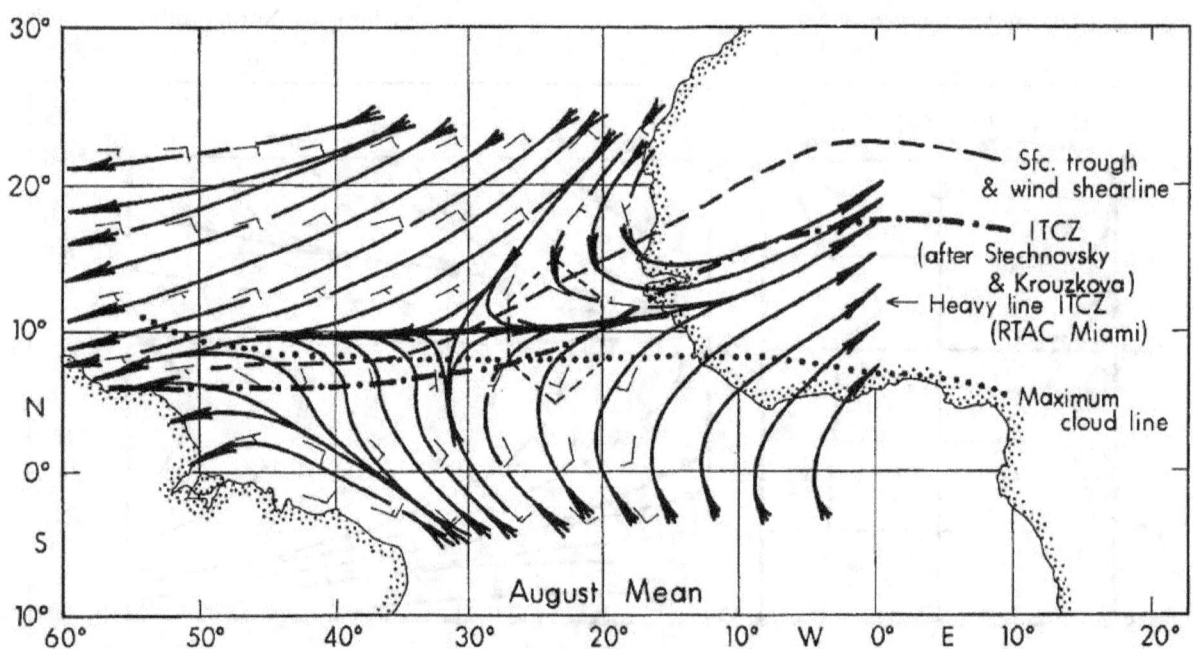

Figure 1-4. Four definitions of the "intertropical convergence zone" (ITCZ) over the tropical North Atlantic (Sadler, 1973).

• *Showers* fall from scattered cumulus or cumulonimbus when wind shear in the vertical and lower tropospheric convergence are both small. In the moist tropics, showers prevail when the weather is generally fair. Diurnal local winds, which develop readily during a showers regime, in turn determine the diurnal variation of the showers.

ResultantWind For climatological purposes, scalar quantities can be simply averaged; their standard deviations indicate variability. On the other hand, wind must be averaged vectorially by first resolving each wind observation into components from north and east, summing over the climatological period, obtaining the averages, and finally reconverting the average north and east components into a single vector—the resultant wind. The steadiness or persistence of the wind is defined as the ratio of the magnitude of the resultant wind to the average speed of the wind without regard to direction. It is usually expressed as a percentage. A resultant of winds all with the same direction would have a steadiness of 100 percent.

South Pacific Convergence Zone (SPCZ) Climatological charts show a zone of maximum cloudiness extending east-southeastward from New Guinea throughout the year. During winter, cloudiness is enhanced by troughs in the upper tropospheric westerlies that tend to become stationary just west of the zone. During summer, surface convergence in the zone is associated with a trough in which cyclogenesis occurs. Mechanical and thermal effects generated by Australia and the maritime continent probably cause the SPCZ. See Figures 6-1 to 6-4, 6-22, 8-41, and 8-42.

Surge Within persistent surface wind regimes that flow toward the heat equator, such as tradewind and monsoons, greater speeds may suddenly appear (surge) and last for a few days. Downwind of the speed maximum of the surge, convergence worsens weather and the associated vertical circulation accelerates. Above, in the upper troposphere, flow in the opposite direction from the surface winds is *divergent* and exports heat to colder latitudes. Surges have been linked to the development of tropical cyclones as well as monsoon depressions, squall lines, equatorial convection, and the onset of El Niño. Their causes seem to be varied and not well-observed or understood. See Figure 8-38.

Tradewind Trough The first edition of AWS TR 240 defined "tradewind trough" as the pressure trough along which the tradewinds of each hemisphere converge. Research aircraft traverses have since shown that the trough is very broad and weak, and that the convergence, which is much narrower and more well-defined, may not coincide with the lowest pressure. This edition, then, replaces "tradewind trough" with the more specific "near-equatorial tradewind convergence" (NETWC).

Tropical Upper Tropospheric Trough (TUTT) See Cold Low.

Walker Circulation Bjerknes (1969) identified a weak climatological feature in the equatorial zonal circulation. In this feature, surface flow from the east was lifted and gave rain over Indonesia, returned from the west in the upper troposphere and sank over the central Pacific. Others have postulated similar circulations associated with the equatorial African and South American thunderstorms. The Walker circulation is much weaker than the Hadley circulation and is often not apparent on the synoptic scale. See Figure 2-7.

1.3 UNITS OF MEASUREMENT

The World Meteorological Organization (WMO) has established a set of units, based on the centimeter-gram-second system, in which to express meteorological measurements. Ideally, this manual should conform. In the United States, however, surface air temperature is still regrettably measured and reported in degrees Fahrenheit (° F) instead of degrees Celsius (° C) and rainfall is measured in inches instead of millimeters (mm) or centimeters (cm). In aviation operations, heights are in feet instead of meters or kilometers (km), distances are in nautical miles instead of kilometers (km), and speeds are in knots instead of meters per second (m s^{-1}).

There has been some progress; upper-air temperatures and dew points in the U.S. are expressed in ° C, potential temperatures in K, and mixing ratios in gm Kg^{-1}.

Whenever U.S. units appear in the text, they are followed by the approximate metric equivalents set in brackets, except for speed; the transfer between knots and m s^{-1} can be readily approximated by applying a factor of one-half. When convenient, figures include both sets of units.

1.4 LOCATOR CHARTS

The map foldouts opposite should help readers in locating the geographical points of reference metnioned in the text. The "Maritime Continent" described on page 1-6 is outlined by the dashed lines on the map of the Eastern Hemisphere, Figure 1-5a.

Chapter 2

PRIMARY PHYSICAL CONTROLS OF THE TROPICAL CIRCULATION

2.1 GENERAL

The tropical circulation and resulting weather patterns are determined by a number of complex and related physical controls. The more observational data acquired, the more the complexity and variability of the tropical atmosphere become evident. These characteristics are vividly displayed by the global coverage of meteorological satellite data. In 1952, Palmer predicted this increase in apparent complexity with increasing data. Referring to new discoveries in tropical meteorology, often made possible in the past by observational requirements of wars or atomic weapons tests, Palmer wrote,

"The new regions thus opened up for exploration only vaguely resemble those preconceived by the theoreticians. It is not only that the griffins and basilisks described by the philosophers are absent, it seems that the country is occupied by creatures of which they have never dreamt."

Since 1952, much progress and many exciting discoveries have been made in tropical meteorology, such as the biennial oscillation in the tropical stratosphere, the detailed structure and influence of oceanic upper-tropospheric troughs revealed by jet aircraft and satellite observations, the preparation of detailed climatologies of the tropical wind field, cloudiness and rainfall, and studies of the tropical general circulation and energy balance, including the role of the monsoons and air-sea interaction during El Niño.

This chapter reviews the major physical factors controlling tropical climate. A grossly simplified description of the earth's heat-energy budget is presented to help in understanding the major driving forces of atmospheric circulations. This is followed by views of the tropical general circulation. Finally, factors controlling the tropical circulation and weather patterns are discussed. Subsequent chapters cover them in more detail. Lack of observations or physical understanding may prevent other influences from being specified.

2.2 EARTH'S HEAT-ENERGY BALANCE

According to Sellers (1965), net radiation at the Earth's surface (R) is given by:

$$R = (Q + q)(1 - á) - I$$

where ($Q + q$) is the sum of the direct and diffuse solar radiation incident on the earth's surface ("insolation"), á is the surface albedo (1 for a perfectly *reflecting* surface and 0 for a perfectly *absorbing* surface), and I is the effective outgoing radiation from the surface.

Over an average year, the terms balance globally. The earth absorbs solar radiation at a rate of about 164 watts m^{-2}. In turn, it radiates at 69 watts m^{-2}, leaving a net surplus of 95 watts m^{-2}. The atmosphere also participates in the exchange. The atmospheric net radiation (R_a) is given by:

$$R_a = (Q_a + q_a)(1 - á_a) - I_a$$

where ($Q_a + q_a$) is the total radiation incident on the atmosphere, $á_a$ is the atmospheric albedo, and I_a is the effective outgoing radiation from the atmosphere. For the earth/atmosphere system to remain in long-term radiation balance, $R = -R_a$, and the components of the atmospheric equation are:

$$-95 = (60 - 155) \text{ watts m}^{-2}$$

Thus, there is a net balance between the gain at the Earth's surface and the loss to the atmosphere. Although the annual global average radiation balance is zero, it will generally not balance (either seasonally or annually) in a given latitude zone. A schematic view of the radiation balance is given in Figure 2-1.

The atmosphere is uniformly a radiation sink at all latitudes, while the Earth's surface, except near the poles, is a heat source. The sum of the two, shown by the solid line in Figure 2-1, determines the radiation balance of the Earth-atmosphere system. To keep the earth from warming and the atmosphere from cooling, there must be energy transfer from the surface to the atmosphere. This vertical exchange occurs mainly by evaporation of water from the surface, as well as by condensation in the atmosphere, conduction of

sensible heat from the surface, and turbulent diffusion (convection) into the atmosphere. In Figure 2-1, since the horizontal axis scale is proportional to the Earth surface area in each latitude band, the area of surplus radiation equals the area of deficit radiation.

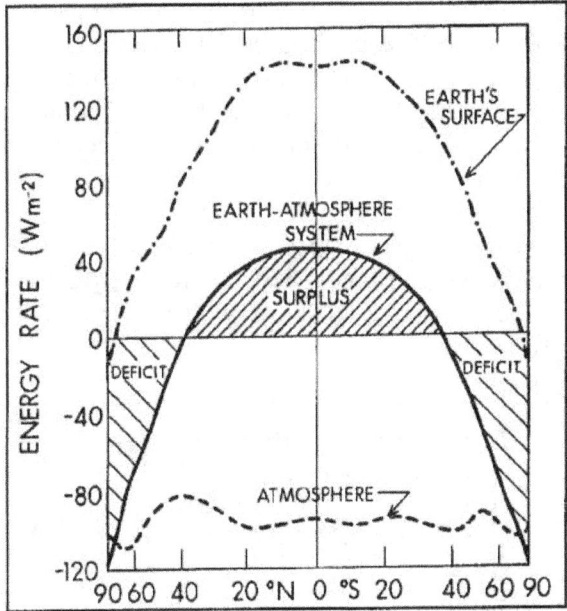

Figure 2-1. Average annual meridional distribution of the net radiation fluxes of the Earth's surface, atmosphere, and Earth-atmosphere system (from Sellers, 1965).

The net radiation is *positive* equatorward of 40° latitude and *negative* poleward of 40°. In order to keep the poles from getting colder and the tropics from getting warmer, energy must be transported from the tropics to higher latitudes. This horizontal heat exchange is carried out by (1) the poleward transfer of sensible heat in the atmosphere, (2) latent heat of condensation subsequently released and carried poleward in the atmosphere, and (3) oceanic circulations. Over the tropical oceans, the latent heat transfer is an order of magnitude larger than the sensible heat transfer. Malkus (1962) therefore concluded that the atmosphere is fueled mainly from below, with 80 percent (global average) of its heat energy initially latent in the form of water vapor. Considerably more than half of this latent heat energy is supplied to the atmosphere by the tropical oceans between 30° N and 30° S.

Figure 2-2 shows the annual water balance in the Earth-atmosphere system. Poleward of about 40° and within about 10° of the equator, precipitation exceeds evaporation; the reverse is true in the subtropical latitudes (about 10 to 40°).

On a long-term basis, regions with excess evaporation must *export* water vapor, while regions with excess precipitation must *import* it. Overall, the oceans lose more water by evaporation than they gain by precipitation; the deficit is made up by runoff from the land areas where precipitation exceeds evaporation.

As Figure 2-2 shows, evaporation is greatest in the subtropics, where it occurs mostly in relatively dry tradewinds. Riehl and Malkus (1957) showed that

regions of oceanic tradewinds provide a significant proportion of the energy in the form of latent and sensible heat needed to drive the atmospheric circulation. This energy, primarily in the form of water vapor, is initially transported equatorward by the relatively steady tradewinds near the Earth's surface. In the equatorial zone, water vapor is lifted and condensed in large, convective cloud systems; they produce sensible heat and potential energy, which are then moved to higher latitudes by the upper tropospheric circulation.

In another study of the heat balance in the equatorial zone, Riehl and Malkus (1958) estimated that only a very small fraction of a 10° latitude belt near the equator needs to be occupied by giant cumulonimbus (Cb) clouds or "hot towers" to maintain the heat

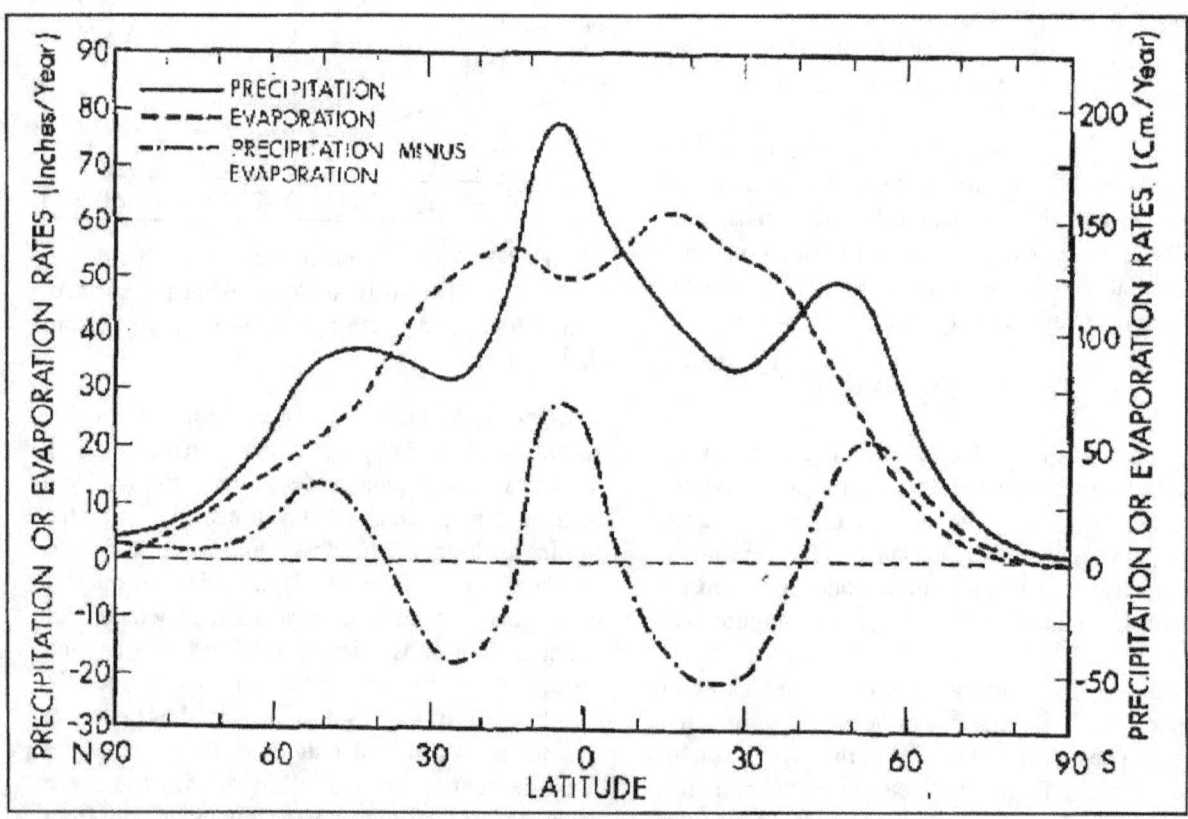

Figure 2-2. Average annual meridional distribution of precipitation, evaporation, and precipitation minus evaporation (from Sellers, 1965).

budget of the equatorial zone and provide for much of its poleward energy transport. They suggested the following descending hierarchy of the fractional area occupied by the various scales of phenomena in the equatorial zone:

- Area of near-equatorial trough zone = A

- Area occupied by synoptic disturbances = 0.1A

- Area occupied by active rain = 0.01A

- Area occupied by undilute towers = 0.001A

Thus, they estimated that at any one time about 30 synoptic disturbances containing a total of about 1,500 to 5,000 giant Cbs is sufficient for heat balance requirements.

Garstang's group at the University of Virginia has been calculating the heat budget for the Amazon basin (see Greco et al., 1990); they suggest that requirements for a global heat balance can be met by only 6-15 synoptic systems and fewer than 1,000 giant Cbs. As mentioned previously, a latitudinal heat-energy balance is achieved by atmospheric and oceanic circulations. Assuming no net heat flux across the poles, Figure 2-3 shows the net annual transport of various components of the heat-energy budget across each latitudinal circle. North of about 5° N, the sensible heat-energy (sensible heat plus potential energy) transport by atmospheric and oceanic currents is northward; south of 5° N, it is southward.

While atmospheric sensible heat-energy transport shows double maxima in each hemisphere, the oceanic sensible heat-energy transport shows a single maximum in the subtropics of each hemisphere. The transport of latent heat energy by atmospheric circulations is more complex. Poleward of 20-25° in each hemisphere, there is net transport of water vapor poleward; equatorward of these latitudes, water vapor is transported toward the near-equatorial tradewind convergence (NETWC), located on average near 5° N. The total poleward heat-energy transport required to

balance the net radiation deficit in higher latitudes combines the three components discussed above, as shown at the bottom of Figure 2-3; it is greatest near 40° N and 40° S. Overall, there is a net energy transport from the tropics by atmospheric and oceanic circulations, and a net import to higher latitudes.

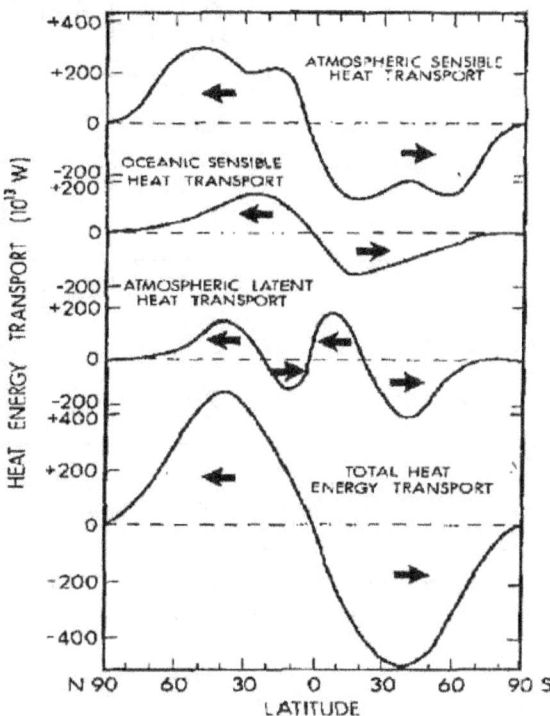

Figure 2-3. Average annual meridional distribution of the components of the poleward heat-energy transport (from Sellers, 1965).

Vonder Haar and Oort (1973) used satellite radiation measurements to estimate total poleward heat transport. They differed little from Sellers (1965), whose estimates are shown in Figure 2-3. But because the tropical oceans absorb more radiant energy (lower albedo) than previously thought, ocean currents transport a larger proportion of sensible heat across the subtropics, as shown in Table 2-1. Over the globe, about 40 percent of the total required meridional heat transport may be accounted for by ocean currents. Hastenrath (1985) discusses heat and water budgets in detail.

TABLE 2-1. Poleward transport of heat energy by ocean currents according to Sellers (1965) and Vonder Haar and Oort (1973), with percentages of the total (air plus ocean) fluxes (units of 10^{13} W).

METHOD	LATITUDE										
	90 N	80 N	70 N	60 N	50 N	40 N	30 N	20 N	10 N	0	10 S
Sellers	0	0	12	34	75	104	152	157	107	-21	-13
% of total	0	0	7	11	17	20	31	47	66	70	60
Vonder Haar/Oort	0	-7	5	5	113	206	244	340	187	26	-13
% of total	0	2	4	2	28	41	48	74	66	61	6

Besides meridional exchanges of heat, *regional* exchanges also occur. Alexander and Schubert (1990) applied the method of Vonder Haar and Oort to data from FGGE; they found that in winter the mid-latitude oceans export heat to the continents, and that the flow is reversed in summer. Not surprisingly, the exchanges are much larger in the Northern than in the Southern Hemisphere.

14

2.3 TROPICAL GENERAL CIRCULATION

The expression "general circulation of the atmosphere" refers to a statistical description of the mean large-scale atmospheric motions over the whole Earth. These statistics are derived from many daily flow patterns and include not only mean conditions but also the variability of the flow resulting from seasonal changes and the effects of transient cyclones and anticyclones (Huschke, 1959).

Many general circulation studies have been made, especially since the worldwide expansion of upper-air networks during and after World War II. Lorenz (1967) made an excellent survey. A group of Massachusetts Institute of Technology (MIT) investigators (Newell et al., 1972) concentrated on the general circulation of the tropics. They processed upper-air data for about 300 stations between 45° N and 30° S for the period July 1957 to December 1964. The tropospheric results were obtained from objective analyses at various pressure levels of the long-term station means and the stratospheric results from latitude-band means, giving equal weight to data from odd and even numbered years to eliminate the effects of the quasi-biennial stratospheric wind oscillation (see Chapter 4).

Mean meridional cross-sections were computed from a linear combination of station values for 10° latitude bands using a weighting scheme to reduce the bias caused by more observations over the continents. These cross-sections present extremely simplified views of the tropical general circulation because of very large east-west departures from the means. For example, during July along 10° N, there are regions of very strong easterly and westerly resultant winds near the surface, but the resultant wind for all meridians is very light. Figure 2-4 shows the mean vertical motion at 500 mb for each season (Kyle, 1970).

In general, motion is upward near the equator and downward in the subtropics. But mean vertical motion varies greatly along latitude circles. Over the continents, the large annual variation is associated with the monsoons. Otherwise, centers of upward motion lie over the continents and centers of downward motion lie over the oceans and the deserts. Remember, too, that time-averaging greatly reduces the intensity of vertical motion. The largest values in Figure 2-4 are less than 20×10^{-4} mb s^{-1}, or 30 mm s^{-1}, two orders of magnitude less than vertical motions inside cumulonimbus.

Where climatological data is sparse, these patterns may have to be revised; for example, the satellite cloud climatologies of Figures 6-1 through 6-4 show cloud along the North Pacific NETWC, whereas Figure 2-4 depicts *sinking* motion.

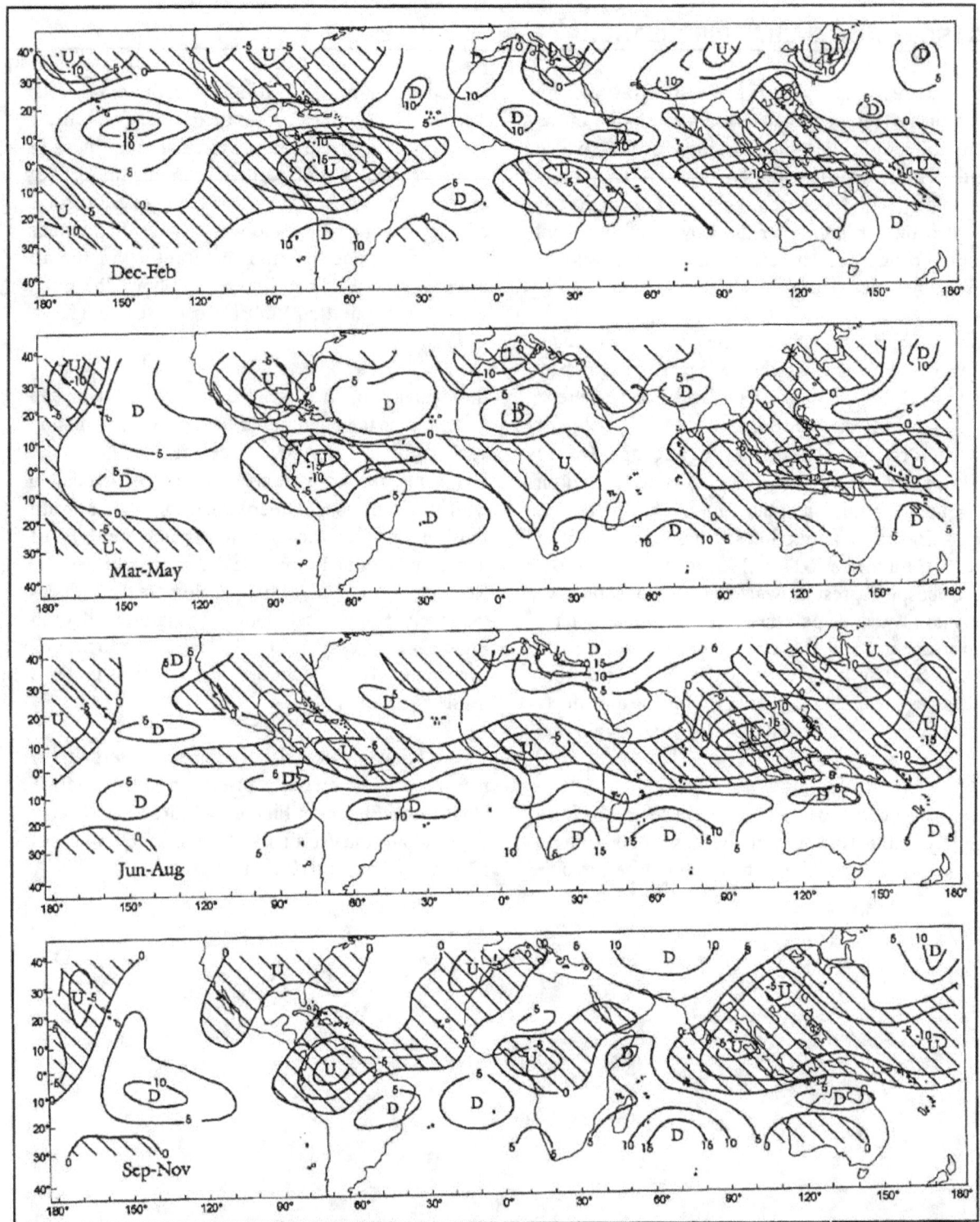

Figure 2-4. Mean vertical velocity in units of 10^{-4} mb s^{-1} or 1.5 mm s^{-1} at 500 mb in the tropics. Areas of upward motion are hatched (from Kyle, 1970).

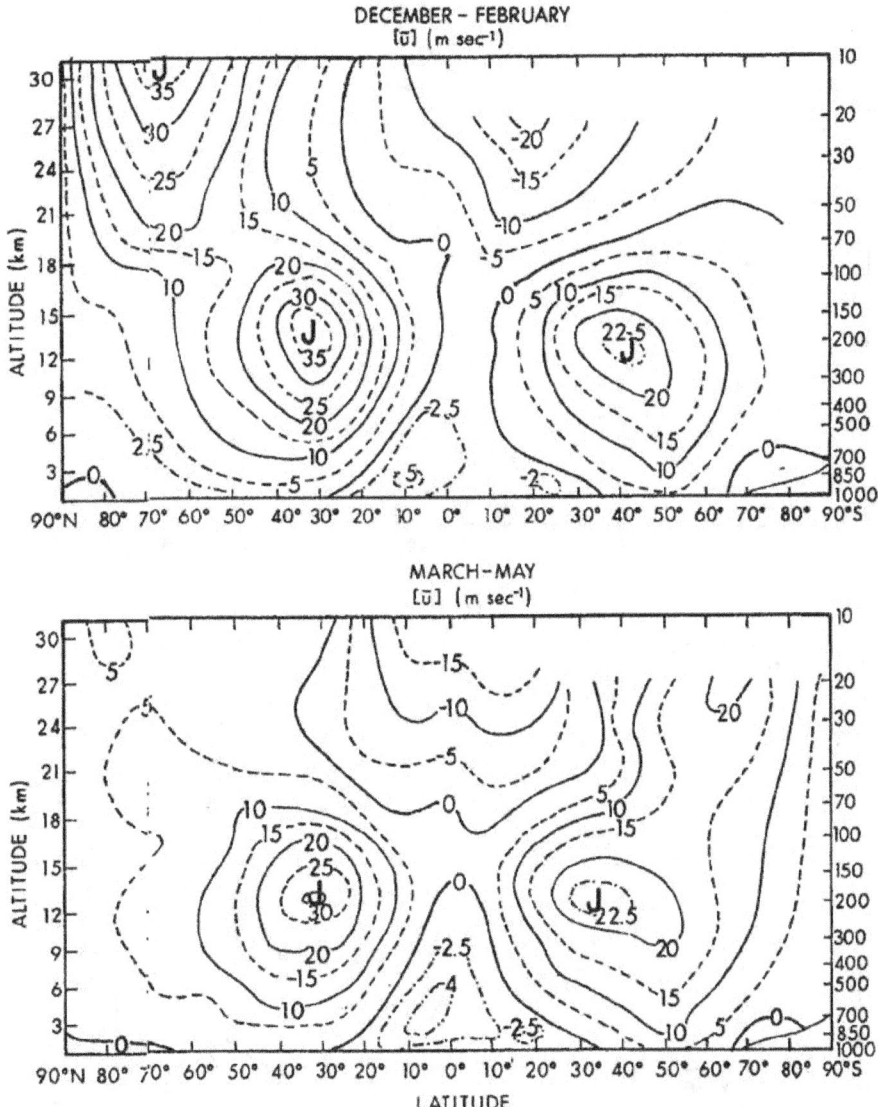

Figure 2-5a. Meridional cross-sections of zonal wind speed (m s⁻¹); westerlies positive; subtropical jets are labeled "J" (from Newell et al., 1972).

Figures 2-5a (above) and 2-5b (next page) show the mean zonal wind cross-sections from the MIT studies for each season . The major features are the regions of westerly flow above the near-surface layer in the subtropics reaching maxima in the subtropical jet streams near 200 mb, and the deep easterly flow in latitudes.

At the surface, the subtropical ridges, which separate easterlies from westerlies, shift about 5° poleward between winter and summer. On average, the ridges are located at about 30° N and 30° S, slightly equatorward of the boundary between heat surplus and heat deficit—see Figure 2-1. Near 200 mb, the ridges are 30° apart from June to August and only 15° apart from December to February.

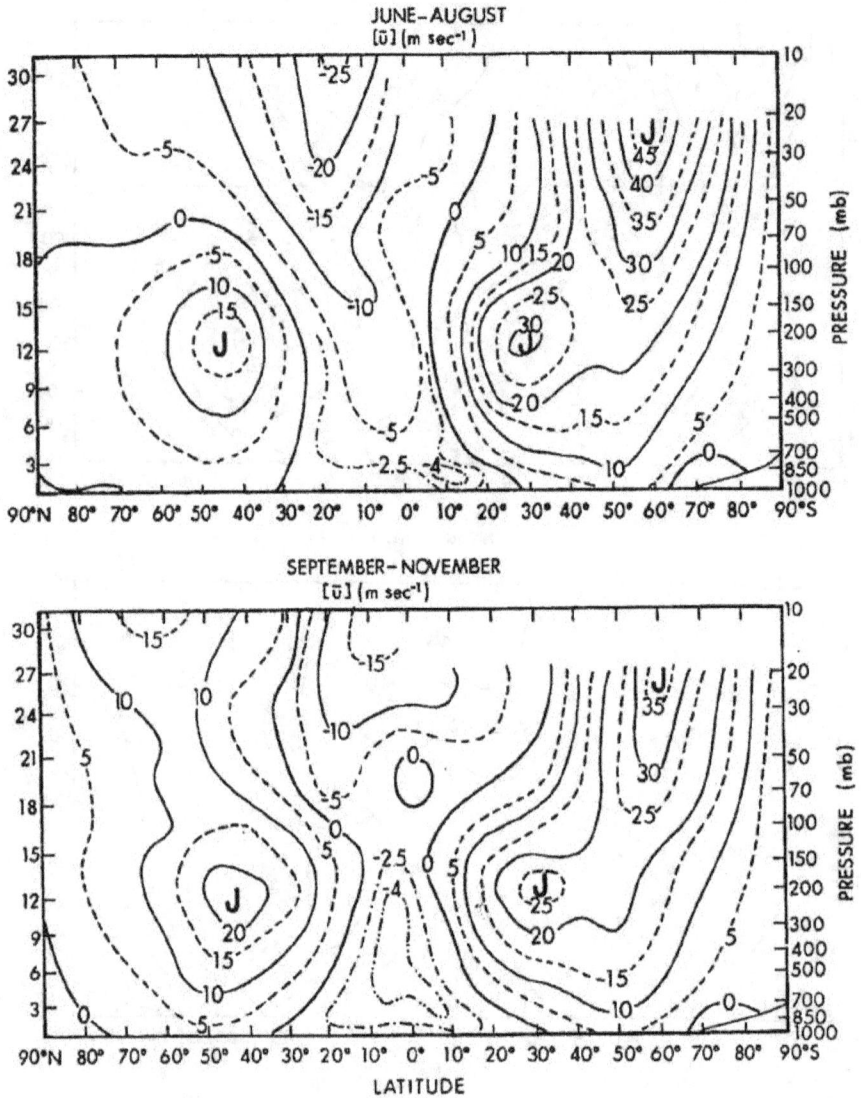

Figure 2-5b. Meridional cross-sections of zonal wind speed (m s⁻¹); westerlies positive; subtropical jets are labeled "J" (from Newell et al., 1972).

By determining the mean meridional motion at various levels, averaging them around latitude circles, and integrating them with respect to pressure between 1,000 and 100 mb (assuming zero vertical motion at the upper and lower boundaries), the mean vertical motion field can be determined from the continuity equation. However, readers should remember that the circulation at any one time may differ greatly from the long-term average.

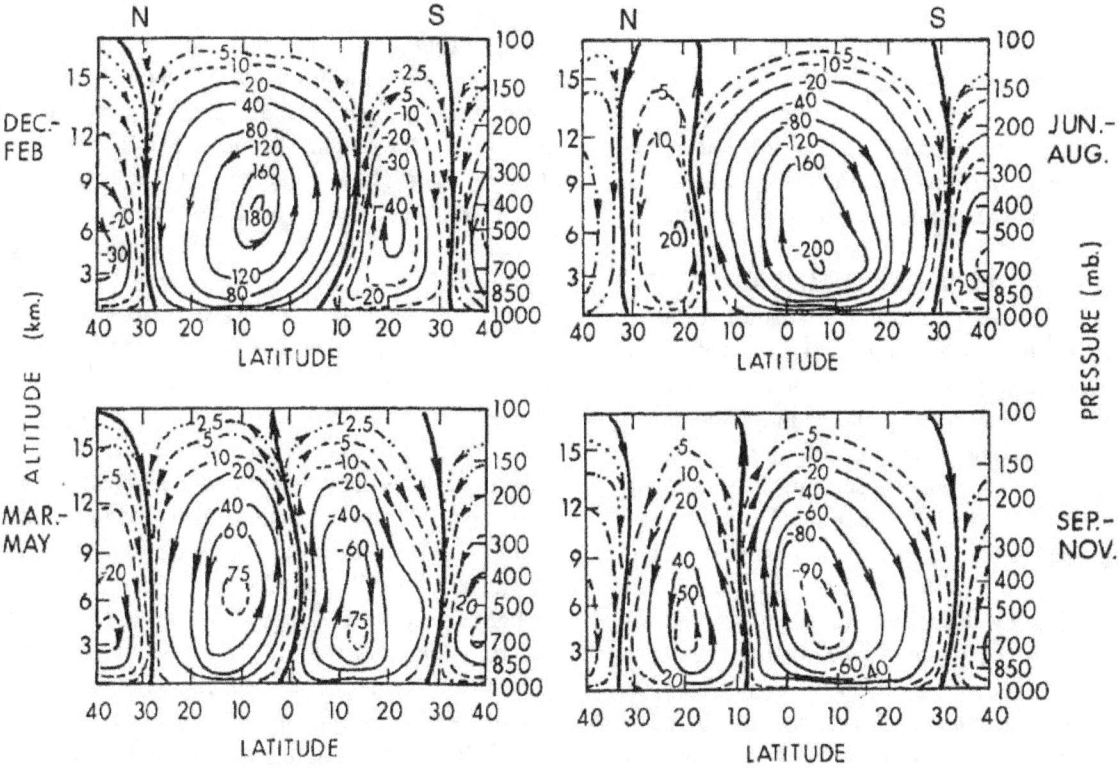

Figure 2-6. Meridional cross-sections of the mean meridional circulation in the tropics in the form of mass-flow streamlines. Units: 10^{12} g s^{-1} (from Newell et al., 1972).

Figure 2-6 depicts the mean meridional circulations and resultant vertical motion fields between 40° N and 40° S in the form of mass-flow streamlines. The spacing of the streamlines is inversely proportional to the strength of the flow; total transport in each cell is given by the labeled values in the center. Mass continuity demands that over the globe the streamlines must be closed. Cellular patterns are evident, with rising motion near the equator and sinking motion in the subtropics. Hadley (1735) postulated these cellular circulations to account for the tradewinds. The mean meridional circulation is strongest in the winter hemisphere, when upward motions of the Hadley cells extend across the equator into the opposite hemisphere. The Hadley cells, in large part, maintain the strong subtropical jet streams of the winter hemisphere through poleward momentum transport in the upper troposphere; they are also responsible for equatorward latent-heat transport in the lower troposphere, as described previously. Hadley cells are

"direct" circulations, with warmer air rising and cooler air sinking. Their primary energy source is the latent heat of condensation released in the rising branches of the cells. Short-period fluctuations, which, when averaged, comprise a Hadley cell, are discussed in 8.4.4.

Mean meridional temperature cross-sections produced by MIT investigators show nearly uniform temperature in the tropical middle troposphere; between 15° N and 15° S, the annual range in mean temperature is only about 1° C. The tropospheric temperature maximum (heat equator), though not pronounced, lies near 5° S during the southern summer and near 20-25° N during the northern summer. This asymmetry results from stronger surface heating over the Northern Hemisphere, with its larger continental areas, and the cooling effect of Antarctica.

2.4 PRIMARY PHYSICAL FACTORS

2.4.1 General. The preceding sections discussed the role of the major energy conversion processes (e.g., radiation, precipitation, and evaporation) in the tropical large-scale circulation. This section covers the more important physical factors controlling or modifying tropical circulation and weather patterns.

2.4.2 Land and Water Distribution. The distribution of land and water affects all scales of motion in the tropics. The large land masses of Africa, Asia, and Australia experience a much larger annual cycle of radiational heating and cooling than do the neighboring oceans, where specific heat is higher. Consequently, pronounced seasonal (monsoonal) wind regimes dominate much of the Eastern Hemisphere tropics. Annual wind changes are less pronounced on the Western Hemisphere land masses. The shifts in these seasonal wind patterns stem from shifts in the mean positions of the low-level monsoon and heat troughs over and near the continents. The frequencies of tropical cyclones, which develop in the troughs, are affected in turn. Over tropical oceans, away from continental influences, the NETWC, where easterly tradewinds from both hemispheres meet, has only a small annual movement.

Although latent heat of *evaporation* is mainly added to air over the ocean, the release of latent heat of *condensation* is greatest over tropical land.

Ramage (1968) has shown that because of frequent thunderstorms, the near-equatorial regions of South America, Africa, and Indonesia generate much more heat for export to higher latitudes than do the low-latitude oceans. The longitudinal distribution of these concentrated heat sources produces large meridional temperature gradients and results in the strong subtropical jet streams shown in Figures 2-5a and b. The heat sources also cause weak vertical circulations along the equator. Bjerknes (1969) identified a cell in which air rose over Indonesia and sank over the western Pacific. He dubbed it "The Walker Circulation." Flohn (1971) extended the concept around the globe, identifying upward branches of the circulation cells over the continents and Indonesia and sinking branches over the oceans. But in the equatorial Pacific, where the Walker Circulation is presumably strongest, data barely supports its existence. At the surface, easterlies extend from 100° W to 150° E, while upper tropospheric westerlies are not found west of 180°. Not surprisingly, others have disagreed with Flohn; Newell's more complicated 1979 view (Figure 2-7) fits the upper tropospheric and surface wind systems, but leaves a gap between 170° E and 130° E, where large interannual variations associated with El Niño occur. The Walker Circulation may be a useful concept and will be revisited in Chapter 6, but forecasters should beware of putting it on a par with Hadley cells.

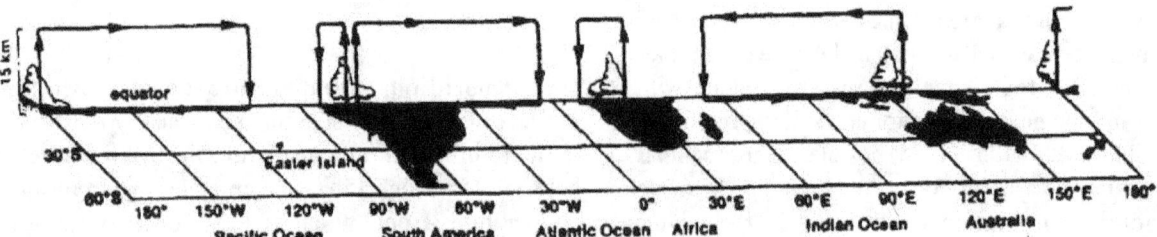

Figure 2-7. Zonal circulation cells along the equator (from Newell, 1979).

The Tibetan plateau is a high-level radiational heat source throughout the year. In summer, with condensation along the Himalayas providing powerful assistance to the effects of radiation, this area plays a significant role in developing and maintaining the high tropospheric anticyclone over Asia and the resulting easterly jet stream south of Asia. In turn, the vertical motion patterns associated with this jet contribute to the extreme aridity over North Africa and Arabia and the heavy rainfall over India and Southeast Asia (see 6.2). The Bolivian Altiplano of South America plays a similar role in initiating, then sustaining, western Amazon basin convection. On much smaller scales, the thermal gradients associated with topography produce local circulations such as land and sea breezes over islands and along coasts, land and lake (or river)

breezes around inland water bodies, and mountain and valley breezes (see Chapter 7). Local topography and distance from moisture sources largely determine rainfall in the tropics.

Some of the heaviest mean annual or mean monthly rains fall at stations on the windward sides of mountains exposed to prevailing flow off a large body of water. The top of Mt. Waialeale, on Kauai, Hawaii (22° N, 159° W) is exposed to the tradewinds at 5,075 feet (1,547 meters) elevation; it gets an average of 449 inches (11,415 mm) a year. Conversely, stations leeward of large mountains are dry, reflecting the rainshadow effect associated with subsiding downslope winds.

2.4.3 Stress-Differential Along Coasts.

According to Bryson and Kuhn (1961), air flowing parallel to a coast is slowed frictionally more by the land than by the sea; as a result, surface winds over land cross isobars toward lower pressure at a greater angle than over the sea. Therefore, if pressure is lower over the land (as it is in all upwelling regions), winds diverge and air subsides at the coast. This is partly why vigorous upwelling borders coastal deserts (see Figure 5 4). If pressure is higher over the land, surface winds converge at the coast; this condition often accompanies continental anticyclones. Figure 2-8 (Fett and Bohan, 1986) shows the processes.

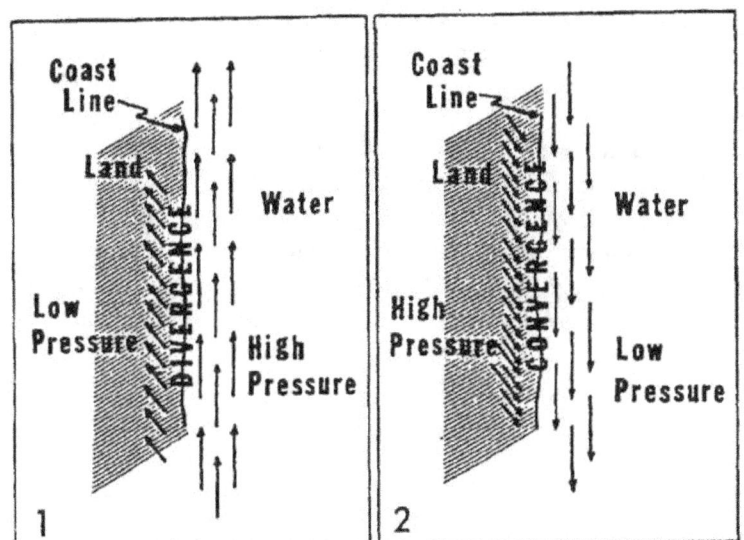

Figure 2-8. Schematic of the effects of coastal friction on low-level wind flow (Fett and Bohan, 1986).

2.4.4 Sea-Surface Temperature.

Sea-surface temperature (SST) has a significant influence on atmospheric circulation and weather systems. Because heat and moisture are exchanged at the air-sea interface, SST largely determines the air temperature and moisture distribution in the surface layer over the tropical oceans. Over warm oceans, latent heat is released by lifting of the surface air in the convective clouds that may precede development

of warm-cored tropical cyclones; some intensify into hurricanes. If hurricanes move into higher latitudes, lower SSTs may deprive them of the latent heat source necessary to maintain the intense circulation. Hurricanes also weaken on moving over land, more the result of the cutoff of their latent and sensible oceanic heat source than from increased surface friction (see 9.6). Intense tropical cyclones modify the distribution of sea-surface temperature significantly. Leipper (1967) studied the effects of Hurricane Hilda (October 1964) in the Gulf of Mexico. Over an area 50 to 100 NM (100 to 300 km) wide near the hurricane path, SST decreased more than 9° F (5° C) when the storm circulation increased mixing and upwelling of cooler subsurface water. The sea surface stayed cooler for at least 3 weeks. More recent measurements have confirmed this effect. Evaporation and conduction beneath early winter cold, dry polar outbreaks can also significantly lower SST.

El Niño, which occurs every few years in the equatorial Pacific, exemplifies the effect of sea-surface temperature on weather. Coastal Peru is normally arid because cold-water upwelling near the coast reinforces the effect of strong atmospheric subsidence. During El Niño, however, when a warm current prevails, copious rain and floods may occur (Caviedes,1975). See 6.3.

2.4.5 Interactions with Mid-latitude Flow. Mid-latitude circulations often affect the tropics. These interactions are vividly illustrated by some weather satellite pictures in which continuous cloud bands extend far into the tropics (see Figure 8-57). Subtropical cyclones develop from upper-level lows which become cut off from the mid-latitude westerlies. In turn, mid-latitudes are affected by recurving tropical cyclones. You may infer from this that "the tropics" cannot be simply defined. The area has been specified variously as:

• Lying between the tropics of Cancer and Capricorn.

• Enclosed by the 68° F (20° C) mean annual isotherm.

• Where average temperature exceeds 20° C in all 12 months.

• Where there is a heat surplus in the Earth-atmosphere system.

These definitions encompass similar areas. Some regions experience a marked annual variation. On the south coast of China, for example, Hong Kong (22° N, 114° E), with a July mean temperature of 83.1° F (28.4° C), indubitably lies in the tropics in summer. However, its January mean temperature of 59.7° F (15.4°C) is scarcely tropical. Suffice it to say that tropical meteorologists should not confine their attention to circumscribed latitudes. They must comprehend the dynamics of mid-latitude circulations and extend their analyses poleward enough to identify and interpret mid-latitude effects on the tropical atmosphere.

2.5 SMALL-SCALE CONTROLS

Tropical circulations are also affected on small scales, primarily those associated with cumulus convection. The role of near-equatorial cumulonimbus has already been stressed. Gray (1970) emphasized their importance for vertical momentum transport and the production of kinetic energy for export to higher latitudes. Cumulus convection can affect a tropical squall line in two ways (Zipser, 1977): For one, highly unsaturated downdrafts are produced by entrainment of cool dry air into middle levels of the cloud line. The leading edge of the downdraft air acts like a cold front, lifting warm moist air and enhancing convection. For another, the downdraft air in the rear of the squall line may eventually become organized over the whole extent of the system and effectively suppress convection there for several hours (see Chapter 8). The direction of movement, environmental wind field, and interacting synoptic-scale influences of these squall lines vary in different parts of the world.

2.6 SUMMARY

The examples given here illustrate interacting atmospheric controls operating at a variety of scales. Although there are some exceptions (such as heat lows), the size and duration of systems affecting the tropics are roughly correlated, as Figure 2-9 suggests.

At one end of the range is the quasi-biennial oscillation (QBO) that covers the equatorial region and follows a 26-month cycle. At the other end are the intense, brief and narrow rain plumes in tropical

cumulus than can be detected only by fast-response instruments (Fullerton and Wilson, 1975).

Complicated interactions between scales must be considered in tropical weather analysis and forecasting. At present, most of the influences can only be evaluated subjectively. Despite some progress, it will probably be many years before sub-synoptic scale motions, which are greatly influenced by convection and topography, can be usefully predicted by numerical methods.

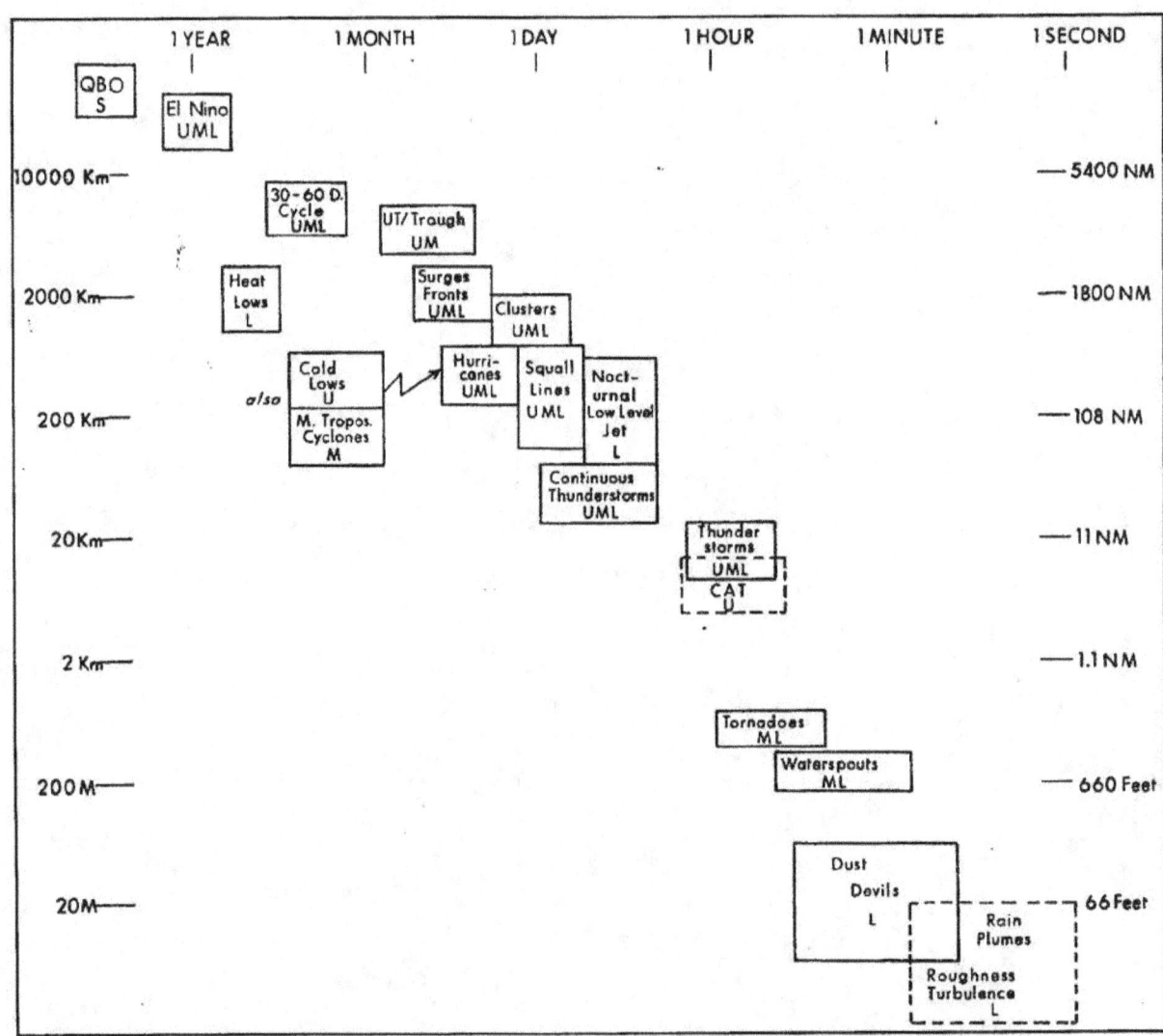

Figure 2-9. Approximate dimension and time scales of tropical systems (after Orlanski, 1975). Key: S -mainly stratosphere; U - mainly upper troposphere; M - mainly middle troposphere; and L - mainly lower troposphere.

Chapter 3

THE OBSERVATIONAL BASIS

3.1 CURRENT DATA SOURCES

A wide variety of data sources is available to tropical meteorologists for analysis, forecasting, and research. This chapter offers a broad review of these sources and their locations. Distribution is a problem. Some areas, such as South Asia, Australia, the western Pacific, and parts of Africa are reasonably well covered with surface and upper-air stations. However, over vast ocean areas and parts of South America, the data is inadequate.

The weather reporting system is organized into six regions by the World Meteorological Organization (1986); the boundaries of those regions are shown in Figure 3-1. Most of the stations in Regions I (Africa), III (South America), and V (Southwest Pacific), and some of the stations in Region II (Asia), and IV (North and Central America), are in the tropics. Therefore, statistics on total number of desired stations and hours of observations in each region given in the following sections will have to be considered with this limitation in mind.

3.1.1 Surface Reports from Land Stations. The current surface synoptic network in the tropics is shown by the stations on the latest available Department of Defense (DoD) Weather Plotting Charts (WPCs). DoD-WPCs 2-15-11 and 2-15-13 are 1:15 million Mercator projections for the region 60° N to 60° S. The hours that surface observations are taken vary, but most stations report on at least 3 of the 4 major synoptic hours (00, 06, 12, and 18Z) and most report intermediate 3-hourly observations.

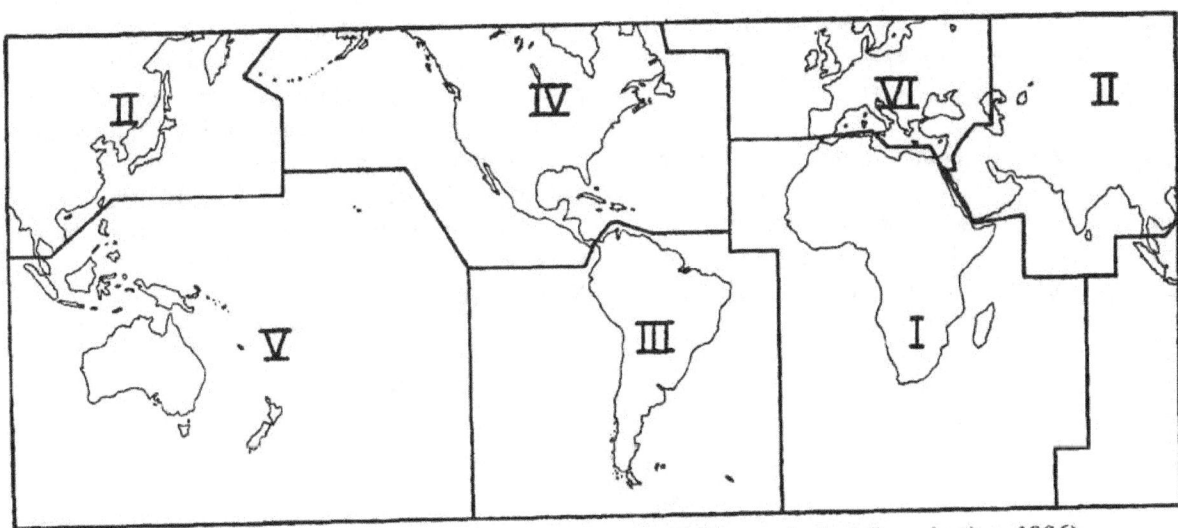

Figure 3-1. Boundaries of the six WMO regions (World Meteorological Organization, 1986).

TABLE 3-1. Implementation of WMO synoptic observing program, 1988 (from WMO, 1989).

Region	Observations Requested	Observations Fully Achieved	Percent
RA I: Africa	704	385	54
RA II: Asia	1172	1091	93
RA III: South America	348	158	45
RA IV: North and Central AMerica	584	421	72
RA V: Southwest Pacific	356	152	43
RA VI: Europe	842	812	96

In many tropical areas there are enough surface synoptic reports for analysis, at least at the 6-hourly synoptic times. However, compared to higher latitudes, smaller weather systems and fewer observations in the tropics limit diagnostic and prognostic information. Table 3-1 compares the number of synoptic observing stations requested by WMO to the number achieving the full observing program in 1988 (WMO, 1989).

There has been little change in the past few years; predominantly tropical regions have been the worst in achieving the WMO's goal for more observing stations. Besides operating first-order surface synoptic stations that transmit real-time data, many tropical countries operate lesser-order synoptic and climatological stations. These observe various meteorological elements (especially rainfall) and periodically send the records to central offices of their national meteorological services.

Climatological stations are operated by agricultural or industrial interests. The non-synoptic data they produce can be used in local-area synoptic studies, in preparing climatological maps, and in other research. In some tropical countries, military weather services and airlines also make observations and transmit them over local military and aviation circuits; these sometimes get into the Air Force's Automated Weather Network (AWN).

3.1.2 Surface Reports from Ships, Moored Buoys, and Automatic Island Stations. In the belt between 30° N and 30° S, about 75 percent of the surface is covered by water. Ship reports are therefore indispensable to tropical analysis and forecasting. Figure 3-2 shows the average daily number of ship reports collected by the National Meteorological Center (NMC) during June 1988 for 4-degree grid areas between 30° N and 30° S. Note the vast empty spaces.

In the Gulf of Mexico, in the waters east of Florida, and around Hawaii, moored weather buoys (shown by asterisks in Figure 3-2) transmit a wide range of meteorological and oceanographic data. Seven land-based automatic weather stations (shown by deltas in Figure 3-2) are now providing important data from remote tropical islands. The Departments of the Navy, Commerce, and Interior plan to install 20 more such stations in the western Pacific. The data resembles that received from moored buoys.

Of the 110,117 radioed ship observations received in June 1988, 30,071 (27 percent) were made between 30° N and 30° S. Of these, 24 percent were made at 00Z, 25 percent at 06 and 12Z, and 22 percent at 18Z.

Circuit overcrowding and deadlines prevent at least 40 percent of ship observations from ever being transmitted on the World Meteorological Organization's Global Telecommunications System (GTS) circuits that interface with the AWN.

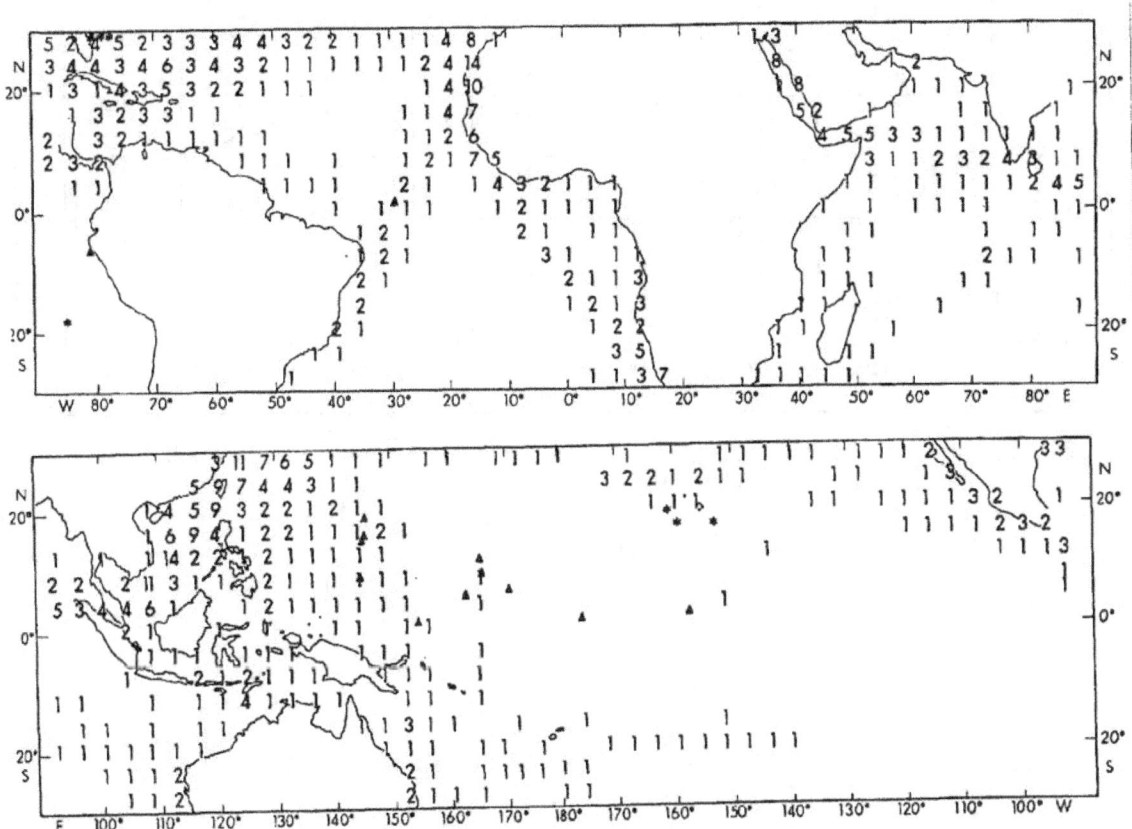

Figure 3-2. Average daily number of shipboard surface weather observations received at the National Meteorological Center in June 1988, totaled for 4-degree latitude/longitude rectangles. Moored weather buoys are indicated by asterisks (*). Deltas (Δ) show locations of automated island weather stations.

About 15 percent of all radioed reports contain errors, most commonly in position and sea-level pressure, the latter due to poor calibration. In critical weather situations, however, analysts can often detect and correct these errors by following individual ships from chart to chart.

Some tropical forecasting units may be required to analyze sea-surface temperature (SST), particularly when longer-range forecasts are made. Because SST changes slowly, SST data from ships and satellites is compiled over several days, then used at NMC and the Fleet Numerical Oceanography Center (FNOC) to prepare their widely disseminated SST analyses.

3.1.3 Surface Reports from Drifting Buoys. In June 1988, WMO (1989) reported that 200 drifting buoys around the world transmitted about 2,000 weather reports over GTS every day. Figure 3-3 shows the distribution of drifting buoys over the tropical Pacific during August 1989 (Climate Analysis Center, 1989). An encouraging number of the buoys is in the Southeast Pacific. Measurements made on the much smaller buoys are more representative of open ocean conditions than measurements made by ships.

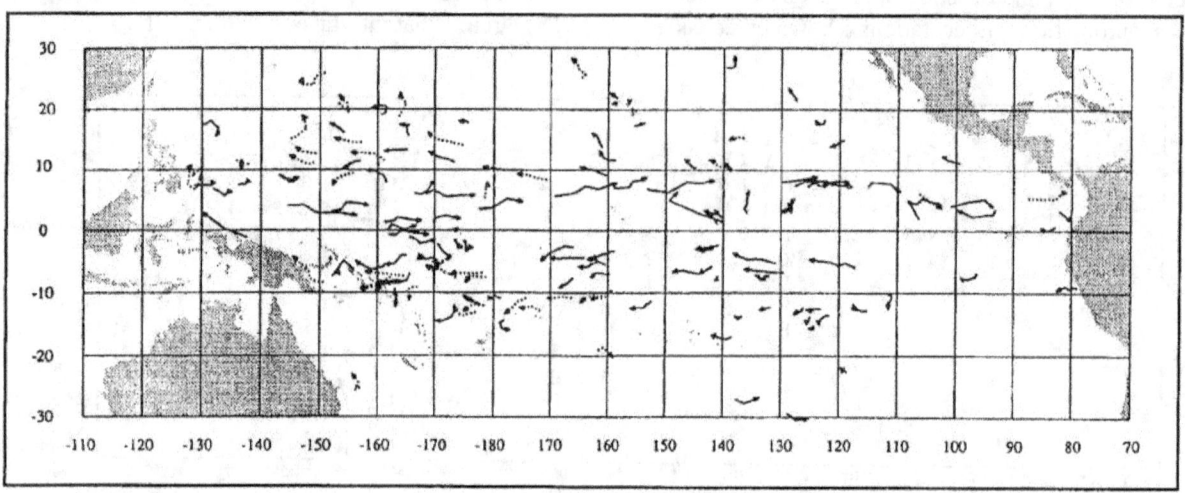

Figure 3-3. Drifting buoy observations in the tropical Pacific during August 1989. Lines indicate weekly displacements (Climate Analysis Center, 1989).

3.1.4 Upper-Air Data. In the tropics, lack of sufficient rawinsonde/PIBAL upper-air observations is often a serious problem. Over the oceans there may not be a single vertical sounding in millions of square kilometers. Worldwide radiosonde observations are made almost exclusively at 00 and 12Z, whereas radiowind and pilot balloon observations are made at 6-hourly intervals at 00, 06, 12, and 18Z. There has been little change in recent years (WMO, 1989). The performance pattern shown in Figure 3-4 resembles that of Table 3-1 and is further reflected in Table 2, page 3-6.

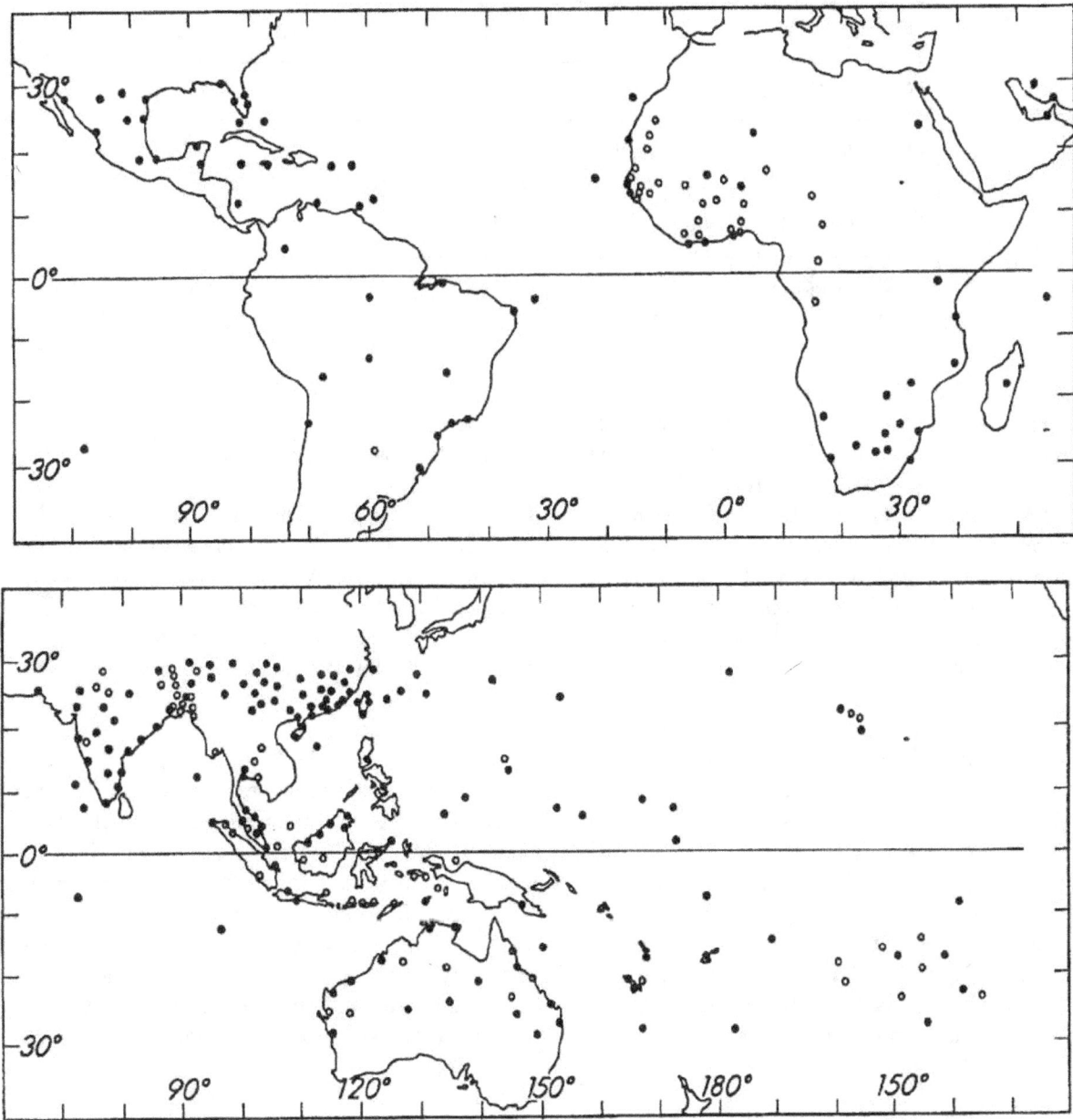

Figure 3-4. Tropical stations making radiosonde (•) and RAWIN/PIBAL (o) observations at the beginning of 1991 (at least one observation a day on two-thirds of the days).

TABLE 3-2. Implementation of WMO radiosonde observing program, 1987 (from WMO, 1989)

Region	Observations Requested	Percent Implemented
RA I: Africa	99	50
RA II: Asia	325	92
RA III: South America	58	56
RA IV: North and Central America	153	95
RA V: Southwest Pacific	97	59
RA VI: Europe	145	92

Commercial aircraft also make valuable contributions with their "AIREP" observations. AIREPs are made as required, not routinely. Figure 3-5 shows the distribution of international commercial airline flights over the tropics. Flights by national airlines greatly outnumber international flights over large countries such as China, Brazil, and Australia and between Hawaii and the US mainland. In most of the oceanic tropics, fewer than five flights a day are made across a given 10° rectangle.

Figure 3-6 (page 3-8) shows the number of AIREPs actually received for each 10° grid area by the Air Force Global Weather Central (AFGWC) over a period of 4 hours. Of the 320 AIREPs, 34 percent were made from just two 10° rectangles (20° to 30° N, and 140° to 160° W); this suggests that requirements other than weather appear to control the pattern of reporting. Twenty-three percent of the reports were made between 20° N and 20° S; in this belt, 78 percent of the 10° rectangles lacked AIREPs. AIREPs are concentrated in the upper troposphere between 300 and 200 mb, where jet aircraft normally fly. Fortunately, these levels are critical for tropical weather diagnosis and prediction.

In 1988 WMO reported that six Aircraft to Satellite data Relay (ASDAR) systems were operating on wide-bodied jet aircraft. With ASDAR, a great many automated wind and temperature measurements are transmitted via satellite to GTS. Operational systems now being deployed will measure turbulence and maximum wind and make vertical soundings during aircraft arrivals and departures.

The Air Force Reserve operates a fleet of 12 hurricane reconnaissance aircraft equipped with the new Improved Weather Reconnaissance System (IWRS), which resembles that used on NOAA research aircraft. Every second it collects position, wind, pressure altitude, absolute altitude, D-value, temperature and humidity data from sensors on the aircraft. Temperature, pressure, humidity and wind beneath the aircraft are sampled by dropwindsondes.

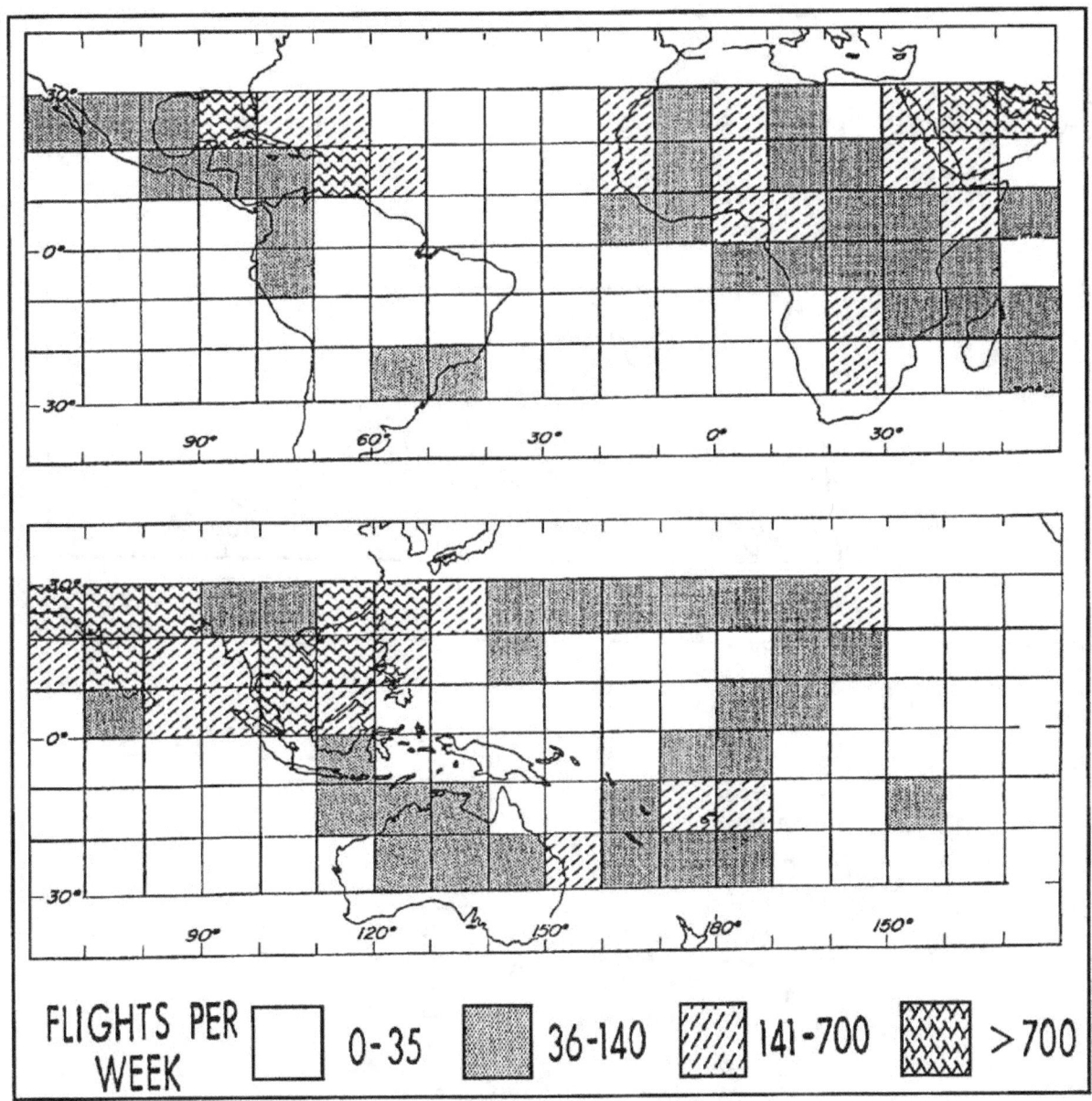

Figure 3-5. Maximum weekly number of international commercial airline flights crossing each 10-degree rectangle between 30° N and 30° S (based on International Civil Aviation Organization charts).

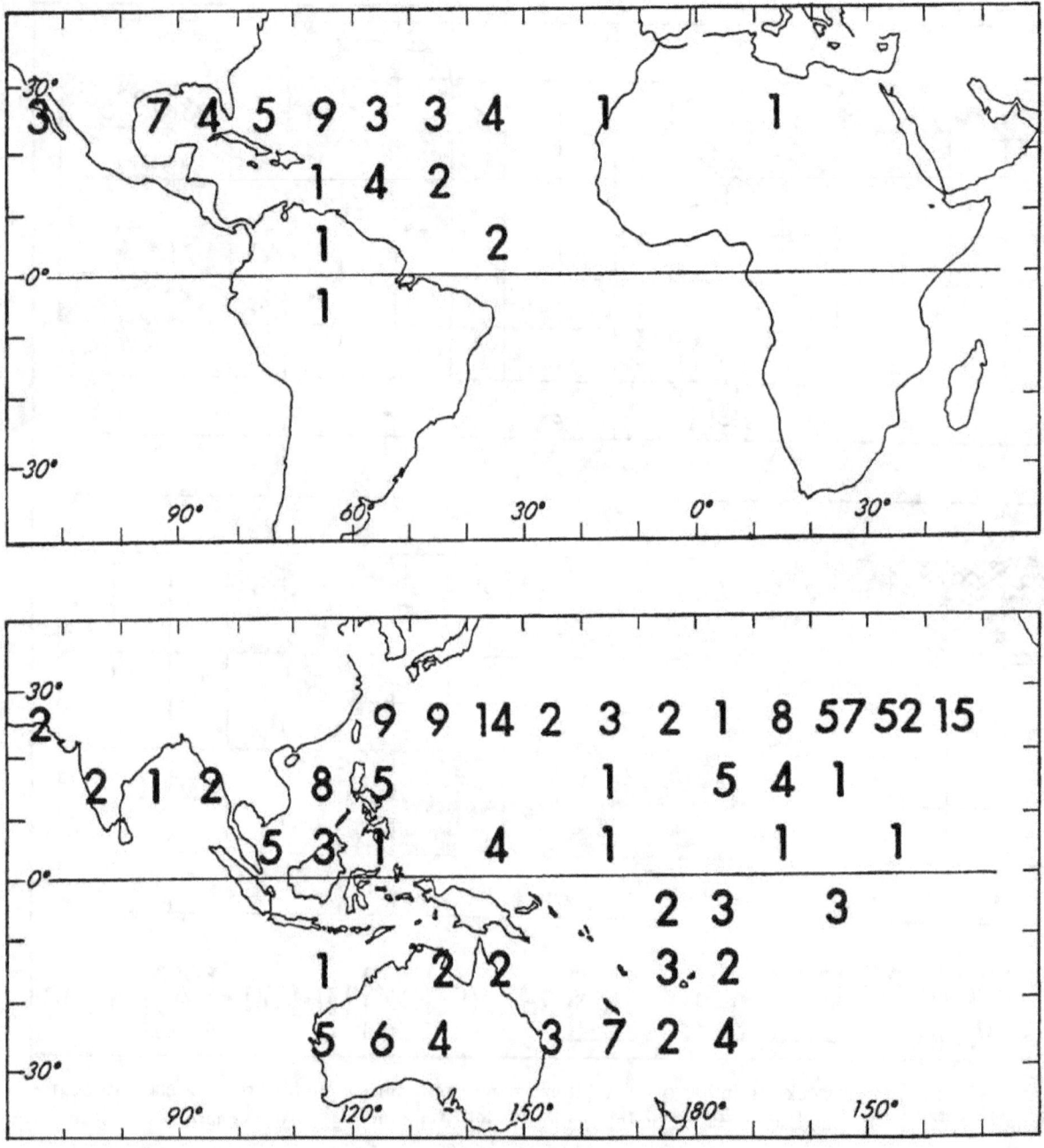

Figure 3-6. Number of aircraft weather observations received at AFGWC for each 10-degree rectangle between 30° N and 30° S from 0000 to 1400Z, 19 December 1990.

3.1.5 Meteorological Satellite data. Satellites have truly revolutionized tropical meteorology. In the mid-latitudes, satellite pictures have led to no great surprises; the typical cloud distributions associated with fronts and extra-tropical cyclones had already been determined from fairly dense surface networks. Apart from a few modifications, satellite data has reinforced the original polar-front concept. But in the tropics, which has always been data-sparse, satellites have enhanced our knowledge of development, movement, and typical distributions of major weather systems. This data often helps the forecaster to relate cloud distributions to circulation patterns, inferring the latter from the former.

The first Television Infrared Observational Satellite (TIROS 1) was launched on 1 April 1960. It was followed by nine more research satellites in the TIROS series, the last of which was launched in July 1965. Based on the knowledge gained from the TIROS series, a National Operational Meteorological Satellite System (NOMSS) became operational in February 1966 with the launches of Environmental Survey Satellites (ESSA) I and II. Besides the TIROS and ESSA series, several of the NIMBUS and Applications Technology Satellite (ATS) series have been used for research and development with some real-time operational applications.

More recently, The National Oceanic and Atmospheric Administration's (NOAA) Polar-orbiting Operational Environmental Satellite (POES), the Defense Meteorological Satellite Program (DMSP), and the Geostationary Operational Environmental Satellite (GOES) systems have provided comprehensive coverage. Each NOAA and DMSP satellite views each part of the earth at least twice daily and contains a suite of sensors that provides conventional satellite imagery as well as vertical and space environmental soundings. DMSP satellites also provide microwave imagery. GOES, anchored at a point over the equator, provides pictures every 30 minutes, and more often on request. Foreign governments also operate weather satellites. Table 3-3 lists satellite capabilities.

TABLE 3-3. Operational weather satellites (AWS/XOOO, 1995).

Program	Agency	Objective
POES (TIROS-N): Polar-oribiting Operational Environmental satellite	NOAA	Meteorological observations, measurements of sea ice and snow cover, assessment of vegetation condition
GOES: Geostationary Operational Environmental Satellite; normally at 75° W and 135° W	NOAA	Operational weather data, cloud clever, temperature profiles, real-time storm monitoring (rapid scan), severe-storm warning, sea-surface temperature
DMSP Block 5D-2: Defense Meteorological Satellite Program; polar orbiter	DoD	Operational weather observations for DoD, tropical cyclone reconnaissance, microwave observations
METEOSAT: Geostationary Meteorological Satellite at 0° E	ESA	Europe operational weather data, cloud cover, water-vapor imagery
GMS: Geostationary Meteorological Satellite at 140° E	Japan	Operational weather data, cloud cover, temperature profiles, real-time storm monitoring (rapid scan), hydrological observations
GOMS/ELEKTRO: Geostationary at 75° E	Russia	Operational weather data, cloud cover, water-vapor imagery, temperature and moisture profiles
METEOR: Meteorological Satellite; polar orbiter	Russia	Meteorological observations, sea-surface temperature, sea ice, snow cover, vegetation condition
INSAT: Geostationary Indian National environmental satellite at 74° E	India	Meteorology, domestic communications, television program distribution
Feng Yun FY-1B: Environmental satellite, polar orbiter	China	Meteorological observations, sea-surface temperature, sea ice, snow cover, vegetation condition

33

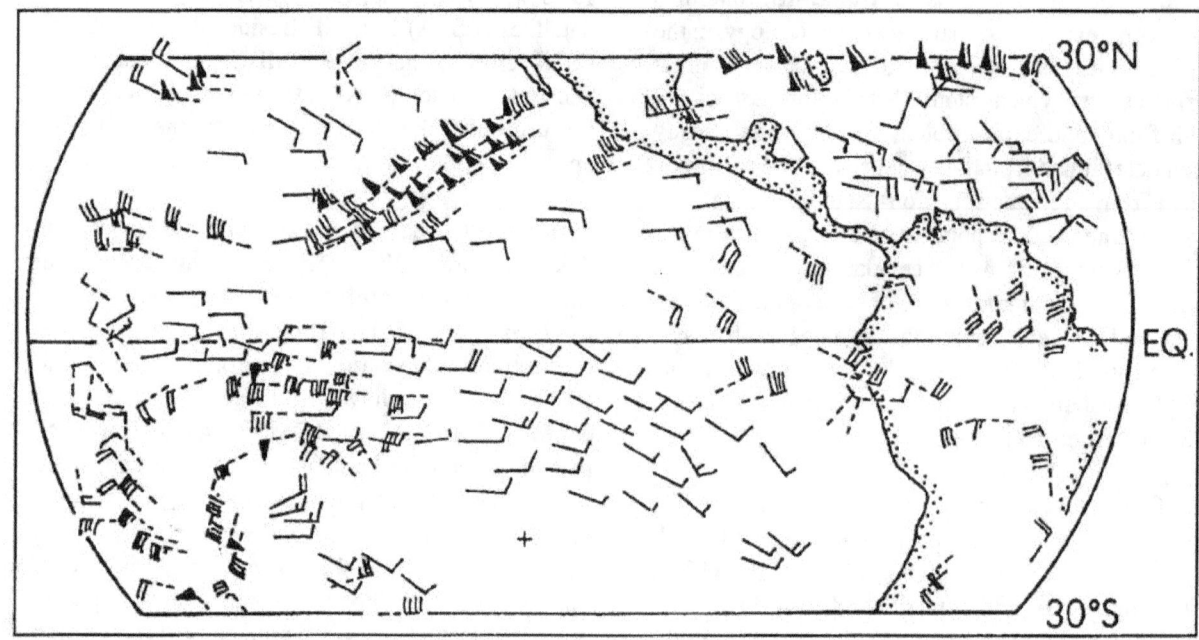

Figure 3-7. Cloud motion vectors (full, low-level; dashed, cirrus-level) obtained from picture sequences made by GOES at 0°, 135° W, on 20 December 1989 (from NESDIS).

Satellite soundings fill large data gaps. They have led to improved global analyses and have benefitted numerical weather prediction. Geostationary satellites are especially valuable to tropical meteorologists. Besides the two operated by NOAA at 75° W and 135° W, a Japanese satellite is located at 140° E. India is responsible for 74° E, while the European Space Agency maintains one at 0° E. The geostationary systems indirectly provide excellent low- and high-level wind data by allowing individual clouds to be tracked (Figure 3-7), again filling many data voids.

Tropical meteorologists use satellite pictures to fix the cloud system centers of tropical cyclones. They determine center movement from successive imagery. In the western North Pacific, satellites have replaced weather reconnaissance aircraft in this role. Cyclone intensity can now be estimated from the appearance of the clouds (Dvorak, 1975). The temperature of the air column in the eye of a tropical cyclone is inversely related to the central surface pressure and directly related to the maximum surface wind. The temperature

difference between the eye and the surrounding region at 250 mb, determined by microwave (54.96 GHz) measurements from NOAA polar-orbiting satellites, has been statistically related to reconnaissance-observed central pressures and winds; standard errors of the estimates amounted to 8 mb and 13 knots for the Atlantic (103 cases) (Velden, 1989) and 13 mb and 17 knots for the western Pacific (82 cases) (Velden et al., 1991).

The IR channel (10.5 to 12.5 μm) that measures cloud top temperatures and its various "enhancements" is especially useful. Tropical meteorologists also use satellite imagery to determine the movement of shear lines, troughs in the TUTT, and the subtropical ridge, as well as the extent and intensity of weather associated with each. Although vapor images (in the 6.7 μm band) of cloudy areas resemble IR images, they can also vertically integrate moisture content where there is no cloud; in so doing, they delineate subsidence and rising motion better than cloud pictures.

The special sensor microwave/imager (SSM/I) on DMSP senses surface wind speed over the ocean where no rain is falling (Holliday and Waters, 1989); this data lets analysts map areas of destructive winds around tropical cyclones. Every 2 weeks, a group at the National Meteorological Center (NMC) combines ship observations with satellite-sensed infrared radiances from the sea to produce excellent global sea-surface temperature charts. Satellite-sensed infrared radiances from cloud tops can be translated into heights. From these, if convective origins are assumed, rainfall can be estimated. As subsequent chapters show, enough satellite data has been accumulated to greatly contribute to tropical climatology.

3.1.6 Weather Radar Data. Weather radar provides the most important short-range local "nowcasting," forecasting, and weather warning aid. It gives forecasters a three-dimensional view of rain areas out to 150 km or more from the observation site. As Table 3-4 shows, radars are scarce in Africa, South America, and the Southwest Pacific. There has been little or no improvement in recent years.

3.1.7 Lightning Detection. In the tropics, lightning detection systems are operating in Australia, Hong Kong, Indonesia, Sri Lanka, and the southern United States (World Meteorological Organization, 1989). Information from these systems is not available on GTS.

TABLE 3-4. Worldwide weather radar station distribution (from WMO, 1989).

Region	Number of radar stations	Wavelength			
		3 cm	5 cm	10 cm	Other
RA I: Africa	54	11	20	16	7
RA II: Asia	193	107	54	32	-
RA III: South America	17	5	3	9	-
RA IV: North and Central America	154	5	81	68	-
RA V: Southwest Pacific	56	2	12	42	-
RA VI: Europe	170	118	39	13	-

3.2 PROBABLE FUTURE DATA SOURCES

3.2.1 Meteorological Satellites. By the end of the 20th century, new sensors will measure tropical winds at several levels, including those in and around tropical cyclones. They will provide the data density needed by the hurricane/typhoon prognostic models and give general support to numerical weather prediction. Other new sensors making accurate, high-resolution measurements in the vertical could greatly enhance analyses used to initialize numerical forecast models. Rainfall measurements from space will greatly help tropical meteorologists. Lightning has been detected at night from polar-orbiting satellites (see Figure 10-2). By the mid-1990s, a new sensor flown on GOES will operate continuously and detect lightning at a 10 km resolution (Christian et al., 1989). Table 3-5 lists proposed new satellite programs. Should they succeed, their products will be made available to operational meteorology.

TABLE 3-5. Proposed satellite missions (AWS/XOOO, 1995). Unless otherwise specified, United States agencies are responsible.

Program	Agency/Status	Objectives
TOPEX/POSEIDON: Ocean	NASA-CNES/Topography Experiment	Ocean surface topography (France)
POES follow-on missions	NOAA-planned for 1996	Advanced capabilities for weather observations
DMSP Block 5D-3	DoD-planned for 2000	Advanced capabilities for meteorological and space weather observations
MOS-2 (Marine Observation Satellite-2)	NASDA/Japan	Passive and active microwave sensing
EOS: Earth-Observing POES	NASA/NOAA	Long-term global Earth research observations
TRMM (Tropical Rainfall Measuring Mission)	U.S./Japan	Total precipitation measurements
NPOESS (National Polar-orbiting Operational Satellite System)	NOAA/DoD/NASA, WITH EUMETSAT coordination-planned for 2006	Weather observations and atmospheric composition; observations of ocean and ice surfaces; land surface imaging; Earth radiation budget; data collection and locations of remote measurement devices; detection and location of emergency beacons; monitoring of space environment

3.2.2 Radar. In the near future, most countries in the tropics will probably maintain the current 3-, 5-, or 10-cm conventional weather radars. However, a revolutionary state-of-the-art weather surveillance radar system (WSR-88D) has been developed under the Next Generation Weather Radar (NEXRAD) program. NEXRAD is a joint Air Weather Service (AWS), National Weather Service (NWS), Federal Aviation Administration (FAA) project that uses Doppler techniques. Not only can a 10-cm Doppler radar detect and locate raindrops, but if a pulsed signal is combined with circular scanning at a fixed elevation, the Doppler shift produced in the reflected signal by horizontal movement of the reflecting particles can be determined; motion toward or away from the radar can be calculated. The principle has been known for a long time (e.g., Probert-Jones, 1960).

A network of Doppler radars is being installed in the United States and at some overseas tropical forecast offices and military installations. This radar will increase the user's ability to detect severe weather while reducing the false alarm rate. The WSR-88D will find and alert the operator to mid-cloud vortexes (e.g., tornadoes, and possibly microbursts), gust fronts, tilted storm axes, and wind shears. It will also allow monitoring of tropical cyclone circulations. Radar operators and severe weather forecasters at major centralized weather facilities will be able to remotely monitor other terminals on the NEXRAD network. Table 3-6 shows where WSR-88Ds are being installed in the tropics (beyond the continental United States). A Doppler radar has been operating at Taipei (25° N, 122° E) since 1987; a network of Doppler radars is planned for Thailand.

The Doppler principle, as applied to a very-high or ultra-high frequency upward-pointing radar that can detect discontinuities in clear air, can be used to determine vertical motion in the radar beam. If two beams are pointed slightly off vertical, the horizontal wind can be estimated through the troposphere above the station. The calculation is based on the two slant-range radial (Doppler) velocity measurements for each level (assuming zero vertical velocity) (Gage et al., 1990). Wind profiles can be measured every 1-2 minutes. NOAA is now installing 30 profilers in the central United States, and maintains research profilers at Saipan (15° N, 146° E), Pohnpei (7° N, 158° E), Christmas Island (2° N, 157°W), Biak (1° S, 136° E), Piura (5° S, 81°W), and Darwin (12° S, 131° E). Data from the Christmas Island profiler is incorporated in routine analyses at NMC and ECMWF. The other stations will also provide data on GTS.

TABLE 3-6. Tropical stations beyond the continental United States at which 10-cm Doppler radars (WSR-88D) are either installed or being installed.

Station	Latitude	Longitude
Kadena AB	25° N	128° E
Anderson AB	4° N	145° E
South Kauai	21° N	158° W
Molokai	21° N	157° W
San Juan	18° N	66° W

3.3 SUMMARY

Data available to tropical meteorologists will increase in the years ahead, and satellite communication techniques will be used to collect and relay the multitude of environmental observations to centralized forecast offices, then disseminate the products to users. The Global Telecommunications System (GTS), already overloaded, should be greatly expanded to cope with present as well as future needs. More detailed satellite data, and new ways of applying, combining, and enhancing multi-spectral images will undoubtedly aid our understanding of tropical meteorology and perhaps enable us to develop better conceptual and numerical models for the tropics.

Chapter 4

PRESSURE AND WINDS

4.1 GENERAL

In this chapter, the major characteristics of tropical pressure and wind fields are illustrated by mean maps, cross-sections, latitudinal profiles, typical station climatology, and circulation models. The wind field is emphasized. In the tropics, important persistent features are better defined by the resultant winds than by pressure and pressure-height fields, which tend to have small gradients. The tropical upper-tropospheric troughs (TUTT) over the Atlantic and Pacific, rarely detectable in the height fields, often show clearly in the resultant wind fields.

In much of the tropics, the two main seasons (wet and dry) are separated by relatively brief transition seasons. Thus, mid-season months (January, April, July, October) describe the annual cycle pretty well. However, local forecasting often demands a more detailed background of monthly or even 5-day (pentad) climatologies. A firm grasp of upper-tropospheric circulation patterns is essential.

Standard deviations (square root of the mean of the squares of the differences between individual values and the mean value) of the surface pressure and other fields provide useful information to the forecaster. Although observational errors contribute, the standard deviation roughly indicates synoptic variability.

4.2 PRESSURE

4.2.1 Mean Sea-Level Pressure. Figures 4-1 through 4-4 give mean sea-level pressure (a) and standard deviation (b) of pressure between 40° N and 40° S for the four mid-season months. Throughout the year, low pressure generally prevails near the equator, while high pressure prevails over the oceanic subtropics. The Southern Hemisphere (SH) subtropical high-pressure ridge migrates over a smaller range of latitude than its Northern Hemisphere (NH) counterpart. In the oceanic ridges, mean pressure is highest in July, reaching 1025 mb; daily values often exceed 1030 mb. Because of extratropical cyclones, standard deviations are much larger in winter than in summer. Although weather is variable equatorward of the ridge, standard deviations of only about 2 mb prevail, not much more than the expected observational error. Most weather disturbances cause pressure fluctuations indistinguishable from noise. It's no wonder tropical meteorologists prefer to analyze weather in terms of the wind field.

In the winter hemisphere, the warm oceanic subtropical ridge and the cold continental ridge form a continuous belt of high pressure around the hemisphere. During summer, as a result of differential heating between land and ocean, heat lows develop over the continents adjacent to the Indian Ocean. The consequent reversal of pressure gradient between winter and summer gives rise to the monsoons (see 6.3.6).

Zonal averages smooth the large departures occurring at various longitudes, but the broad-scale meridional variations remain. Figure 4-5 (on page 4-7) shows the zonally averaged sea-level pressure for January and July, computed from Figures 4-1 and 4-3. Arrows show the north-south movement of the subtropical ridges and the low-latitude trough between January and July. The average trough position varies from 5° S to 6° N. The mean ridge in the NH ranges over 7 degrees (30° to 37° N), compared to 5 degrees (33° S to 28° S) for the SH ridge. Because the trough moves more than the ridges, the trade winds are much broader in winter than in summer. Pressure in the zonal mean subtropical ridge rises from 1015 mb in summer to slightly over 1020 mb in winter, reflecting the summer heat lows and cold highs of the continents.

JANUARY

SEA LEVEL PRESSURE (MBS) - MEANS

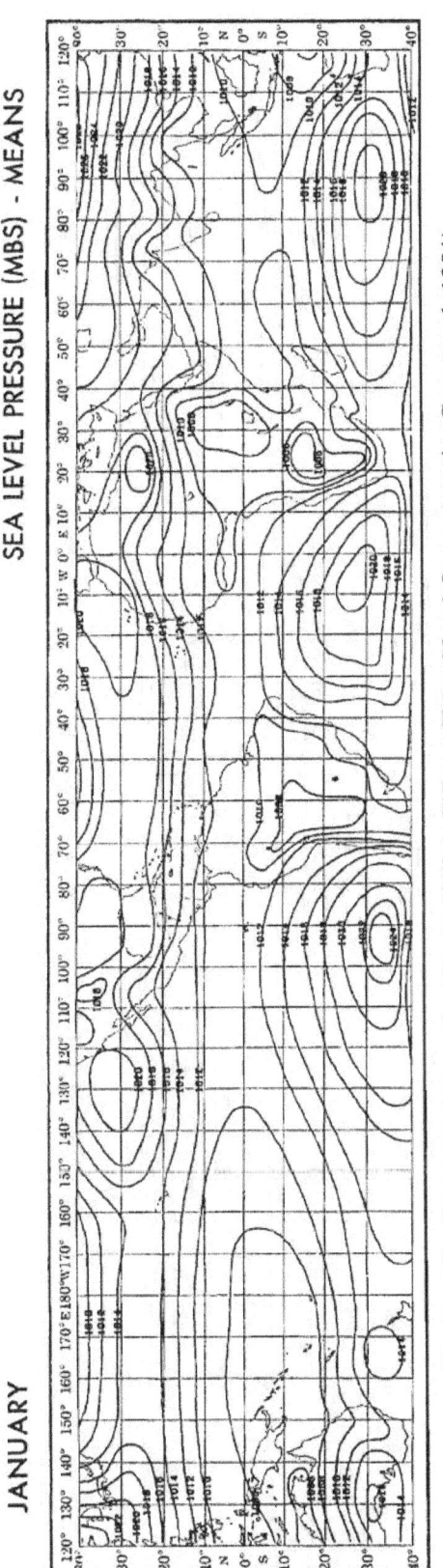

Figure 4-1a. January mean sea-level pressure in millibars (From Naval Oceanography Command, 1981).

JANUARY

SEA LEVEL PRESSURE (MBS) - STANDARD DEVIATIONS

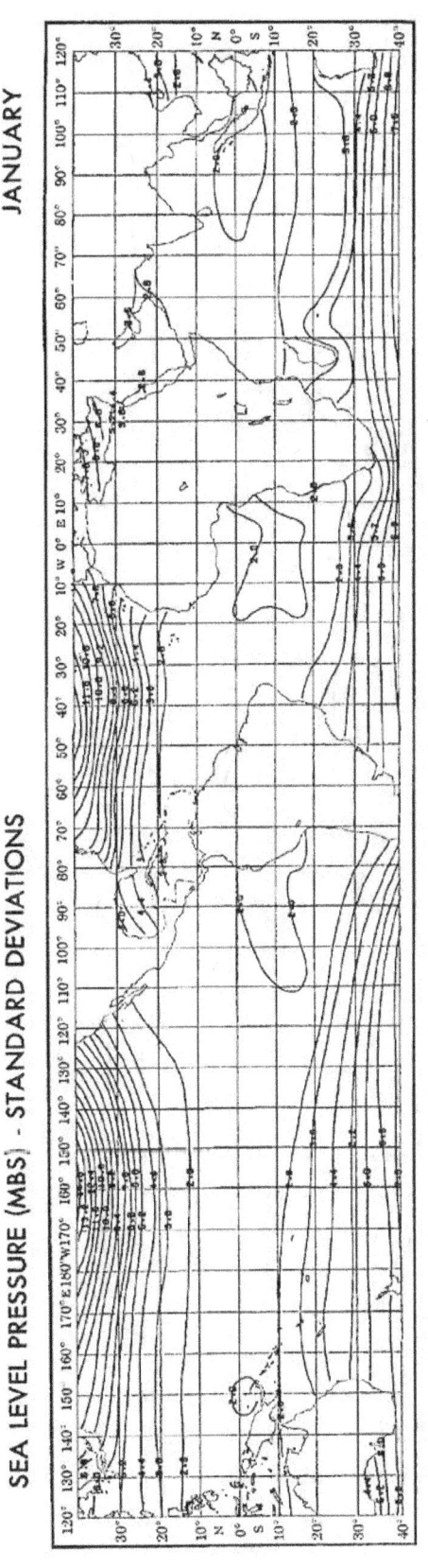

Figure 4-1b. January standard deviation of sea-level pressure in millibars (From Naval Oceanography Command, 1981).

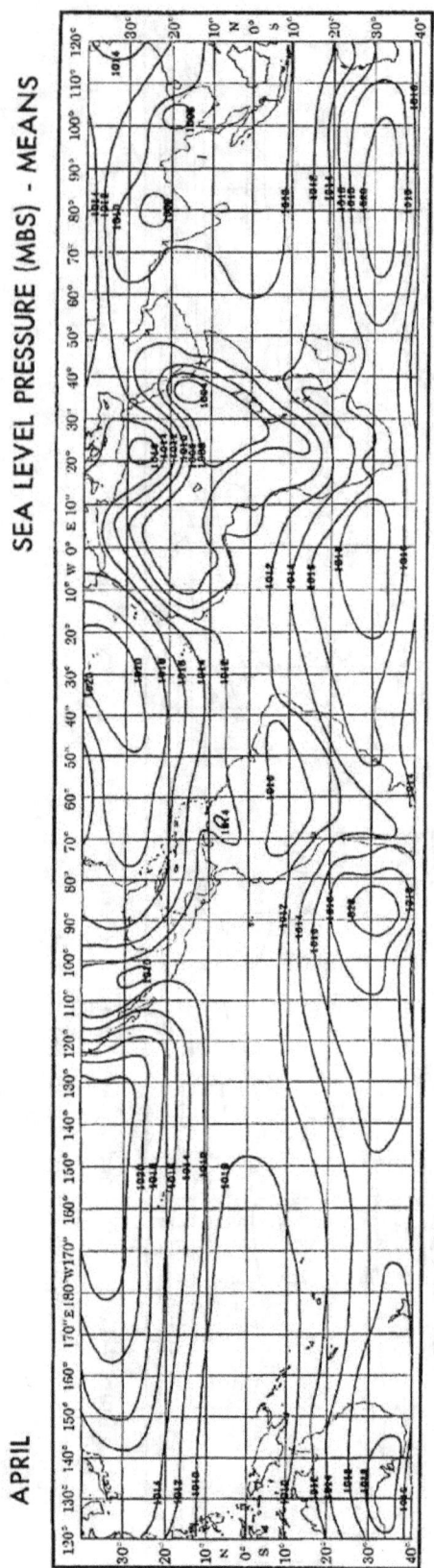

Figure 4-2a. April mean sea-level pressure in millibars (From Naval Oceanography Command, 1981).

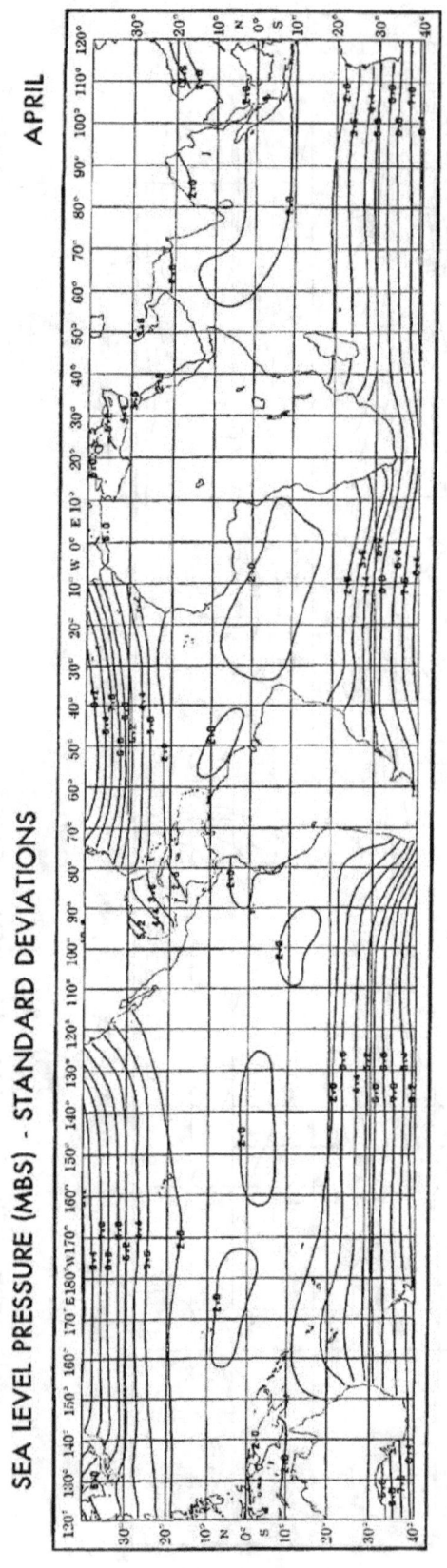

Figure 4-2b. April standard deviation of sea-level pressure in millibars (From Naval Oceanography Command, 1981).

42

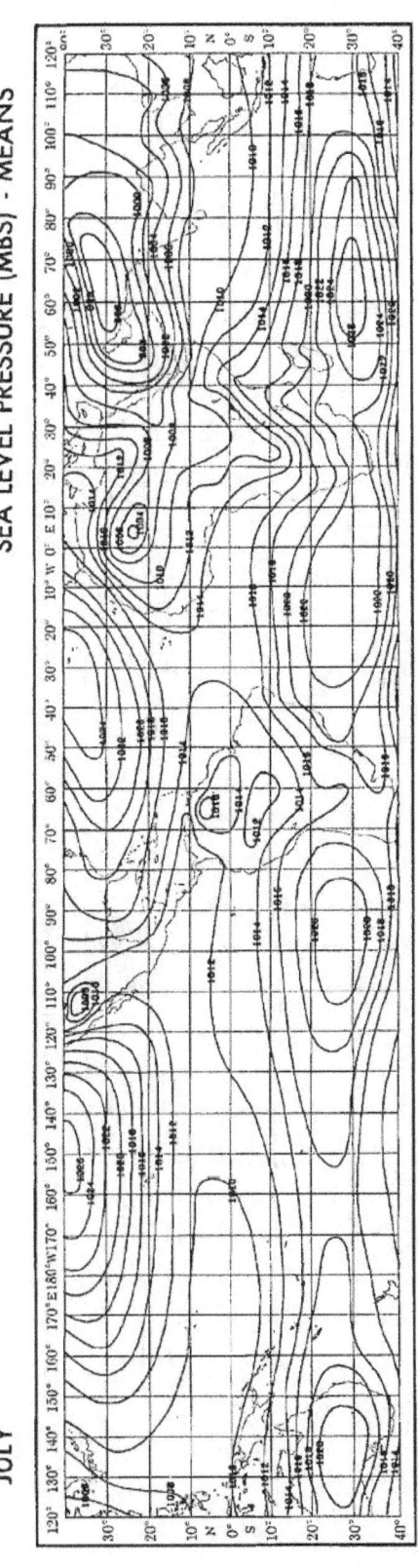

Figure 4-3a. July mean sea-level pressure in millibars (From Naval Oceanography Command, 1981).

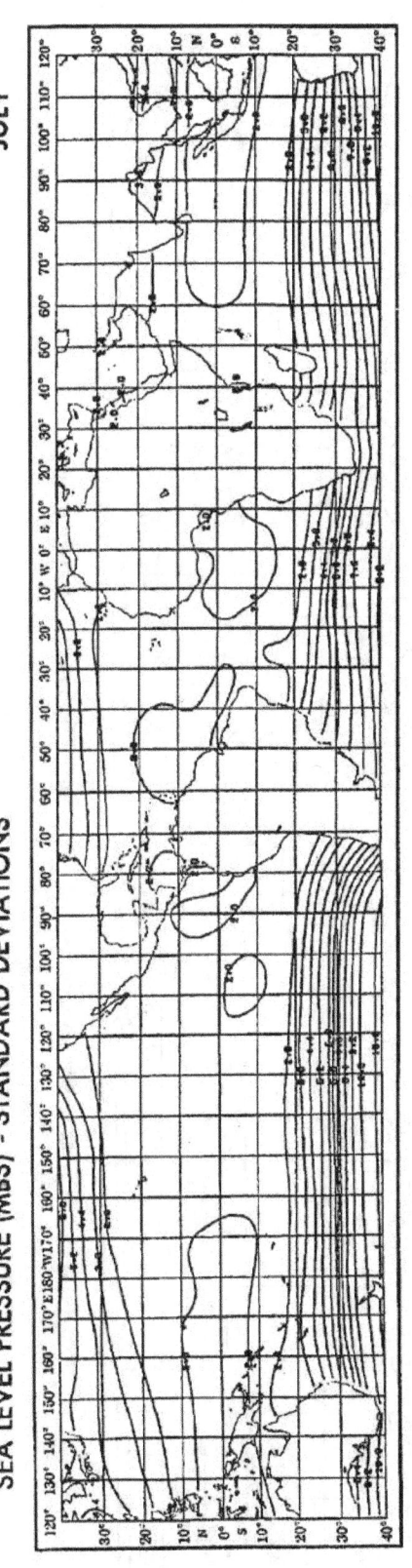

Figure 4-3b. July standard deviation of sea-level pressure in millibars (From Naval Oceanography Command, 1981).

Chapter 4

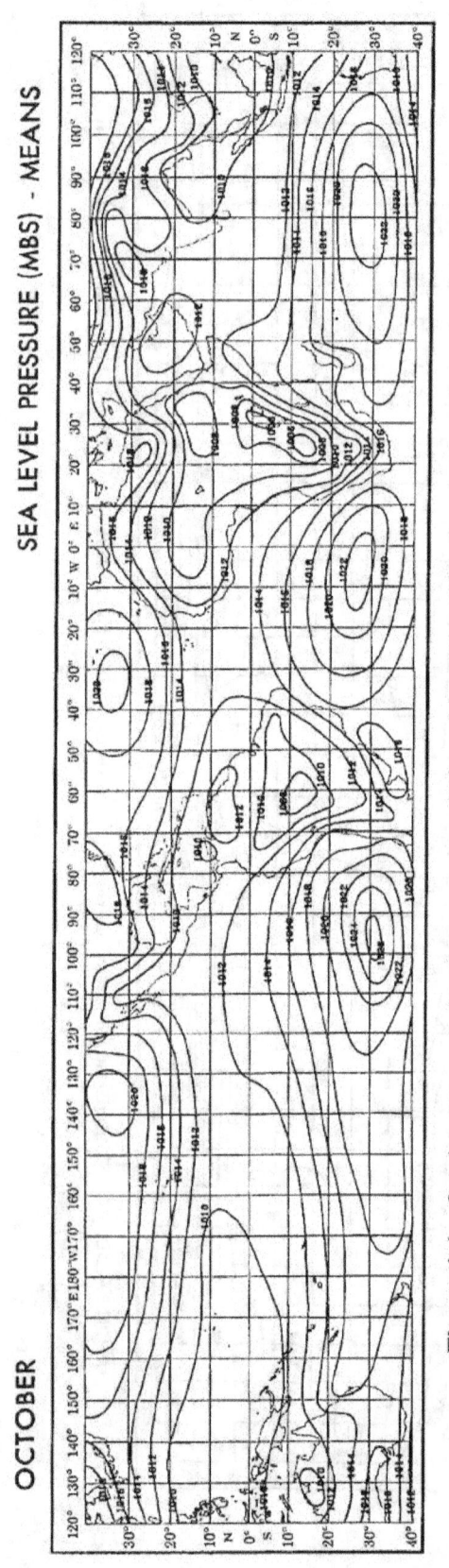

OCTOBER

SEA LEVEL PRESSURE (MBS) - MEANS

Figure 4-4a. October mean sea-level pressure in millibars (From Naval Oceanography Command, 1981).

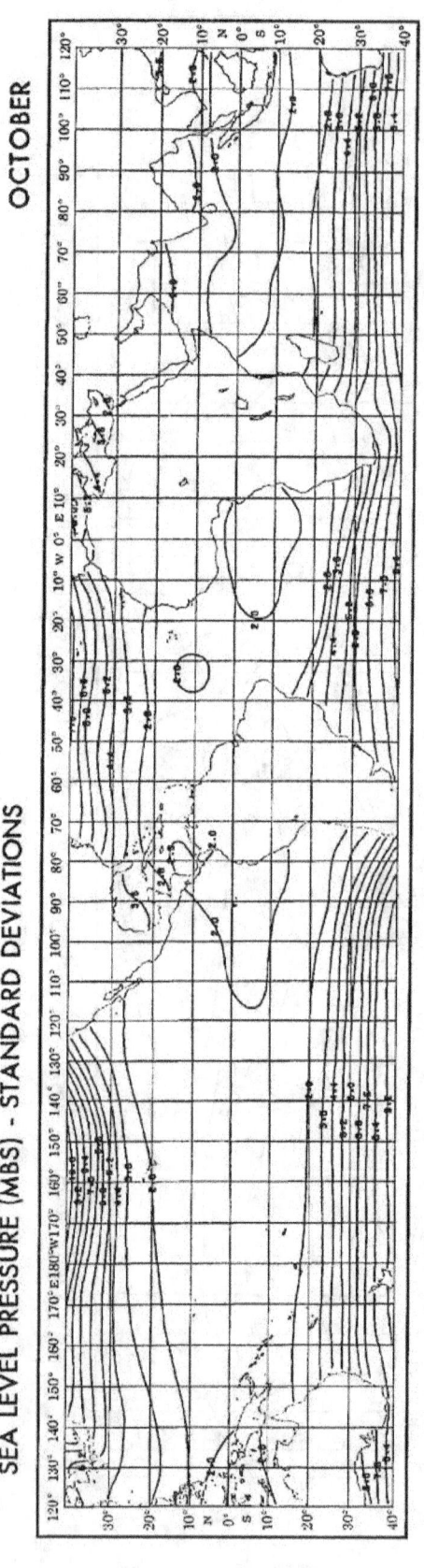

OCTOBER

SEA LEVEL PRESSURE (MBS) - STANDARD DEVIATIONS

Figure 4-4b. October standard deviation of sea-level pressure in millibars (From Naval Oceanography Command, 1981).

44

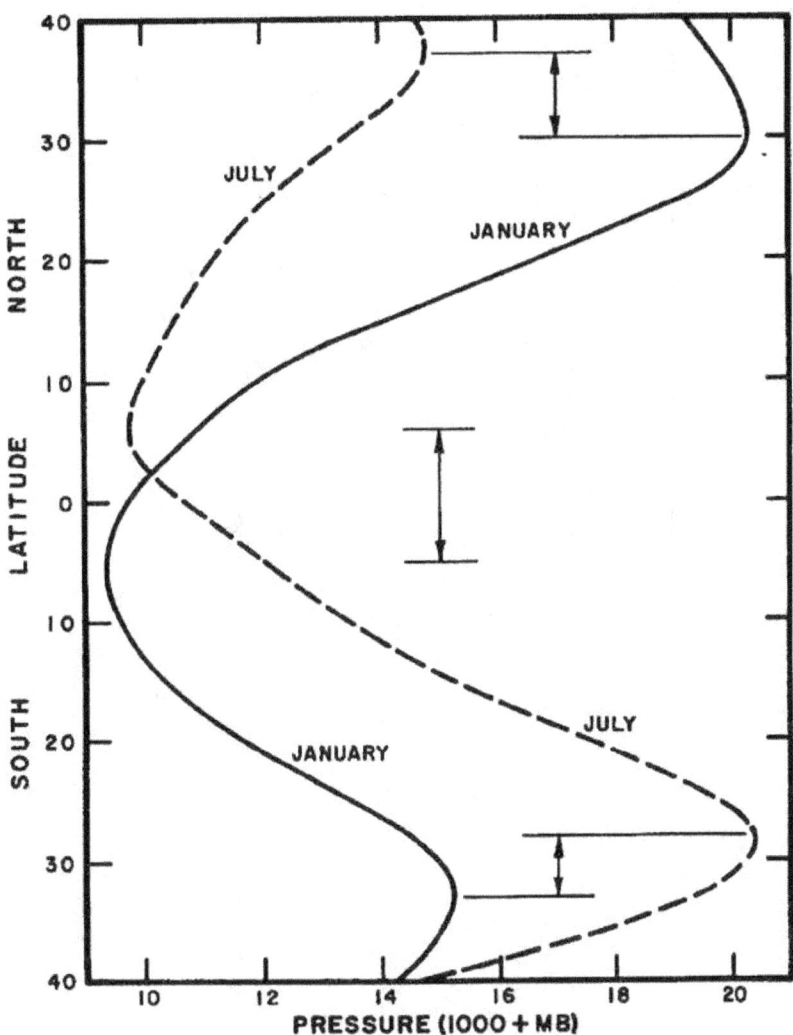

Figure 4-5. Average annual meridional distribution of sea-level pressure for January and July (derived from Crutcher and Davis, 1969).

4.2.2 Large-Scale Pressure Changes.
Simultaneous widespread pressure variations lasting a few days and of unknown origin have been observed over India (Eliot, 1895) and the Caribbean (Frolow, 1942). Figure 4-6 (on the next page) shows simultaneous pressure anomalies for three Pacific-area stations over a 15-day period. The diurnal and semidiurnal variations have been removed by using departures from the monthly means of hourly values. In the tropics, shear lines may be over 2,000 NM (3,700 km) long, while vortexes are rarely more than 1,000 NM (1,900 km) across. Thus, although the stations in Figure 4-6 are too far apart for the same synoptic system to affect all three at once, their broad-scale pressure changes are well correlated. In a case

studied by Palmer and Ohmstede (1956), pressure over the entire tropical Pacific fell almost uniformly by 2.5 mb in 48 hours. Sadler et al. (1968) reported an even more massive event over the western Pacific and East Asia–see Figure 4-7, next page.

More recently, Lander (1991) studied a sequence of widespread pressure fluctuations over the Pacific during May 1986. As Figure 4-8 shows, a north-south oscillation covered most of the basin and implied corresponding fluctuations in surface winds. According to Guard (1985) such changes accompany wind surges; in the forward parts of these, weather deteriorates (see 8.4.4).

45

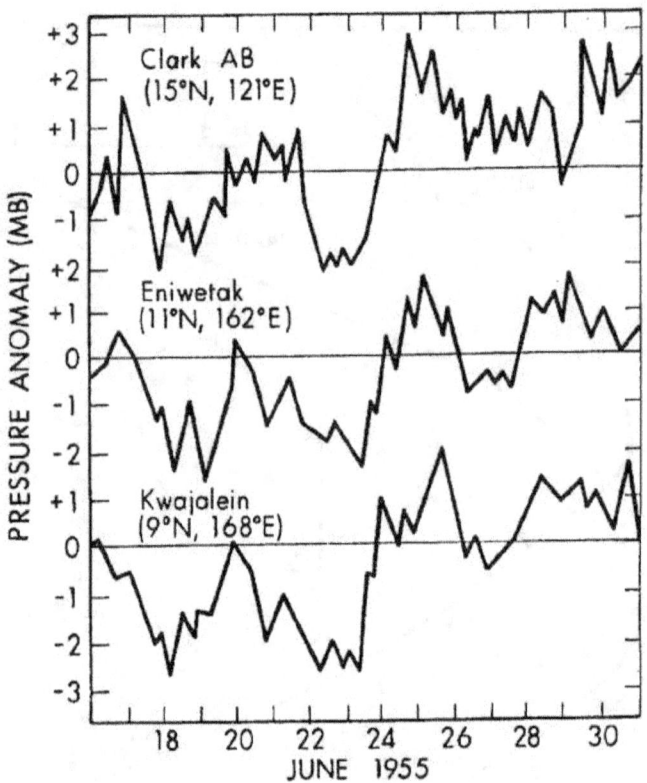

Figure 4-6. Simultaneous hourly surface pressure anomalies for three Pacific stations (from Brier and Simpson, 1969).

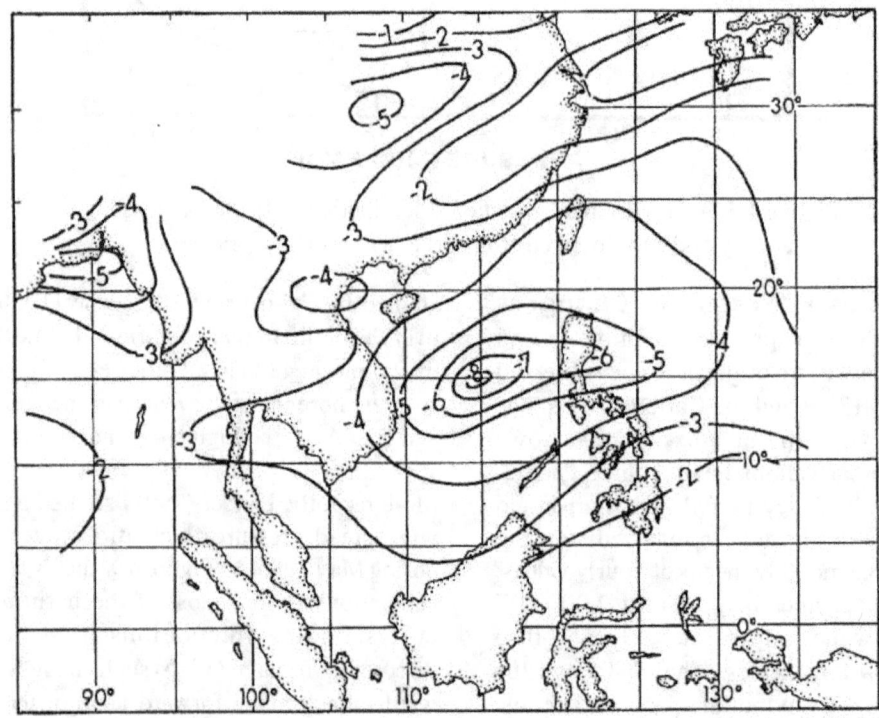

Figure 4-8. Correlations between daily surface pressure at Tarawa (1° N, 173° E) and other Pacific stations for May 1986 (from Lander, 1991).

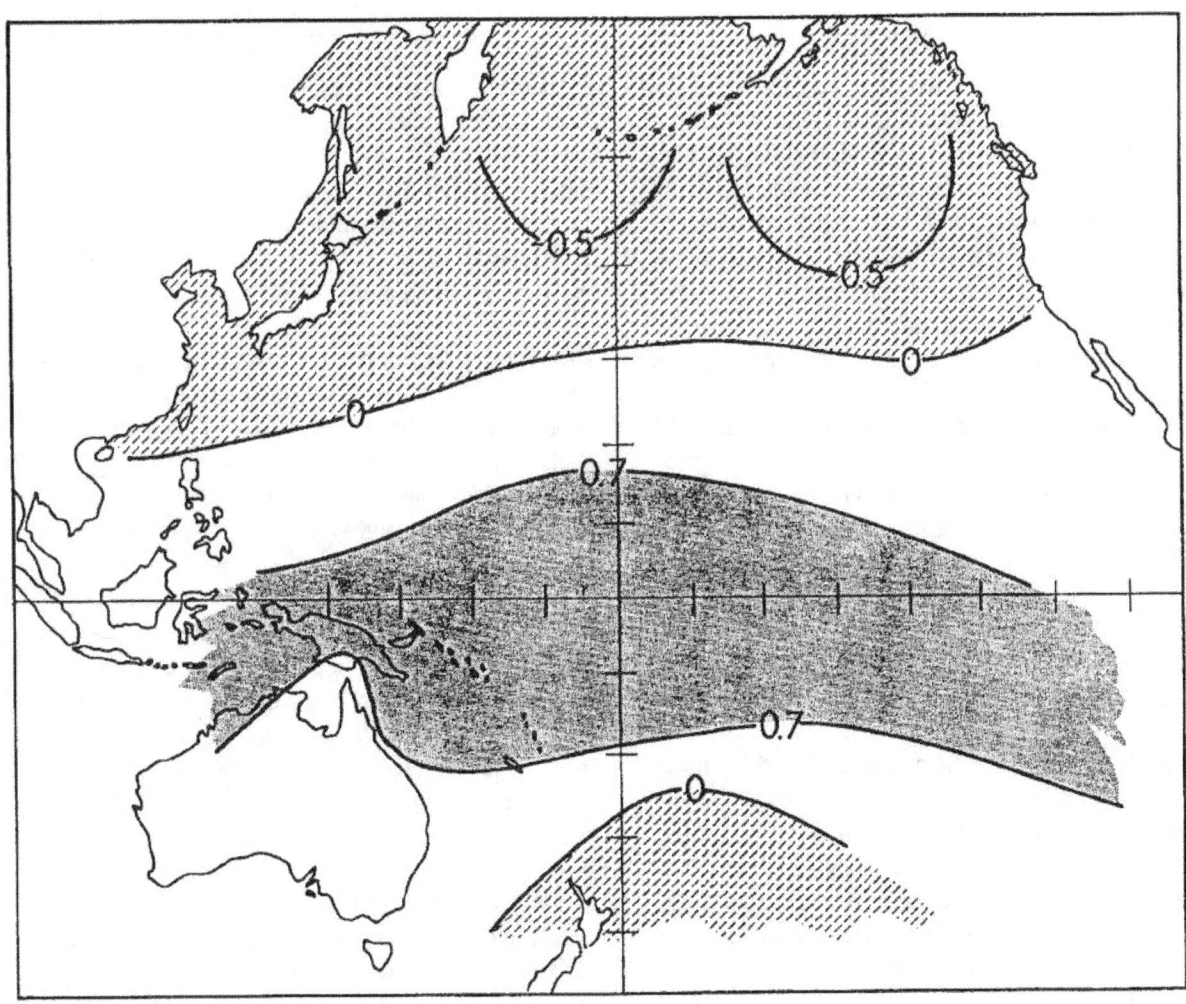

Figure 4-7. Surface pressure change (mb) between 00Z 21 July and 00Z 23 July 1966 (from Sadler et al., 1968).

4.3 WINDS

4.3.1 General. The most analyzed levels in the tropics are the surface/gradient-level and 200 mb, primarily because the most observations available are for those levels. This is fortunate, since the lower and upper tropospheres control the vertical motion fields that produce most tropical weather.

The regions of upward motion shown in Figure 2-4 generally coincide with troughs (convergence) at the gradient level (Figures 4-9 to 4-12) and with anticyclones (divergence) at 200 mb (Figures 4-13 and 4-14). Mid-tropospheric analyses delineate mid-tropospheric cyclones and help in predicting tropical cyclone movement; unfortunately, data only comes from rawin and rawinsonde stations (refer to Figure 3-4). This section presents the resultant wind climatology for the lower and upper levels for the mid-season months. Local winds are discussed in Chapter 7.

4.3.2 Resultant Gradient-Level Winds. Charts for mid-season months shown in Figures 4-9 to 4-12 were extracted and reduced from Atkinson and Sadler, 1970; their Air Weather Service Technical Report 215 contains monthly charts on a 1:20 million Mercator map projection. The "gradient level" is defined as the lowest level at which predominantly friction-free flow occurs. Over most of the tropics, this is about 3,000 feet (1 km). For stations below 1,000 feet elevation, winds at 3,000 feet were used when available. For stations with data only at standard pressure levels, the 850-mb winds were used and adjusted by comparing with vertical wind profiles at neighboring stations. For stations above 1,000 feet, resultant winds at 5,000 feet (1.5 km) were generally used. Over the oceans, resultant surface winds from ship reports formed the basis of subjective estimates of the resultant gradient-level winds.

The wind field is represented on the charts by kinematic analyses (Palmer et al., 1955). Full lines depict streamlines (arrowheads show flow direction),

dashed lines show the resultant isotachs drawn for 5-knot intervals, and shaded areas represent resultant speeds less than 5 and greater than 15 knots. To retain patterns, streamlines were drawn through large mountain barriers, such as the Andes in South America; remember, however, that gradient-level flow has little or no meaning in these regions.

Resultant winds give no direct information on variability. Steadiness serves the same purpose with winds as standard deviation does with scalar quantities such as pressure and temperature. The long-term resultant wind speed is quite highly correlated with steadiness and roughly indicates variability (see Figure 11-15).

The major features shown by the mean sea-level pressure fields (subtropical high-pressure ridges and low-latitude troughs) are equally evident in the gradient-level wind fields. This coincidence arises from the fact that for long-term means, gradient-level winds parallel isobars to within about 3° from the equator. Along the equator, where Coriolis force is zero, a balance is struck between the mean pressure gradient and mean friction. The mean resultant wind, then, blows perpendicular to the isobars, from high to low pressure. This is also true from day to day if the direction of the pressure gradient remains unchanged.

Over the western Indian Ocean in January (July), winds blow from the north (south) across the equator. Easterlies prevail in an east-west pressure gradient over the central and western parts of the equatorial Pacific and Atlantic, and over northern South America—refer to Figures 4-1 to 4-4.

During El Niño, the pressure gradient over the equatorial Pacific often reverses to west-east and westerlies then replace easterlies (see 6.3.5). The central North and South Atlantic and the central North and South Pacific are least influenced by the large continents. Over these regions, tradewinds prevail.

The near-equatorial tradewind convergence (NETWC), where the northeast and southeast trades merge, lies in a broad, very weak, trough. In the summer hemisphere and over the oceans, the gradient-level wind patterns closely resemble those of the surface winds. But over the continents in winter, large thermal gradients produce significant differences. For example, mean sea-level pressure in January shows an intense anticyclone over East Asia north of 40° N (Figure 4-1a) while the gradient wind field in Figure 4-9 shows a small anticyclone near 30° N.

Discussions of the major features shown on the resultant gradient-level wind charts for the mid-season months are provided on the following pages opposite the charts.

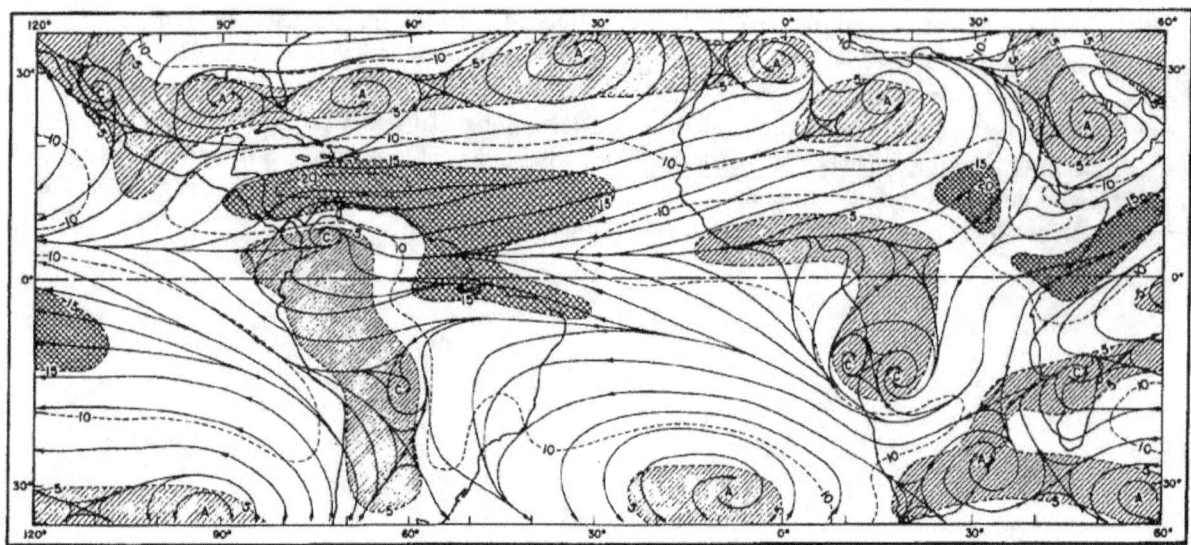

Figure 4-9a. Resultant gradient winds (knots) for January, 120° W to 60° E (from Atkinson and Sadler, 1970).

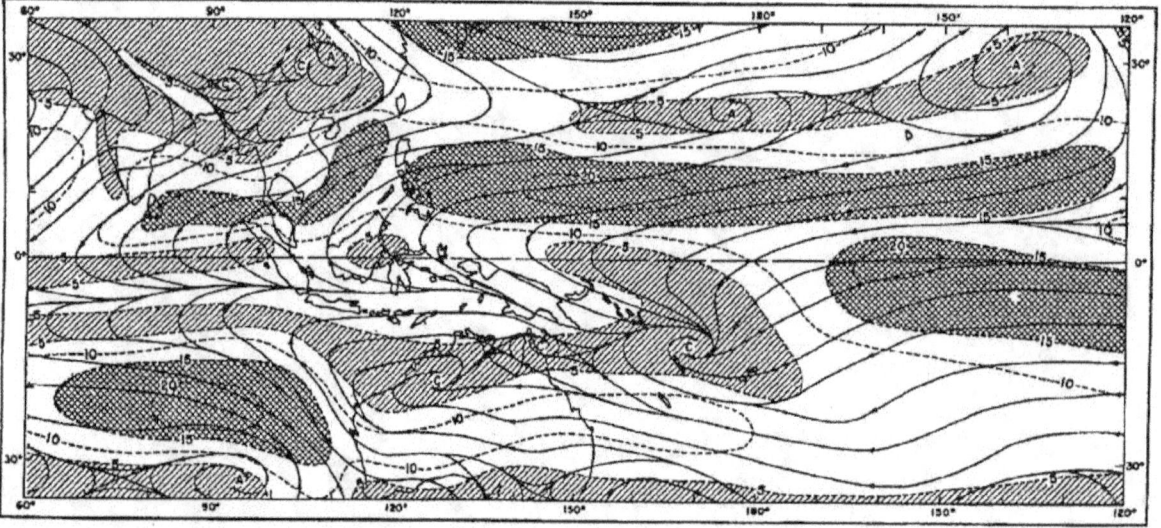

Figure 4-9b. Resultant gradient winds (knots) for January, 60° E to 120° W (from Atkinson and Sadler, 1970).

Major features shown in Figure 4-9, opposite:

Northern Hemisphere

• The tropics are dominated by the oceanic subtropical ridges, which merge with the continental anticyclones.

• The northeasterlies south of this ridge system are strongest near 10° N.

• The tradewinds converge near 5° N. Winds blow across the equator from Africa to 150° E with a near-equatorial trough over Indonesia.

• Anticyclones prevail over the winter (NH) continents; heat lows prevail over the summer (SH) continents. The strongest northerlies along the equator occur over and near Africa, reflecting the large pressure gradient between a winter high in the north and a summer heat low to the south.

Southern Hemisphere

• A trough extends from a low near 13° S, 170° E, westward across northern Australia, where it becomes a monsoon/heat trough. West of Australia, the trough lies near 10° S. Near 50° E it trends poleward to link with the monsoon/heat trough over southern Africa. Most tropical cyclones develop in the oceanic parts of this trough, which is 8,000 NM (15,000 km) long.

• The trough extending southeastward from the low near 13° S, 170° E, is associated with tropical and extratropical synoptic systems and relative maxima of cloud and rain along what is known as the "South Pacific Convergence Zone." It persists throughout the year and has no NH counterpart.

• The broad subtropical anticyclone over the South Atlantic also persists throughout the year, as does the trough east of the Andes.

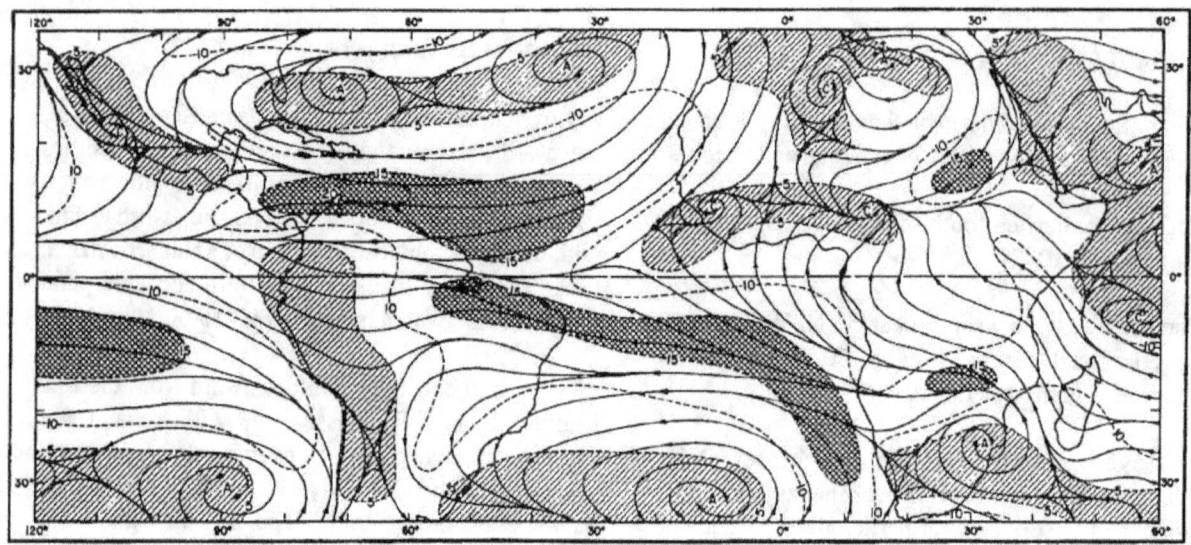

Figure 4-10a. Resultant gradient winds (knots) for April, 120° W to 60° E (from Atkinson and Sadler, 1970).

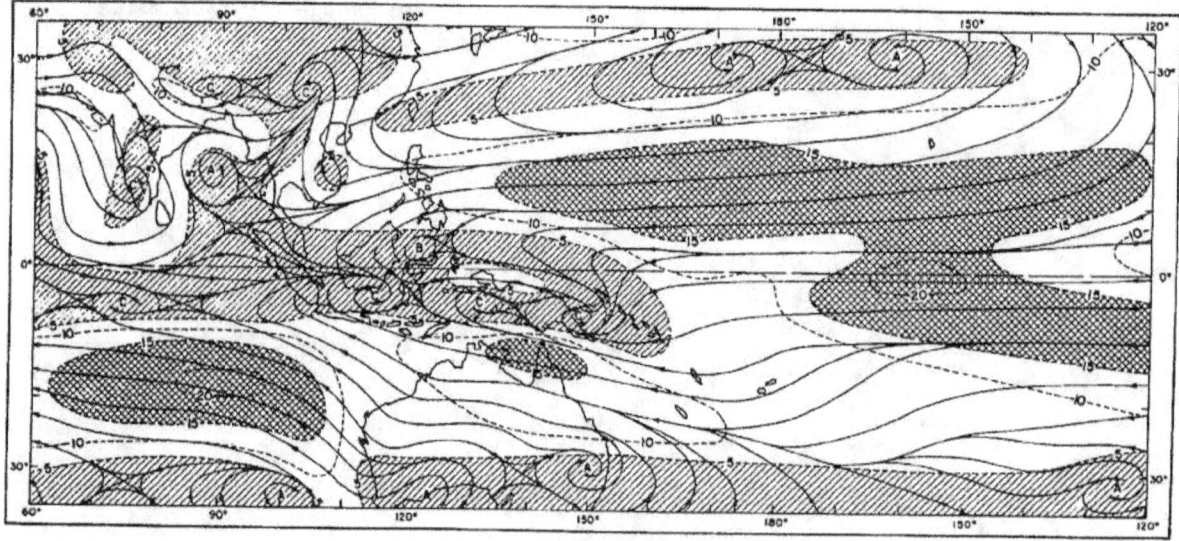

Figure 4-10b. Resultant gradient winds (knots) for April, 60° E to 120° W (from Atkinson and Sadler, 1970).

Major features shown in Figure 4-10, opposite:

Northern Hemisphere

• The oceanic subtropical ridges, the near-equatorial trough and the near-equatorial tradewind convergence lie close to their January positions.

• Northeasterlies south of the ridge have diminished, especially over the South China Sea, where the winter monsoon is almost over.

• Heat lows develop over China, southern India, and western North Africa, replacing the ridges of January.

• The low over northern Africa stems from extratropical cyclogenesis east of the Atlas Mountains (Gleeson, 1954).

• Anticyclonic circulations persist over the Arabian Sea and the Bay of Bengal, but by May, they give way to the southwest monsoon.

Southern Hemisphere

• The monsoon trough has moved north; it now extends from 50° E to 150° E. Over Indonesia and the Indian Ocean, the near-equatorial troughs of both hemispheres are nearly equidistant from the equator. This structure, typical of spring and autumn, favors "twin" tropical cyclone formation—one in each hemisphere.

• The subtropical ridge over the oceans remains poleward of 30° S.

• The heat lows over central South America, southern Africa, and northern Australia are replaced by tradewinds; anticyclones start to move over southern parts of the continents.

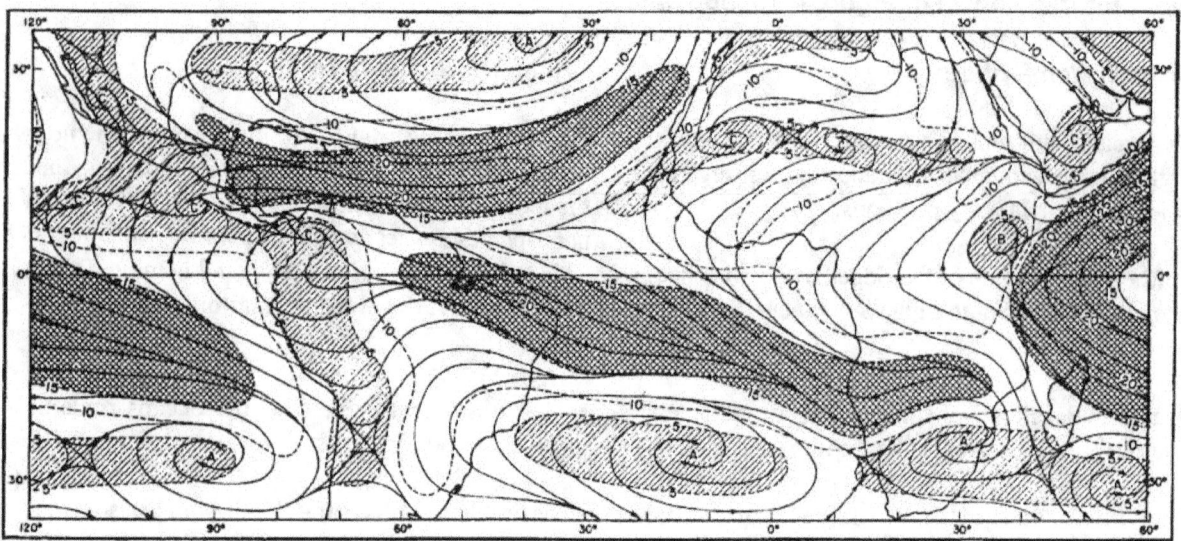

Figure 4-11a. Resultant gradient winds (knots) for July, 120° W to 60° E (from Atkinson and Sadler, 1970).

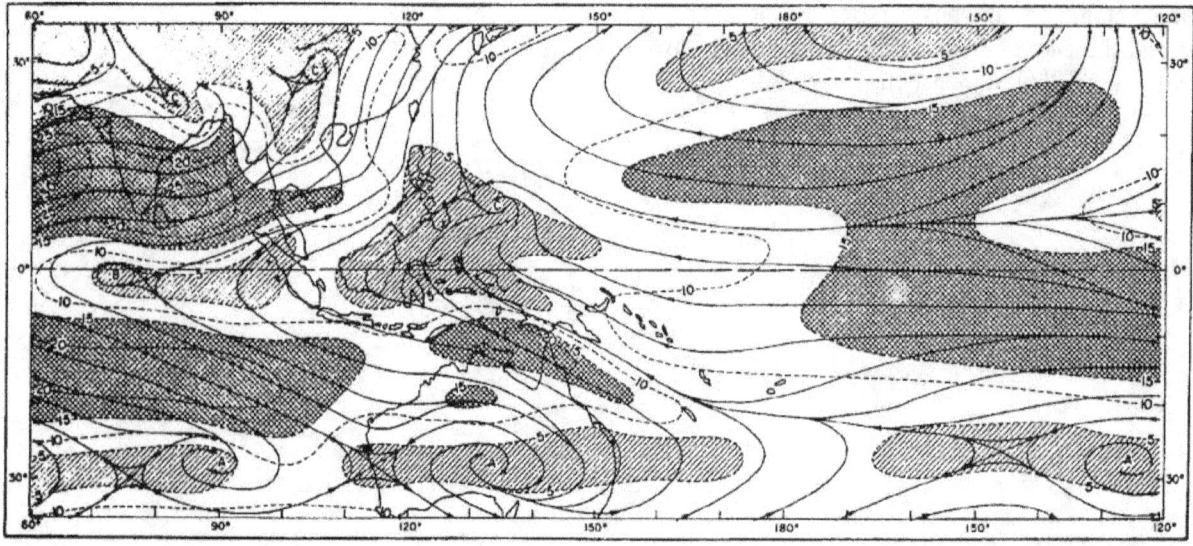

Figure 4-11b. Resultant gradient winds (knots) for July, 60° E to 120° W (from Atkinson and Sadler, 1970).

Major features shown in Figure 4-11, opposite:

Northern Hemisphere

• The oceanic anticyclones have moved well north and expanded. South of the anticyclones, tradewinds are strongest between 15° N and 20° N.

• The heat/monsoon trough is well-established over southern Arabia and North Africa and over south and southeast Asia; westerlies exceed 35 knots in the Arabian Sea and 25 knots in the Bay of Bengal. The cyclone east of the Philippines reflects a concentration of tropical cyclone tracks. Over the Philippines, southwesterlies are weak because the monsoon trough there ranges over about 10° of latitude.

• A heat low has developed over Mexico.

• The cool-season NETWC over the northeast Pacific has been replaced by a trough extending westward from northern South America.

Southern Hemisphere

• The weak monsoon trough is now confined to the Indian Ocean.

• Strong tradewinds dominate the tropics.

• The SPCZ now reflects passage of extratropical systems.

Figure 4-12a. Resultant gradient winds (knots) for October, 120° W to 60° W (from Atkinson and Sadler, 1970).

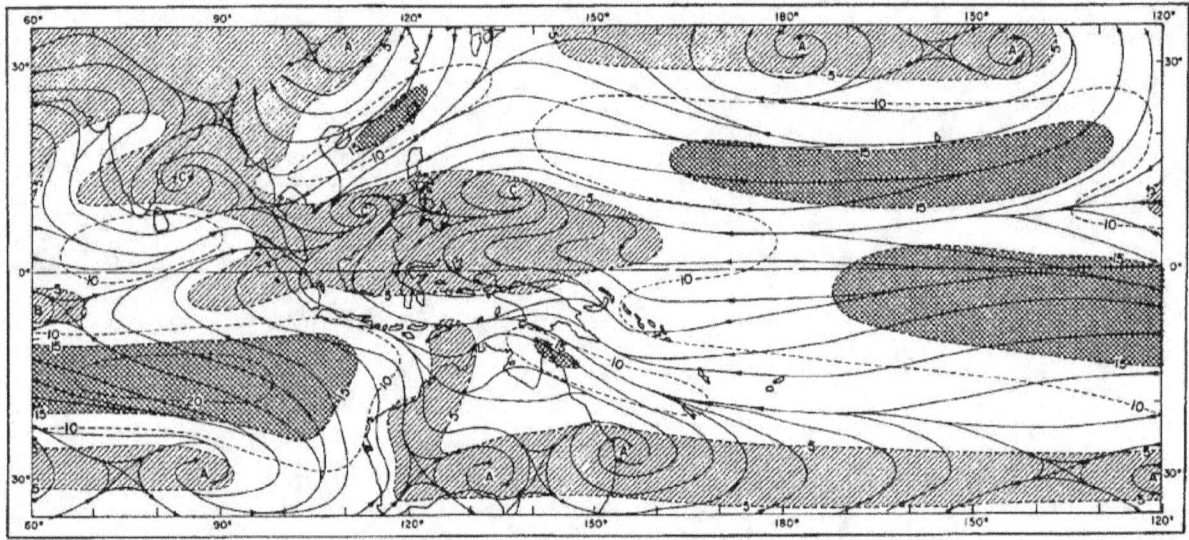

Figure 4-12b. Resultant gradient winds (knots) for October, 60° E to 120° W (from Atkinson and Sadler, 1970).

Major features shown in Figure 4-12, opposite:

Northern Hemisphere

• Except for a minor heat low over the Sahara, cyclonic circulations are gone from the continents.

• The southern Asia heat/monsoon trough now lies across the central Arabian Sea and Bay of Bengal. It is linked to the western Pacific near-equatorial trough, which has remained quasi-stationary. Tropical cyclones may form in this east-west trough, which now extends to Africa.

• Monsoon westerlies have abated over North Africa, replaced by a trough between 5 and 10° N.

• In the eastern North Pacific, the near-equatorial trough has shifted slightly south. Tropical cyclones may form in it on both the Pacific and Caribbean sides of Central America.

Southern Hemisphere

• The broad band of transequatorial winds east of Africa and north of Australia have abated. Otherwise, the tradewinds have not changed much.

• A heat low has formed over southern Africa in advance of the rainy season, and troughs have appeared over central South America and northern Australia where they anticipate the heat lows of summer.

4.3.3 Resultant 200-mb Winds. These fields are shown in Figures 4-13 through 4-16 on the following pages. A brief discussion accompanies each of the charts, which are based on AIREP as well as rawin and rawinsonde averages. Sadler (1975b) gives more detail. In the longitudes where only a single ridge is found in the tropics, it is termed the "subtropical" ridge. When two ridges occur, the poleward one is termed "subtropical." The equatorward ridge is termed "subequatorial," and the intervening trough is known as the "tropical upper tropospheric trough" (TUTT— see Figure 8-20).

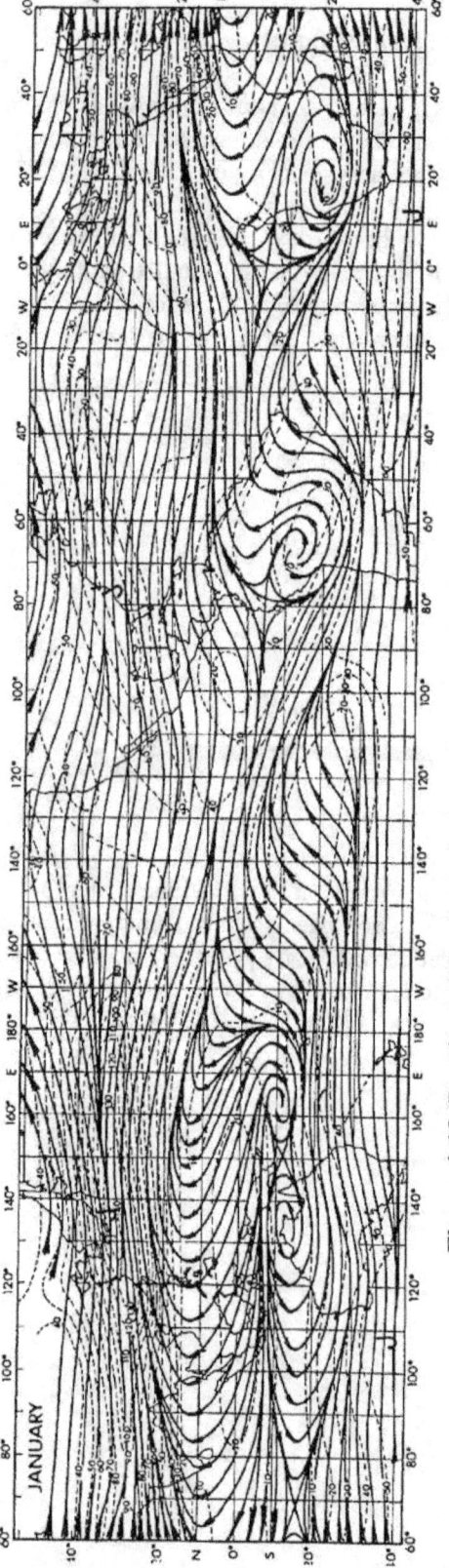

Figure 4-13. Resultant 200-mb winds (knots) for January (from Sadler and Wann, 1984).

Northern Hemisphere

• After being established in November, the mean subtropical jet stream (STJ) reaches 143 knots just south of Japan.

• The subtropical ridge near 10° N extends from 0° eastward to 170° E.

Southern Hemisphere

• The STJ generally lies south of 35° S.

• The subtropical ridge, lying between 10 and 20° S, extends from 0° eastward to 180°.

• In the SH, TUTTs extend to the equator, where westerlies prevail. The TUTT axes are oriented southeast-northwest. In the Pacific, the axis extends from 30° S, 105° W to 175° W and the equator; in the Atlantic, from 30° S, 15° W across South America to 75° W and the equator. The troughs are separated by an anticyclone over western South America.

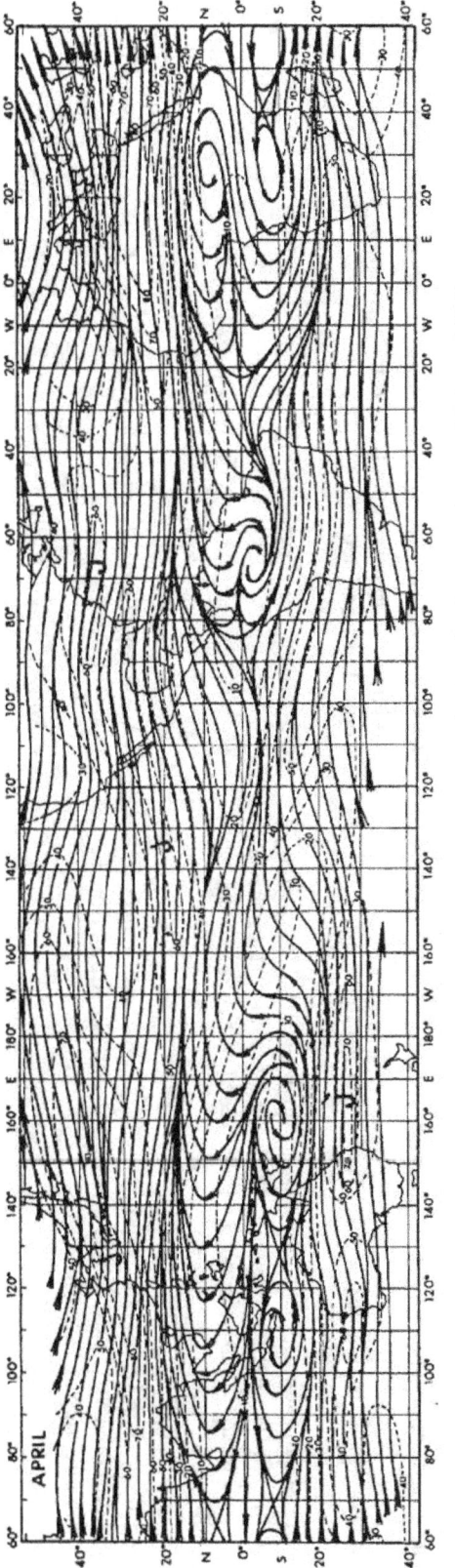

Figure 4-14. Resultant 200-mb winds (knots) for April (from Sadler and Wann, 1984).

Northern Hemisphere

• The STJ shifts north of 35° N and weakens.

• The subtropical ridge moves slightly south to near 8° N.

• A narrow band of equatorial easterlies extends from 170° E westward across the Indian Ocean and Africa into the Atlantic.

Southern Hemisphere

• The SH STJ has moved north to about 25 or 30° S. Maximum resultant speeds exceed 70 knots east of Australia.

• The subtropical ridge has moved north to between 10 and 15° S.

• TUTTs are much weaker.

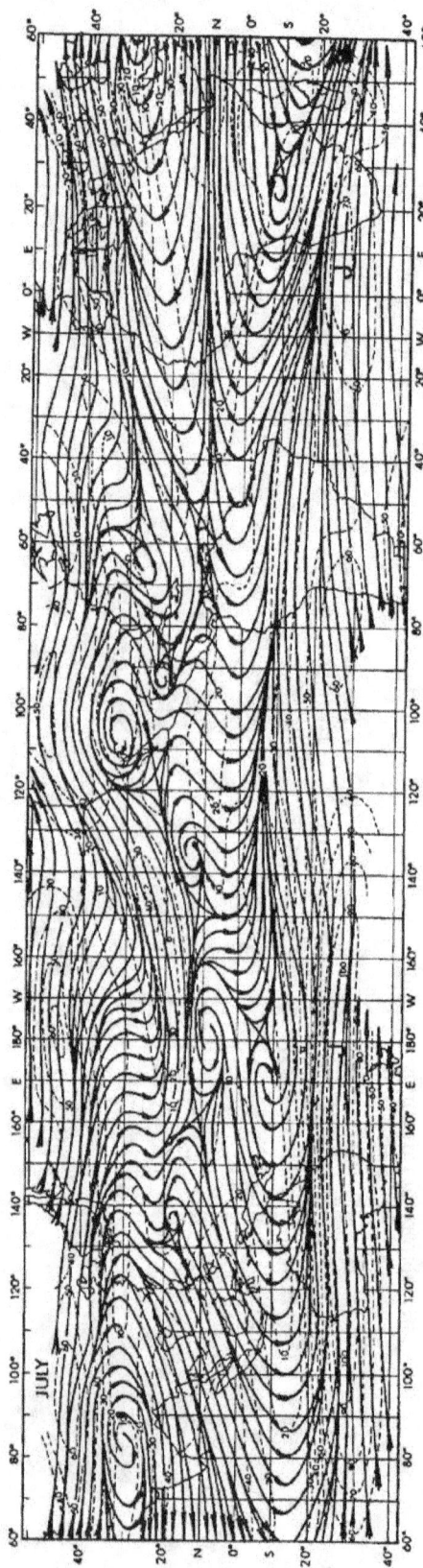

Figure 4-15. Resultant 200-mb winds (knots) for July (from Sadler and Wann, 1984).

Northern Hemisphere.

• The circulation changes dramatically from April to July.

• The STJ disappears and the polar jet stream moves north of 35° N.

• An anticyclone is established near 32° N over North America.

• Over Asia, the subtropical ridge moves from near 10° N in April to 30° N in July. To the south a tropical easterly jet stream is established over the north Indian Ocean with mean winds exceeding 50 knots over southern India.

• Over the Indian Ocean, the strong southerly low-level monsoon flow across the equator (Figure 4-8) is matched by equally strong flow in the opposite direction at 200 mb.

• TUTTs oriented southwest-northeast have appeared over the North Atlantic and North Pacific. The mean mutes day-to-day shifts of the trough and the transient effects of embedded cold-core cyclones.

• A subequatorial ridge extends from 130° E eastward to 50° W. Anticyclones in the eastern and western Pacific parts of the ridge overlie tropical cyclone activity.

Southern Hemisphere

• The STJ near 28° S has moved only slightly south since April but has strengthened to 105 knots over Australia.

• A subtropical ridge, lying between 5 and 15° S, extends westward from 180° to 20° W.

• Over the South Pacific, the TUTT is no longer apparent.

60

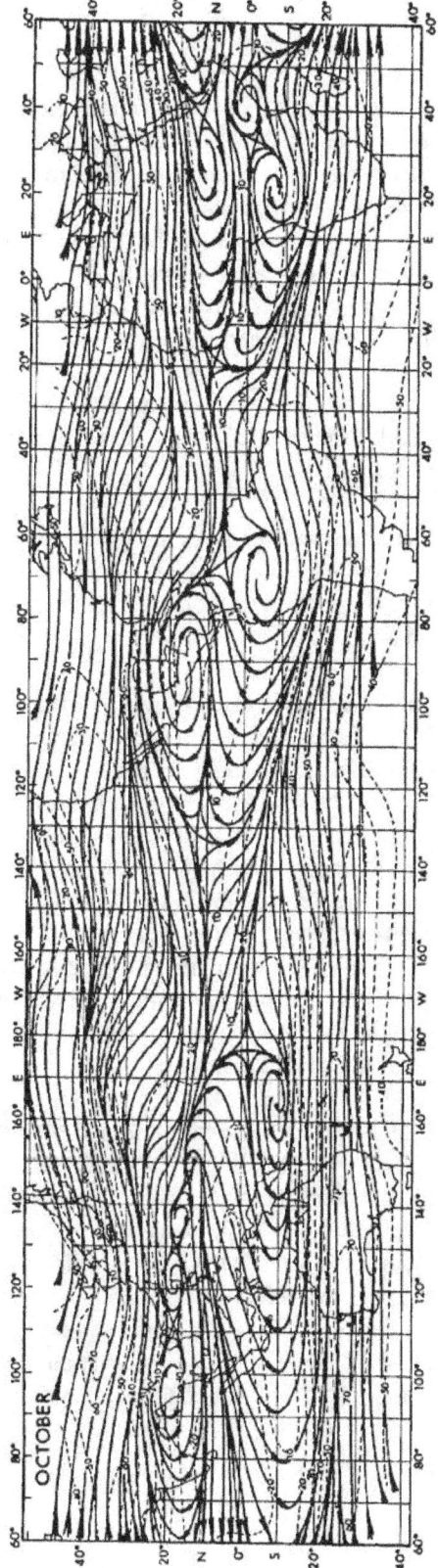

Figure 4-16. Resultant 200-mb winds (knots) for October (from Sadler and Wann, 1984).

Northern Hemisphere

• The NH ridge over Asia and Africa has moved south to between 20 and 10° N. The North American anticyclone has shifted south to near 15° N; except where equatorial westerlies intercede, the subtropical ridge is continuous around the NH.

• The tropical easterly jet has abated.

• Equatorial westerlies have reappeared over the eastern Pacific and Atlantic, but small resultant speeds indicate large daily variability.

• The TUTTs have weakened and the subequatorial ridge has gone.

Southern Hemisphere

• The STJ has moved only slightly from July, having weakened to about 75 knots over Australia.

• The ridge lies between 5 and 10° S with anticyclones over the rain forests of South America and Africa.

Studies have been made of the mean wind field at various levels for specific regions of the tropics. An upper-air atlas prepared by Ramage and Raman (1972) in conjunction with the International Indian Ocean Expedition (IIOE) presents resultant wind analyses for standard pressure levels from 850 to 100 mb and representative meridional cross-sections for the area from 45° N to 50° S and from 20° E to 155° E. Chin and Lai (1974) produced charts for 850, 500 and 200 mb for the area from 46° N to 14° S and from 66° E to 147° E.

Global cross-sections can provide only the broadest view. They are particularly uninformative in the monsoon regions. For example, although the strongest mean surface winds blow from the southwest across the Arabian Sea in summer (Figure 4-11), Figure 2-5 shows the global average to be *easterly* in those latitudes.

4.3.4 Tropical Jet Streams.

4.3.4.1 Subtropical Jet Streams (STJs). As shown in Figures 4-13 and 4-16, STJs are climatological features of the tropical general circulation. Over the NH in winter, the STJ has a three-wave pattern with ridges and maximum speeds over eastern North America, the Middle East, and south of Japan. The mean latitude ranges from 20 to 35° N. Steadiness exceeds 90 percent and speeds may occasionally exceed 200 knots. Over the SH, the winter STJ exceeds 100 knots only over and to the east of Australia, where it lies near 27° S.

4.3.4.2 Tropical Easterly Jet (TEJ). This summer feature is much broader than the STJ. At 150 mb it extends from 5 to 15° N and across the Indian Ocean and Africa. Average speeds exceed 70 knots. Between 1956 and 1962, 151 knots was once measured at 59,000 feet (18 km) over Bombay (Flohn,1964). The TEJ will be discussed again in Chapter 6.

4.3.4.3 Low-Level Jet Streams (LLJs). Two types of LLJ (seasonal or year-round) occur in the tropics.

LLJs flow where persistent, large surface temperature gradients enhance surface pressure gradients, and where vertical motion above about 6,500 feet (2 km) is inhibited by subsidence. Favored locations include desert littorals, especially where coastal waters are upwelling, and over equatorial upwelling (Figure 5-4). LLJs also border heat troughs or sharply defined zones of heavy rain. In all of these cases, the thermal wind opposes the wind in the surface layer. Grossman and Friehe (1986) and others have pointed out two processes that define and sharpen the seasonal LLJ:

• Surface friction over the sea and a nocturnal surface inversion over land ensure that wind increases with height above the surface.

• The horizontal temperature gradient, by creating a thermal wind opposite to the surface wind, causes the wind above the surface friction layer to decrease rapidly with height.

The largest and most intense LLJ prevails over the western Indian Ocean during the NH summer, extending from east of Madagascar across eastern Somalia to Peninsular India. At its core height of 5,000 feet (1.5 km), speeds may exceed 60 knots (see Figure 4-11). Over eastern Somalia, the jet is enhanced by the temperature gradient across upwelling coastal waters and the coastal desert. Also, it is confined and accelerated by a subsidence inversion and by mountains to the west that parallel the flow. LLJs have been observed along the west coasts of South America south of the equator, and in Namibia.

At somewhat higher levels (between 5,000 and 15,000 feet/ 1.5 to 4.5 km), northern South America experiences LLJs along about 15° N during summer. They occur in spring and early summer over west Africa and south of the Mei-Yu front over south China (see 6.3.6.4). These jets seldom exceed 25 knots and seem to result from temperature gradients caused partly by release of latent heat of condensation. All seasonal or annual LLJs appear to accelerate in response to surges from higher latitudes.

Nocturnal LLJs develop over land when fresh synoptic winds coincide with clear skies. Radiational cooling, by creating a surface inversion and reducing dissipation by friction, allows the winds between about 1,000 and 3,000 feet (300 and 900 meters) above the surface to increase; a bordering mountain range and an overlying inversion narrow the jet layer and further increase the speed. The jet is strongest near dawn. Along the coastal plain of northeastern Saudi Arabia between 24 and 29° N, a nocturnal jet in the prevailing summer northwesterlies may exceed 50 knots between 800 and 1,500 feet (250 and 450 meters) (Vojtesak et al., 1991). At Daly Waters (16° S, 133° E) in Australia, a nocturnal LLJ occurs on 19 percent of all winter nights. It is generally 100 or 200 NM wide and about 500 NM (1,000 km) long (Brook, 1985).

Over land, daytime surface heating that leads to vertical mixing weakens and may dissipate all LLJs. This process is discussed in 7.4.5.2.

4.3.5 Tropical Stratospheric Winds. A quasi-biennial oscillation (QBO) of winds in the tropical stratosphere was discovered by McCreary (1959). This phenomenon has been intensively investigated by Reed et al. (1961) and by many others. The oscillation between easterly and westerly wind regimes occurs on an average of every 26 months, but the period is irregular, varying between 21 and 30 months (hence the name "quasi-biennial").

The maximum of the oscillation occurs at 75,000-85,000 feet (23-26 km) above the equator and decreases downward and poleward. The easterly or westerly wind regime, which first appears above 98,000 feet (30 km), sinks at about 3,000 feet (1 km) a month without losing amplitude in the first 23,000 feet (7 km). But below 75,000 feet (23 km), it attenuates sharply and is barely noticeable at 100 mb, the average level of the tropical tropopause.

Figure 4-17 (next page) illustrates the QBO in the equatorial stratosphere from 1953 through 1984 (Naujokat, 1986). The annual cycle of stratospheric wind change is small at the equator, where the QBO is most prominent. In the troposphere and at more poleward stations, the larger annual cycle makes the observed wind pattern more complex. For example, the 21-km mean monthly zonal winds at Darwin (12° S, 131° E) from 1958 to 1968 reflect both the annual and quasi-biennial cycles shown in Figure 4-18 (Hopwood, 1968). The average amplitudes of the two cycles are shown in Figure 4-19.

Along the equator, the QBO dominates; poleward of about 20°, the annual oscillation is much larger. Although the QBO has been well-described, its exact cause is still debated (Wallace, 1973). The climatology of tropical stratospheric winds cannot be satisfactorily represented in standard format such as resultant winds on constant pressure surfaces or by vertical cross-sections. Instead, the mean winds for a given month and year can be anticipated by extrapolating into the future the recent QBO cycle for the station and levels of interest. At most, the QBO affects the tropospheric circulations only marginally (Trenberth, 1980); it affects weather not at all.

63

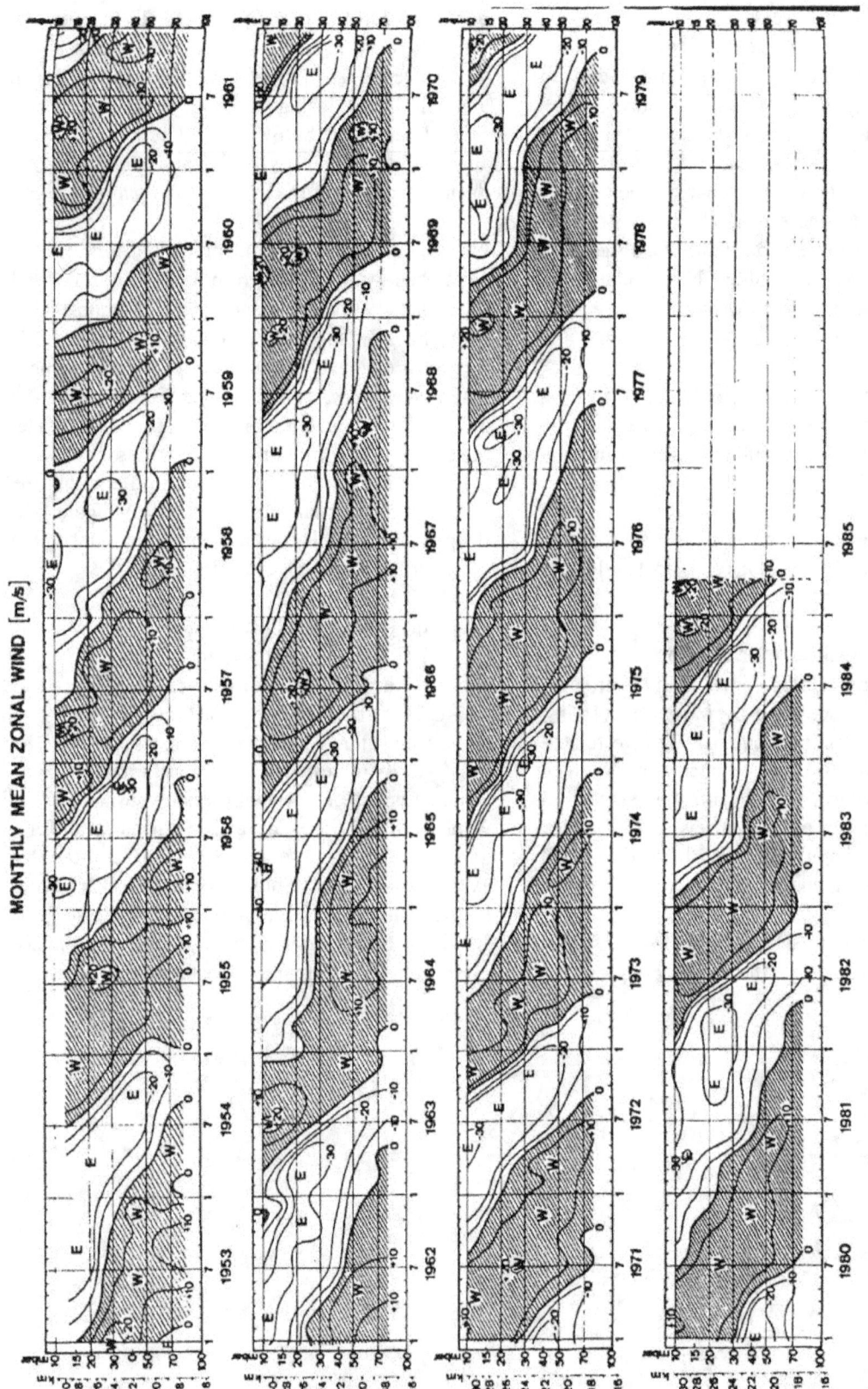

Figure 4-17. Time-height section of monthly mean zonal wind components (meters/sec) at Canton Island, 4° S, 174° W (January 1953 to August 1967); Gan (1° S, 73° E (September 1967 to December 1975); and Singapore, 1° N, 104° E (January 1976 to April 1985) (Naujokat, 1986).

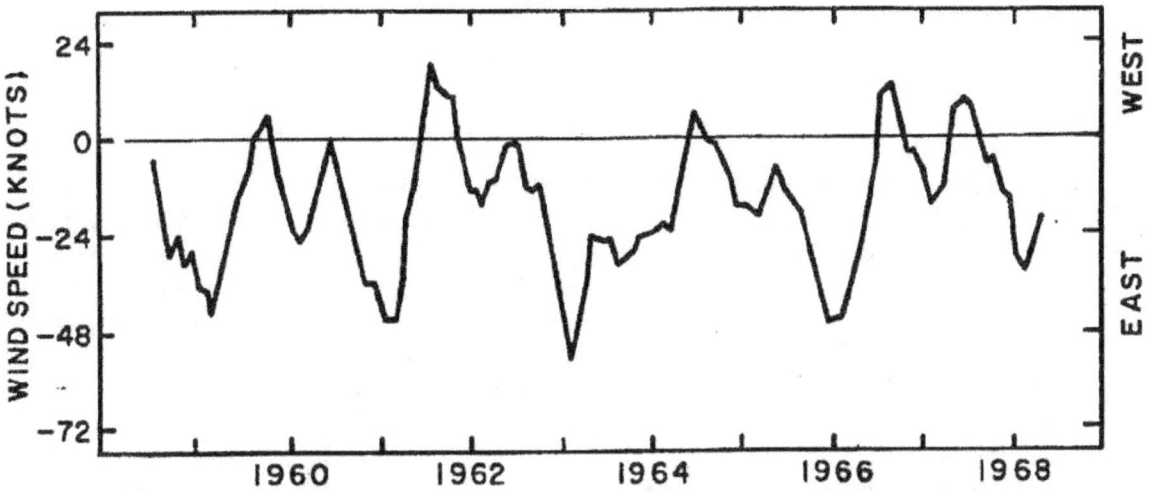

Figure 4-18. Mean monthly zonal wind speed at Darwin (12° S, 131° E) at 70,000 feet (from Hopwood, 1968).

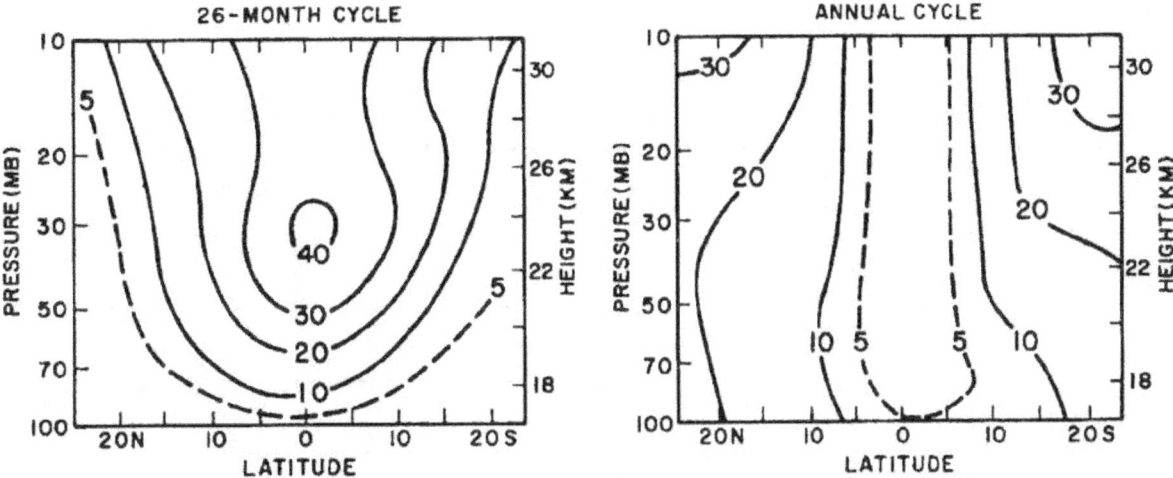

Figure 4-19. Average amplitude (knots) of the quasi-biennial and annual cycles of the mean monthly zonal wind speed in the tropical stratosphere (from Reed, 1964).

Chapter 5

TEMPERATURE AND WATER VAPOR

5.1 General

This chapter examines the distributions and variations of temperature and water vapor in the global tropics. Water, in its atmospheric vapor and liquid forms (weather), dominates almost every topic from here on.

Most of the tropical atmosphere is characterized by small horizontal temperature gradients. Near the surface, differential heating of land and water (or cold ocean waters upwelling near coasts) may cause large temperature gradients; however, the effects of these disappear or are greatly reduced several hundred feet (a few hundred meters) above the surface. Even the strongest cold fronts, which may penetrate deep into the tropics, can scarcely be found in the surface temperature field over the oceans below 20° latitude. Although temperatures vary little, water vapor varies perceptibly throughout the troposphere, the result of annual changes in the large-scale tropical circulation and from day-to-day changes caused by synoptic weather systems.

As Table 5-1 shows, performances of all electrooptical devices can be seriously affected by atmospheric water and dust. Forecasters should memorize this table and use it to assess the effects of tropical weather on electrooptics.

TABLE 5-1. Weather effects on electrooptical devices (Seagraves, 1989).

Weather Type	Visible & Near-IR Imagers	Thermal Imagers	Lasers	Guidance Devices
Clouds	Severe	Severe	Severe	Severe
Fog	Severe	Moderate	Severe	Moderate/Severe
Rain/snow				
Light	Moderate	Moderate	Moderate	Moderate
Heavy	Severe	Severe	Severe	Severe
Smoke/dust/sand				
Hvy density	Severe	Moderate	Severe	Moderate/Severe
Mdt density	Moderate	Moderate	Moderate	Moderate
High humidity	Low	Moderate	Low	Low/Moderate

5.2 Surface Air Temperature and Water vapor

5.2.1 Means. Mean surface air temperatures and standard deviations for January, April, July, and October are shown in Figures 5-1a-d. Because of their size, these maps can show only gross temperature features–they should not be used to obtain values for specific locations. Temperature, humidity, and rainfall climatologies for individual stations are readily available in many publications, such as Crutcher and Davis (1969), numerous USAF Environmental Technical Applications Center (USAFETAC) documents, and the eleven volumes of the U.S. Navy Marine Climatic Atlas of the World.

Air over the tropical oceans has a mean temperature of about 82° F (28° C) throughout the year. In contrast, summer means over the subtropical continents exceed 90° F (32° C). Over the open tropical oceans, surface air temperature is high and the standard deviation small. This is especially so over the western oceans, where upwelling is insignificant.

Large temperature gradients near the east coasts of North America and Asia result from nearby warm ocean currents such as the Gulf Stream and the Kuroshio Current; the latter flows northeastward from Taiwan along Okinawa and the coast of Japan as far as 35° N. Small shifts in wind direction over these currents may lead to considerable heating or cooling of the air by the sea. Large values of the standard deviation result.

Persistent oceanic anticyclones cause equatorward winds to blow along the west coasts of the continents (see Figures 4-1 to 4-4 and 4-9 to 4-12) where cold sub-surface waters upwell and the air is cooled. In general, mean air temperature differs little from mean sea-surface temperature (SST), being noticeably cooler only off coasts that experience frequent polar outbreaks. The heating (or cooling) of subtropical continents closely follows the poleward (equatorward) movement of the sun. Regions are warmest when the sun is most nearly overhead and coldest when it is farthest away. On the other hand, because of the sea's large specific heat, SST changes lag the sun's movement by 1 to 2 months. Over tropical lands, the mean temperature distribution is influenced by land/sea contrasts along coasts and lake shores, elevation differences, cloudiness, and other local effects.

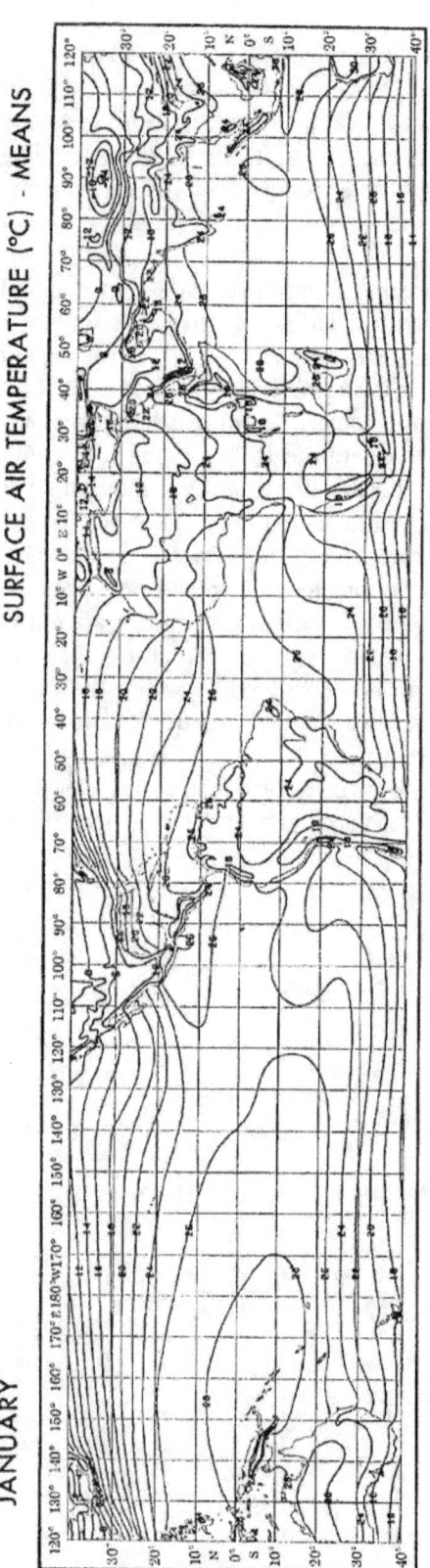

Figure 5-1a. Surface air temperature means (° C) for January (from Naval Oceanography Command, 1981).

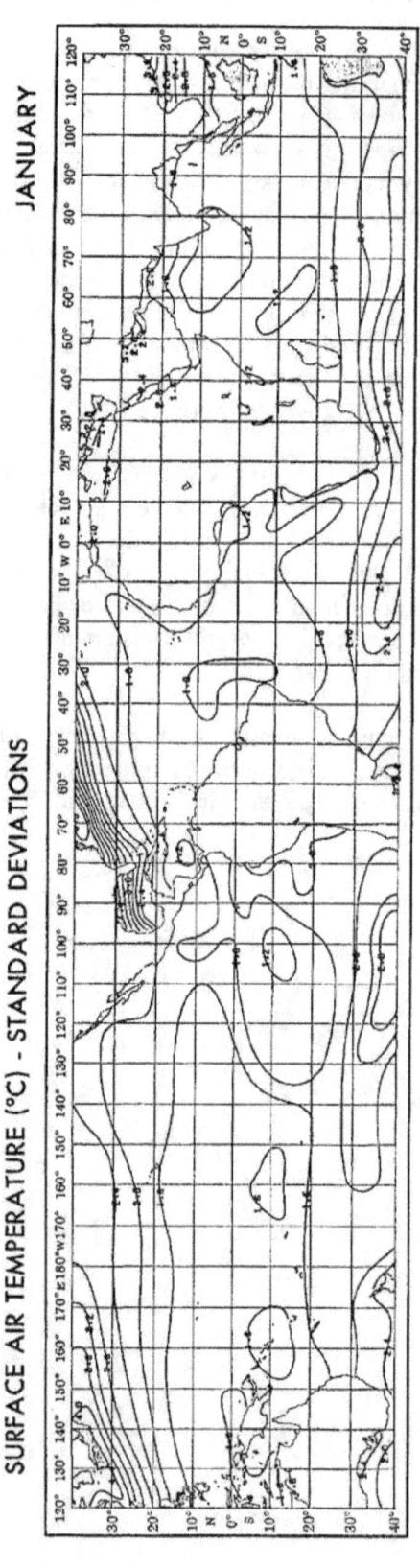

Figure 5-1b. Surface air temperature standard deviations (° C) for January (from Naval Oceanography Command, 1981).

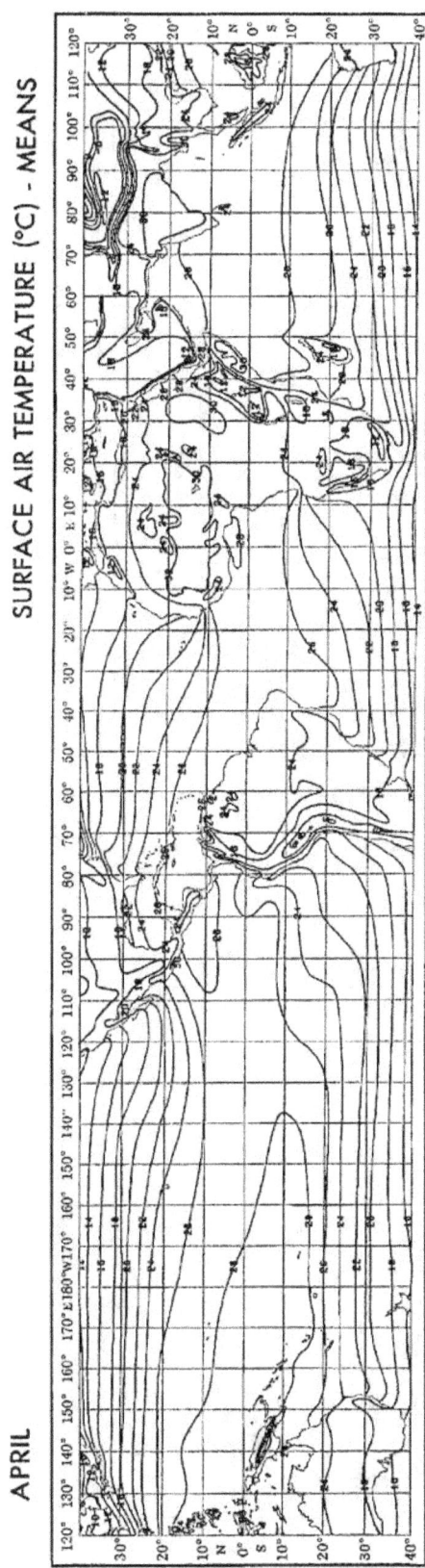

SURFACE AIR TEMPERATURE (°C) - MEANS

APRIL

Figure 5-1c. Surface air temperature mean (° C) for April (from Naval Oceanography Command, 1981).

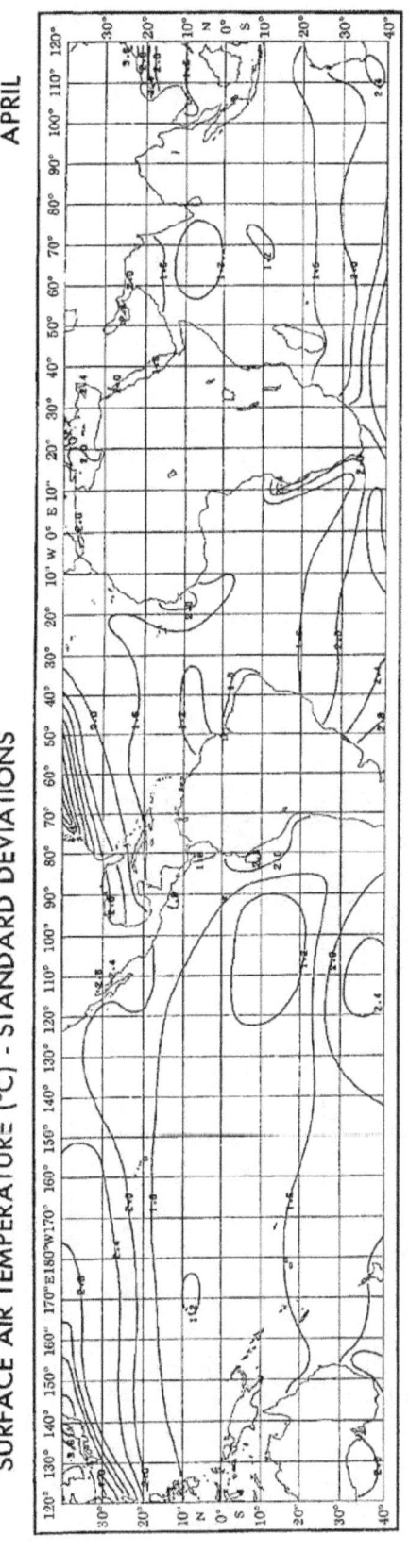

SURFACE AIR TEMPERATURE (°C) - STANDARD DEVIATIONS

APRIL

Figure 5-1d. Surface air temperature standard deviations (° C) for April (from Naval Oceanography Command, 1981).

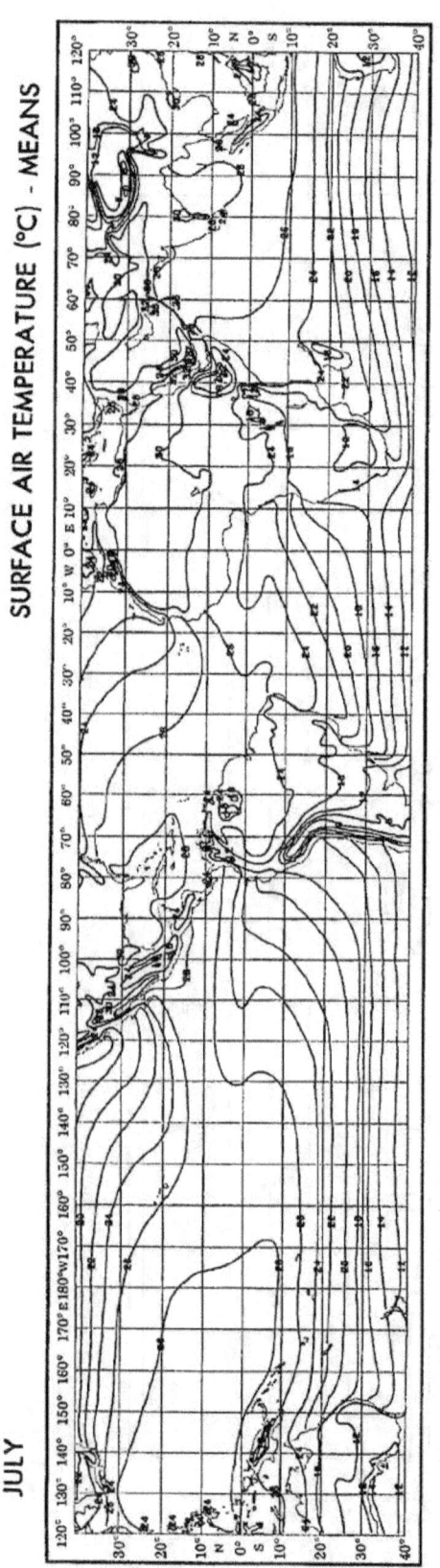

Figure 5-1e. Surface air temperature means (° C) for July (from Naval Oceanography Command, 1981).

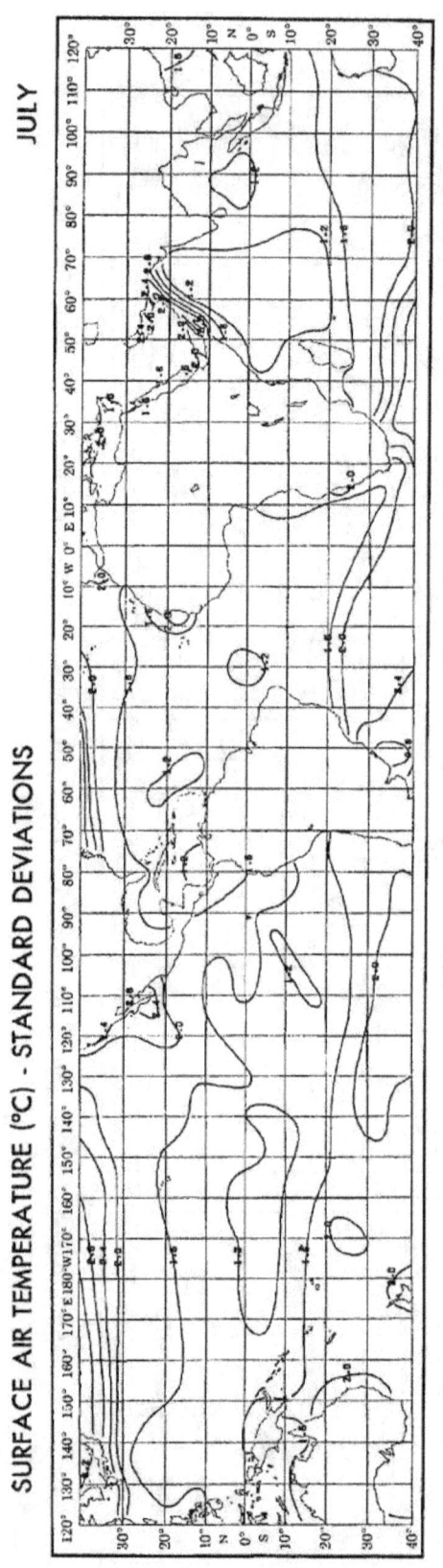

Figure 5-1f. Surface air temperature standard deviations (° C) for July (from Naval Oceanography Command, 1981).

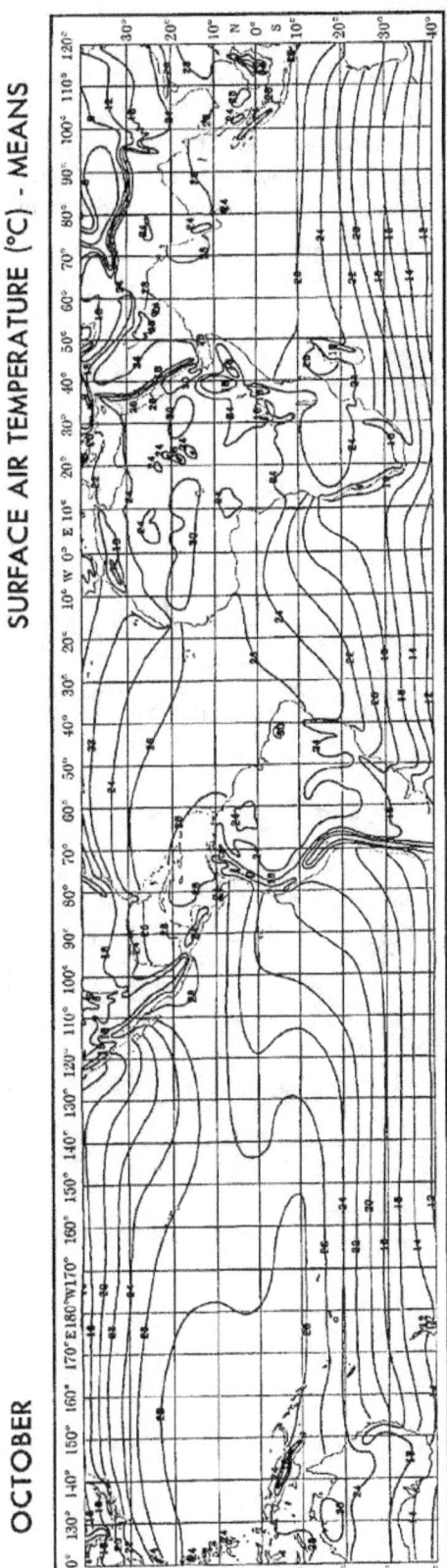

Figure 5-1g. Surface air temperature means (° C) for October (from Naval Oceanography Command, 1981).

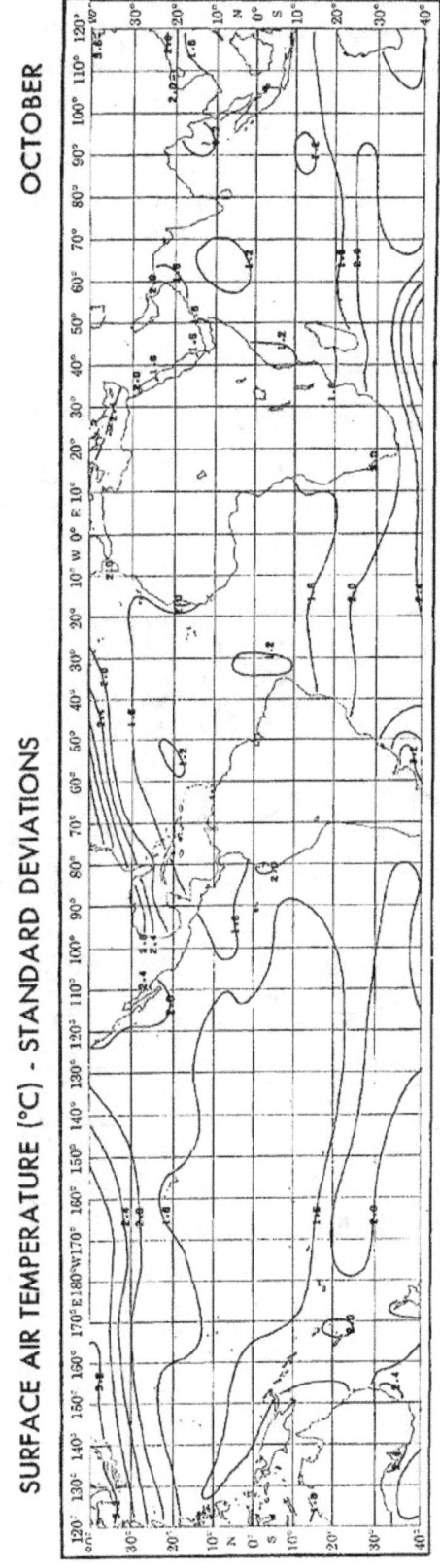

Figure 5-1h. Surface air temperature standard deviations (° C) for October (from Naval Oceanography Command, 1981).

To illustrate typical surface temperatures and annual variations over tropical continents, Figure 5-2 gives mean monthly temperatures for representative stations. The stations lie along approximately meridional profiles across Africa, East Asia/Australia, and the Americas. The three regions are alike in that the annual range is large in the subtropics and small within 10° of the equator, amounting to 42° F (23° C) at Shanghai and 2° F (1° C) at Singapore.

In higher latitudes, the annual variation approximates a sine curve, with the extremes generally occurring in January and July. The curves are more complex in lower latitudes, with double maxima and minima at some stations. This results from the sun passing directly overhead twice a year and from corresponding changes in humidity and cloudiness. In low latitudes, highest temperatures are experienced during the sunny months just before the rainy season starts.

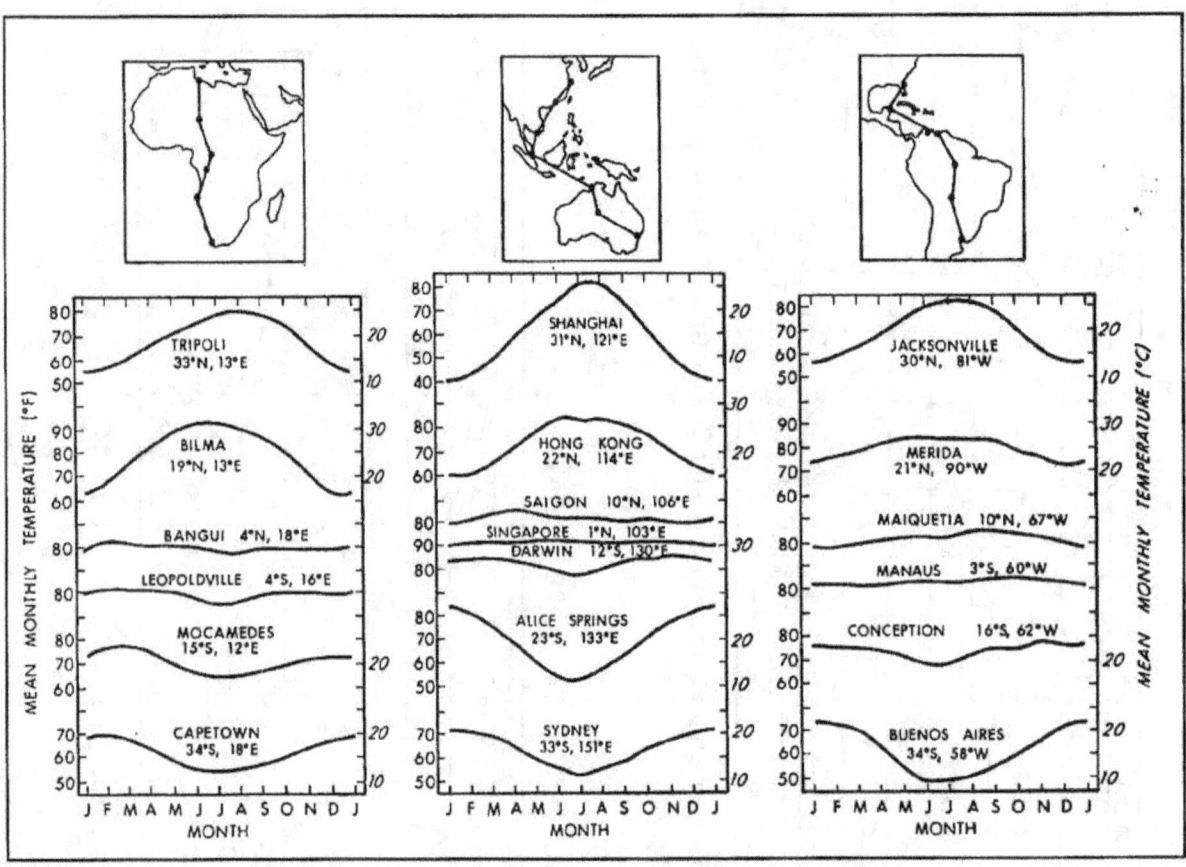

Figure 5-2. Annual march of mean monthly temperatures at selected tropical stations along meridians through the continents.

5.2.2 Upwelling/Downwelling.

Wind blowing across open water produces a stress on the surface (see Sadler et al., 1987) that causes a current to flow in a direction turned 90° anticyclonically from the wind direction. In the NH there is a surface stress to the *right* of the wind; in the SH, the surface stress is to the *left* of the wind. Therefore, when the wind blows parallel to a coast, surface water is forced either away from or toward the coast, depending on the sense of the wind. Outward flow causes water to upwell along the coast, as shown in Figure 5-3. The upper cross-section shows the effect of a surface wind directed into the paper. Note that the wind stress has raised the surface water level off the coast and lowered it along the coast. The lower cross-section is a plan view. The arrow with the wavy shaft denotes a geostrophic surface current generated by the change in surface water level.

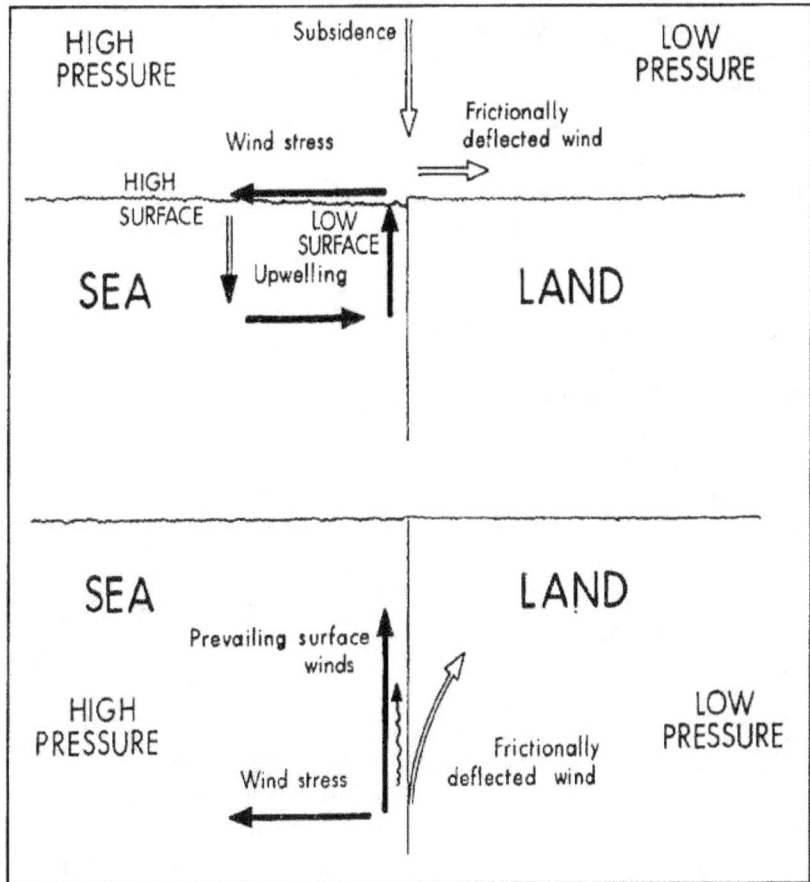

Figure 5-3. Schematic showing the effects of surface wind blowing parallel to the west coast of a southern hemisphere continent such as South America or southern Africa.

If upwelling extends below the surface mixed layer and into the thermocline, cooler water will be brought to the surface. Thus, a north wind along the west coast of North America, or a south wind along the east coast of Somalia, will produce cold-water upwelling. The wind stress associated with upwelling also causes water to accumulate off the coast and increase the ocean pressure there. In an analogy to the atmosphere, a geostrophic ocean current develops roughly parallel to the wind. Therefore, in almost all coastal upwelling, the geostrophic current enhances surface cooling by advecting water from higher latitudes. The exception is off Somalia in summer where the southwest monsoon upwells cold water, but the geostrophic current brings warmer water from the south. Even so, the net result is significant surface cooling. The large thermal gradient that develops across the Somali coast causes a low-level jet stream to develop; the

surface wind increases, further enhancing the upwelling. The net result of this positive feedback is the strongest, steadiest surface wind in the tropics (see Figure 4-11).

If the wind flow were reversed, the resulting cyclonically turning current would produce warm water *downwelling*, which occurs along the coast of China in winter during the northeast monsoon. However, the strong geostrophic current generated by the wind brings cold surface water from the north and causes net cooling at the surface.

Coriolis force changes sign across the equator. Thus easterly winds, blowing along the equator, are divergent (Gordon and Taylor, 1970); wind stress forces surface water away from the equator and upwelling results. Westerlies cause downwelling.

73

Figure 5-4 shows upwelling regions and coastal deserts. The winds driving coastal upwelling undergo frictionally induced divergence along the coast (see 2.4.3). Over these same regions, large-scale subsidence, associated with subtropical anticyclones or with the Asian summer monsoon in the case of the western Indian Ocean, further contributes to relatively dry weather (Lydolph, 1973) and helps maintain coastal deserts. Off western Australia, upwelling fails to lower the SST, probably because warm sub-surface water drifts southward around Northwest Cape (22° S, 114° E).

Upwelling persists through the year off the west coasts of the Americas and southern Africa, and also off the north coast of South America. Since the water in these regions is normally colder than the land, the annual reversal of the temperature gradient between land and sea, needed for monsoons to develop, cannot occur.

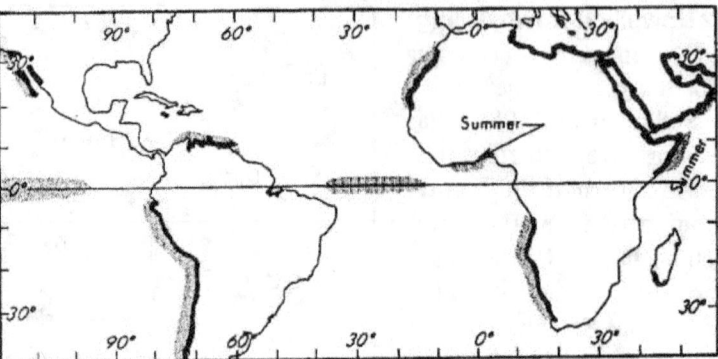

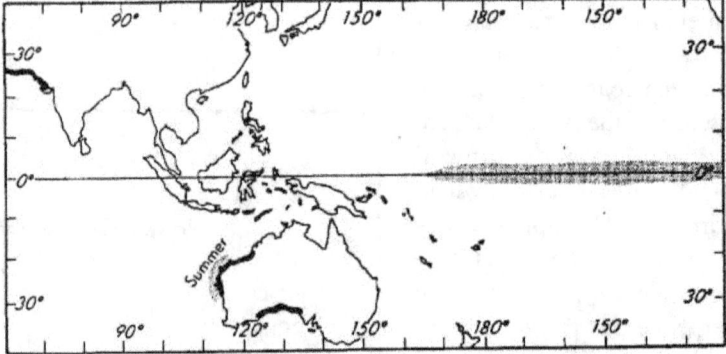

Figure 5-4. Upwelling regions (stippled). The littoral (coastal region) bordering upwelling is always relatively dry (heavy lines) and may sometimes be desert (Meigs, 1973).

Along the west coast of South America during El Niño (see 6.3.5), cold water in the thermocline is too deep to be reached by upwelling, and SST becomes anomalously high.

5.2.3 Elevation Effects. Over tropical continents away from the coasts, mean monthly surface temperatures show a regular decrease with elevation of about 3° F per 1,000 feet (4.5° C per km). With elevation, dew point decreases less than temperature. This results in higher mean relative humidities and more fog for stations at higher elevations. When combined with rugged terrain, this can pose serious problems for aircraft operations. Near the coast, the patterns of surface temperature and dew point related to elevation may be influenced by land and sea breezes, marine inversions, frictionally-induced divergence, and other local effects.

5.3 UPPER-AIR TEMPERATURE AND WATER VAPOR

5.3.1 Means. The mean structure of the upper-air temperature and humidity fields is illustrated by constant-pressure charts and mean vertical soundings. Figure 5-5 shows the mean 700-mb temperature and relative humidity for winter and summer.

In the NH winter, the 700-mb temperature exceeds 8° C over most of the tropics below 20°. Mean horizontal temperature gradients are small, except over East Asia. The 700-mb 8° C isotherm corresponds to a freezing level of about 14,000 feet (4,270 meters). In the NH subtropical high-pressure

ridges, 700-mb relative humidities are below 30 percent. Mean relative humidities above 70 percent are confined to bad weather areas near the equator.

In the NH summer, the isotherm pattern shifts northward with the sun, and warm centers (exceeding 12° C at 700 mb) appear above pronounced surface heating over southern Asia, North America, and North Africa. The 700-mb relative humidities are highest in regions of deep cloud, and lowest in the subtropical ridges.

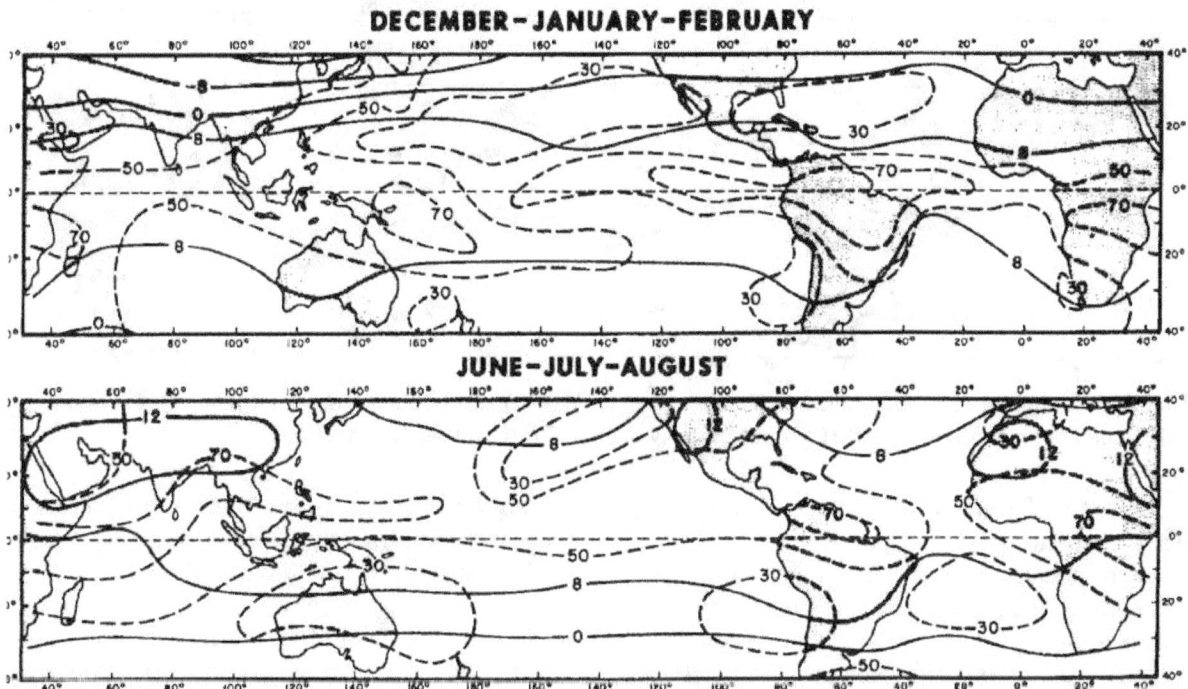

Figure 5-5. Mean 700-mb temperature (° C), solid lines) and relative humidity (percent, dashed lines) for winter (top) and summer (bottom) (after Crutcher and Davis, 1969).

Temperature and relative humidity changes between winter and summer are best illustrated by Figure 5-6. Large temperature changes occur over the NH mid-latitude continents, especially East Asia. Equatorward of 20°, the temperature changes are less than 4° C and often less than 2° C. Since patterns of mean monthly relative humidity resemble those of mean monthly cloudiness and rainfall over much of the tropics, changes in all three patterns occur in phase. In general, 700-mb relative humidities decrease in the SH tropics and increase in the NH tropics from January to July.

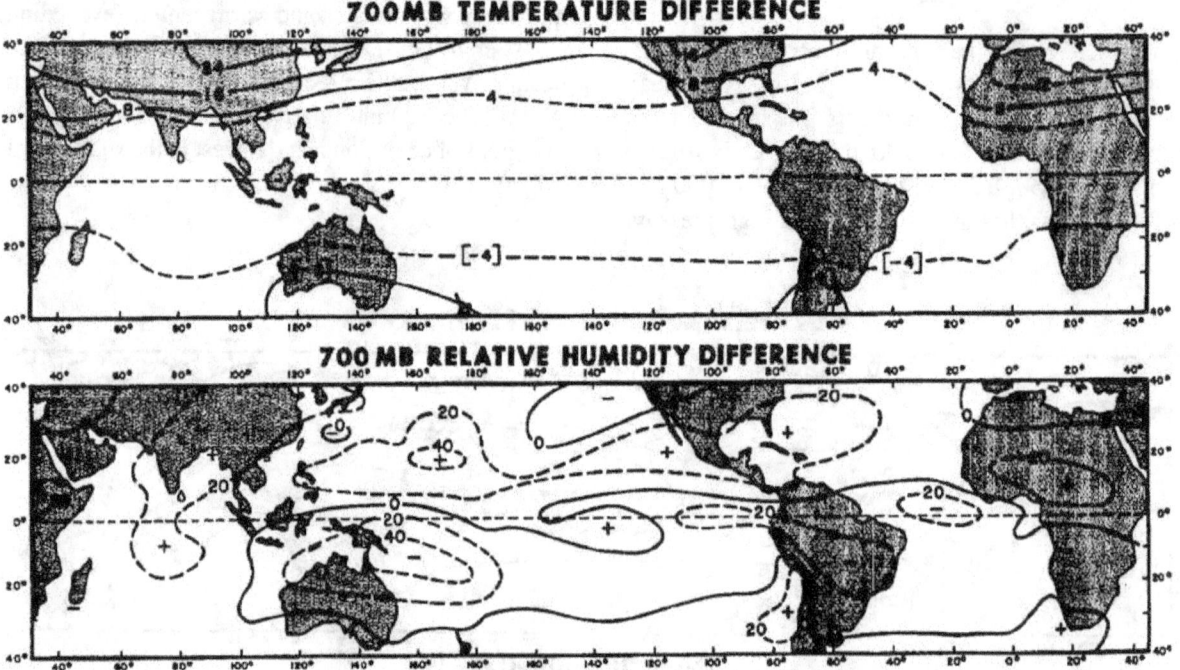

Figure 5-6. Annual changes of mean 700-mb temperature (° C) and relative humidity (percent), NH summer (SH winter) minus NH winter (SH summer) (derived from Crutcher and Davis, 1969).

5.3.2 Mean Precipitable Water. "Precipitable water" is defined as the depth of liquid water that would be obtained if all the water vapor above a unit area of the earth's surface were condensed. It therefore indicates total moisture content within the column of air having a cross-sectional unit area. Tuller (1968) prepared mean monthly and annual maps of the world distribution of precipitable water based on the years 1964 through 1966; his maps for January and July are shown in Figure 5-7, next page.

In January, mean values in the tropics range from less than 0.50 inches (13 mm) over North Africa to more than 1.75 inches (45 mm) over much of the low-latitude tropics, especially in the SH. Large gradients of precipitable water and elevation coincide, as along the western Andes, the coast of eastern Africa, and the southern Himalayas. Also, note the large meridional gradient over western Africa.

In July, the area with mean precipitable water over 1.75 inches (45 mm) is largely confined to the NH. Values exceed 2.5 inches (60 mm) near Bangladesh. Lowest values are found over Australia and southern Africa. The monthly maps of mean precipitable water show a meridional shift that accompanies the sun over the continents, and lags it by about 2 months over the oceans. The annual range is small over the equatorial oceans, high mountains, and higher latitudes. The greatest annual range, more than 1.50 inches (38 mm), occurs over the Bay of Bengal.

76

JANUARY

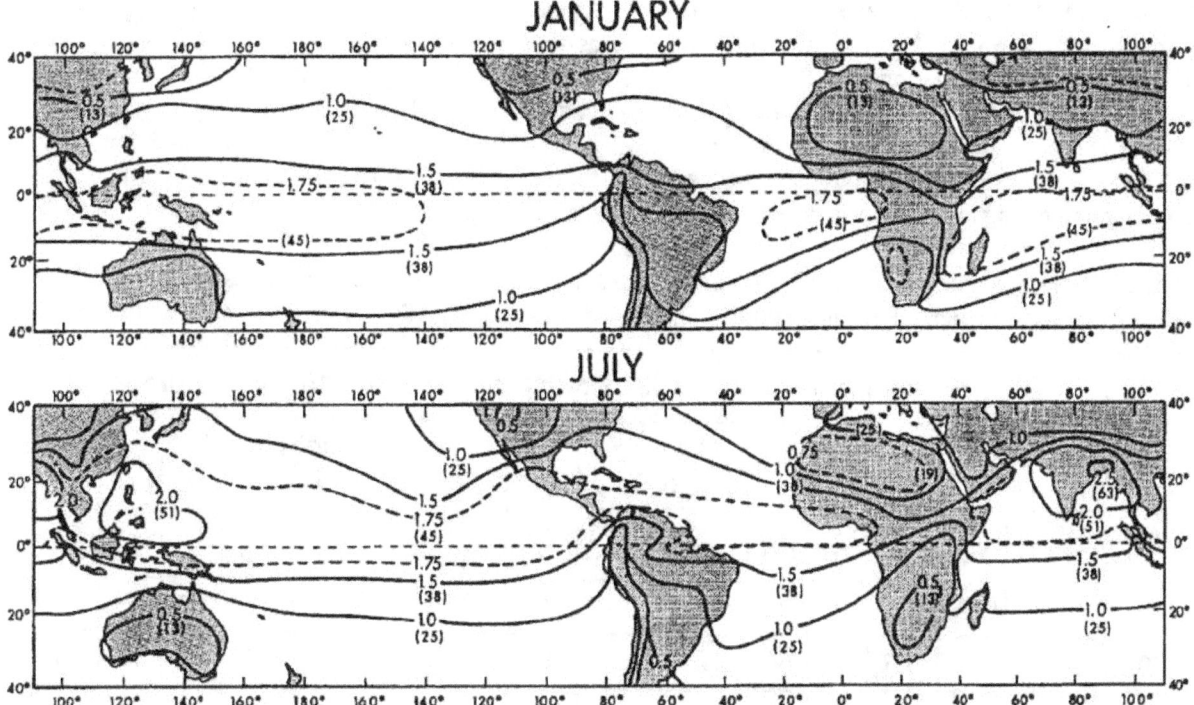

JULY

Figure 5-7. Mean precipitable water (inches and mm) for January and July (after Tuller, 1968).

Stephens (1990) established a statistical link between mean SST and mean precipitable water over the ocean–see Figure 5-8.

Since the relationship is almost linear for SST greater than 20° C, the patterns of ocean surface air temperature in Figure 5-1 and of precipitable water in Figure 5-7 are alike. If they know the SST, forecasters can use Figure 5-8 to estimate precipitable water.

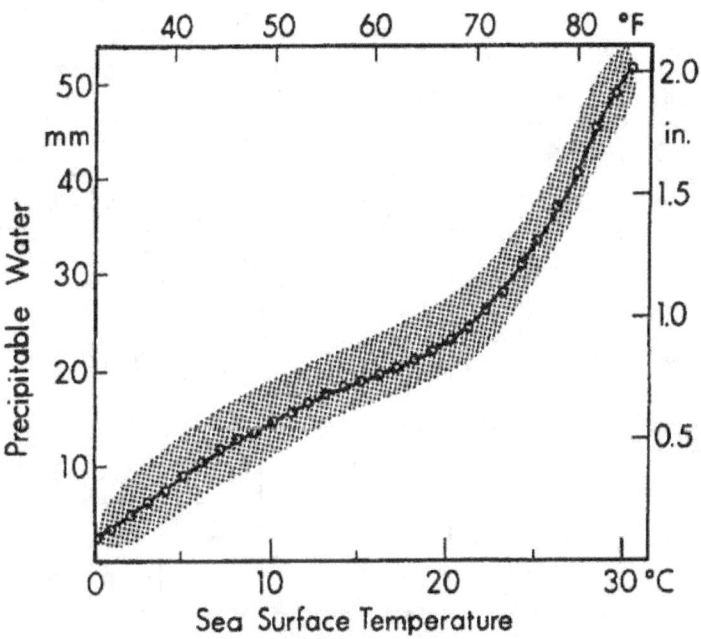

Figure 5-8. Annual mean relationship between precipitable water and SST. The shading ranges over one standard deviation above and below the averaged data points (Stephens, 1990).

5.3.3 Mean Vertical Soundings. The "Standard Atmosphere" adopted by the International Commission for Air Navigation in 1924 (List, 1963), formed the basis for "normal" atmospheres (ESSA et al., 1967) for 15 and 30° N. These atmospheres are representative of the tropics and subtropics. They were derived by computing the zonal averages of pressure, temperature, and humidity at standard pressure levels, as shown in Table 5-2. The months of January and July are given for 30° N. At 15° N, where annual variation is small, mean annual values suffice. The mean vertical profiles of temperature and dew point for the surface to 100 mb are shown in Figure 5-9, on the next page.

TABLE 5-2. Normal atmosphere for 30° N (January and July) and 15° N (Annual). P = pressure in mb; Height X 10^3 = geopotential height; T = temperature (° C); Td = dew point (° C); RH = relative humidity (percent).

P	30° North January					30° North July					15° North Annual				
	Height X 10^3					Height X 10^3					Height X 10^3				
	Ft	km	T	Td	RH	Ft	km	T	Td	R	Ft	km	T	Td	RH
1000	0.57	0.17	13.5	12	80	0.39	0.12	27	23	78	0.39	0.12	26	23	75
850	5.1	1.5	9.5	2	60	5.1	1.5	18	10	60	5.1	1.5	17	13	75
700	10.3	3.1	1	-10	45	10.4	3.2	9	1	60	10.4	3.2	9	-5	35
500	18.9	5.7	-17	-29	30	19.3	5.9	-6	-17	40	19.3	5.9	-9	-21	35
400	24.1	7.3	-27	-39	30	24.9	7.6	-18	-28	40	24.9	7.6	-19	-33	30
300	31.5	9.6	-40			31.8	9.7	-33	-44	30	31.7	9.7	-33		
250	34.4	10.5	-48			36.1	11	-42			35.9	10.9	-43		
200	39.4	12	-57			40.7	12.4	-52			40.6	12.4	-53		
150	45.9	14	-65			46.6	14.2	-62			46.5	14.2	-64		
100	53.5	16.3	-69			54.5	16.6	-69			52.7	16.1	-80		
70	61.3	18.7	-69			62	18.9	-64			61.3	18.7	-71		
50	67.3	20.5	-64			68.9	21	-59			67.8	20.7	-63		
30	76.4	23.3	-57			79.7	24.3	-52			78.1	23.8	-54		
20	86	26.2	-51			88.6	27	-47			87.5	26.7	-48		
10	100.1	30.5	-43			101	30.8	-39			101.7	31	-37		

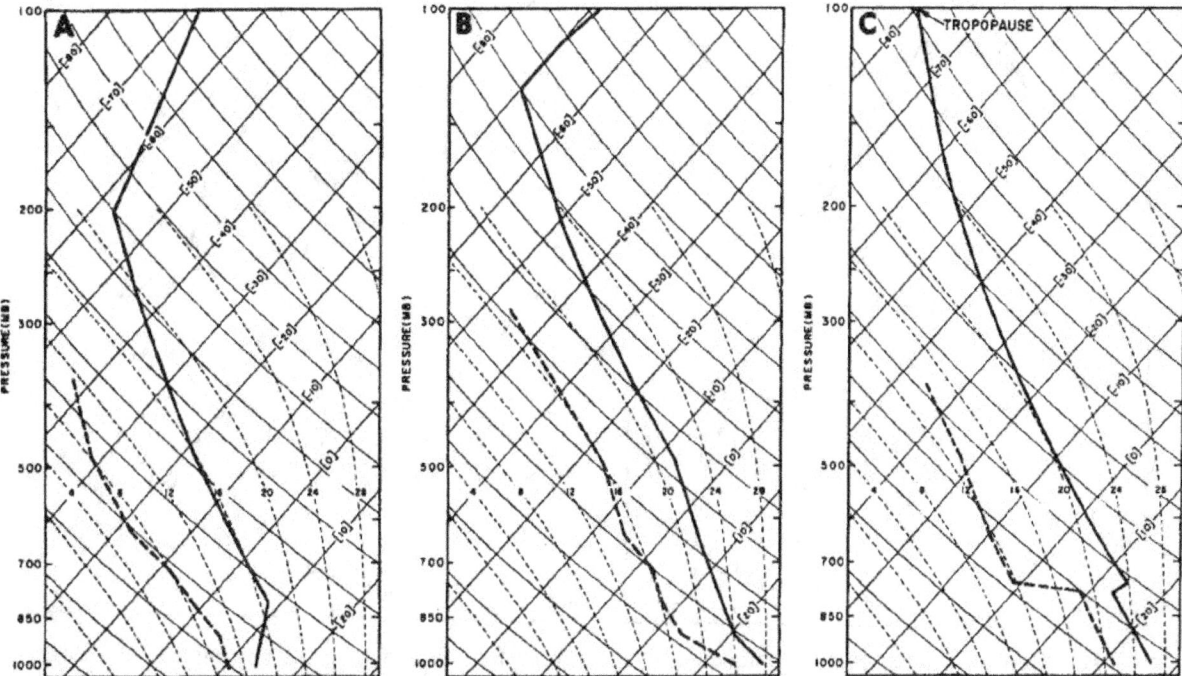

Figure 5-9. Standard atmosphere adiabatic diagrams for the tropics. (A) 30° N January, (B) 30° N July, and (C) 15° N annual. Solid lines are temperature; dashed lines, dew point (ESSA, et al., 1967).

In the 30° N normal atmosphere for January, shown in Figure 5-9(A), a fairly stable lapse rate extends from the surface to 800 mb, giving way to a lapse rate near that of the U.S. Standard Atmosphere (USSA) above 800 mb. The mean tropopause lies near 200 mb. The dew point decreases a little with height below 900 mb and more above.

The 30° N normal atmosphere for July in Figure 5-9(B) is much warmer than in January. From the surface to 250 mb it is 14 to 17° C warmer than the USSA. The temperature lapse rate is large below 900 mb, then slightly less than the USSA between 900 and 500 mb, and slightly more above 500 mb. The mean tropopause height is near 130 mb. The July dew point lapse rate is large below 900 mb and less above.

The 15° N normal atmosphere in Figure 5-9(C) shows a temperature inversion between 800 and 750 mb that is typical of the tradewinds. The inversion (or marked stable layer) separates the moist, well-mixed lower layer from drier air aloft; dew point therefore decreases markedly with height through the inversion or stable layer. Because the height, thickness, and intensity of inversions often change quickly, they would normally be lost in monthly and zonal averaging at constant heights. This is avoided by averaging daily values of the inversion height, thickness, and intensity, and then synthesizing the typical profile shown in Figure 5-9(C).

Gutnick (1958) studied the climatology of the tradewind inversion in the Caribbean, including the seasonal, spatial, and diurnal variation of the inversion frequency, height, strength, and thickness. He found that above the inversion, the mean temperature lapse rate was slightly greater than in the normal atmosphere. The mean tropopause height at 15° N is 100 mb, typical of the tropics equatorward of 20°.

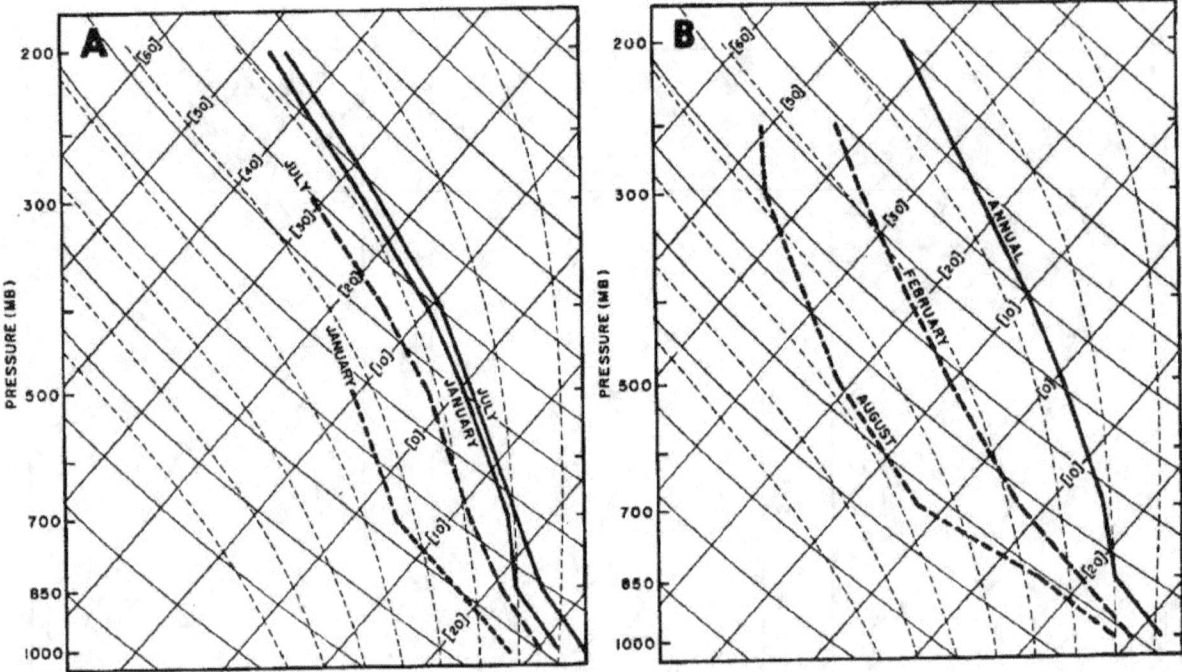

Figure 5-10. Mean vertical temperature and dew-point soundings for (A) Bangkok, and (B) Papeete, during the dry and wet season. Solid lines are temperature; dashed lines, dew point.

Bangkok, Thailand (14° N, 100° E) and Papeete, Tahiti (18° S, 150° W) are representative of the continental and oceanic tropics. Figure 5-10 shows their mean vertical temperature and dew-point soundings for dry- and wet-season months. Above 850 mb at Bangkok, July is only 1 to 2° C warmer than January, but the dew points at all levels are much higher. In July, the average dew-point spread from the surface to 300 mb is only 4 to 6° C. Since Papeete's mean monthly temperatures are within 1° C of the annual mean, only that sounding is shown. As in Bangkok, dew points are much lower at Papeete in the dry season (August) than in the rainy season (February).

Upper-air dew points vary much more from day-to-day than temperatures. Over most of the tropical troposphere equatorward of 20°, the standard deviation of temperature in any month is generally less than 2° C. Much of this stems from occasional cold advection from mid-latitude systems in winter, from cyclonic cells in the TUTT in summer, and from differential radiation effects under cloudy and clear skies. Dew-point standard deviations are two to three times larger. This results from the upward and downward air motions associated with meso- and synoptic-scale changes acting on the relatively large vertical gradients of dew points.

5.3.4 Equivalent Potential Temperature. The equivalent potential temperature (θ_e) has been used by various investigators to describe the structure of the tropical troposphere. This element, which combines the effects of temperature and moisture variations, is conservative with respect to dry- and pseudo-adiabatic processes and proportional to the total static energy in the atmosphere. The equivalent potential temperature is defined as the potential temperature that an air parcel would have after undergoing the following processes, which are rarely realized in the real atmosphere:

• Dry adiabatic expansion until saturation.

• Pseudo-adiabatic expansion until all moisture is precipitated out.

• Dry-adiabatic compression to 1,000 mb.

The atmosphere is defined as "potentially unstable" if θ_e *decreases* with height, and "potentially stable" if θ_e *increases* with height. Typical lapse rates of θ_e show that the lower part of the tropical troposphere is almost always potentially unstable.

Garstang et al. (1967) studied soundings made aboard a research vessel in the tropical Atlantic. The soundings were grouped into *disturbed* (cloudy with rain) and *undisturbed* (fair weather) as determined from hourly weather observations and cloud photographs. The mean temperature profiles differed little, but the dew-point lapse rates showed large differences that in turn resulted in large differences in θ_e lapse rates—see Figure 5-11.

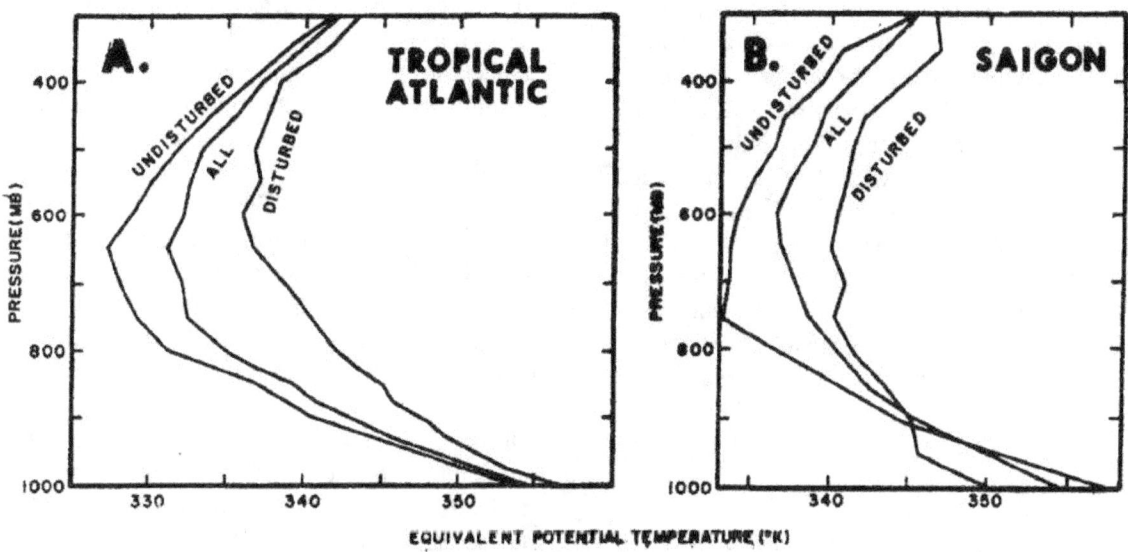

Figure 5-11. Vertical structure of equivalent potential temperature (θ_e) for disturbed and undisturbed days on the tropical Atlantic (after Garstang, et al., 1967) and Saigon (after Harris and Ho, 1969).

The mean θ_e lapse rate for all days is typical of much of the tropics; in the lower troposphere, a large lapse rate reaches a θ_e minimum between 600 and 700 mb; θ_e then increases with height. During disturbed days, mid-tropospheric θ_e becomes larger due to the increased moisture aloft. Conversely, θ_e becomes *smaller* on undisturbed days due to *decreased* moisture aloft. Thus, the lower troposphere is more potentially unstable (there is a larger decrease of θ_e with height) on *undisturbed* days.

In a similar study for Barbados (13° N, 59° W), where the soundings were divided into five synoptic groups depending on the percentage of stations reporting rainfall during a 24-hour interval, Garstang et al. (1967) also found large differences in mid-tropospheric θ_e values between greatly disturbed days (more than 85 percent of stations reporting rain) and greatly suppressed days (less than 16 percent of stations reporting rain).

Harris and Ho (1969) studied the virtual equivalent-potential temperature (θ_{ev}) for stations in southeast Asia as a measure of atmospheric structure during periods of increased or suppressed convection. Compared to θ_e, θ_{ev} more accurately assigns temperature to each pressure level. θ_e and θ_{ev} have essentially the same vertical distributions.

The degree of convection was determined by a radar index that represented the percentage of the area (within a 50-NM/100-km radius of the station) covered by radar echoes. Daily values of this index defined 4 days with increased convection and 4 days with suppressed convection at Saigon/Ho Chi Minh City (11° N, 107° E) during July 1966. When the θ_{ev} lapse rates were compared with the monthly mean θ_{ev} lapse rates shown in Figure 5-11, the results confirmed those of Garstang et al. The above studies show that θ_e varies characteristically with height and time in association with synoptic disturbances. Parodoxically, the tropical troposphere is most potentially unstable when convection is suppressed, becoming more stable as convection increases. The associated (and distinctly different) rainfall regimes will be discussed in Chapter 6.

The cause of the mid-tropospheric increase in θ_e is the upward transport of sensible and latent heat. Thus, changes in θ_e lapse rates (determined mainly by moisture variations in the mid-troposphere) and stability are often the result rather than the cause of the increased convection associated with tropical disturbances. This partially explains why conventional stability indexes developed for mid-latitudes are of limited use in forecasting convection over much of the tropics.

Soundings made even in disturbed conditions show θ_e increasing above 600 mb. How, then, can heat energy be exported from the surface to the high troposphere to fuel the Hadley cell? The "hot tower" hypothesis of Riehl and Malkus (1958) (see 2.2) provides the answer. Within the central region of a huge cumulonimbus undiluted by entrainment (see Figure 6-10), θ_e remains nearly constant with height; heat is transported upward. This very reasonable hypothesis is still unconfirmed, because there have so far been no accurate soundings in the core of a giant tropical cumulonimbus.

Gray (1968) determined the average vertical distribution of θ_e over the tropical oceans for about 15 years of the two warmest summer months and the two coldest winter months. Figure 5-12 (next page) shows the distributions for the regions where tropical cyclones form. Gray defined the term "potential buoyancy" of cumulus in the lower troposphere as the difference in θ_e between the surface and 500 mb. The average potential buoyancy is large over all tropical oceans, decreases with increasing latitude, becomes negative in higher latitudes, and is considerably less in winter than in summer. As a result, winter cumulonimbus is rarer and less intense.

The difference in θ_e between the top and bottom of an inversion is noteworthy. Where there is a solid stratus overcast, θ_e increases with height. A change to a decrease with height from the bottom to the top of the inversion indicates instability and precedes stratus breakup and eventual development of scattered cumulus.

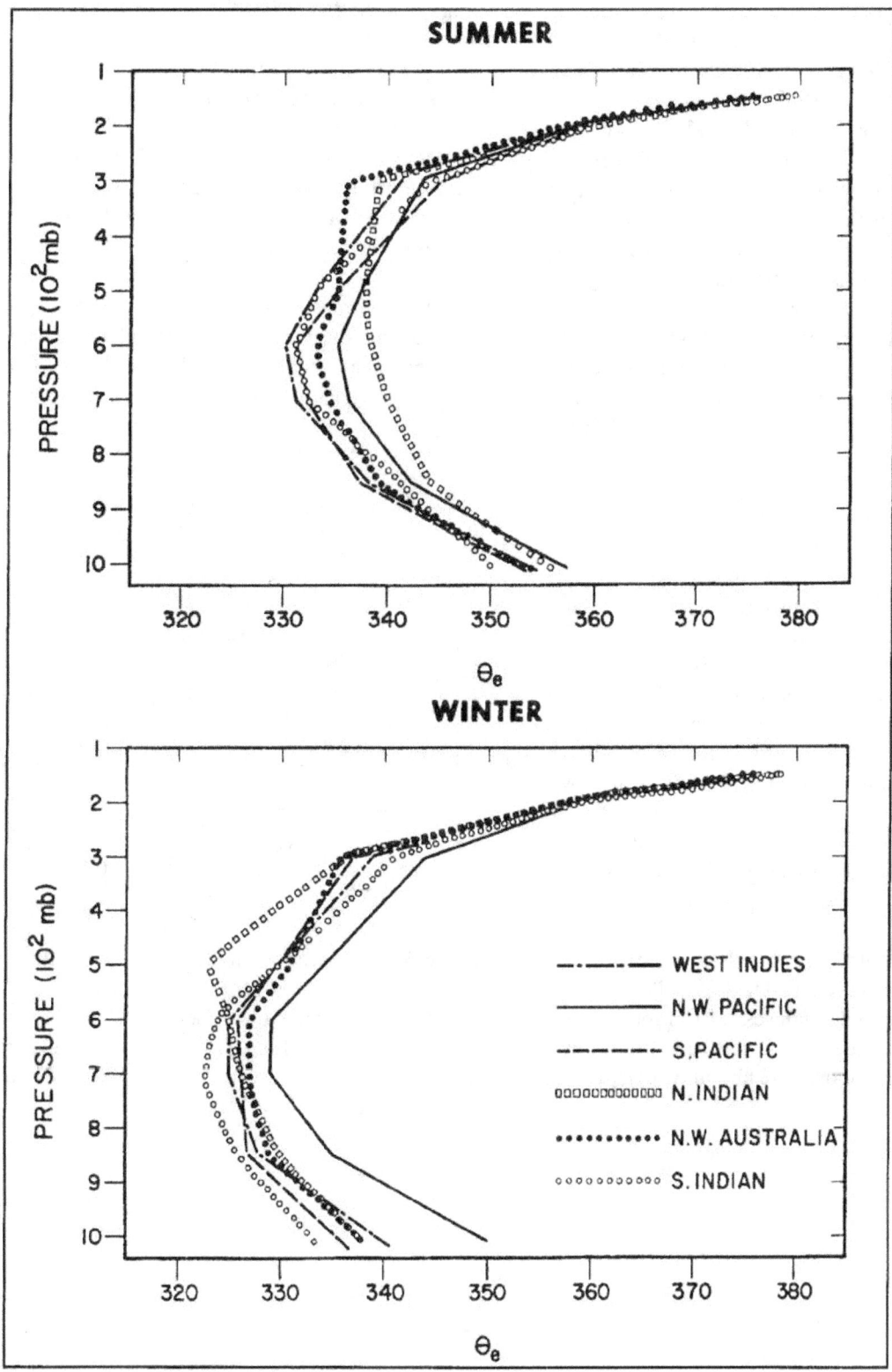

Figure 5-12. Average vertical distribution of equivalent potential temperature (θ_e) during the two warmest and two coldest months (after Gray, 1968).

5.3.5 Tradewinds. The tradewind inversion has been extensively investigated since it was first mapped over the Atlantic Ocean by the Meteor expedition (von Ficker, 1936). Riehl (1979a) summarized his and others' work. The inversion marks the lower boundary of air subsiding in the oceanic subtropical Anticyclones. Figures 4-9 to 4-12, 5-1, and 6-1 to 6-4 reveal that surface air in the anticyclones first flows toward the western coasts of Africa and the Americas, where it is cooled and stabilized by the upwelling water. The inversion may exceed 5° C. If it stays above the surface, cooling by the sea may reduce the surface air temperature to the dew point and cause sea fog to form. At times, the inversion may reach the surface; skies are cloud-free or scattered in the dry, subsiding air.

Farther downstream, winds first parallel the coast, then head out to sea. During this time, SST increases along the trajectory, and the air receives considerable heat and moisture from below. The resulting convection raises the inversion to between 1,500 feet (500 meters) and 3,000 feet (1 km). Since this is above the condensation level, extensive stratus or stratocumulus develops at the inversion base. An average aerological sounding for the southern California coast in summer (see Figure 5-13) shows not only an intense inversion, but θ_e increasing from 312 K at the base to 323 K at the top, indicating strong potential stability. No exchange takes place through the inversion and the cloud top is uniformly flat.

Wylie et al. (1989) found that in July 1987, stratocumulus increased 500 to 1,300 NM (1,000 to 2,500 km) west and southwest of California when the air was being heated and moistened from below. Deep precipitating clouds are absent (Garcia, 1985).

Farther into the North Pacific, when the air reaches about 20° N, the inversion has been raised above 5,000 feet (1,500 meters); heating by the sea surface has diminished. The dashed line in Figure 5-13 is a typical sounding for Lihue (22° N, 159° W); it shows θ_e decreasing from 319 to 315 K through the inversion. This potential instability enables cloud tops to penetrate the inversion and evaporate; dry air then subsides through the inversion.

Tradewind cumulus, typical of this regime, is scattered; tops are at various heights and often wispy. The inversion height changes little as the tradewinds approach the equator (Kloesel and Albrecht, 1989).

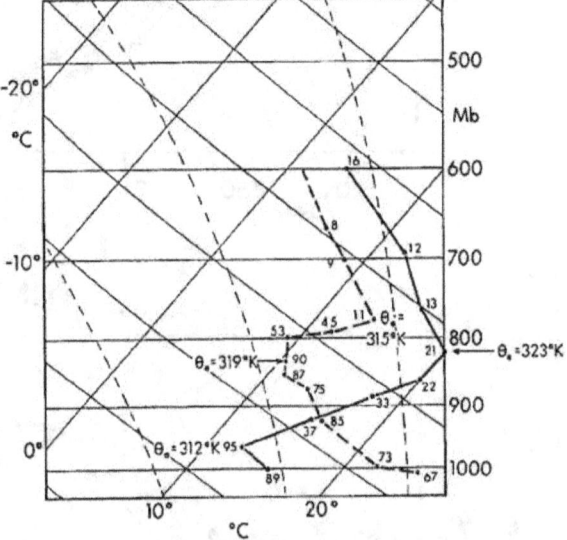

Figure 5-13. Average sounding for the southern California coast in summer (solid line, Neiburger et al., 1961) and a typical summer sounding (dashed line, 19 June 1976) at Lihue, 22° N, 159° W. Relative humidities (percent) and equivalent potential temperatures (K) are also shown.

84

Riehl has pointed out that the inversion is not an impermeable boundary. Convection from below and subsidence from above combine to mix moist rising air with dry sinking air. Thus, in summer, surface relative humidity over the Central Pacific is about 75 percent, compared to 85 percent at Hong Kong (22° N, 114° E) where, in the prevailing southwesterlies, subsidence is weak or absent. Consequently, cloud bases there are about 500 feet (150 meters) lower than in the tradewinds.

As the tradewinds approach the cooled upwelled water along the equator over the southeast tropical Pacific and Atlantic, cooling and divergence stabilize the air. This causes shears in the vertical to reach 16 knots in the lowest 300 feet (100 meters) (Wallace et al., 1989), while lowering the inversion and reducing clouds. The southeast tradewinds then cross the equator and converge with the northeast tradewinds between 5 and 10° N, where a persistent cloud band (the near-equatorial tradewind convergence, NETWC) is found. The character of the NETWC will be discussed in Chapters 6 and 8.

A satellite picture of the Eastern Pacific (Figure 5-14) shows typical tradewind clouds. Figure 5-13 showed that over coastal upwelling the stratus cloud top has about the same temperature as the sea surface, and can seldom be distinguished on satellite IR images (see also Figure 6-7).

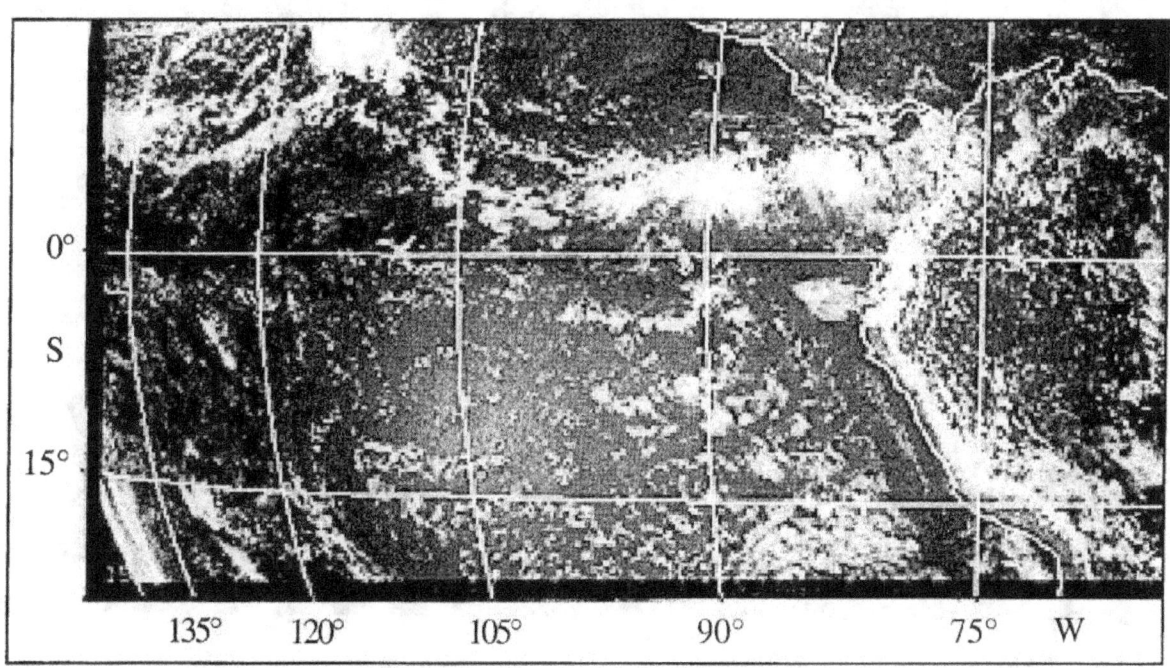

Figure 5-14. GOES visible image of the Eastern Pacific, about 1500L, 19 January 1984.

Since the tradewind inversion limits vertical motion, it also affects horizontal motion when the tradewinds impinge on mountains that are as high or higher than the inversion. Satellite images of cloud swirls that develop leeward of Guadeloupe Island (29° N, 118° W) have often been reproduced (see for example, Dvork and Smigielski, 1990). As an island obstacle slows the flow, there is convergence some distance upwind and acceleration around the edges of the obstacle; swirls form to leeward .

Figure 5-15. Tradewind clouds photographed from the Space Shuttle in June 1985. The view is across the island of Hawaii toward the east.

Figure 5-15 is a space shuttle photograph that looks eastward across the island of Hawaii. It shows lines of cumulus paralleling the windward coast and cyclonic and anticyclonic vortexes to leeward. This pattern accounts for very light tradewinds along the windward coast, strong winds around the northern and southern ends of the island, and a well-developed sea breeze along the leeward coast (see Figures 7-6 and 7-16).

The Central American mountain chains also block the leeward tradewinds, except where they are funneled through three isthmuses, as will be shown in Figures 8-35 and 8-36. In the strait between Taiwan and China, the northeast monsoon is similarly funneled. In winter (as mentioned previously), the subtropical Oceanic anticyclones combine with the continental anticyclones to form a continuous high-pressure ridge. Thus the northeast winter monsoon across the South China Sea and the southeast winter monsoon across the Timor Sea are, in all important aspects, tradewind regimes with subsidence aloft and inversions between 5,000 and 10,000 feet (1.5 and 3 km).

5.3.6 Low-Latitude Westerlies. Equatorward of heat troughs and monsoon troughs, surface winds with a westerly component prevail. Most notable is the southwest monsoon of the NH summer, extending from West Africa to east of the Philippines and between 5° N and 25° N over the Indian Ocean. Away from the trough, surface westerlies generally converge and underlie upper-tropospheric divergence. Most of the troposphere is moist (see Figure 5-10A). Where orographic lifting occurs, as over the western Ghats and the Khasi Hills of India, and the Arakan Mountains of Burma, summer rainfall exceeds 200 to 300 inches (5,000 to 7,500 mm).

Year-long heat lows cause persistent surface westerlies in equatorial West Africa and western Colombia. On windward mountain slopes, annual rainfall may reach 500 inches (12,500 mm) (Walters et al., 1989).

Chapter 6

CLOUDINESS AND RAINFALL

6.1 GENERAL

Tropical cloudiness and rainfall have been the subjects of intense investigation recently by university groups at Texas A & M, Florida State, Hawaii, Colorado State, and others. The results of these investigations are summarized and included in this chapter.

The distributions and variations of cloudiness and rainfall are presented for various scales from the large-scale global patterns to the convective-scale patterns associated with individual rainstorms.

6.2 CLOUDINESS

6.2.1. Mean Cloudiness. Various investigators have used meteorological satellite data to derive mean cloudiness over the tropics. Sadler et al. (1984) used operational nephanalyses prepared by the National Environmental Satellite Service (NESS) to determine mean monthly cloudiness between 30° N and 30° S (Figures 6-1 to 6-4). The following introduction to the broad-scale features of tropical cloudiness should be supplemented by careful study of the monthly charts in the original publication, and a comparison of those charts with mean circulation features, especially at the gradient level.

Over most of the tropics, cloudiness and rainfall are directly related. Figures 6-1 to 6-4 show that regions of average deep convection are also cloudy. As mentioned in Chapter 5, this is not true over the eastern parts of the oceanic subtropical anticyclones. There, where the tradewinds first blow along the coast and then head out to sea, they are heated and moistened from below; extensive stratus and stratocumulus form beneath the tradewind inversion. The cloudiness maximum extending southeast from New Guinea persists throughout the year. The largest gradients are found east of Tibet, south of the North African deserts, and near upwelled cold water.

Clouds are orographically enhanced to windward, and depleted to leeward of the Philippines, Madagascar, and Central America (see Figure 7-37).

6.2.2. Annual Cloudiness Variation. The largest annual variation occurs over south Asia between 70° E and 110° E, where desert-like conditions in winter give way to the wet summer monsoon. The oceanic cloudiness minima associated with the subtropical highs and near-equatorial maxima move meridionally following the Sun, with a lag of about 3 months. Except where the tradewinds converge, maximum cloudiness coincides with surface westerly winds on the equatorward side of a surface pressure trough. Clouds are relatively fewer along the trough, which is anchored by a surface temperature maximum. Over the continents, this summer heat trough is replaced in winter by an anticyclone that is also almost cloud-free—hence, the deserts (see Figure 6-29).

Apart from a brief "Indian summer" in August and September, south China remains cloudy (see Figure 6-29). In the eastern Pacific, an east-west cloudiness maximum is often observed along 5° S between February and April. The effect, caused by convergence within the southeast tradewinds, may extend to South America. The fact that there are very few ship observations and no research flights in this area has inhibited study.

Climatologically, the cloudiness of tropical cyclones is counteracted by subsidence-generated clear skies on their peripheries. Hence, average cloudiness is not affected.

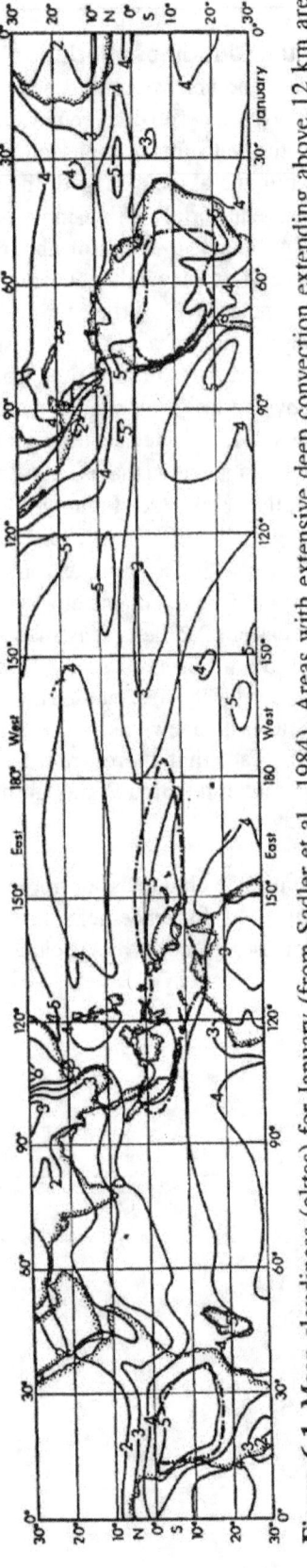

Figure 6-1. Mean cloudiness (oktas) for January (from Sadler et al., 1984). Areas with extensive deep convection extending above 12 km are enclosed by dot-dashed lines (from Janowiak et al., 1985, and Duvel, 1989).

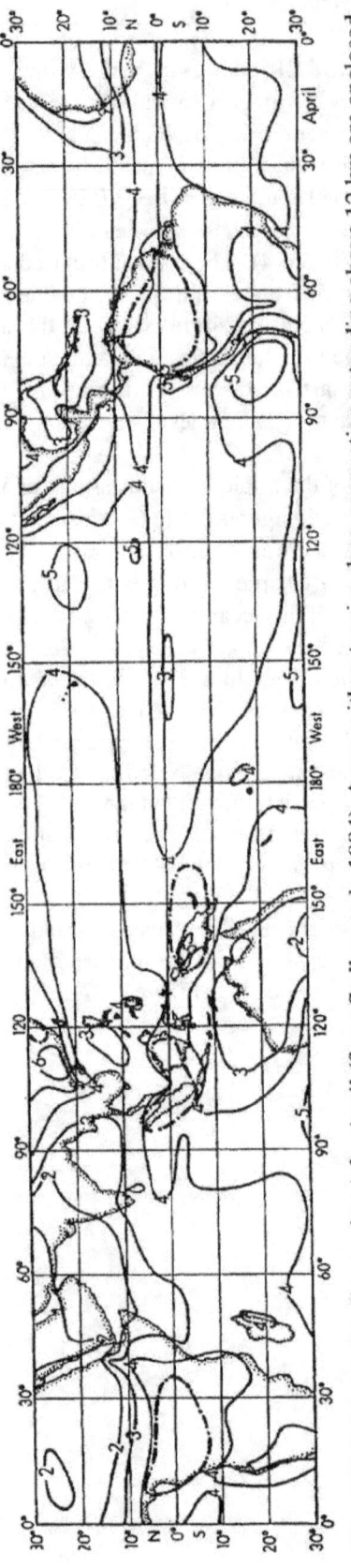

Figure 6-2. Mean cloudiness oktas) for April (from Sadler et al., 1984). Areas with extensive deep convection extending above 12 km are enclosed by dot-dashed lines (from Janowiak et al., 1985, and Duvel, 1989).

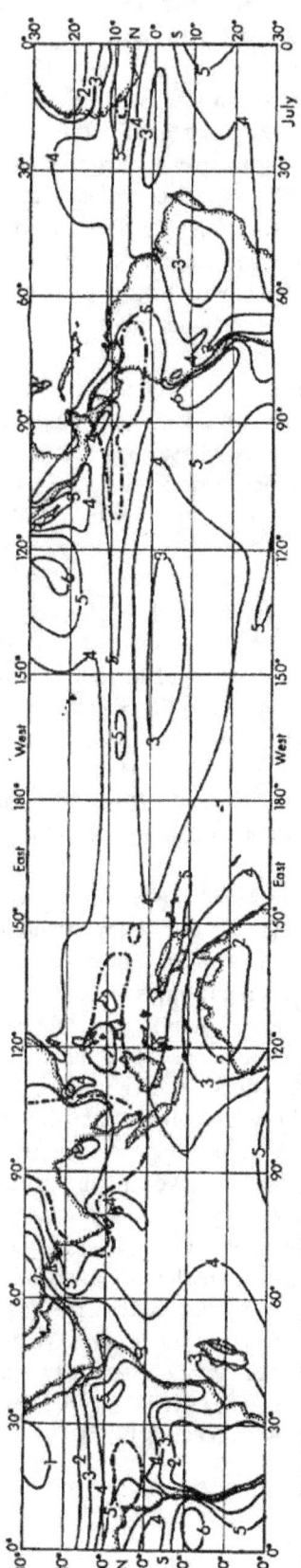

Figure 6-3. Mean cloudiness (oktas) for July (from Sadler et al., 1984). Areas with extensive deep convection extending above 12 km are enclosed by dot-dashed lines (from Janowiak et al., 1985, and Duvel, 1989).

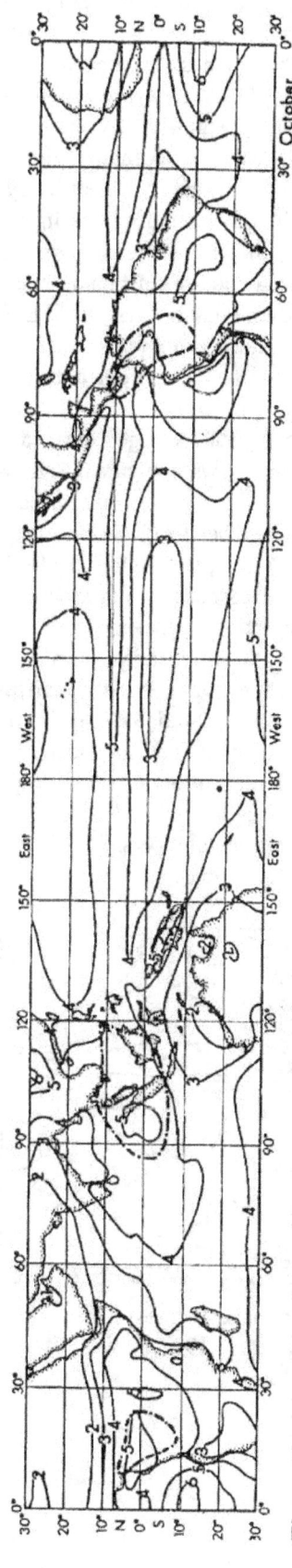

Figure 6-4. Mean cloudiness (oktas) for October (right) (from Sadler et al., 1984). Areas with extensive deep convection extending above 12 km are enclosed by dot-dashed lines (from Janowiak et al., 1985, and Duvel, 1989).

91

6.2.3 Year-to-Year Cloudiness Variations.
Sadler (1969) found that, for any calendar month, the general locations of cloudiness maxima and minima varied little from year to year; however, year-to-year cloudiness may change significantly *within* these regions. In a year when some maximum regions have above-normal values, there is a tendency for compensation by below-normal values in nearby minimum regions; this suggests changes in the intensity of the vertical motion cells linking maxima and minima, a condition that occurs during El Niño (see 6.3.5). There is little year-to-year change in total average cloudiness for the whole tropical belt.

6.2.4 Zonal Averages.
Averages derived from Sadler et al. (1984) are shown in Figure 6-5. Between 5° N and 5° S, cloudiness changes little throughout the year. Values are higher to the north along the oceanic NETWC. A relative minimum along the equator reflects increased stability due to cool upwelling in the Atlantic and eastern Pacific. In the NH summer, the south Asian monsoon moves the cloudiness maximum north and intensifies it. Compared to the NH, annual SH variation and meridional gradients of cloudiness are much less, but total cloudiness is more. In this ocean hemisphere, more moisture is available, the ocean buffers the annual cycle, and stratocumulus persists in the eastern parts of the tradewind regions.

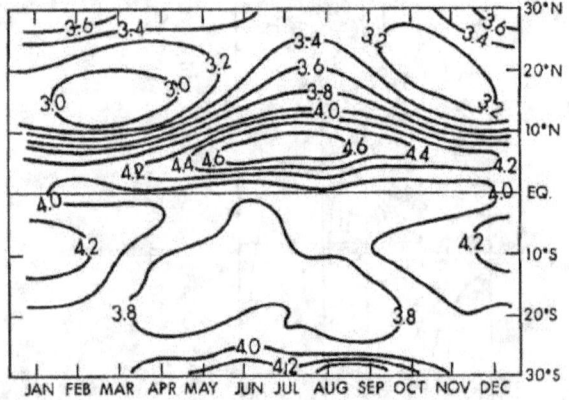

Figure 6-5. Mean zonal cloudiness in oktas for each month (from Sadler et al., 1984).

6.2.5 Large-scale Organization of Deep Tropical Cloud Systems.
A special Study Group on Tropical Disturbances established as part of the Joint Organizing Committee of the Global Atmospheric Research Program (GARP, 1969) made a census of cloud systems over the tropics. Daily satellite pictures for 1967 were used to determine the typical large-scale organization of deep clouds in various parts of the tropics. Three major types were identified and labeled "cloud clusters," "monsoon clusters," and "popcorn" cumulonimbus.

A cloud cluster covers an area of 100 to 650 NM (200 to 1,200 km) on a side, and consists of numerous cumulonimbus (Cb) cells whose tops are seen as bright patches from which cirrus streamers emanate. Clusters are usually arranged in bands thousands of NM long.

Monsoon clusters are larger than cloud clusters; they extend from 250 to 500 NM (500 to 1,000 km) from north to south and 250 to 1,000 NM (500 to 2,000 km) from east to west; they are distributed over the land and adjoining seas of southern Asia during the summer (southwest) monsoon. Most monsoon clusters are not aligned in bands. Bright elongated areas of Cb activity occur within monsoon clusters.

Distinguishing between cloud clusters and monsoon clusters is not easy and may not be important. The term "cluster" contains little information on the nature or cause of cloud aggregation, and reflects the meteorologists' tendency to substitute labels for explanations. The satellite image for 19 July 1985 in Figure 6-6 shows clusters of various sizes and orientations. Over the whole area of the image, a moderate southwest (summer) monsoon prevailed. The cloud system oriented southwest-northeast across southeast China and the Ryukyus might comprise "cloud clusters," while that covering the Philippines could be termed a "monsoon cluster." Apart from a very weak 500-mb low over the Ryukyus, the clusters were not readily related to synoptic features. This was also true of a group of Atlantic clusters between 8° and 14° N, studied during the GARP Atlantic Tropical Experiment (GATE) (Martin et al., 1984).

Figure 6-6. GMS infrared (IR) image for 0900Z 19 July 1985, showing cloud clusters over and east of Southeast Asia.

"Popcorn" cumulonimbus comprises a few Cb cells that cover an area of about 3,000 NM² (10,000 km²). The cells undergo pronounced diurnal variation; in their afternoon maturity, a collection of these clusters resemble (in satellite pictures) randomly distributed kernels of popcorn. This distinct type occurs mainly over tropical continents where there is sufficient low-level moisture and weak wind shear in the vertical; it has been labeled as "showers" (see 1.2). Over northern South America, in the relatively undisturbed conditions shown in Figure 6-7, popcorn cumulonimbus developed between 1100 and 1600L.

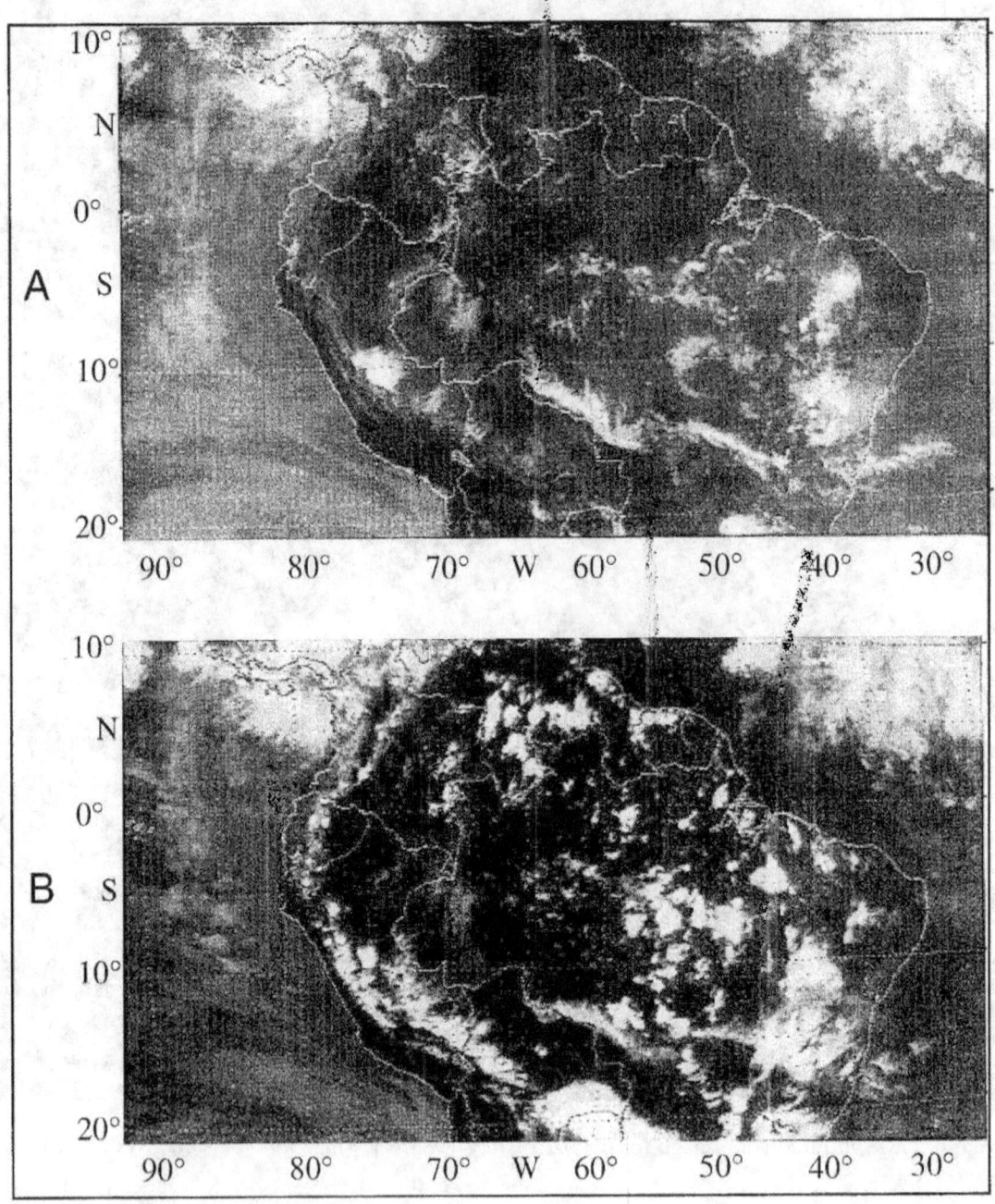

Figure 6-7. GOES IR images for: (A) 1100L 25 October 1988 and (B) 1600L 25 October 1988, showing development of "popcorn" cumulus over northern South America.

In the tropical Pacific, most deep clouds occur as clusters of numerous cells. They make up the near-equatorial tradewind convergence that is clearly revealed in maps of monthly average cloudiness (Figures 6-1 to 6-4). The mean NH cloud band is prominent throughout the year (Figure 6-5); it is nearest the equator over the central Pacific and farthest from it over the western and eastern Pacific. The SH near-equatorial mean cloud band is prominent only from October through March. Within each band, the space between clusters is about two to three times the size of an individual cluster. Clusters in the eastern and western North Pacific often show a vortical structure. Throughout the year in the South Pacific, a prominent band of cloud clusters (the South Pacific Convergence Zone) extends from near 10° S, 160° E to 30° S, 120° W. In this band, the space between cloud clusters is less than in equatorial regions. See Chapter 8 for a more detailed discussion of these bands.

Martin and Karst (1969), in a detailed statistical study of deep cloud systems over the tropical Pacific (25° N to 25° S and 160° E to 100° W) used satellite photographs from March 1967 to February 1968. The photos identified four types of cloud systems: *oval, line, wave,* and *spiral-vortex.* The cloud systems were categorized on the basis of satellite images; accompanying circulations were almost never identified. Only the spiral vortexes could, with some confidence, be associated with tropical cyclones. The oval type occurred more than 500 times a year; the other types, between 100 and 200 times a year. From near zero at the equator, frequencies of all types increased to their maxima at 5-10° latitude. Ovals, waves, and vortexes were two-thirds more common in the NH than in the SH. The vortex and oval types extended over about 150,000 NM^2 (500,000 km^2), while the wave and line types were about 50 percent larger. The vortex type lasted longest (about 6 days) and the oval type was the briefest (about 2 days).

Near the equator, the cloud systems generally moved westward, but poleward of 20°, they moved eastward, with some poleward component. There were significant annual and areal differences in the frequency, distribution, and movement of the cloud systems, especially in the NH. Differences were larger and more common in the western part of the area studied.

Martin and Suomi (1972) studied cloud clusters over the tropical Atlantic for the summers of 1969 and 1970. The average cloud cluster lasted longer in the Atlantic than in the Pacific, possibly because the Atlantic was studied for only two summers, while the Pacific was examined for the whole year. Clusters were larger in the east and central Atlantic than in the west, perhaps because many clusters originated well within Africa as squall lines before moving west across the ocean.

Hayden (1970) measured cloud-cluster dimensions and spacing in the tropical North Pacific. He used a computer program to objectively analyze satellite-derived brightness fields for July and October 1967 and January and April 1968. He concentrated on the 500-NM- (1,000-km-) wide zone of maximum brightness between 110° W and 130° E where the predominant cloud-cluster width ranged from about 150 NM (280 km) in winter to about 250 NM (460 km) in summer. Clusters were most often either 360 to 480 NM (670 to 890 km) apart or 600 to 720 NM (1,100 to 1,330 km) apart, with little annual variation. For comparison, Hayden applied the technique between 30° N and 60° N in the Pacific for October 1967. The mid-latitude clusters were usually about twice as large as the tropical clusters. Assuming a relationship between cloud-cluster width and the size of associated circulation features, Hayden concluded that a grid size suitable for tropical analysis should be about half that used in mid-latitude analysis.

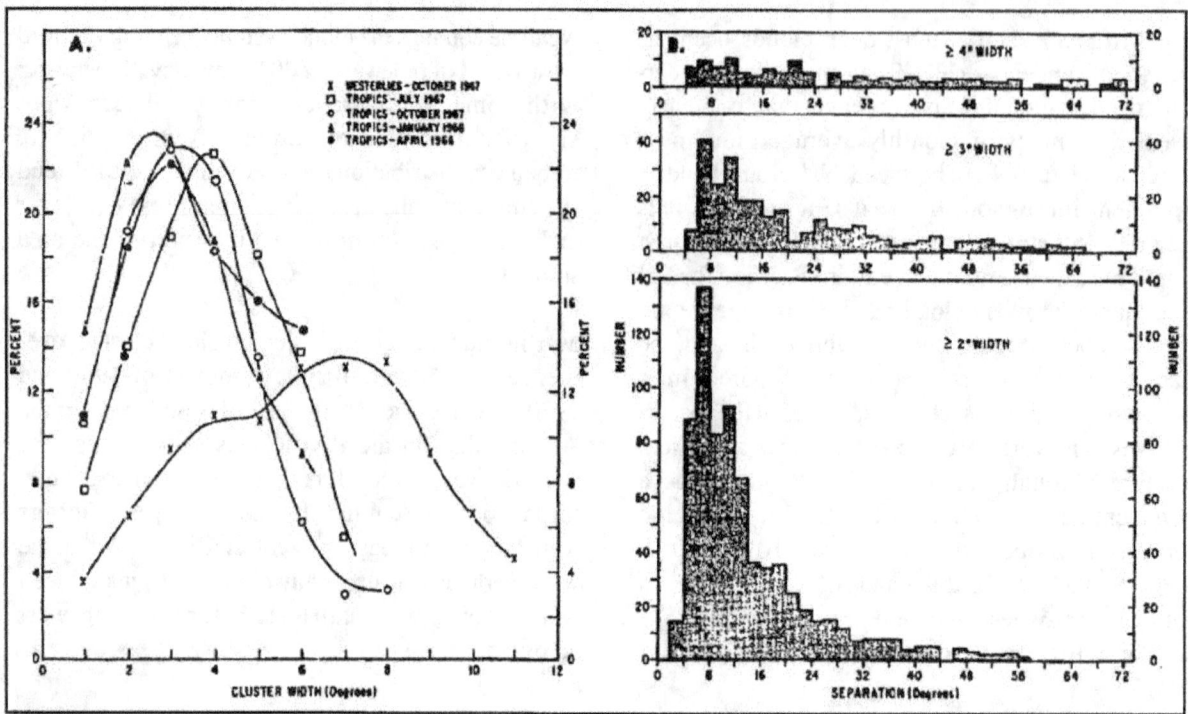

Figure 6-8. Cloud-cluster frequency distributions of width and separation distance for months shown in the tropical North Pacific (adapted from Hayden, 1970). (A) Frequency distributions (percent) of cloud-cluster widths for the 10° latitude band of maximum brightness. October 1967 frequency distributions for the mid-latitude North Pacific are included for comparison. (B) Frequency distributions of cloud-cluster separation distances as (bottom to top) smaller clusters are excluded. Distributions are an aggregate of the 4 months for the 10° latitude band used for (A).

Figure 6-8 shows frequency distributions of cloud-cluster widths and separation distances found by Hayden. Relative dominance (expressed as a percentage) is computed by weighting the frequency of a cluster width by its area and dividing by the total area of all cloud clusters.

According to Hayden, the 600-720 NM separation distance corresponds to the principal synoptic scale of the tropics, while the shorter 360-480 NM separation corresponds to a faster-moving secondary scale enhanced by interaction with the larger scale.

In a different approach, Williams and Gray (1973) classified the stages of cloud-cluster development after analyzing the characteristics of the wind, thermal, and moisture fields surrounding satellite-defined mesoscale cloud clusters in the tradewinds

of the western North Pacific (0 to 30° N, 125° E to 160° W) for October 1966 to October 1968. They identified five categories of cloud clusters (*pre-storm, developing, conservative, non-conservative* or *developing-dying*, and *dying*) 3 to 6 degrees of latitude wide. A sixth category (clear areas at least 10 degrees of latitude wide) was added. Altogether, 1,257 individual clusters and 553 clear areas were identified.

For each category, Williams and Gray composited Rawinsonde Data relative to the category center for 16 levels from the surface to 30 mb. From that, they derived fields of wind, vorticity, divergence, kinetic energy, temperature, moisture, stability, and isobaric heights. The results support physical reasoning. For example, the composited wind fields showed significant low-level cyclonic shear across the cloud clusters. Pre-storm clusters had the largest shear and there was anticyclonic shear across the clear areas. Since little curvature was evident in the composited low-level wind field, the relative vorticity was produced primarily by the horizontal wind shear. This suggests the absence of easterly wave-like disturbances (see 8.2.1).

In the clusters, net convergence in the lower troposphere (below 400 mb) was overlain by net divergence in the upper troposphere (resulting in upward motion); the reverse was true for the clear areas. Over the cloud-cluster centers, average vertical wind shear (900 to 200 mb) was relatively light (less than 10 knots). Because the authors did not distinguish between clusters associated with lower-tropospheric circulation systems and those associated with upper-tropospheric systems, the composite environmental fields surrounding each category of cluster could not be distinguished. *Twenty-seven percent of all the clusters observed on any one day disappeared within 24 hours.*

Although the organization of tropical Atlantic cloud systems resembles those of the Pacific, an SH near-equatorial band is much less evident. From its position near the equator in NH winter, the NH band shifts to near 10° N in summer; circulations associated with clusters in the North Atlantic occasionally develop into hurricanes. In the South Atlantic, a curved band of cloud clusters, similar to the one in the Pacific, extends southeastward from near 20° S, 40° W, into the mid-latitudes.

Over the Amazon Basin, cloud clusters and monsoon clusters predominate in the January-May wet season (Greco et al., 1990); in the July-October dry season, popcorn cumulonimbus is common. Randomly distributed over the plains and mountains, popcorn Cb varies diurnally. Cumulus clouds develop in the forenoon, become thunderstorms during the afternoon, and dissipate in the evening.

Over tropical Africa, clouds are organized into large clusters similar to those of the Atlantic and Pacific, usually accompanying squall lines (see Figure 8-28). Generally, the clusters first appear over the East

African mountains and move westward at about 250 NM (500 km) a day. During the NH summer, the clusters travel between 5° N and 10° N. They may develop into vortexes before emerging into the North Atlantic. Although the vortexes generally weaken as they move across the Atlantic, some redevelop into hurricanes in the western North Atlantic. During the SH summer, cloud clusters are found over Africa between 5° S and 10° S.

Over the Indian Ocean and southeast Asia, cloud systems both resemble and differ from those in other areas. In the South Indian Ocean they are organized into large clusters similar to those in the Pacific. They are most active in the SH summer, moving westward along 10° S to 12° S. The accompanying circulations cannot be identified unless the clusters become vortical. Clusters also occur during the SH winter near 5° S.

During the NH winter, intense cloud clusters develop over the maritime continent, generally in response to surges of the winter monsoon. They undergo strong diurnal variation and are characterized by extensive stratiform clouds, the tops of which appear as cirrus shields in satellite pictures. Rain comes about equally from embedded deep convective cells and from the stratiform clouds (Williams and Houze, 1987). A similar distribution has been observed elsewhere in the tropics, as well as in the middle latitudes (Houze, 1989).

During the NH spring (April and May), cloud clusters are generally confined to between 5°N and 10°N. Onset of the southwest monsoon is accompanied by a dramatic change. In late May and early June, cloud clusters lie north and south of the equator, occasionally straddling it. By mid-June, one or two large cloud systems extend west to east to cover most of the area between 70° and 90° E and 10° and 20° N. These large cloud systems, called "monsoon clusters," occasionally reach as far east as 110° E. Smaller monsoon clusters sometimes occur over the Arabian Sea (50° E to 70° E, 5° S to 15° N). Many of the cumulonimbus tops embedded in these clusters appear in satellite pictures.

Based on the results of the GARP Study Group, a highly idealized picture of typical cloud organization in the tropics emerges. Figure 6-9 shows the large-scale deep tropical cloud systems on a typical summer day in each NH and SH region. Many mesoscale cloud systems would occur within each of the larger systems. Smaller, independent cloud systems, which may appear in daily satellite pictures, have been omitted. In view of the great variety of daily patterns and the prevailing westward motion of low-latitude cloud systems, the complexity and variability of the tropical atmosphere is not surprising.

Tropical forecasters have long been aware of the large, organized cloud systems of tropical cyclones and monsoon depressions, but tended to

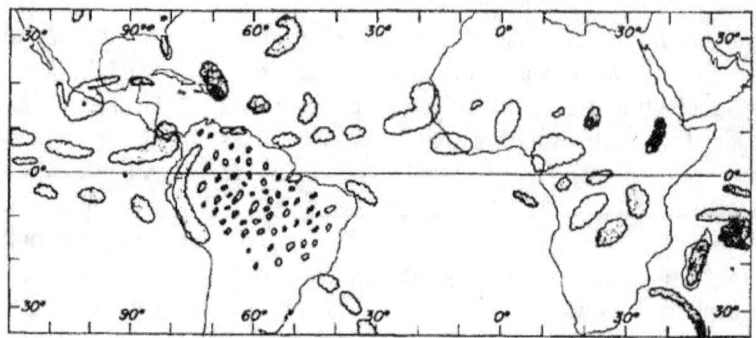
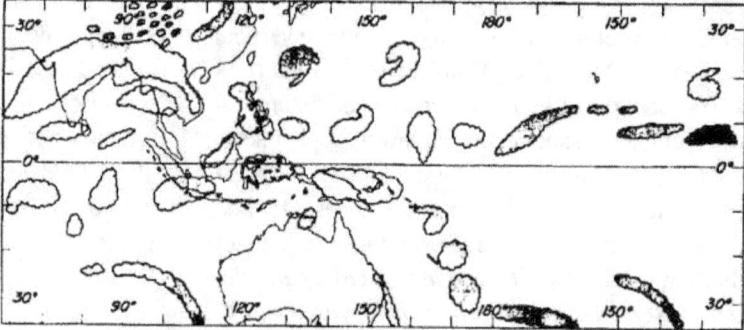

Figure 6-9. Idealized view of the larger-scale tropical cloud systems on a typical summer day in each hemisphere.

think that other rain-producing clouds were rather scattered. Weather satellites have now shown that rain from clouds in clusters contributes more than 80 percent of the total amount, that scattered convection is relatively unimportant, and that fair weather usually covers a large area.

6.2.6. Tropical Cloud Types. In this section, tropical cloud types, their characteristics, and their relative frequencies are discussed. Much of the information is condensed from Palmer et al. (1955, 1956), with additional information from more recent studies. The cloud definitions are taken from the WMO International Cloud Atlas (1956a). Pictures of individual cloud types are readily available in the Atlas and are not presented here.

The international cloud classification is applicable worldwide. However, the relative frequency of various cloud species does differ between the tropics and higher latitudes and some sub-varieties found in high latitudes may not occur in the tropics (Malkus and Riehl, 1964). While all cloud species are found over both land and sea in the tropics, there is a clear

distinction between oceanic and orographic clouds. Far from land, diurnal variation is relatively small. Over continents and large islands, diurnal variation is pronounced and must be carefully considered in forecasting (See Chapter 7). Operational satellite images are not detailed enough to allow individual cloud elements to be readily identified.

6.2.6.1 Cumulus. Cumulus predominates in the tropics. As discussed earlier, it transports much of the energy required to drive the atmosphere. The following discussion of the physics of cumulus is based on Simpson and Dennis (1972).

Atmospheric movement is driven by the Sun, but indirectly; heat is absorbed at the Earth's surface, from which it is transmitted upward, mainly as water vapor. When cumulus develops, water vapor condenses into liquid cloud droplets, turning latent heat of evaporation into sensible heat of condensation. The process is made irreversible if rain falls out of the cloud; then the sensible heat can be used to drive air motion.

Figure 6-10 illustrates a hierarchy of tropical cumulus according to size. The bubble-like small cumulus or the plume or jet-like larger cumulus are driven upward by condensation and the resultant buoyancy of the cloud with respect to the environment. For growth to be maintained, a small supersaturation is needed to sustain condensation in the face of vapor removal by the growing droplets.

Measurements have shown that the temperature and liquid water in a cloud are less than what would evolve were the cloud to be composed only of isolated parcels of air that had risen from the Earth's surface. As shown in Figure 6-10, unsaturated air is drawn (entrained) into the cloud from its surroundings. Compared to small clouds, the buoyant central core of large clouds is less affected by entrainment, and the clouds last longer.

An added complication is introduced by shear (Figure 6-11). Air entrains on the upshear side, where there is a component of wind into the cloud, and detrains on the downshear side of the cloud. The larger clouds of Figure 6-10 are more slowly eroded by entrainment than the smaller clouds, since their central cores are better protected from invasion by dry air.

Condensation alone cannot produce raindrops (of 0.5 mm or more in diameter), but rain does fall. Two processes may ensure this; the more important of the two is growth by droplet collision, causing coalescence. The other occurs in the parts of a cloud colder than about -5° C. Because saturation vapor pressure is lower over ice than over water, the ice particles in the cloud grow faster than water droplets and sometimes grow large enough to fall.

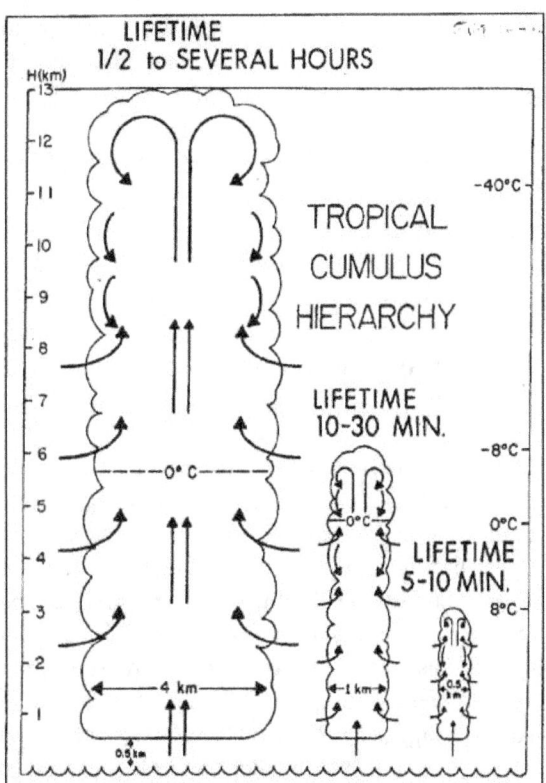

Figure 6-10. Schematic hierarchy of tropical cumulus (from Simpson and Dennis, 1972).

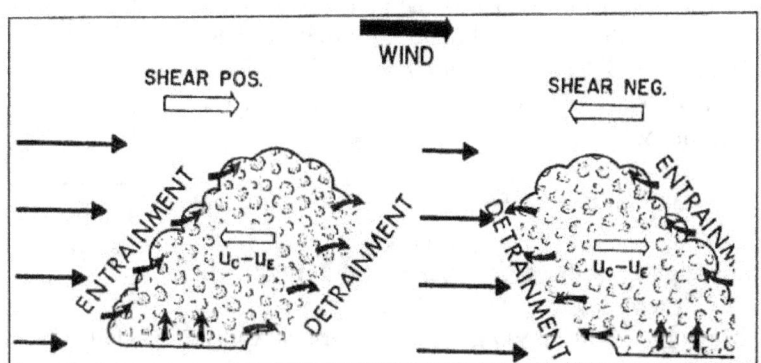

Figure 6-11. Cumulus leaning with shear vertically. On the left, cloud velocity (U_c) is *less than* wind velocity (U_e). In the cloud on the right, the reverse is true.

When drops grow large enough to fall from a cloud and do not evaporate in the air beneath the cloud, they reach the surface as rain. Within the cloud, the falling drops scavenge a significant fraction of all cloud droplets. The relative humidity in the cloud is thus lowered, most of the remaining droplets evaporate, and the cloud "rains out." By the time the rain stops, only a few wisps of the cloud from which the rain originated may remain.

Although the mechanism is not clear, only the presence of both ice and water in a cloud can generate the large electric gradients that cause lightning. In the tropics, where a cloud must extend above 16,000 feet (4 km) for ice to form, thunderstorms are confined to only very deep convective clouds that generally extend above 30,000 feet (9 km).

When rain falls from a cloud, it may, through friction and evaporation, induce a cold downdraft of air that reaches the surface. There, cooling due to cloud shade might be further enhanced and the chance of new convection reduced. Or the cold air, spreading outward and converging with surrounding air (the gust front), may trigger upward motion and new convection.

Cumulus clouds are detached, generally dense, with sharp outlines developing vertically in the form of rising mounds, domes, or towers, of which the bulging upper part often resembles a cauliflower. The sunlit parts of cumulus are brilliantly white; the shaded base is relatively dark and nearly horizontal. Four species of cumulus are defined in order of increasing vertical development: *fractus,* or irregular shreds of cloud having a ragged appearance; *humilis,* of slight vertical extent; *mediocris,* of moderate vertical extent; and *congestis,* of great vertical extent.

The most striking variation among cumulus is the extent of lean (or shear) of the cloud (see Figure 6-11). On this basis, Palmer et al. (1955) divided cumulus into "doldrum" cumulus, with little or no lean or shear, and "trade" cumulus, in which shear is significant.

The doldrum class is typical of the light wind and small wind-shear regions in lower latitudes. Bases are slightly wider than tops, at about 3,000 feet (900 meters) and 2,000 feet (600 meters), respectively. Bases are usually 1,500 feet (450 meters), lowering to 1,300 feet (400 meters) or below in showers. Tops vary between 6,000 and 12,000 feet (1,800 and 3,600 meters). Winds are light and there is little shear below 15,000 feet (4,500 meters).

As its name implies, trade cumulus is most common in the tradewinds between 10 and 30° latitude. Wind shear through the cloud layer causes the cloud axis to lean, as was shown in Figure 6-11. In the tradewinds, the wind is usually strongest near the cloud base and decreases with height. This is in response to a temperature gradient directed to the right (left) of the tradewind flow in the NH (SH). The resulting thermal wind diminishes the tradewind. This effect, combined with frictionally reduced speed at the surface, produces the typical tradewind profile, first explained by Shepard and Omar (1952) and shown in Figure 6-12.

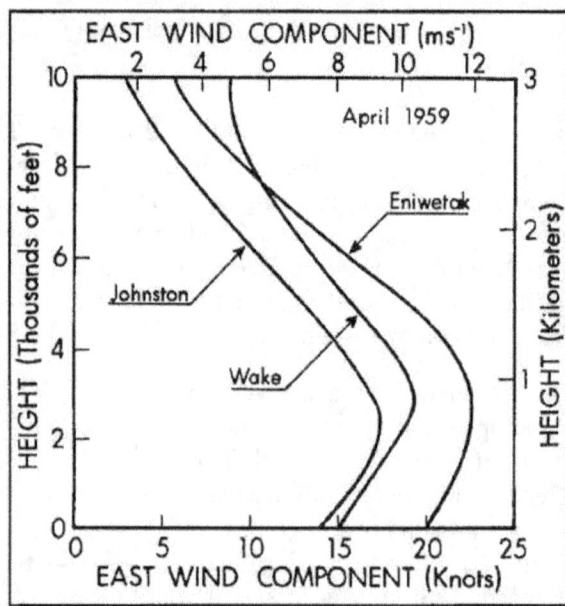

Figure 6-12. Mean zonal tradewind speed in the lowest 3 km for three tropical Pacific stations (derived from Adams, 1964).

Daily shear often exceeds mean shear. Bases of trade cumulus are usually near 2,000 feet (600 meters). Tops vary with atmospheric structure and synoptic situation; they generally evaporate within the tradewind inversion. Mean tops are near 7,400 feet (2.3 km) but commonly much lower–see Figure 6-13.

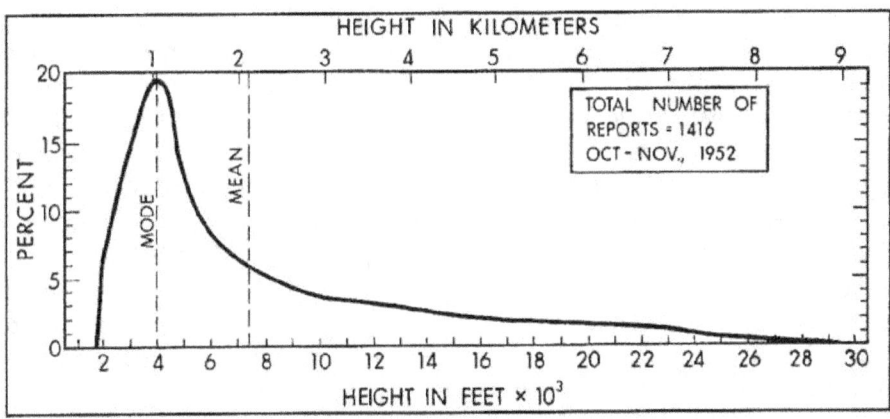

Figure 6-13. Percent frequencies of cumulus top heights for the area 2.5° N to 25° N and 155° E to 170° W (after Palmer et al., 1955).

6.2.6.2 Cumulonimbus (Cb) is a heavy, dense, very tall cloud in the form of a mountain or huge tower, usually accompanied by showers, thunder, and lightning. The upper portion is usually smooth, fibrous, or striated cirrus. The almost flattened Cb tops often spread out in the form of anvils or vast plumes. Satellite pictures sometimes show plumes from individual Cbs extending for hundreds of miles. Over tropical oceans, Cbs are almost always restricted to synoptic or mesoscale disturbances. They are much more common over land; some stations report thunderstorms on more than half the days of the year (see Figures 10-1, 10-2, and 10-3). Over the oceans, Cb tops usually range between 30,000 and 40,000 feet (9 and 12 km). But in the walls and feeder bands of intense tropical cyclones, they can reach 50,000 to 60,000 feet (15 to 18 km). Tops are somewhat higher over land than over oceans. A radar study (Ghosh, 1967) of thunderstorms over India during the southwest monsoon showed mean heights at maximum development to be 43,000 feet (13 km), with an extreme of 56,000 feet (17 km). Studies over Florida and Africa gave similar results. Cbs are typically from 1 to 10 NM (1.5 to 15 km) wide. Despite their size, individual Cbs have short lives; from development to dissipation they seldom last more than 2 hours. Tornadoes and waterspouts occasionally accompany Cb in the tropics, but appear to be less severe than in mid-latitudes.

6.2.6.3 Stratus and Stratocumulus prevail over the eastern tropical oceans, which are dominated by inversions and where the air is being warmed and moistened from beneath. Stratus is a greyish, uniform cloud with a horizontal base. It usually occurs in extensive layers, but may be in ragged patches (fractostratus) caused by cooling and condensation when rain falling from above saturates the air. Stratocumulus is a grey or whitish cloud occurring in patches, sheets, or layers. Its rounded or roll-shaped elements may or may not be merged and are usually arranged in orderly groups, lines, or undulations. Stratocumulus is more common over the oceans with moderate low-level winds and associated mixing. Stratus is more often observed along coasts at night and in the early morning with light winds. Stratus is also common along coasts where warm moist air is advected over cool coastal waters. It may reach the surface and become fog.

6.2.6.4 Fog prevents most visual and near infrared electrooptical devices from operating. Conditions favoring fog are the same in the tropics as in middle latitudes.

• *Radiation Fog.* Over flat open ground, nocturnal cooling in the tropics rarely lowers surface air temperature to the dew point, and fog does not form. However, under clear skies, air that has a local moisture source and that is trapped in river valleys may often be cooled enough at night for fog to form. The Congo and Amazon river basins are the most susceptible regions in the tropics, experiencing 60 to 120 days of fog a year. A different mechanism tends to keep very wide rivers clear of fog (see 7.4.5.2).

• *Advection Fog.* When warm, moist air moves over water that is colder than the dew point, condensation usually occurs and fog forms. This occasionally happens northwest of the Gulf Stream and the Kuroshio Current. Over the cold coastal waters of south China and the Gulf of Tonkin, surface air in the northeasterlies of the winter monsoon is heated and moistened from below, and stratus predominates. During spring, when the monsoon weakens and surface winds veer, the air may be sufficiently cooled from below for fog to form. Hong Kong (22° N, 114° E), where fog occurs on a quarter of the days in March and April, once experienced a week of fog. Periods of fog or low stratus are known locally as "crachin."

Forecasters can anticipate fog when either an anticyclone over China moves eastward, or when the subtropical anticyclone extends westward across the Philippines and the South China Sea, and coastal winds turn from northeast to east or southeast. With time, surface air parcels pass over increasingly warmer seas and pick up more moisture before reaching the coast. Consequently, Hong Kong's dew points rise about 4° F (2° C) a day. If the sequence lasts long enough, the dew point rises to the SST, and fog forms. Crachin is confined to coastal waters and accounts for the large gradient of mean cloudiness there (see Figure 6-2). By the end of April, warming coastal waters end the chance of fog (Ramage, 1954).

The advection fog mechanism leads one to expect fog over cold coastal upwelling. Fog usually develops if sea-surface temperature drops sufficiently along the air trajectory, and if dry subsiding air does not reach the surface. It has been observed along the Ivory Coast in summer, the Angola coast in winter, and along the northern part of the southeastern Arabian coast in summer.

Even within upwelling regions, low stratus (rather than fog) may form if surface air is locally warmed along its track. For example, at Guayaquil (2° S, 80° W), visibilities below 0.5 NM (1 km) have almost never been observed (Edson and Condray, 1989) because SST is about 6 to 8° F (3 to 4° C) higher there than at the upwind entrance to the Gulf of Guayaquil. At Lima (12° S, 77° W) on the coast of Peru, ceilings below 300 feet (100 meters) are rare. In fact, since surface winds generally blow up the SST gradient of the upwelled water and parallel to the coast, fog is likely only when the wind direction shifts to onshore and the air is cooled from below.

During winter, a cold anticyclone over the plateau region of southern Africa occasionally extends to the west coasts of southern Angola and northern Namibia. Hot and dry downslope easterlies, funneled by the valleys, then displace the prevailing coastal southerlies and the associated stratus or fog. As in a California Santa Ana, temperatures reach 105° F (40° C). The dust-laden, gale-force easterlies may blow intermittently for up to 10 days; they tend to weaken at night (D. and J. Bartlett, 1992). Similar wintertime interruptions probably occur along the upwelling coast of northwest Africa.

Research vessel measurements along the east coast of Somalia in summer 1964 (see Figure 6-14, next page) never recorded visibilities below 5 NM (10 km). There were only scattered low clouds, even though air moving over the cooling surface had, only a day before, a dew point 16° F (9° C) higher than the temperature of the upwelled water. Aerological soundings revealed that dry subsiding air had reached the surface and prevented cooling from causing saturation. This is typical; summer fog has never been reported off Somalia (Meteorological Office, 1949).

102

Figure 6-14. Measurements made along about 50° by R.V. Discovery between 16 and 21 August 1964 (Ramage, 1971).

The illustration opposite shows Discovery's route and a kinematic analysis of surface winds. The middle cross-section shows isotherms (° C, solid lines), isohumes (percent, dashed lines), and the inversion layer (shaded).

The bottom cross-section shows air temperature at deck level (solid line), sea-surface temperature (dashed line), and relative humidity at deck level (dot-dashed line).

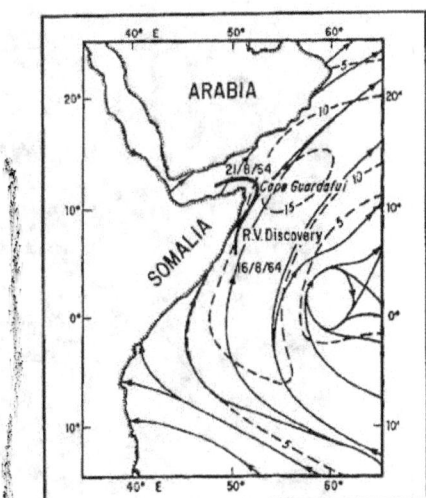

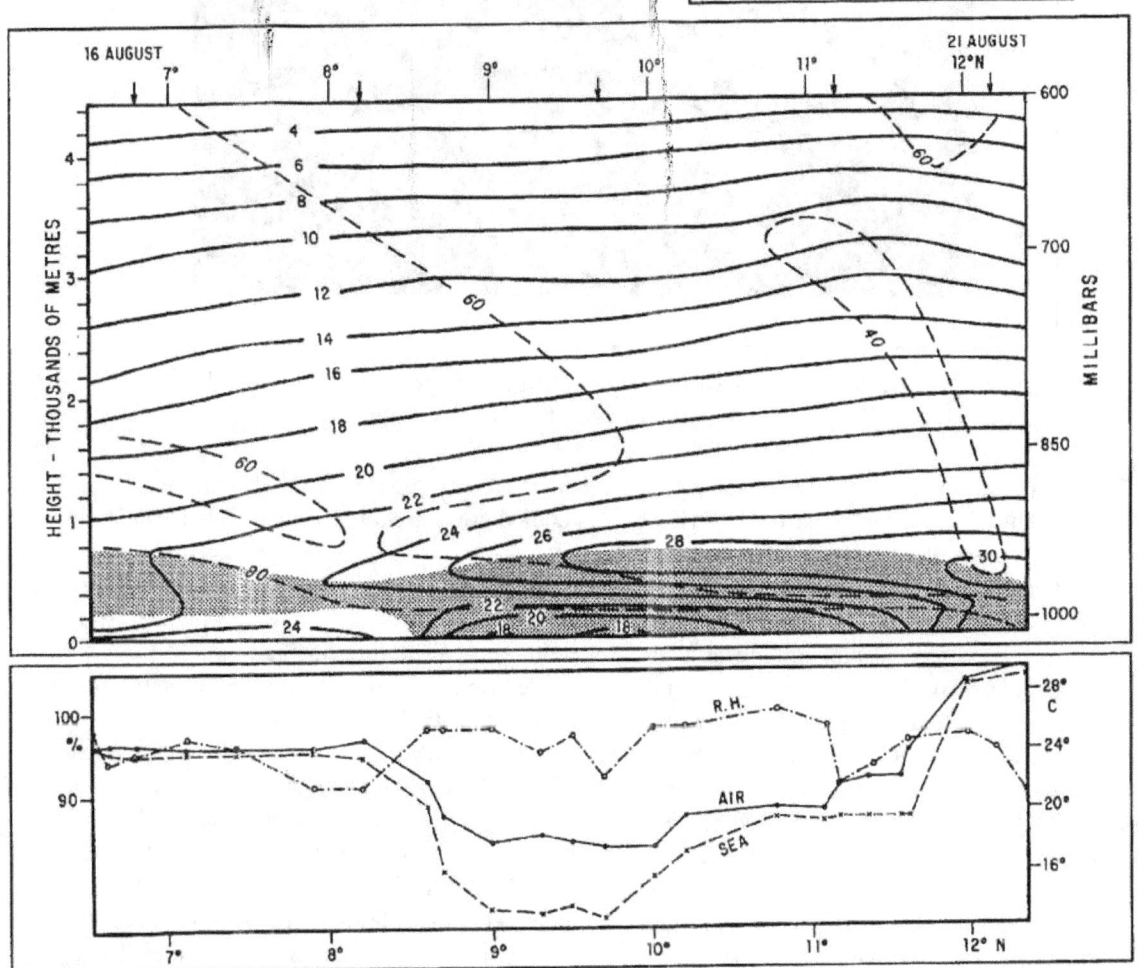

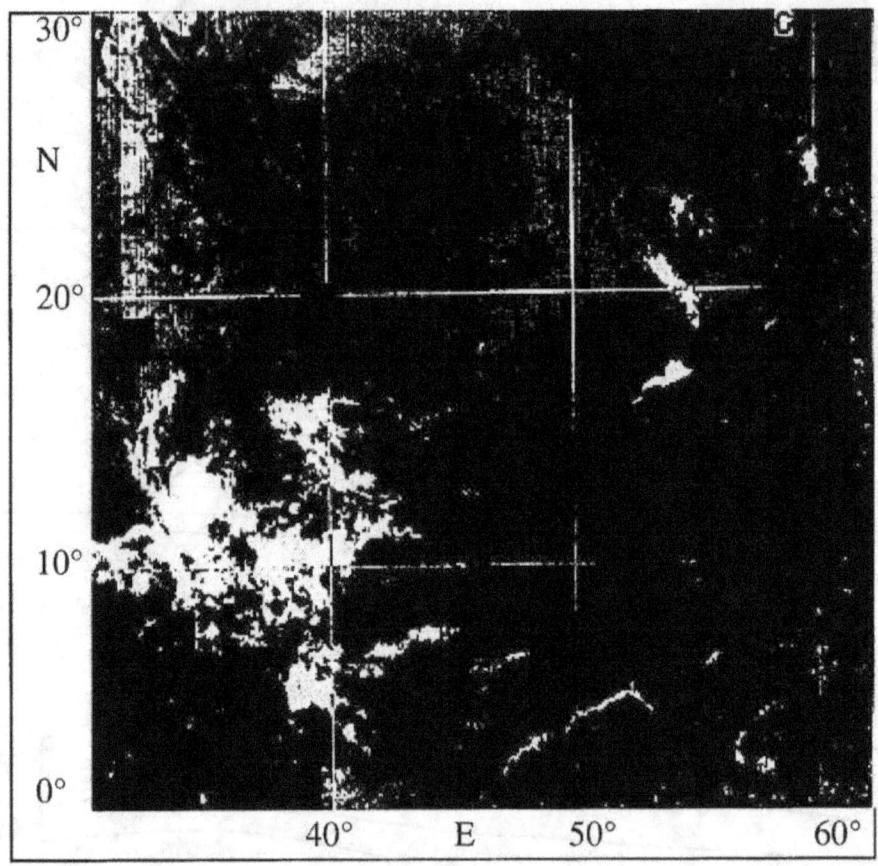

Figure 6-15. DMSP visible image for 7 August 1986 showing clear skies off Somalia and a patch of fog or stratus along the southeast coast of Arabia.

Figure 6-15 shows the typically clear summer skies off Somalia. There are two reasons for this phenomenon. On a small scale, the difference in surface stress between land and sea along upwelling coasts always favors divergence and causes the air to sink (see 2.4.3). On a larger scale, sinking is also produced upwind of a climatological speed maximum in the surface southwesterlies (Figure 4-11). When combined with absence of convection as SST decreases downstream, these two effects bring dry subsiding air down to the surface. Similar effects occur off other upwelling coasts (see Figure 5-14).

To the northeast, along the coasts of Arabia and Pakistan, climatological convergence in the summer southwesterlies (see Figure 4-11) lifts the subsidence inversion above the surface; morning stratus and

drizzle are common. Fog can develop where the air is cooled downstream (see Figures 6-15 and 8-16).

In fog-prone areas, subtle differences in surface temperature gradients, low-level divergence, inversion height, vertical mixing, insolation, and local winds may determine whether the cloud lies on the surface as fog, or a little above the surface as stratus. Forecasting the difference is important because surface-based visual and near infrared electrooptical devices fail in fog, but can function under a low stratus deck. Surface observations can distinguish between fog and stratus, but satellite pictures cannot. One cannot generally infer the difference between stratus tops and the ocean surface from satellite IR images and, of course, they give no hint of ceilings. However, multispectral imaging may soon enable stratus tops

104

to be distinguished from the ocean surface. Forecasters can find the stratus in visible images. Using centrally-estimated and disseminated SST charts, they then need to predict surface wind direction and estimate down-trajectory surface temperature change before considering the chance of fog. A sounding can determine the base of the subsidence inversion. Remember that even small topographical features less than 150 feet (50 meters) high may prevent sea fog from spreading inland.

6.2.6.5 Altocumulus and Altostratus. In the tropics, these clouds are typically associated with disturbed weather. They may persist as the residue of "rained out" cumulonimbus. Altocumulus resembles stratocumulus in structure and color, but has smaller elements and appears at higher altitudes (by definition, from 6,500 to 25,000 feet/2 to 7.5 km). Altostratus, in the form of grey or bluish sheets or layers of striated, fibrous, or uniform appearance, may be formed from spreading of cumulus or cumulonimbus; it may contribute as much rainfall as the active convection. Extensive alto-type clouds usually accompany tropical cyclones. Independently formed alto-type clouds are usually found east of upper tropospheric cyclones or troughs. *Nimbostratus*, a thick grey cloud rendered diffuse by fairly continuous rain, may develop from altostratus under these conditions.

6.2.6.6 Cirrus and Cirrostratus consist of ice particles and may be formed from Cb activity (remnant anvils) or independently. Cirrus is a fibrous cloud composed of detailed elements in the form of white filaments, patches, or narrow bands. Cirrostratus is denser and more extensive than cirrus, appearing as a whitish veil that may totally cover the sky, often producing halo phenomena. Thin cirrus is common in the tropics. Independently formed Cirrostratus is generally found with upper-level cyclones and on the south side of subtropical jet streams. *Cirrocumulus* appears as thin white patches of cloud without shadows. It is composed of very small elements in the form of grains or ripples that may be merged, separate, or be more or less regularly arranged. Cirrocumulus is rare in the tropics.

Cirriform clouds are usually much higher in the tropics than in mid-latitudes. Using aircraft reconnaissance flights over India during the southwest monsoon, Deshpande (1965) found that the height of cirriform cloud bases averaged 38,000 feet (11.5 km); 73 percent ranged between 35,000 and 45,000 feet (10.5 and 13.7 km). Tops averaged 41,000 feet (12.5 km); 50 percent were between 40,000 and 45,000 feet (12.2 and 13.7 km). Both bases and tops were 4,000 to 5,000 feet (1.2 to 1.5 km) higher in the southwest monsoon than in other seasons. These findings generally agree with double-theodolite observations made at Manila and Djakarta during the International Cloud Year 1896-97 (Stone, 1957). These observations showed cirrus bases to range between 36,000 and 41,000 feet (11 and 12.5 km), and to be slightly higher in summer than in winter.

Sadler and Lim (1979) compared cirrus motion vectors calculated from sequential geostationary satellite pictures to winds measured by radiosondes. For the tropical eastern hemisphere, differences were least at 200 mb and 150 mb. This suggests that the cirrus usually lay between those levels (41,000 to 47,000 feet (12.4 to 14.2 km), in agreement with the earlier findings. In view of this, and because of innumerable reports from jet aircraft, those meteorologists who still assign the mid-latitude altitude of 25,000 feet to tropical cirrus should desist from doing so.

6.2.7 The Relationship of Summer Cloudiness to Other Elements Over the Tropical Atlantic.
Warren et al. (1989) analyzed the frequency and distribution of cloud types over the oceans, using ship data for the period 1952-1981. Extracts from their charts for June-August over the tropical Atlantic are given in Figure 6-16a. To help interpret the cloud data in Figure 6-16a, see Figure 6-16b, which provides August data for related elements such as long-term means of sea-level pressure, resultant surface wind, sea surface temperature, outgoing long-wave radiation, frequency of highly reflective clouds, and cloudiness. Circulation changes little between June and August; comparisons are possible.

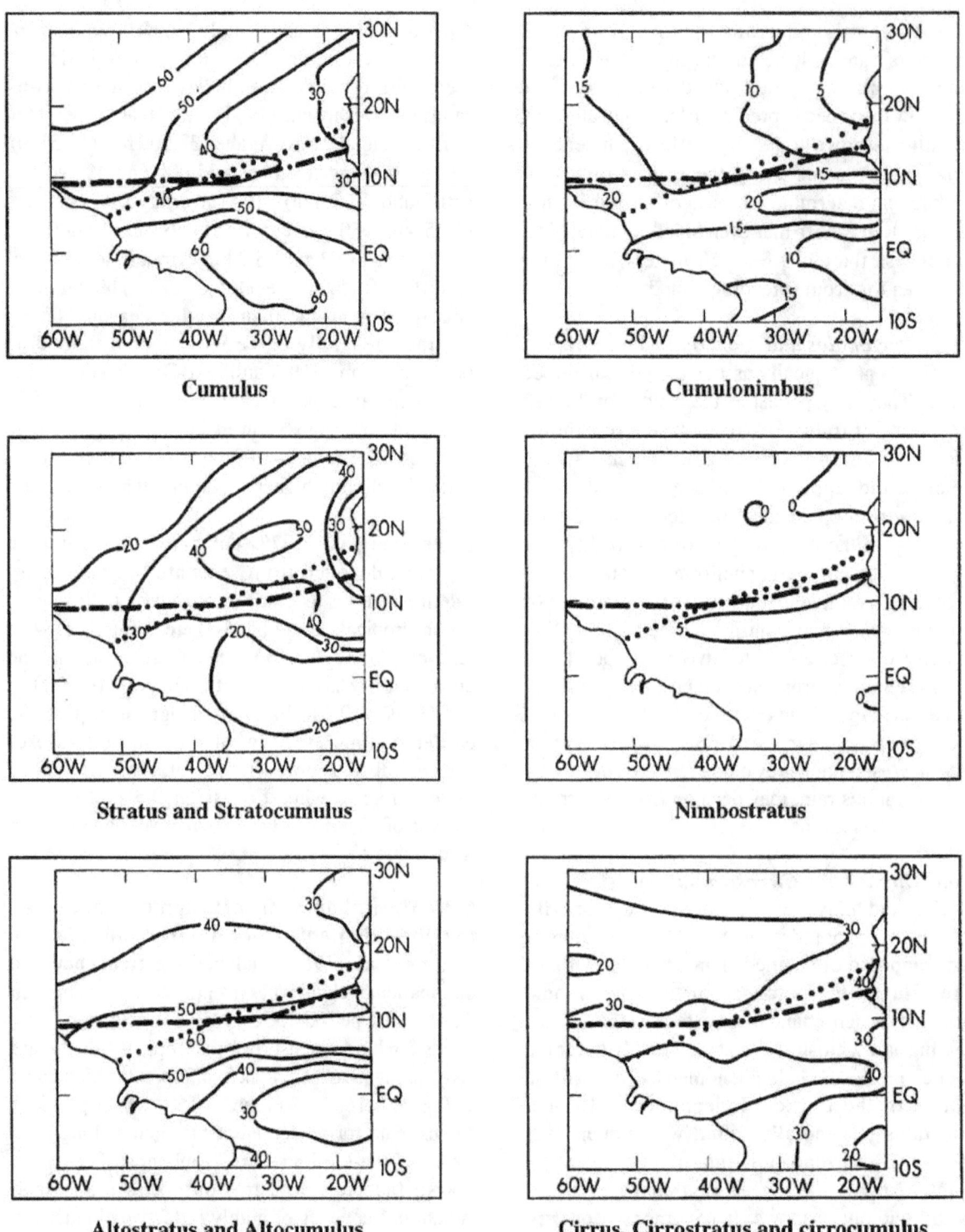

Figure 6-16a. Mean summer (June-August) percent frequency of cloudiness by type over the tropical Atlantic. The surface trough is shown by the dotted line; the surface convergence Asymptote, by the dot-dashed line (from Warren et al., 1989).

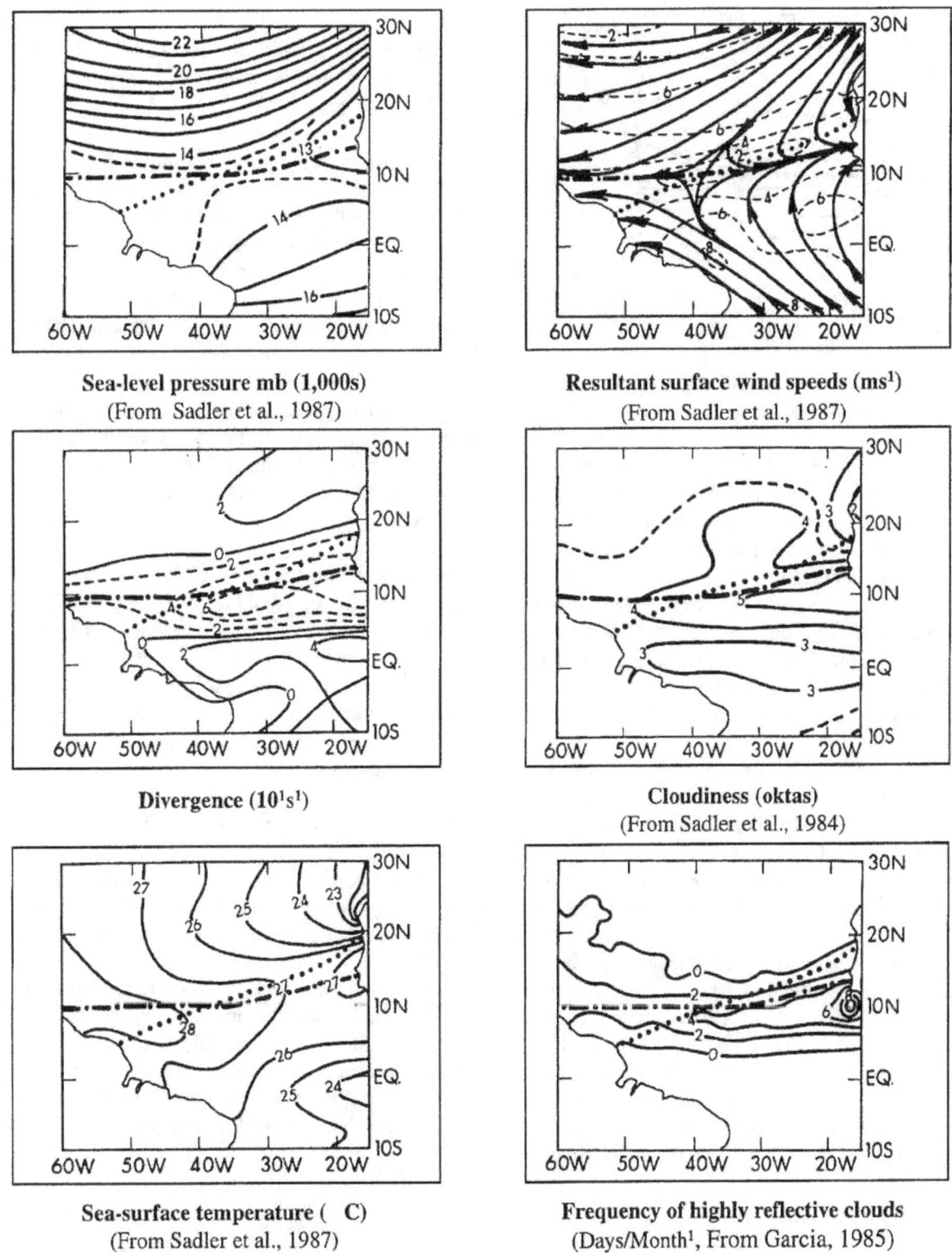

Sea-level pressure mb (1,000s)
(From Sadler et al., 1987)

Resultant surface wind speeds (ms[1])
(From Sadler et al., 1987)

Divergence (10[1]s[1])

Cloudiness (oktas)
(From Sadler et al., 1984)

Sea-surface temperature (°C)
(From Sadler et al., 1987)

Frequency of highly reflective clouds
(Days/Month[1], From Garcia, 1985)

Figure 6-16b. August mean values of various elements used to assist in interpreting the data in Figure 6-16a. See continuation, opposite.

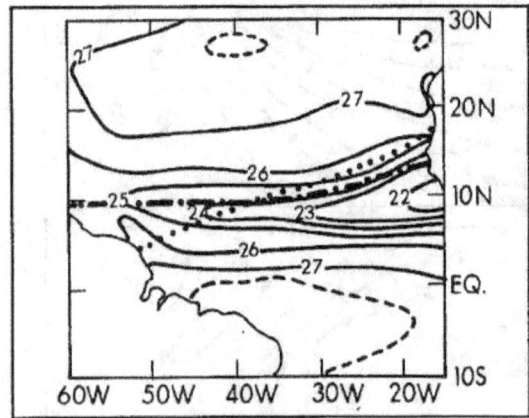

Outgoing longwave radiation (10s of W m)
(From Janowiak et al., 1985)

Forecasters should note that the surface trough and asymptotes of convergence shown by the resultant surface wind diagram in Figure 6-16b lie in a broad, weakly-defined area of low pressure shown in the sea-level pressure diagram. Note also that the low-pressure area is roughly located where the sea surface is warmest. Although flow is convergent along the asymptotes in the divergence diagram, maximum convergence (and the worst weather) coincide where the winds decrease downstream.

As shown in Figure 6-16a, cumulus is the most common cloud; it is observed more than half the time in the heart of the tradewinds. Clouds are scattered, with tops limited by the tradewind inversion and divergent flow. Outgoing longwave radiation exceeds 270 Wm⁻², signifying tops below 15,000 feet (4.5 km).

Stratus and Stratocumulus predominate in the band of fresh tradewinds after they leave the cool upwelled water along the North African coast and are heated from below. Subsidence limits the tops to about 3,000 feet (1 km). Cumulonimbus, altostratus and altocumulus, and nimbostratus are most common (total cloud exceeds 5 oktas) above a band of convergent westerlies south of the monsoon trough.

Highly-reflective cloud is most common here; outgoing longwave radiation is less than 240 Wm⁻². Farther west, where the tradewinds converge, the effect is less pronounced.

Cirrus, cirrostratus, and cirrocumulus are often obscured by lower clouds in the bad weather region and not reported. They appear, therefore, to have a lower frequency. Upper-tropospheric east-northeast to northeast winds (see Figure 4-15) would carry high cloud generated in the bad weather toward the southwest, where lower clouds are scattered and high clouds can be observed more often.

Along the equator, where surface winds diverge most, cloudiness is least. Cumulus predominates and highly-reflective clouds are absent. The southeast tradewinds near the equator are cooled from below and convection dies down.

Vortexes moving westward from Africa are generally concentrated between 10 and 15° N (see Figure 9-12), north of the bad-weather clouds that tend to be confined to the low-level westerlies south of the vortex centers. The cloud charts agree well with studies based on satellite pictures. For example, Duvel (1989) analyzed three summers, using the METEOSAT geosynchronous satellite.

Figures 6-16a and b offer examples of the range of tropical climatological charts now available to forecasters. Growing interest in climate and climate change is largely responsible for the development of these charts, but synoptic meteorology is benefitting.

Although this discussion has been restricted to a relatively small area and a single season, its findings can be applied to other tropical oceans for which climatological charts are available and that experience upwelling, tradewinds, tradewind convergence, monsoon troughs, and low-level westerlies.

6.3 RAINFALL

In the absence of shadowing by orography, most significant rain in the tropics falls in either of two circumstances:

• From deep nimbostratus with embedded cumulonimbus. Wind shear in the vertical and lower tropospheric convergence are both strong and, although rain intensity may fluctuate considerably, skies remain predominantly overcast. Lapse rates are small (see Figure 5-10).

• From scattered towering cumulus or cumulonimbus, when vertical wind shear and lower tropospheric convergence are both weak. Lapse rates are relatively large. The term "rains" is assigned to the former, and "showers" to the latter.

6.3.1 The Character of Significant Rain in the Tropics.

6.3.1.1 Thunderstorm Frequency and Rainfall. Heavy rain falls from thunderstorms. The average thunderstorm lasts about the same time everywhere—seldom over 2 hours. Consequently, where thunderstorms are overwhelmingly the predominant ⁻ain producers, the average number of thunderstorm days should be directly related to the average rainfall. Figure 6-17, derived from Portig (1963), shows large areal variation in this relationship. It ranges from 16 days over western Africa and southeastern Tibet to less than 2 over China, southwestern India, and the ocean. At the ends of the range, Portig identified two types of rain regimes: the West African regime, in which frequency of thunderstorm days and rainfall increase to maxima in midsummer, and the western Indian regime, in which frequency of thunderstorm days decreases as rainfall increases. Indian meteorologists have long known that periods of maximum thunderstorm frequency precede and follow the summer rains (Ram, 1929). The summer rains, particularly along the west coast of India, seldom result from thunderstorms. Between Bombay and northern India, where four Rawinsonde stations are located, the thunderstorm days/rainfall ratio spans both of Portig's categories shown in Figure 6-17. The gradient can be studied by comparing monthly averages of the ratio of thunderstorm frequency to rainfall, along with the wind shear and lapse rate between 850 and 300 mb (see Figure 6-18, next page). At the moist adiabatic lapse rate, virtual temperature differences range between 33.5° C ($T_{850} = 29°$ C) and 43.5° C ($T_{850} = 20°$ C).

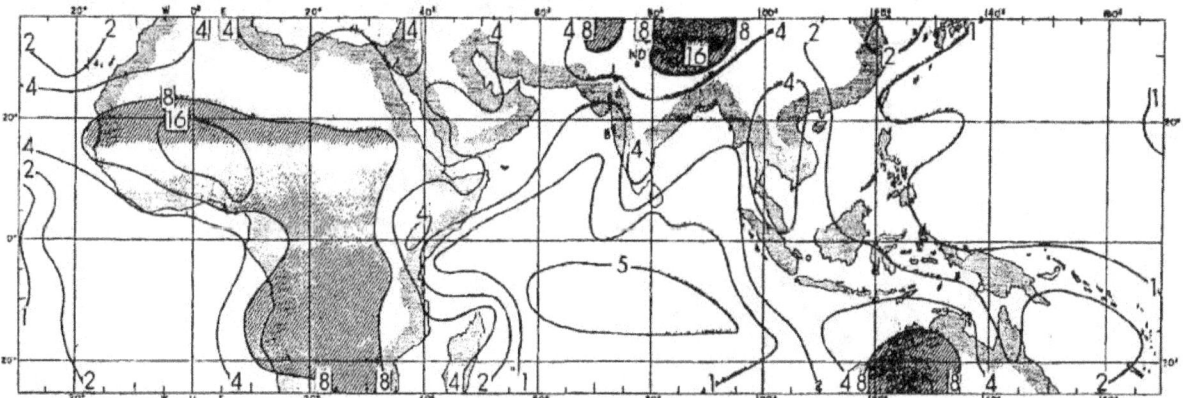

Figure 6-17. Ratio of the mean number of thunderstorm days to the mean annual rainfall in decimeters. Areas with ratios greater than 8 are shaded.

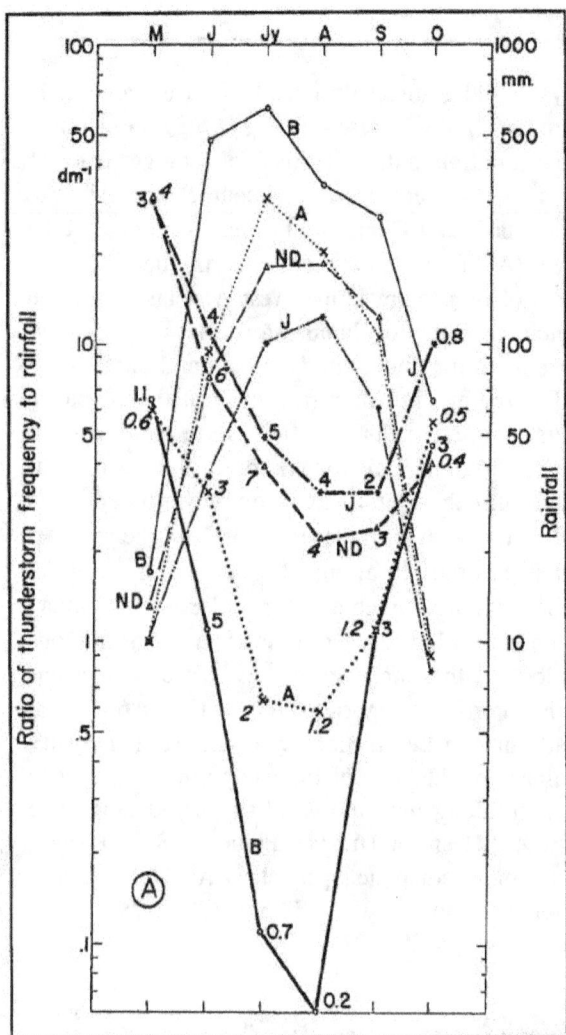

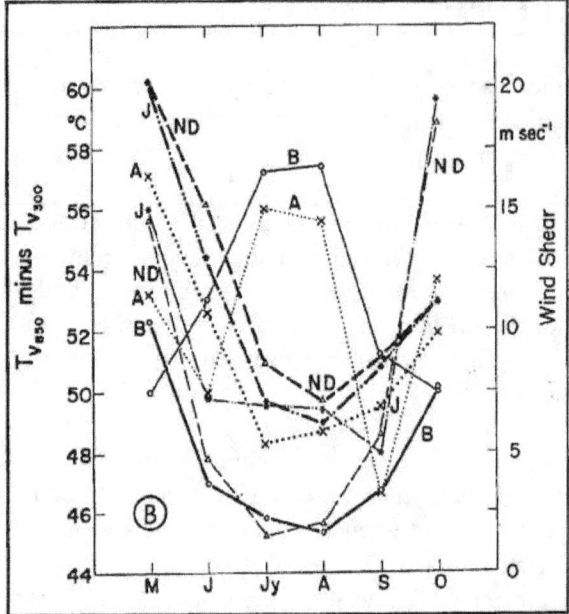

Figure 6-18. Monthly means of thunderstorms, rainfall, lapse-rate, and wind shear in the vertical at four stations in western India. B-Bombay, solid lines (19° N, 73° E); A-Ahmedabad, dotted lines (23° N, 73° E); J-Jodhpur, dot-dashed lines (26° N, 73° E); ND-New Delhi, dashed lines (29° N, 77° E). In (A), the heavy lines represent the ratio of thunderstorm frequency to rainfall in decimeters, the numbers denote thunderstorm frequencies, and the light lines denote rainfall. In (B) the difference in virtual temperature between 850 and 300 mb is shown by heavy lines; wind shear between 850 and 300 mb is shown by the light lines.

Jodhpur and New Delhi, with most of their thunderstorm days in July, are in Portig's West African category. At Bombay and Ahmedabad, typical of the western Indian category, thunderstorm days are fewest in August. However, at all four stations, the thunderstorm days/rainfall ratios are least in midsummer when rainfall is greatest. From May to August at Bombay, where the rains are heavy, the ratio decreases from 6.5 to 0.06, whereas at Jodhpur, on the edge of the Great Indian Desert, the decrease is an order of magnitude less. Thus, in both categories, the trend toward mid-summer is the same. The average amount of rain falling from a thunderstorm is unlikely to increase spectacularly as the season advances. Perhaps, then, a different mechanism begins to produce rain, becoming quite important in the north and overwhelmingly predominant in the south.

As summer advances, the monsoon circulation develops and intensifies over the two southern stations. Strong and generally convergent westerlies, underlying strong and generally divergent easterlies, lead to large-scale rises through the middle troposphere, increased humidities, decreased lapse rates to near moist adiabatic, and great cloud sheets with embedded cumulonimbus from which considerable rain falls. The reduced insolational heating of the surface, near-stable lapse rate, and large shear of wind in the vertical all inhibit thunderstorm formation and a rains regime prevails.

Over the two northern stations, the intensifying monsoon deepens the moist layer, decreases lapse rate, and hinders thunderstorm development. However, Jodhpur and New Delhi, which lie near the heat trough and upper-tropospheric ridge axes, experience decreasing shear of wind in the vertical, a trend that favors thunderstorms. During summer, the upper troposphere is relatively dry, the lapse rate is conditionally unstable, shear is small, and insolation heats the surface; all these conditions favor thunderstorms, but not extensive rain. A prevailing showers regime only rarely gives way to rains. Meridional profiles over western Africa resemble those of India.

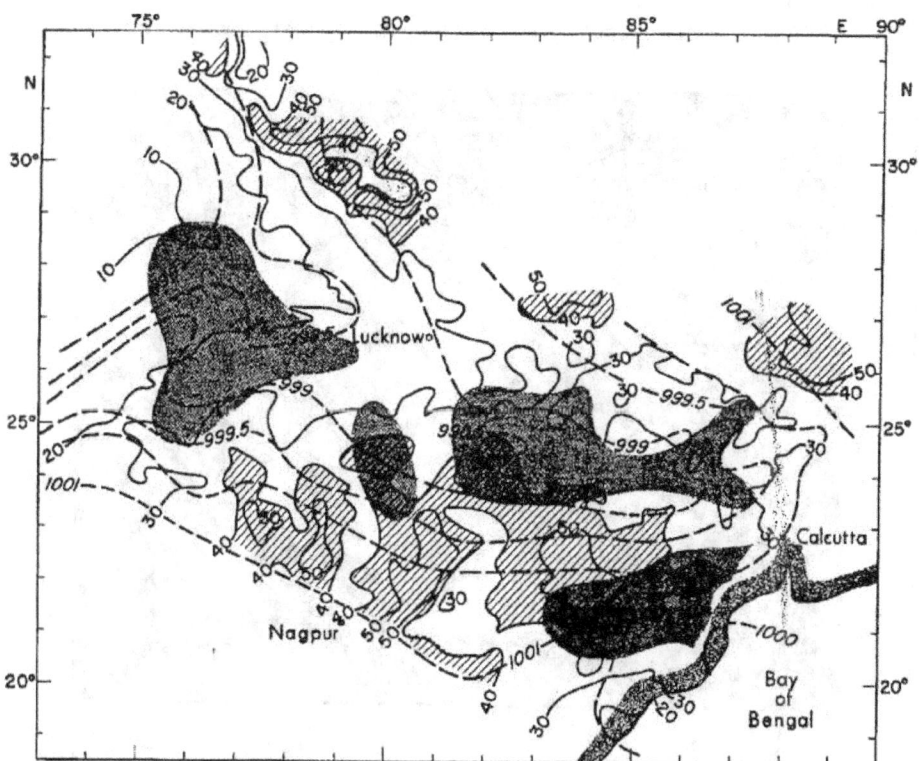

Figure 6-19. Comparison of July mean rainfall and mean sea-level pressure over northern India. (Data from the India Meteorological Department.)

In the summer monsoon trough of central India (see Figure 6-19), rainfall is less than on either side of the trough. In the figure, solid lines are isohyets (in cm); they are not shown above 50 cm (20 inches). Hatching denotes areas where the rainfall average exceeds 40 cm (15 inches). The dashed lines are isobars (in mb). The shaded areas are those with an average of more than 10 thunderstorm days a month.

Thunderstorms, which would be expected to be more frequent over the higher ground on either side of the trough, are more frequent along the trough. Thus, during the wet season, extensive fair weather is also thunderstorm weather. Beneath scattered convective cells, showers may be heavy, but average rainfall per unit area is only a fraction of that caused by the rains.

Gentle subsidence favors fine weather since the accompanying inversion or isothermal layer limits convection. However, an inversion can also enhance convection (Fulks, 1951). Air beneath the inversion is moist and is heated during the day; air above the inversion may cool radiationally and potential instability throughout the troposphere increases. Under these conditions, if some thermal or orographic agent ruptures the inversion, thunderstorms may suddenly develop. Lines of thunderstorms (squall lines) generally last much longer than individual thunderstorm cells. They are discussed in 8.4.2.

6.3.1.2 Continuous thunderstorms are rare, but they occur when rains fall from thick nimbostratus. A huge, stationary and persistent thunderstorm develops within the cloud mass, apparently initiated and sustained by upper-tropospheric divergence above geographically anchored surface convergence (Sourbeer and Gentry, 1961; Daniel and Subramaniam, 1966; Chen, 1969). Surface wind is light and variable, with no gusts or squalls. Radiation and heat divergence in the upper troposphere must dissipate the heat released by condensation in the rising air so rapidly that the lapse rates necessary for a thunderstorm are maintained for hours. These systems have never been successfully probed by radiosondes. More than 10 inches (250 mm) of rain may fall in a day, and floods result.

Hawaii, India, Malaysia, south Florida, Puerto Rico, eastern Indochina, the Philippines, and South China have all experienced continuous thunderstorms. On 2 and 3 December 1978, 20 inches (512 mm) of rain fell on Singapore (1° N, 104° E) during a continuous

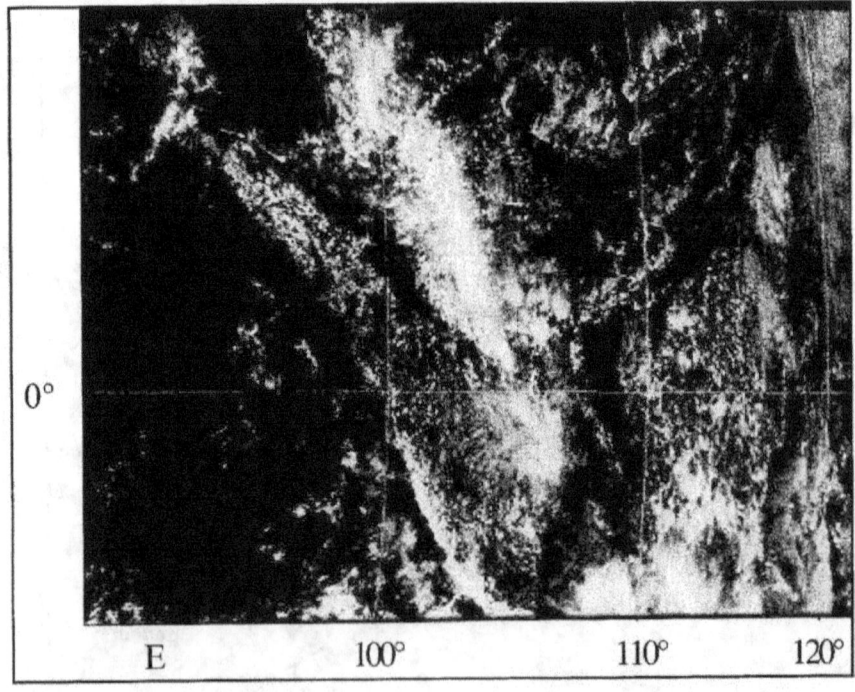

Figure 6-20. Satellite view of storm cloud over Singapore (1° N, 104° E) at 1300L, 2 December 1978.

thunderstorm. This was the heaviest rainfall since observers started keeping records in 1869. It amounted to 21 percent of the annual average. Seven people died and 3,000 were evacuated. The weather satellite photo in Figure 6-20 shows an intensely reflecting blob of cloud over the island. Objective kinematic analyses of the 950 and 200 mb winds at 0000Z on 2 December are shown in Figure 11-23. The torrential rain started at about 0300Z, 3 hours later. At 200 mb over Singapore, the NH and SH Hadley cell circulations were diverging. At the surface, flow converged into a weak low.

Over south China, continuous thunderstorms associated with Mei-Yu fronts (see 6.3.6.4) have produced heavy 24-hour rainfalls (Tao and Chen, 1987). Thunderstorms over south China are most common in May or June during the Mei-Yu season. On 29 May 1973, 33.5 inches (850 mm) fell at Yangjiang (22° N, 111° E). On 31 May 1977, 34.8 inches (884 mm) fell at Haifeng (23° N, 115° E). Over western Malaysia, development tends to shift south from 22 November to 11 December. In the cases studied, the area enclosed by the 10-inch (250-mm) isohyet ranged from 436 to 4,360 NM² (1,500 to 15,000 km²).

Continuous thunderstorms are likeliest in spring or autumn. They often develop beneath upper-tropospheric poleward divergent flow (typical of winter) adjoining equatorward divergent flow (typical of summer). Surface conditions give few clues. Convergence into a weak low is generally all that can be seen; this is a very common occurrence to accompany such a rare event.

In some respects, continuous thunderstorms resemble the mid-latitude mesoscale convective complexes described by Maddox (1983), but they are generally

much smaller. Giant persistent thunderstorms do not fit the thunderstorm model (Byers and Braham, 1949). Continuous thunderstorms can be categorized as neither rains nor showers. They are so rare that the best a forecaster can do is to issue a warning if an active thunderstorm develops within thick nimbostratus.

6.3.2 Significant Tradewind Rainfall. Over the open ocean, where tradewinds and the winter monsoon prevail, monthly average rainfall seldom exceeds 2 inches (50 mm). However, on mountain ranges exposed to the tradewinds, monthly totals of more than 30 inches (750 mm) are not uncommon. This was explained in terms of orographic uplift. Doubts arose, however (Ramage, 1978b), when analysis of tradewind rainfall at the top of Mt. Waialeale on Kauai (5,240 feet/1,598 meters at 22° N, 159° W, annual average 433 inches (11,000 mm), showed that strong tradewind days with rainfall less than .02 inches (5 mm) were as common as those with rainfall greater than 2 inches (50 mm). Similar tradewind inversions were present in both categories. Only light mountain rain falls in an environment of scattered fair-weather cumulus, while persistent, generally moderate, rain falls when the mountain lies beneath an area or line of clouds extending to the upwind side. Then, cloud droplets grow along the air trajectory, unhindered by dry air entrainment. When this cloudy air is lifted along the mountain face, prolonged light or moderate rain falls at the mountain top.

On 14 and 15 October 1976, 7.32 inches (186 mm) of uninterrupted rain fell at Waialeale (see Figure 6-21, below). On the east coast, 10 NM (20 km) away, only 0.79 inches (20 mm) fell. Tradewinds were fresh and the inversion height was 8,000 feet (2.5 km). The cumulus cloud band that caused the rain was aligned

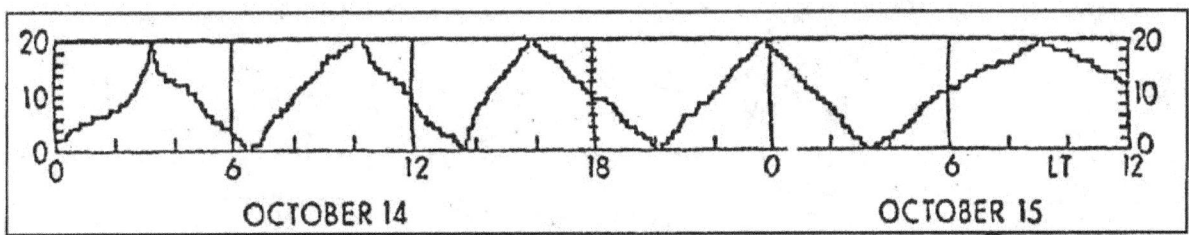

Figure 6-21. Rainfall record at Mt Waialeale (22° N, 159° W) for 14 and 15 October 1976. The tipping bucket activated the recorder at 1-mm intervals. The total was 186 mm.

parallel to the wind. It may have been a shear line (see 8.4.3), and presumably marked slight local convergence. Mountains exposed to the tradewinds or to the winter monsoon in the Pacific, the South Indian Ocean, the Philippines, Indochina, southeastern Indonesia, the Caribbean, and Central America, share this pattern. Maximum rain occurs along the mountain slope about 2,000 to 5,000 feet (600 to 1,500 meters) below the inversion base.

The NH and SH tradewinds converge over the central and eastern Pacific and the central and western Atlantic (see Figure 6-16), and occasionally during spring and autumn over the western Indian Ocean. Cloud persists along the convergence zone, where lower-tropospheric flow parallels the zone axis. Thus, cloud particles grow steadily by accretion; lifting raises the inversion and increases cloud depth, though not to the level of equatorial continental clouds (next section, Stowe et al., 1989). Rain can be light or moderate and longlasting. See Chapter 8 for more discussion.

6.3.3 Mean Rainfall. Rainfall varies greatly over short distances; even where dense rain-gauge networks exist, constructing accurate isohyetal charts is demanding and time-consuming (Meisner, 1979).

Over the open ocean, rainfall measurements at coral islands are probably representative, but too scarce to allow confident interpolation. We can turn to the continuous field measurements made from meteorological satellites to obtain a fair general picture of tropical rainfall. Since deep clouds rain more than shallow clouds, outgoing longwave radiation (OLR), which is inversely related to the height of the emitting surface, is also inversely related to rainfall. Figure 6-22 shows the annual mean OLR for about 10 years. Where there are good raingauge networks, OLR less than 220 Wm^{-2} has been found to correspond to annual rainfalls greater than 100 inches (2,500 mm) and OLR greater than 260 Wm^{-2} to rainfalls less than 20 inches (500 mm).

Rain is heaviest close to the equator over Africa, Indonesia, and South America. The tradewind regions and subtropical Africa, Australia, and southwest Asia are dry. India is not particularly wet since the annual total combines a very wet summer with a desert-like winter. Although the eastern tradewinds are as cloudy as the continental equatorial regions (Figures 6-1 to 6-4), the inversion stops clouds from growing deep enough to give significant rain. The near-equatorial tradewind convergence (NETWC) over the central and eastern North Pacific and the North Atlantic are sharply defined in the OLR, but radiate 20 Wm^{-2} less than the continental equatorial maxima (see 8.4.8).

East of Hawaii, OLR is less than what is recorded in other tradewind regions. Sheets of higher cloud, unconnected to convection or rain, are common here in winter (Morrissey, 1986) and account for the difference (see 8.4.6).

The elevated cold surfaces of Tibet and the Andes may contaminate the data, suggesting higher cloud tops and more rain than might actually fall there.

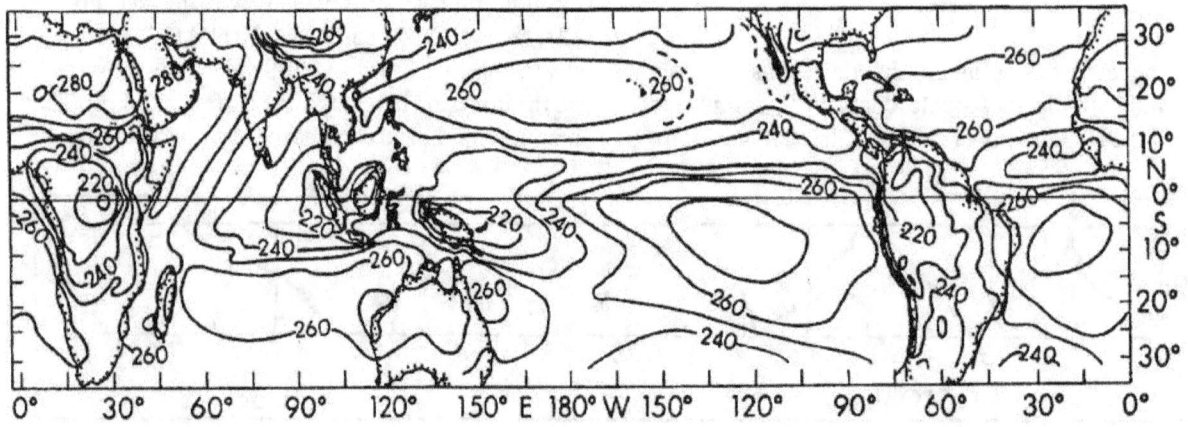

Figure 6-22. Mean annual outgoing long-wave radiation in wm^{-2} (from Janowiak et al., 1985).

Figure 6-23 compares the mean annual zonal OLR and cloudiness. The OLR maximum and cloudiness minimum coincide near 20° N, where the cloud-free deserts are more extensive than the stratocumulus decks of the eastern ocean tradewinds. Although the coldest clouds straddle the equator (as was shown in Figure 6-22), the Pacific and Atlantic NETWCs and equatorial fine weather to the south shift the global minimum of OLR and maximum of cloudiness to 6° N and secondary turning points to 1° S. Near 20° S, the OLR maximum and cloudiness minimum coincide. In contrast to the NH, the much larger cloudiness minimum reflects dominance of eastern ocean stratocumulus and near absence of deserts with clear skies.

Summing up, cloudiness can give a rough idea of rainfall, except over the eastern tropical oceans where a persistent inversion allows extensive, but shallow, stratocumulus to develop over a heating ocean.

6.3.4 Rainfall Variability is inversely proportional to rainfall (see Figure 6-37). Variability tends to be directly proportional to OLR. Consequently, droughts are uncommon in wet (low OLR) regions; in the deserts, almost all rain falls in rare but brief and intense storms that cause severe erosion.

6.3.5 Interannual Variation and El Niño. In the tropics, most significant variations have been associated with the phenomenon known as "El Niño"

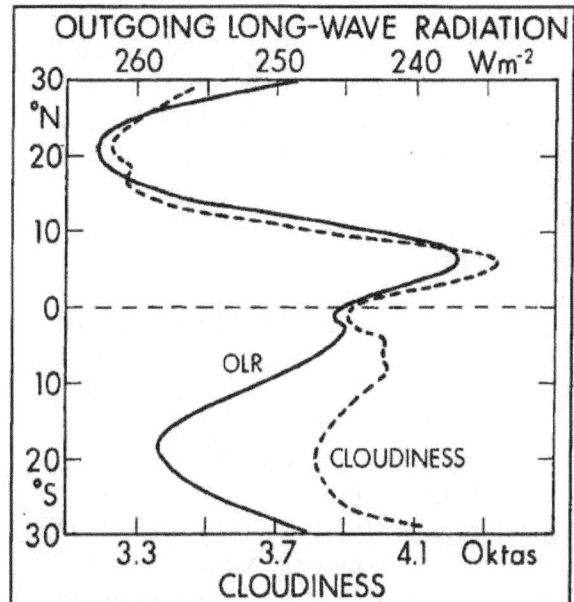

Figure 6-23. Mean annual zonal cloudiness in oktas (from Sadler et al., 1984) and outgoing longwave radiation in Wm⁻² (from Janowiak et al., 1985).

(Enfield, 1989). At irregular intervals ranging from 2 to 12 years, the equatorial Pacific and waters along the coasts of Ecuador and Peru become anomalously warm—see Table 6-1. Usually, this ocean change first appears in the east in the SH summer and extends westward through the year, to weaken and disappear at the end of the next SH summer.

TABLE 6-1. Starting years of El Niño and anti-El Niño events (1877-1988) (from Kiladis and Diaz, 1989).

El Niño (warm) events											
1877	1880	1884	1888	1891	1896	1899	1902	1904	1911	1913	1918
1923	1925	1930	1932	1939	1951	1953	1957	1963	1965	1969	1972
1976	1982	1986									
Anti-El Niño (cold) events											
1886	1889	1892	1898	1903	1906	1908	1916	1920	1924	1928	1931
1938	1942	1949	1954	1964	1970	1973	1975	1988			

In El Niño events, complex interactions between air and sea cause the sea-surface temperature to change. The sequence appears to start along the equator in the western Pacific when a surge of anomalous westerly surface winds, often accompanied by tropical cyclones to north and south (see Figure 9-15), generates an equatorial internal oceanic (Kelvin) wave that travels eastward.

The Kelvin wave, which is trapped along the equator because the Earth's rotation confines the water motions to the near-equatorial zone, moves warm water eastward. This heats the overlying air and allows surface westerlies to shift eastward. A Kelvin wave takes 2 to 3 months to reach the coast of South America, where it deepens the thermocline. Successive bursts of equatorial westerlies generate further Kelvin waves that maintain and enhance El Niño and give it a characteristic period of more than a year.

As El Niño progresses, cool upwelled water disappears off South America (and subsequently in the central equatorial Pacific) as westerlies replace easterlies and shallow scattered clouds give way to deep convective systems. The maritime continent becomes relatively dry. In the most intense Niño on record (1982-83), active convection extended all the way to South America. Over the much narrower Atlantic, similar interactions between ocean and atmosphere have much shorter periods and therefore appear as noise in the annual cycle.

El Niño is one manifestation of a basin-wide oscillation between normally high surface pressure in the southeast Pacific and normally low surface pressure over Indonesia—see Figure 6-24. When southeast Pacific minus Indonesia pressure is abnormally small (that is, when the east-west pressure gradient is small and the tradewinds are relatively weak), El Niño occurs. This "Southern Oscillation" is further discussed in 12.4.3.3.

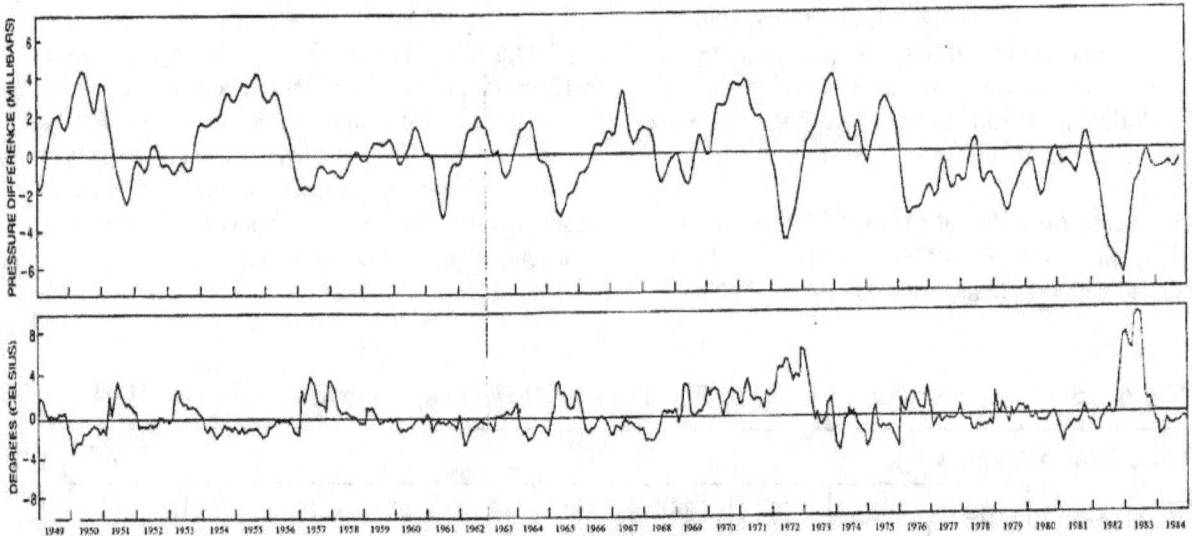

Figure 6-24. The top graph shows the Southern Oscillation Index (SOI), calculated by subtracting surface pressure (mb) at Darwin (12° S, 131° E) from surface pressure at Easter Island (27° S, 109° W), then smoothing and plotting departures from the long-term mean difference. The bottom graph shows departures of monthly sea-surface temperatures from long-term means at Puerto Chicama (8° S, 79° W).

Since the Niño of 1972-73, extensive research has sought to explain why the phenomenon starts, why it persists, and how to forecast it. Increasingly sophisticated air-sea interacting numerical models are now being used. These have generally failed, as have statistically based forecasts, probably because the historical record is too short to encompass the wide variability among El Niño events. Despite this, better observations have usually made it possible to recognize the early signs of El Niño (though false alarms are not uncommon). For example, in mid-December 1990, a surge of surface westerlies along the equator in the western Pacific was followed by tropical cyclones developing to north and south (Figure 9-15). The Kelvin wave moved across the Pacific and resulted in anomalously high SSTs off Peru in March 1991, possibly presaging El Niño. However, the westerly surge proved to be isolated, and SST in the eastern Pacific returned to normal a few weeks later.

Statistics have linked rainfall anomalies in various parts of the tropics to the stages of El Niño and to "cool" periods between Niño events (see Table 6-1). Hastenrath (1990a) used the distribution of highly reflective clouds (Garcia, 1985) to determine relationships of tropical rainfall anomalies to the Southern Oscillation. His findings agree with those of Ropelewski and Halpert (1987,1989) shown in Figure 6-25, opposite.

Hastenrath concluded that the equatorial convection centers that lie over South America, Africa, and occasionally, the maritime continent, tend to vary in unison. Thus, long-range forecasting is sometimes possible. Over northern Peru and Ecuador, where the data is too sparse to permit the statistical analyses depicted in Figure 6-25, El Niño may have a massive effect on rainfall. The coastal plains are normally deserts, but at the start of El Niño, during the SH summer, they can be five to seven times as wet as usual (Cobb, 1967, 1968). In the "Super Niño" of 1982-83, coastal rainfall was more than 40 times normal (Goldberg et al., 1987). As Table 6-2 shows, summer over India tends to be drier during El Niño, though this is not always true. India experienced a severe drought in 1979 when El Niño was absent.

TABLE 6-2. Relationship of moderate or strong Niño to same-year Indian summer monsoon rainfall, 1875-1980 (Ramage, 1983).

	NUMBER OF SEASONS	
	Niño	Non-Niño
Rainfall above normal	5.5	50.5
Rainfall below normal	15.5	34.5

The Hawaiian Islands are also affected. Seventy-five percent of the cool season months at the end of El Niño are drier than usual; conversely, during anti-Niño, 68 percent are relatively wet. In the central Pacific, Christmas Island (2° N, 157° W) and Canton Island (2° S, 172° W), normally dry, receive an order of magnitude more rain during El Niño.

The typhoon season that is usually dying out by mid-November is prolonged toward the end of El Niño. Development occurs farther east than usual, probably because warmer water there encourages persistent surface troughs. In 1982-83, phenomenal activity in the SH (shown in Figure 6-26) gave French Polynesia as many tropical cyclones as in the previous century!

Although El Niño is not duplicated over the equatorial Atlantic, the rainfall regimes of coastal Angola and coastal Peru are similar. Both coasts are bordered by cold upwelling water and are very dry (see Figure 5-4). March and April (the Angolan "wet" season) are dry when coastal upwelling and equatorial easterlies are strong, and wet when coastal upwelling and equatorial easterlies are weak (Hirst and Hastenrath, 1983). As with Peru, the interannual variation of Angolan coastal rainfall is large.

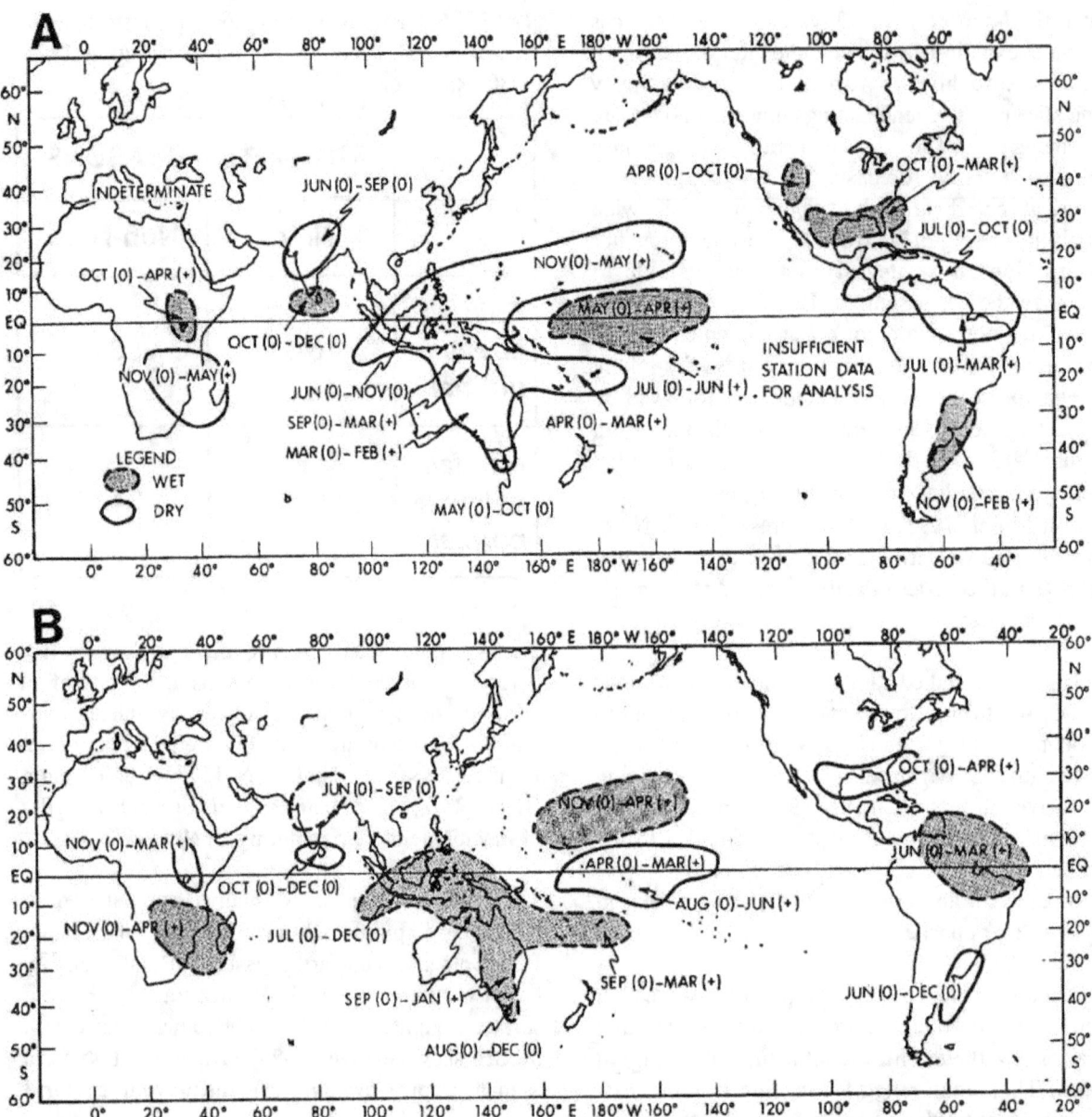

Figure 6-25. Schematic representation of principal rainfall anomalies found to accompany El Niño and anti-Niño years. (A) shows anomalies associated with El Niño (Ropelewski and Halpert, 1989), while (B) shows anomalies associated with anti-Niño or a cool year (Ropelewski and Halpert, 1989). Parenthesized zeros (0) indicate the months of a Niño or cool year from January through December. Parenthesized plus signs (+) indicate the months of the following year.

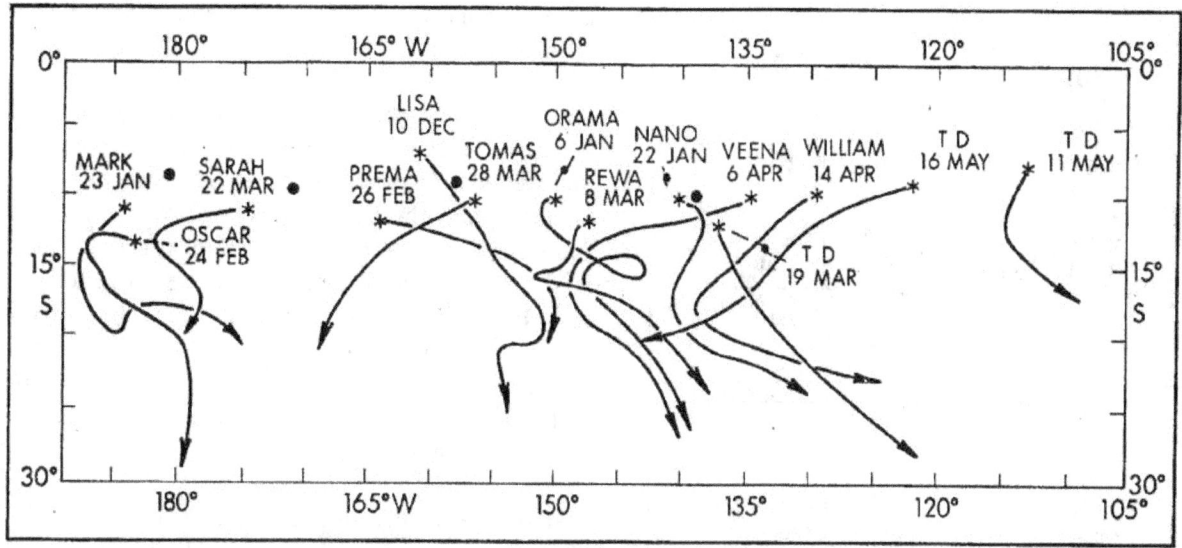

Figure 6-26. Tracks of South Pacific tropical cyclones from tropical depression stage onward for December 1982 through May 1983 (Sadler and Kilonsky, 1983).

6.3.6 Seasonal Distribution—The Monsoons

6.3.6.1 Definition and Extent of the Monsoons. The monsoons blow in response to the annual change in the difference in air pressure over land and sea, which in turn is caused by the difference in temperature between land and sea. The annual change is greatest where continents border oceans. When the Sun moves north of the equator in the NH summer, southern Asia, because of its relatively low heat capacity and the presence of the Himalayas and Tibet, is rapidly warmed. On the other hand, the northern Indian Ocean and the western Pacific store the Sun's heat within their deep surface layers. Consequently, the land gives off heat more readily than the sea; the air over the land becomes warmer, and the air pressure lower, than over the neighboring ocean.

Thus it is that during summer, air flows from the Indian Ocean toward lower pressures over southern Asia. The air ascends as it is heated over the land until it reaches a level at which the pressure gradient is reversed, whereupon it flows on a return trajectory from land to sea. There it descends, and is once more taken up by the landward-directed pressure gradient–

see Figure 6-27, summer. As long as the land is significantly warmer than the sea, this great circulation persists.

In winter, the reverse occurs—see Figure 6-27, winter. The low heat capacity of Asia relative to the northern Indian and western Pacific Oceans ensures that over the land, air is colder than over the sea; the winter monsoon prevails. At low levels, air flows out from the continent over the sea where it rises and returns in the middle and upper troposphere to the land, sinks to the surface, and resumes the cycle.

African monsoons differ from those in Asia. During the NH summer, the Deserts of North Africa heat rapidly and, as over Asia, cause pressure to fall. South of the equator during the SH winter, cooling causes a pressure rise. A pressure gradient is therefore established from south to north across Africa, setting up a massive flow of air, also from south to north across the equator. Because Coriolis force changes sign at the equator, air flow is more directly from high to low pressure there than at higher latitudes. Africa influences the circulation as far as 400 NM (800 km) to the east, merging with the influence of Asia to the north.

119

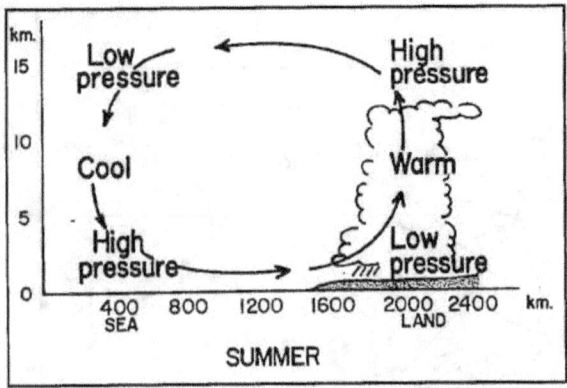

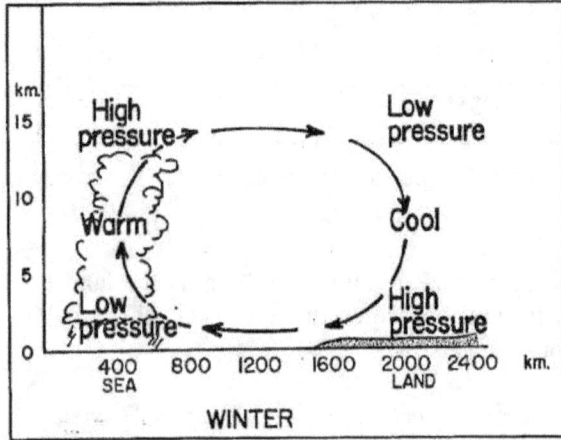

Figure 6-27. Schematic representation of the vertical circulation associated with the summer (top) and winter monsoons.

In the NH summer, wind circulates in a huge gyre that blows from the southeast around the northern edge of the South Indian Ocean anticyclone toward the coast of Africa, then swings to south across the equator, and then to southwest to parallel the African, Arabian and Asian coasts. It finally sweeps across India, Burma, and the Indochina-Thailand peninsula as the southwest or summer monsoon (see Figure 4-11). Six months later, a complete reversal takes place. Northern Africa is cold and southern Africa is warm—the winds now blow from the north across the

equator in the western Indian Ocean and over East Africa.

Australian monsoons are less intense, but persistent. Monsoon tendencies have been identified in many other regions; for example, Mexico, the southwestern United States, the Caspian Sea, and even parts of Europe. Since an annual shift in circulation direction is the fundamental property of the monsoons, monsoon regions encompass January and July circulations in which:

• Prevailing wind direction shifts by at least 120° between January and July.

• The average frequency of prevailing wind directions in January and July exceeds 40 percent.

• The mean resultant winds in at least one of the months exceed 6 knots.

• Fewer than one cyclone-anticyclone alternation occurs every 2 years in either month in a 250 x 250 NM (500 x 500 km) square.

These criteria are met in a contiguous region extending from western Africa to Indonesia with southward protrusions to Madagascar and northern Australia–see Figure 6-28, next page.

Weather is not a monsoon criterion. There is no simple connection between the direction of the surface pressure gradient and rainfall. However, most rain falls in summer or fall, except for a near-equatorial band that has a double maximum shown in Figure 6-28. The summer maximum is associated with heat trough intensification, although the heat troughs themselves are dry. The double maximum is probably associated with temporary intensification of near-equatorial troughs in the transition seasons.

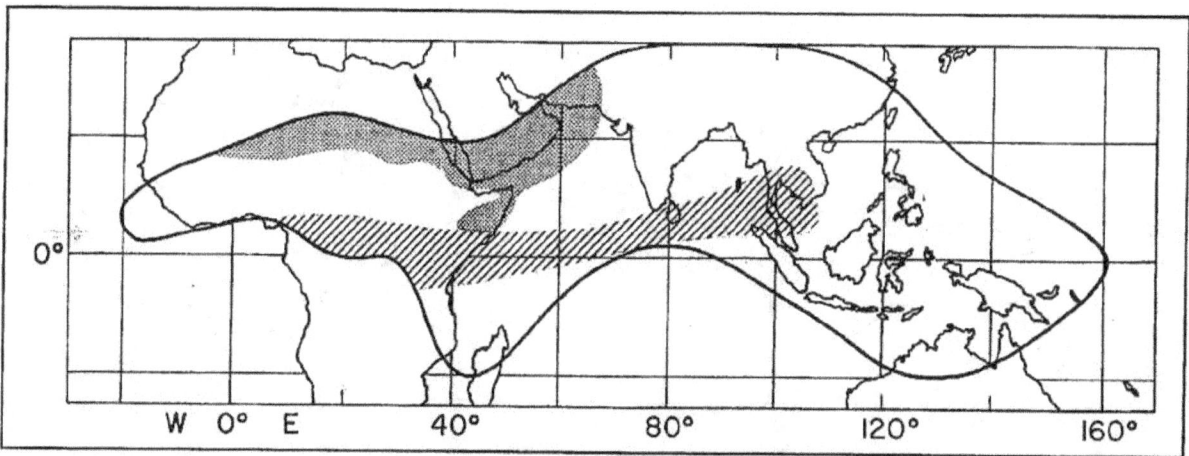

Figure 6-28. The area satisfying monsoon criteria is enclosed by the solid line. Within the enclosed area, deserts (average annual rainfall less than 250 mm) are shaded. The area with a bimodal annual rainfall variation is hatched. Except for a small portion of eastern peninsular India, the remainder of the monsoon area experiences a summer or fall rainfall maximum. As pointed out in 5.2.2, upwelling along the north and west coasts of South America, and along the west coast of southern Africa, prevents monsoons from developing there.

6.3.6.2 Role of the Himalayan-Tibetan Massif in the Monsoons. By protecting the land to the south from cold polar outbreaks, the chain of mountains extending from Turkey to west China produces a sharp discontinuity in surface monsoon characteristics along about 100° E. Thus, in summer, the south Asian monsoon is stronger (with a more intense heat trough) than the east Asian monsoon, whereas in winter, the east Asian monsoon is the stronger (with a more intense polar anticyclone).

Figure 6-29 reveals a large cloudiness gradient between northwest India and western China, a distribution that prevails throughout the year and is reflected in the rainfall. This weather discontinuity derives from the mountain-plateau mass of the Himalayas and Tibet which at times may also influence SH monsoons. In the broadest sense, the NH monsoons comprise three parts:

• East of Tibet, where spring and summer are wet and significant winter precipitation falls.

• West of Tibet, where the climate is desert-like.

• South of Tibet, where summers are wet and winters are desert-like.

These distributions stem from the combined mechanical-thermal effect of Tibet. During summer, central and southeastern Tibet is a strong radiational heat source; with the aid of condensation along the Himalayas, it anchors an upper-tropospheric subtropical ridge across southern Tibet (see Figure 4-15). The heating increases the N-S temperature gradient of the summer monsoon and causes a speed maximum in the upper-tropospheric easterlies (the easterly jet) south of India. Thus, upstream to the east of about 70° E, rising motion beneath divergent easterlies favors clouds and rain.

Downstream to the west of 70° E, subsidence beneath convergent easterlies keeps skies clear. Karachi (25° N, 67° E) and Bombay (19° N, 73° E) (see Figure 8-1) typify the two regimes. Both are coastal stations upwind of mountain chains. Both are exposed to the fresh southwesterlies of the summer monsoon, which are slightly moister at Karachi. June-September rainfall averages 6.00 inches (152 mm) at Karachi, and 56.72 inches (1,441 mm at Bombay. Figure 6-30 gives July circulation.

Central and southeast Tibet are almost snowfree during fall, winter and spring; these areas persist as high-level radiational heat sources. Pressure surfaces

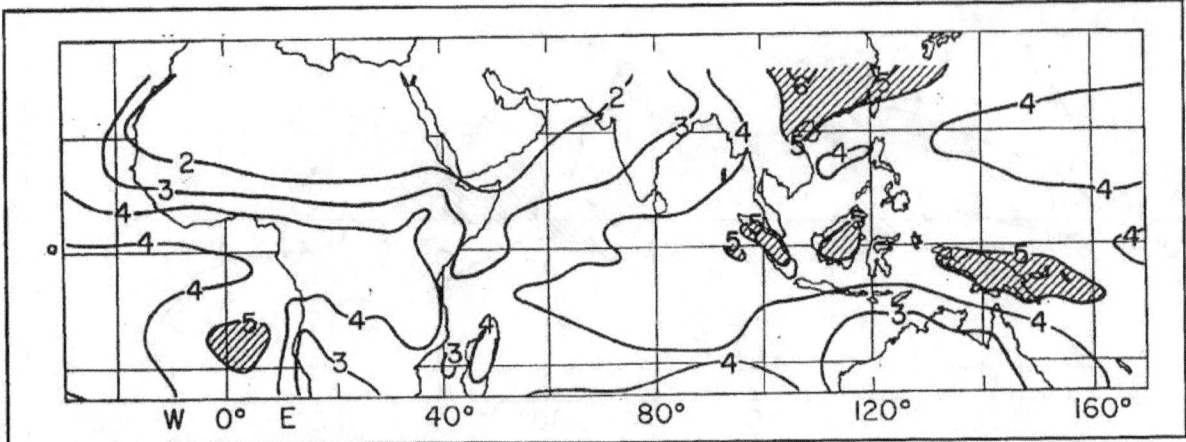

Figure 6-29. Mean annual cloudiness in oktas, based on 3 years of weather satellite photographs (from Sadler, 1969).

are raised there, diminishing the S-N temperature gradient and the strength of the subtropical jet to the south (see Figure 4-13). Immediately east of the plateau, subsiding air is warmed by compressional heating. Over central China it flows alongside very cold air which has swung around the northern edge of the massif; here, the very large temperature gradient produces an exceptionally strong jet stream. With speeds increasing downstream east of 100° E, upper-tropospheric divergence favors large-scale upward motion and increased cloudiness.

West of 100° E, generally convergent upper-tropospheric westerlies over the Middle East and North Africa favor sinking and less cloud. See Figure 6-31, next page, for January circulation.

Year-long subsidence persists over the great deserts of southwest Asia and North Africa, extending to the surface in winter anticyclones and to the lower troposphere in summer heat lows.

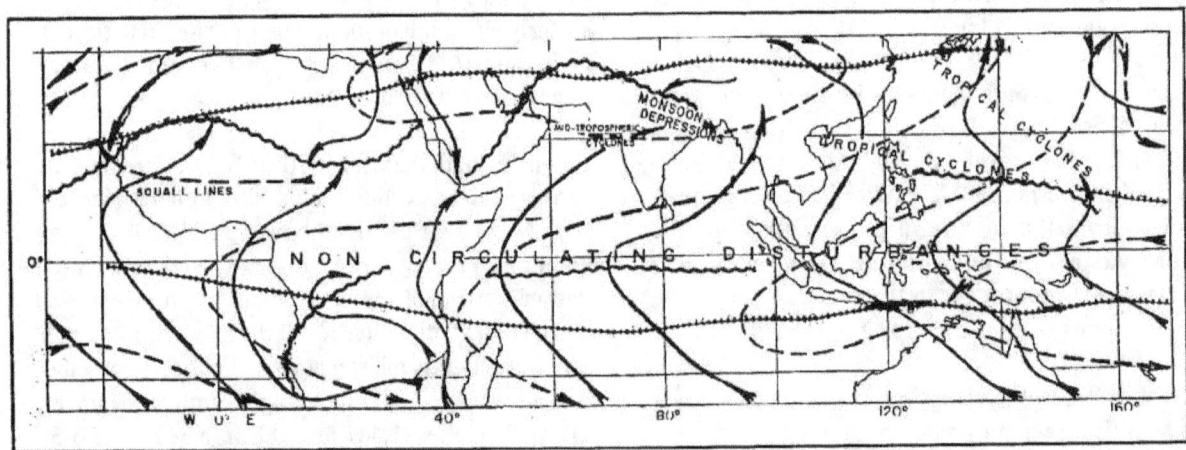

Figure 6-30. July monsoon circulation. Solid lines (thickened where speeds exceed 14 knots) represent resultant winds at the surface. Major pressure troughs are indicated by wriggly lines. Dashed lines (thickened where speeds exceed 40 knots) represent resultant winds at 200 mb. Major pressure ridges are denoted by lines of joined diamonds. Preferred regions for synoptic systems are labeled.

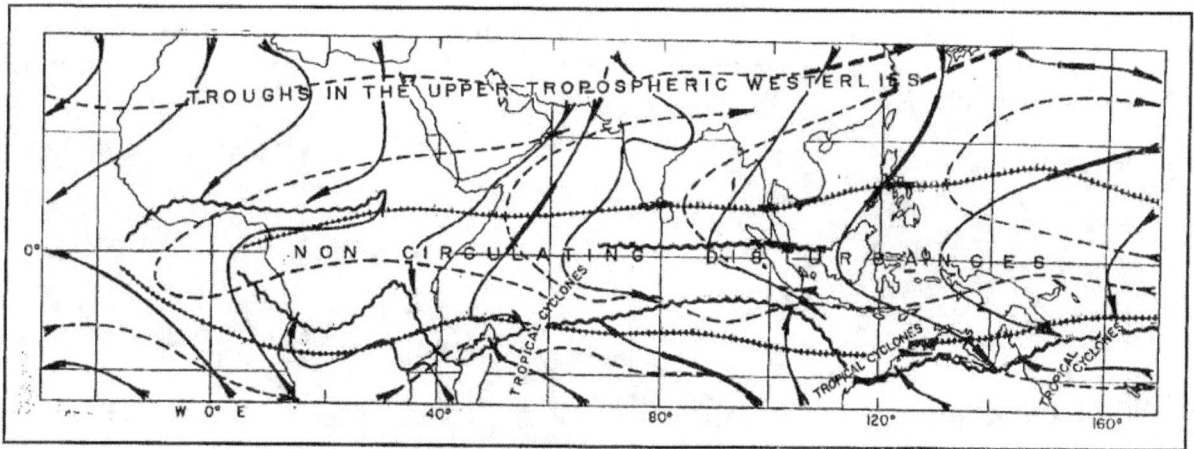

Figure 6-31. January monsoon circulation; same representation as in Figure 6-30. The thickened dashed lines here denote 200-mb wind speeds greater than 120 knots.

6.3.6.3 Rain During the Monsoons. The monsoons, as modified by Himalaya-Tibet, set the stage for weather. Over and near the continents, the cold dry air of the winter monsoon usually predisposes the weather to be fine, while the warm moist air of the summer monsoon favors weather that is unsettled. Although moving surface disturbances during the monsoons are rare and surface winds are notably steady, rainfall on a scale from days to weeks is surprisingly variable. Synoptic changes and intensification and decay of quasi-stationary bad-weather systems are responsible.

The circulation components of the monsoons shown in Figure 6-32 are discussed in detail in Chapter 1 (heat lows), Chapter 8 (monsoon depressions, West African cyclones, mid-tropospheric cyclones and upper cyclones) and Chapter 9 (tropical cyclones). Figures 6-30 and 6-31 show the regional distribution of some of these synoptic components. The circulations can be separated into the three groups in the monsoon region. Predominantly **summer** systems are shown in **boldface**; predominantly *winter* systems, in *italics*:

• Circulations in which downward motion and fine weather predominate: *polar* and *subtropical anticyclones*, and **heat lows**.

• Circulations in which upward motion and wet weather predominate: **tropical cyclones, monsoon depressions, West African cyclones**, and **mid-tropospheric cyclones** (Rao, 1976).

• Other systems with marked weather gradients: *troughs in the upper-tropospheric westerlies, the Polar Front*, near-equatorial troughs, quasi-stationary non-circulating disturbances, surface trans-equatorial flow, and **squall lines**.

Tropical cyclones, monsoon depressions and squall lines interrupt the surface monsoon circulation as they develop and move. However, other synoptic systems may move very little and do not significantly change surface wind directions. Thus, quite variable weather may occur in what is superficially a steady monsoon. As described earlier, most rain falls in the form of rains or showers.

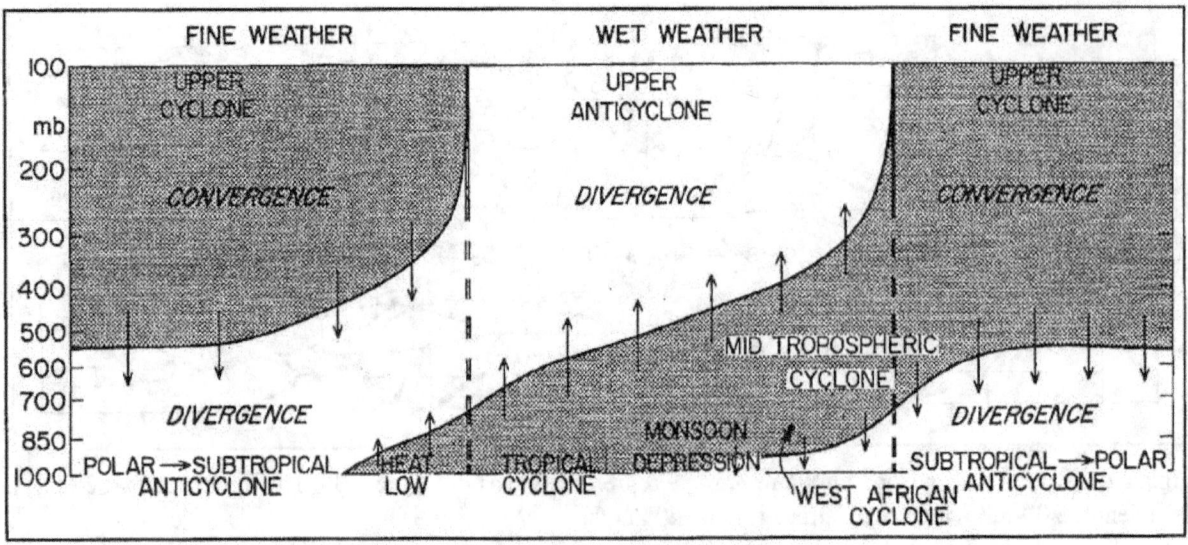

Figure 6-32. Circulation components of the monsoon, arranged schematically according to weather, divergence, and vertical motion. Solid lines denote levels of non-divergence.

6.3.6.4 March of the Seasons. During spring and fall (the transition seasons between winter and summer monsoons), complex changes affect the large-scale circulations and the mixtures of synoptic systems. The changes differ within the monsoon area and from one year to the next—Indian droughts are well-known. The spring transition lasts about twice as long as the fall transition, because in spring the normal equator-pole temperature gradient is reversed, while fall marks return to the normal gradient. For example, over West Africa the heat trough takes 5 months to move from 9° N to 22° N, and only 3 months to return to 9° N.

Since the monsoons arise from differing heat capacities of land and ocean, circulations in a transition season change first in the surface layers. After several weeks, the change has spread through the troposphere and the transition season is over.

In the Indian Ocean-south Asia region, there is an east-west trough throughout the year in the tropics of each hemisphere. In the NH summer, the heat trough lies above 25° N and there may be a very weak secondary trough close to the equator (Figure 6-33). As summer gives way to winter, the primary trough

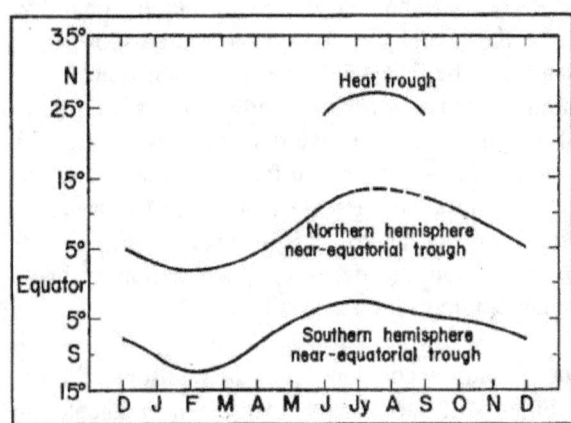

Figure 6-33. Annual latitudinal variation of lower-tropospheric pressure troughs over the Indian Ocean.

weakens and dissipates and the secondary trough becomes predominant. The sequence reverses from winter to summer. The transition season "jumps" are caused by the rapid fall cooling and spring heating of the Deserts of northwest India and Pakistan. This accounts for the usual absence of a double rainfall maximum between 10° and 25° N. South of the equator, despite considerable day-to-day fluctuations, a single trough follows the annual march of the Sun.

124

Over Africa, a weak heat trough is sometimes found between 5° and 10° latitude in each hemisphere during the transition seasons. As the year advances, the trough in the winter hemisphere disappears and the trough in the summer hemisphere intensifies and follows the Sun poleward. The sequence then reverses through the following transition season.

Over the Indian Ocean west of 50° E, northeast and southeast tradewinds occasionally blow at the same time. Climatologically, they meet between the equator and 5° N in April and November, and near 10° S in March and December. In April and November, the winds accelerate downstream to produce net divergence and a cloud minimum; in March and December, downstream deceleration produces net convergence and cloud and rainfall maxima in a near-equatorial tradewind convergence (NETWC) (see 8.4.8.1). In some transition seasons, a strong NETWC persists for several weeks (March 1972 is an example); in others, it is hard to find. Near 10° S in March and December, corresponding coastal effects tend to be masked, since December represents the start and March the end of the summer monsoon.

The Indian Ocean-south Asian regime is weakly duplicated in the northern Australian region.

In late spring and early summer, East Asia experiences an unusual transition season. North of the monsoon trough, and separated from it by a weak ridge, a very active frontal trough (the "Mei-Yu front") dominates south China and Taiwan. It was studied during the Taiwan Mesoscale Experiment (TAMEX) from 1 May to 29 June 1987 (Kuo and Chen, 1990). As the front moves slowly northward between May and July, it may stop and even shift a little southward at times. The front separates relatively cool air in the north from moist, Conditionally unstable air in the south. Along the front, weak mesoscale depressions form and drift slowly eastward. When one is overlain by vigorous upper tropospheric divergence, very heavy rain (and occasionally a continuous thunderstorm) may develop within the depression.

Using satellite images for 1983-85, Miller and Fritsch (1991) identified many of these systems as mesoscale convective complexes (MCCs) that resembled those described for the United States by Maddox (1983). In the 850- to 700-mb layer, a low-level jet reaching 25 knots sometimes precedes and accompanies heavy rain to the north.

In Figure 6-34, opposite, a Mei-Yu front extends from northern Indochina to southern Taiwan. Hong Kong (22° N, 114° E), lying under the northern part of an extremely cold cloud mass, experienced 8 inches (200 mm) of rain in 7 hours. Absence of shear in the deep clouds suggests that they extended from a surface trough to an upper-tropospheric ridge. As the Mei-Yu front moves northward, it is followed by a brief period of better weather under a weak ridge. This, in turn, gives way to the monsoon trough and the typhoon season proper in July.

The satellite helps forecasting. A very cold, non-shearing mass of cloud, many times larger than a single thunderstorm and possibly anchored and enhanced by orography, can give prolonged heavy rain. The Mei-Yu front is not duplicated elsewhere in the tropics, probably because only over China does persistent cloudiness hinder a heat trough from forming, and the absence of east-west mountain ranges allows cool, mid-latitude air to reach the tropics even in early summer. Over the ocean, troughs near the equator spawn cloud clusters; poleward, tropical cyclones may develop in the troughs. Over land west of 70° E, fine weather prevails in the heat troughs, but on their equatorward sides hybrid cyclones (see 8.3.1) may develop.

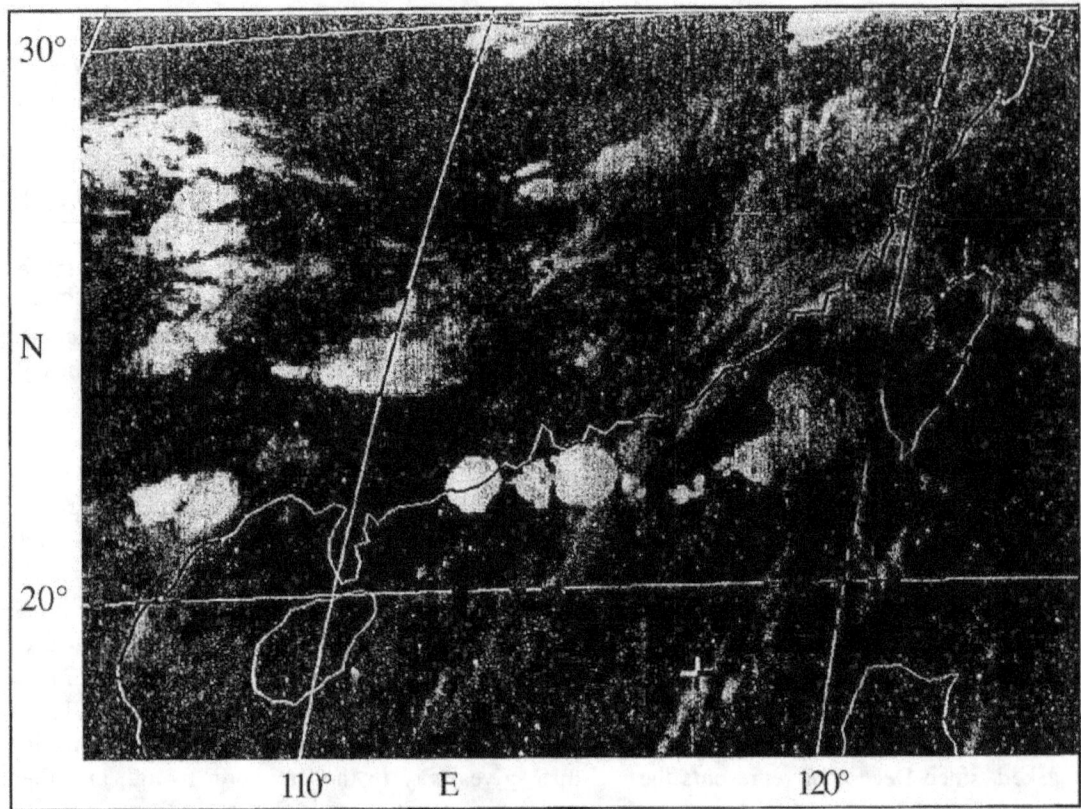

Figure 6-34. GMS IR image of a Mei-Yu front extending from northern Indochina to Southern Taiwan at 1800Z, 28 May 1982.

6.3.6.5 Summary. Although surface disturbances during summer and winter are rare, middle- or upper-tropospheric disturbances are not. In the transition seasons, differences between the monsoon area and surrounding areas are least. Thus, latitude for latitude, the range and complexity of annual variations are greater in the monsoon area than beyond.

The general character of the monsoons and their interregional variations reflect the juxtaposition of continents and oceans. However, without the great mechanical and thermal distortions produced by the Himalayas and the Tibetan Plateau, the vast NH deserts would be moister; Central China would be much drier and no colder in winter than India. Within the monsoon area, annual variations are seldom in phase. The climatological cycles merely determine necessary conditions for certain weather regimes; synoptic changes then control where and when rain will fall and how heavily, and whether or not winds will be destructive. The great vertical circulations comprising the monsoons (see Figure 6-27) often undergo wide-ranging, nearly simultaneous accelerations (surges) or decelerations, apparently triggered by prior changes in the cold parts of the circulations (see 8.4.4).

Monsoon weather, on the scale of individual clouds, seems to be determined by changes occurring successively on the macro- and synoptic scales. Rains set in—not when cumulonimbus gradually merge, but when a synoptic disturbance develops, perhaps in response to change in a major vertical circulation. Showers are also a part of the synoptic cycle. Individually intense, but collectively less wet, they succeed or precede rains as general upward motion diminishes.

126

6.3.7. Monthly Distribution. The tropics are divided into the four regions shown in Figure 6-35 (Asia/Australia and the Americas) and Figure 6-36 (the Pacific and Africa). The bar graphs show mean monthly rainfalls. Average rainfall is given by the number beneath the station location. Graphs are not shown for stations with an average annual rainfall of 5 inches (125 mm) or less. Most of the stations shown have 30 or more years of data.

Some general deductions can be made from the rainfall graphs. Most tropical stations have wet and dry seasons or significantly more rain in some months than in others. Near the equator, some stations have a double maximum, while more poleward stations generally experience a single maximum during or just after the high-sun period. The rainy season, as shown by the monthly means in Figure 6-35, starts abruptly in Bombay, India, and San Salvador, El Salvador. The rainy season begins more gradually at other stations, such as at Addis Ababa (9° N, 39° E), Ethiopia—see Figure 6-36. Similar variations occur at the end of the rainy season.

At some stations, rain may vary considerably within the rainy season. For example, Kingston, Jamaica (18° N, 77° W), as shown in Figure 6-35, has a relative maximum in May and June and a marked decrease in July, followed by an increase to the yearly maximum during October.

Several station pairs illustrate the effects of orography, elevation, and exposure to the prevailing wind. For example, Honolulu (21° N, 158° W), on the lee side of the island of Oahu, receives an average of 22 inches (559 mm) compared to 137 inches (3,480 mm) at Hilo, (20° N, 155° W) on the windward side of the island of Hawaii (see Figure 6-36). Jamestown (1° S, 6° W), at 40 feet (12 meters) elevation on the coast of the island of St. Helena in the South Atlantic receives 5 inches (125 mm) a year compared to 32 inches (813 mm) at Hutts Gate, which is at 2,062 feet (855 meters) near the center of the island. For further discussions, the reader is referred to climatology textbooks such as Trewartha (1981), and special treatises on the climates of particular areas.

ASIA/AUSTRALIA

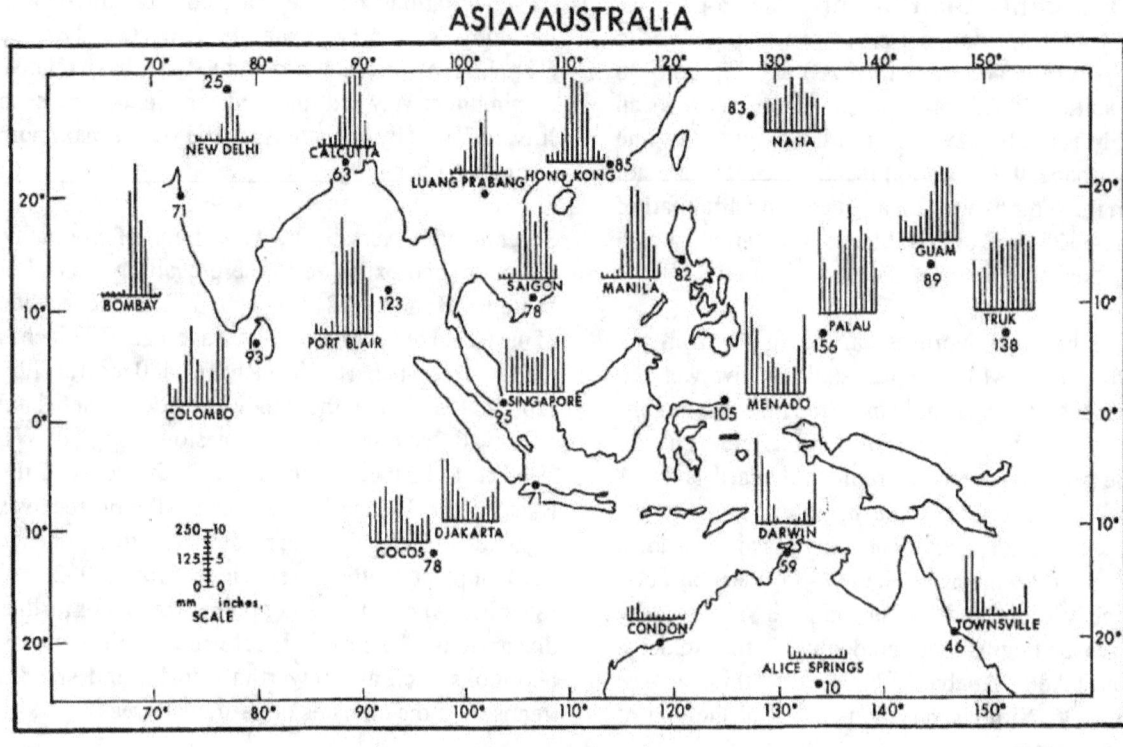

AMERICAS

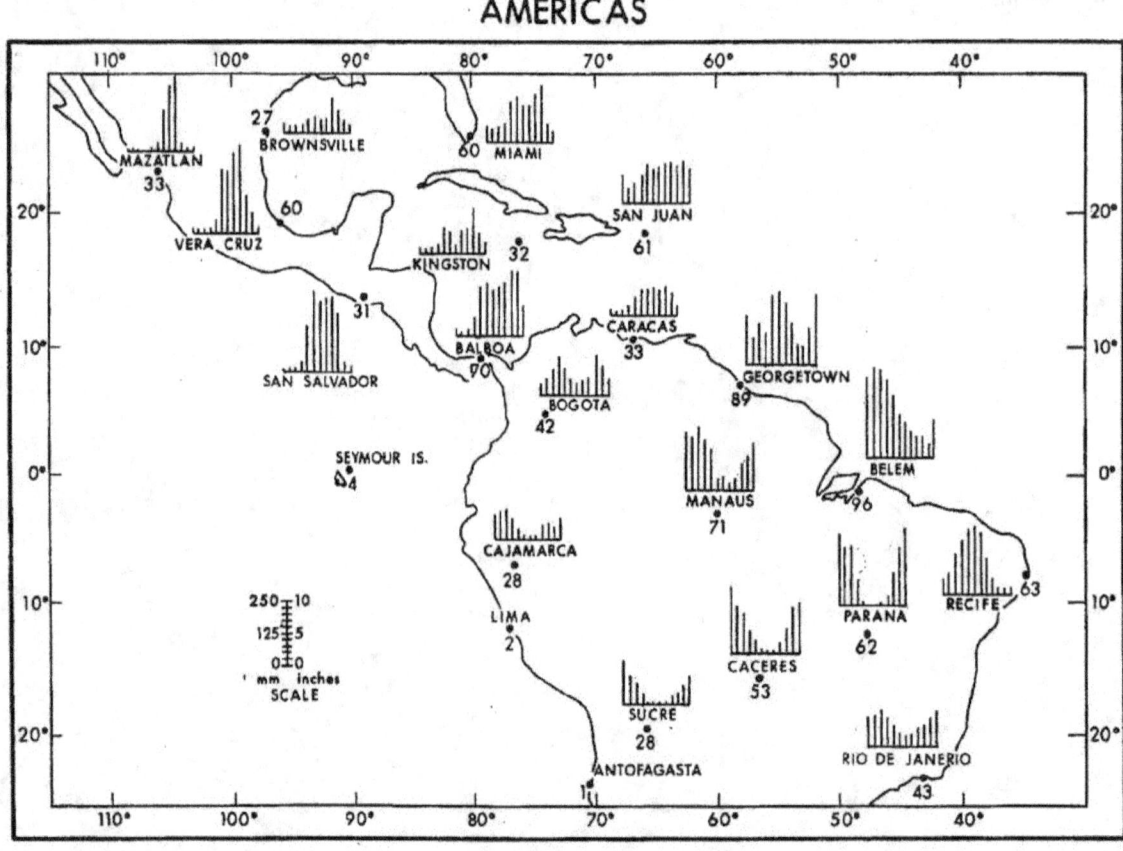

Figure 6-35. Mean monthly rainfall for selected stations in Asia/Australia and the Americas.

PACIFIC

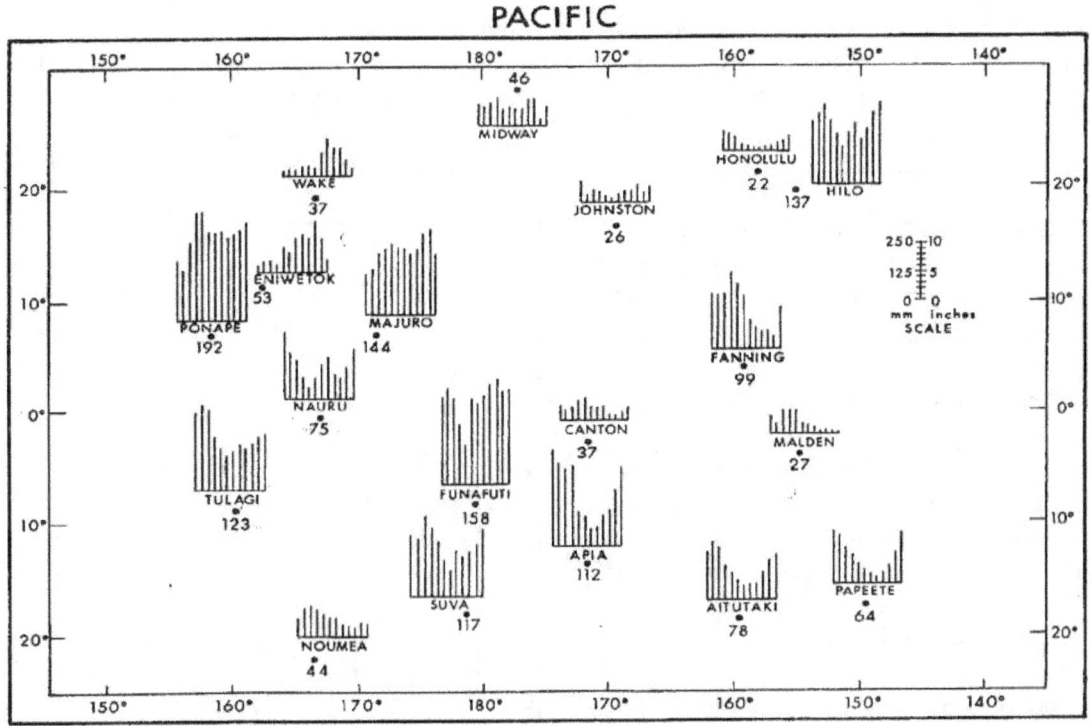

AFRICA

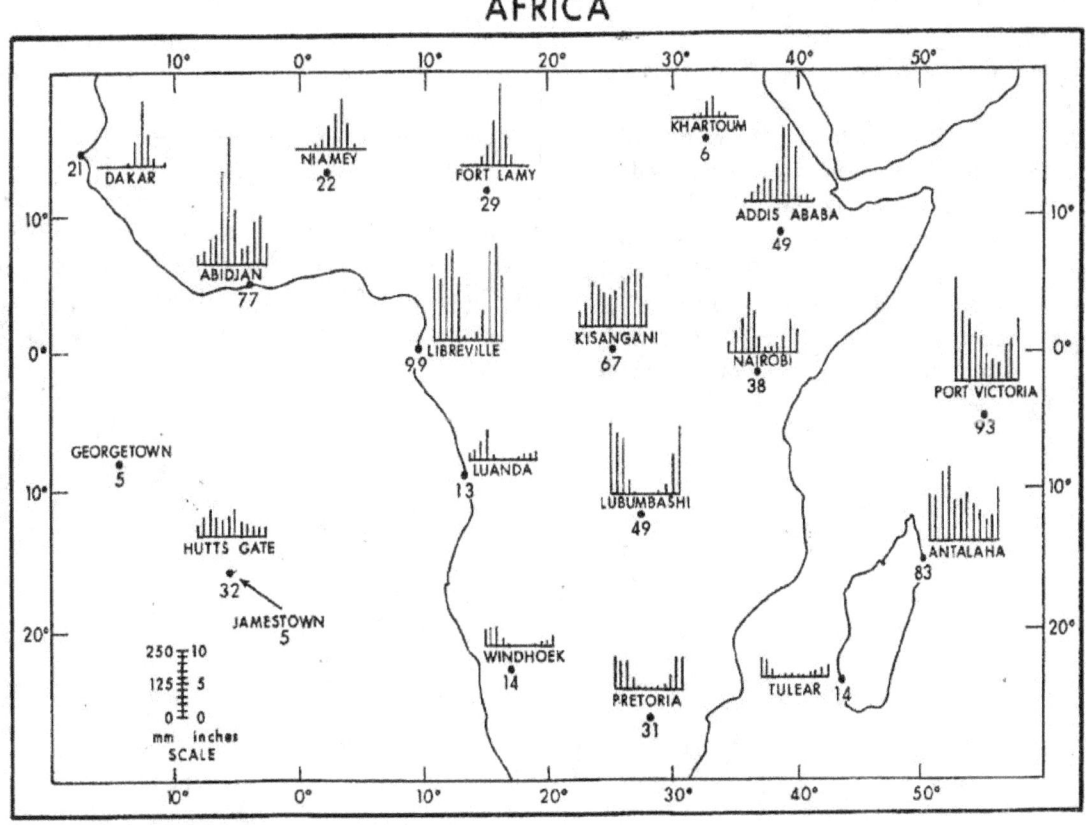

Figure 6-36. Mean monthly rainfall for selected stations in the Pacific and Africa.

129

6.3.8 Monthly Variability. The coefficient of variation (C_v) is defined as the ratio of the standard deviation of the monthly amounts S to the mean monthly amounts X, times 100 percent:

$$C_v = (S/X)100$$

In Figure 6-37, the coefficient of variation is plotted against mean monthly rainfall at Indian stations during the rainy season (Mooley and Crutcher, 1968). The solid line is fitted by Eye to the points; however, the individual station months vary considerably about this line. The dashed line is from a similar study by Coligardo (1967) for stations in the Philippines. For large mean monthly rainfalls, both curves approach a limiting C_v value of about 40 percent. Variability increases rapidly as amounts drop below 8 inches (200 mm). Below 2 to 3 inches (50 to 75 mm), standard deviations of the rainfalls exceed the means; i.e., C_v greater than 100 percent.

In the Sahel of North Africa, south of the summer heat trough, C_v is very large. As Osman and Hastenrath (1969) pointed out, it was unusually wet in July 1958 when the heat trough lay to the north along about 19.5° N, and unusually dry in July 1966 when the trough stayed south of 16° N. Nicholson (1981) confirmed this finding for the Sahel as a whole. Various statistical distributions have been used to describe annual and monthly rainfall variability. The normal (Gaussian) distribution approximates the frequency distributions at stations with large annual or monthly rainfalls. For drier stations, distributions that are skewed to the right, such as log-normal, square-root normal, or gamma distributions, are more appropriate. Although the gamma distribution is more complex than the other types, it generally fits rainfall series better under a wide range of conditions (Mooley and Crutcher, 1968). Further details on these statistical distributions and their applications are available in various textbooks (Brooks and Carruthers, 1953).

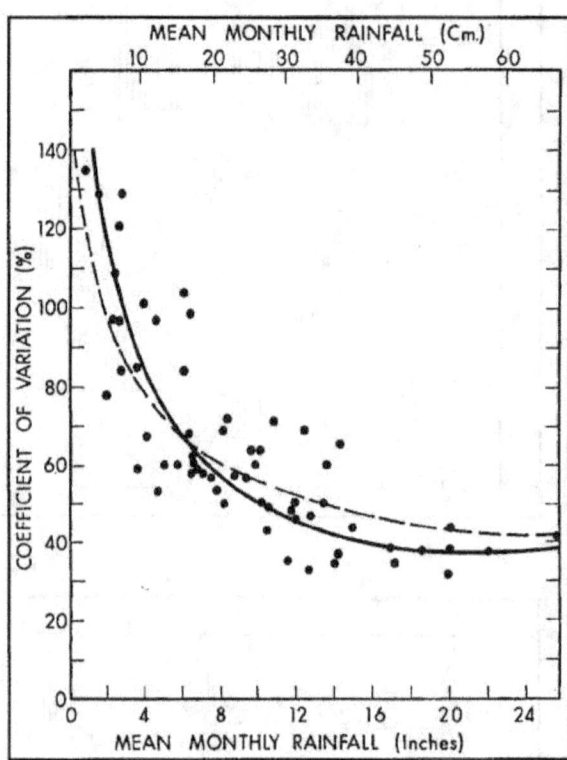

Figure 6-37. Mean monthly rainfall versus coefficient of variation for stations in India (derived from Mooley and Crutcher, 1968). The dashed line is from a similar study for stations in the Philippines (after Coligardo, 1967).

6.3.9 Pentad Distribution and Beginning and Ending of Rainy Seasons.

Some tropical stations compile pentad (5-day) rainfall averages that often reveal climatic features hidden in monthly averages. Over south China (Figure 6-38), the Mei-Yu rains ease off and are followed by a relatively dry spell from 1-15 July, generally as a weak high-pressure ridge moves over from the south. The second rainy period coincides with more frequent tropical cyclones. This sequence cannot be found in the monthly averages for June and July. Japanese stations to north and east follow the pattern with about a 2-week delay (Ramage, 1952b).

Figure 6-39 shows long-term pentad rainfall for selected Indian stations (Ananthakrishnan and Pathan, 1970). Missing from the bar graphs of monthly means (Figure 6-35) is the secondary minimum near the 46th pentad (mid-August) at all stations north of 16° N. More frequent "breaks" in the monsoon rains between 10 and 20 August (Ramamurthi, 1969) may be the cause. Were pentad rainfall averages more generally available, climatological regimes could be much better defined.

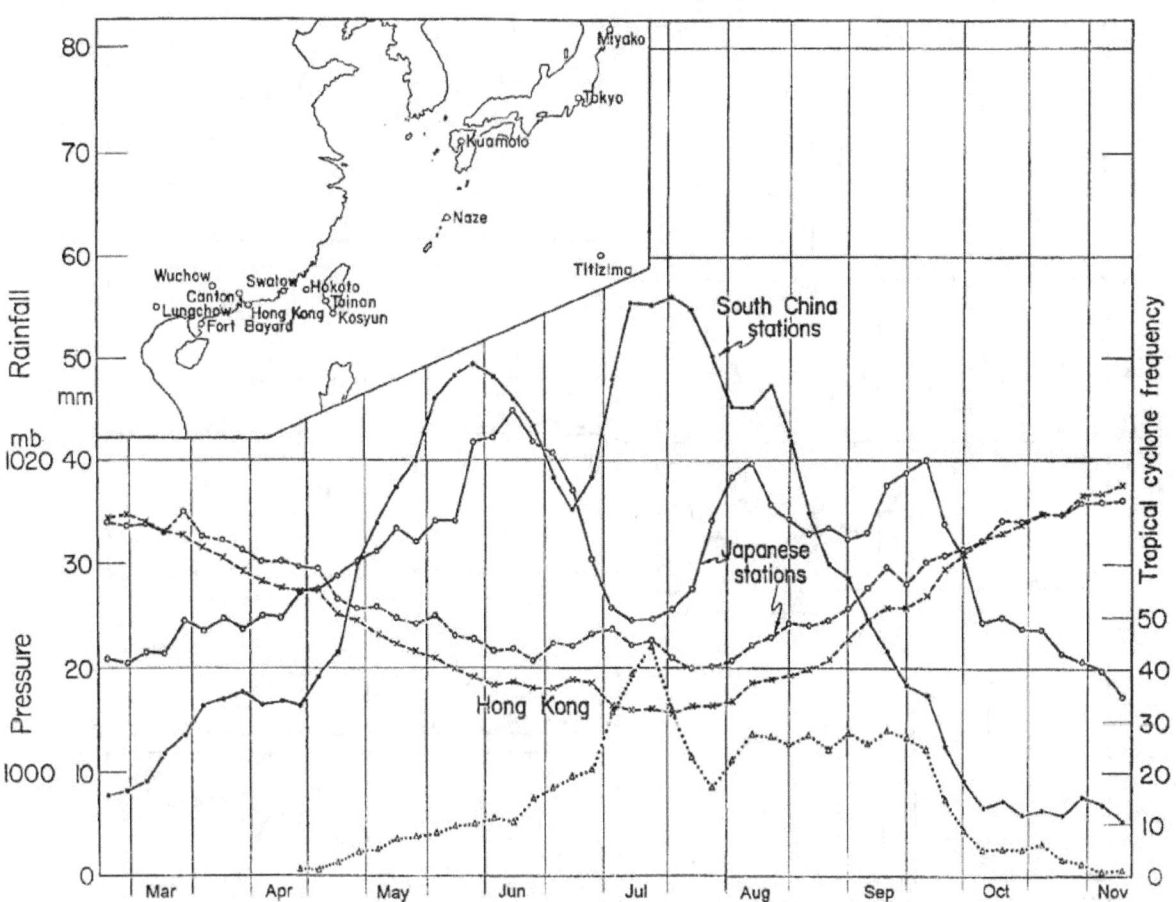

Figure 6-38. Pentad rainfall means (solid lines) for nine stations in southern China and five stations in or south of Japan. The dashed lines denote pressure for Hong Kong and the five Japanese stations. The dotted lines are 50-year tropical cyclone frequencies for southern China.

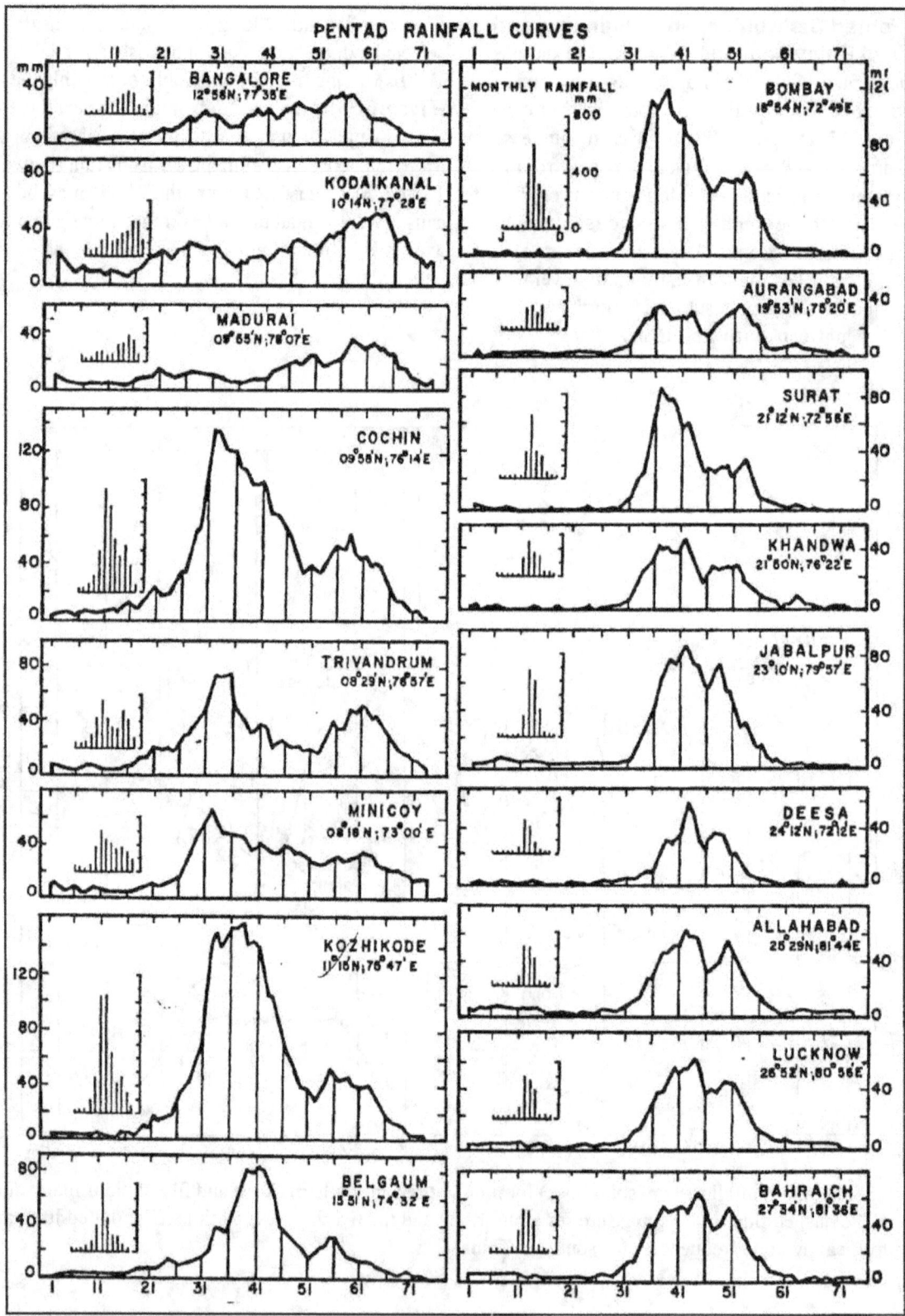

Figure 6-39. Pentad mean rainfalls and monthly means (bars) for selected Indian stations (Ananthakrishnan and Pathan, 1970).

132

Where the tropics experience distinct dry and wet seasons, as in the monsoons, there is great interest in knowing when the rainy season will start and finish. Staff weather officers supporting tropical operations are often urged by field commanders to forecast rainy season beginning and ending dates. Estimates often depend on gross climatological data (such as mean monthly rainfalls) without a firm understanding of what is being forecast. The rainy season is not easy to define. The remainder of this section reviews definitions and delineations of the rainy season in various parts of the tropics.

Pentad rainfalls have helped define the start and end of the rainy season over India (Rao, 1976), although the task is challenging. Experienced Indian forecasters settled on the following six objective rules for declaring the onset of the monsoon over Kerala (in the southwest corner of India) (Ananthakrishnan et al., 1967).

1. Beginning from 10 May, if at least five out of the seven stations report 24-hour rainfall of 1 mm or more for 2 consecutive days, the forecaster should declare on the second day that the monsoon has advanced over India.

2. Thereafter, the daily rainfall distribution should be watched; if it is found that three or more stations out of seven report no rainfall for 3 consecutive days, the forecaster should indicate on the third day that the monsoon has receded from Kerala. The recession of the monsoon will thus be preceded by weak monsoon conditions for at least a day or two. (There is nothing wrong in saying that the monsoon has receded in the early stages of its onset if we bear in mind the pulsatory character of the monsoon. As a matter of fact, such announcements have been made in the daily weather bulletins of the past.)

3. An important point to be borne in mind in the practical application of Rule 2 is that this rule can be applied *only* if the monsoon has not advanced into Konkan (the province north of Kerala) and is still confined to south of 13° N. If the monsoon has advanced north of this latitude it is illogical to recede it from Kerala on the first rainfall criteria given. In this case, we can only say that the monsoon is weak over Kerala.

4. After stating that the monsoon has receded on the basis of Rules 2 and 3, forecasters should continue to keep a watch on the rainfall of the seven stations; when Rule 1 is again satisfied, they should declare that "the monsoon has revived over Kerala," or alternatively "a fresh advance of the monsoon has taken place over Kerala."

5. Rules 2 and 3 can again be applied if required.

6. The date of permanent onset of the monsoon for the purpose of records may be taken as "that date after which it does not become necessary to recede the monsoon over Kerala."

As mentioned in the earlier discussion of the monsoons, rain associated with large-scale circulation changes falls at the whim of synoptic events; at any location, wet weather is interrupted by fair-weather intervals. Table 6-3 shows the normal dates the summer monsoon rains start in various Indian subdivisions, as well as the earliest and latest dates. The apparent progression from south to north is seldom realized in a single season. The rains set in during synoptic pulses and may occasionally start earlier in the north.

Figure 6-40 traces the onset dates at Bombay (19° N, 73° E) between 1879 and 1975. The terms "monsoon" and "rainy season" must be clearly differentiated. "Monsoon" applies to prevailing surface seasonal winds; i.e., the northeast (winter) monsoon and the southwest (summer) monsoon. Lay people, including those who use weather forecasts, equate the summer monsoon with increased rain. However, the start or finish of the rainy season may differ significantly from the changes in prevailing surface winds. Along the west coast of India, westerlies generally set in at least a month before the rainy season starts.

TABLE 6-3. Advance of Indian summer monsoon (normal starting dates for summer monsoon rains).

Area	Normal	Earliest	Latest
Coastal Karnataka (13-15° N)	4 June	19 May 1962	14 June 1958
North Konkan (17-20° N)	8 June	29 May 1956	25 June 1959
West Bengal (22-25° N)	7 June	27 May 1962	23 June
Vidarbha and most of Madhya Pradesh (18-25° N)	12 June	First week of June	Last week of June
Bihar (22-27° N)	12 June	6 June	1 July
East Uttar Pradesh (25-28° N)	15 June	5 June	3 July
West Uttar Pradesh (25-30° N)	25 June	10 June	9 July

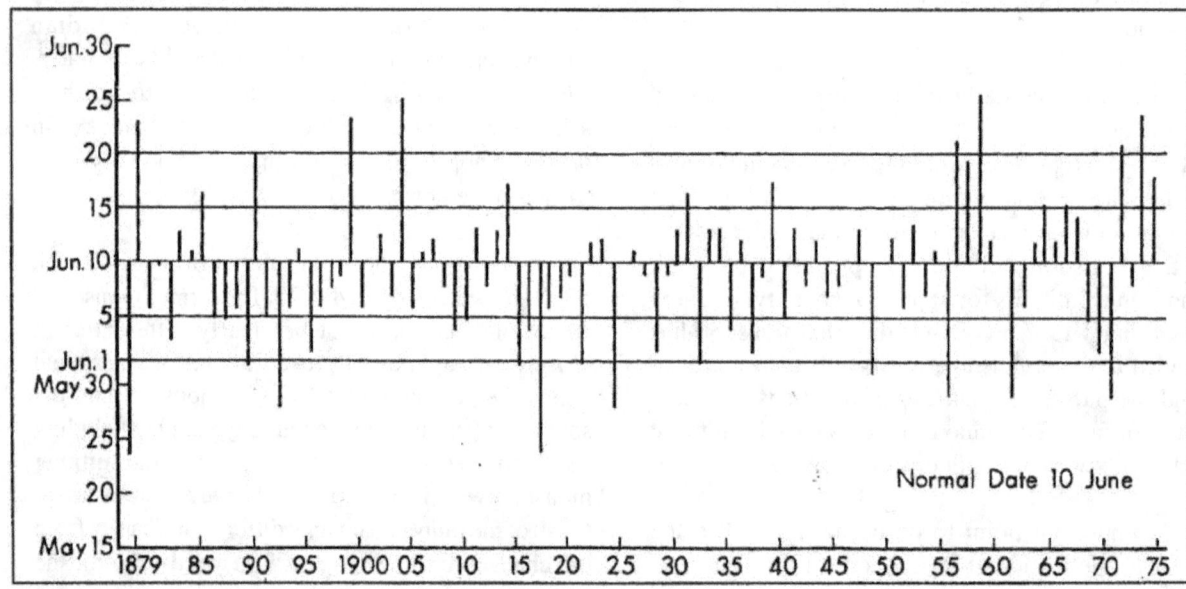

Figure 6-40. Onset dates of southwest monsoon rains over Bombay (19° N, 73° E) from 1879 through 1975 (Rao, 1976).

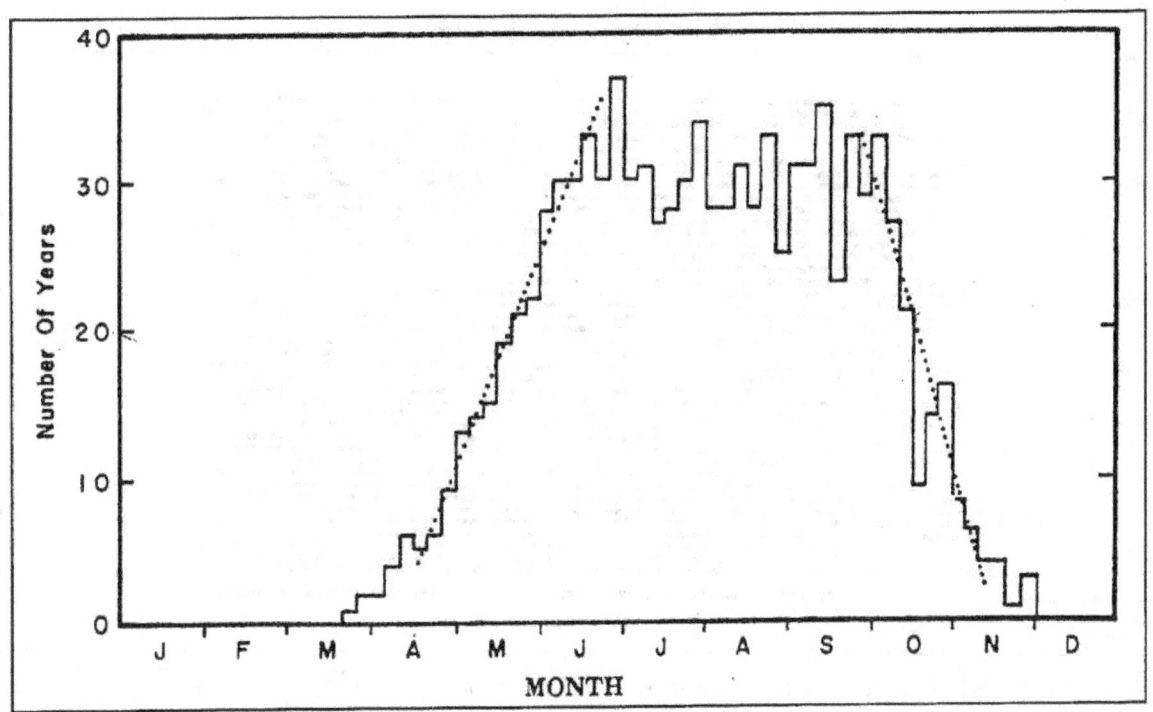

Figure 6-41. Number of years each pentad received more than 25 mm of rain at San Salvador (14°N, 89° W) (40-year POR). The dotted line shows linear best-fit (after Griffiths, 1964).

Griffiths (1964) tried to define the beginning and ending of the rainy season at San Salvador (14° N, 89° W) in two ways. In a climatological approach, he summed the daily rainfall over 40 years by pentads. In a "rainy" pentad, more than 1 inch (25 mm) of rain falls. Figure 6-41 shows the number of years on which each pentad was rainy and gives a subjective estimate of any pentad being rainy (dotted lines). For example, after about 20 May and before about 15 October, there is over a 50-percent chance of any pentad being rainy, with the probability much less before 20 May and after 15 October.

Since such a straight climatological approach gives no direct measure of year-to-year variation, Griffiths also used the "practical approach." He computed the soil-water budget from daily rainfall and an assumed average daily loss of .02 inches (5 mm) due to runoff and evaporation. Figure 6-42 shows the periods each year at San Salvador when the soil-water budget exceeded 1 inch (25 mm). This representation may be useful to agriculture and to the military for trafficability estimates. In some years, short spells of rain precede the main rainy season, while in other years the rains continue once they start.

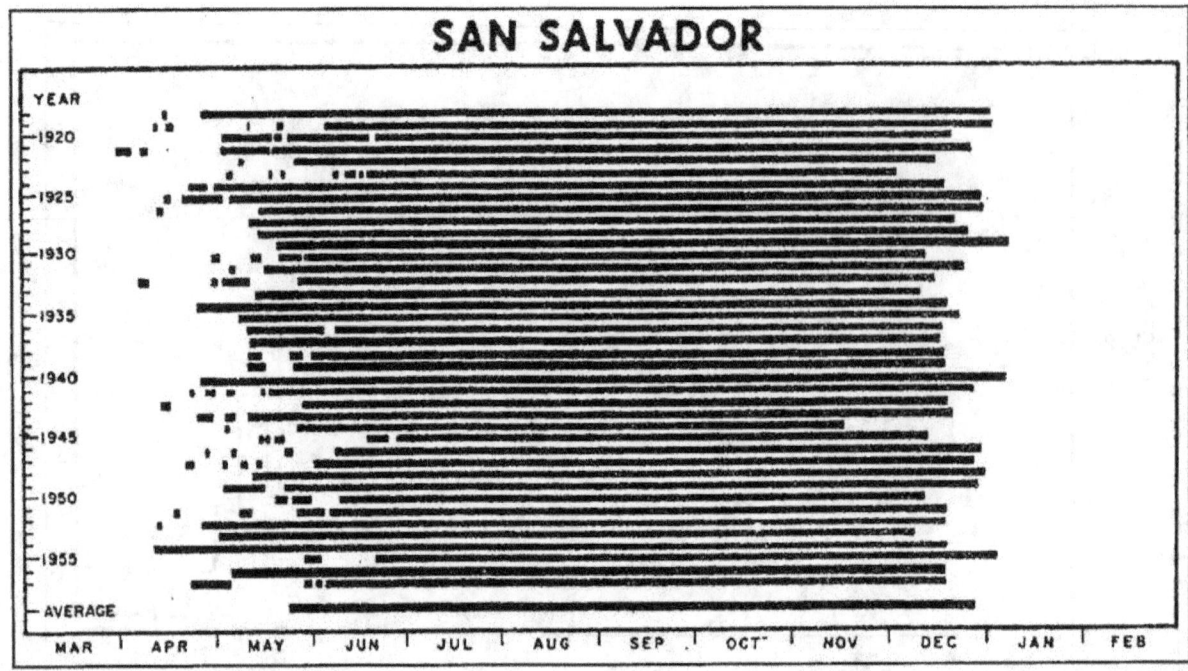

Figure 6-42. Periods each year 1918-57 at San Salvador when the soil-water budget exceeded 25 mm (after Griffiths, 1964).

Using the soil-water budget definition and ignoring breaks of less than 5 days once the main rains set in, Griffiths calculated the average onset date at San Salvador as 25 May (with a range from 11 April to 26 June) anu the average ending date as 2 January (with a range from 21 November to 12 January). Note that the average beginning date shown in Figure 6-42 agrees with that found by the climatological approach, but that the ending date is much later. This lag results from the slow drying of the soil after the rains have ended.

Gramzow and Henry (1972) first defined a "rainy" pentad as having more than 1 inch (25 mm) of rain. They then delineated the wet season over Central America as being encompassed by the first and last rainy pentad. Although some stations experienced a relatively dry spell in mid-summer, the method worked well--see Figures 6-43 and 6-44. The complex patterns are affected by the prevailing low-level easterlies.

The rains start first on the Caribbean coasts of Panama and Costa Rica, move into Nicaragua, and progress across the isthmus. Uplift on the El Salvador mountains causes an early start there. Progress is then northward across Honduras. The rains end first in the mountains of El Salvador, Honduras, and Nicaragua, and much sooner on the Pacific coast than on the Caribbean.

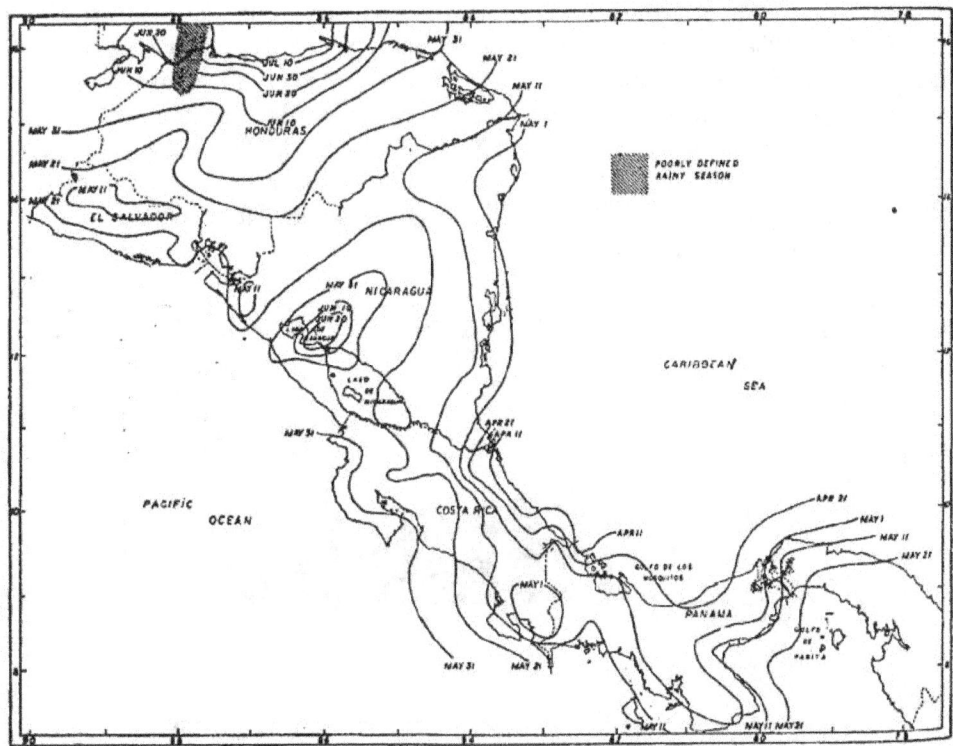

Figure 6-43. Mean date of start of the rainy pentads (more than 1 inch/25 mm) over Central America (Gramzow and Henry, 1972).

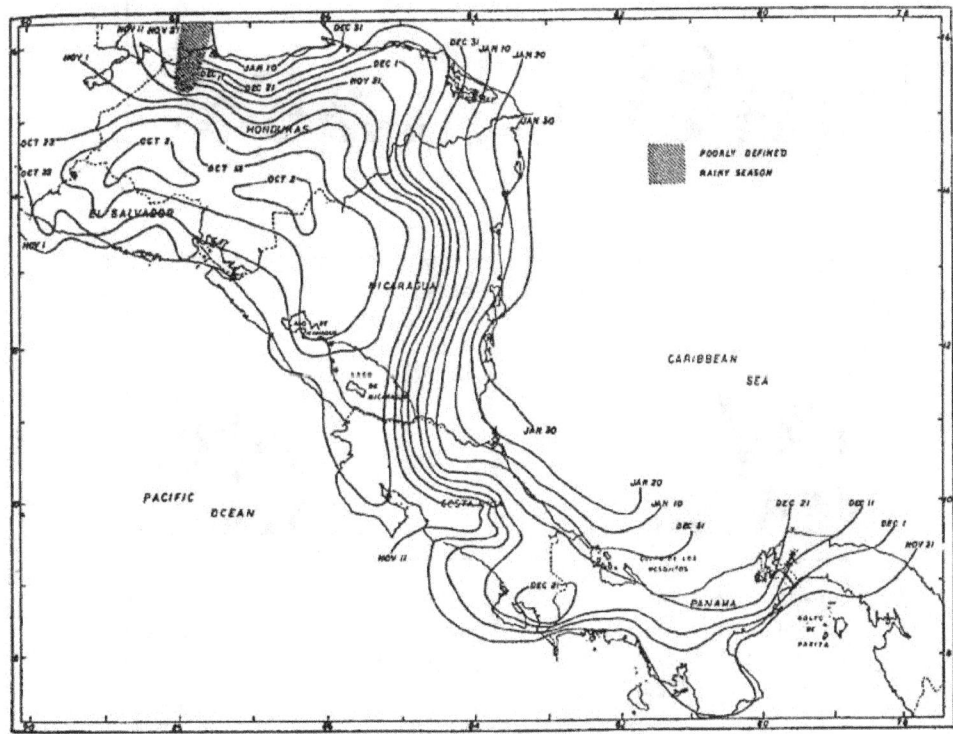

Figure 6-44. Mean date of end of the rainy pentads (last pentad with more than 1 inch/25 mm) over Central America (Gramzow and Henry, 1972).

Horel et al. (1989) used pentad averages of outgoing longwave radiation for 1974-1988 to define the rainy seasons of Central America and the Amazon basin. For each, Horel calculated the *fractional low radiance* (FLR), defined as the ratio of the number of 2° by 2° latitude/longitude grid areas that have OLRs of less than 200 Wm⁻² divided by the number of grid areas in the enveloping 10° by 10° box (usually 25).

High FLR combines frequent and/or intense convection over a 5-day period. As Figure 6-45 shows, deep convection is confined to the summer of each hemisphere, with the equator more like the Amazon. There is little evidence of deep convection passing back and forth across the equator during the transitions (Figure 6-46, next page).

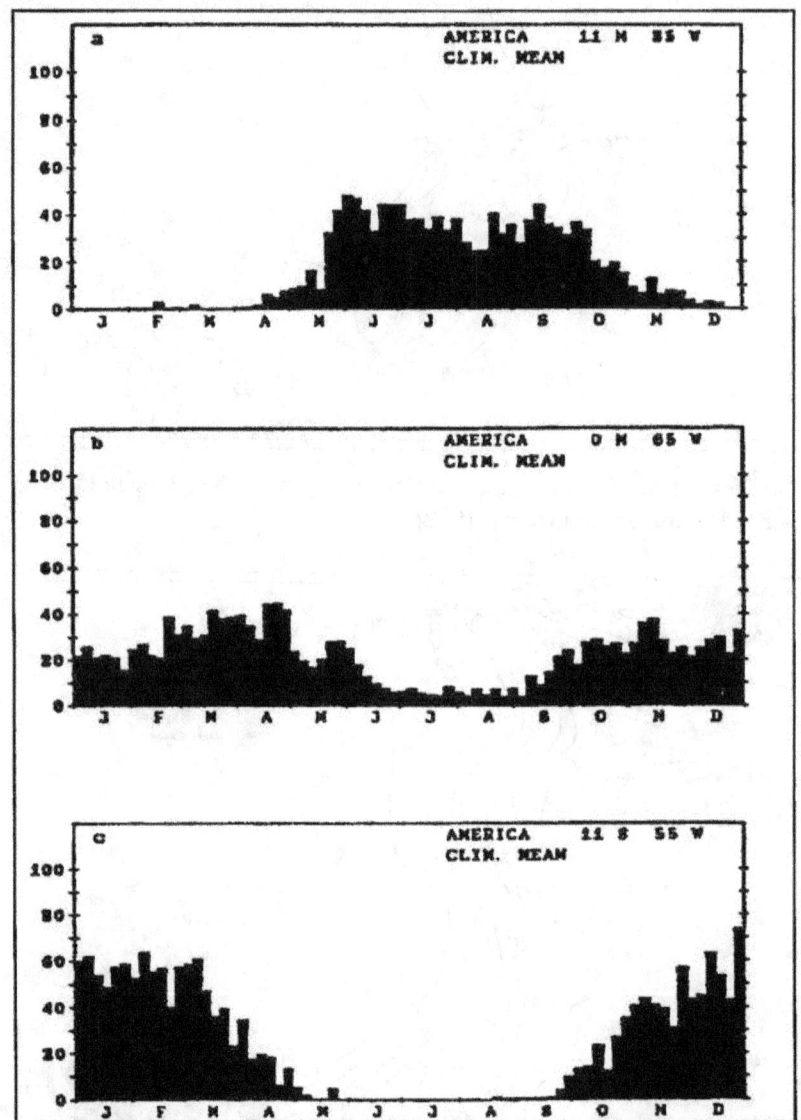

Figure 6-45. Histograms showing the climatological annual progression of fractional low radiance (FLR) based on OLR Data for 1974-88. Box A is centered on 11° N, 85° W; Box B, on the Equator, 65° W; and Box C, on 11° S, 55° W. (Horel et al., 1989.)

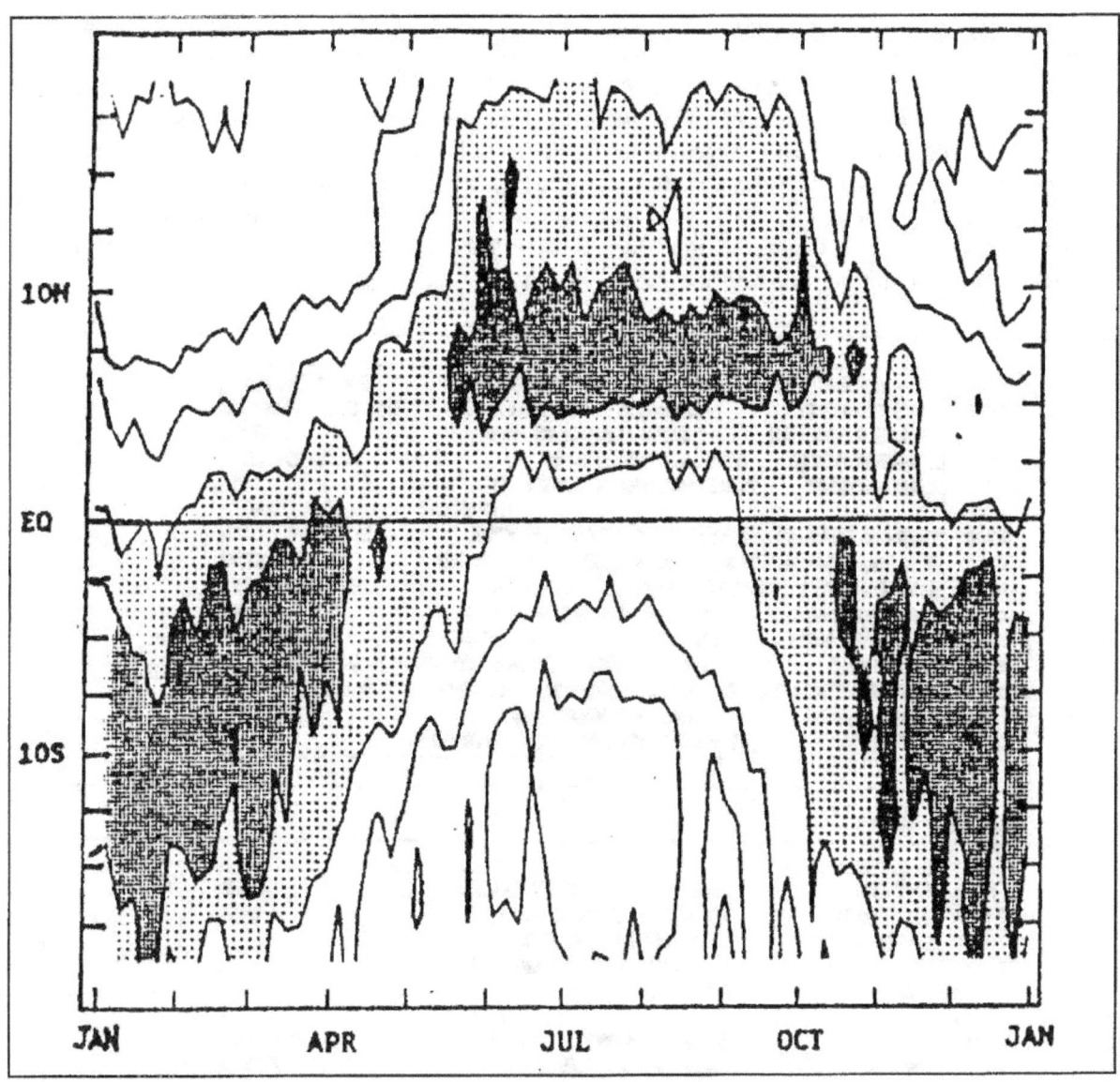

Figure 6-46. Time-latitude diagram of the 8-year (1980-87) Climatology of pentad OLR across the Americas. Areas with OLRs less than 200 Wm^{-2} are heavily shaded; those with OLRs 220-200 Wm^{-2} are lightly shaded. Contour interval is 20 Wm^2 (Horel et al., 1989). Transition seasons are brief, especially during October.

As shown in Figure 6-47, Horel et al. next defined wet-season durations in each hemisphere for the tropical Americas and Africa. Their "wet seasons" started with the first of five pentads during which the maximum FLR is in the opposite hemisphere. This definition fails to take account of heavy convection in the same pentad in both hemispheres, and does not allow any time overlap. Over Central America, the mean wet-season duration agrees with the dates determined by Griffiths (1964) and Gramzow and Henry (1972). The Amazon wet season and the Central American dry season start as the 200-mb Bolivian anticyclone develops and cease as it weakens (refer to Figures 4-13 and 4-16). Africa's southern wet season is longer than in the north, where the dryness might be partly due to downstream subsidence induced by Himalaya/Tibet (see 6.3.6.2).

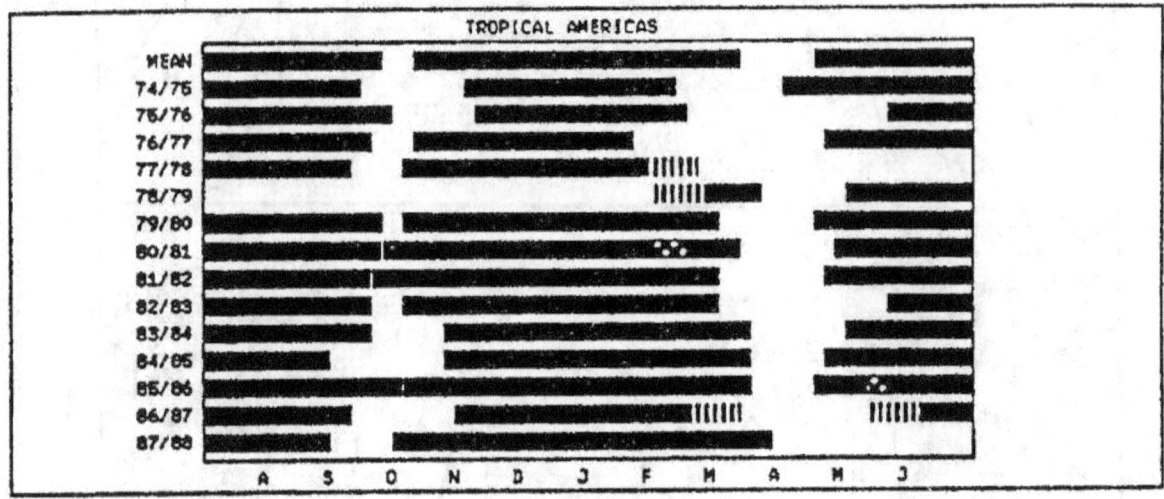

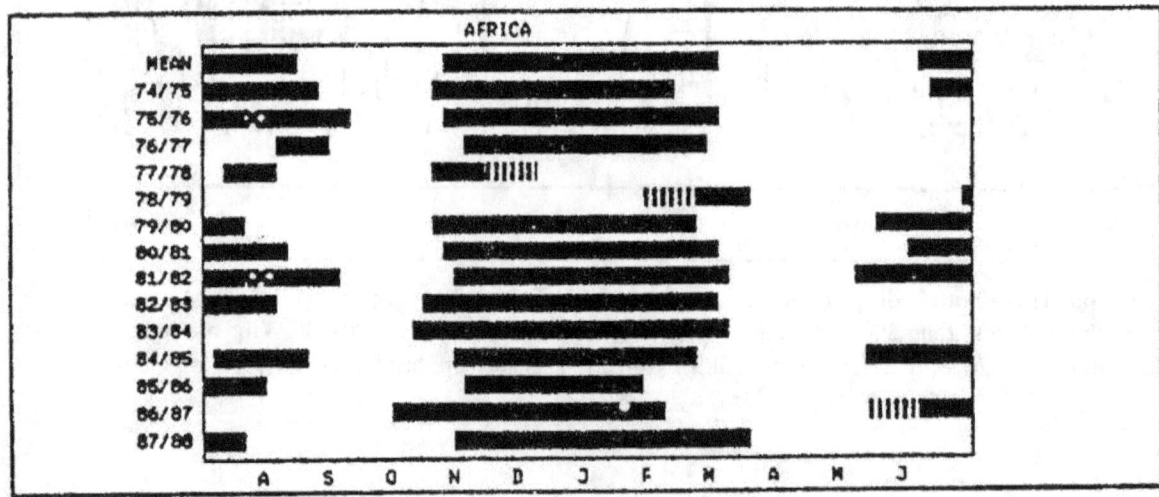

Figure 6-47. The durations of wet seasons over the tropical Americas (top) and Africa (bottom). The summer wet season in each hemisphere is represented by a solid bar. Transition seasons, when a wet season is not established, appear as gaps in the bar. A striated bar represents the first and last months of periods without data. Hatched regions within a bar denote "break" periods, when the maximum FLR shifts briefly into the opposite hemisphere (Horel et al., 1989).

For the area around Ho Chi Minh City/Saigon (11° N, 107° E), Fukuda (1968) defined the end of the rainy season as the first day of the first 14-day period when the average daily rainfall at seven stations in the Saigon area was less than .08 inches (2 mm); that is, when the average total of the daily rainfalls over a 2-week period was less than 1.10 inches (28 mm). This criterion was selected because after 1 October, 2 weeks of little rain is unlikely to be followed by significant falls.

Figure 6-48 shows the 3-day average of the daily rainfalls. The arrows indicate the dates the rainy seasons ended. End dates ranged from 23 October

(1962) to 24 December (1966) with an average date of 15 November.

In some years, the rains ended abruptly, while in others they tapered off. Interannual variability was large between 1966 and 1968.

In summary, tropical meteorologists trying to make long-range forecasts of a rainy season's beginning or ending should express them in probability terms based on long-period data. The operational user must clearly understand how the rainy season is defined. Despite user pressure, dogmatic forecasts such as "the rainy season will start on 15 May" must be avoided.

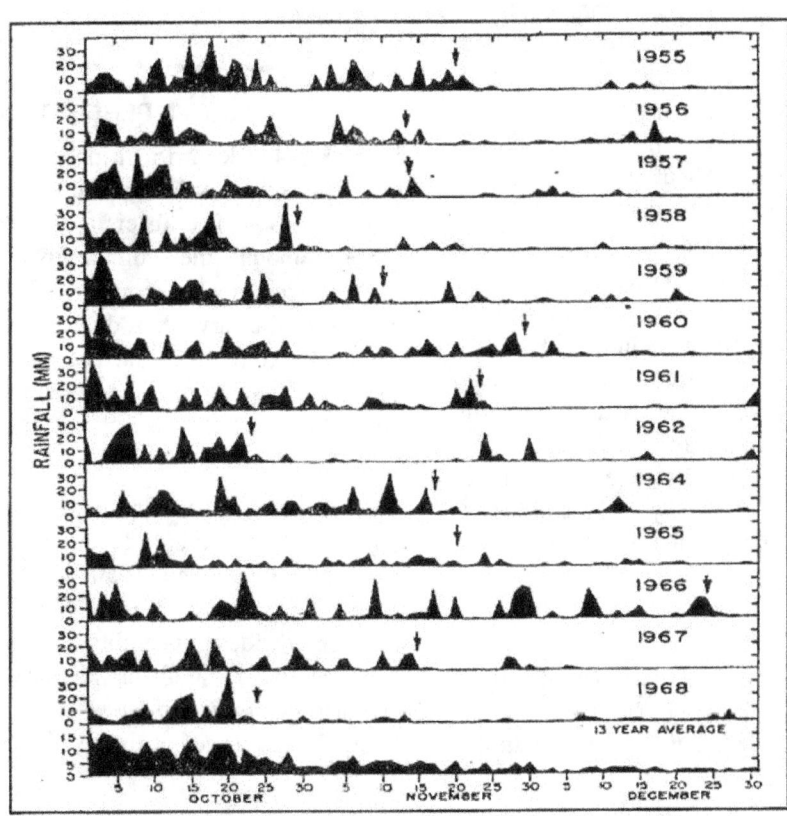

Figure 6-48. Consecutive 3-day means of the total rainfall at seven stations near Ho Chi Minh City (11° N, 107° E) for each year from 1955 to 1968 (less 1963) and the 13-year average. The arrows show the ending dates of the rainy season by Fukuda's criterion (Adapted and expanded from Fukuda, 1968).

6.3.10 Daily Distribution. As in the mid-latitudes, synoptic-scale disturbances are primarily responsible for creating the environment that favors most tropical rainfall. Within these disturbances, rain from convective and cloud-scale processes lasts from tenths of hours to hours (Garstang, 1966) and accounts for most of the accumulation (Henry, 1974). Rain falls mostly within a small fraction of the total periods with rain. On the island of Barbados (13° N, 59° W), half of the total rain falls in 10 percent of the time intervals with rain, whether 24-hour, 1-hour, or 0.1-hour periods are used as the basic interval (LaSeur, 1964b). Only on mountains exposed to the tradewinds is this distribution unlikely (refer to Figure 6-21).

Garstang (1966) expanded LaSeur's single-station amount-frequency analysis to Florida, where 10 percent of the rainy days accounted for more than 70 percent of the areal rainfalls. Similarly, at stations in southeast Asia half the rain fell in about 10 percent of the hours (Riehl, 1967). About 80 percent of the area encompassed by a synoptic disturbance lacked rain at any instant. In another study, Riehl (1949) showed that over central Oahu, most rain fell during synoptic disturbances. On average, ten rainstorms a year with a median rainfall exceeding 0.75 inches (19 mm) produced over two-thirds of the total annual rainfall. Riehl also showed that as the rainstorm intensity increases, orography becomes less important. Riehl et al. (1973) came to a similar conclusion for Venezuela. Many more studies of tropical daily rainfall gave similar results. They can be described by composite tropical rainfall frequency vs. amount curves.

To construct such a curve, order the amounts. Then start the plot at zero with the smallest amount and finish at 100 cumulative percent frequency with the largest amount. A curve derived by Martin (1964) for 30 years of daily rainfall at five stations in North and Central America is shown in Figure 6-49.

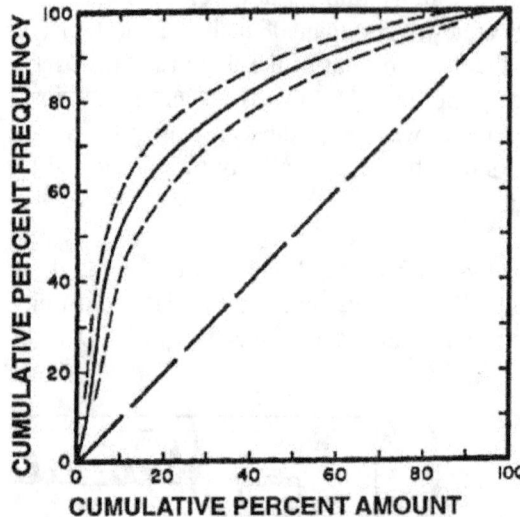

Figure 6-49. Relationship between cumulative rainfalls and cumulative rainfall frequency (after Martin, 1964). The abscissa gives the cumulative percent amount; the ordinate gives the cumulative percent frequency of rainfall. The dashed lines paralleling the curve show departures of two standard deviations from this mean. The dashed diagonal line shows the distribution if it always rained the same amount for each time period considered.

The curve also fits 20 additional stations in the region and closely resembles frequency-amount curves derived by Olascoaga (1950) for Argentina, and by Riehl (1950) for San Juan (18° N, 66° W). The curve's apparently wide applicability led Martin to call it the "universal daily rainfall distribution curve." The curve can also be applied to other time intervals, such as hours. The data used is given in Table 6-4, next page.

TABLE 6-4. Data for the universal rainfall distribution curve (from Martin, 1964).

Cumulative % Amount	Cumulative % frequency	Standard Deviation
0	0	0
10	50.6	5.1
20	66.1	3.9
30	75.4	3.2
40	82.1	2.7
50	87.2	2.1
60	91.2	1.7
70	94.3	1.3
80	96.7	0.9
90	98.7	0.4
100	100	0

The universal curve can be used to estimate the frequency distribution of daily rainfalls at a tropical station where only the mean monthly rainfall and mean monthly numbers of days with rain are known. For simplicity, assume a station during a given month has an average of 20 days with rain and a mean total of 10 inches (250 mm). Over 10 years, 200 days with rain, totaling 100 inches (2,500 mm), could be expected during a given month.

Table 6-5 gives estimated frequency distribution of daily rainfalls at this station during a 10-year period based on the universal rainfall distribution curve values given in Table 6-4. The percent frequencies corresponding to each 10-percent rainfall are determined by subtracting subsequent numbers in the percent frequency column in Table 6-4. In this hypothetical case, 60 percent of the total falls during days reporting 1 inch (25 mm) or more.

TABLE 6-5. Examples of applying the universal rainfall distribution curve to determine distribution of daily rainfall. A given month has an average of 20 days with rain and a mean rainfall of 10 inches (250 mm). The table gives estimates of various daily rainfalls during a 10-year period.

Rainfall		Number of days required to produce this amount (rounded to nearest whole day)	Average rainfall per rainy day	
inches	mm		inches	mm
10	250	50.6m X 200 = 101.2 (101)	0.1	2.5
10	250	15.5 X 200 = 31.0 (31)	0.32	8.1
10	250	9.3 X 200 = 18.6 (19)	0.52	13.2
10	250	6.7 X 200 = 13.4 (13)	0.77	19.6
10	250	5.1 X 200 = 10.2 (10)	1	25.4
10	250	4.0 X 200 = 8.0 (8)	1.25	31.7
10	250	3.1 X 200 = 6.2 (6)	1.66	42.2
10	250	2.4 X 200 = 4.8 (5)	2	50.8
10	250	2.0 X 200 = 4.0 (4)	2.5	63.5
10	250	1.3 X 200 = 2.6 (3)	3.33	84.6
100 in	2500mm	200 days		

Cobb (1968), on applying Martin's universal curve to daily rainfalls for southeast Asian stations, found significant differences In the annual mass-distribution curves between one group of stations influenced primarily by the *southwest* monsoon and another group influenced by the *northeast* monsoon. He concluded that describing rainfall distributions at a variety of tropical stations might require a family of curves. Similar conclusions were reached by Wexler (1967) in a study of rainfall in Thailand. Martin's universal curve estimates the distribution of daily rainfalls fairly well. Better estimates demand individual daily rainfalls, or frequency distributions of daily rainfalls such as those given in USAFETAC's "Surface Observation Climatic Summary" series.

6.3.11 Rainfall Variation with Elevation.
Orography is undoubtedly a major factor in determining rainfall distribution. In a comprehensive discussion, Barry (1981) pointed out that many local or regional complications prevent generalizing. Figure 6-50 (from Lauer, 1975) illustrates this. In West Africa, the south side of Mt. Cameroon (4° N, 9° E), exposed to the summer monsoon, has a rainfall maximum at its *base*; on its northeast side, where tradewinds dominate, the maximum lies at 5,000 feet (1,500 meters). Convective heating over inland plateaus may further raise the maximum above 6,500 feet (2,000 meters). In general, the maximum zones tend to rise as annual rainfalls diminish.

That Lauer has by no means included all profiles is confirmed by two from the Hawaiian Islands (Giambelluca et al., 1986) added to Figure 6-50.

The summit of Mt. Waialeale (22° N, 169° W) lies about 1,600 feet (500 meters) below the tradewind inversion; the inversion does not seriously impede vertical motion. However, on the island of Hawaii, where peaks reach 13,000 feet (9,000 meters), the inversion hinders rising air and the tradewinds flow mostly around the obstacle. Thus the rainfall

maximum at 2,000 feet (600 meters) is lower than on Waialeale.

Cobb (1966) lists five factors that influence terrain effect:

- Shape, size and roughness of the mountains.

- Direction and distance from the moisture source.

- Intervening terrain between moisture source and the orographic barrier.

- Wind velocity.

- Thickness and stability of the moist layer.

- Height and intensity of any subsidence inversion. All these factors must be considered by meteorologists forecasting for mountainous areas in the tropics, as well as by climatologists preparing isohyetal maps.

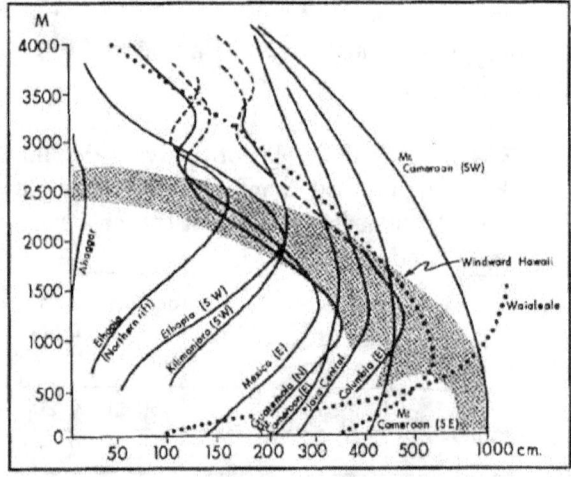

Figure 6-50. Generalized profile of mean annual precipitation (cm) vs. altitude in the tropics. The zone of maximum is shaded (from Lauer, 1975). The dotted lines are from Giambelluca et al., 1986.

6.3.12 Mesoscale Distribution. As shown in the discussion of daily rainfalls, a few days contribute most of the total and, as Riehl (1979a) pointed out, most rain comes from mesoscale disturbances measuring from about 13.5 to 54 NM (25 to 100 km) across. They develop within favorable macroscale systems and are suppressed when the macroscale is unfavorable (Henry, 1974).

The size of mesoscale disturbances can be inferred by relating daily rainfalls between station pairs at various distances apart, as shown in Table 6-6, below. Henry (1974) studied mesoscale rainfall systems in Central and South America using the contingency index (CI) computed from daily rainfalls as follows: The 1-inch (25-mm) rainfall isopleth delineates the extent of mesoscale rainfall systems. The actual frequency of occurrence in each of the nine blocks is given by the a's. The column and row totals are given by the x's and y's, respectively, and the total number of days is given by T. A similar table can be constructed showing the number of occurrences expected in each block by chance, indicated by b's.

$$b_1 = \frac{x_1 y_1}{T}, \quad b_2 = \frac{x_2 y_2}{T}, \quad \dots \quad b_9 = \frac{x_3 y_3}{T}$$

The contingency index is then computed as follows:

$$CI = \frac{(a_1 + a_5 + a_9) - (b_1 + b_5 + b_9)}{T - (b_1 + b_5 + b_9)}$$

TABLE 6-6. The relationship of daily rainfall between station pairs (Henry, 1974).

		RAINFALL AT STATION A			
		Zero	**<25 mm**	**>25 mm**	**Total**
RAINFALL AT STATION B	**Zero**	a_1	a_2	a_3	y_1
	<25 mm	a_4	a_5	a_6	y_2
	>25 mm	a_7	a_8	a_9	y_3
	Total	x_1	x_2	x_3	T

The numerical values of CI range from 1.0 to -1.0, with 1.0 indicating a *perfect relationship* between amounts at the stations (i.e., all values in blocks a_1, a_5, or a_9). A CI of zero indicates *no relationship* between the stations, and a CI of less than zero indicates an *inverse relationship*. In other words, the CI value indicates the tendency for no rain, light rain, or heavy rain to occur on the same days at two stations. CI calculations for hundreds of station pairs in Central and South America led to a typical profile of CI versus station separation (Figure 6-51).

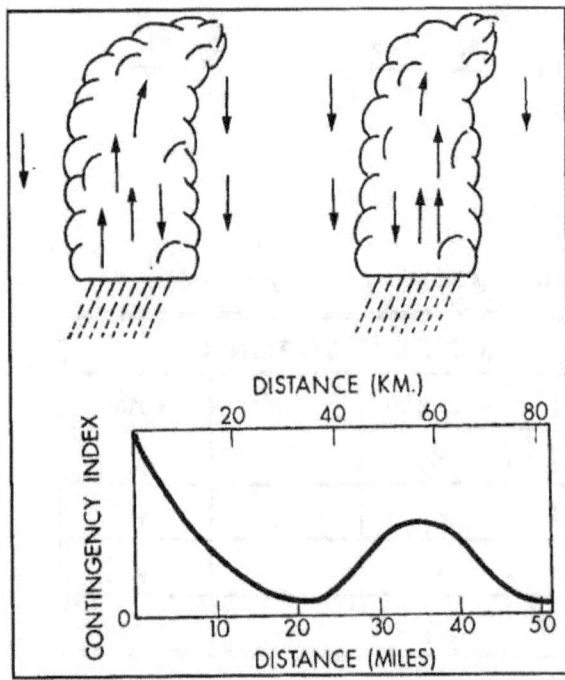

Figure 6-51. Composite curve of contingency index (CI) as a function of distance between Central American rainfall stations. Mesoscale clouds and circulations drawn atop the distance scale illustrate physical relationships (after Henry, 1974).

The CI falls off with distance to a minimum at 19 NM (35 km), then increases to a maximum at about 35 NM (65 km); this suggests that mesoscale rain systems are about 13 NM (25 km) wide and that their centers are about 30 NM (55 km) apart. Rainbird (1968) obtained a similar result for two summers over

the SeSan sub-basin of the Mekong River. Henry (1974) found that storms are smaller during the dry season, but that they have larger CI than in the wet season. Rainfall per storm is about the same. Schroeder (1978) concluded that, over the Hawaiian Islands, orography so dominated CI that no indisputable mesoscale structure could be detected. It is hard to understand why orography over central and South America did not also mask the mesoscale.

From hourly rainfall data for stations in Thailand and Cambodia during the southwest monsoon, Henry (1974) determined the typical rainfall associated with a mesoscale convective disturbance. In his analysis, rainshowers of less than 0.4 inches (10 mm) were omitted; this eliminated 50 percent of the rainy days and about 15 percent of the total rainfall. Henry constructed a composite rainstorm for the months May through October for 13 stations in Thailand and Cambodia (Figure 6-52); individual rainstorms, of course, depart significantly from this composite.

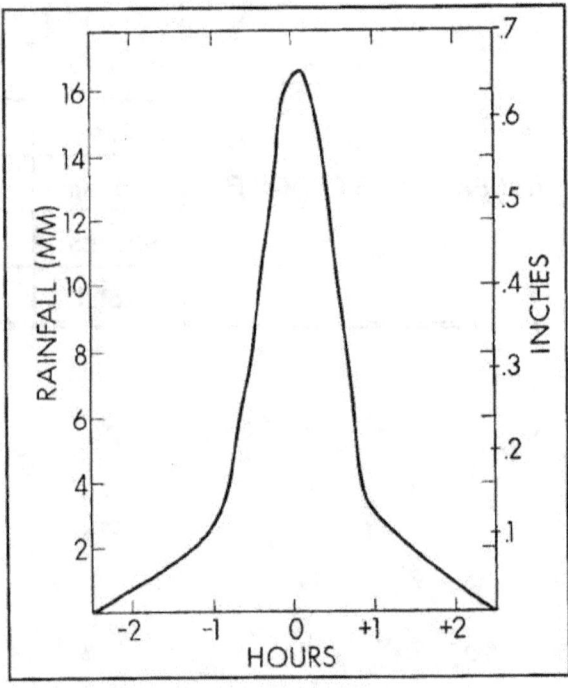

Figure 6-52. Typical rainfall associated with a single mesoscale convective system (from Henry, 1974).

In any hour there may be many showers. Garstang (1959) used 10 years of record to derive the frequency distributions of shower amounts and durations at Piarco (11° N, 61° W) during the wet and dry seasons (Figure 6-53).

Most showers total less than 0.16 inches (0.4 mm) in the dry season and less than 0.4 inches (1 mm) in the wet season. Showers usually last 12 minutes or less and rarely persist for more than 24 minutes.

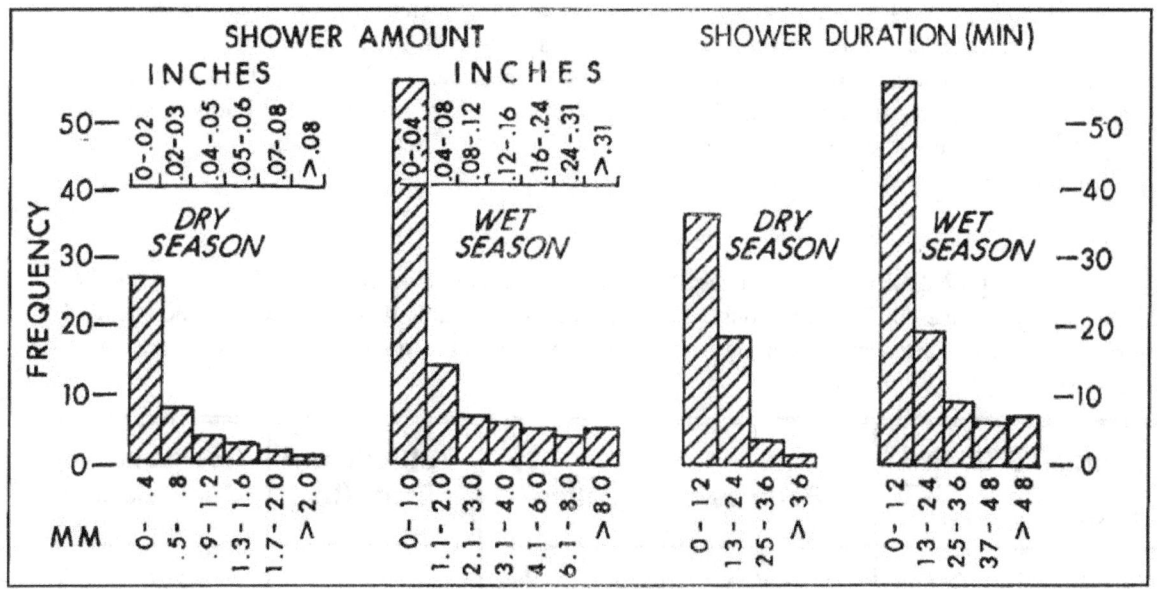

Figure 6-53. Frequency distribution of rainshower amounts and durations (minutes) at Piarco, Trinidad (11° N, 61° W) during dry and wet seasons (after Garstang, 1959).

Winner (1968) applied empirical relationships in estimating from the mean annual rainfall how often various hourly rainfalls are equaled or exceeded at tropical stations. He used hourly rainfall data for 32 stations in the Panama Canal Zone.

When applied to 13 stations in Thailand and Cambodia, the estimated and observed frequencies agreed well. The graphs in Figure 6-54 give the mean number of clock-hours per year that various hourly rainfalls (shown in inches next to each line) are equalled or exceeded.

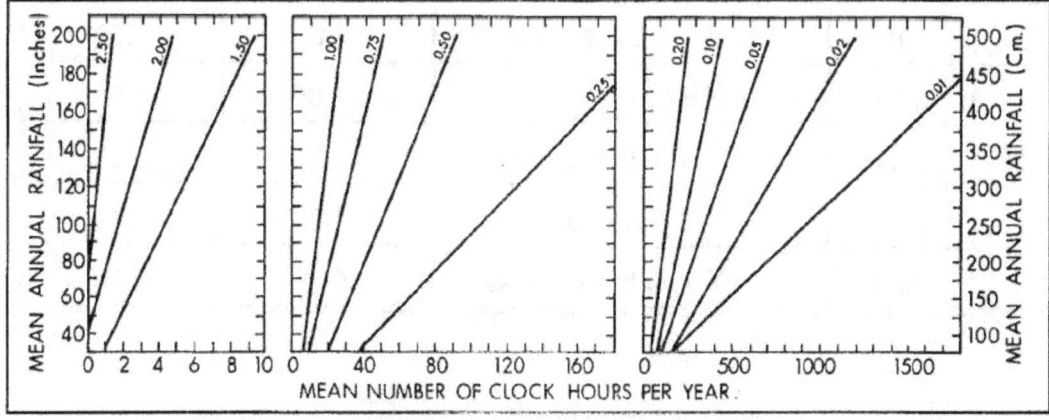

Figure 6-54. Nomogram based on mean annual rainfall to determine mean number of clock-hours that a specified rainfall rate is equaled or exceeded (after Winner, 1968).

147

Clock-hour rainfall rate is the total amount of rain measured during a 1-hour period. (The term "clock-hour" is used because observations are usually made on the hour; that is, at 0700, 0800, etc.). If, during this time, it rained a total of 0.05 inches in only 5 minutes, the clock-hour rainfall rate would be 0.05 inches h^{-1}. This would be true even though instantaneous rates may have ranged from zero to some value greatly exceeding the clock-hour rate. Thus, a station with a mean annual rainfall of 100 inches (2,500 mm) could expect approximately 350 clock-hours per year, with an hourly rate of 0.05 inches (1.3 mm) or more, 49 hours of 0.5 inches (12.7 mm) or more, and 2 hours of 2 inches (50 mm) or more. Winner's regression equations for various hourly rainfalls are given in Table 6-7, along with the standard error of the estimate and the correlation coefficients between the estimated and observed frequencies. The best relationships were for rates from 0.02 to 0.5 inches (0.5 to 12.7 mm) h^{-1}. Forecasters can use Table 6-7 to estimate the chances that flooding or rain-reduced visibility would interfere with or stop airport operations.

TABLE 6-7. Empirical relationships between the mean annual rainfall given by X and the mean number of hours per year given by Y that various clock-hour rainfalls are equalled or exceeded. The standard errors of estimate (in hours) and the correlation coefficient between the estimated and observed frequencies are also shown (after Winner, 1968).

Rainfall Rate (h^{-1})		Regression Equation	Standard Error (h)	Correlation Coefficient
Inches	mm			
0.01	0.3	Y = -219.8 + 0.45X	219.4	0.79
0.02	0.5	Y = -42.1 + 0.24X	65.0	0.92
0.05	1.3	Y = -6.8 + 0.14X	32.1	0.94
0.10	2.5	Y = 4.2 + 0.09X	16.8	0.96
0.20	5.1	Y = 5.3 + 0.05X	9.3	0.96
0.25	6.3	Y = 5.7 + 0.04X	7.0	0.96
0.50	12.7	Y = 6.9 + 0.016X	6.0	0.87
0.75	19.1	Y = 2.9 + 0.008X	4.6	0.78
1.00	25.4	Y = 2.5 + 0.005X	3.6	0.64
1.50	38.1	Y = -0.6 + 0.002X	2.2	0.53
2.00	50.8	Y = -1.1 + 0.0011X	1.3	0.44
2.50	63.5	Y = -0.6 + 0.0004X	0.7	0.36

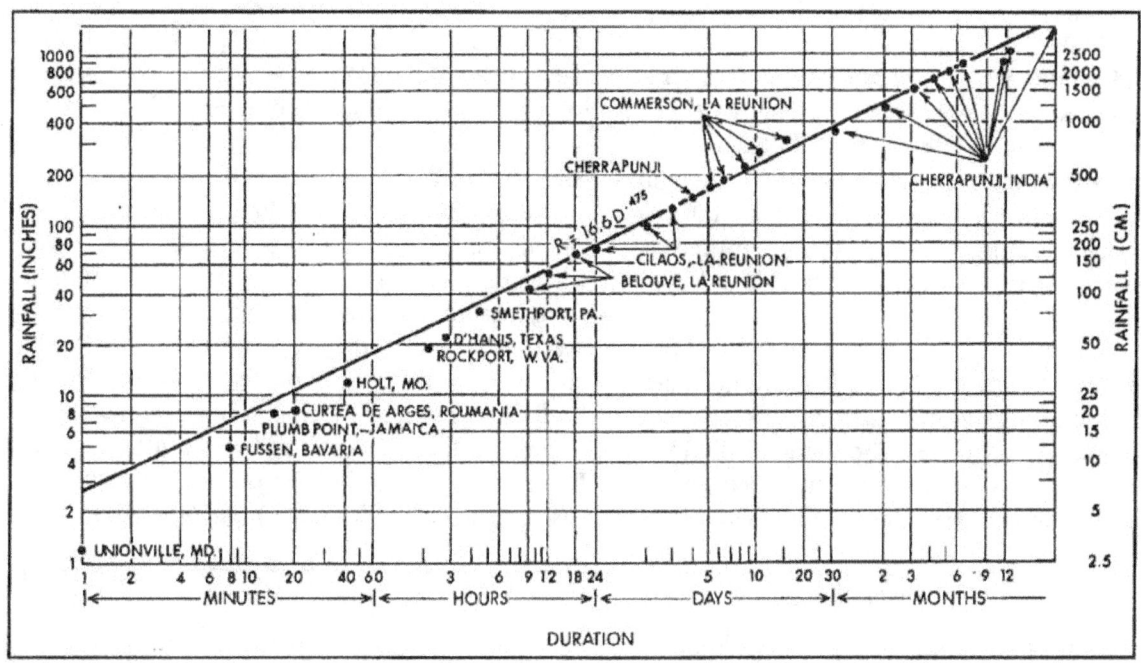

Figure 6-55. World record observed point rainfalls for various time periods (after Paulus, 1965; Dhar, 1977; and Chaggar, 1984).

6.3.13 Rainfall Extremes. The tropics experience extremely heavy rain much more often than in the higher latitudes. Figure 6-55 gives the world's greatest observed point rainfalls for various time periods. Cherrapunji (25° N, 92° E, at 4,300 feet/1,313 meters) holds all the records for periods of 4 days, as well as for 30 days and above. Three mountainous stations on Reunion Island (21° S, 56° E) hold the record for 9 hours to 16 days (except for the 4-day record held by Cherrapunji).

The envelope curve for these extremes is given by $R = 16.6D^{0.475}$ (where R is rainfall in inches) and $R = 42.16D^{0.475}$ (where R is the rainfall in centimeters). D is the duration in hours. A similar formula was developed earlier by Fletcher (1950) who showed that extreme rainfall (inches) is proportional to the square root of duration (hours), according to the formula $R = 14.3D^{0.5}$. The equation for centimeters is $R = 36.32D^{0.5}$.

Many of these tropical daily rainfall extremes have been associated with intense tropical cyclones; examples are the world record 24-hour fall of 73.6 inches (186.9 cm) at Cilaos, Reunion; 49.1 and 97.0 inches (124.7 and 113.4 cm) at two stations in the mountains of Taiwan (Paulus, 1965); and 42.2 inches (107.2 cm) at Kadena Air Base, Okinawa (Jordan and Shiroma, 1959). The latter event occurred as Typhoon Emma passed the island on 8 September 1956; it was somewhat surprising, since (compared to Reunion and Taiwan) Kadena is near sea-level on the fairly low-lying island of Okinawa. Jordan and Shiroma pointed out that for a long time the station lay under the wall cloud and a feeder band of the typhoon.

On the other hand, a coastal station on northern Kauai in Hawaii recorded more than 38 inches (96.5 cm) in a 24-hour period during a winter storm. But all records for 6 to 15 days were broken at Commerson, Reunion (Chaggar, 1984), when, from 14 to 28 January 1980, tropical cyclone Hyacinthe moved slowly clockwise around the island, the track looping twice. The 15-day record of 252 inches (643.3 cm) here almost equalled the 30-day record held by Cherrapunji.

Extremely heavy rains can also fall in relatively innocent-looking situations. For example, over 20 inches (50 cm) were recorded over parts of southeast Florida in less than 24 hours on 21 January 1957. Sourbeer and Gentry (1961) found little synoptic-scale justification for the downpour, even though upper-tropospheric divergence overlaid lower-tropospheric convergence. The continuous thunderstorms described in Section 6.3.1.2 are other examples.

Since daily amounts are the most commonly available rainfall data, most studies of extreme rainfall have been concentrated there. The Gumbel double-exponential distribution is the statistical model most often applied to extreme values (Gringorten, 1960; Gumbel, 1958). It fits well the annual extremes of wind speed, rainfall, temperature, pressure, and other meteorological elements. It can sometimes be applied to extremes within a month or season. To determine the Gumbel rainfall distribution for a station, find for each year of the record the rainfall for the wettest day, then order these from lowest to highest. The cumulative probability will proceed by steps = $1/n$, where n = the number years of record. Tests have shown that for a variety of stations, the points fall on a straight line if they are plotted on extreme probability paper with a double exponential grid. Probable return periods, longer than the period of record, can then be estimated by extending the line. Remember that this estimate is based only on what has gone on before. It gives a *probability forecast*.

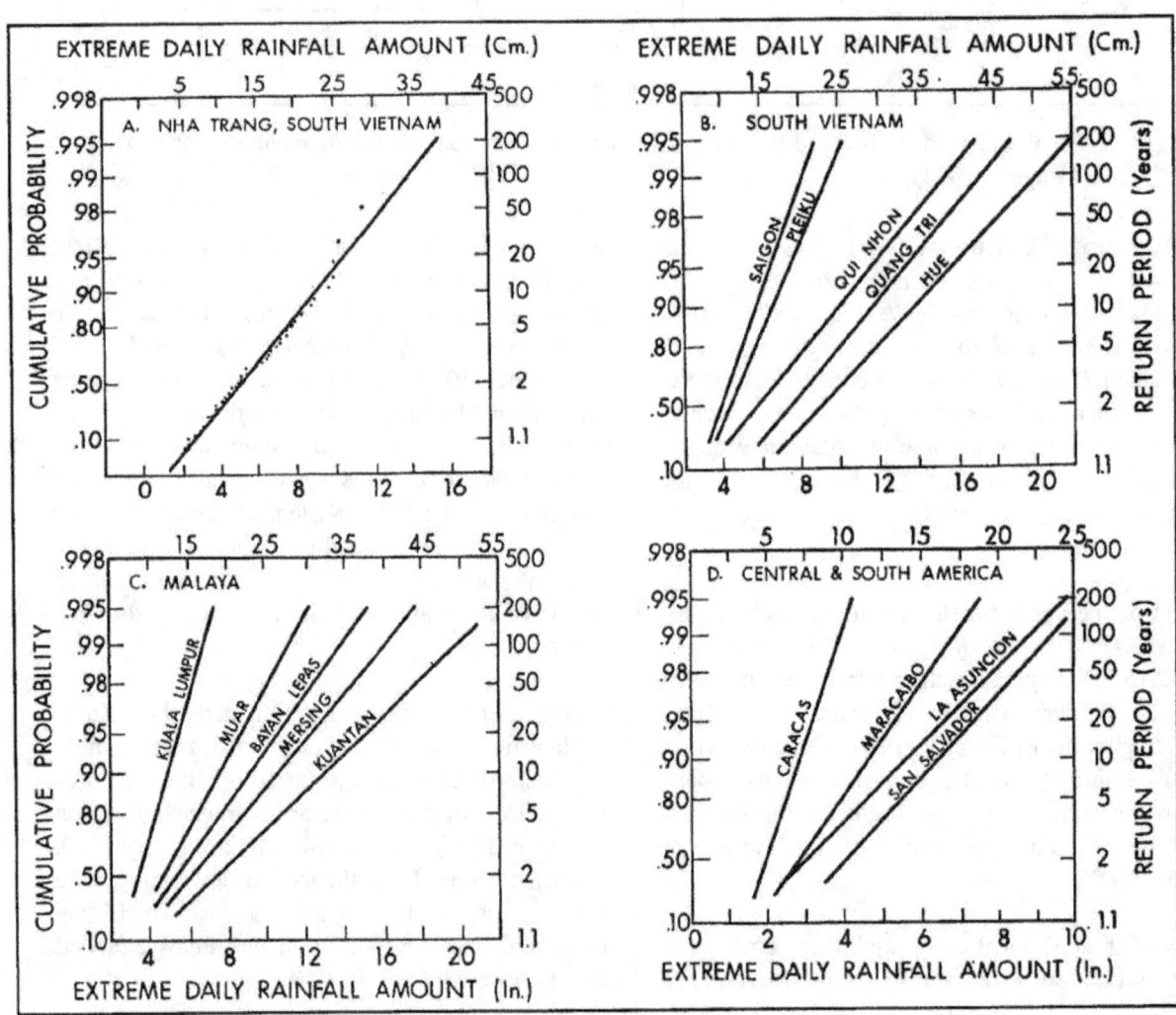

Figure 6-56. Probable extreme daily rainfall for selected tropical stations (adapted from Atkinson, 1968; Lockwood, 1967; and Griffiths, 1967).

The Gumbel distributions of extreme daily rainfalls at some tropical stations are presented in Figure 6-56, on the previous page. The figure shows (in the upper left-hand panel) the distribution for Nha Trang (12° N, 109° E) For 49 years of record. The solid line is the double-exponential distribution fitted to the extremes. The abscissa is the annual extreme daily rainfall, the left-hand ordinate is the cumulative probability (F), and the right-hand ordinate represents the cumulative probability expressed as a return period (R), based on the relationship $R = 1/(1-F)$. The plot shows that on average, once every 2 years Nha Trang can expect daily rainfall to reach about 5 inches (125 mm), while during a 100-year period, one daily rainfall of slightly over 14 inches (355 mm) can be expected.

In the graphs for stations in Vietnam and West Malaysia (Lockwood, 1967), Figure 6-56 shows that East coast stations are wettest because from September through December they may be hit by typhoons or experience continuous thunderstorms. The graph for Central and South America shows that Caracas (10° N, 67° W), in a high, protected valley inland from the north coast, differs from La Asuncion (11° N, 64° W) on an offshore island. The wetter San Salvador (14° N, 89° W), in contrast to the Venezuelan stations, is affected by developing Pacific tropical cyclones.

For other tropical locations, Fletcher (1949) analyzed extreme rainfalls over Venezuela, and the U.S. Weather Bureau studied extreme rainfalls In Panama (1943), the Hawaiian Islands (1962), and Puerto Rico and the Virgin Islands (1961). These studies were made primarily to support hydrological engineering, but they are also valuable in meteorology.

6.3.14 Rainfall Associated with Tropical Cyclones. Extremely heavy rain falls beneath the eyewalls and the rainbands of intense tropical cyclones (see 6.2.13 and Figure 6-55). Although measurements are scanty over the open ocean, Gray (1978) reported averages of rain gauge readings in the vicinity of tropical cyclones over Florida (Miller, 1958) and the west Pacific (Frank, 1976), while Riehl (1979a) gave measurements from Hurricane Donna

(1960) as well as computations based on moisture convergence toward the center. Figure 6-57 shows that the four disparate procedures give similar results between 60 and 250 NM (100 and 600 km) from the cyclone center. Rainfall is an order of magnitude greater at the inner radius than at the outer radius. Between 60 NM (100 km) of the center and the eyewall, hurricane winds prohibit accurate measurement. Riehl's calculations suggest a further order of magnitude increase in rainfall over this distance.

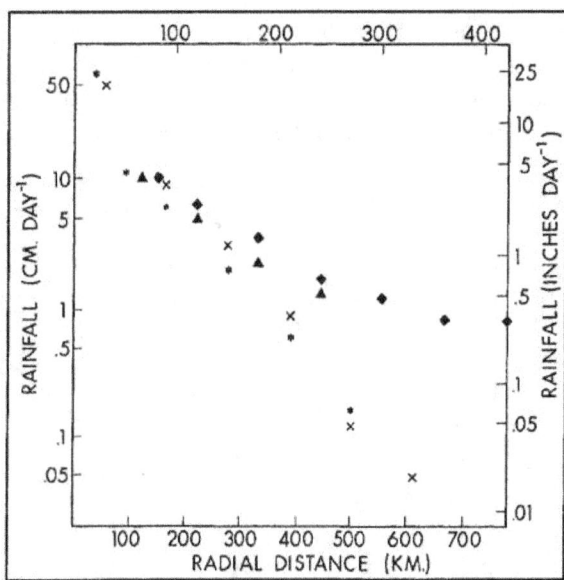

Figure 6-57. Rainfall at radial distances from tropical cyclone centers, as computed from moisture convergence (asterisks), Riehl, 1979a; as measured in Hurricane Donna (x's), Riehl, 1979a; as measured in Florida around tropical cyclones (deltas), Miller, 1958; and as measured on west Pacific Islands around typhoons (diamonds), Frank, 1976.

Goodyear (1968) investigated the rainfall of 46 tropical storms and hurricanes crossing the low-lying United States Gulf Coast from 1940 through 1965. Figure 6-58 shows the rainfall with respect to the cyclone center positions for those 46 storms averaged from 24 hours before to 24 hours after landfall. The maximum of 6 inches (150 mm) occurs slightly inland at 20 to 40 NM (40 to 80 km) to the right of the storm track, where the wind is blowing onshore. This fits the data in Figure 6-57 quite well. In the successive 12-hour subperiods of the 48-hour period (not shown), the center of maximum rainfall moves from offshore in the first period to 43 to 62 NM (80 to 116 km) inland in the fourth period. The maximum 6-hour point rainfall averaged 4.2 inches (107 mm) and ranged from 0.95 to 11.0 inches (24 to 280 mm). For 12 hours the average was 5.9 inches (150 mm) and the range 1.47 to 15.55 inches (37 to 395 mm). Where rugged terrain lifts the strong onshore winds, the average maximum rainfall can be much greater.

Figures 6-57 and 6-58 can be used in anticipating areal rainfall distribution from a tropical cyclone. When meteorologists are asked to make a point forecast, however, they should keep the following data from Kadena Air Base (26° N, 128° E) in mind (Detachment 8, 20th Weather Squadron. 1988):

Between 1945 and 1988, 31 typhoons gave maximum sustained winds of 50 knots or more at Kadena. From three typhoons whose eyes passed over the station, rain amounted to 3.89 inches (98 mm) from one, and over 10.46 inches (266 mm) from each of the others. On every bearing of the eyes from Kadena and an eye approach distance of zero to 140 NM (260 km), at least one total exceeded 10 inches (254 mm), but many others did not. One center that passed 7 NM (13 km) to the northwest gave only 2.63 inches (67 mm). When Kadena lay under a rainband for several hours, totals were large; when the station lay mostly *between* rainbands, they were small.

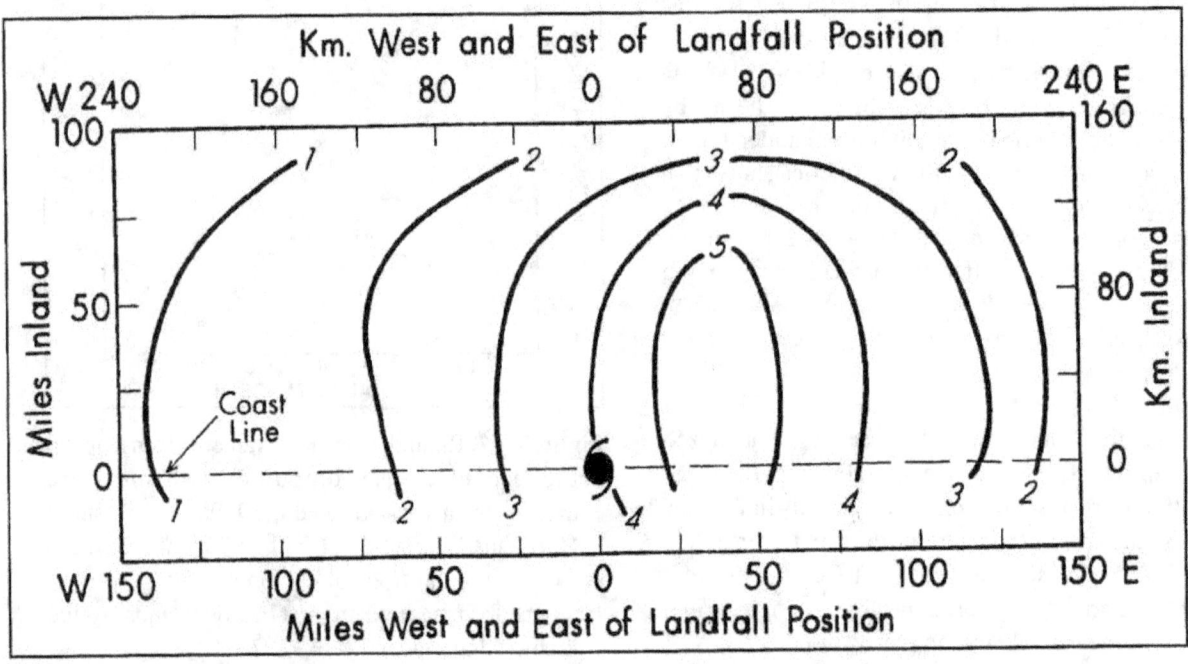

Figure 6-58. Average 48-hour rainfall (inches) for 46 Gulf Coast hurricanes (after Goodyear, 1968).

Chapter 7

DIURNAL VARIATIONS

7.1 GENERAL

In the tropics, gradients of surface pressure, wind, and temperature (except for tropical cyclones) are small. In fact, the standard deviations of these elements (Figures 4-1 and 5-1) are about the same as the variations imposed by the Sun's diurnal cycle. *Forecasters must always understand and take account of diurnal variations because they may often overwhelm changes accompanying synoptic or mesoscale weather systems.* For example, The diurnal cycle is so powerful that during the summer between 1000 and 1100L, stations along the east coast of West Malaysia never record rain.

7.2 DIURNAL TEMPERATURE VARIATION

The diurnal variation in temperature, following the daily march of the Sun, is generally largest at the surface. By affecting air density, it determines to a great degree the nature of the diurnal variations of pressure, wind, humidity, cloudiness, and rainfall (Buchan, 1883). These in turn, by modifying insolation and outgoing longwave radiation, help determine the amplitude of the temperature cycle. The mean diurnal temperature variation at any station depends on a number of factors, which include humidity, cloudiness, wind speed, and topography.

As shown in Figure 7-1, temperature was measured in two instrument shelters on Willis Island (16° S, 150° E; 1,300 by 500 feet (400 by 150 meters); elevation 30 feet (9 meters) during tradewinds. As shown in Figure 7-2, at the upwind end of the island (beach site), diurnal variation amounted to 1.3° F (0.7° C); at the middle of the island, variation was 5.4° F (3° C).

Measurements made on a buoy-mounted mast, 26 feet (7.8 meters) above the surface in the North Atlantic tradewinds (Figure 7-2, Prümm, 1974), showed an average daily range of 0.47° F (0.26° C), almost the same as the diurnal variation of the sea surface temperature. Thus, an island about the size of a supertanker can heat the air significantly, while shallow water and occasionally exposed reefs can triple the diurnal range observed over the open ocean.

The diurnal range is 10° F (5° C) or less on small islands or along coasts with a prevailing onshore flow, but 10 to 20° F (5 to 10° C) at coastal stations affected by land breezes and at interior stations during the rainy season. The range is greater than 20° F (10° C) at interior stations during the dry season. Figure 7-3 (opposite) illustrates the mean diurnal temperature and dew-point changes at a tropical continental station (Luang Prabang, 20° N, 102° E) during dry (March) and wet (August) months and on a coral island (Eniwetak, 11° N, 162° E). The annual curves at Eniwetak are typical of any month.

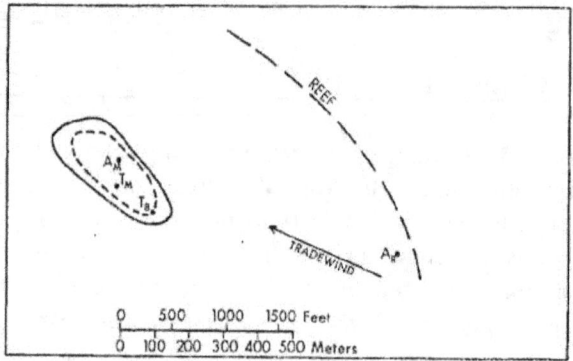

Figure 7-1. Instrument sites on Willis Island. Thermometers at T_M (main site) and T_B (beach site); anemometers on island (A_M) and on reef to windward (A_R) (Neal, 1973).

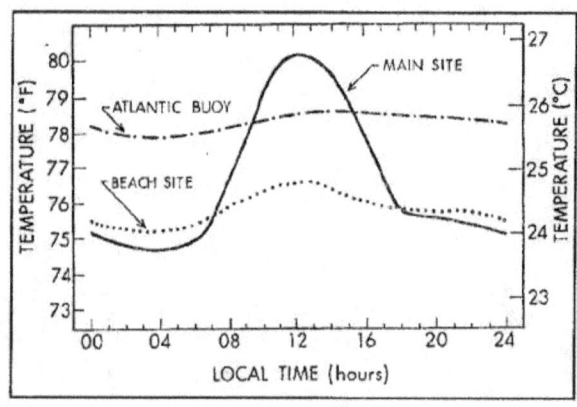

Figure 7-2. Diurnal variation of surface air temperature at a buoy in the North Atlantic tradewinds average between 4 and 13 Feburary 1969 (Prümm, 1974), and at Willis Island (main site and beach site) for 59 undisturbed tradewind days in July-September 1964 (Neal, 1973).

In general, the diurnal curves show a minimum near sunrise and a maximum in the afternoon; however, they depart significantly from a sine curve. Hence, the average of the mean daily maximum and minimum temperatures will slightly overestimate the mean temperature determined by averaging all the mean hourly values.

The diurnal range of 25° F (14° C) at Luang Prabang in March is more than double the range of 12° F (6.7° C) in August when rainfall is an order of magnitude more. In contrast, the mean diurnal range at Eniwetak is only 5° F (2.8° C), but still ten times as much as over the open ocean. For these stations, the dew point varies in phase with the temperature, but the mean diurnal range is much less. At Luang Prabang, dew point increases rapidly from 0800 to 1000L, then stays relatively constant until after sunset, when it begins to decrease. This "square-wave" effect may result from the Sun's heat dissipating a nocturnal inversion.

Over humid continents and on the windward sides of mountainous islands, dew point is highest in the afternoon and lowest in the early morning, similar to the curves shown in Figure 7-3. But over dry regions and on the lee sides of mountainous islands, this diurnal cycle is reversed (Palmer et al., 1955).

Over land, turbulence is greatest in the afternoon. In the tradewinds, this increases downward leakage of dry air through the tradewind inversion and causes an afternoon dew-point minimum.

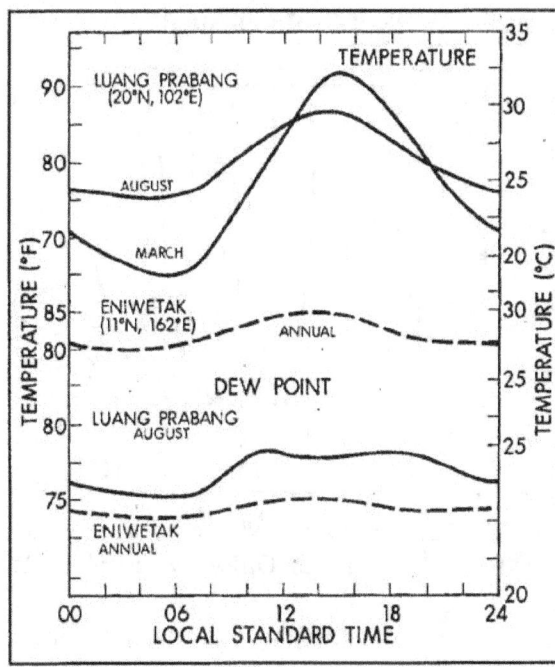

Figure 7-3. Mean diurnal August and March temperature and August dew-point curves at Luang Prabang (20° N, 102° E). Annual temperature and dew-point curves at Enewetak (11° N, 162° E).

7.3 DIURNAL PRESSURE VARIATION

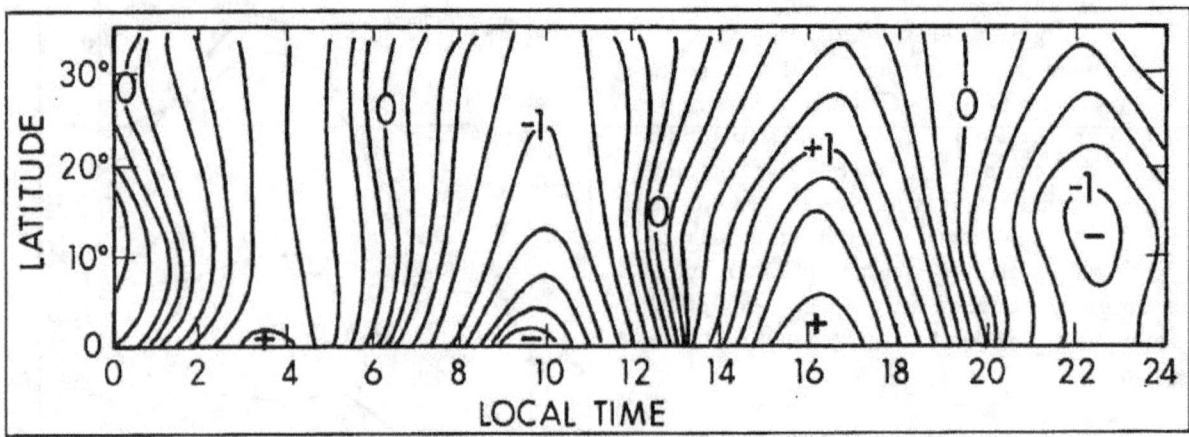

Figure 7-4. Adjustments needed to eliminate the diurnal variation of surface pressure (mb) over the tropical oceans (from Meteorological Office, 1968; Jenkins, 1945). Isobars are labeled -1, 0, and +1.

Mean hourly pressures roughly follow a systematic sequence, characterized by two minima near 0400 and 1600L and two maxima near 1000 and 2200L. Figure 7-4 shows that amplitudes are greatest over the equator and about half as large at 30° latitude. Tidal forces generated by the Sun, and the alternate heating and cooling of the air, account for most of the diurnal variation. Thus the magnitude of the mean diurnal pressure range is closely related to the mean diurnal temperature range, which in turn depends on the mean water vapor content, cloudiness, and surface moisture that control solar heating and cooling by terrestrial radiation. Figure 7-5 illustrates the effect of cloudiness at Hong Kong (22° N, 114° E). The clear-day maximum is about 0.5 mb higher and the minimum about 0.3 mb lower than on cloudy days.

Pressure tendencies can be used in extrapolating motion of pressure systems. In the tropics, the usually small pressure gradients can be recognized locally only if the diurnal variation is first removed from the station data. A further adjustment can be made for cloudiness.

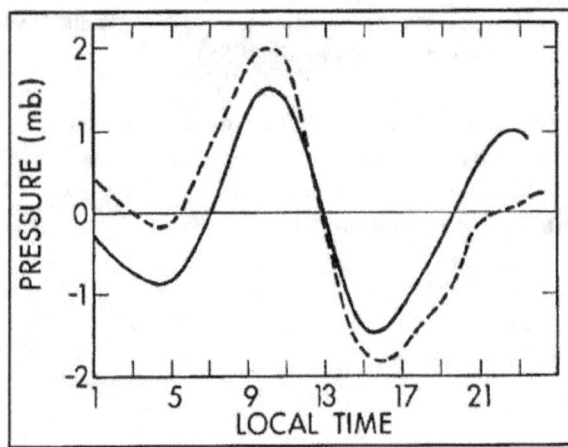

Figure 7-5. Mean diurnal January variation of surface pressure for Hong Kong (22° N, 114° E). Solid line: days with cloudiness greater than 95 percent; dashed line: days with cloudiness less than 20 percent.

7.4 DIURNAL WIND VARIATION

7.4.1 Open Ocean. The research vessel "Challenger" cruised the world's oceans from 1873 to 1876, spending most of the time in the tropics. Over the open ocean, winds underwent a very small diurnal variation; the maxima occurred about midday and midnight, with a range of less than 1 knot (Buchan, 1883). Subsequent measurements over the tropical oceans and upwind of very small coral islands (see Figure 7-2 for an example) produced similar results. Magnitude and timing of the variation strongly suggest that the diurnal variation of pressure is responsible for the diurnal variation of wind.

7.4.2 Land-Sea Breezes. In the tropics, lower-tropospheric diurnal wind changes are most pronounced near coast lines and over the land. Above this layer, wind changes are relatively small. Land-sea breezes are the most familiar local circulations. According to Defant (1951), along tropical coasts and the shores of large lakes a diurnal cycle of onshore and offshore winds occurs in response to large temperature variations over land and very small temperature variations over water. As the land warms with respect to the sea, air over the land becomes less dense than air over the sea and a surface pressure gradient directed from sea to land is established. Since pressure in the land-anchored warm air column falls off with height more slowly than in the cool air column over the sea, above about 3,000 feet (1 km) above the surface the pressure gradient is reversed and directed from land to sea. Thus, air flows from sea to land in the surface layer and from land to sea above the surface layer. This sea-breeze circulation sets in a few hours after sunrise, continues during daylight and dies down after sunset. Later, as radiation cools the land with respect to the sea, the surface temperature and pressure gradients are reversed; a seaward-directed land breeze begins to blow in the surface layer with a return flow from sea to land above. The land-breeze circulation persists until after sunrise.

Although the scale is much larger, the vertical circulations associated with the monsoons (Figure 6-27) resemble the land-sea breeze circulations; the summer monsoon corresponds to the sea breeze and the winter monsoon corresponds to the land breeze.

The similarity can be extended. Year-long upwelling of cold water along a coast, by maintaining higher pressure over the sea than over the land, not only prevents monsoons, but inhibits the land breeze and enhances the sea breeze. The effect is lessened by nocturnal downslope winds generated by mountains near the coast. Land-sea breezes are weak or absent along swampy coasts such as those of the Guyanas of South America (Edson and Condray, 1989). Temperature differences across the shore line are small, especially when the prevailing flow is onshore. The sea breeze may extend inland up to 25 to 50 NM (50 to 100 km), but the seaward range of the land breeze is much smaller. In the vertical, the sea breeze may reach altitudes of 4,300 to 4,600 feet (1,300 to 1,400 meters) in tropical coastal areas, and is strongest several hundred feet above the surface. In contrast, the nocturnal land breeze is usually shallow. The sea breeze typically reaches speeds of a few knots.

7.4.2.1. Observations. Many studies of the land-sea breeze have been based on surface wind and cloud observations. At stations some distance from the equator, the sea breeze or land breeze usually persists long enough for the Earth's rotation (Coriolis force) to cause the wind direction to turn clockwise in the NH and anticlockwise in the SH. The lower the latitude (and the smaller the Coriolis force), the farther inland the sea breeze is likely to penetrate. As the sea breeze "front" moves inland, it is accompanied by a distinct wind shift, a pronounced drop in temperature, and an increase in relative humidity. Convergence at the leading edge may cause convection; conversely, nighttime convergence may accompany the leading edge of the land breeze. Sometimes stratus or fog is carried in from the sea. Some studies have included observations of the vertical structure of the circulations.

Leopold (1949) studied interaction of land-sea breezes with the prevailing easterly tradewinds and resulting cloud systems in the Hawaiian Islands. He listed three factors that influence the interaction:

• The size of the land area.

• The heights and shapes of mountain ranges and their relation to the height of the subsidence inversion.

• The aspect of the area over which the land-sea breeze develops (i.e., windward or leeward exposure to the tradewinds).

To these factors, we should add the following:

• The shape of the coastline.

• The strength of the tradewinds.

• The vertical profile of tradewind speed.

Leopold developed models of four major types of interaction, each named for the geographic feature of the islands where each type is most common. The original paper gives details.

Diurnal variations of surface winds for Hilo (20° N, 155° W) on the windward and Waikoloa Beach (20° N, 156° W) on the leeward coast of the island of Hawaii are shown in Figure 7-6, below. The hodograph for each of the three stations represents the distribution of hourly wind vectors on a polar diagram. For example, a wind from south (east) is plotted on the radius labeled 360° (270°) with distance from the origin proportional to the speed. The curve connects the ends of all the hourly wind vectors. The island of Hawaii penetrates the tradewind inversion, and diverts flow around its corners. This swirl, aided by more leeward sunshine, produces a stronger sea breeze at Waikoloa (against the tradewind) than at Hilo (with the tradewind), while the Hilo land breeze is stronger than its sea breeze (see also Figure 7-16).

Garrett (1980) used mountain slope measurements to develop a vertical-plane model of the Hilo land-sea breeze—see Figure 7-7(opposite), which shows the return flow branches

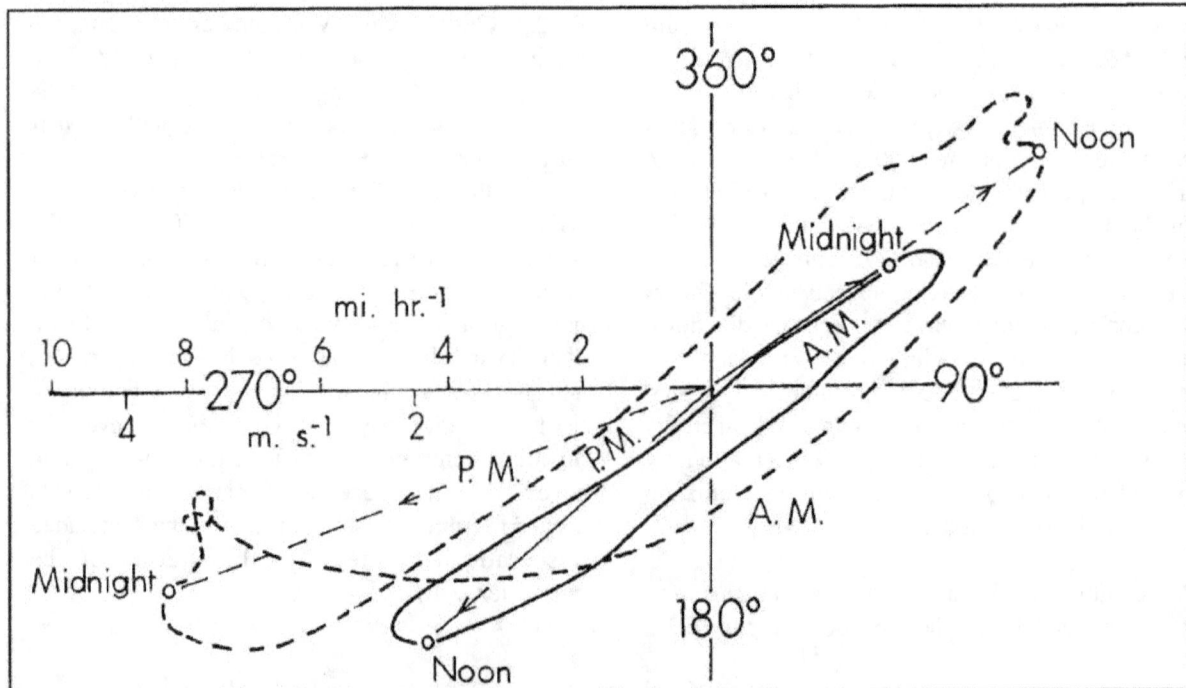

Figure 7-6. Hodographs of summer diurnal variation of surface winds at Hilo (20° N, 155° W) on windward Hawaii (solid line) and at Waikola Beach (20° N, 156° W) on leeward Hawaii (dashed line) (Ramage, 1978a).

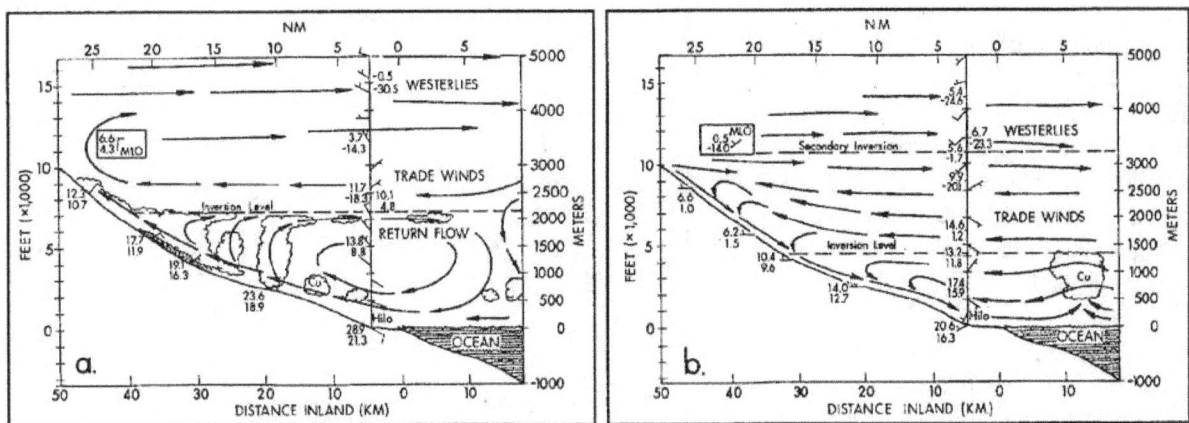

Figure 7-7. Artist's conception of east-west land- and mountain-sea breeze circulations during summer over the eastern part of the island of Hawaii. At left, daytime upslope-sea breeze; at right, nighttime drainage-land breeze. At each surface station and in the upper-air sounding above Hilo, dry-bulb temperatures (° C) are plotted above dew points. On the wind arrows, one full feather denotes 10 knots. The Mauna Loa Observatory (MLO) (20° N, 156° W) data is boxed because terrain there slopes north-south rather than east-west (Garrett, 1980).

An early study of the vertical structure of land-sea breezes was made by van Bemmelen (1922), who used pilot balloon observations for Djakarta (6° S, 107° E), at the head of a bay. Figure 7-8 shows speed of the onshore or offshore winds at various levels throughout the day. The surface sea breeze blows from about 1000 to 1900L, becoming strongest in mid-afternoon. Because of surface frictional effects, the sea-breeze appears earlier and is much stronger about 500 feet (150 meters) above the surface than at the surface. It deepens to about 4,300 feet (1,300 meters) near 2100L after it has ended at the surface. Above the sea breeze, a pronounced return flow from land to sea is centered near 6,500 feet (2 km). Near the surface, the nighttime land breeze and its overlying return flow from the sea are much weaker than the sea breeze system.

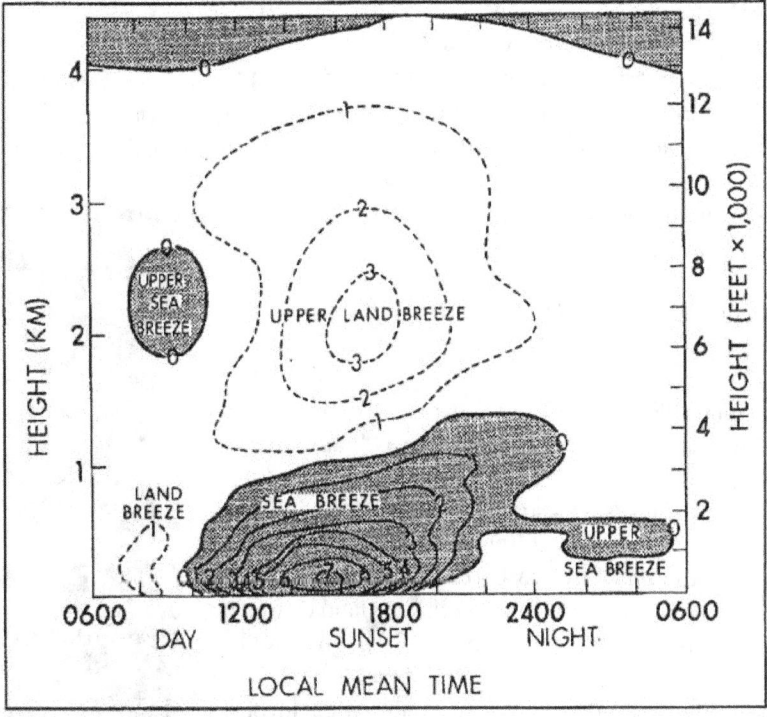

Figure 7-8. Isotachs (m s⁻¹) for the land-sea breeze at Djakarta (6° S, 107° E). Winds with a sea-breeze component are shaded (after van Bemmelen, 1922).

Hsu (1970) described a University of Texas field study of the Texas Gulf Coast land-sea breeze, based on a mesoscale station network near Galveston (29° N, 95° W), during the summers of 1965 to 1967. Using only those days with very weak or zero synoptic flow, Hsu synthesized a model of the land-sea breeze system; Figure 7-9, next page, is a simplified form of the model.

The lower portion of the onshore flow is the sea breeze and that of the offshore flow the land breeze. The levels of maximum wind in each current and their approximate heights are depicted by arrows. The elliptical shapes enclose lower and upper parts of the circulation systems. The approximate level of convective condensation (900 mb) is shown by the horizontal dashed lines.

At 0900L, the air is still cooler over the land than over the sea and the land breeze is still blowing. By 1200L, the land has become warmer than the water and the circulation has reversed. At this time a line of small cumulus may mark the sea-breeze front. At 1500L, the sea breeze is fully developed and showers may be observed at the convergence zone marking the sea-breeze front 15 to 20 NM (30 to 40 km) inland. Low-level divergence near the coast causes subsidence there. At 1800 and 2100L, the sea breeze is gradually weakening. By 0000L, there is no wind at the surface, the sea breeze is barely evident a little above the surface, and a temperature inversion (occasionally accompanied by fog) appears over the land.

With the land now cooler than the water, a land breeze becomes well developed by 0300L and strongest near 0600L. A weak land-breeze convergence with a line of cumulus develops offshore near sunrise. The land breeze continues until mid-morning, when the sea breeze resumes. In this model, the land breeze in the near-surface layer is as strong as the sea breeze. But because of day-night differences in stability and friction, the daytime sea breeze at the surface is considerably stronger than the nighttime land breeze. These differences also account for the vertical circulation being deeper for the sea breeze, with the return flow above the convective condensation level.

Figure 7-10 shows a well-developed sea breeze along a relatively straight coasts bordering a flat coastal plain. As with the Texas Gulf Coast model in Figure 7-9, clouds show the sea breeze up to 30NM (55 km) inland. Air sinks, keeping skies clear along the coast and up to 40NM (75km) offshore.

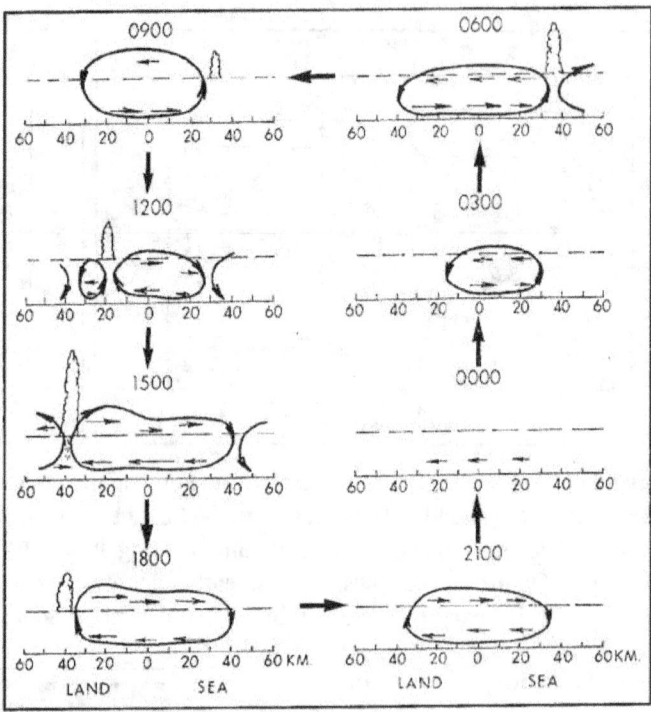

Figure 7-9. Synthesized empirical model of the land-sea breeze along the Texas Gulf Coast. Arrow lengths are proportional to wind speeds (modified from Hsu, 1970).

Figure 7-10. Western Bight of Benin (Atlantic coast of Equatorial Africa) during the afternoon (NASA photo). The arrow points north.

160

In general, the larger the land mass and the higher the mountains, the stronger the sea and land breezes. Sea breezes are enhanced and land breezes are retarded along windward coasts, but mountains penetrating the tradewind inversion slow the tradewinds and inhibit the sea breeze, forcing air to flow around the obstacle (see Figures 5-15, 7-6, and 7-16). Along leeward coasts land breezes are usually enhanced and sea breezes retarded. Across a bay, sea breezes diverge and land breezes converge; across a headland or isthmus, sea breezes converge and land breezes diverge.

7.4.2.2 Numerical Models. Estoque (1962) derived numerical models of the sea breeze for different synoptic situations. The models are applied to a 6,500-foot (2-km) thick layer across a straight, flat coast; they are integrated to obtain time and space variations of winds and temperatures. The driving force for the circulation is a sine function temperature wave applied to the earth's surface initially at 0800L.

Figure 7-11 shows the results of the integration at 1100 and 1700L for three initial geostrophic winds: zero, offshore at 10 knots, and onshore at 10 knots. The horizontal component of the wind vectors is proportional to wind speed, as shown by the scale legend, but the vertical motion scale is greatly exaggerated.

With zero initial wind, the sea breeze forms near the shore and moves about 15 NM (30 km) inland by

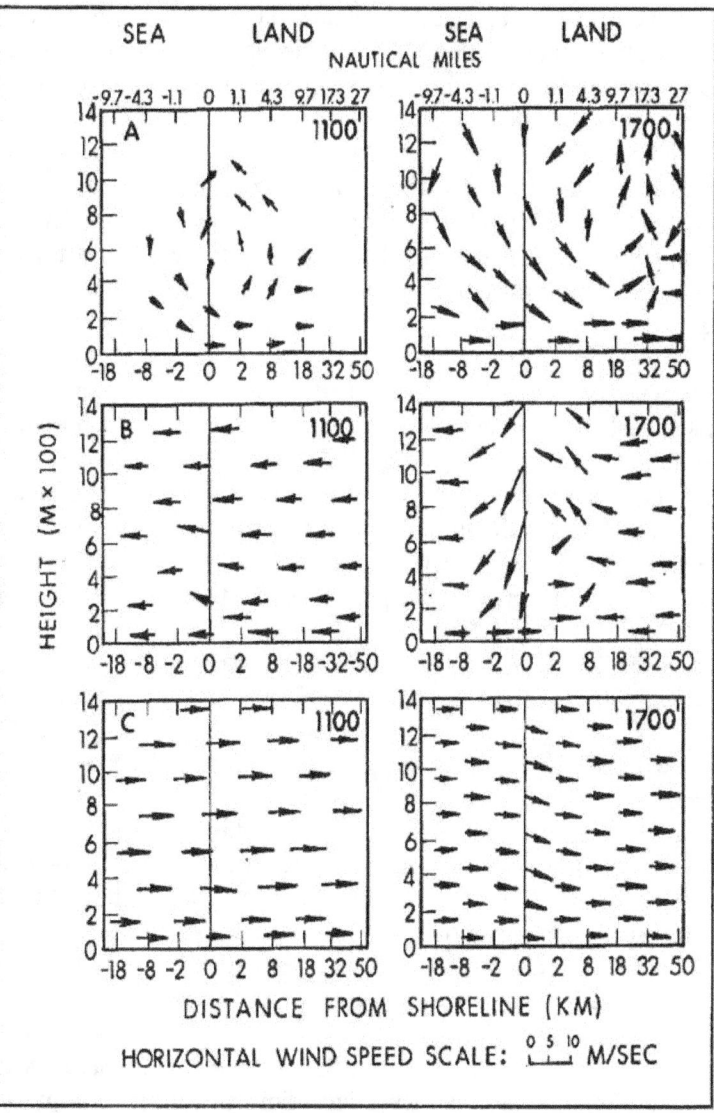

Figure 7-11. Numerical model of the sea-breeze (1100 and 1700L) with three prevailing synoptic flows: (A) calm, (B) wind offshore 10 knots, and (C) wind offshore 10 knots (from Estoque, 1962).

1700L. With an offshore initial wind, the sea breeze is absent at 1100L, but by 1700L it has formed and penetrated about 5 NM (10 km) inland.

With an onshore initial wind, no pronounced vertical motion develops up to 25 NM (50 km) inland since the onshore advection inhibits the temperature from rising there. Note that the resultant vertical motions are most pronounced in the zero and offshore prevailing wind cases. By afternoon, this typically results in a line of cumulus inland and clearing over the shoreline. The model underlines the importance of synoptic flow in determining the character of the sea breeze.

Bechtold et al. (1991) used a numerical model to simulate interaction between an advancing sea breeze and an opposing synoptic current. They concluded that wherever the leading edge of the sea breeze is brought to a standstill, upward motion (intensity) is greatest. McPherson (1970) expanded Estoque's model to three dimensions and applied it to the Texas Gulf Coast sea breeze. He simulated Galveston Bay (29° N, 95° W) by using a square indentation of an otherwise straight coast.

In the numerical results, the sea breeze convergence and the associated vertical motion fields are distorted. The greatest upward motion first develops northwest of Galveston Bay and later shifts to its northeast. The distortion caused by the bay becomes damped as the convergence zone moves inland. The preference for convergence and convection northwest and northeast of Galveston Bay has been verified by observations (Hsu, 1970).

Pielke (1974) used the eight-level three-dimensional primitive equation model to describe the life-cycle of the south Florida sea breeze during late spring and summer (Figure 7-12, opposite). The arrows show horizontal wind velocity at 50 meters 3, 5, 8, and 10 hours after simulated sunrise. The solid lines are axes of convergence. Although the roughness difference between land and sea was included, Pielke found that

the difference in heating between land and water was chiefly responsible for the sea-breeze convergence patterns. One must remember that most land surfaces are much rougher than south Florida's. In this model, the sea breeze was confined to below 11,800 feet (3.6 km). With the most common synoptic flow from southeast, convergence first develops along the east coast and moves slowly inland. Convergence along the west coast sets in about 2 hours later, and remains almost stationary.

Confirming the model, Miami often has early morning showers; skies clear as cumulonimbus develops to the west, and by late afternoon, heavy thunderstorms dominate the western horizon. With southwest synoptic winds, the outcome is reversed as the west coast becomes clear during the day and deep convection occurs in the east. Once more, observations agree.

Another numerical model simulation (Xian and Pielke, 1991) showed that convergence between sea breezes blowing across a neck of land is greatest when the neck is 80 NM (150 km) wide and there is no synoptic wind. No model treated the complete land-sea breeze cycle until Mak and Walsh (1976), who found that greater stability in the near-surface layer at night accounted for the land breeze being weaker than the sea breeze.

Numerical models have realistically simulated sea breezes between the ocean and flat, low land. Complexities introduced by rugged coastlines and mountains have been addressed with a two-dimensional model (Mahrer and Pielke, 1977). Observations support their conclusions that the combined land-sea breeze and mountain circulations are stronger than either acting alone, and a mountainous coast aids inland penetration of a sea breeze.

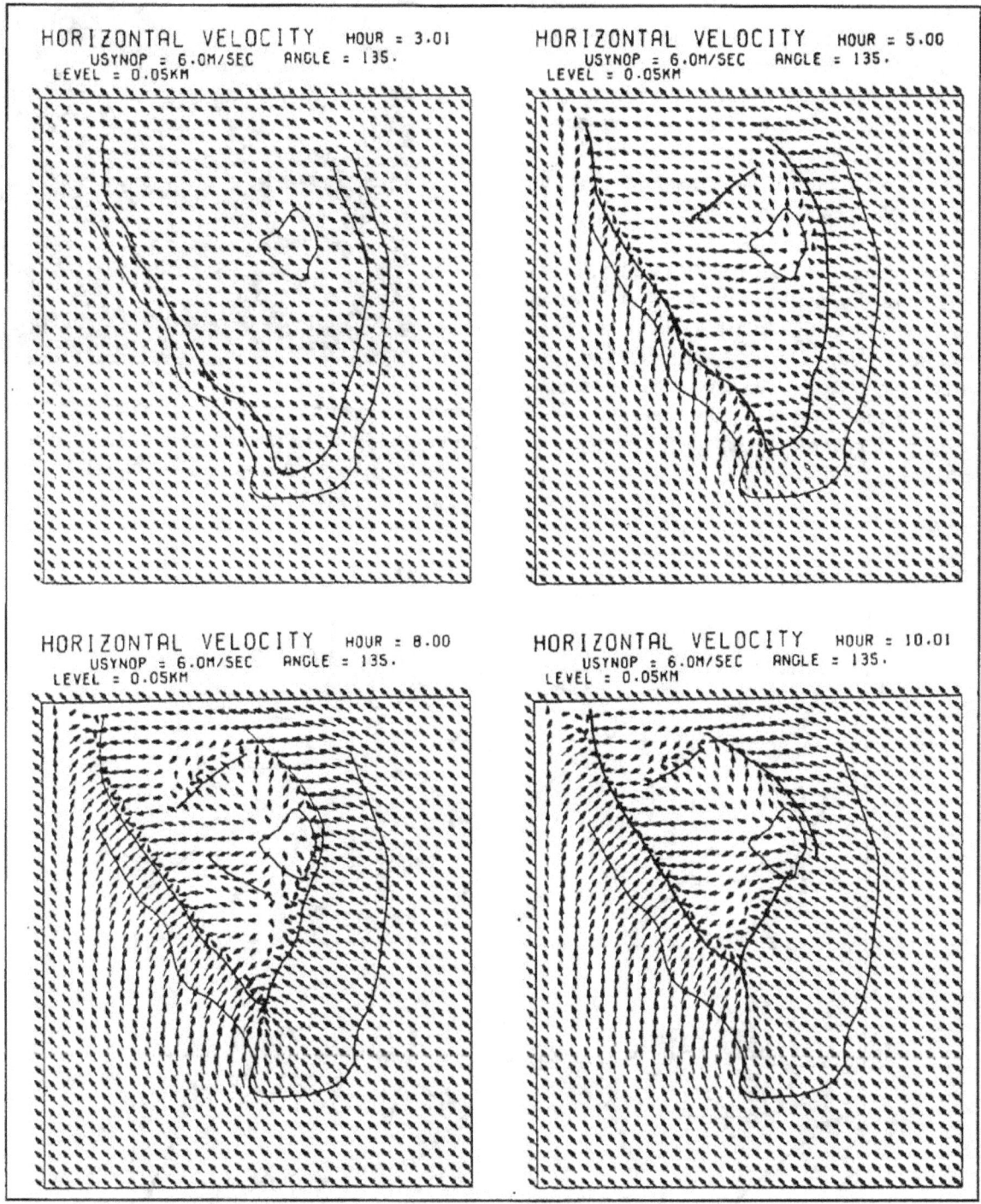

Figure 7-12. Numerical model of south Florida sea breeze (from Pielke, 1974).

7.4.3 Lake Winds. Large tropical lakes experience pronounced land-lake breeze circulations. Over Lake Victoria (2° S, 32° E), converging land breezes cause the air to rise at night, while diverging lake breezes cause subsidence and clear skies in the afternoon—see Figure 7-13). The much smaller Lake Okeechobee (27° N, 81° W) in Florida exerts a similar effect, as shown by Pielke's model in Figure 7-12.

Diurnal variation over Lake Victoria (Leroux, 1983) is modified by the environmental wind, as shown in Figure 7-14. Prevailing southeasterlies are stronger in April than in July. On early April mornings they sweep across the lake, but converge with a land breeze in July. In the afternoon, lake breezes prevail everywhere. Similar patterns have been reported over other lakes.

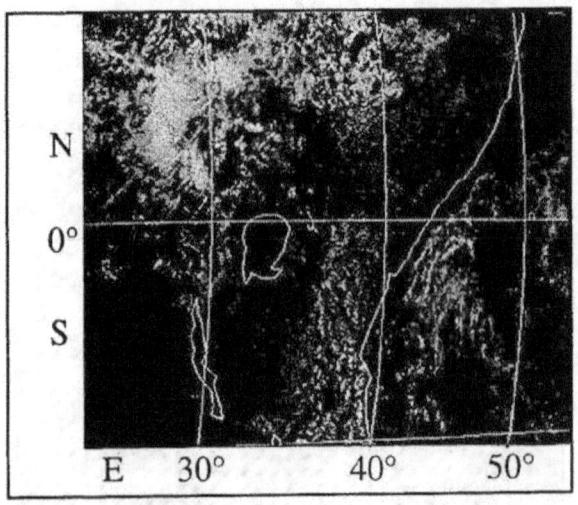

Figure 7-13. Meteosat IR image of Lake Victoria (2° S, 32° E) at 1400L 26 July 1990, showing effects of lake winds on clouds.

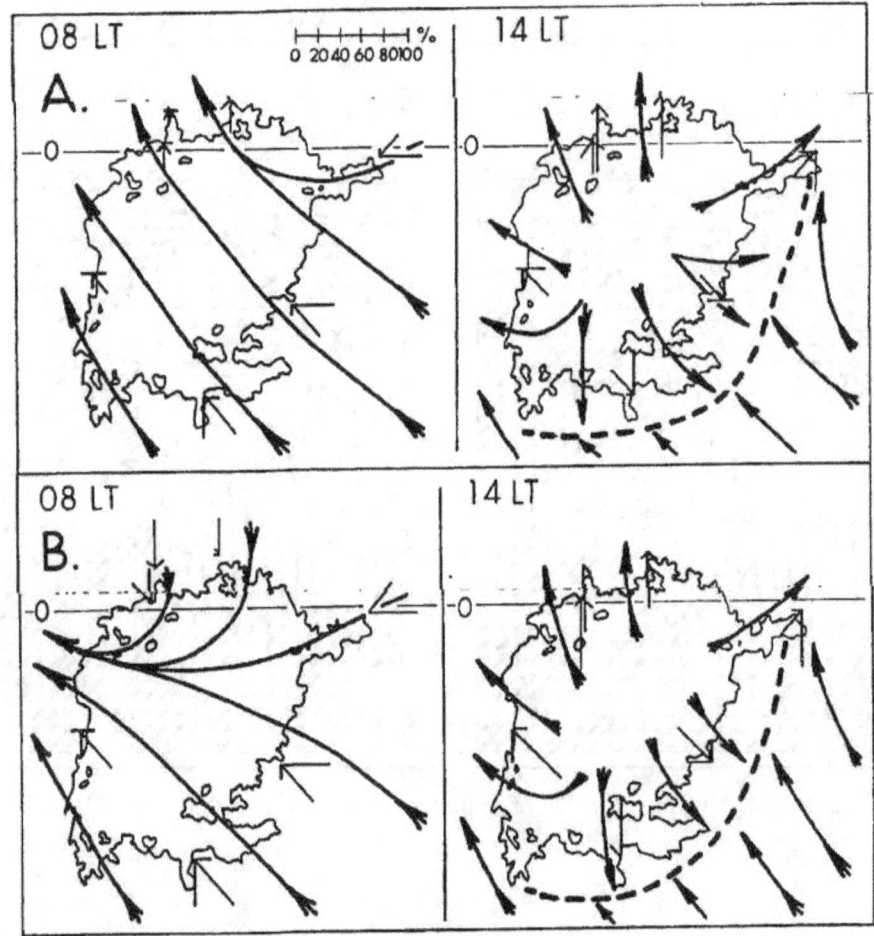

Figure 7-14. Surface circulation over Lake Victoria. (A) April, (B) July with 0800L on the left, 1400L on right (from Leroux, 1983).

Figure 7-15 shows that even the small bays and tributaries of Lake Volta (7° N, 1° W) are cloud-free during the afternoon, suggesting a very weak environmental wind. The larger the lake, and the weaker the environmental wind, the greater the amplitude of the land-lake breeze cycle. Even extensively irrigated surfaces may produce the same effect as lakes (Hammer, 1970).

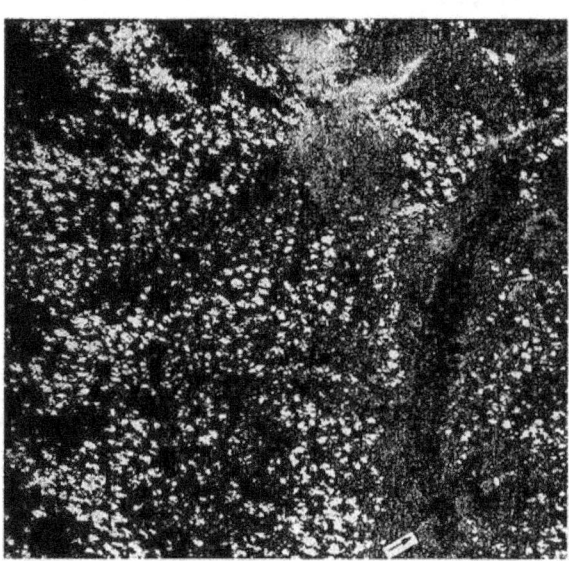

Figure 7-15. Lake Volta in the afternoon. Cumulus is evenly distributed over land, but all the water areas are clear of cloud (NASA).

7.4.4 Mountain-Valley Winds. These have been described by Defant (1951) and others. They are important local circulation features in the tropics where insolation is great and synoptic-scale wind regimes are weak. The temperature difference between heated mountain slopes and air at the same altitude over nearby valleys causes air to rise (*anabatic wind*) along the slopes in daytime. These circulations are well developed on sun-facing slopes and weak or absent on shady slopes. At night, radiational cooling of the mountain surface generates a reverse circulation and downslope flow (*katabatic wind*). The upslope wind layer is generally 300 to 600 feet (100 to 200 meters) thick with characteristic speeds of 4 to 6 knots. Nocturnal downslope wind is shallower and weaker. Figure 7-7 showed the slope winds on the island of Hawaii; a more detailed study was made during the Hawaiian Rainband Project (HARP) in 1990.

Figure 7-16 shows measurements at automatic stations near dawn and in the afternoon during a day on which a well-developed inversion lay near 6,500 feet (2 km). The two major mountains, Mauna Kea and Mauna Loa, penetrated the inversion, while the saddle between them was close to the inversion, which remained near 2 km.

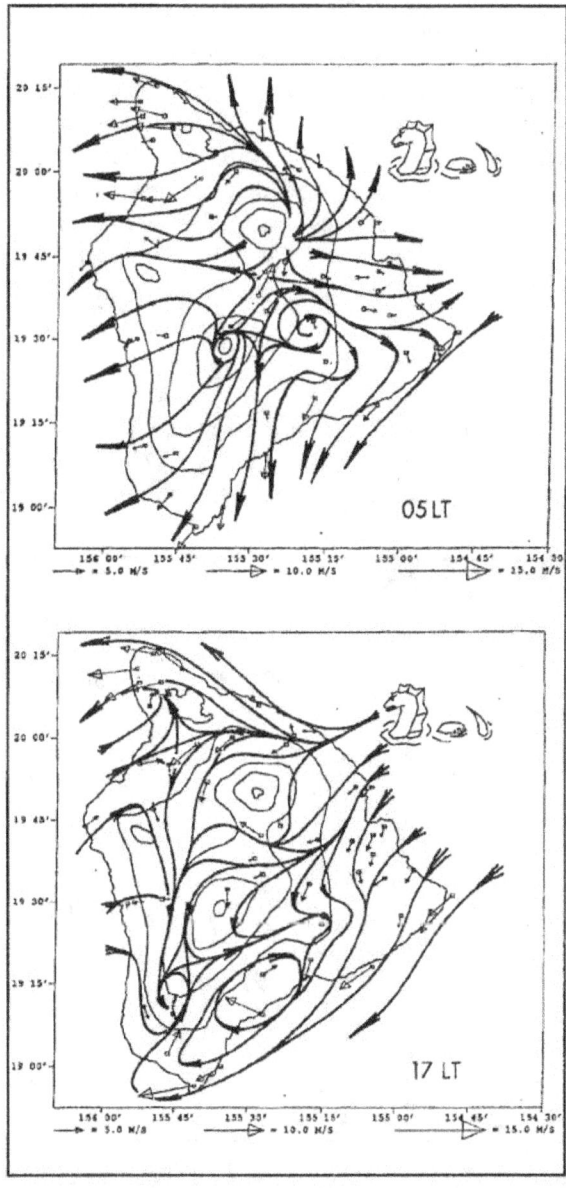

Figure 7-16. Surface winds on the island of Hawaii at 0500 (top) and 1700L (bottom) on 2 August 1990. Land contours are at 1 km intervals.

165

In contrast, the other trans-island saddle, north of Mauna Kea, lay well below the inversion at 3,500 feet (1,060 meters). Land breezes prevailed near dawn, except near the inversion, where calms were common. The strongest breezes were in the direction of the tradewinds around the northern and southern ends of the island, and downwind from the northern saddle. Downslope winds surrounded Mauna Loa and Mauna Kea and converged in the saddle between them. In the afternoon, the island blocked and diverted the tradewinds, which swirled and accelerated around the northern, eastern, and southern corners of the island. Along the west coast, consequently, a cyclonic circulation developed in the north, an anticyclonic circulation developed in the south, and the sea breeze was enhanced, while winds circulated anticyclonically over the southeastern part of the island. The two trans-island saddles, particularly the northern one, funneled the tradewinds. Winds were light and variable near and above the inversion.

Figure 7-17 gives an example of the mountain-valley wind in the Mount Kenya (Equator, 37° E) area. During the day, divergence between upslope flows

causes the valleys to be clear; convergence results in clouds over the mountains. At night, downslope flows *converge* in the valleys, where cloud is likely, and *diverge* from the mountain tops, which are usually clear. In an afternoon weather satellite image of South America (Figure 6-7), lines of cumulonimbus along the western and eastern Andes are separated by relatively clear skies over the Altiplano. No separation can be seen where the Altiplano is narrow. Consequently, Quito (Equator, 78° W; 10,226 feet (3,120 meters)) experiences a sharp afternoon cloudiness and rainfall maximum.

A somewhat different effect (Lyons, 1979) is produced by Mt. Haleakala on the island of Maui (21° N, 156° W); since the peak penetrates the tradewind inversion by 3,000 feet (1 km), maximum afternoon cloud rings the summit, as shown in Figure 7-18. At night, convergence between the land breeze and the tradewind makes the windward coast cloudiest.

Land-sea breezes and upslope-downslope wind regimes reinforce one another. For example, in the Red Sea area (Flohn, 1965) steep escarpments rise abruptly close to the coast, and generally fair skies allow maximum surface heating.

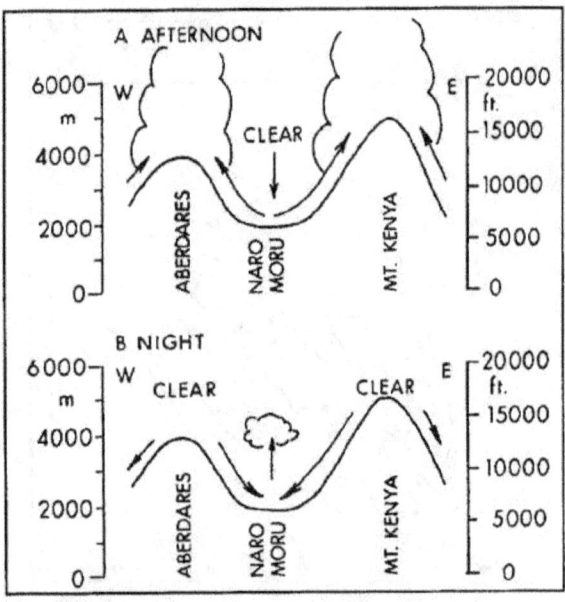

Figure 7-17. Diurnal circulation and cloudiness pattern in the Mt. Kenya area(0°, 37° E). Vertical exaggeration about 1:25 (Hastenrath, 1985).

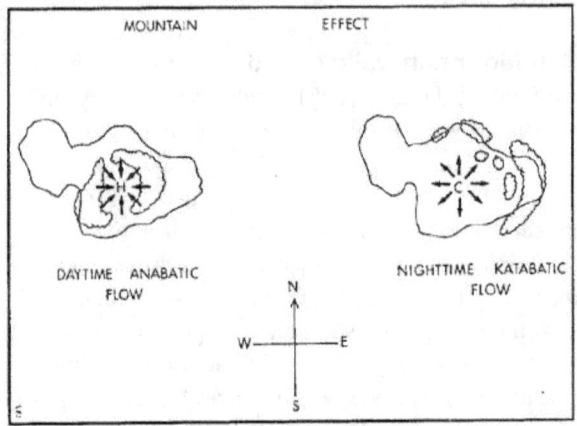

Figure 7-18. Cloud pattern associated with the effect of Mt. Haleakala (21° N, 156°W; 3,11 meters altitude) showing anabatic (upslope) and katabatic (downslope) winds. "H" indicates heating and "C" indicates cooling (Lyons, 1979).

Figure 7-19 shows the resultant surface wind directions for two daily periods (0600 to 0800L and 1200 to 1500L) during July and August. During summer, the resultant gradient-level wind blows from the northwest over the Red Sea and from the west-southwest over the Gulf of Aden. In the morning, downslope-land breezes prevail, causing convergence over the water; by afternoon, upslope-sea breezes cause divergence over the water and local convergence over land. These circulations affect rainfall and vegetation patterns significantly. The west coast of Guatemala and the Mediterranean coast of Asia Minor have similar very strong combined wind systems. As the morning advances and mountain wind and sea breeze develop together, and blow in the same direction, a sea breeze "front" and its associated convergence effects can seldom be detected.

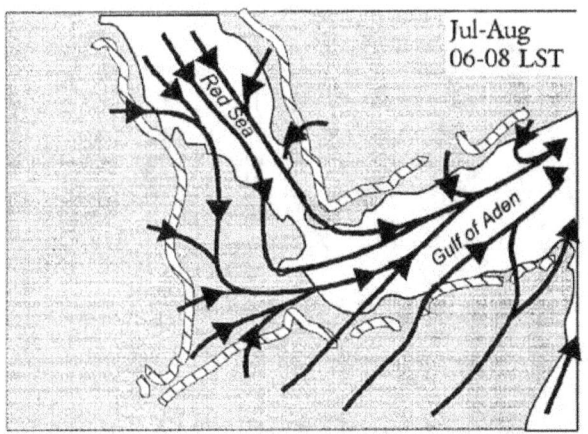 

Figure 7-19. Resultant surface streamlines over the Red Sea/Gulf of Aden in the morning (left) and afternoon (right) during July and August (modified from Flohn, 1965).

7.4.5 Vertical Mixing.

7.4.5.1. General. From the surface to the gradient level, friction decreases and wind increases with height. In the absence of local winds, the vertical gradient of wind speed is controlled by vertical mixing, which in turn is controlled by the lapse rate. Over the sea, lapse rate undergoes little diurnal variation, but over land, surface cooling at night increases stability and reduces vertical mixing, while daytime surface heating *decreases* stability and *increases* vertical mixing. Thus, with no land-sea or mountain-valley winds, surface winds are weakest at night and strongest during the day. With light tradewinds at Honolulu, a classical land-sea breeze cycle prevails (Figure 7-20A). With typical tradewinds, no sea breeze appears; in fact, tradewinds freshen during the afternoon as island heating increases vertical mixing, brings stronger tradewinds to the surface, and reverses the sea-breeze effect. (Figure 7-20B).

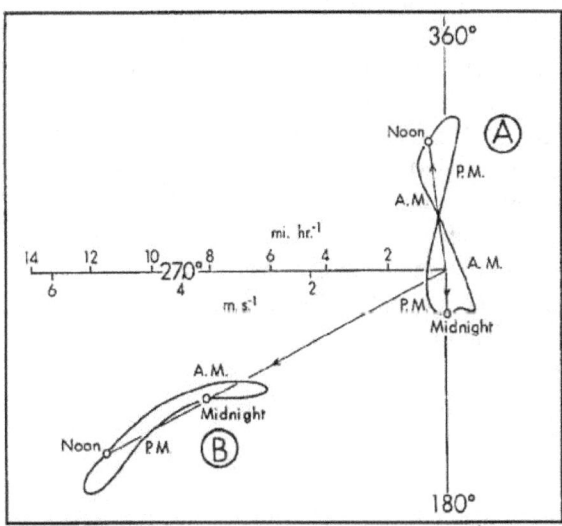

Figure 7-20. Hodograph of diurnal variation of surface winds at Honolulu Airport (21° N, 158° W). (A) shows tradewinds averaging 4 knots or less and winds from other directions. (B) shows all tradewinds (Ramage, 1978a).

167

Observers have often reported sea breezes extending more than 50 NM (100 km) inland by midday—impossibly far. On most of these occasions, synoptic wind and sea breeze blow in the same direction. Inland, vertical mixing after sunrise freshens the surface wind; the direction may suggest sea breeze, but its origin is local.

7.4.5.2 Low-Level Jets (LLJs). The LLJ poses some problems for forecasters. Beneath an LLJ, airports may experience gusty winds during the day; at night, aircraft on approach may let down through a layer with strong wind shear. The rapidly diminishing wind causes a sudden and sometimes disastrous loss of lift.

Beneath strong LLJs over land, the effects of vertical mixing often dominate other local winds. Typically, at Baledogle (3° N, 45° E) beneath the Somali jet, surface winds average less than 10 knots around dawn, but gust up to 25 knots in the afternoon (Walters, 1988). Conversely, the jet is strongest in early morning and weakest in early afternoon.

Between 23 June and 3 July 1977, 3-hourly pilot balloon observations were made at Obbia (5° N, 48° E; 400 feet (120 meters elevation) on the east coast of Somalia, and at Burao (9° N, 45° E; 3,420 feet (1,043 meters) elevation) on the plateau in northwestern Somalia (Ardanuy, 1979). In presenting composite 3-hour profiles of the Obbia and Buran wind speeds, Figure 7-21 shows the Somali jet to be strongest at night near 2,450 feet (750 meters) (much lower than the regional average of about 5,000 feet/ 1.5 km). Speeds reached 49 knots at Obbia and 39 knots at Burao. At Burao, turbulent mixing during the day reduced the jet and increased the surface wind. By late afternoon, speeds were nearly constant with height.

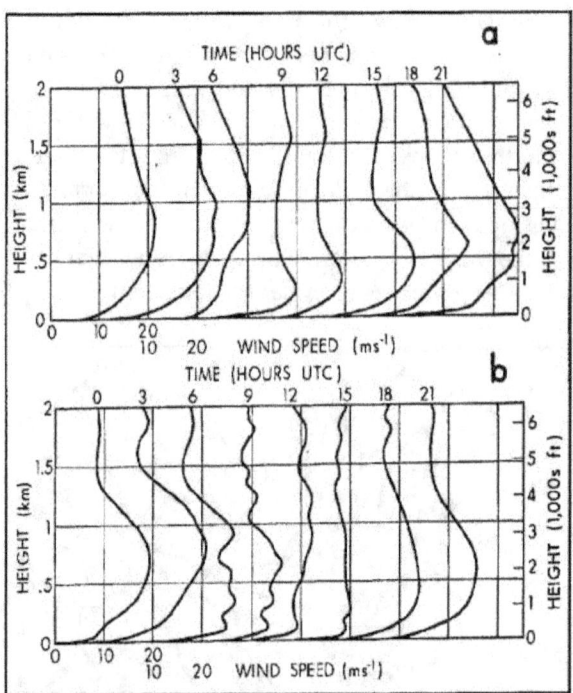

Figure 7-21. Composite 3-hour profiles of wind speeds at Obbia (A) and Burao (B). The X-axis shows a staggered scale, beginning at zero for each profile (from Ardanuy, 1979).

The sequence differed at Obbia, on the east coast, where there was less daytime turbulence. The jet maximum moved down to 800 feet (250 meters) during the morning, then rose. At the same time, overland surface temperature increased by 32 degrees F (18 degrees C) and caused the thermal wind to increase. This very large daytime temperature difference between dry land and cold upwelled coastal waters fails to generate a sea breeze at Obbia. As at Honolulu during fresh tradewinds (see Figure 7-20B), the downward mixing that is in control maintains surface southwesterlies.

During relatively clear weather over the Amazon Basin, large rivers (those wider than 8 NM/15 km) affect the local circulation (Garstang et al., 1990).

Large rivers are 14 to 16 degrees F (8 to 9 degrees C) warmer than the forest canopy at night and 4 to 6 degrees (2 to 3 degrees C) cooler by day. At night, as shown in Figure 7-22(A), a surface inversion develops over the cooling forest. Reduced frictional drag leads to the development of a jet at 1,000 to 2,300 feet (300 to 700 meters) above the surface; its speed reaches 30 to 40 knots. Over the warm river, convective heating prevents a jet; winds are weaker than those above the forest, and convergence develops inland from the upwind bank.

By day, since vertical mixing is greater over the heated forest, flow accelerates over the river, leading to convergence inland from the downwind bank (Figure 7-22B). These processes cause relatively little rain along the banks of large rivers; satellites show less cloud over rivers. The Congo River of central Africa should produce similar effects.

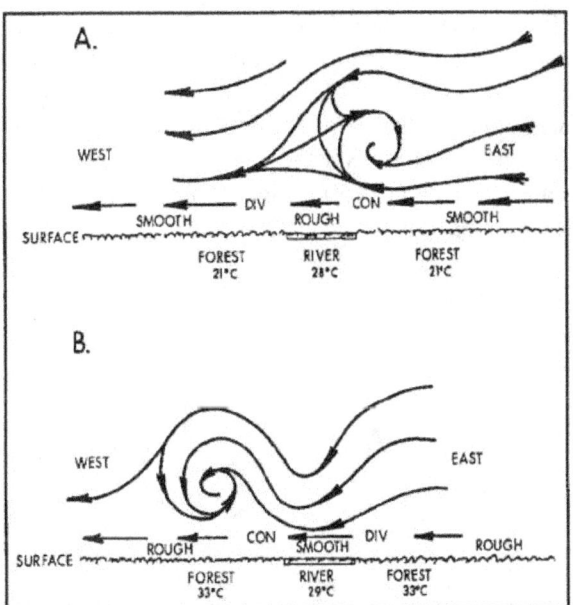

Figure 7-22. Schematic diagram of circulations over and near large rivers of Amazonia: (A) night, and (B) day (Garstang et al., 1990).

7.4.5.3 Interactions Along Coasts. Even more complex interactions occur along coasts, as shown in the following examples.

• *Onshore synoptic wind.* In his study of the Texas land-sea breeze (7.4.2.1), Hsu (1970) selected 9 days that lacked any significant synoptic wind. For the 22 remaining days of the periods studied, fresh onshore synoptic flow prevailed, and the strongest sea breeze occurred at midnight (Eigsti, 1978). On these nights, an LLJ was present, with maximum speeds near dawn at about 3,000 feet (1 km). The LLJ, in the same direction as the sea breeze, seems to have been responsible for increasing the weakening sea breeze after mid-afternoon. Eigsti suggested that, at the coast, onshore flow would hinder nocturnal cooling enough to prevent a near-surface inversion from developing there. Thus, the surface would be linked to the nocturnal jet, as shown in Figure 7-23. A low-level jet coinciding with the boundary layer top makes the sea breeze persist into the night. The jet, becoming supergeostrophic a short distance inland and causing pressure to fall downstream, would increase the onshore surface wind (the sea breeze) at the coast. At Bombay (19° N, 73° E) in May, westerlies prevail. Their associated LLJ enhances the sea breeze, which increases through the evening.

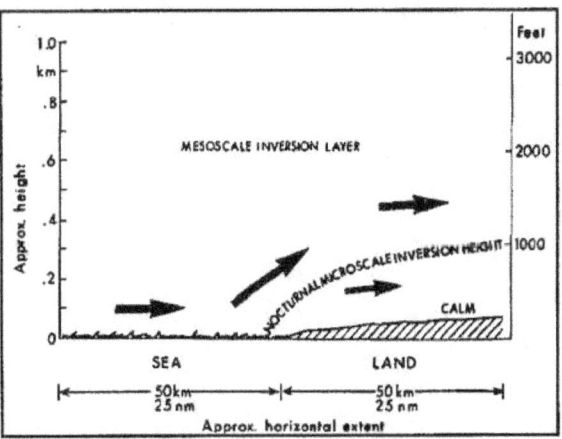

Figure 7-23. Nocturnal ocean-land interface with *onshore* prevailing wind.

• **Offshore synoptic wind.** When the synoptic wind blows *away from* the coast (as shown in Figure 7-24), surface wind speed along the coast is in phase with LLJ speed; the land breeze is strongly enhanced, reaching a maximum near dawn. The sea breeze is weaker than the land breeze. The north coast of Somalia provides an example; the Somali jet enhances the land breeze and opposes the sea breeze (Vojtesak et al., 1990).

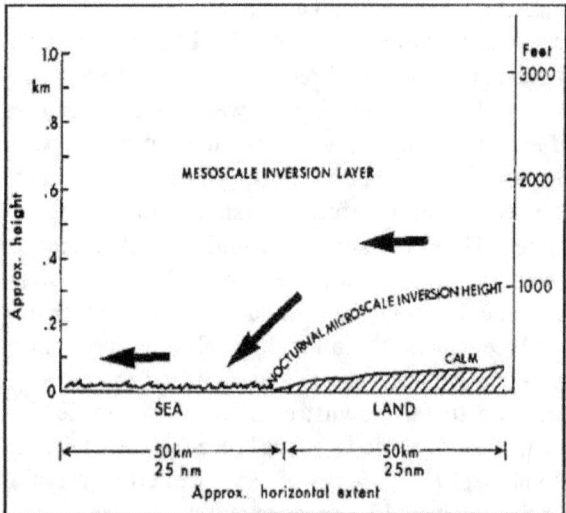

Figure 7-24. Nocturnal ocean-land interface with *offshore* prevailing wind. The low-level jet reaches the surface along the coast and just offshore, enhancing the land breeze (from Hsu, 1979).

• **Synoptic wind parallel to coast.** In this event, the land-sea breeze is not affected much at the shoreline, but the LLJ may cause complications a few miles inland, where the diurnal heating/cooling cycle is significant. Detailed observations are needed. In summer, LLJs parallel the shores of the Persian Gulf.

7.4.6 Summary. So far, discussion has concentrated on diurnal wind changes at the surface and in the lower troposphere driven by the diurnal thermal cycle. Wallace and Hartranft (1969) used data from the surface to 100,000 feet (30 km) to extend our knowledge of diurnal wind variations in the free atmosphere. They computed annual mean 12-hour wind differences between 0000 and 1000Z (or 0300 and 1500Z) at 27 pressure levels for 105 radiosonde stations throughout the world. In polar latitudes, a simple pattern of flow across the poles from the daytime to the nighttime hemisphere is observed. This agrees well with atmospheric tidal theory. In low and middle latitudes, however, the wind differences are strongly related to topography; land-sea contrasts and terrain slope appear to be in control. Above 900 mb at tropical stations, the average 12-hour wind differences were less than 4 knots. This about equals the accuracy of the observations and is smaller than hour-to-hour changes or differences over a few tens of kilometers. Therefore, it is doubtful that these slight diurnal wind changes in the free atmosphere could be detected or applied in routine tropical analysis and forecasting.

7.5 DIURNAL CLOUDINESS VARIATION

The previous chapter made passing reference to clouds associated with diurnal wind variations, and Figures 7-7, 7-10, and 7-12 to 7-18 should be kept in mind. In this chapter, attention is focused on non-precipitating clouds. Diurnal variation of deep convective clouds will be discussed in 7.6 (on rainfall), since the former implies the latter.

7.5.1 Open Ocean Tradewind Cloud.
Chapter 2 reported that over the open ocean, surface winds vary semi-diurnally by about 1 knot in response to the semi-diurnal pressure wave. Brier and Simpson (1969) claimed that the resulting divergence of the surface wind could account for diurnal variations of cloudiness and rainfall at Wake Island (19° N, 167° E). The small divergence (5 X 10^{-8} s^{-1}) corresponds to a vertical motion of only about 0.02 cm h^{-1}, which is very unlikely to cause variations in cloud or rain.

Brill and Albrecht (1982) used data from the Atlantic Tradewind Experiment (ATEX) to determine diurnal variations of the sub-cloud mixed layer and the base of the tradewind inversion. This was highest at dawn and lowest in mid-afternoon with a range of about 260 feet (80 meters), somewhat less than estimated by Holle (1968) for the tropical Atlantic (13° N, 55° W) in August and September 1963, and in line with what Kraus (1963) found. The difference between cloud base and inversion base (a measure of the depth of tradewind cloud) was greatest (2,800 feet/ 860 meters) at 0400L. Diurnal variation of the divergence of the surface winds measured in the ATEX array equalled 1 X 10^{-6} s^{-1} with a maximum at 0700L and a minimum at 1900L. Brill and Albrecht then inserted the data into a time-dependent one-dimensional model of the tradewind boundary layer, while specifying long-wave radiative cooling. The model simulated the diurnal variation of tradewind cloudiness rather well. Changing the external variables confirmed what Refsdal (1930) had suggested—that the diurnal variation of radiational heating and cooling at the cloud top contributes most to the diurnal variation in cloud depth and cloudiness and that divergence due to the semi-diurnal pressure wave has a negligible effect. During GATE and ATEX, cloud depth and cloudiness were generally greatest

near 0800L and least near 1600L, with a range of 14 percent (Figure 7-25). The values given here average rather diverse data. They generally apply to shallow tradewind cloud. Preponderant evidence supports a morning maximum and an afternoon minimum and control by radiation.

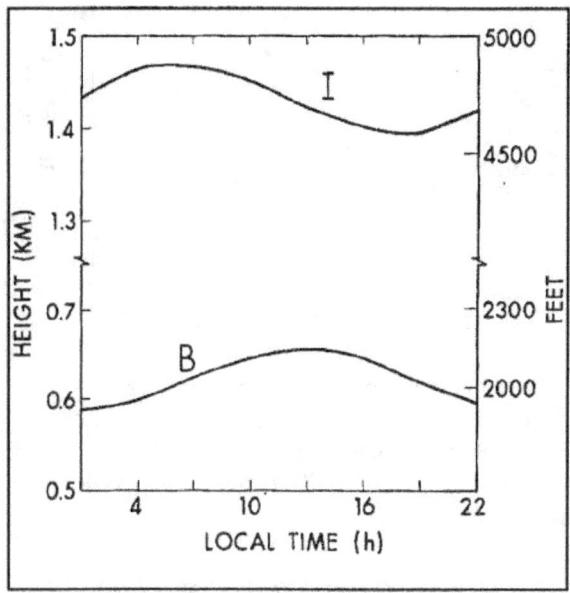

Figure 7-25. Average diurnal variation of the inversion height (I) and the cloud base height (B) for a period of undisturbed tradewinds during ATEX (Brill and Albrecht, 1982).

7.5.2 Open Ocean Stratocumulus.
These extensive cloud sheets in the eastern Pacific and Atlantic tradewinds (see Figures 6-1 to 6-4) are capped at about 3,000 feet (1 km) by a strong inversion. Minnis and Harrison (1984) analyzed GOES data for November 1978 and found very uniform diurnal variation of cloudiness for all the stratocumulus regions. The data shown in Figure 7-26 (next page) is typical, with a dawn maximum and an afternoon minimum. It resembles the tradewind cloudiness variation shown in Figure 7-25, but the range (43 percent) is much greater. The range of 2.8 K in the temperature of the cloud top is equivalent to a range of 1,000 feet (300 meters) in the height of the inversion base. Turton and Nichols (1987) have reproduced the

171

same diurnal variation of cloudiness in a numerical model that emphasizes radiation. Similar results were obtained by Short and Wallace (1980), who found cloud tops to be relatively cold (high) in the morning and warm (low) in the evening in both stratocumulus and tradewind cloud regimes.

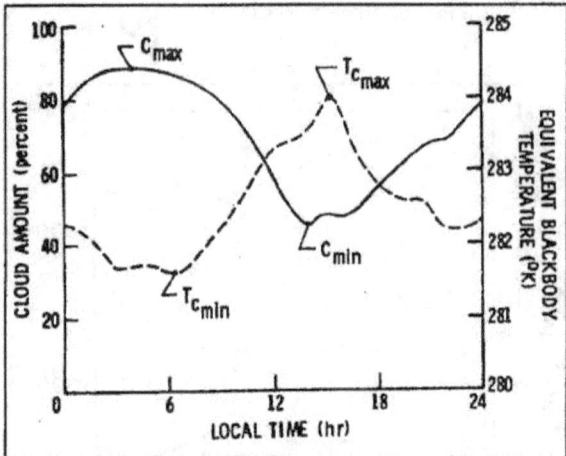

Figure 7-26. Mean November 1978 cloudiness (C) and cloud top temperature (T_c) for a 250 x 250 km² region at 21.4° S, 86.3° W (Minnis and Harrison, 1984)

7.5.3 Open Ocean and Coastal Fog. When warm moist air moves over cold water, generally near the coast, and the subsidence inversion is some distance above the surface, air temperature may be lowered to the dew point (see 6.2.6.4). As with ocean stratocumulus, nocturnal radiation from the top makes it deepest and densest near dawn. Daytime heating of the top reverses the process. Fog may extend into coastal bays and river mouths and over low-lying shore lines; it is common over coastal south China in spring. Fog dissipates quickly over land as the Sun region warms the surface and the air temperature rises above the dew point. At the same time, the fog thins over coastal waters, but the surface temperature is unchanged. Besides daytime heating of the fog top, the vertical circulation of the sea breeze may contribute. Sinking motion over coastal waters (see Figure 7-11) brings drier air to the surface and causes

the dew point to fall below the sea-surface temperature, thus dissipating the fog (Ramage, 1971).

7.5.4 Overland Stratocumulus and Fog. Along flat exposed coasts, stratocumulus is fairly common and follows the oceanic pattern; radiative cooling from the top of a moist layer favors nocturnal development, while warming from above and convective mixing dissipate the cloud during the morning. At Lima (12° S, 77° W), drizzle ("garua") falls between 2300 and 0900L from stratus that deepens during the night. This accounts for almost all the rainfall, averaging 0.39 inches (10 mm) a year (Prohaska, 1973). Similar weather prevails along the west coast of South America from 8° S to 25° S.

Over the Amazon basin during the SH winter, surface air within a cold surge (see 8.4.4.1) is heated and moistened from below and is capped by an inversion at 3,000 feet (1 km). As in the tradewinds over the eastern oceans, extensive nocturnal stratus develops, sometimes giving drizzle. The cloud generally breaks up during the morning (Walters et al., 1989).

In river valleys, cooling under clear skies, coupled with surface moisture and nearly stagnant conditions, may lead to fog at night. As with stratocumulus, dissipation follows sunrise. Radiation fog may develop over flat coastal areas bordering deserts. Strong sea breezes bring moist air inland during the day. At night, surface air cools markedly under clear skies; since the flat terrain hinders land breezes, fog may form in the previously moistened air. The fog burns off quickly after sunrise. The coasts of Pakistan, the Persian Gulf, northern Angola, and eastern Somalia all experience such fogs. The Guyanas are also affected during the September-October dry season.

Duvel (1989), analyzing METEOSAT pictures, found extensive altostratus and altocumulus decks at about 16,000 feet (5 km) over tropical Africa and the Atlantic; he suggested that they may be not uncommon elsewhere in the tropics. As with stratocumulus, coverage is greatest around dawn.

7.5.5 Popcorn Cumulus. Figure 6-7 showed popcorn cumulus over the Amazon region. This is a regular feature in the absence of disturbances such as squall lines, It develops in the forenoon, sometimes becomes thunderstorms in the afternoon, and dissipates at night. Small terrain gradients, leading to weak local wind regimes, account for the random distribution of cloud elements. The diurnal cycle is directly controlled by radiation; upward motion and condensation result from daytime heating of the ground; radiational cooling at night favors subsidence, and that evaporates clouds.

7.6 DIURNAL VARIATION OF RAIN

7.6.1 Over Open Oceans. Amounts of cloud that rarely produce any rain respond to the daily cycle of radiation; that is, greatest in early morning and least in the afternoon. Ruprecht and Gray (1976) analyzed 13 years of cloud clusters over the tropical west Pacific and found that over twice as much rain fell on small islands from morning (0700 to 1200L) clusters as from evening (1900 to 2400L) clusters. The heaviest rain fell when it was part of an organized weather system and when diurnal variation was most pronounced. Fu et al. (1990) used satellite IR images over the tropical Pacific to confirm and refine these findings. Deep convective cloudiness was greatest around 0700L and least around 1900L. Gray and Jacobson (1977) explained the observations in terms of radiational differences between cloudy disturbed areas and their surroundings.

Figure 7-27 shows that the clear surroundings cool more by night than by day. The only large diurnal variation in the disturbance is at the cloud top, where daytime heating contrasts with large nighttime cooling. Lower tropospheric inflow and upper tropospheric outflow maintain upward motion in the disturbance and pressure surfaces slope accordingly.

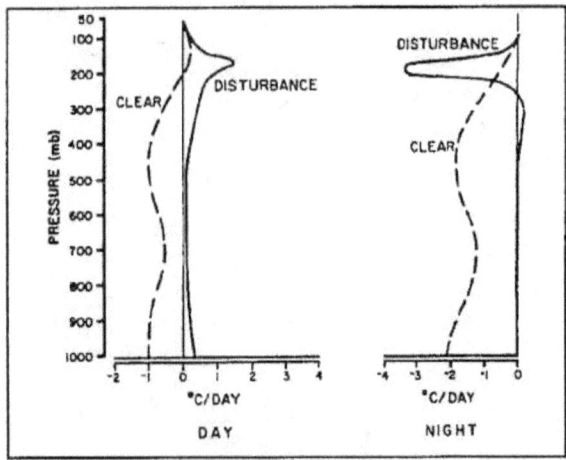

Figure 7-27. Estimated rates of net radiational warming within a tropical disturbance and in the surrounding clear or mostly clear regions (Gray and Jacobson, 1977).

In Figure 7-28, the day-night radiation differences enhance nocturnal low-level inflow and upper-level outflow and so intensify the disturbance. The atmospheric response lags the radiational forcing by 1-2 hours.

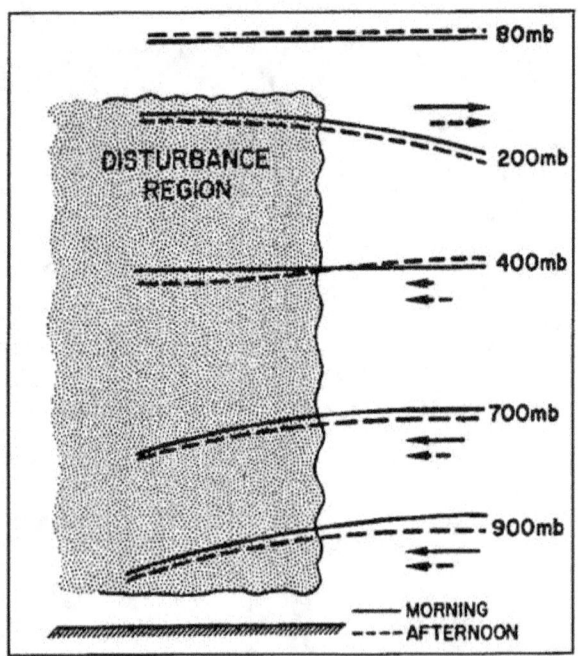

Figure 7-28. Idealized slopes of pressure surfaces between a disturbance and the surrounding clear air during the day (dashed lines) and at night (solid lines), along with the resultant night vs. day inward-outward radial wind patterns (Gray and Jacobson, 1977).

The same processes were found in typhoons (Muramatsu, 1983; Zehr, 1987), Atlantic tropical cyclones (Steranka et al., 1984), and tropical cyclones in the Australian region (Lajoie and Butterworth, 1984). Over tropical cyclones, the area covered by very cold cirrus (less than -65° C) is greatest in early morning. The cirrus, advected outward from the cyclone center, is warmed and thinned by sinking. Thus, the area covered by cirrus is greatest in late afternoon (Browner et al., 1977).

Figure 7-29 from Zehr (1987) shows diurnal variations in the areas of "cold" and "warm" cirrus. Infrared images made through the 11.2 μm window do not delineate the two diurnal variations, but enhanced IR images allow the area of "cold" cirrus to be determined; after the diurnal variation has been removed, increases or decreases in this area precede tropical cyclone intensification or weakening.

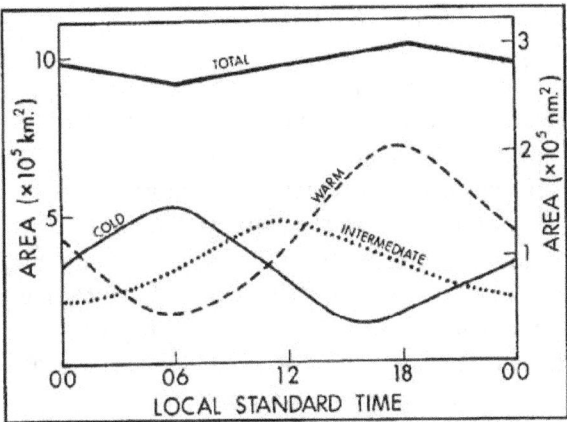

Figure 7-29. Diurnal variation of satellite-observed radiating cloud surfaces associated with tropical cyclones. Areas of "cold" (<-65° C), "intermediate" (-40° C to -65° C), "warm" (-15 to 40° C), and "total" (<15° C) (from Zehr, 1987).

A variety of numerical models have confirmed that diurnal variations in radiative forcing of the disturbance and its surroundings could account for the observed diurnal variation in rainfall (Fingerhut, 1978; Dudhia, 1990; Tao et al.,1991).

7.6.1.1 Squall Lines. During the Line Islands Experiment (LIE) and GATE, when attention was concentrated on the tradewinds, squall lines were common rain producers; they formed the basis of the model described in 8.4.2 (Zipser, 1977). In the GATE array of research vessels centered near 10° N, 23° E, the air was least stable at around 0300L (Albright et al., 1985). When significant convection, generally in the form of squall lines, occurred, it was usually under

way about then. Since the squall lines took several hours to develop fully, maximum rainfall was measured on the research vessels at about 1400L (McGarry and Reed, 1978). The maximum surface convergence that occurred at this time is consonant with squall line intensity.

Disturbances resembling squall lines may also have been responsible for diurnal variations reported by other investigators; examples are Reed and Jaffe, 1981; Augustine, 1984; Albright et al., 1985; Duvel, 1989; and Shin et al., 1990. Over the Atlantic and Pacific near equatorial convergence, cold cloud tops were more common between noon and mid-afternoon, while the South Pacific convergence zone (SPCZ) (see 8.4.5) experienced dawn and mid-afternoon maxima, as did the Pacific tradewinds.

7.6.1.2 Summary. The morning rainfall maximum associated with western Pacific cloud clusters and the early morning instability in the tradewinds both originate from the nocturnal radiational cooling of cloud tops. In the western Pacific cloud clusters, which remain under radiational control, rainfall diminishes in the afternoon. In the tradewinds, middle tropospheric dryness favors disturbances resembling squall lines. These form in the early morning, but in the several hours they take to intensify, they are not under radiational control; consequently the rainfall maximum is pushed into the afternoon.

Sharma et al. (1991) estimated oceanic rainfall from 10 months of SSM/I data obtained between August 1987 and July 1988. The satellite crossed the tropics at about 0600 and 1800L, allowing differences between early morning and evening to be determined. The mornings were slightly wetter that the evenings.

Majuro Atoll (17° N, 171° E) is generally dominated by the tradewinds; it experiences little diurnal variation in thunderstorms, which are likely only as weather disturbances pass; diurnal variation on the atoll is masked—see Figure 7-30.

175

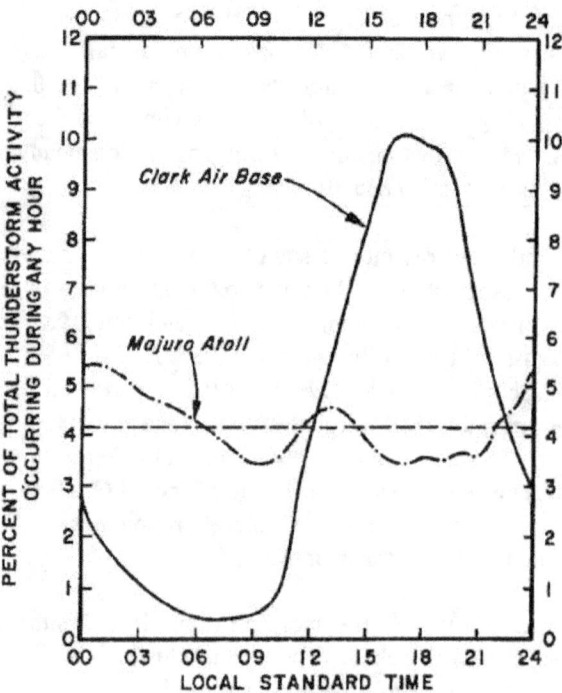

Figure 7-30. Percent of the total thunderstorm activity occurring durning any hour at Majuro Atoll (7° N, 171° E), and Clark AB (15° N, 121° E).

7.6.2 Over Land.

7.6.2.1 General. So far in this chapter (except for atmospheric tides and near-shore fog), diurnal variations of pressure, temperature, winds, cloudiness and rain have been directly forced by the diurnal variation of the Sun's radiation. Now as we move from small islands to large islands to continents and from offshore to interior, the effect of the radiation cycle on local winds (land-sea breezes, mountain-valley winds) determines the distribution of divergence and vertical motion, and that in turn can complicate, counteract, and sometimes overwhelm direct radiational forcing of cloud development and rain. ***Understanding diurnal variation of rainfall demands knowing the local wind regimes.*** They may be powerful enough to blank out synoptic or mesoscale influences, and often dominate weather forecasting.

7.6.2.2 Synoptic-Scale Weather Disturbances. Under cloudy skies, surface temperature does not vary much;

diurnal variation of local winds is damped out. The radiation cycle, following the Gray-Jacobson (1977) model described earlier, favors an early morning rainfall maximum and an afternoon minimum. This happens when the summer monsoon is active over India (Prasad, 1970), and is also true during the months of greatest rainfall over southeast Asia (Figure 7-1, opposite), except in the extreme south of Vietnam, where less cloudiness and converging sea breezes cause an afternoon maximum.

In the wet monsoon regions, when disturbances weaken and cloudiness diminishes, radiational heating of the ground could cause a second maximum of rainfall in the afternoon (none of the curves of Figure 7-31 goes to zero during the afternoon). This transitional regime is typical of summer over east China and southwest Japan (Ramage, 1952a). During a break in the monsoon rains, a showers regime gives scattered thunderstorms and most rain falls in the afternoon and evening (Prasad, 1970).

Over the flat Korat Plateau of eastern Thailand, Ing (1971) used weather satellite pictures to stratify summer days into "weak monsoon" (scattered clouds) and "strong monsoon" (generally overcast). The diurnal rainfall variations were very different, as shown in Figure 7-32. An afternoon maximum for the weak monsoon arose from convection induced by surface heating, while a nocturnal maximum for the strong monsoon was caused by destabilizing cooling of the cloud top, enhancing convection.

7.6.2.3 Squall Lines. Over the Sahel and northern South America, squall lines account for most of the rain (see 8.4.2). They tend to develop during mid-afternoon, when convective heating is greatest; off the northeast coast of South America nocturnal convergence between land breeze and tradewind may also be responsible. For many hours, and sometimes for more than a day after development, intensity and movement of a squall line are not controlled by the diurnal cycle. Consequently, Leroux (1983) failed to find any pattern in the diurnal variation of rainfall over West Africa; the same is probably true for northern South America.

176

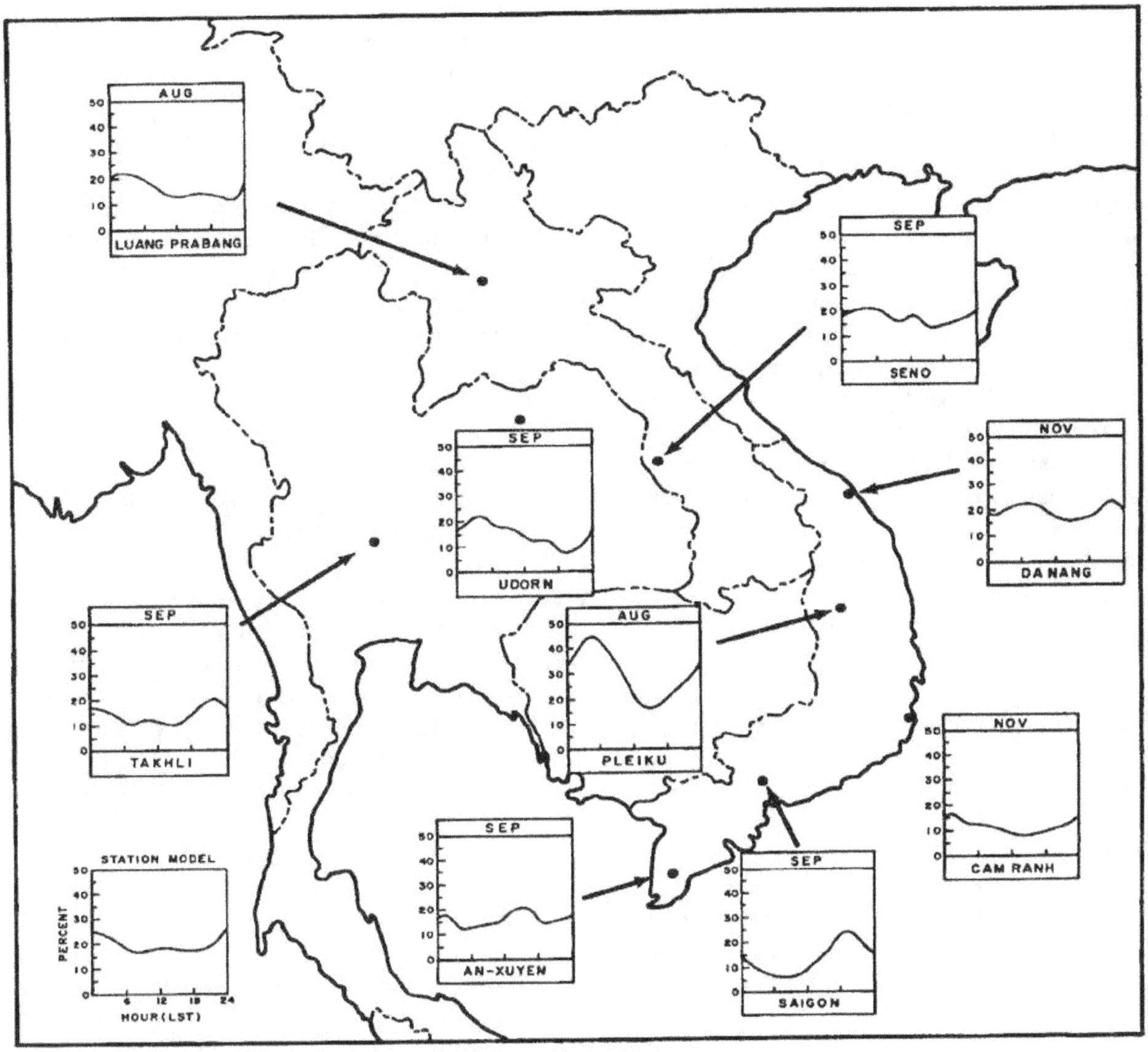

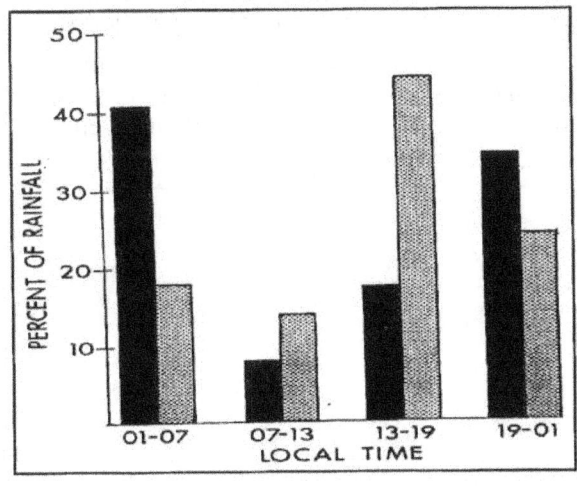

Figure 7-31. Diurnal variation of rainfall frequency during the month of maximum rainfall at selected stations inn southeast Asia.

Figure 7-32. Mean percent of total rainfall for 6-hourly periods at 11 stations on the Korat Plateau, Thailand (June-August 1967 and 1968). Solid bars: during strong monsoons; shaded bars, during weak monsoons (from Ing, 1971).

7.6.2.4 Tradewind Regimes. Diurnal variations in rainfall frequency on the island of Hawaii (Figure 7-33, opposite) cover a wide range. Along the northeast coast, exposed to the tradewinds, rainfall is most common at night, when a land breeze converges with the trades (see Figures 7-6, 7-7 and 7-16), and is rarest around noon, when the coast lies in the sinking branch of the sea breeze circulation. On the sheltered west coast, the night is dry as the land breeze sweeps out to sea (Figure 7-6); the sea breeze produces an afternoon maximum. Along the southeast coast of the island, parallel to the tradewinds, land breeze/tradewind convergence always lies offshore; a sharp rainfall maximum stems from sea breeze/mountain wind lifting. At all inland higher-level stations, the moist mountain-sea breeze produces an afternoon maximum, and the dry, sinking downslope flow produces a nocturnal minimum (note the dew-point differences between day and night in Figure 7-7).

The data in Figure 7-33 reflects responses to a strong diurnal heating-cooling cycle. When upper cloud covers the island of Hawaii ("moist-dry-moist sandwich," see 8.4.6.1) nocturnal rain increases along the east coast. Reduced surface cooling, rather than rain falling from the upper layer, is the cause. This diminishes the land breeze and brings its convergence with the tradewind inshore from its normal offshore position (Chen and Schroeder, 1986), enhancing coastal rain.

Kousky (1980) reported similar patterns for northeast Brazil—see Figure 7-34 on Page 7-28. This area usually lies south of the NETWC and of the squall lines, and is relatively dry. The coasts exposed to easterly onshore flow experience a nocturnal maximum from convergence with the land breeze. 50 to 150 NM (100 to 300 km) inland, rain is greatest during the day as the sea breeze advances inland and convective heating develops. Farther inland, mountain-valley winds dominate (see Figure 7-17) and, since most of the stations are in valleys, they experience a nocturnal maximum.

7.6.2.5 Dry Regimes. The diurnal variation of monsoon rainfall at 17 stations in the Sudan was analyzed by Pedgley (1969). Only one of the stations was coastal; nevertheless, a wide variety of types was found, possibly because of squall lines. An afternoon or early evening maximum dominates only far from the Ethiopian highlands. At inland stations near the highlands, rain is more evenly distributed through the day; however, at most places a weak afternoon maximum resulting from radiational surface heating can still be detected. A weak secondary maximum in the early morning is also evident at many stations. The Sudan is topographically rather uniform.

7.6.2.6 Lake Regimes. Since a lake is warmer than its surroundings at night and cooler during the day, convergent land breezes cause nocturnal showers over the lake. By day, diverging lake breezes are accompanied by a rainfall minimum. Thus the local winds over and around Lake Victoria in Africa (see 7.4.3) and Lake Okeechobee in Florida (see Figure 7-12) determine the diurnal variation of rainfall. Similar effects have been observed over Lake Nicaragua (12° N, 85° W) and Lake Maracaibo (10° N, 72° W) and can be expected over other tropical lakes. Bingham (1911) experienced rainy season nocturnal thunderstorms over Lake Titicaca (16° S, 69° W) in the Andean Altiplano.

Where lakes are surrounded by mountains, there are very complex diurnal patterns. On the northern shore of Lake Victoria (Figure 7-14), Entebbe (Equator and 32° E) experiences an early morning rainfall maximum in April, when it is part of the central lake regime, and an afternoon maximum in July after the lake breeze sets in. At Kisumu (Equator and 35° E), northeast of the lake on an inlet perpendicular to the environmental flow, the land breeze always lies offshore. The afternoon rainfall maximum accompanies the lake breeze (Lumb, 1970).

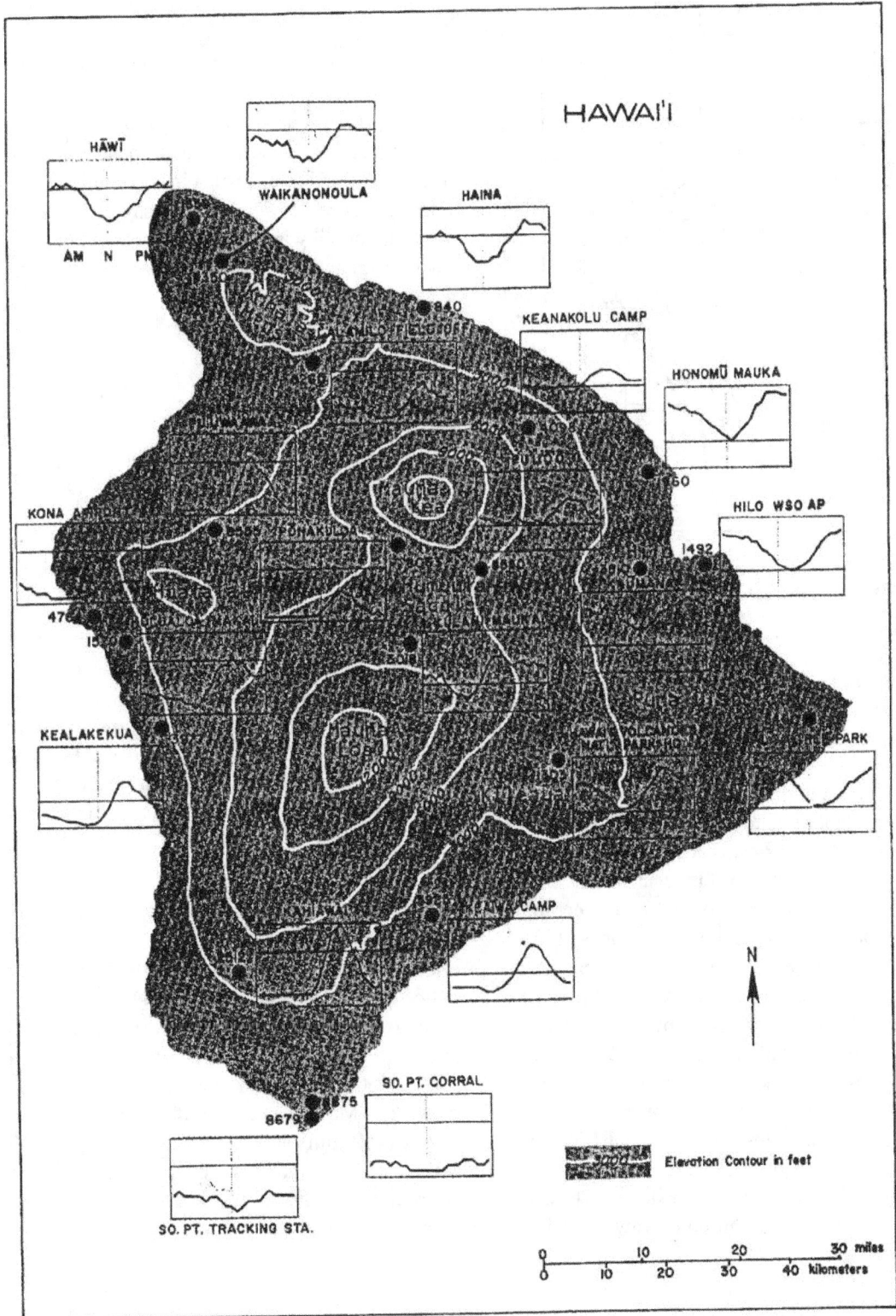

Figure 7-33. Diurnal variation in frequencies of rainfall at stations on the island of Hawaii. Noon local time is marked by a very light vertical line down the middle of each box; the 10-percent frequency is marked by the horizontal line (Schroeder et al., 1978).

179

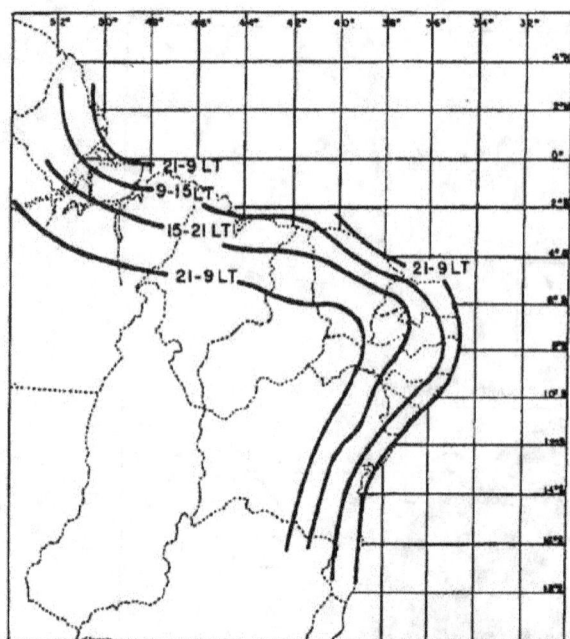

Figure 7-34. Positions of the rainfall maxima over northeast Brazil by observation times (Kousky, 1980).

7.6.2.7 *Mountain Slopes.*

In the equatorial western Andes, maximum rainfall occurs first (1400 to 1900L) *above* the condensation level, and later (1900 to 2400L) *below* the condensation level (Walters et al., 1989). Similar separation has been observed over the Ethiopian highlands (Vojtesak et al., 1990) and on the island of Hawaii. This somewhat surprising sequence can be explained in terms of the local winds. After sunrise, the upper slopes heat first and generate a stronger anabatic wind than the lower slopes. Rapid condensation produces cloud close to or on the ground, and rain does not evaporate before reaching the surface. The cloud and rain, by cooling the air and interrupting insolation, weaken and may even *reverse* the upslope wind. Air stops rising and rain eases off during the afternoon. Katabatic flow from this level now converges with the anabatic wind (possibly reinforced by a sea-breeze component) generated over the still sunny lower slopes. Thus, the center of rising motion is shifted downward and the

rainfall maximum is displaced to the evening. The lowest slopes of the eastern Andes experience an even later maximum, extending to dawn (Johnson, 1976).

7.6.2.8 *Moist Near-Equatorial Regimes.*

Over west Malaysia, extensive weather disturbances are rare and a weak Coriolis force results in less complicated land and sea breezes. In summer (Ramage, 1964), diurnal variation of the wind leads to a cycle of convergence and divergence over the peninsula (Figure 7-35).

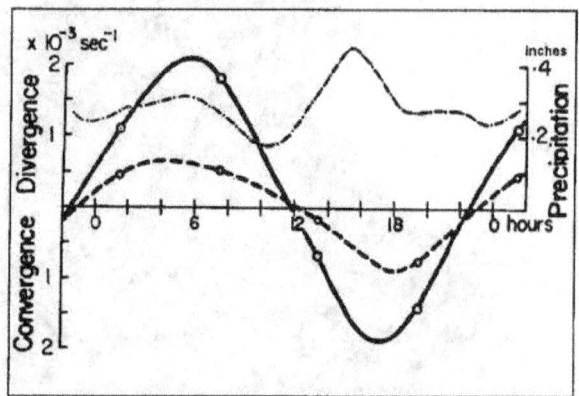

Figure 7-35. Mean horizontal divergence over West Malaysia for 25 days in August 1957 and 1958 derived from coastal pilot balloon observations. Solid lines, 300-meter elevation; dashed lines, 600-meter elevation; dot-dashed lines, 3-hourly running means of rainfall at 11 uniformly distributed stations (Ramage, 1964).

There is no simple relationship to the diurnal variation of rainfall, but it can be explained in terms of interactions between local winds, the prevailing synoptic southwesterly wind and orography. West Malaysia is a roughly elliptical peninsula about the size and shape of Florida. The western and eastern ranges rise to 6,500 feet (2 km) and are oriented along the peninsula. Mean August rainfall reaches about 10 inches (250 mm) along the seaward slopes of both ranges. Figure 7-36 shows the locations of five diurnal rainfall regimes over West Malaysia in August; each will be described in turn.

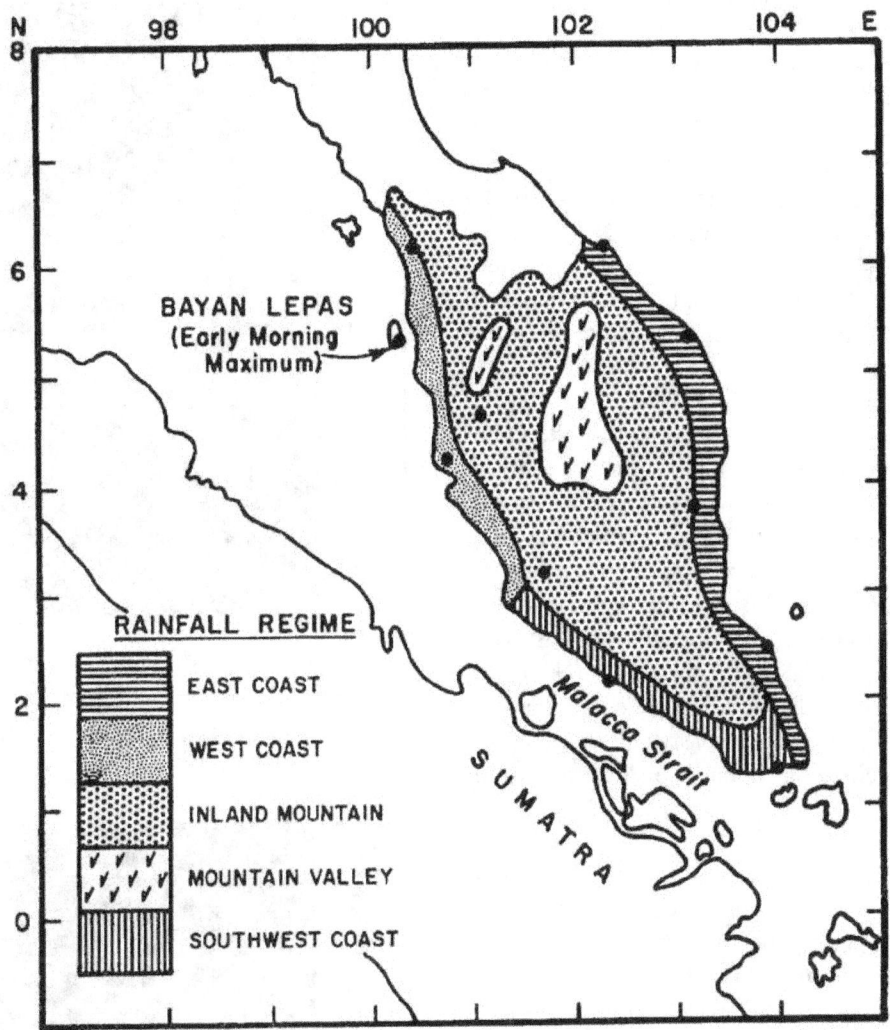

Figure 7-36. Areal extent of diurnal rainfall regimes over West Malaysia in August (adapted from Ramage, 1964).

• *East Coast Regime.* In the east, a mountain slopes regime prevails. Anabatic winds, interacting with the southwest synoptic wind, cause an early afternoon maximum inland. As the upslope flow dies down beneath the rain and cloud, it is replaced by the synoptic wind. Convergence between it and the sea breeze moves eastward to give a late afternoon rainfall maximum along the coast.

• *West Coast Regime.* There is little interaction between the Malaysian and Sumatran land-sea breezes to either intensify or weaken the low-level synoptic-scale convergence. Consequently, diurnal rainfall variation is slight.

• *Inland-Mountain Regime.* During the day, convective heating is reinforced near the coast by the sea-breeze convergence to result in an afternoon rainfall maximum.

• *Mountain-Valley Regime.* No hourly rainfall measurements were made in this area; however, downslope winds converging in the valley at night should result in a nocturnal maximum here.

• *Southwest Coast (Malacca Strait) Regime.* At night, land breezes from west Malaysia and Sumatra converge and cause upward motion and rainfall in the Malacca Strait (the "Sumatra"). Since the prevailing winds are from the southwest, the Sumatra lies near the Malay coast, causing a nighttime and early morning rainfall maximum there. A similar effect is observed at Bayan Lepas (5° N, 100° E) on the east coast of the island of Penang where downslope winds from the island converge with land breezes from the mainland to produce an early morning maximum.

Over the entire Malaysian peninsula, local circulations are so powerful that as much rain falls to leeward (the NE coast) as to windward (the SW coast). In winter, under prevailing northeasterly winds, diurnal variation in the interior remains unchanged, but the patterns are reversed along the coasts; the maximum is in the early morning along the east coast and in the afternoon along the west coast (Nieuwolt, 1968). Similarly, the Pacific coast of Central America, leeward of year-round northeast tradewinds, has a sharp early evening maximum (Condray and Edson, 1989).

In western Colombia, prevailing winds blow onshore throughout the year. Along the coast, most rain falls between midnight and 0400L, and in the afternoon and evening farther inland (West, 1957).

In northern Taiwan, summer flash floods occur mostly at night as downslope flow converges with the prevailing southwesterlies (Kuo and Chen, 1990). At Clark Air Base (15° N, 121° E) on the central Luzon plain, thunderstorms are about 30 times as likely between 1600 and 1700L as between 0700 and 0800L (see Figure 7-30).

These examples illustrate the role of the synoptic wind. By enhancing the land breeze, an offshore synoptic wind keeps coastal skies clear at night; during the afternoon and evening it spreads convective clouds, generated over the mountains, to the coast. An onshore synoptic wind converges along or near the coast with the land breeze and results in a nocturnal sea/mountain breeze keeps convection away from the coast.

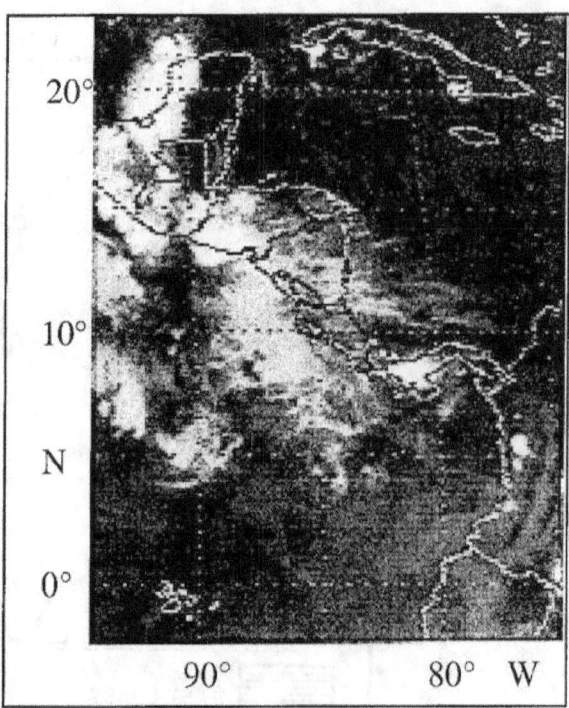

Figure 7-37a. 1900L GOES IR image, Central America, 6 July 1985.

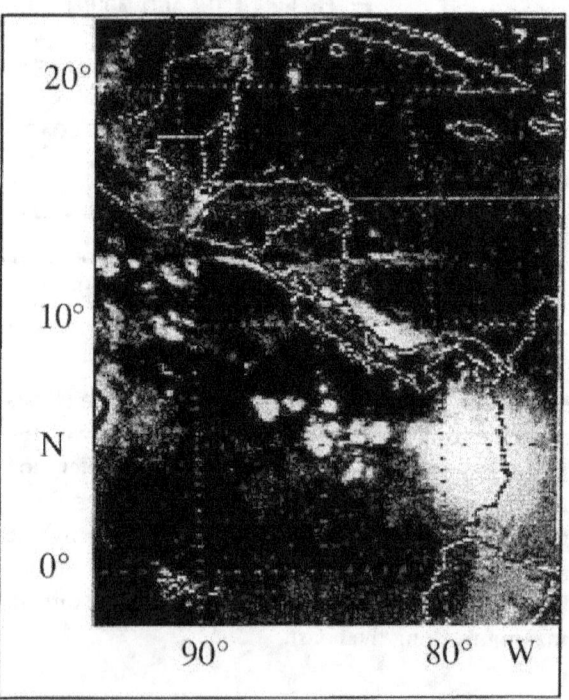

Figure 7-37b. 0700L GOES IR image, Central America, 7 July 1985.

Satellite views of central America for July 1985 show that at 1900L (Figure 7-37a, on the previous page) in the tradewind region north of Panama, mountain convection has spread to the west coast of Yucatan and to the southwest coasts of Guatemala and El Salvador; it is confined inland, however, over western Colombia, where surface westerlies prevail. Twelve hours later, at 0700L (Figure 7-37b), the leeward tradewind coasts are clear and showers are well offshore. Nocturnal convection over the Mosquito Gulf, however, has spread inshore, as has a massive cloud system south of the Gulf of Panama.

The diurnal variation of rainfall is probably more marked over and around the Indonesian island of Sulawesi than anywhere in the world. The island, which comprises four long mountainous peninsulas and three enclosed bays, is shown below in Figure 7-38 as a three-dimensional representation.

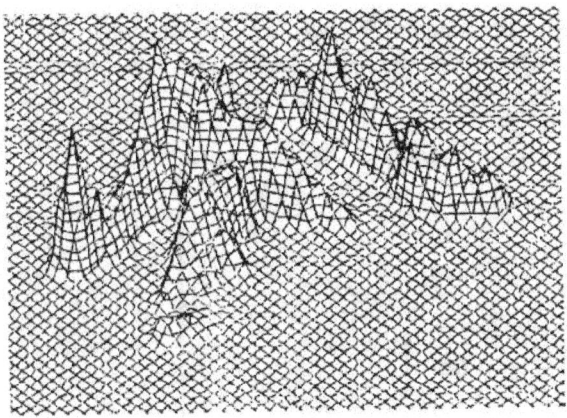

Figure 7-38. Three-dimensional presentation of the island of Sulawesi, as viewed from the southeast.

Figure 7-39 (next page) shows Cb centers over and around Sulawesi plotted for (A) 0500 and (B) 1700L; the data is from geostationary satellite photos taken during 15 days of the wet season. Early morning offshore thunderstorms are found where land breezes converge, but there are few over land, where the land breezes diverge. In the afternoon, prevailing sea breezes converge over land to cause thunderstorms. They diverge over the bays; skies are almost clear.

In Figure 7-40 the per unit area ratio of the number of sea to land thunderstorms reaches a maximum of 9.5 at 0600L; numbers are the same near local noon; and at 1700L the ratio of land to sea thunderstorm numbers (8.5) is greatest. Figure 6-22, showing the mean annual outgoing longwave radiation values over Sulawesi are about 10-15 W m^{-2} lower than over the neighboring water, suggests that daytime convection is more intense and deeper (colder cloud tops) than at night. Although Figures 7-39 and 7-40 were based on December data, such intense local effects should prevail throughout the year.

The satellite images for 20 July 1985 (Figure 7-41, on page 7-33) confirm this. In the early morning, eastern Borneo and most of Sulawesi were clear of convection, which was especially intense over the Macassar Strait and the Gulf of Tomini. Twelve hours later, heavy convection dominated eastern North Borneo and the northern peninsula of Sulawesi, while the seas were generally clear. Over southern Sulawesi mesoscale stability inhibited the diurnal cycle. Murakami (1983) analyzed GMS pictures for December 1978 and January 1979 and found the Sulawesi picture repeated (though less strongly) throughout Indonesia and the surrounding seas, while Keenan et al. (1989) found similar patterns over eastern Indonesia, the Timor Sea, and northern Australia during the summer monsoon.

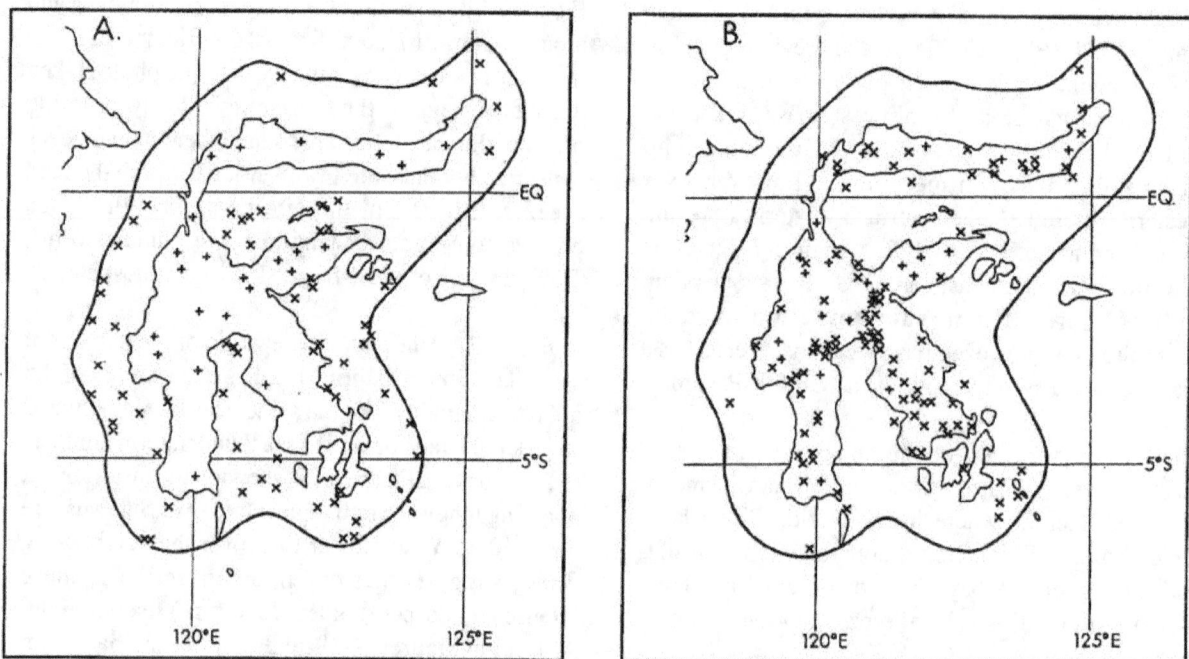

Figure 7-39. Centers of cumulonimbus activity over and around the island of Sulawesi on 1-15 December 1978 are marked by the "x" symbols. (A) 0500L; (B) 1700L. The "+" symbols mark peaks above 1,800 meters. Based on GMS infrared photographs.

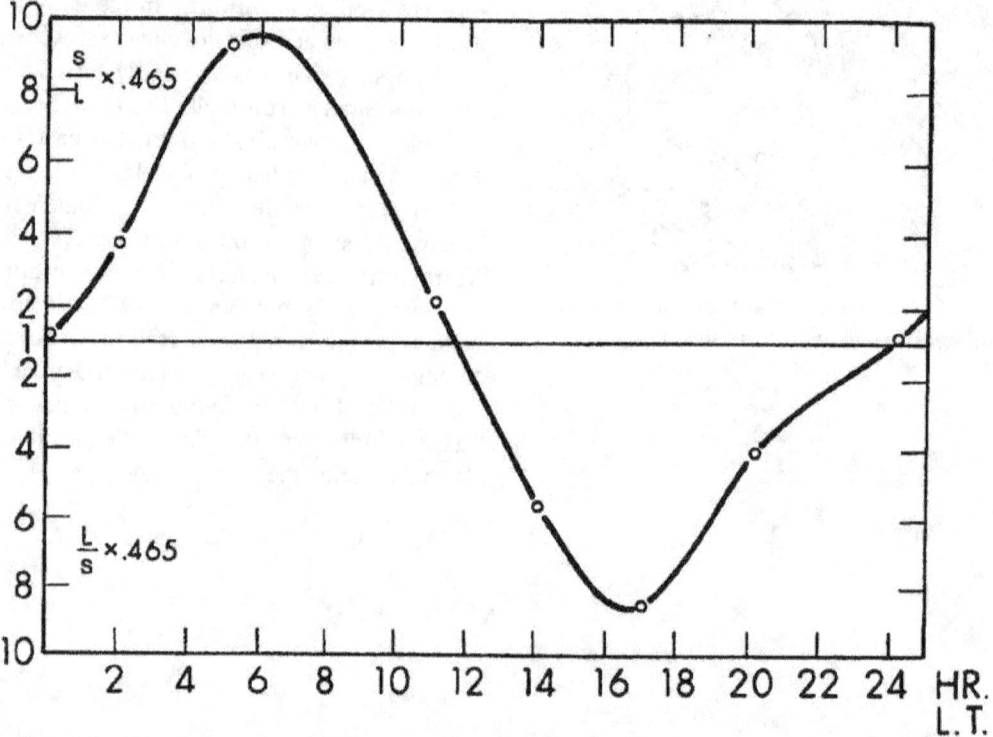

Figure 7-40. Diurnal variation of ratio (normalized by area) between numbers of cumulonimbus centers over Sulawesi and the neighboring sea, based on 3-hourly GMS IR photos. As the curve passes through L/S x 0.465 = 1, the ratio is reversed to emphasize the symmetry of the variation.

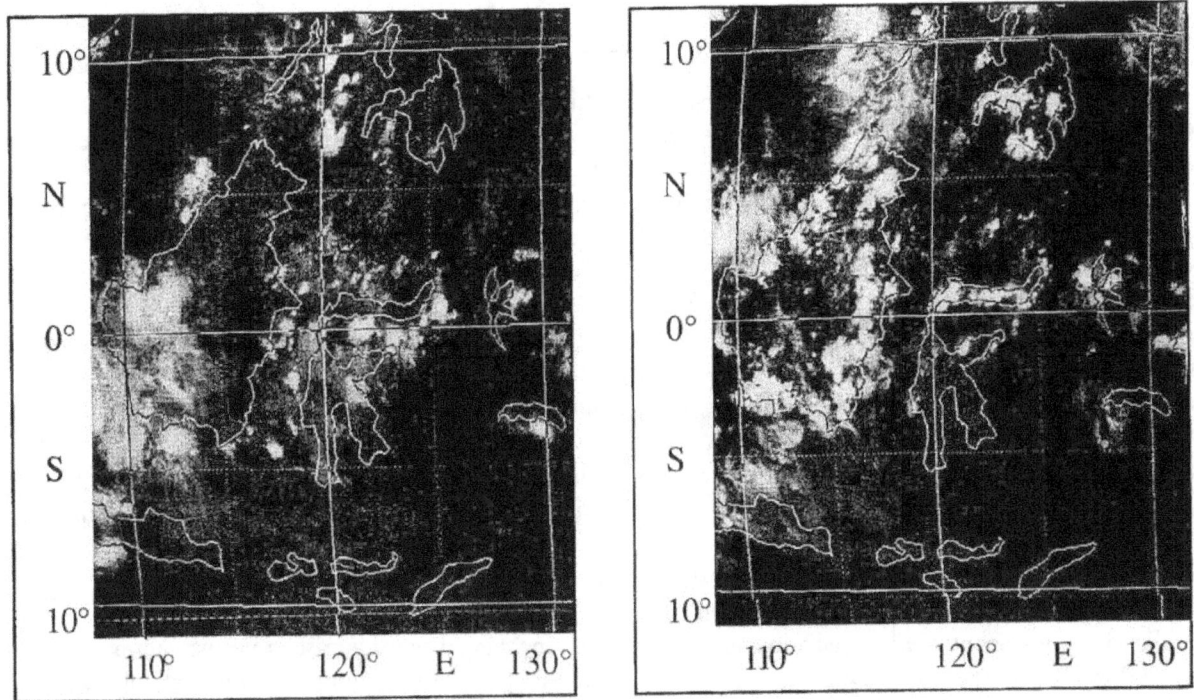

Figure 7-41. GMS infrared images of eastern Borneo and Sulawesi on 20 July 1985. (A) 0500L, (B) 1700L.

7.7 SUMMARY

At sea, diurnal variation of surface temperature is insignificant, but over land in the tropics, forecasters should first predict synoptic scale cloudiness. If scattered clouds are expected, then a strong surface heating-cooling cycle will affect the local winds and vertical mixing, and through them control where and when rising and sinking motions and convection will develop. The Sun will not heat a damp or flooded surface as fast as a dry surface. Thus, recent rain can postpone development of afternoon convection. By initially shielding the surface, early morning stratocumulus or fog can have the same effect; even cirrus can reduce the amplitude of the convective cycle. If the forecaster expects cloudy skies, the surface heating-cooling cycle and local wind developments will be muted; diurnal variation of rainfall, now largely controlled by radiation at cloud top, will favor a nocturnal maximum.

In regions where the mid-troposphere is relatively dry, squall lines usually develop when the diurnal cycle favors convection. However, squall lines move, intensify, and dissipate largely independently of the diurnal cycle, and so tend to blur or shift the normal diurnal variation of rainfall. Since diurnal rainfall patterns in tropical lands stem from many factors, they cannot be simply categorized. Therefore tropical meteorologists should obtain climatological data for the stations of interest. For most first-order tropical stations this is available in USAFETAC standard summaries (see, for example, Edson and Condray, 1989) or from various national meteorological services. Remember, almost all first order stations are located on coasts or in valleys, and that a short distance away, diurnal variations may be quite different. Once the climatology of diurnal rainfall variation is known, physical explanations of the causal mechanisms can be sought.

Lacking rainfall data, forecasters can use geostationary satellite pictures to develop a useful climatology. In the deep tropics, a relatively short period can provide useful information, as was true for Sulawesi. Very high resolution IR images from polar-orbiting satellites can pick up nocturnal fog or stratus, and the first signs of early morning convection.

Chapter 8

TROPICAL SYNOPTIC MODELS

8.1 GENERAL

A synoptic model attempts to specify the idealized or generalized space distribution of meteorological elements such as clouds, rainfall, wind, temperature, or pressure in a distinct type of atmospheric system (Huschke, 1959). It also attempts to condense the general results of extensive empirical investigations so as to describe the essential kinematic, dynamic, and thermodynamic aspects of the particular type of atmospheric system.

The most important mid-latitude synoptic model is the extratropical cyclone (along with its associated air masses and frontal systems) that was developed by the Norwegian school of meteorology during and just after World War I. With minor variations, it has stood the test of time as the basic synoptic model for mid-latitude weather systems, and was little changed by cloud imagery from meteorological satellites.

By the early 19th century, tropical cyclones were known to rotate about a center of low pressure and to travel east to west before recurving parabolically eastward and poleward. During the first half of the 20th century, tropical meteorologists tried to apply mid-latitude models. Out of this came various concepts including tropical and intertropical fronts, and three fronts pinwheeling around a tropical storm. None of these satisfied the mid-latitude test of air-mass discontinuity, even though cold fronts may penetrate below 15° latitude during winter.

Before the advent of satellite meteorology, tropical meteorologists thought that moving systems caused changes in tropical weather. The key to a good forecast was to know what was upstream. After the satellite arrived, weather forecasting did not improve significantly; changes within a weather system may affect a particular location much more than changes produced there by movement of the weather system. Extrapolating intensity changes is harder than extrapolating movement; this has hindered the

development of useful tropical models. Also, models adequate for one region may not be suitable elsewhere; the reader should keep an open mind and allow operational experience to determine which concepts are most useful to analysis and forecasting. As LaSeur (1964a) said in a survey of tropical synoptic models,

"In the literature one encounters such diverse opinions and interpretations, often of essentially the same data, plus considerable speculation and unwarranted generalization or extrapolation of concepts and results of limited validity, that the resulting impression is one of disquieting confusion."

In surveying the origin of tropical cyclones, Gray (1968) made a similar observation, as follows:

"There has been too much qualitative and incomplete reasoning concerning the physical processes of development. General conclusions have been drawn from atypical case studies. Theories of development have been advanced without supporting data or plausible physical substantiation. Numerical model experiments have been made where initial assumptions are not realistic."

Despite this confusion, it is evident that a variety of tropical synoptic systems are associated with areas of disturbed weather. One can have more confidence in a model that has been found to work in more than one region.

This chapter surveys various synoptic models. The organization follows LaSeur's (1964a) division of synoptic models into three broad classes: *waves*, *vortexes*, and *linear disturbances*. The models will also be classified according to *weather*, *divergence* and *vertical motion* (see Figure 6-32).

8.2 WAVES

8.2.1 Waves in the Easterlies. During World War II, Riehl (1948) postulated the easterly wave theory. Frank (1969) proposed a variation called the "inverted V." Neither of these helped the tropical forecaster. Past weather analyses from many tropical stations show regular successions of easterly waves across an entire ocean; however, even careful post-analysis cannot maintain adequate time-continuity for individual waves.

Any report of showers in the tradewinds was used to justify introducing a wave axis near the report, just as "cold fronts" were called on during World War II; for many years, easterly waves conveniently "explained" most tropical weather. Evidence is now lacking for the notion that tropical cyclones form from easterly waves. Among weather-producing systems in the tropics, the "classical" easterly wave, if it exists, is much less common and more regionally restricted than once believed. For example, Simpson (1969) thought that easterly waves are confined to the North Atlantic, whereas Aspliden et al.(1965-67) could find no evidence for them there in satellite or conventional data.

Although Reed and Recker (1971) claimed to find waves in the western Pacific when the region was influenced by a train of typhoons, careful analysts have been unable to identify easterly waves anywhere outside of the North Atlantic. Even here, associated weather lay sometimes east and sometimes west of the wave axis, thus further limiting its usefulness as a synoptic model. For an up-to-date, rather conservative opinion, the reader is referred to Riehl (1979a).

It is now recognized that many waves in the low-level easterlies stem from upper-level cyclones (cold lows); therefore, the National Hurricane Center at Miami uses the more general term "tropical waves" (Simpson et al., 1968) in lieu of "easterly waves." A tropical wave is defined as "a trough, or cyclonic curve maximum, in the tradewind easterlies." The wave may reach maximum amplitude in the low or middle troposphere or may be the reflection from the upper troposphere of a cold low (Figure 8-22) or equatorward extension of a mid-latitude trough.

In addition to these mechanisms, scanty observations have allowed the poleward extensions of low-level, low-latitude vortexes to be mistaken for easterly waves. Since there has always been more data in the subtropical tradewinds than in lower latitudes (especially in the Atlantic), it has often been easier for meteorologists to track the waves than the low-level vortexes with which they may have been associated. Similarly, it could not be asserted that upper cold lows caused low-level tropical waves until wind data from jet aircraft and satellite cloud pictures became available.

8.3 VORTEXES

8.3.1 General. Weather satellite data shows that cyclonic cloud and circulation patterns (hereafter termed "vortexes") are much more common in the tropics than was previously thought. Some weak vortexes develop into tropical storms (maximum wind 34 to 63 knots) or even hurricanes (maximum wind greater than 63 knots), while others eventually dissipate without reaching this intensity. Although all vortexes can be detected in weather satellite pictures, pictures alone do not reveal the tropospheric level of maximum cyclonic vorticity. To determine this, tropical meteorologists need to use all available conventional data and have a thorough knowledge of the regional and seasonal climatology of various types of vortexes.

As Figure 6-32 showed, tropical vortexes can be schematically arranged in terms of vertical motion. In fine weather vortexes (anticyclones, heat lows, and upper-tropospheric cyclones/cold lows), middle-tropospheric air subsides. But in bad weather vortexes (tropical cyclones, monsoon depressions, West African cyclones and mid-tropospheric cyclones), middle-tropospheric air *rises*. Thus, in their associated weather, heat lows and anticyclones are more alike than heat lows and monsoon depressions.

Heat lows are described in 1.2, "Terminology," and tropical cyclones are treated in Chapter 9. Circulations in the other vortexes are strongest and pressure gradients largest some distance above the surface: 3,000-5,000 feet/1-1.5 km for monsoon depressions, 6,500-10,000 feet/2-3 km for west African cyclones, 13,000-16,000 feet/ 4-5 km for mid-tropospheric cyclones, and 33,000- 39,000 feet/10-12 km for cold lows.

Following Sanders (1984), the first three (bad weather) types are classified as "hybrid" cyclones. All form near or equatorward of monsoon/heat troughs; their environments are (for the tropics) strongly baroclinic, with marked easterly shear. Most rain falls equatorward of the center. These cyclones differ significantly from tropical cyclones, which are most intense at the surface, are embedded in a barotropic environment (weak horizontal wind shear in the vertical) and have a much more uniform distribution of rain around the center. Douglas (1987) suggested that hybrid cyclones may resemble pre-hurricane depressions, well before an Eye has formed. They have also been observed in summer over the South China Sea and in the north Australian region (McBride, 1987); Davidson and Holland, 1987; Zhao and Mills, 1991).

8.3.2 Monsoon Depressions. During the Indian summer monsoon the monsoon trough normally lies over land (see Figure 6-19). It occasionally extends southeastward over the northern Bay of Bengal where monsoon depressions may develop in it after lower tropospheric southwesterlies freshen and weather deteriorates over peninsular India (Ramanathan and Ramakrishnan, 1932).

In an extensive discussion, Rao (1976) pointed out that monsoon depressions cause most of India's summer rain. They move between west and northwest to central India before weakening or filling. Most rain falls in the southwest quadrant and is heavy at times.

Table 8-1 gives the average frequency of monsoon depression development. For the season, the range is from 3 to 13. Depressions usually move west-northwest at 2 to 4 knots east of 85° E and about twice as fast to the west. All storms weaken to depression intensity as they cross the coast. As shown by Figure 8-1, the track envelope is very narrow. Monsoon depressions usually last between 3 and 4 days.

TABLE 8-1. Average monthly and seasonal frequency of Bay of Bengal monsoon depressions for the period 1891-1970 (Rao, 1976). Depressions have winds up to 33 knots, storms from 33 to 62 knots.

	June	July	August	September	Total
Depressions	0.9	1.3	1.7	1.8	5.7
Storms	0.4	0.5	0.3	0.4	1.6
TOTAL	1.3	1.8	2	2.2	7.3

Half of the time, monsoon depressions form in a southeastward extension of the monsoon trough over the northern Bay of Bengal. In 15 percent of cases, development starts from a cyclonic circulation between 850 and 500 mb. The remainder originate from weak lows that have moved westward across Burma. Many of these may be remnants of South China Sea typhoons. According to Rao, techniques for forecasting development have not been established. Fortunately, since development is restricted to the north Bay of Bengal, forecasters can concentrate on looking for early signs there.

During summer MONEX, a monsoon depression that lasted from 5 to 8 July 1979 was intensively observed and attempts were made to model it. Sanders (1984) called it a "hybrid" cyclone that depended on baroclinic processes (strong shear in the vertical between surface westerlies and upper tropospheric easterlies) for its development and maintenance. Thus, it lay between tropical cyclones that *avoid* shear and mid-latitudes cyclones that depend entirely on shear. Douglas (1987) suggested a schematic of the vertical circulation of the MONEX depression (Figure 8-2). Upward motion is concentrated just south of the surface low center, most of which is clear of cloud. Douglas ascribed the asymmetrical cloud field to the large shear of wind in the vertical.

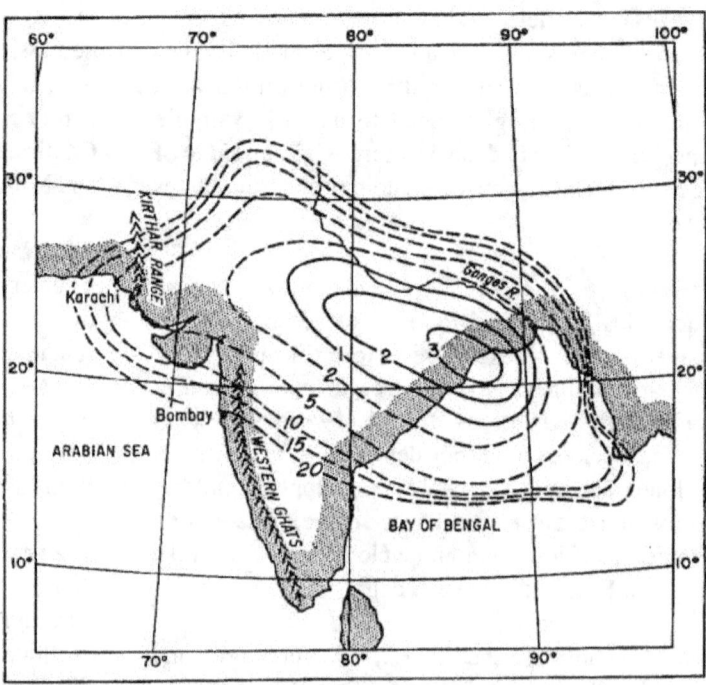

Figure 8-1. Summer. Mean areal frequency of surface monsoon depressions. Isopleths in times per season (full) and in years between occurrences (dashed) (Ramage, 1969).

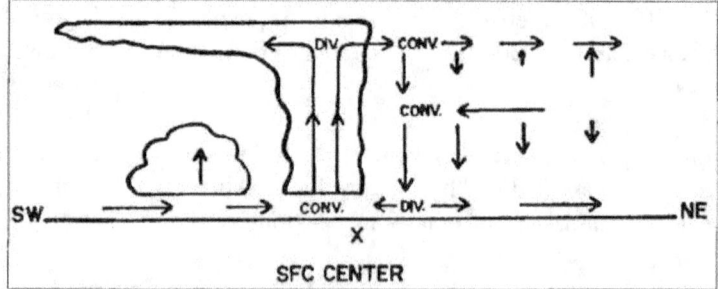

Figure 8-2. Schematic of the vertical circulation across a monsoon depression (Douglas, 1987).

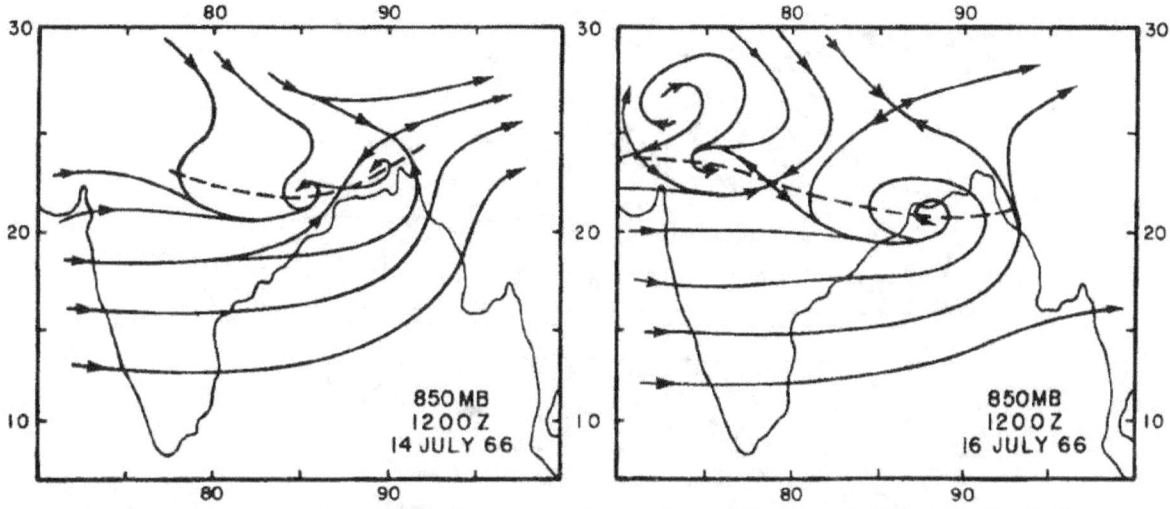

Figure 8-3. Streamline analysis for 850 mb over the Bay of Bengal for 14 and 16 July 1966. The monsoon trough (dashed) migrated south before a depression formed on 17 July (after Sadler, 1967).

Figure 8-3 illustrates the formation of a Monsoon depression. On 14 July 1966, the monsoon trough lay near its normal position over land. Two days later it extended out over the Bay of Bengal. The following day (17 July) a depression of 990 mb formed in the trough and moved inland.

The satellite image in Figure 8-4 shows a Monsoon depression forming over the northwestern Bay of Bengal. Almost all the cloud lies southwest of the center and may have accompanied a surge that moved

northeastward 3 days earlier (see Figure 8-16). After moving west-northwestward, the Monsoon depression weakened and became indistinguishable from the monsoon trough on the 16th.

During the transition months (April-May, October-December) the monsoon trough is farther south. Lacking the strong baroclinicity of summer, vortexes that develop in it differ from Monsoon depressions and often reach tropical storm or hurricane strength before striking land.

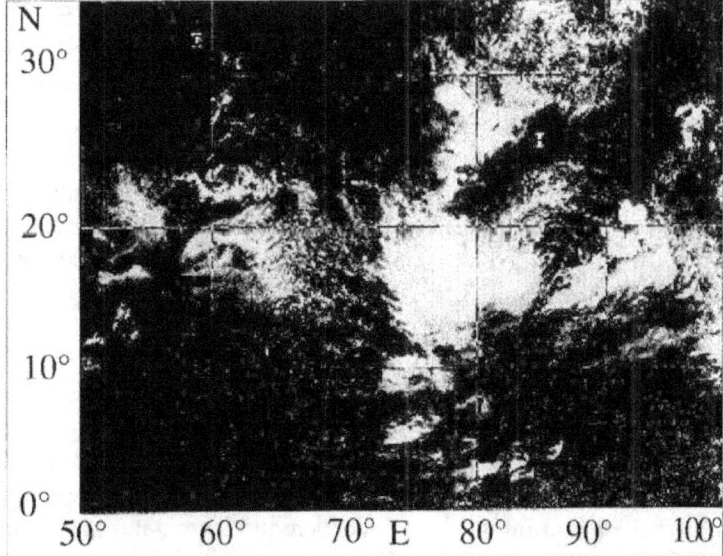

Figure 8-4. DMSP visible image of a Monsoon depression developing over the northern Bay of Bengal on 12 August 1986. A mid-tropospheric cyclone is filling over the northern Arabian Sea.

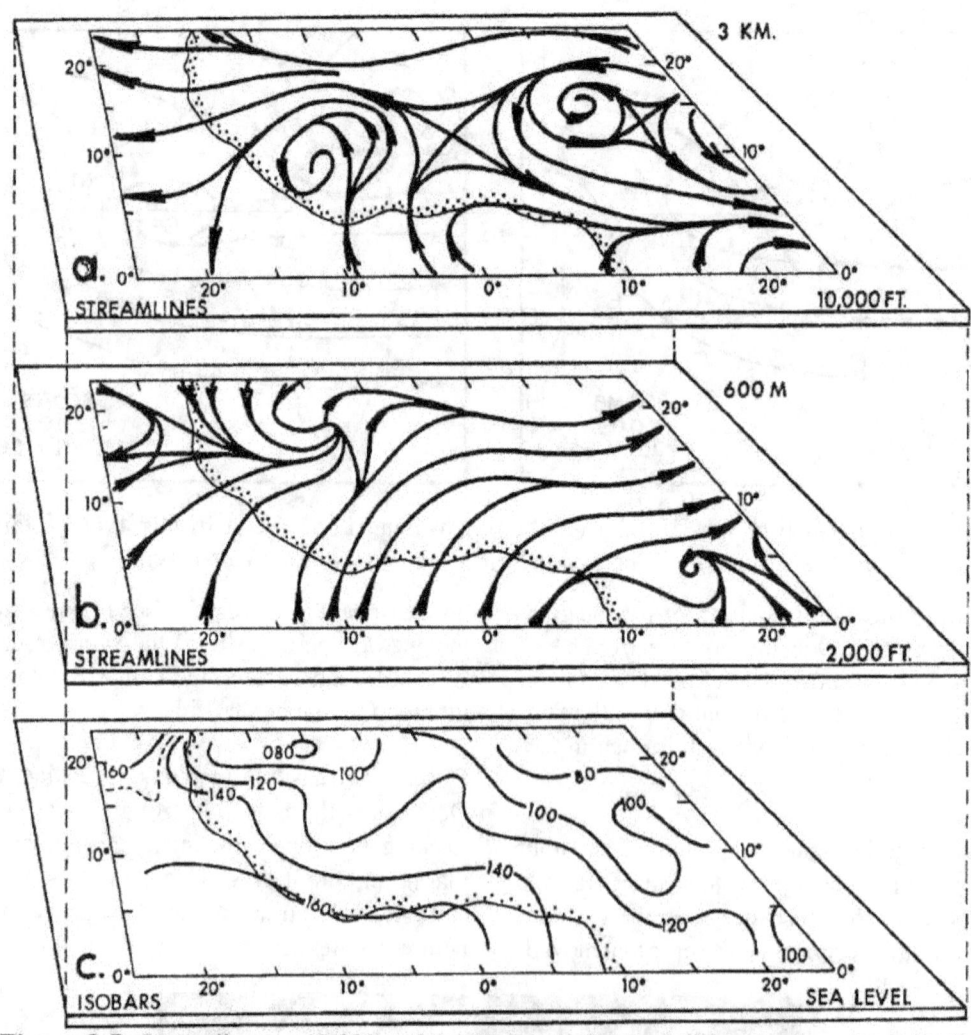

Figure 8-5. Streamlines at 10,000 and 2,000 feet (3 km and 600 meters) and sea-level pressure analysis over west Africa for 1200Z, 24 August 1967 (adapted from Carlson, 1969a).

8.3.3 West African Cyclones. Cyclones may originate over West Africa, but not in the surface heat trough, which is controlled by subsidence. Cyclones develop in a trough 6,500 to 10,000 feet (2 to 3 km) above the surface, about 500 NM (1,000 km) south of the surface trough. As with the mid-tropospheric cyclones discussed later, the heat trough may exert an indirect effect on cyclone development. The cyclones, moving west, generally do not have an associated surface-wind vortex until they near the coast. Carlson (1969a) tracked a number of disturbances across Africa during August and

September 1967. The 700-mb level vortexes appeared to originate over Africa between 10 and 15° E and move westward at 12 to 20 knots. Figure 8-5 shows 10,000- and 2,000-foot (3-km and 600-meter) streamline and sea-level pressure analyses for a day in August 1967. The vortexes become strongest over the west coast of Africa. Perhaps enhanced instability of the mid-tropospheric jet accentuates growth here. Convection is concentrated within and ahead of the vortex trough and, by releasing latent heat of condensation, promotes growth over West Africa.

In a later paper, Carlson (1969b) expanded his synoptic analyses to the period 5 July to 18 October 1968. He used these and the earlier analyses in synthesizing a model (Figure 8-6) of a typical disturbance near the coast during the warm season. Mean cloudiness is generally less than 10 percent near 20° N, but over 30 percent near 5° N, and slightly more along the 10,000-foot (3-km) wave axis through the vortex. In individual disturbances, convection is most intense between 11 and 15° N; farther south, clouds are mainly dense and persistent stratocumulus. As the disturbances approach the coast, a vortex (separate from the string of vortexes often found in the surface heat trough near 20° N) often extends down from 10,000 to 2,000 feet (3 km to 600 meters). The jet instability produced by the sub-Saharan baroclinic zone and the accentuated convection are both greater over Africa than over the ocean.

Leroux (1983) in his extensive treatise on the climate of Africa, failed to find any evidence for the vortexes described by Carlson, except off the west coast; the case studies he reproduced bear him out. He classified the westward-moving weather systems as squall lines; as these pass, surface easterlies replace monsoonal westerlies. In any case, as the vortexes start to cross the Atlantic, they weaken. Reed et al. (1977) composited data as far west as 30° W from 20 August to 23 September 1974 (Phase III of GATE). Their vortex model differed little from Carlson's and confirmed an earlier finding by Burpee (1975).

8.3.4 Mid-Tropospheric Cyclones. In certain tropical regions and seasons, cyclones reach their greatest intensity in the middle troposphere. Two types have been studied: the cool season subtropical cyclones over the eastern North Pacific and North Atlantic, and the Arabian Sea cyclones, found near the west coast of India during the summer monsoon.

8.3.4.1 Subtropical Cyclones. Simpson (1952), in the first detailed study of subtropical cyclones, showed them to be a major circulation feature of the subtropics during the cool season. These cyclones, known as "Kona storms" in the Hawaiian Islands, originate when a closed upper low is cut off from the main

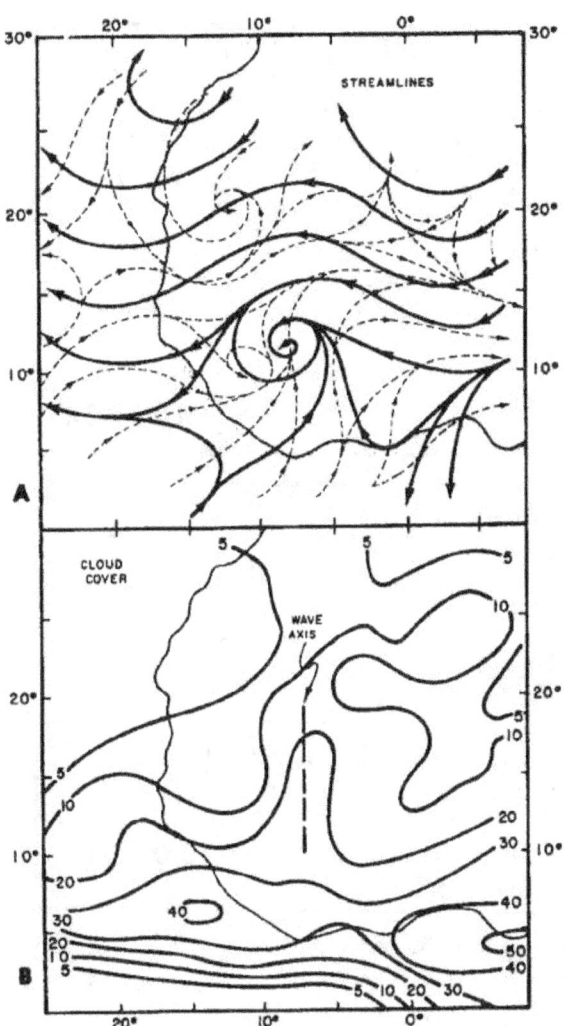

Figure 8-6. Disturbances over west Africa. (A) Schematic flow pattern at 10,000 feet (3 km) (solid lines) and 2,000 feet (600 meters) (dashed lines). (B) Mean percentage cloudiness with respect to wave axis (after Carlson, 1969b).

stream of the upper-level westerlies. Simpson identified two sources for 76 Kona storms that were observed over the North Pacific from November through March for a period of 20 years. Two-thirds developed from occluded cyclones trapped in low latitudes after being blocked by a warm high. The other third resulted from baroclinic development of mid- and upper-tropospheric cut-off lows which gradually extended their circulations to the surface.

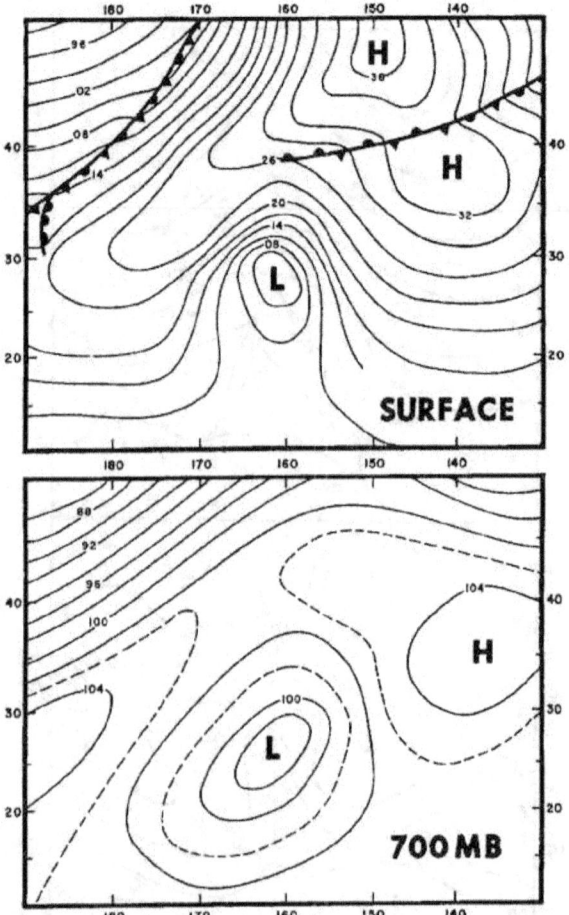

Figure 8-7. Surface and 700 mb pressure-height charts for 7 Jan. 1949 illustrating a subtropical cyclone in the N. Pacific (after Simpson, 1952).

Figure 8-7 shows the surface pressure and 700-mb contour charts for a subtropical cyclone that formed from an occluded frontal wave. In the North Pacific, subtropical cyclones move erratically but are usually confined between 15° N and 35° N and 174° E and 140° W, as shown in Figure 8-8. Simpson also found evidence of subtropical cyclones in the North Atlantic (15° N to 35° N and 30° W to 60° W) most often from November through January (16 during a 10-year period). Simpson also developed a composite model of Pacific storms. Maximum wind and rainfall occur in a quadrant 300 to 800 km east from the center, depending on the size of the storm. Except for the few subtropical cyclones that develop tropical cyclone characteristics, small pressure gradients and light winds prevail in the central region (within 160 km of the center). In their final stages, subtropical cyclones generally move to where frontogenesis is favored and

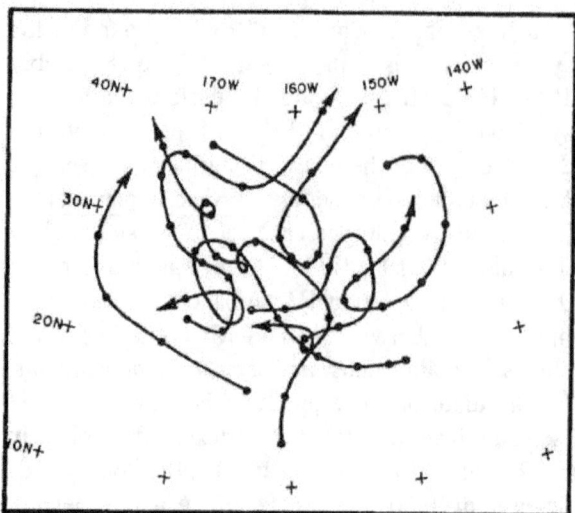

Figure 8-8. Typical tracks of North Pacific subtropical cyclones (adapted from Simpson, 1952).

generate a tropical wave. Occasionally, however, latent heat of condensation, released near the center, causes them to acquire the warm-core properties of tropical cyclones before they recurve and are absorbed into the polar westerlies.

Ramage (1962), using conventional data and early TIROS satellite pictures, also studied the wind and weather patterns of subtropical cyclones. Figure 8-9 composites the surface wind distribution around a 4-5 April subtropical cyclone.

Figure 8-9. Composite surface winds (knots) around a North Pacific subtropical cyclone on 4-5 April 1960 (adapted from Ramage, 1962).

The strongest winds, largest pressure gradients and greatest convergence occurred in the mid-troposphere between 400 and 600 mb; however, there wasn't enough upper-air data for a composite to be constructed. Figure 8-10 is a schematic radial cross section of the clouds, weather, and vertical motion in this storm. The tropopause and the low-level subsidence inversion are also shown. Horizontal convergence is greatest near 600 mb, resulting in general upward motion above this level and sinking beneath, except for the near-surface layer. The cyclone eye appears to be rather large, with scattered clouds and little weather. Significant weather is confined to an annulus from 85 to 270 NM (160 to 500 km) from

the center. This model differs from Simpson's, which locates most rain east of the center. This may have resulted from Simpson not distinguishing between true subtropical cyclones which are completely cut off from the polar westerlies, and large-amplitude troughs in the polar westerlies. In the latter, bad weather is generally concentrated east of the trough axis. Subtropical cyclones may sometimes last several weeks. Because of their energy-producing character and the insignificant effect of surface friction, subtropical cyclones do not decay in situ but are eventually absorbed by large-amplitude troughs in the polar westerlies.

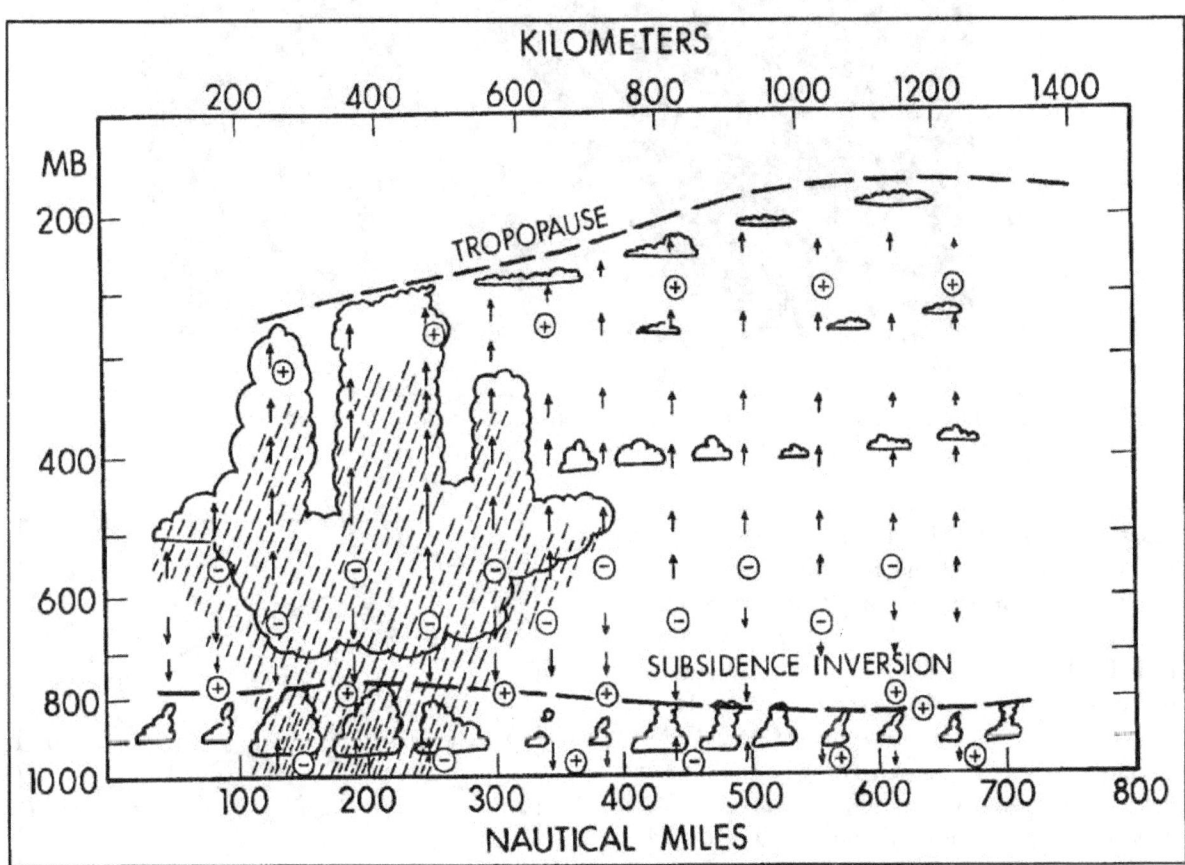

Figure 8-10. Schematic radial cross section of clouds, weather, and vertical motion in a symmetrical subtropical cyclone. "+" denotes horizontal divergence; "-" denotes convergence (from Ramage, 1962).

Figure 8-11 is a GOES visible image of a North Pacific subtropical cyclone that extended to the surface. On the previous day it had been cut off in the southern end of a sharp trough in the upper-tropospheric westerlies. Three days later, it moved rapidly northeastward ahead of a new trough approaching from the west.

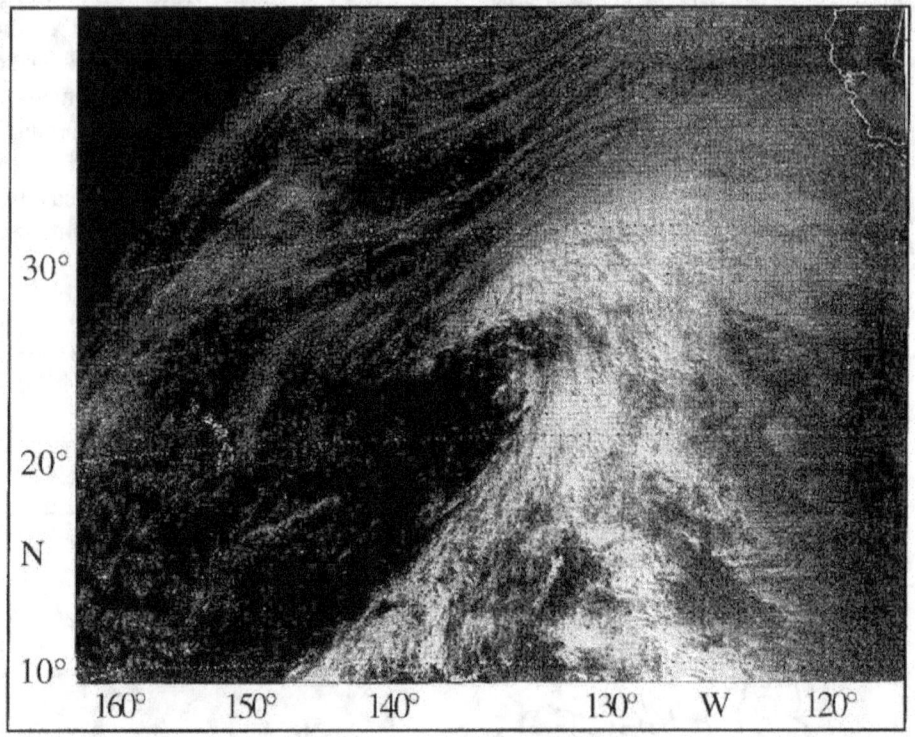

Figure 8-11. GOES visible image of a subtropical cyclone centered near 23° N, 139° W, on 1 January 1986.

8.3.4.2 Arabian Sea Cyclones. These systems develop between 700 and 500 mb in the monsoon trough lying across the northeast Arabian Sea; they are the major producers of rain along the west coast of India. Figure 8-12 shows the mean July position of the monsoon trough in the surface layer and at 500 mb as determined from the resultant-wind field. Note the large slope of the trough over western India and Pakistan compared to eastern India, where Monsoon depressions predominate. Above 400 mb, steady easterlies prevail.

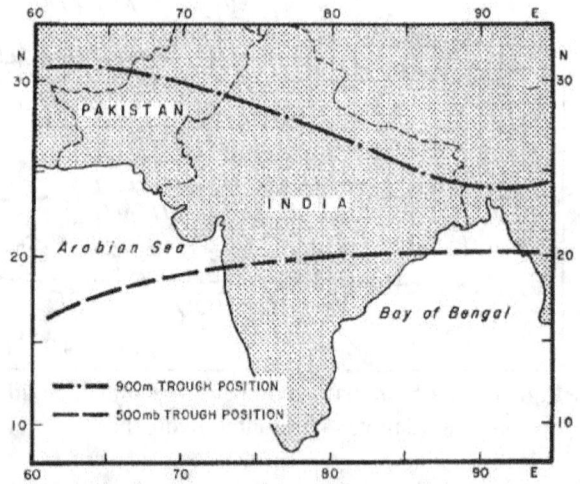

Figure 8-12. Mean July positions of the monsoon trough over India at 3,000 feet and 500 mb (adapted from Miller and Keshavmurthy, 1968).

Figure 8-13 is a meridional cross section of July resultant zonal winds along western India. A subtropical ridge line overlies the surface monsoon trough near 30° N. The trough slopes from there to the 500-mb level near 20° N. Baroclinicity is greatest south of 20° N where the tropical easterly jet overlies strong low-level westerlies.

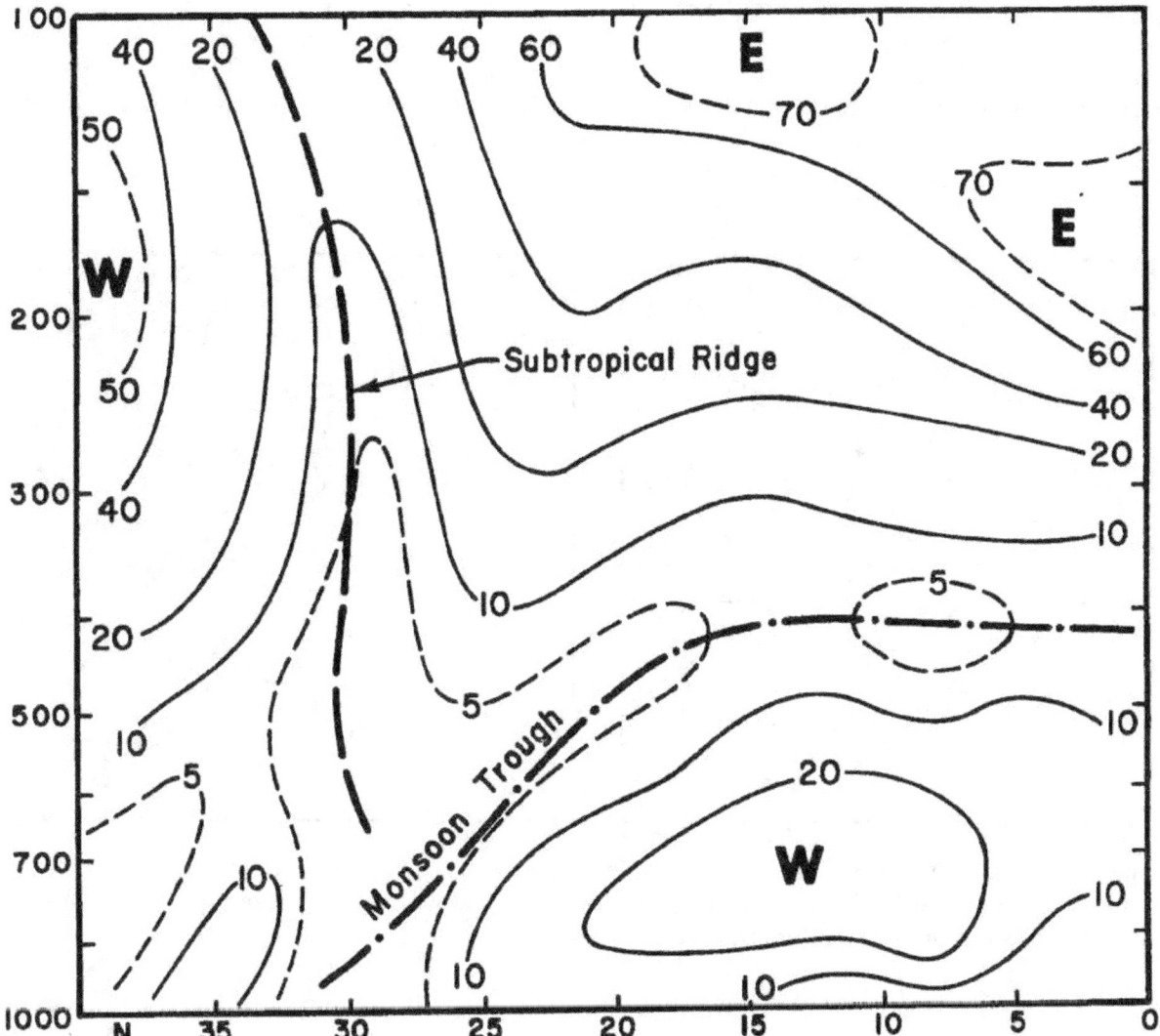

Figure 8-13. Meridional cross section along western India of the July mean zonal winds (knots) (adapted from Miller and Keshavamurthy, 1968).

Miller and Keshavamurthy (1968) studied an Arabian Sea mid-tropospheric cyclone that was probed by research aircraft between 26 June and 10 July 1963. Composite analyses of wind, pressure, temperature, moisture, and weather were prepared using the daily cyclone center at 500 mb as the origin of a moveable coordinate system. Figure 8-14 shows the composite kinematic analyses for the near-surface and 600-mb levels. A weak trough near the coast is the only evidence of a surface disturbance, while at 600 mb a cyclone is well developed. The composite temperature fields show the cyclone to be colder than the environment at 700 mb and warmer at 500 mb. This resembles the thermal structure of subtropical cyclones.

Figures 8-15a and b are composites of the vertical-motion and cloud distribution associated with the cyclone. The cumulonimbus symbols show that convection was best developed slightly west and well south of the cyclone center.

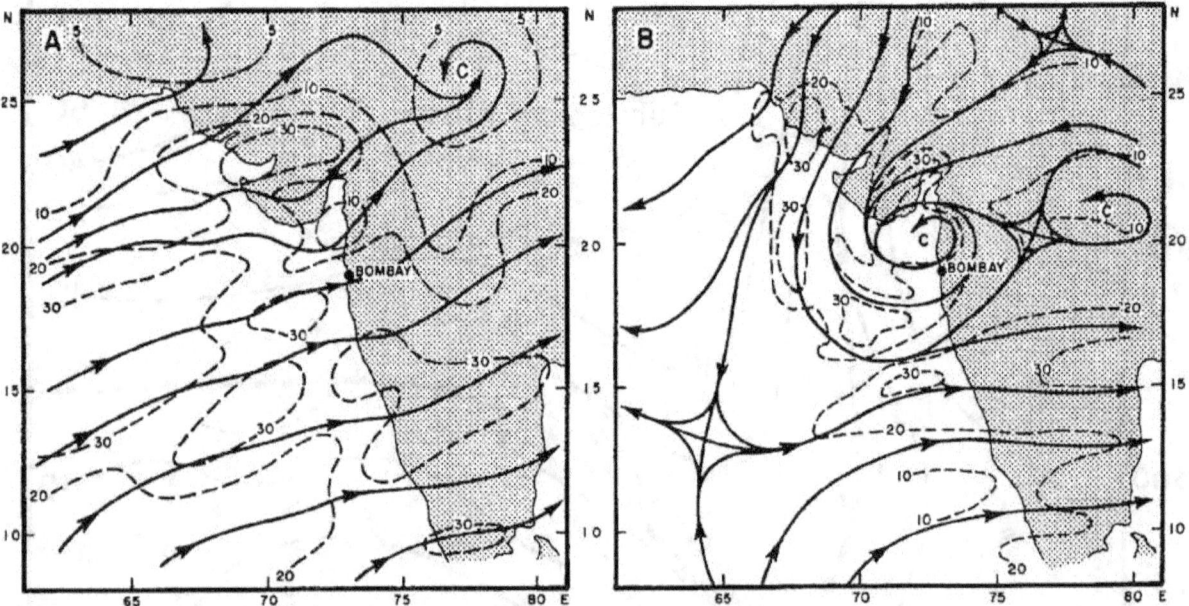

Figure 8-14. Composite kinematic analyses (knots) for 1-10 July 1963. (A) shows a near-surface layer (500 to 900 meters). (B) is the 600-mb level, showing a well-developed mid-tropospheric cyclone over western India (adapted from Miller and Keshavamurthy, 1968).

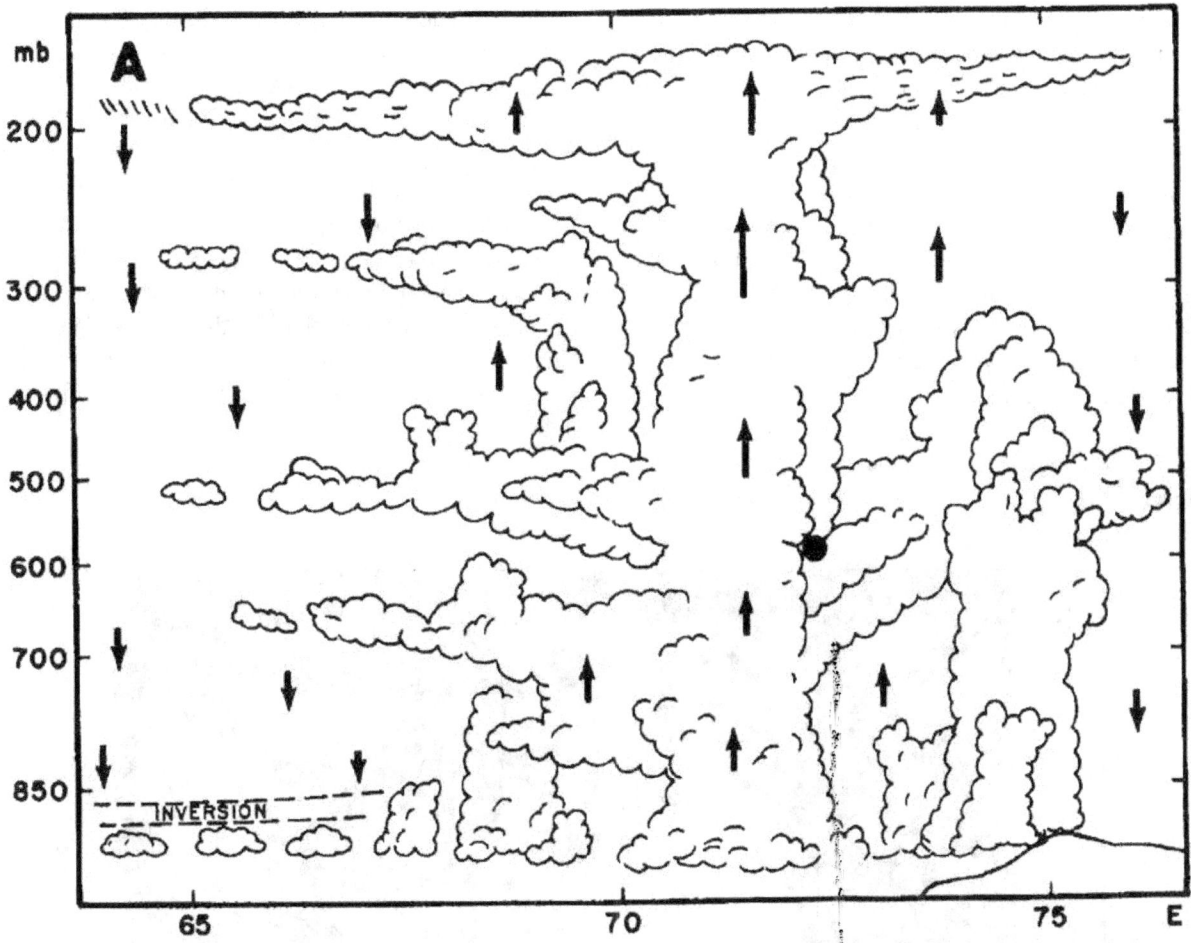

Figure 8-15a. Vertical east-west cross-section of the mid-tropospheric cyclone shown in Figure 8-14. The large dot locates the cyclone center at 600 mb. The arrows depict relative vertical motion computed from composite kinematic analyses.

Figure 8-15b. Horizontal cloudiness and rainfall distribution of the cyclone shown in Figure 8-14. Hatching delineates rainfall of more than 4 cm day[-1] and the shaded areas are broken to overcast middle or high cloud (adapted from Miller and Keshavamurthy, 1968).

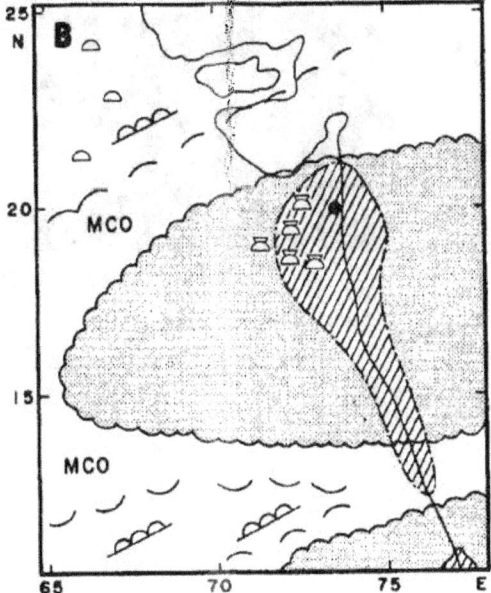

199

Ramage (1966), applying Petterssen's (1956) views on large-scale thermodynamic influences, concluded that export of excess cyclonic vorticity generated in the heat low (monsoon trough) to the north could develop and maintain the mid-tropospheric cyclone. Subsequently, numerical diagnostic models were applied to Miller and Keshavamurthy's data (see for example Carr (1977) and Brode and Mak (1978). Although Ramage's hypothesis could not be confirmed, the models suggested that instability in baroclinic flow is a prerequisite for mid-tropospheric cyclone development.

Forecasters should be aware that these cyclones often develop after a Monsoon depression has moved inland to the east and started to fill. The heat low may also have intensified and increased export of cyclonic vorticity to the south. Lack of aircraft data in the middle troposphere often handicaps analysis.

Figure 8-16 is a DMSP visible image of a mid-tropospheric cyclone over the northern Arabian Sea on 8 August 1986. The cyclone moved slowly westward and became very weak by the 12th (see Figure 8-4). Note orographic clouds along the Western Ghats and relatively little cloud along the monsoon trough over central India. The cloud mass south of India moved northeast during the next day. By the 12th, it lay southwest of a developing monsoon depression.

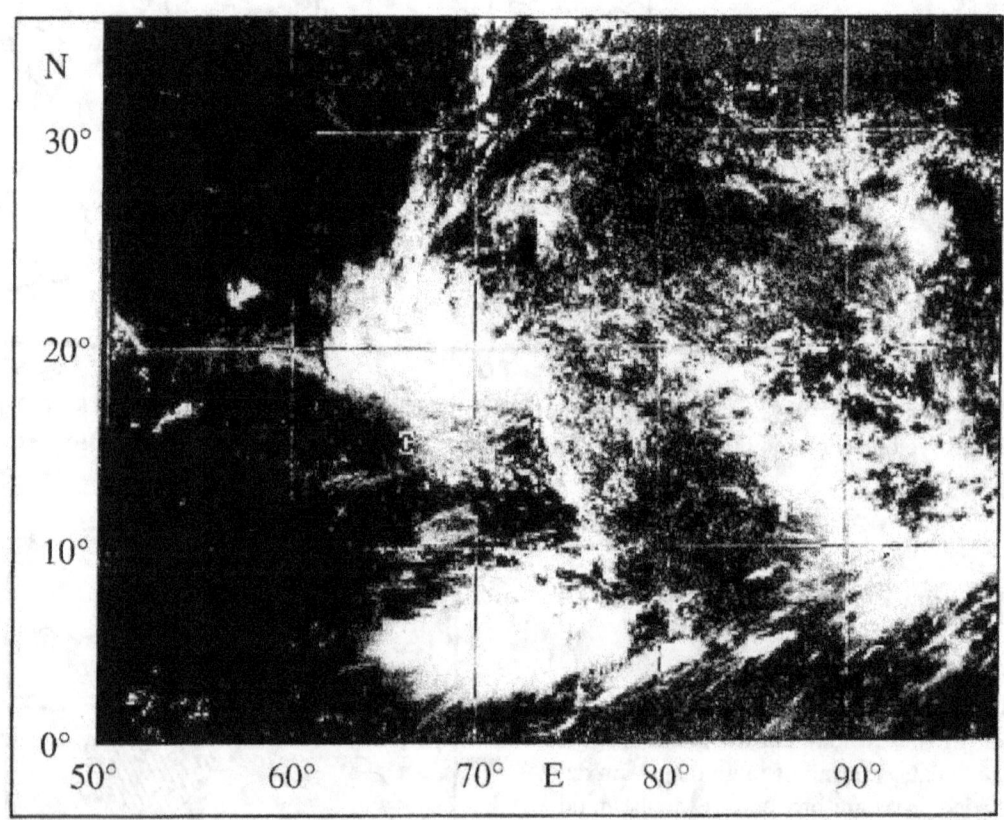

Figure 8-16. DMSP visible image of a mid-tropospheric cyclone over the northern Arabian Sea on 8 August 1986.

If the circulation of a hybrid cyclone extends to the surface over the ocean, the surface wind distribution can mislead a forecaster who has a tropical cyclone in mind. As Figure 8-9 showed, a large central region of light winds may be surrounded by an annulus of stronger winds. Applying a tropical cyclone model to ship observations or SSM/I estimates in the strong wind zone, a forecaster may overestimate wind speed nearer the cyclone center. Although the satellite image in Figure 8-16 does not show an eye, the cloud can sometimes look like a tropical cyclone. Reconnaissance aircraft are rarely available to settle the question, and no rules of thumb have been developed. Thus, hybrid cyclones have been mistaken for tropical cyclones, especially when the only means of detection and identification are satellite images. Forecasters should err on the safe side, but be prepared for a few surprises.

Hybrid cyclones are the predominant vortexes equatorward of vigorous, quasi-stationary heat or monsoon troughs. Large meridional temperature gradients favor development. The gradient over the northern Arabian Sea in summer is usually so large that the cyclonic circulation may not extend to the surface. Hybrid cyclones have also been observed in summer over the South China Sea and in the north Australian region (McBride, 1987); Davidson and Holland, 1987; Zhao and Mills, 1991). Rare flights by reconnaissance aircraft suggest that hybrid cyclones develop or are maintained north of cold upwelled equatorial waters over the eastern North Atlantic and eastern North Pacific. In these regions, some tropical cyclones may have started out as hybrid cyclones.

In other parts of the tropics, where the atmosphere is usually barotropic, baroclinicity may temporarily increase and favor hybrid cyclogenesis. For example, a typhoon filling over South China, by greatly warming the air, increases baroclinicity to the south and enhances the chance of a hybrid cyclone developing over the South China Sea.

The monsoon and West African depressions, presumably influenced by the stationary monsoon/low-latitude trough, move rather steadily westward. Elsewhere, hybrid cyclones move slowly and irregularly. The best prediction for them might be "no movement."

8.3.5 Upper-Tropospheric Cyclones. Since Palmer (1951b) first reported tropical upper-tropospheric cyclones (cold lows), they have puzzled meteorologists. We know that they are most intense between 200 and 300 mb and that their tangential winds may exceed 50 knots. Since the lows are colder than the environment, they weaken downward. According to Frank (1970) 60 percent do not reach 700 mb, and most of these ("dry lows") have little direct effect on surface weather. As shown in Figure 8-17, a cold low northwest of Hawaii on 8 July 1979 was enclosed by a cirrus annulus. The cyclonic circulation neither reached nor affected the lower troposphere, where undisturbed bands of low cloud extended across the center.

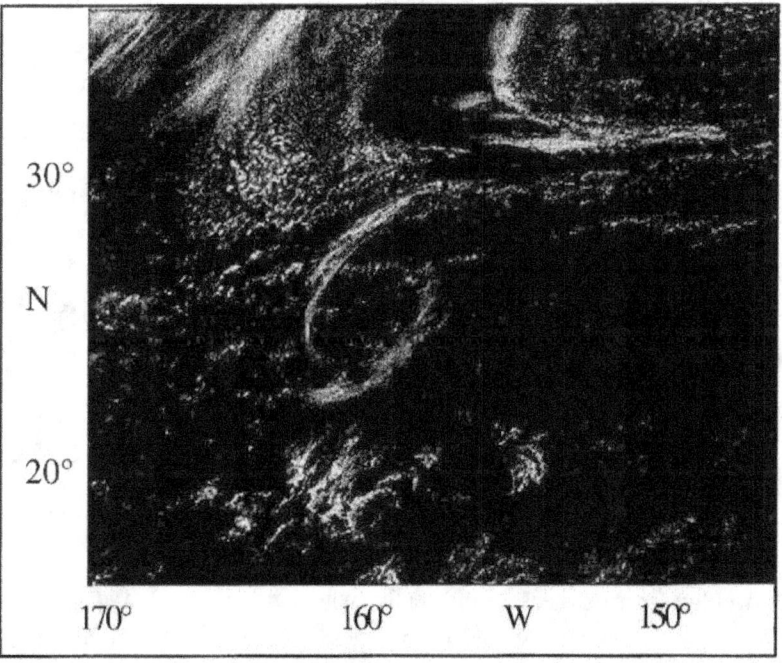

Figure 8-17. GOES visible image of a "dry" upper-tropospheric (cold) cyclone over the central North Pacific on 8 July 1979.

Thin cirrus, on the periphery of most dry cold lows, allows them to be readily recognized in satellite images. Occasionally, deep, short-lived cumulonimbus develops in the center of the low. Once convection is triggered near the surface, weak vertical wind shear and potential instability favor development. However, dry air entraining into a cloud shortens its life.

The cloud systems of cold lows that reach the surface ("wet lows") are composited in Figure 8-18. While the central core is relatively cloud-free, a cloudy ring (heavy dashed line) surrounds the center at a radius of about 135 NM (250 km). The eastern part of the ring is twice as cloudy as the western part. In a Caribbean cold low studied by Carlson (1967), rain was confined south and east of the center, in agreement with Frank's composite.

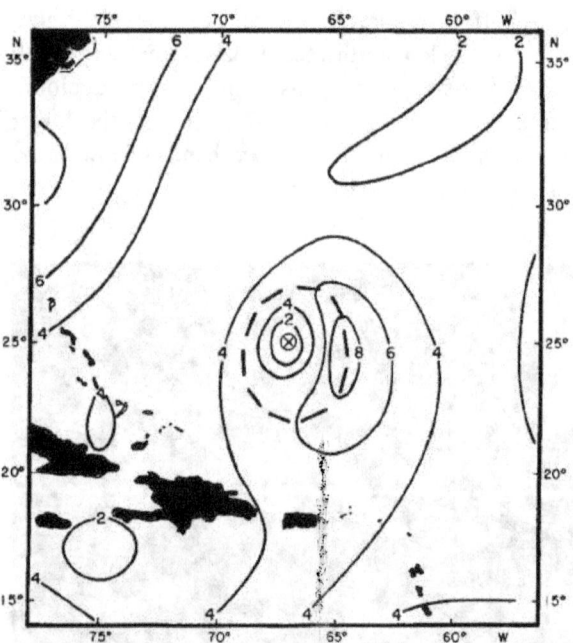

Figure 8-18. Composite mean cloudiness (tenths) derived from 13 cases of "wet" upper-tropospheric (cold) cyclones in the N Atlantic, 1961-66 (adapted from Frank, 1970).

In wet lows, air rises east of the center (where cloud and rain are concentrated) and sinks west of the center, similar to the distribution accompanying a trough in the upper-tropospheric westerlies (Carlson, 1967). Mostly, layers of middle and high clouds predominate; these tend to inhibit convection, which is usually confined to the southeast quadrant. The dry subsiding air in the cold-low center is sharply delineated in weather satellite moisture channel pictures.

Figure 8-19 shows a cold low that had moved slowly westward across Hawaii. Deep convection in the eastern semicircle gave rainfalls of up to 8 inches (200 mm) over eastern Hawaii and eastern Maui. After calculating kinetic energy production for both dry and wet lows, Frank concluded that both were "direct" systems, in which cold interior air sank and warm peripheral air rose. Cold lows, usually unconnected to higher latitude energy sources, can persist for several days. Especially for dry lows, where condensation heating is absent, the sinking, warming cold air must be radiationally cooled more than the environment if the horizontal temperature and pressure gradients are to be maintained. The drier sinking core should cool more, but Riehl (1979a) doubted if that were enough to ensure that a cold low persists.

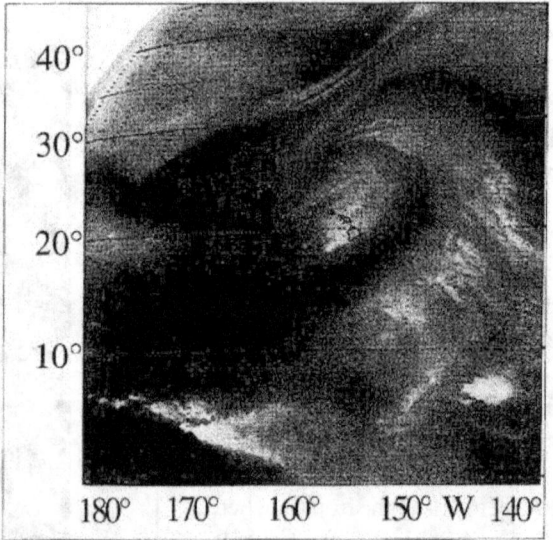

Figure 8-19. GOES moisture channel image of a cold low centered west of Hawaii on 24 May 1984. Note moist air east of the center and dry air in the rest of the circulation.

The tropical upper-tropospheric trough (TUTT) is a well-known summer feature over all oceans except the North Indian Ocean. The TUTT incorporates a series of cold lows, appears in response to a heating continent, and is most intense just after mid-summer. It is usually oriented WSW-ENE in the NH and WNW-ESE in the SH.

As a short-wave trough in the upper tropospheric westerlies passes poleward of the TUTT, a jet-like surge of cold air west of the trough sweeps westward and equatorward around the TUTT, where the increase in cyclonic vorticity causes a cold low to develop and may result in an apparent split in the TUTT. On the eastward side of the low, a smooth, sharp-edged sheet of cirrostratus develops, while deep convection, with

frequent thunderstorms, occurs beneath and east of the cirrostratus. In the NH, the TUTT starts developing in late April and is a persistent feature by June. It is most intense in August (Figure 8-20) and becomes rare by mid-November.

Sadler (1976a) has made many detailed studies of cold lows and the TUTT over the Pacific. From them, he developed a model showing interaction with low-level flow. Divergence 100 to 250 NM (200 to 500 km) east of an upper cold-low center induces a surface wave-like disturbance that may be mistaken for an easterly wave. On rare occasions, the disturbance becomes a typhoon. Figure 8-21 (next page) is a three-stage schematic of the model.

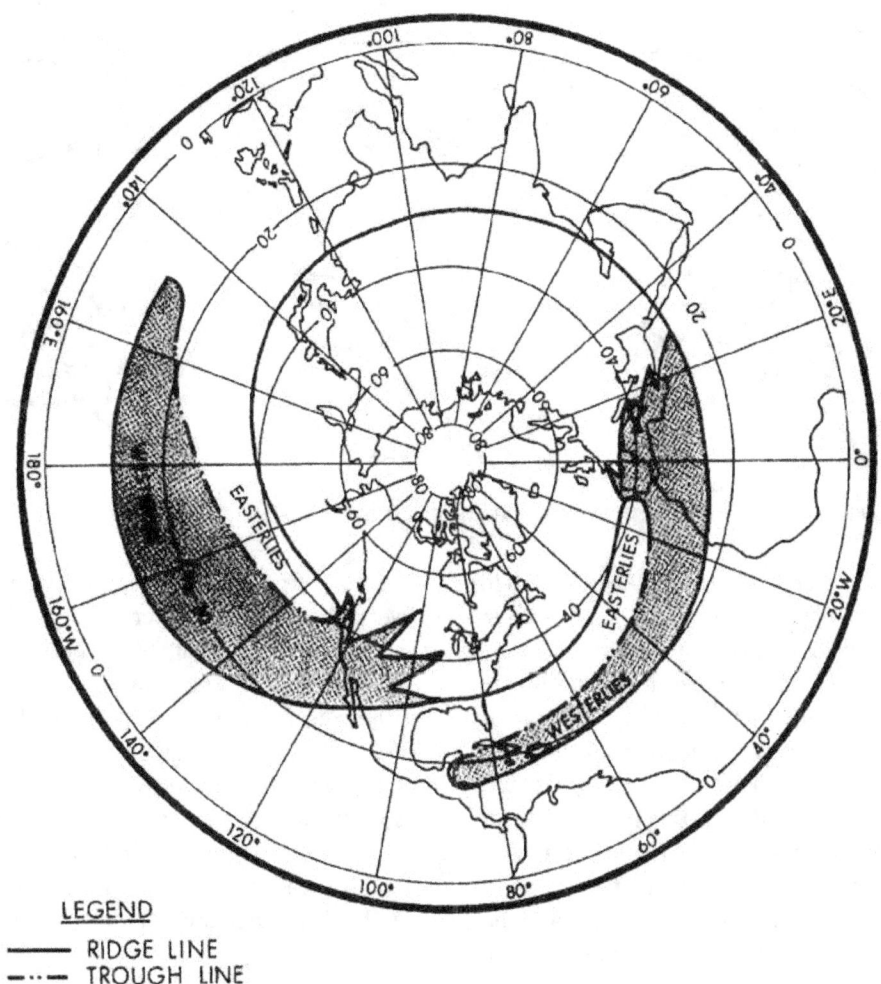

LEGEND
——— RIDGE LINE
—·—·— TROUGH LINE

Figure 8-20. Mean locations of 200-mb ridges and troughs over the Northern Hemisphere in August (after Sadler, 1964).

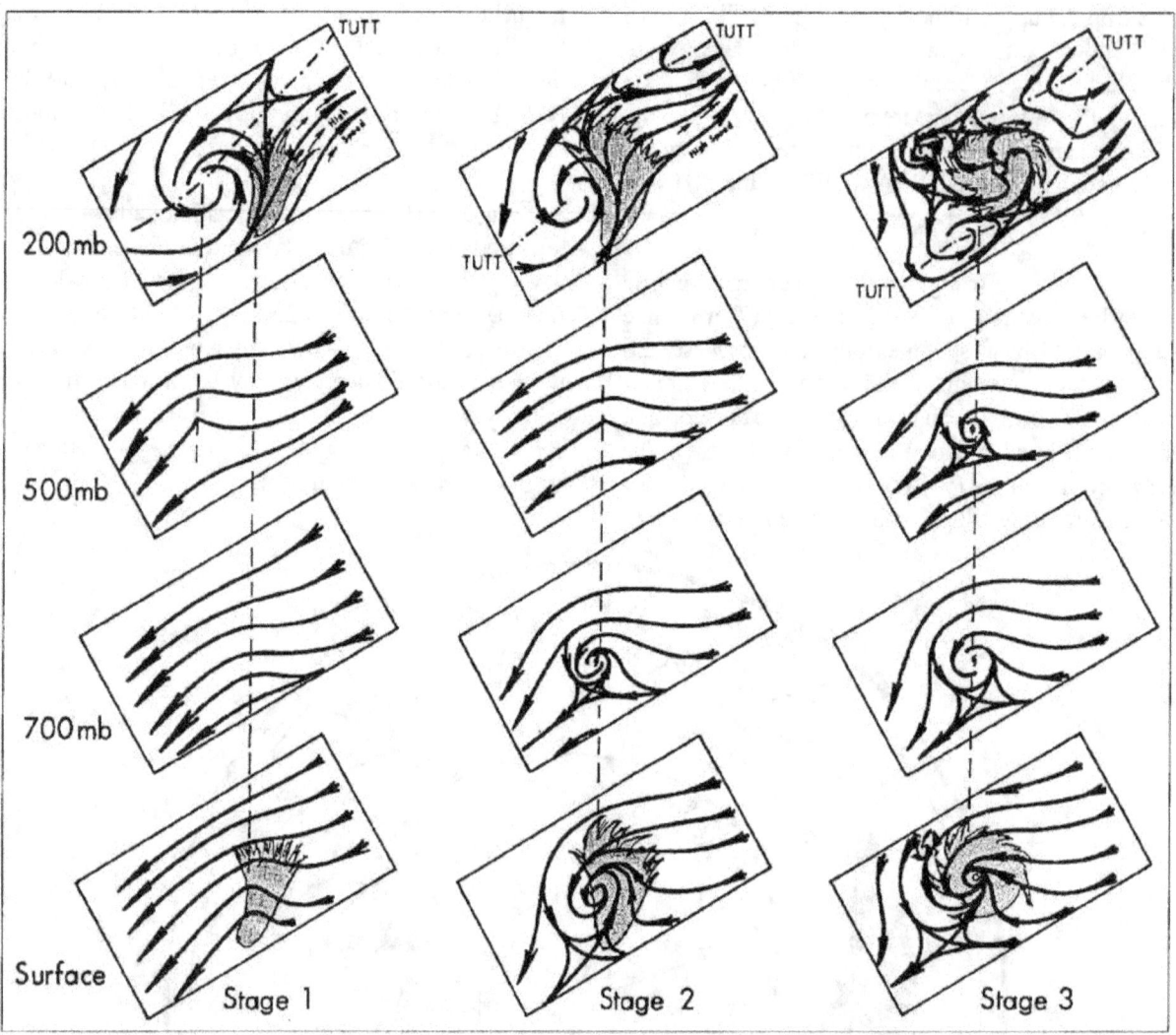

Figure 8-21. Schematic model of a tropical cyclone initiated by an upper-tropospheric low (Sadler, 1976a).

Stage 1. In the upper cell, 500 to 1,000 NM (1,000 to 2,000 km) across, winds reach 40 knots or more in all quadrants. Flow diverges east and southeast of the center, generally accelerating downstream toward the strong southwesterlies of the TUTT to the east. Beneath it, the divergence induces a trough that extends up to 700 mb. This is where organized convection is first observed. Locating the clouds with respect to the low-level trough is difficult because, at this stage, the trough is weak. Whether the extensive convection, initiated from aloft, develops prior to the surface trough cannot be determined from available observations.

Stage 2. Organized convection beneath the upper divergence east of the upper cell expands significantly. This in turn alters the upper-level flow as the heating due to condensation builds a sharp ridge east-northeast of the cell, forces the cell northward, and splits the TUTT. At the same time, the cold low shrinks and no longer penetrates down to the 500-mb level. In low levels, the wave develops into a weak depression with a closed vortex below 500 mb. The main convective cloud lies east of the low-level vortex. Other clouds may be distributed around the circulation.

Stage 3. Upper-tropospheric flow becomes more complex as increased convection intensifies the ridge or even builds an anticyclone whose outflow interacts with and further distorts the TUTT. In the western end of the northern branch a small cyclone often forms, with its own cloud system. The model depicts it only for stage 3, but it may occur in stage 2. In the lower layers, the depression becomes a tropical storm extending up through 500 mb and capped by an anticyclone. The upper outflow is not impeded to the east where it merges with the larger scale westerlies south of the TUTT, and so the cirrus shield is skewed toward the east. North of about 15° N, storms that are triggered by cold lows in the upper trough cannot tap an upper-level outflow channel to the south—into large-scale easterlies. Perhaps this is why these storms seldom intensify into strong typhoons.

Between 27 and 31 July 1970, an upper cold low moved westward across Midway Island (28° N, 177° W). In the time-altitude section (Figure 8-22) (Sadler, 1976a), easterlies prevailed below 700 mb. The upper low centered near 250 mb changed little in intensity. It was 4 to 5° C colder than its surroundings. Beneath it, the weather was fine and the air dry. Humidity rose after its passage, as an induced trough in the tradewinds (135 NM (250 km) east of the cold low center) approached Midway, where showers developed. The low-level trough was capped by an anticyclone at about 250 mb. As the trough passed Midway, winds veered from east-northeast to southeast. This cross section fits stage 1 of Figure 8-21. Lacking upper-tropospheric data, a forecaster could easily label the trough in the tradewinds an easterly wave.

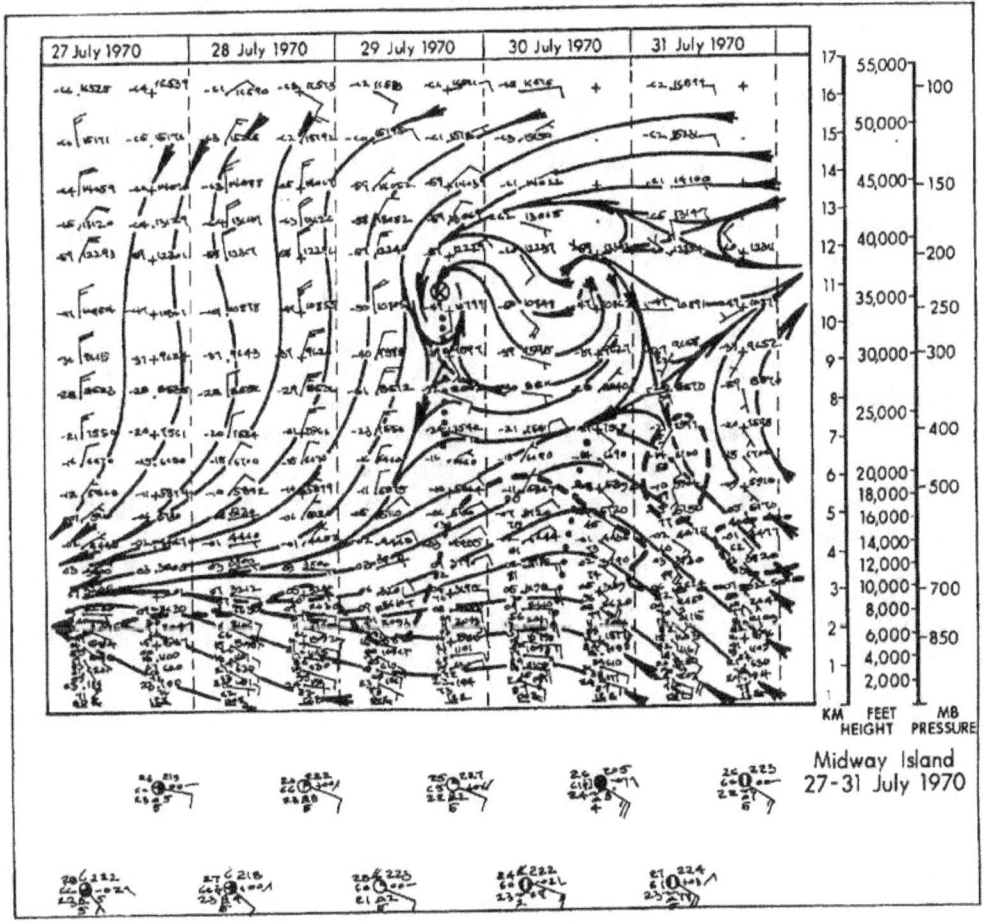

Figure 8-22. Time-altitude section and 12-hourly surface observations for Midway Island (28° N, 177 W°), 27-31 July 1970. Dotted lines are troughs; dashed lines, 50-percent relative humidity (Sadler, 1976a).

205

A diagnostic study of cold lows over the western North Pacific (Kelley and Mock, 1982), supported Sadler's model of a cold low. From 24 July through 7 August 1988, a cold low persisted over southwest Texas. Whitfield and Lyons (1992) computed mean wind, vorticity, and temperature fields for this period, and determined the rainfall distribution. The Sadler model, based on scanty over-ocean data, was once again validated, this time over a data-rich region.

Cyclones develop during the warm season in the TUTT over the South Pacific; see Figure 4-13 for its January position. As over the NH tropical oceans, these cold lows occasionally trigger tropical cyclones. Since the low-level monsoon trough in the South Pacific seldom extends east of the dateline, it seems likely that most of the rare tropical storm developments east of this meridian stem from cold lows (except in the 1982-83 Niño—see Figure 6-26).

Kousky and Gan (1981) described a South Atlantic summer TUTT, closely resembling the NH systems. Although the embedded cold lows move erratically, those between 10 and 15° S tend to move westward and significantly affect the distribution and intensity of convection over eastern Brazil. Lyons (1991) has identified a rather weaker TUTT over the southeast Indian Ocean, suggesting that it is a channel through which mid-latitude waves can affect the weather of equatorial Africa. Jet aircraft observations generally allow cold lows to be analyzed and tracked. Lows that extend downward into the middle troposphere tend to persist and to move steadily; shallow lows are usually transient and move erratically. After cold lows and the TUTT are identified in the wind field, their cloud signatures can be recognized in satellite images, especially the moisture channel. Since the lows are mostly free of cloud, the images alone cannot provide unequivocal identification. The pictures reveal the dry central area as well as outbreaks of deep convection east of the center; this is early evidence that the low is modifying weather and circulation in the lower troposphere.

Hybrid storms and cold lows are puzzling. We do not understand why interaction between higher latitudes and the tropics sometimes results in cold lows and at other times in subtropical cyclones, nor do we understand what causes a cold low to intensify enough to produce bad weather on its eastward/equatorward periphery. Why do Arabian Sea mid-tropospheric cyclones rarely reach the surface, while those in the South China Sea often do, and in that sense, resemble subtropical cyclones? And what causes mid-tropospheric cyclones very rarely to intensify into hurricanes, and in at least one instance, for a cold low to be so transformed (9.5.5).

All these cyclones are direct and energy-exporting; they do not readily dissipate. In contrast to tropical cyclones, they are not locally dependent on heat and moisture for their survival, and so often persist over land. Forecasters can use satellite images to help identify hybrid cyclones. They lack eyes and generally spiral convective bands, usually appearing as amorphous masses of dense cirrostratus. Cold lows may possess an annulus of cirrus around a clear center in which a few cumulonimbus may be scattered. More often, the cirrus is in one quadrant or semicircle only, with a very sharp inner edge lying along a jet stream. Usually, by differentiating cloud top heights, IR images are more informative than visual images, while the moisture channel gives the best picture of subsiding dry air in the centers of cold lows.

8.3.6 Temporal Storms of Central America. A phenomenon that may be variously caused by tropical depressions, hybrid cyclones, or cold lows is the temporal of the Pacific coast of Central America and the north coast of Honduras (Portig, 1976). Each location on the Pacific coast experiences one or two temporals a year. They are most likely in September and October, with a secondary maximum in June. They resemble hurricanes in their rainfall and clouds; however, the associated low-level winds are relatively weak. Pallman (1968) studied temporals using conventional and satellite data, but was hampered by lack of upper-air data over the eastern Pacific. His model of the low-tropospheric contour pattern and weather distribution (Figure 8-23) indicates extensive altostratus on the east side of the storm with embedded cumulonimbus and continuous rain in the northeast quadrant. Thunderstorms are very rare. Since temporals often remain almost stationary for many days, they can be responsible for catastrophic rain and major flooding, accentuated by orographic effects.

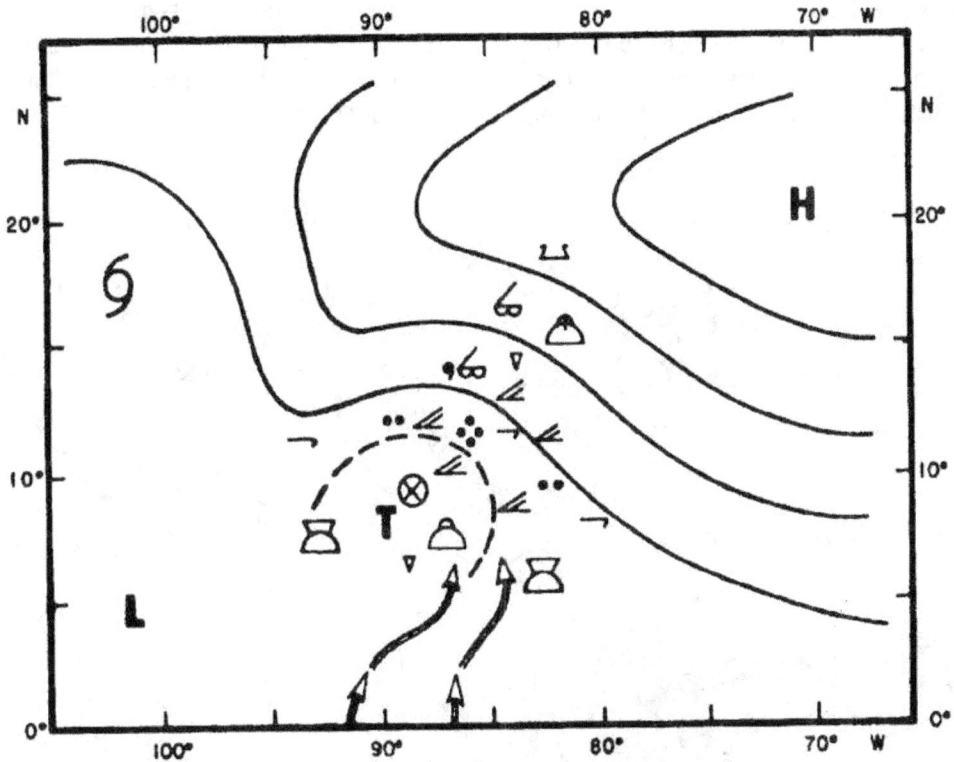

Figure 8-23. Model of the lower-tropospheric contour and weather patterns of a temporal (adapted from Pallman, 1968).

8.4 LINEAR DISTURBANCES

8.4.1 General. A linear disturbance is a synoptic system in which the vorticity or divergence, or both, tend to be concentrated in a zone whose length is much greater than its width (LaSeur, 1964a). Through the use of satellite imagery, these disturbances are much better understood. Before satellites, tracking them over the oceans was difficult. Conventional data are sparse and differences in temperature, moisture and wind are subtle. As a line passes, winds, clouds and rain increase. Often, during World War II, such weather was ascribed to fronts, although no air-mass discontinuity could be detected. Later, easterly waves were blamed, but satellites showed how short-lived many of the disturbances were. Despite the daunting variety, especially between 10° N and 10° S, the following line disturbances occur rather often: squall lines, remnant or reactivated cold fronts, shear lines, tradewind or monsoon "surges," trough superpositions, and near equatorial tradewind convergence.

8.4.2 Squall lines. These are generally defined as non-frontal lines of active thunderstorms from several to some tens of nautical miles wide and hundreds of nautical miles long; they last much longer than the lifetimes of the component cumulonimbus elements. In monsoon regions, they mainly occur during spring and autumn. A typical tropical squall line is shown in Figure 8-24.

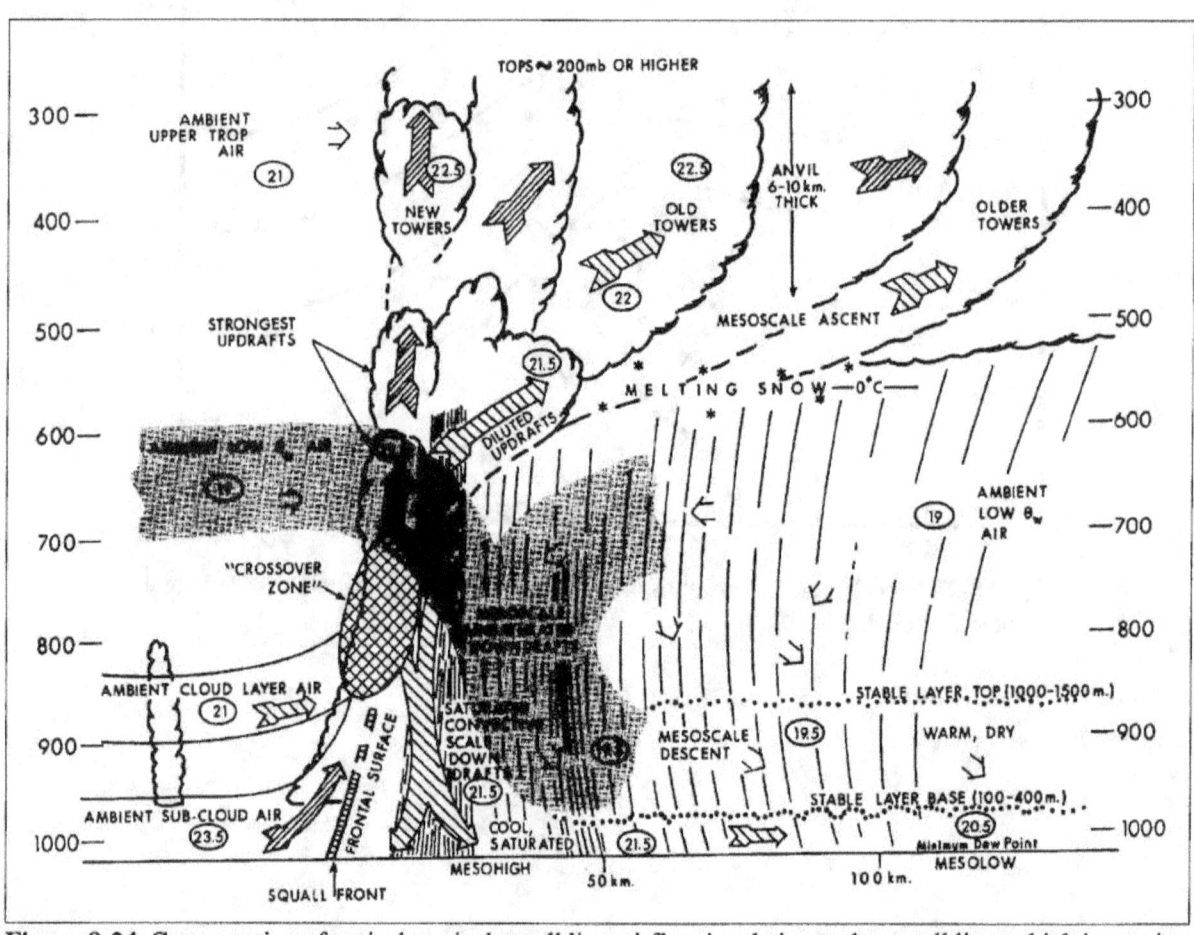

Figure 8-24. Cross-section of typical tropical squall line; airflow is relative to the squall line, which is moving from right to left. Circled numbers are typical values of θ$_w$ in ° C (Zipser, 1977).

Based on squall lines he investigated during the Line Islands Experiment (LIE) in 1967 and GATE in 1974, Zipser (1977) described an archetypal squall line for the tropics (Figure 8-24).

Zipser's model is similar to the tropical squall line model conceived independently by Houze (1977). To quote from Zipser:

"1. Relative winds are directed into the squall line from the front at all levels, with a minimum relative flow often noted about 700-600 mb. Well behind the squall, the relative flow is strongly outward from the squall near the surface, still outward — but much less so — just above the mixed layer, and from either direction in the 600-900 mb layer, with indications that the flow from the rear may increase with time.

2. New cumulonimbus cloud growth takes place above the squall front along the leading edge, with older towers joining the anvil mass trailing behind. The anvil is 20,000 to 33,000 feet (6-10 km) thick, and rain falls from the anvil base for 50 NM (100 km) or more to the rear in a region having no significant clouds below the anvil base, which is near 13,000 to 16,000 feet (4-5 km). [About half of the total rain of a squall line falls from the anvil].

3. At 500 feet (150 meters) in altitude, the region from 5 to 15 NM (10 to 30 km) behind the squall front, where almost all active cumulonimbus clouds are located, has very cool, near-saturated air. The remainder of the system to the rear has air varying widely in θ_w and in relative humidity. [On a Skew-T diagram, the wet bulb potential temperature (θ_w) can be found by first plotting the wet bulb temperature, and then following the saturation adiabat from that point to 1000 mb, and reading the temperature at the intersection. As with θ_e and θ_{ex} [5.3.4] an atmospheric layer is potentially unstable (stable) if θ_w decreases (increases) with height.]

4. Soundings taken in the rear portion of the squall system show cool, near-saturated air of intermediate θ_w [21.5 C] near the surface [the mixed layer]; a deep layer of relatively warm air with low relative humidity and low θ just above that; and near-saturated conditions again near anvil base.

5. The stable layer which separates the two lowest layers (4, above) has its base, which marks the top of the mixed layer, from 130 to 1600 feet (40 to 500 meters), with 330 to 1,300 feet (150 to 400 meters) rather common.

6. The winds in the surface layer and in the mixed layer behind the squall line are divergent in the range of 1 to 5 X 10^{-4} s^{-1} over distances of 15-55 NM (30-100 km). On the same scale, the mesoscale sinking just above the mixed layer is in the 5-25 cm s^{-1} range at 1600 feet (500 meters).

7. The dewpoint at the surface drops with squall-front passage, but often drops still further in the anvil rain area, reaching an absolute minimum about 50-100 NM (100-200 km) behind the squall line.

8. A mesoscale high-pressure region at the surface tends to accompany the squall line itself, followed by a mesoscale low-pressure region about 50-100 NM (100-200 km) behind the squall line, but with considerable variations in individual cases.... If the relative flow is from the front at 700 mb, much of the air must pass between active cumulonimbus towers to arrive in the heavy rain area behind the squall and eventually in the anvil rain area....because the cumulonimbus towers are not continuous either in distance along the squall line or in time....If the relative flow is from the rear, the air enters the anvil rain area directly. There is evidence that air in some squall lines enters from both front and rear, converging near the heavy-rain area.... Evaporation of rain into this air is still believed to be the most important factor allowing some of it to sink several kilometers to within a few hundred meters of the surface."

209

As mentioned in 2.5, convection is generated along the leading edge of the cold downdraft air, while the downdraft air to the rear of the squall line may suppress convection. As a result, the squall line moves forward. For squall lines to develop, persist and propagate, they require not only moist surface air, but also relatively dry (low θ_w) middle tropospheric air, which, on being cooled by evaporation from rain, sinks and establishes a cold front at the leading edge of the squall line. Zipser (1969), in a study of a cloud system that passed across the Line Islands 31 March to 1 April 1967, described how unsaturated convective downdrafts in the rear of a squall line suppressed convection. The squall line formed rapidly on the evening of the 31st, was most intense around midnight, and then quickly dissipated during daylight of the 1st.

Since it developed where wind shear in the vertical was strong (westerlies replacing easterlies above 600 mb), Zipser concluded that mid-tropospheric air with low equivalent-potential temperature was entrained into the mesoscale system and carried to the surface in organized downdrafts that thereupon inhibited further cumulus production—see Figure 8-25. In Figure 8-25C, the disturbance is dissipating and the downdraft is maintained primarily by rain falling from the extensive cloud shield. At this time satellite pictures revealed a thin line of cumulus moving outward from the main cirrus cloud mass.

Successive positions of this surface outflow boundary are shown in Figure 8-26, which is based on satellite, surface, and aircraft observations. The solid lines show high confidence, while the dashed lines are more conjectural. Satellite photos suggested that the bordering dashed lines marked the limit of the downdraft area. The main cloud system, which had appeared as a bright mass on the morning satellite picture, had almost

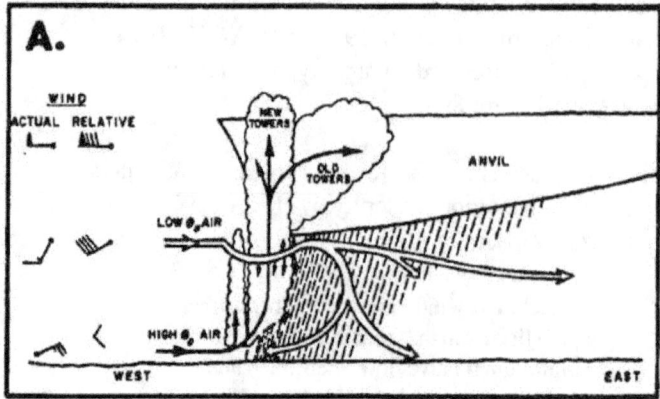

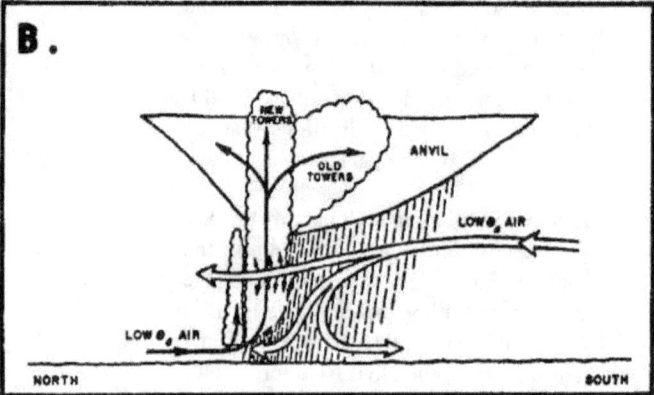

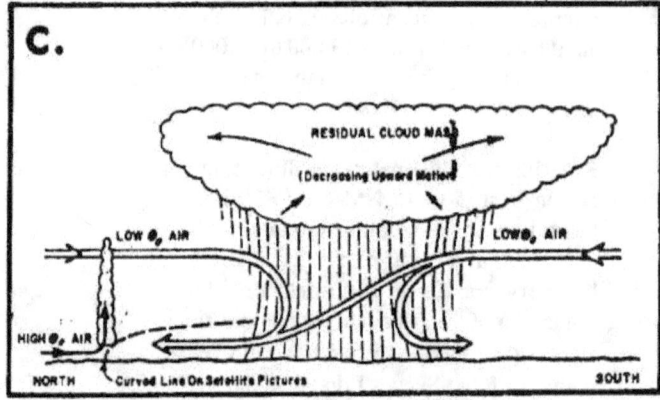

Figure 8-25. Schematic streamlines of airflow relative to convective disturbances in the Line Islands area. (C) represents the N-S section of the dissipating phase (after Zipser, 1969).

completely gone by 1711L. Sequences of satellite pictures have shown that where two outflow boundaries intersect, large cumulus is likely to develop. Had the Zipser squall line been close to a coast, where orography could be involved, intensification, rather than dissipation, might have ensued.

Over West Africa, where squall lines are responsible for most of the rain, and even for generating summer duststorms (Leroux, 1983) (10.5.3), they may develop at the leading edges of anticyclonic surges from mid-latitudes (see 8.4.4). The surges move in from the northeast and displace the monsoon westerlies. The initiating impulse may quickly dissipate. However, in a numerical simulation (Rennick, 1976), instability in the 700-mb easterly jet along about 14° N favored squall-line development. At the western edge of the squall line, low-level cold outdraft continually regenerates the convection; since squall lines are embedded in an environment with easterly vertical shear, they may move westward faster than the environmental flow (Eldridge, 1957). Most rain falls in the low-level westerlies, accelerated toward the squall line and lifted by the cold air. As shown in Figure 8-27 (Leroux, 1983), a mountain chain affects weather some distance ahead of the line, generating showers to windward and subsident drying to leeward.

Squall lines are most likely to develop over the southern Sudan, generally during the afternoon.

Although some West African squall lines last several days, 80 percent die out in less than 24 hours (Cochemé and Franquin, 1967). They are most common from June through October and best developed between 10 and 15° N. By combining several studies, the following rough statistics were produced:

•Time interval between squall line passages: 2.5 to 5 days

•Distance between squall lines: 1,000 to 2,000 NM (2,000 to 4,000 km)

•Westward speed: 6-7 degrees of longitude a day, but a few are much faster.

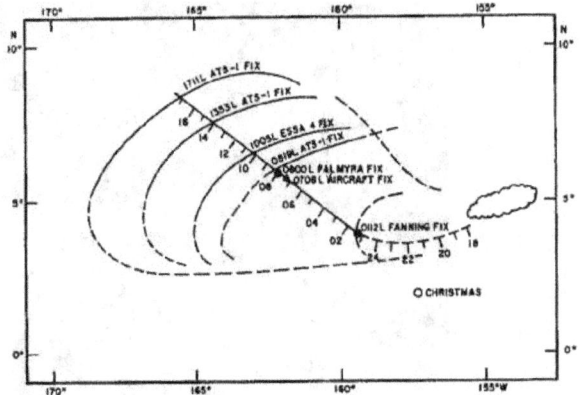

Figure 8-26. Isochrones of the leading edge of the downdraft air from a disturbance in the Line Islands area on 1 April 1967 (after Zipser, 1969).

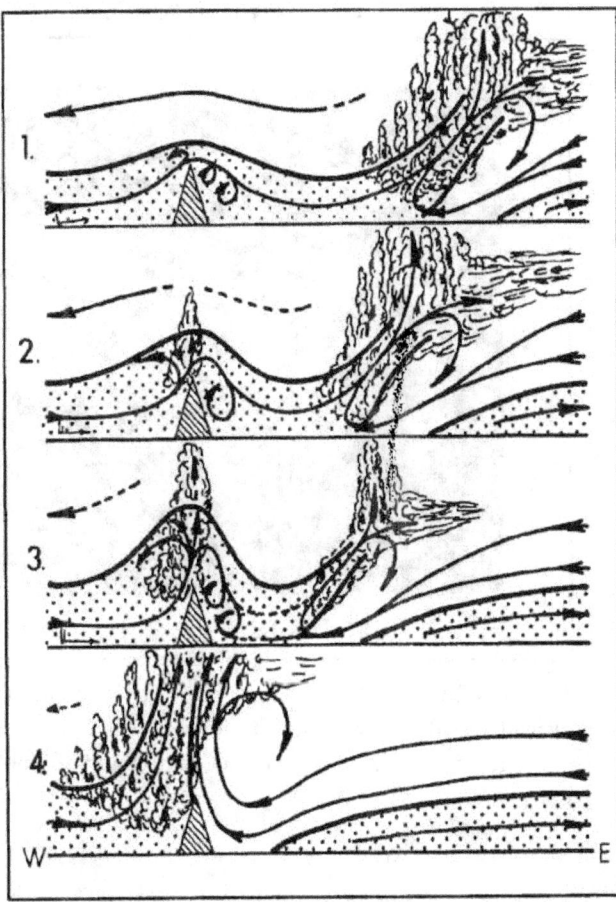

Figure 8-27. Schematic of cloud development over a ridge west of an approaching west African squall line (from Leroux, 1983).

211

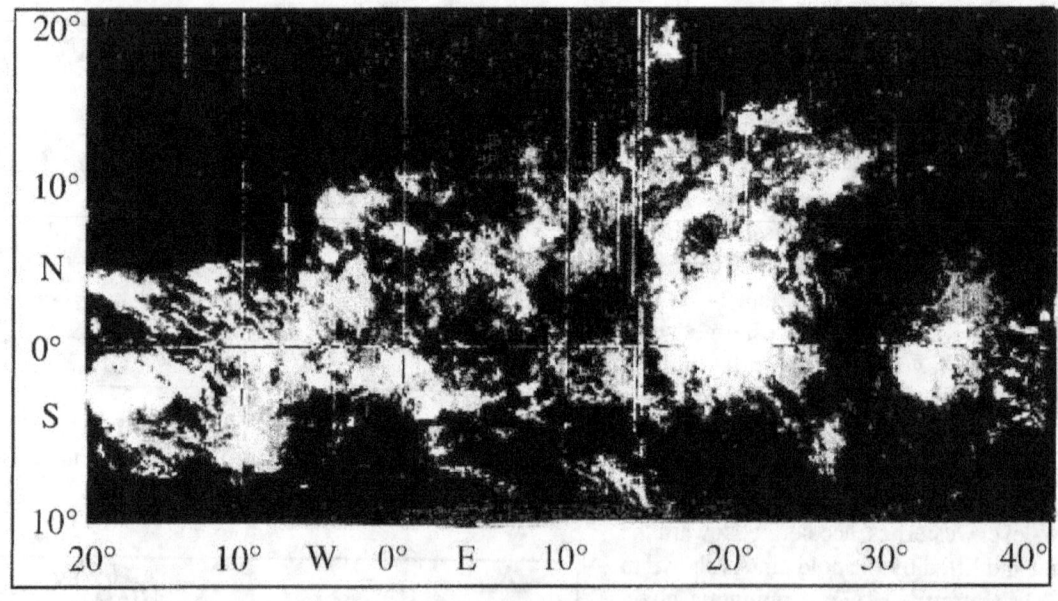

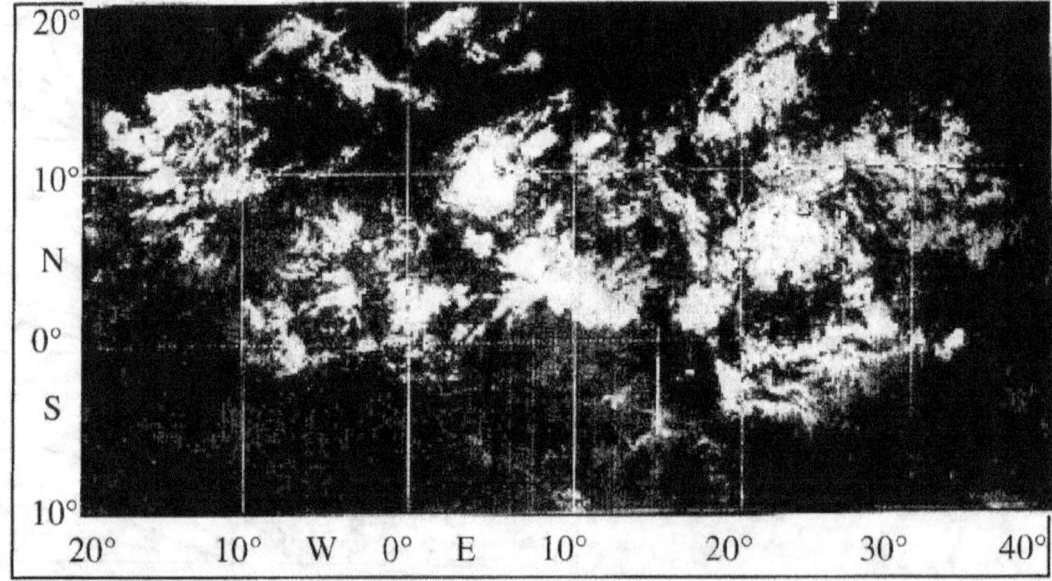

Figure 8-28. DMSP IR images of west Africa on 18 July (top) and 19 July (bottom) 1987. A curved squall line extended from about 11° N to the equator. At 7° N it moved westward (from 15° E to 3° E) in 24 hours, at a speed of about 30 knots. Note how other clouds changed between images.

212

Figure 8-28 (opposite) shows successive 24-hour IR images of a West African squall line that was typically convex toward the west. It moved west at about 30 knots.

Over tropical South America, east of the Andes, squall lines resembling those of West Africa (Fernandez, 1982) are responsible for most of the rain (8.4.8.2). This is also true over the Australian tropics when easterly winds prevail (Drosdowsky and Holland, 1987). Scattered synoptic observations made every 6 hours, or twice-daily pictures from polar-orbiting satellites, are seldom frequent or detailed enough to allow the forecaster to maintain analytical continuity, let alone predict squall lines. Continuous radar coverage or hourly geostationary satellite pictures are essential if the rapid intensity changes are to be monitored or extrapolated. Even with these aids, forecasts rarely succeed beyond a few hours.

Disturbances are likely to become squall lines wherever a moist lower troposphere is overlain by a relatively dry middle troposphere; they are therefore favored in the tradewinds, in the general neighborhood of heat troughs, and during the transition seasons in monsoon regions. Squall lines are extremely rare in the deep, moist air of the summer monsoon and within 5 to 10° of the equator.

The impact of squall lines on the diurnal variation of rainfall is discussed in 7.6.2.3; for their association with destructive winds, see 10.5.2.

8.4.3 Surface Cold Fronts and Shear Lines.
Cold fronts often penetrate to low latitudes over tropical continents during winter. Even though currents of polar air are usually rather shallow by the time they reach the subtropics, surface temperature and dew-point discontinuities and a shear line can be detected to low latitudes over land because of repeated nocturnal cooling in the clear, dry air mass behind the front (Stone et al., 1942). Over the oceans, satellite imagery and continuity facilitate tracking fronts. The front's leading edge is usually marked by a line of active convection; a series of convective lines may parallel the front on its poleward side. Cloud tops average about 10,000 to 15,000 feet (3 to 4.5 km).

Associated low ceilings and rain may be enhanced by orography. Frontal shear lines occasionally reaching Guam in winter cause extended periods of low ceilings and poor visibility.

Figure 8-29 schematically relates the shear line to a cold front farther poleward. After becoming stationary, the shear line coincides with an Asymptote of convergence within the tradewinds, and the mechanism described in the discussion of near-equatorial tradewind convergence ensures growth of cloud droplets and sometimes rain. In shear lines, stratiform clouds generally reach 8,000 to 12,000 feet (2.4 to 3.7 km) with embedded convection. Where shear lines intersect mountain slopes significant rain may fall on windward slopes (see 6.3.2).

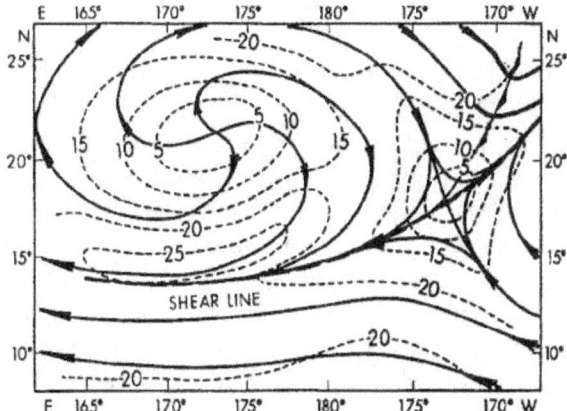

Figure 8-29. Model of the surface kinematic pattern associated with a cold front extending into a tropical ocean shear line. Isotachs (dashed) labeled in knots (adapted from Palmer et al., 1955).

On 31 December 1987 and 1 January 1988, a shear line, along which fresh northeasterlies converged with weak east-southeasterlies, lay stationary across the island of Oahu in Hawaii. Orographic lifting by the Koolau Range, aided by a trough to the west in the upper tropospheric westerlies, built cloud tops to 19,000 feet (5.8 km). Although thunderstorms were absent, continuous rain amounting to up to 25 inches (635 mm) in 24 hours fell over and to windward of the mountains on southeastern Oahu (National Research Council, 1991).

Figure 8-30, for 0000Z 25 December 1985, is a satellite view of a cold front crossing Hawaii, where winds shifted from southwest to northwest. Below 15° N the front merged into a shear line, with cloud tops ranging between 7,000 and 12,000 feet (2 and 3.5 km); tradewinds freshened as it passed. To the east, shear lines had formed in deep southwesterlies beneath a subtropical jet stream.

During the normally dry months of February and March, a shear line originating from a NH cold front may reach the coastal mountains of Venezuela and give continuous rain that is heaviest between 500 and 1,500 meters on the north-facing windward slopes (Goldbrunner, 1963). This phenomenon, referred to as the "invierno de las chicarras" strikingly resembles the tradewind rainfall periods on Mt. Waialeale (6.3.2).

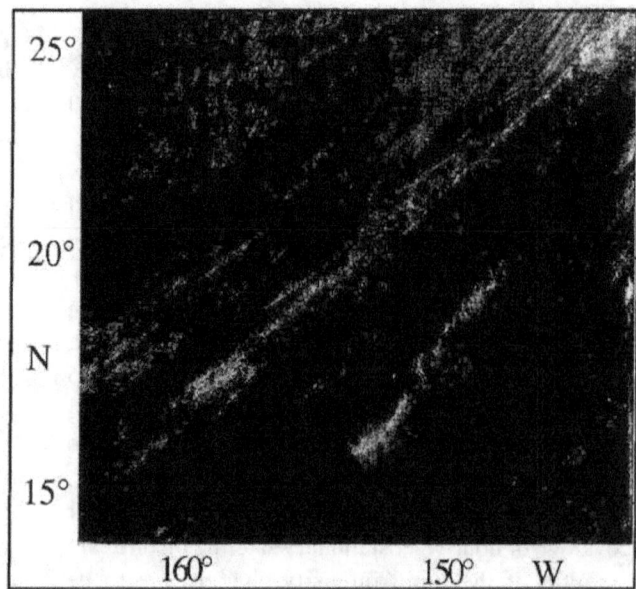

Figure 8-30. GOES visible image at 0000Z 25 December 1985 showing a cold front crossing Hawaii. South of 15° N, the front merged with a shear line.

Shear lines are sometimes very hard to identify, even with satellite pictures; eventually, they dissipate. As described below, a cold front may suddenly disappear within a strong equatorward surge.

8.4.4 Surges. For more than 50 years, tropical meteorologists have often observed sudden accelerations within major surface wind currents such as the tradewinds and the monsoons. Downwind of the acceleration maximum, convergence and increased cyclonic shear to the *left* of the maximum axis in the NH (to the *right* of the axis in the SH) enhanced convection and sometimes favored cyclogenesis. Lack of data hindered finding any development continuity, and isolated speed increases were often discounted because meteorologists were looking for downstream movement of acceleration centers; no *local* generating force could be identified.

There is now no doubt that throughout the tropics, and in all seasons, wind currents locally accelerate and decelerate, usually without apparent cause, but often with significant effect on weather. The best-understood surges originate in the winter hemisphere

and will be discussed first. They will be followed by the more mysterious, but no less important, surges in the summer monsoon and the tradewinds. Oppositely directed upper-tropospheric surges almost certainly accompany surface surges, but are seldom observed or reported.

8.4.4.1 Winter Surges. In the winter hemisphere, meridional temperature gradients are largest and branches of the Hadley cell best developed across the continents. The vertical circulation of a Hadley cell branch may suddenly accelerate in response to a deep trough in the upper-tropospheric westerlies (Ramage, 1971). Southeast Asia and the South China Sea experienced such an event in mid-January 1967—see Figure 8-31.

A cold front crossed Hong Kong (22° N , 114° E) on the 14th; on the same day, surface winds increased across the South China Sea and heavy rain began over North Borneo. Apart from a remnant moving southward along the coast of Vietnam (Chang et al., 1979), the cold front had disappeared as the northerly winds south of the front freshened simultaneously.

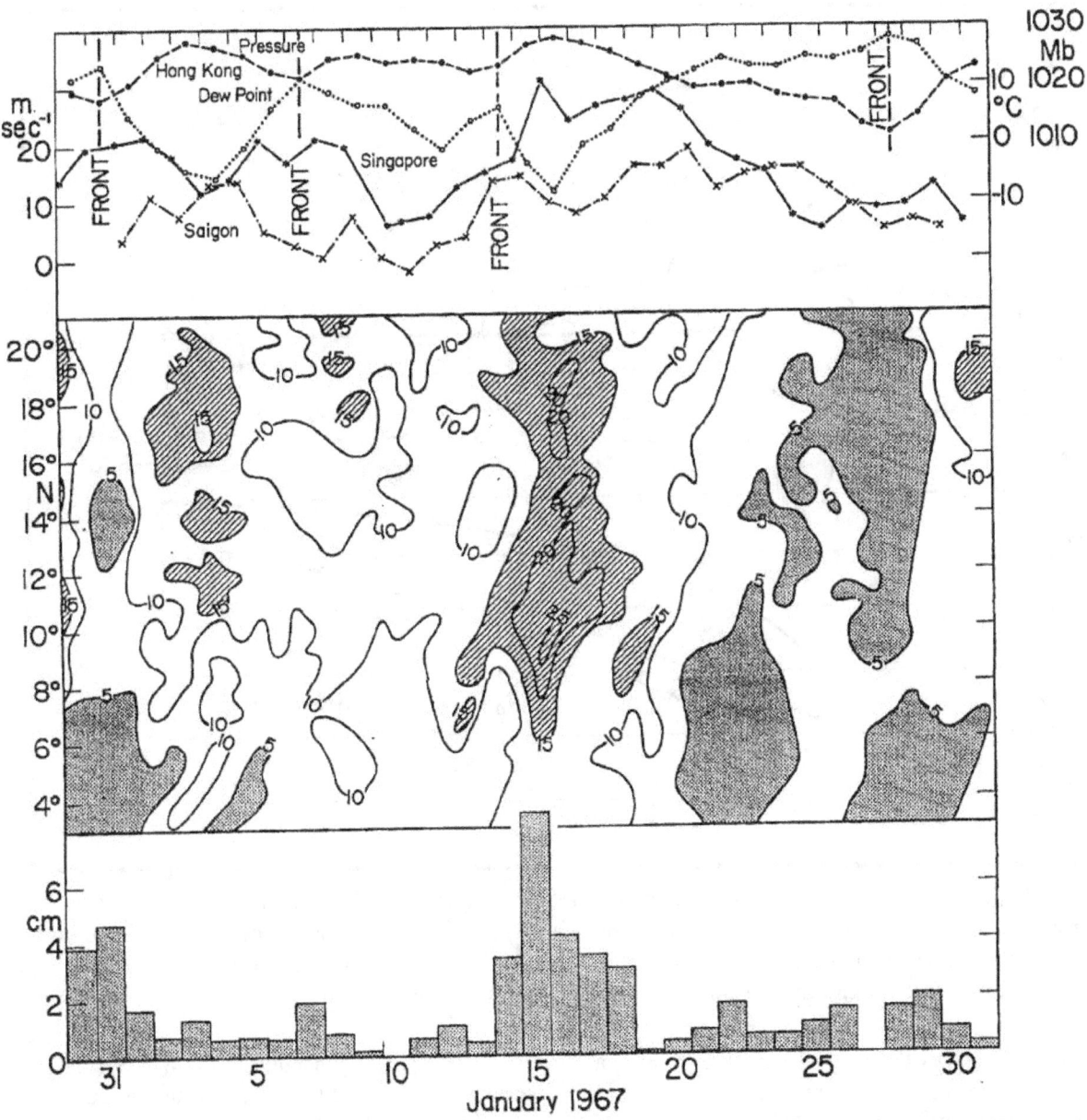

Figure 8-31. Daily values for January 1967 over the South China Sea. The graph at the top shows Hong Kong sea-level pressure (dashed line) and dew points (dotted line); Saigon southerly components of the 200-mb wind (dot-dashed line); and Singapore 200-mb winds (solid line). The chart in the middle shows surface isotachs (ms⁻¹) averaged zonally between 110 and 118° E. The bar graph at the bottom is a histogram of rainfall averaged for five coastal stations in North Borneo.

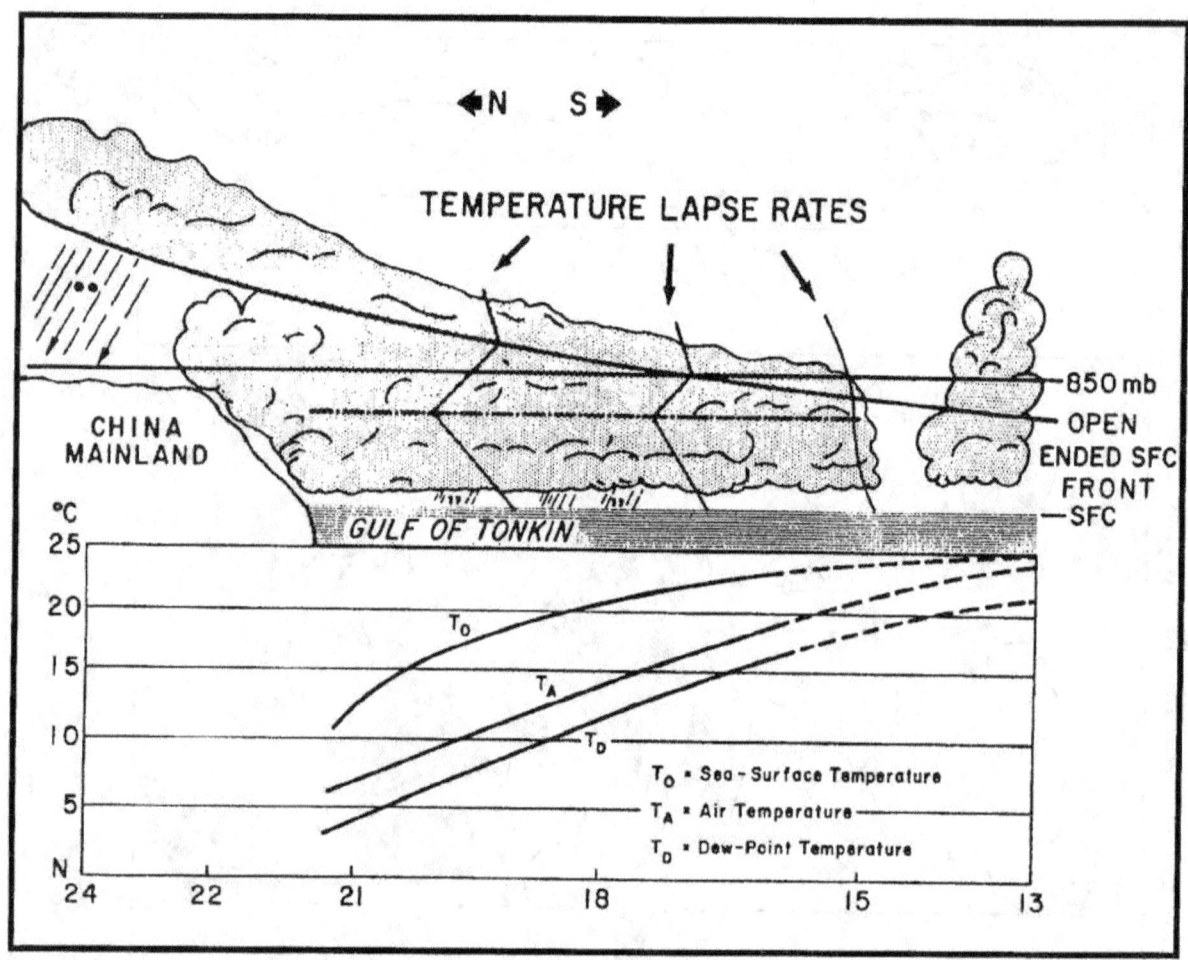

Figure 8-32. Schematic meridional cross section over the South China Sea during a surge of the winter (northeast) monsoon. At the top is the cloud, weather, and temperature profile; the dot-dashed line lies at the base of the frontal inversion. The bottom half of the figure shows the latitudinal averages of sea-surface, air, and dew-point temperatures (adapted from Navy Weather Research Facility Staff, 1969).

Figure 8-32 shows the vertical structure of such a system. In the middle and upper troposphere, cold air also swept southward and subsided on the eastern side of a ridge; on the southern and western sides of the ridge, easterlies accelerated over Singapore (1° N, 104° E) and southerlies over Saigon (11° N, 107° E).

Rather weak surges generated during Winter MONEX travelled south at 80 knots across the northern and central South China Sea, and appeared to have the character of synoptic-scale gravity waves (Chang et al, 1983). During surges, the cross-isobaric component of the surface wind is much greater than normal. Once established, the area of stronger winds associated with a surge may remain almost stationary for a few days.

normal. Once established, the area of stronger winds associated with a surge may remain almost stationary for a few days.

On 9 December 1985, a cold front passed Hong Kong. At about the same time, northeasterly winds increased by about 10 to 15 knots over the South China Sea. Prior to the surge, the satellite showed scattered convection near 10° N, north of a surface trough (Figure 8-33). Everywhere to the north clouds were limited by a subsidence inversion. Twenty-four hours later (Figure 8-34) shallow low cloud had reached about 20° N, but to the south the convection near 10° N had intensified.

During this time, the jet stream south of Japan increased from 110 to 150 knots. Subsequent surges further increased winds over the South China Sea, and led to a Tropical depression developing near 8° N on the 15th.

Vigorous cold outbreaks channeled by north-south mountain ranges also affect

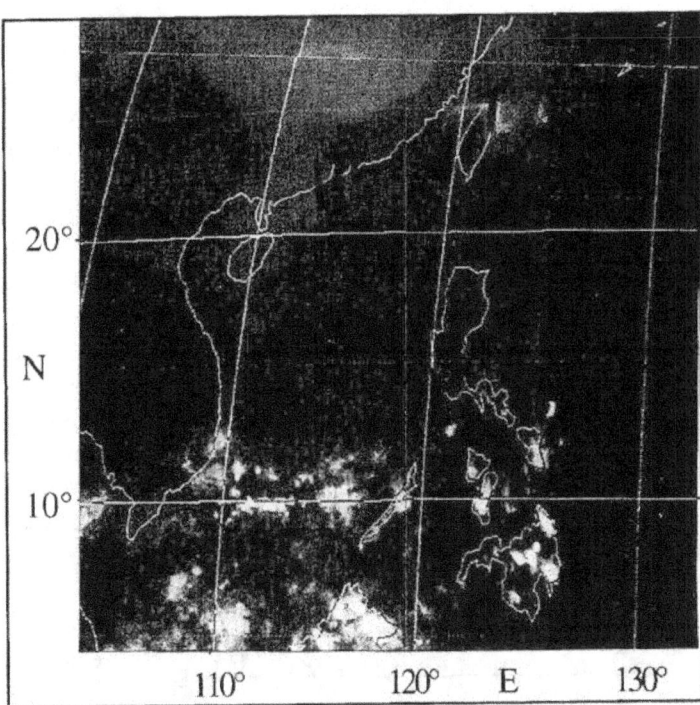

Figure 8-33. GMS IR image for 0900Z 9 December 1985, prior to a cold surge across the South China Sea.

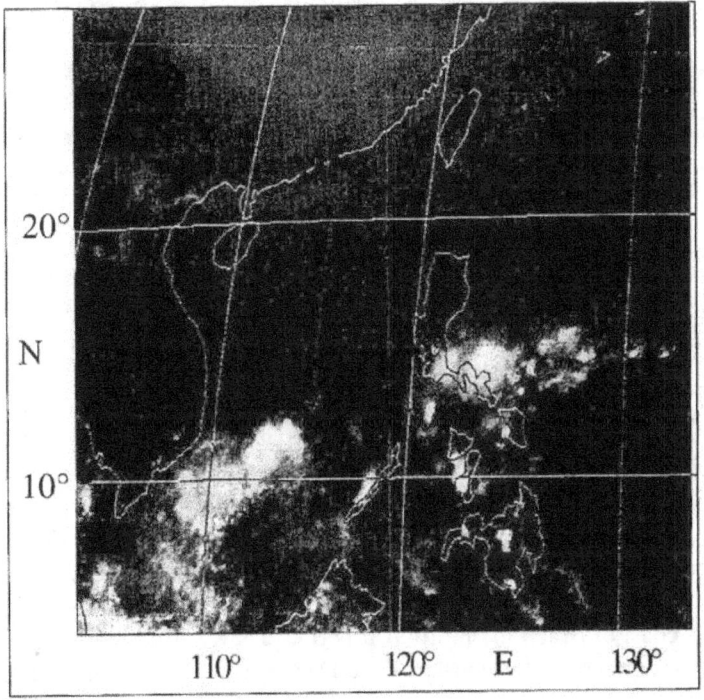

Figure 8-34. GMS IR image for 0900Z 10 December 1985, following a cold surge across the South China Sea.

Vigorous cold outbreaks channeled by north-south mountain ranges also affect Central America and the Caribbean. "Typically, increase of wind speed and shift in direction are observed several hours prior to the arrival of cold air" (Hastenrath, 1985). These "Northers" funnel through the mountain passes of southern Mexico, Nicaragua, and Panama, as shown in Figure 8-35.

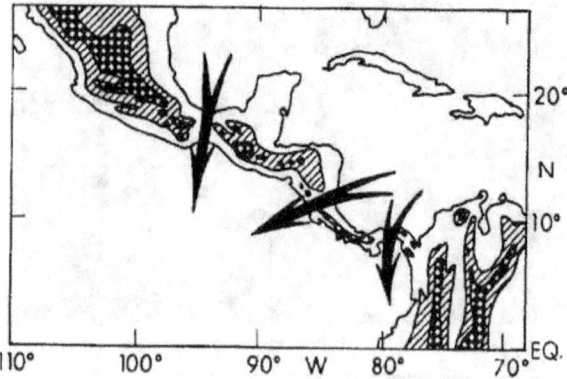

Figure 8-35. Topography of Central America. Land above 600 meters is hatched; above 1,800 meters, checkered (Sadler and Lander, 1986).

Each winter, the Isthmus of Tehuantapec experiences an average of 20 Northers; the Canal Zone, only three (Trewartha, 1981). In the Pacific, hundreds of nautical miles downstream from the passes, the resulting high winds, cool upwelled water, and a cloud minimum are features of the long-term ship data climatology for January (Figures 8-36a, b, c, and d) (Sadler and Lander, 1986).

Significant surface pressure rises at Hong Kong (22° N, 114° E) have been statistically related to heavy rain along the east coast of West Malaysia within a day or so (Gan, 1963). Similarly, Brooks (1987) used increases in the pressure gradient between Houston (30° N, 95°W) and Merida (21° N, 90°W) to forecast the onset of a Norther in Honduras within 48 hours. Kousky (1979) counted 152 cold fronts at 18° S on the coast of Brazil in one 10-year period. They were most common between March and December.

Occasionally, the fronts reached 5° S and enhanced rainfall. Surface pressures varied nearly simultaneously between 18° S and 4° S. Parmenter

(1976) used satellite pictures in tracking a cold front across South America in July 1975. When the front reached about 15° S, winds turned southerly and accelerated as far north as 5° S. By the following day, they had reached 10° N. Equatorial convection increased as well. These observations again suggest near-simultaneous accelerations of the South American Hadley cell branch.

In winter, when a vigorous cold front starts moving south over North Africa, the strong winds may cause a severe dust storm (10.5.3). However, once over the desert, the front is hard to follow. Hamilton and Archbold (1945) and Johnson (1964) noted that freshening northeast winds often start to blow dust well ahead of where the front might reasonably have been expected. Did a branch of the Hadley cell suddenly accelerate? Figure 8-37 uses the leading edge of a North African dust storm to define a cold front position. This implies that the front moved at more than 30 knots over the previous 24 hours, a surprising rate for such low latitudes. The analysis is probably wrong. This rather common error can be corrected by dropping the front from analysis below 25° N and marking the advancing edge of the duststorm with a surge line.

Dvorak and Smigielski (1990) used satellite pictures to identify outbreaks over the tropical Americas in March 1986 and over North Africa in October 1978. In both cases, divergence beneath the upper-tropospheric low-latitude ridge west of the outbreak overlay intense convection. Walters et al. (1989) reported that over central America, equatorward-advancing upper tropospheric "cold pools" often accompany surface surges. Once again, these observations suggest near- simultaneous accelerations of a branch of the Hadley cell.

During winter over southern Africa and Australia, cold fronts may be preceded by "leader fronts" up to 300 to 400 km ahead of the main cold front, often with greater wind and temperature changes than at the main front itself (Taljaard, 1972). Similar sequences are possible every 3 to 5 days during the SH winter along and off the east coast of Africa. In a case study by Cadet and Dubois (1981), a deep mid-latitude low moved eastward to the south of the continent, as

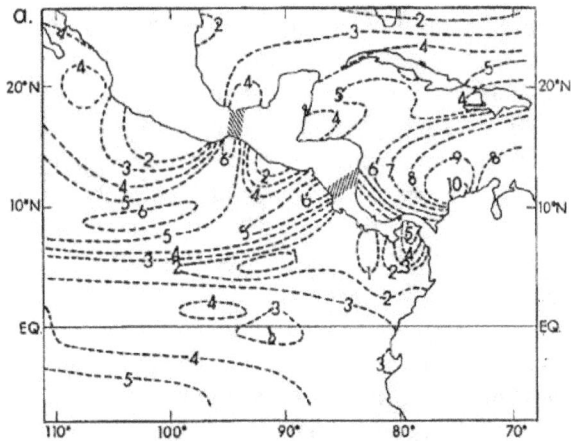

Figure 8-36a. January isotachs—ship data climatology (Sadler and Lander, 1986).

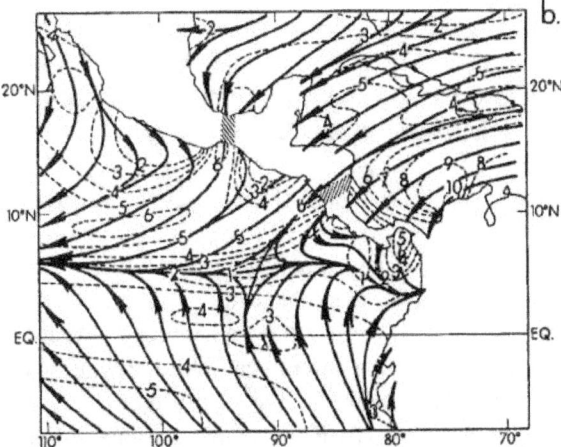

Figure 8-36b. January kinematics—ship data climatology (Sadler and Lander, 1986).

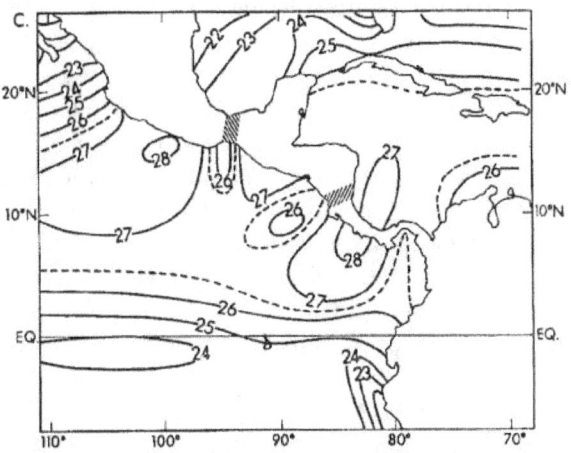

Figure 8-36c. January sea-surface temperature—ship data climatology (Sadler and Lander, 1986).

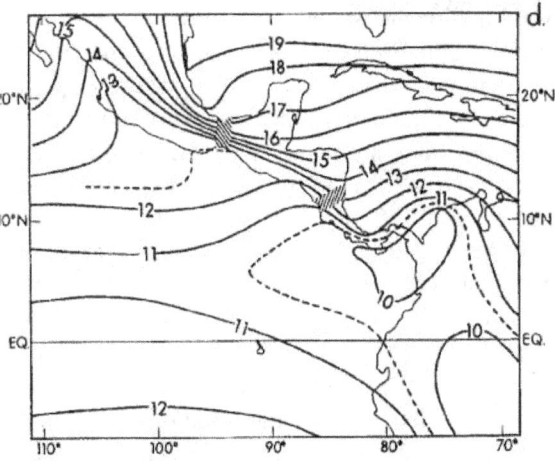

Figure 8-36d. January sea-level pressure—ship data climatology (Sadler and Lander, 1986).

Figure 8-37. Synoptic sequence over North Africa for a 48-hour period centered on 23 February 1943. Previous and following day frontal positions indicated by dashed lines. Shading shows area of dust-restricted visibility (adapted Solot, 1943). The analysis is discussed in the text.

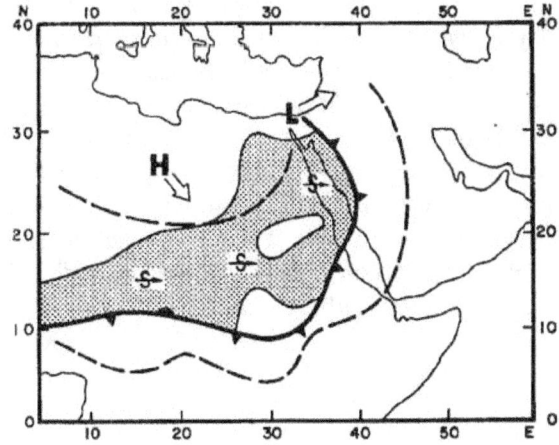

219

southerlies increased in the Mozambique Channel, well ahead of the surface cold front. Within a day or two, the surge had crossed the equator, intensified the Somali jet and increased convection over Somalia.

Although surges are best observed at the surface, several investigators (e.g., Chang and Lau, 1980, 1982) confirmed that upper-tropospheric poleward flow diverges from the enhanced convection in the deep tropics. Downstream from where the poleward flow merges with the subtropical jet, the westerlies accelerate. Cause and effect are hard to ascribe. Is the sequence started in the tropics or in higher latitudes? That everything seems to happen at once suggests that a surge cycle goes as follows: Initially, as a surge develops, convergence and cyclonic vorticity downwind of the speed maximum enhance upward motion and so increase rainfall. In the upper troposphere (above the rain), heat accumulating from condensation is exported by the divergent flow to colder regions. Thus convective instability is maintained. Stability returns when the surface surge weakens and heat is no longer exported aloft (no longer needed?). Convection then warms the upper troposphere, and by increasing stability, inhibits further convection. Once more, everything seems to happen at once.

Over the maritime continent, generally fair weather prevails when the northeast monsoon is weak. When circulation strengthens, rising motion increases, but not consistently. For example, in January 1967 (see Figure 8-31), the mid-month surge was much wetter over North Borneo than the surge at the start of the month, whereas the reverse was true over eastern West Malaysia, which experienced floods in the earlier period. Rainfall streakiness probably also typifies the equatorial regions of Africa and South America. Thus, when a Hadley cell branch is weak, local forecasts of relatively fair equatorial weather can be made with some confidence. When a surge develops, figuring just where the heavy rain will fall is much more difficult, although rain is more likely where surface winds curve cyclonically.

Analysts should watch carefully for these regional accelerations and be prepared to eliminate fronts from their analyses of the strongest cold outbreaks. They should expect to retain fronts in analysis poleward of 20 degrees latitude, and for longer times during spring, (or in the mid-ocean tradewinds,) than equatorward of the continents in the depth of winter. They should remember that the Himalayas prevent central Asian cold outbreaks from reaching the Indian Ocean.

8.4.4.2 Other Surges. As mentioned in 8.3.2, a surge in the southwest monsoon over Peninsular India precedes development of a monsoon depression over the northern Bay of Bengal. Case studies (Davidson and Holland, 1987; Zhao and Mills, 1991) have shown that surges also precede development of monsoon depressions over Australia. Surges generated in the winter hemisphere apparently crossed the equator and contributed to the generation of tropical cyclones in the South Pacific (Arakawa, 1940) and in the eastern North Atlantic (Morgan, 1965). Attempts to link the surges to fronts failed. In the Australian region, when a tropical cyclone develops in the monsoon trough, low-level westerlies north of the trough and low-level easterlies south of the trough usually have already increased (McBride and Keenan, 1982). Here, and in the western Pacific (Douglas, 1987), tropical cyclones in their early stages resemble monsoon depressions.

Lander (1990) studied simultaneous development of tropical cyclones on either side of the equator in the western Pacific, and found that equatorial surface westerlies freshened prior to cyclogenesis (see Figure 9-15) and may send an oceanic Kelvin wave toward South America (6.3.5). Chu and Frederick (1990) suggested that equatorial surges might stem from pressure rises to the west originated by surges out of East Asia. The case they studied lasted 11 days, but 3 to 5 days is more usual. In line with this, Love (1985a,b) concluded that surges originating in the winter hemisphere, by increasing equatorial pressure, could induce surges in the surface westerlies of the summer hemisphere and so favor cyclogenesis there.

Jeandidier and Rainteau (1957) emphasized that forecasting westerly monsoon surges is prerequisite to forecasting weather in the Congo Basin.

What else might cause the pressure gradient to increase and the summer monsoon to freshen? Guard (1985) thinks that the pressure falls accompanying tropical cyclones sometimes generate upstream surges, even in the opposite hemisphere, while the widespread pressure changes described in 4.2.2 could also be responsible; about two a year develop east of the Philippines. According to Guard (1985), these pressure falls might originate from intensifying upper cyclones (8.3.5). Minor surges might also result from the heat/monsoon trough deepening under clear skies.

In Figure 8-38, Guard has categorized summer monsoon surges. In a weak surge, the monsoon is shallow, the upper level return flow from northeast is weak, and a showers regime prevails. In a strong

surge, the southwesterlies are deep and vigorous and are overlain by strong northeasterlies. This large vertical wind shear accompanies a rains regime.

Surges in the tradewinds seem to comprise high-pressure pulses originating on the eastern sides of subtropical Anticyclones and may occur simultaneously in the northeast and southeast trades (Krishnamurti et al., 1975). The surges generally lack fronts. They may affect patterns of convergence and convection in the NETWC, while on the equatorward side of a speed maximum, cyclonic shear favors development of a cloud line.

Over Africa in the warm part of the year, anticyclonic pulses/surges can reach the tropics between the intense heat lows centered over central West Africa and southwest Asia (see Figure 4-3). This intervening zone stretches across Libya, Egypt, and Sudan. According to Leroux (1983) surges trigger the

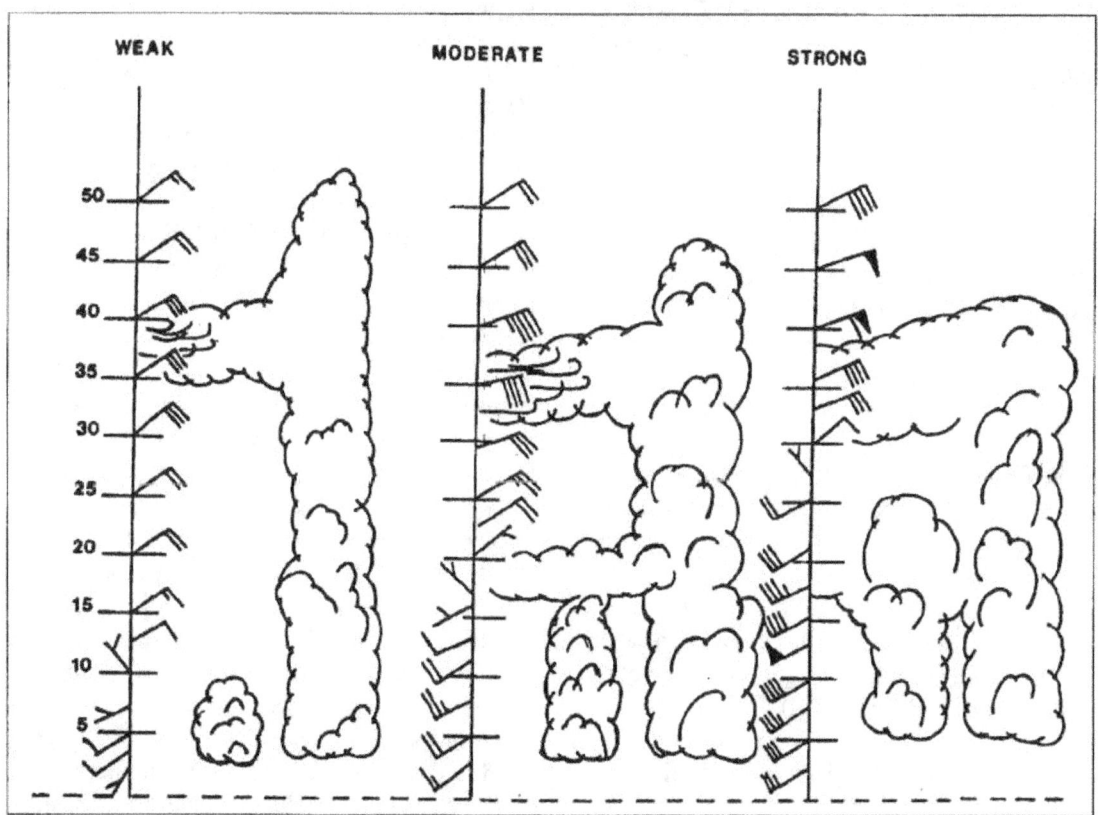

Figure 8-38. Schematic vertical profile of winds and clouds associated with weak, moderate, and strong surges of the southwest (summer) monsoon. Heights in thousands of feet (Guard, 1985).

haboobs of Khartoum (16° N, 33° E) (10.5.3.3), and farther south, where more moisture is available, the squall lines of West Africa (8.4.2). The time-sequence is shown in Figure 8-39, with dust storms predominating in the dry regions around and north of the heat trough, and squall lines forming where moist southwesterlies prevail. The schematic incorporates Leroux's view that westward progression of the anticyclonic pulse causes the squall line/dust storm also to move west. Strong environmental easterly shear is a much more likely cause of this. Similar surge-induced squall lines move westward across tropical South America (8.4.2).

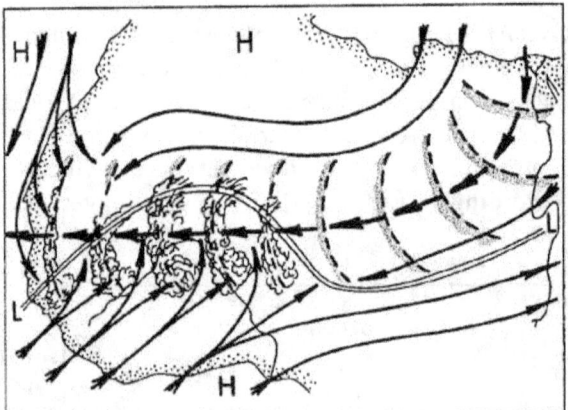

Figure 8-39. Schematic time-sequence (according to Leroux, 1983) of a surge (dashed line) progressing from Egypt to west Africa during the warm season. Streamlines depict surface flow. North of the heat trough (double line) the surge raises dust (stippling); near the coast, influx of moist, maritime air inhibits this effect. South of the heat trough, the surge causes squalls.

During the SH summer, upper-tropospheric poleward outflow from enhanced tropical convection is accompanied by a strengthened subtropical jet stream. Hurrell and Vincent (1990) observed this over the southwest Pacific and south Indian Oceans. In a Pacific case study they reported that convection came from a developing tropical cyclone in the South Pacific Convergence Zone (SPCZ).

Perhaps the continuous thunderstorms described in 6.3.1.2 are sometimes triggered by surges from both north and south, directed toward a weakly cyclonic circulation in the surface layers, and away from an overlying upper tropospheric anticyclone.

8.4.4.3. Summary. Rapid wave-like propagation with convergence in the surface winds flowing toward the heat equator, matched by divergence and heat export in the oppositely directed upper tropospheric flow, causes surges to suddenly "appear" in the tropics; once there, the surface speed maximum and bad weather downstream of the maximum may not move much for several days. This could account for stationary bad weather in the low-level easterlies over East Africa (Johnson, 1962) and near-absence of moving disturbances over Indonesia (Braak, 1921-29). Upstream of the speed maximum, weather is better than usual; in the winter monsoon and the tradewinds, the inversion lowers.

In surges, heat is exchanged between the tropics and the winter hemisphere; sometimes, between the tropics and higher latitudes in the summer hemisphere. Heat export in the upper-tropospheric part of a surge causes instability and contributes to convection along the forward edge of the surface surge. Cause and effect are hard to determine; convergence in low levels, divergence aloft, and enhanced convection coincide and arise from distant causes; the surprising variability of tropical weather (Lau and Lau, 1990) may reflect this. Surges contribute to climatology. They are most common during winter in longitudes dominated by continents; Africa, the Americas, and East Asia/Australia. There, low-latitude rainfall is heaviest and upper-tropospheric flow to colder latitudes (sometimes across the equator) is greatest and the subtropical jets are strongest (Yang and Webster, 1990). Even in summer, upper-tropospheric outflow from the Sudan into a jet maximum over Turkey fits the pattern (Figure 4.15).

Surges range widely in strength, extent, duration, and weather effects. Except with intense polar outbreaks, they can seldom be anticipated. Regional and annual variations are poorly understood. But long-time evidence from disparate sources confirms their importance. Their effects on tropical weather demand to be investigated; as pointed out by Keenan and Brody (1988) for the Australian summer monsoon, these effects may well have been overemphasized. In the meantime, tropical forecasters should watch for changes in surface wind speeds within extensive monsoon and tradewind streams. An isolated report of stronger wind associated with a thunderstorm, needs careful evaluation before being dubbed a surge.

Forecasters must avoid repeating the error of their predecessors with fronts in the 1930s and easterly waves in the 1950s; not all mysterious weather is surge-induced!

8.4.5 South Pacific Convergence Zone (SPCZ). The charts of average cloudiness (Figures 6-1 to 6-4) show a persistent maximum extending east-southeastward from New Guinea. It also appears as a relative minimum in annual average OLR (Figure 6-22), though not as well-marked south of 10° S as the cloudiness maximum. The clouds are deepest from October to March.

8.4.5.1 Winter. Hill (1964) analyzed two periods of extensive altostratus between Australia and Fiji in June of 1962 and 1963. Light rain fell. In both cases, a remnant cold front was moving northeastward beneath divergent flow east of a trough in the upper-tropospheric westerlies. The divergence maximum was created by passage of an isotach maximum. In the lower troposphere, sinking dried the air. Hill suggested a model for this decoupling between upper and lower troposheres (Figure 8-40) that may also account for the "sandwich" described in 8.4.6.1. Middle- and upper-tropospheric upward motion responded to the jet-stream maximum, while lower-tropospheric sinking accompanied outflow from the surface anticyclone.

Upper troughs that slow down and often become stationary cause the climatological cloudiness maximum. At 1600Z 2 July 1985 (Figure 8-41, next page), a well-developed SPCZ extended southeastward from New Guinea. It lay in divergent northwesterlies east of an upper trough that had recently crossed the Australian coast. The western edge of the SPCZ cloud sheet coincided with sharply shearing upper-tropospheric winds west of a strong jet stream that increased (diverged) downstream from 55 knots over southeastern New Guinea to 150 knots at 27° S, 160° E. For a few days, the system was almost stationary.

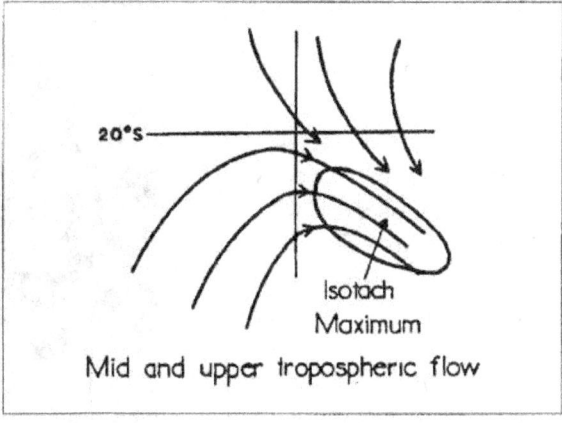

Mid and upper tropospheric flow

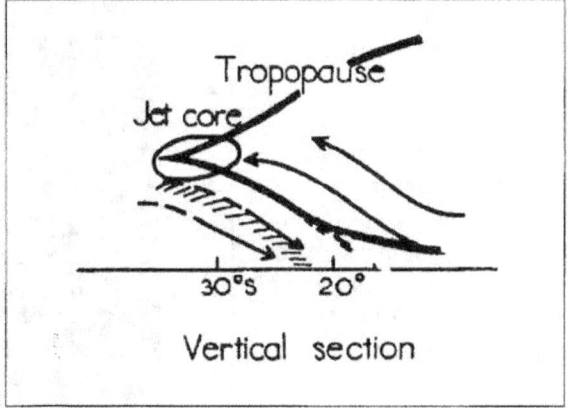

Vertical section

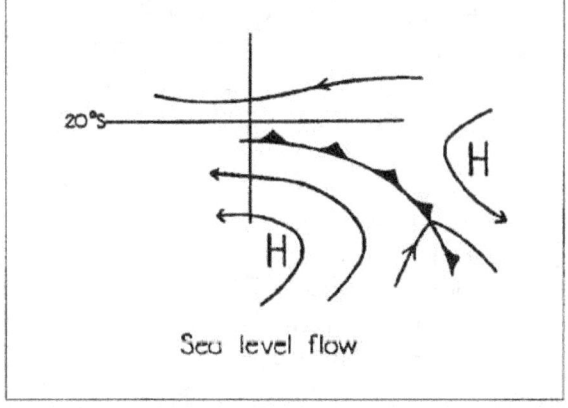

Sea level flow

Figure 8-40. Three-dimensional synoptic model of an altostratus layer overlying a surface front in the region of the southwest Pacific convergence zone (SPCZ) during winter. The north-south line usually lies near 160° E (Hill, 1964).

At the surface in the SPCZ, southeasterly tradewinds prevailed, reaching 25 to 30 knots beneath the jet stream. A weak depression remained near 10° S, 160° E. Observers reported skies overcast with middle and high clouds, little or no convection, and scattered, generally light rain. The jet stream and its accompanying bad weather moved east on 4 July, along with the upper-tropospheric trough; the tradewinds moderated.

The Hill model seems to describe spring and autumn situations east of Asia (Guard, 1990). Frequently, in the region of cold fronts, dense cloud below 20,000 feet (6 km) underlies dense cirrus above 26,000 feet (8 km), with little cloud between. Anticipating this distribution helps in planning air refueling.

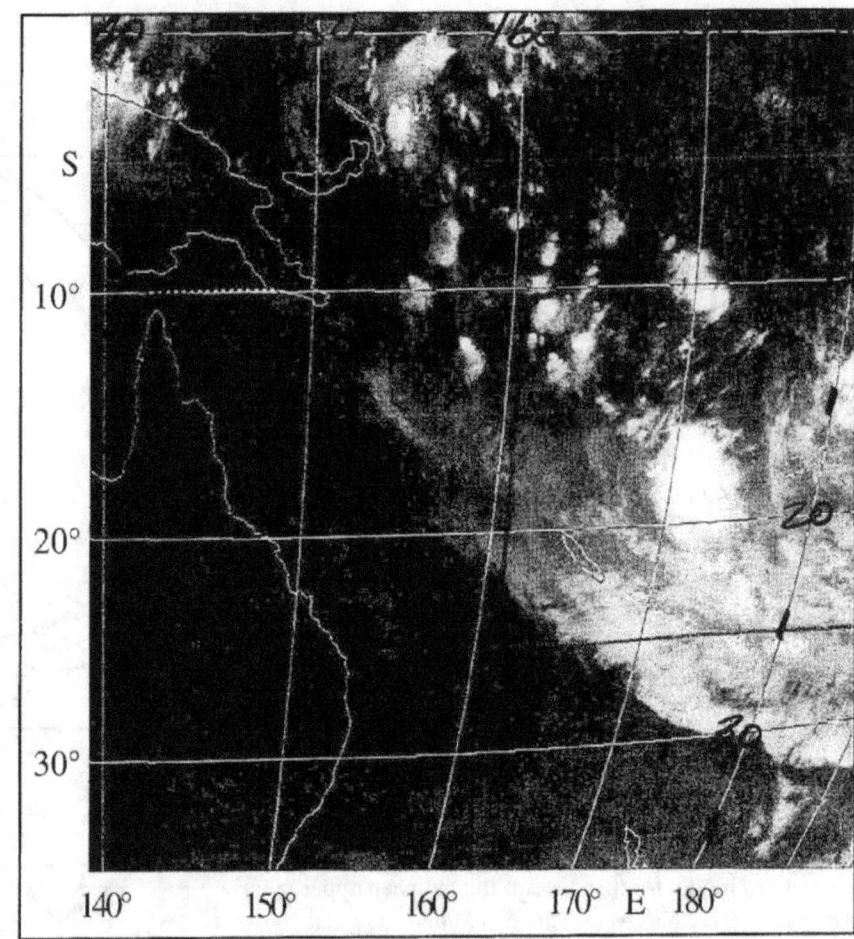

Figure 8-41. GMS IR image at 1600Z 1 July 1985 of an SPCZ in its normal climatological position.

8.4.5.2 Summer. In summer, the convergence zone lies between surface northeasterlies to the north and south-easterlies to the south, and includes a climatological low east of New Guinea. As in winter, the SPCZ is almost stationary. Vincent (1982, 1985) analyzed the SPCZ for the period 10-18 January 1979—see Figure 8-42. Along an upper trough, west of the surface trough, upper divergence, lower convergence, and a minimum in OLR coincided. Three Tropical depressions formed and moved east or southeast. Such developments are common during summer; depressions may remain almost stationary in the convergence zone for many days before moving away. Some quickly intensify into tropical storms or hurricanes. Forecasting is difficult. Unnecessary warnings may be given if early slight intensification is extrapolated. On the other hand, a low that has remained stationary for days can catch a forecaster by surprise if it suddenly intensifies and starts to move. Sequences of satellite pictures can help.

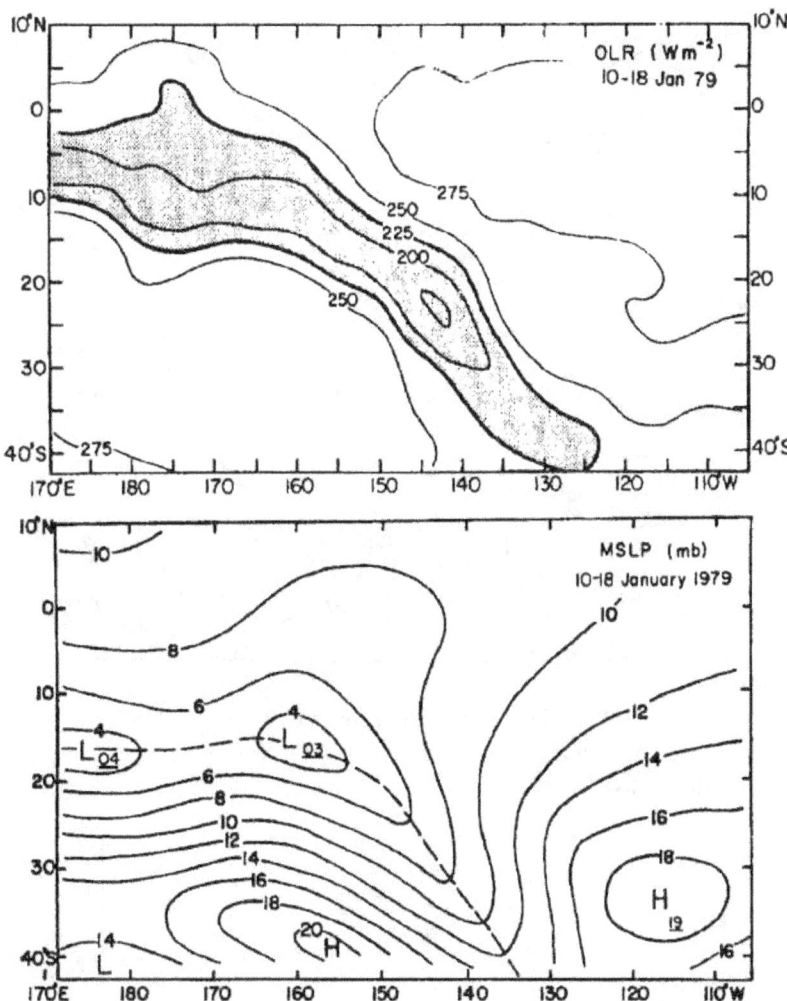

Figure 8-42. 10-18 January SPCZ in the southwest Pacific. Top: Average outgoing longwave radiation (W m⁻²); the area < 225 W m⁻² is shaded; it probably contained deep convection. Bottom: Average mean sea-level pressure (Vincent, 1985).

8.4.6 Troughs in the Upper-Tropospheric Westerlies. In middle latitudes, rising motion ahead of these eastward-moving troughs intensifies fronts and may cause wave cyclones to form. Except over Asia and North Africa in summer, the troughs often extend into the tropics and may even reach the equator. Figure 8-43, on the next page, (Ramage, 1971) shows a wintertime trough almost stationary over Indochina after having moved eastward south of the Himalayas. A front lay across the northern part of the South China Sea, with fresh surface northeasterlies to the north.

As upper-level convergence caused strong subsidence down to 930 mb (Figure 8-44), skies were clear west of the upper trough (Figure 8-45). East of the upper trough, upper-level divergence produced ascent and extensive cloud. In the south, upglide along the front (SSW winds), and heat and moisture added to the cold surface air by coastal waters, almost saturated air below and above the front. Visibilities were poor; little rain fell. Wang et al. (1985) ascribed springtime heavy rain over Taiwan to a similar interaction between surface front and upper trough.

225

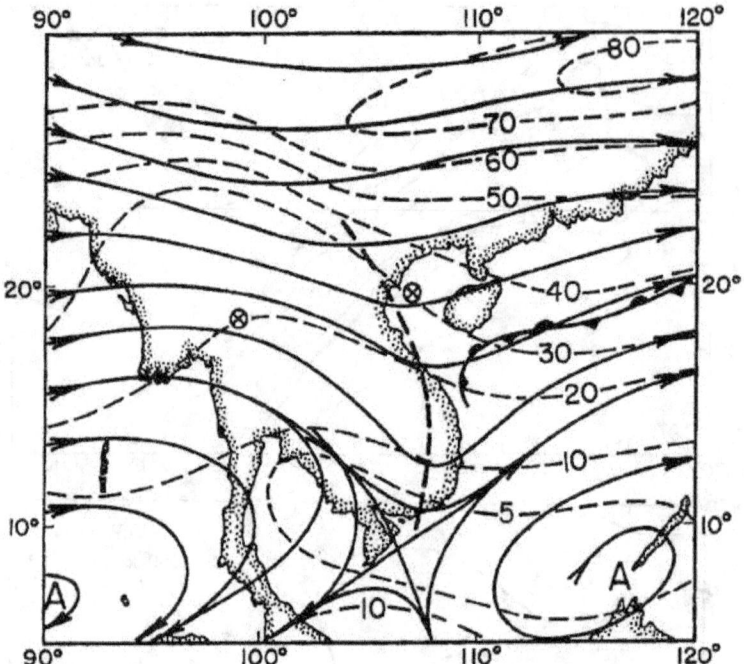

Figure 8-43. Circulation at 300 mb for 0000Z 6 Feb 1968. Isotachs in m s⁻¹. The surface front is shown; the soundings from Figure 8-44 are shown as circled X's.

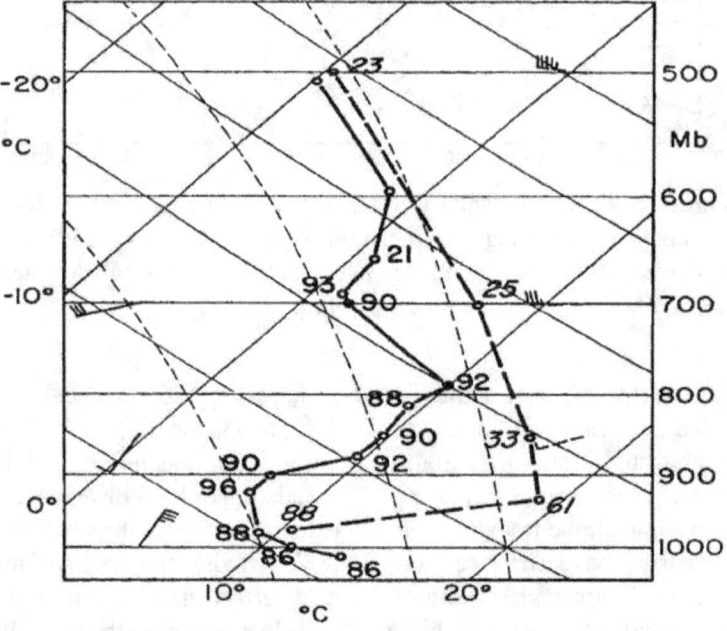

Figure 8-44. Soundings, relative humidities, and winds (one full barb denotes 10 knots) at 0000Z 6 February 1968. Dashed line: West of the trough in the subtropical westerlies (Chiangmai, 19° N, 99° E). Solid line: East of the trough in the subtropical westerlies (USS Belnap, 19° N, 107° E) (Navy Weather Research Facility Staff, 1969).

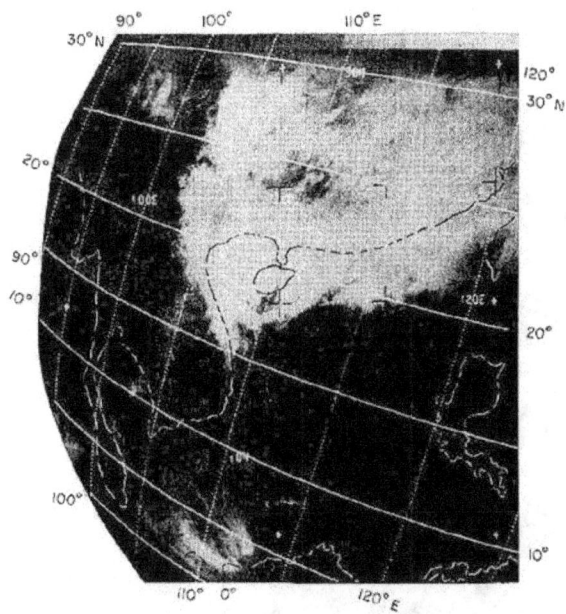

Figure 8-45. Clouds photographed by ESSA 3 at 0500Z 6 February 1968.

During the dry cool season, upper-tropospheric troughs affecting Senegal (13 to 16° N), and very rarely, other parts of tropical North Africa (Seck, 1962) resemble the cool-season troughs of the SPCZ (8.4.5.1). Skies east of the trough become overcast with altostratus; the underlying anticyclonic easterlies are even drier than the South Pacific tradewinds; very little rain reaches the ground. Australia (Southern et al., 1970) and Southern Africa (Fox, 1969) experience similar situations. At the start and end of the cool season, when more moisture is available, frontal thunderstorms and even hail may precede the trough; as shown in Figure 8-46, the cloud zone was moving northeastward ahead of a sharp trough in the upper-tropospheric westerlies. Rain from showers and thunderstorms occasionally exceeded 50 mm.

In Zambia (8 to 18° S) the rainy season lasts from December through March. Kumar (1979) analyzed five rainy seasons (1972-73 through 1976-77), finding that an average of five to seven upper-tropospheric westerly troughs extended into Zambia each month. Rainfall increased east of about 80 percent of the troughs and decreased after they had passed.

Over the Hawaiian Islands, mean resultant 200-mb winds blow from slightly north of west in winter and slightly south of west in summer (Figures 4-13 and 4-15). Troughs in the upper westerlies can occur throughout the year. The most intense, giving heavy rain east of the trough axis, are confined to the cool season. On 19 April 1974, northeastern (windward) Oahu experienced up to 20 inches (250 mm) of rain as it lay east of a stationary trough in the upper-tropospheric westerlies (Schroeder, 1977). Although the tradewinds continued to blow, the inversion disappeared (Figure 8-47, page 8-43) and a "continuous" thunderstorm developed, anchored by the mountains that were also orographically lifting the tradewinds.

If low-level convergence is already concentrated east of the upper trough, say along a front, and if the trough is stationary or slow-moving and surface flow is light or moderate, upward motion east of the trough may then act long enough on a volume of air for deep convection and severe weather to develop. However, rain is not always enhanced east of an upper trough. Although Han concluded (1970) in his study of 300-mb troughs that daily rainfall greater than 0.10 inches (2.5 mm) over northern Oahu is more likely with the trough near to or west of the island than in other locations, average trough-day rainfall (0.15 inches (3.8 mm)) was not much more than average non-trough-day rainfall (0.10 inches (2.5 mm)).

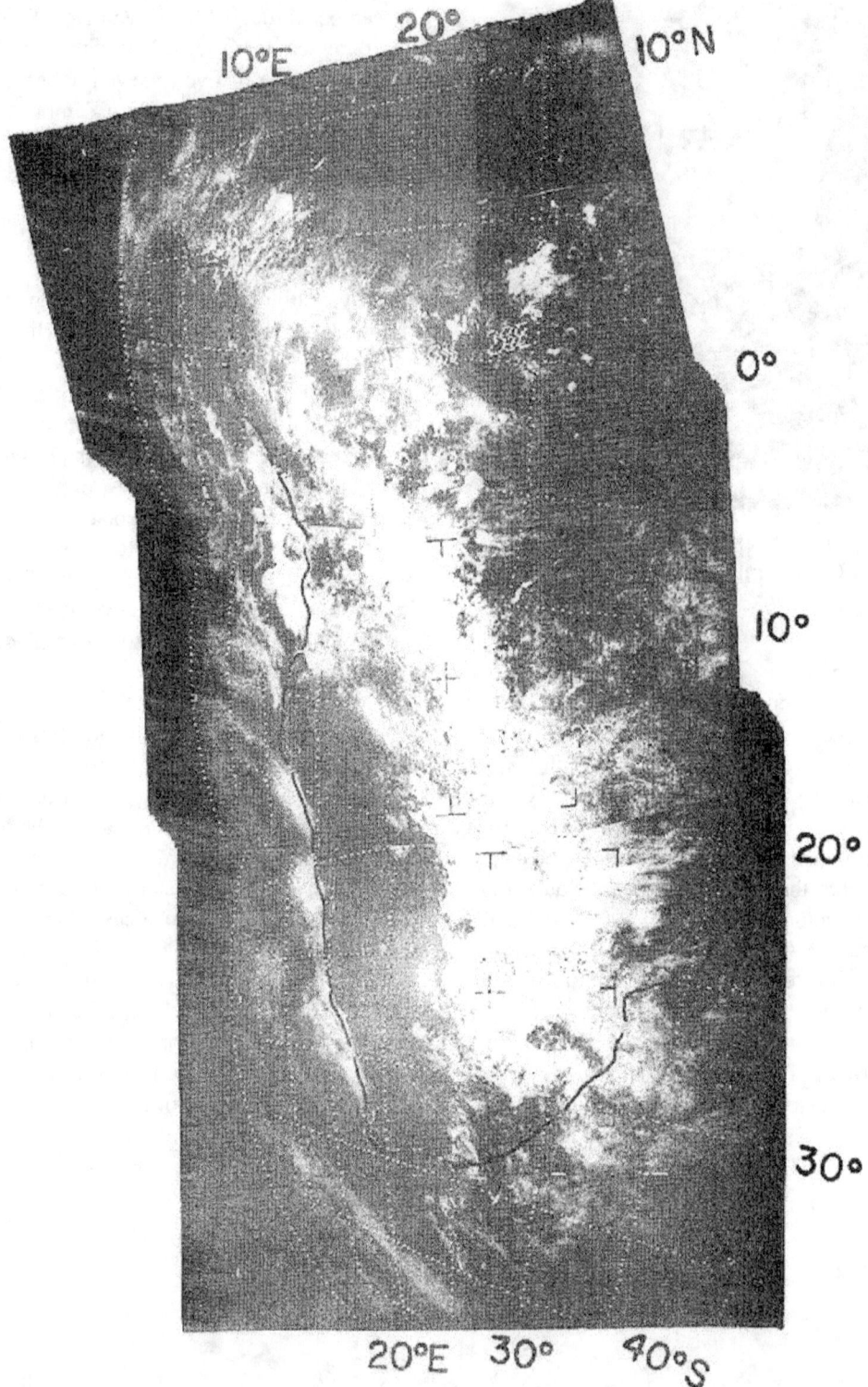

Figure 8-46. Clouds over Southern Africa photographed from ESSA 3, 17 April 1968. (Fox, 1969).

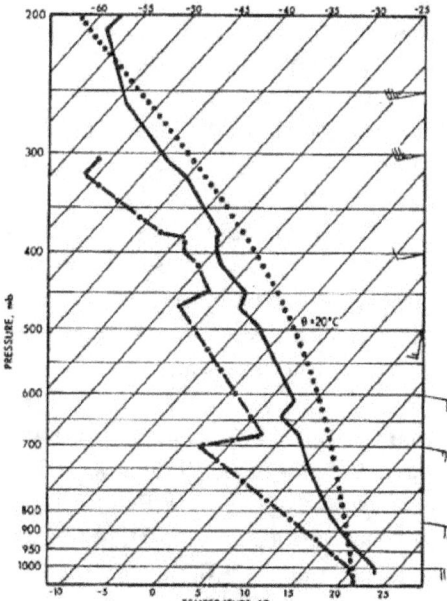

Figure 8-47. Sounding at Lihue (22° N, 159° W) at 0000Z 19 April 1974. Temperature, solid line; dew point, dot-dashed line.

In the tradewinds, the relationship between the height of the subsident dry layer and the direction of the 200-mb wind could indicate vertical motion. For 1976 at Lihue (22° N, 159° W), the median height of the dry layer base was 800 mb with northwest winds at 200 mb and 785 mb with southwest winds at 200 mb.

Figure 8-48 shows that dry-layer bases above 650 mb are much more likely with 200-mb southwesterlies but on two-thirds of occasions the difference between the effects of northwest and southwest flows is negligible.

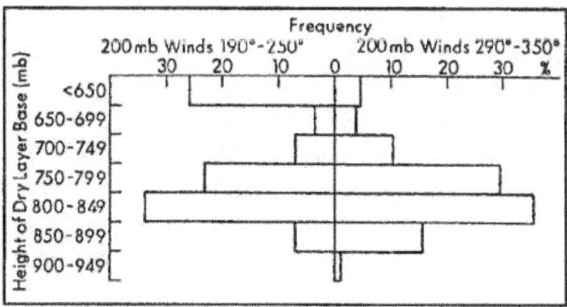

Figure 8-48. Frequency distribution of heights of the dry layer base as a function of 200-mb wind direction, Lihue, 1976.

Often, east of an upper-tropospheric trough, an extensive sheet of altostratus is separated by a subsiding dry layer from the tradewind inversion and the moist tradewinds beneath. This condition may last several days. The dry layer prevents any rain falling from the altostratus from reaching the surface. Rising motion east of the trough is apparently decoupled from the lower troposphere by the widespread subsidence associated with the subtropical anticyclone. This, along with another possible cause of "moist-dry-moist sandwiches," are discussed in the following paragraphs.

8.4.6.1 Moist-Dry-Moist "Sandwiches." In the region of Hawaii on about 20 percent of days east of an upper-tropospheric trough, a moist-dry-moist sandwich is observed. The sounding shown in Figure 8-49, made during a prolonged rain spell on the top of Mt. Waialeale (see Figure 6-21) is typical. Moderate moisture convergence caused by rising motion east of the trough causes middle cloud to develop within a normally dry zone, but why rising motion fails to extend to the surface is puzzling. More puzzling is the fact that moist-dry-moist sandwiches are just as common beneath upper-tropospheric northwesterlies, but generally unaccompanied by middle cloud.

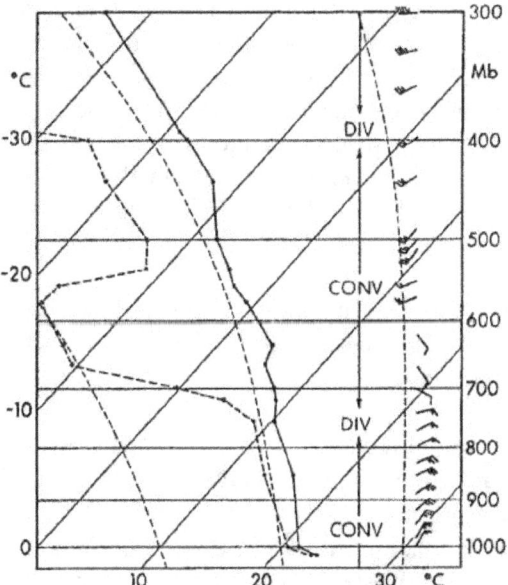

Figure 8-49. Lihue Sounding at 0000Z, 15 October 1976. Solid line: temperature; dashed line, dew point.

During the Line Islands Experiment (LIE), 70 percent of the soundings at Fanning Island (4° N, 159° W) had sandwiches. Of 1,200 dropsonde observations made over the eastern tropical Pacific in 1979, 68 percent had sandwiches (Kloesel and Albrecht, 1989), suggesting that in convectively suppressed conditions two distinct cloud layers may exist. The lowest OLR of any tradewind area, recorded east of Hawaii (Figure 6-22), as well as dryness above 700 mb east of the upper trough in Figures 8-44 and 8-45, could also be evidence of the phenomenon.

In the absence of middle-tropospheric moisture convergence *east* of an upper trough, how does moisture accumulate in subsiding air? Kloesel and Albrecht (1989) postulated the nearby existence or pre-existence of deep cumulus, which moisten levels above the inversion by advection, and whose downdrafts bring relatively drier air to the top of the inversion. Detailed observations are needed to test this hypothesis.

Decoupling of upper and lower tropospheres also occurs in winter over the southwest Pacific (see 8.4.5.1) and North Africa, and may be not uncommon elsewhere in the tradewinds and the winter monsoon. Forecasters must often choose between little rain or a storm east of an upper- tropospheric trough. Upper winds often fail to reveal whether upper and lower tropospheres are coupled (Figure 8-47) or decoupled (Figure 8-49). Weather satellites may not help much, since they view only the upper cloud deck, even though in a storm, this will be denser and colder than in the decoupled mode.

Interaction between a trough in the upper westerlies and the near-equatorial tradewind convergence is discussed in 8.4.8.

8.4.7 Superposition of Tropical and Extratropical Disturbances. Between 15 and 25° latitude, disturbances of tropical origin are typical of summer; whereas in winter, troughs in the upper-tropospheric westerlies may appear. Rising motion

east of these troughs is often insufficient to break down a tradewind-type inversion and bad weather seldom results (8.4.5 and 8.4.6). Occasionally, a remnant tropical vortex may be moving westward at a low latitude as an upper-tropospheric trough approaches from the west. As the two systems superpose, the relatively deep moist layer accompanying the vortex is further deepened east of the upper trough, and an unseasonable wet spell may result. Then, as the two systems draw apart, weather becomes seasonally normal again.

On 1 December 1962 (see Ranganathan and Soundarajan, 1965; Ramage, 1971), a weak tropical vortex moving across southern Peninsular India became stationary off the west coast, just east of an almost stationary trough in the upper-tropospheric westerlies. From the 2nd through the 4th, both systems intensified, and widespread rain resulted (see Figures 8-50 and 8-51).

At Bombay (19° N, 73° E) (Figure 8-52), southwesterlies prevailed above 500 mb from the 3rd through the 5th. In the lower troposphere, pressure fell on the 3rd and 4th as the tropical vortex intensified, and southerlies set in. Moist air had spread through the troposphere by the 4th. Then as the upper trough moved eastward, subsidence dried the air above 700 mb and the rain ceased. The southern vortex had moved westward and both systems quickly weakened. At Bombay, 1.9 inches (48 mm) of rain fell in just over 24 hours on the 4th and 5th. The Bombay December average is 0.08 inches (2 mm); four out of five Decembers in Bombay are completely dry.

Superposition can also occur in a transition season. In September 1970, a late monsoon depression was moving west over India. Instead of weakening, it intensified as it came under the influence of three early waves in the middle- and upper-tropospheric westerlies. It stalled before recurving eastward. Rainfalls reached 11.8 inches (300 mm) (Mishra and Singh, 1977).

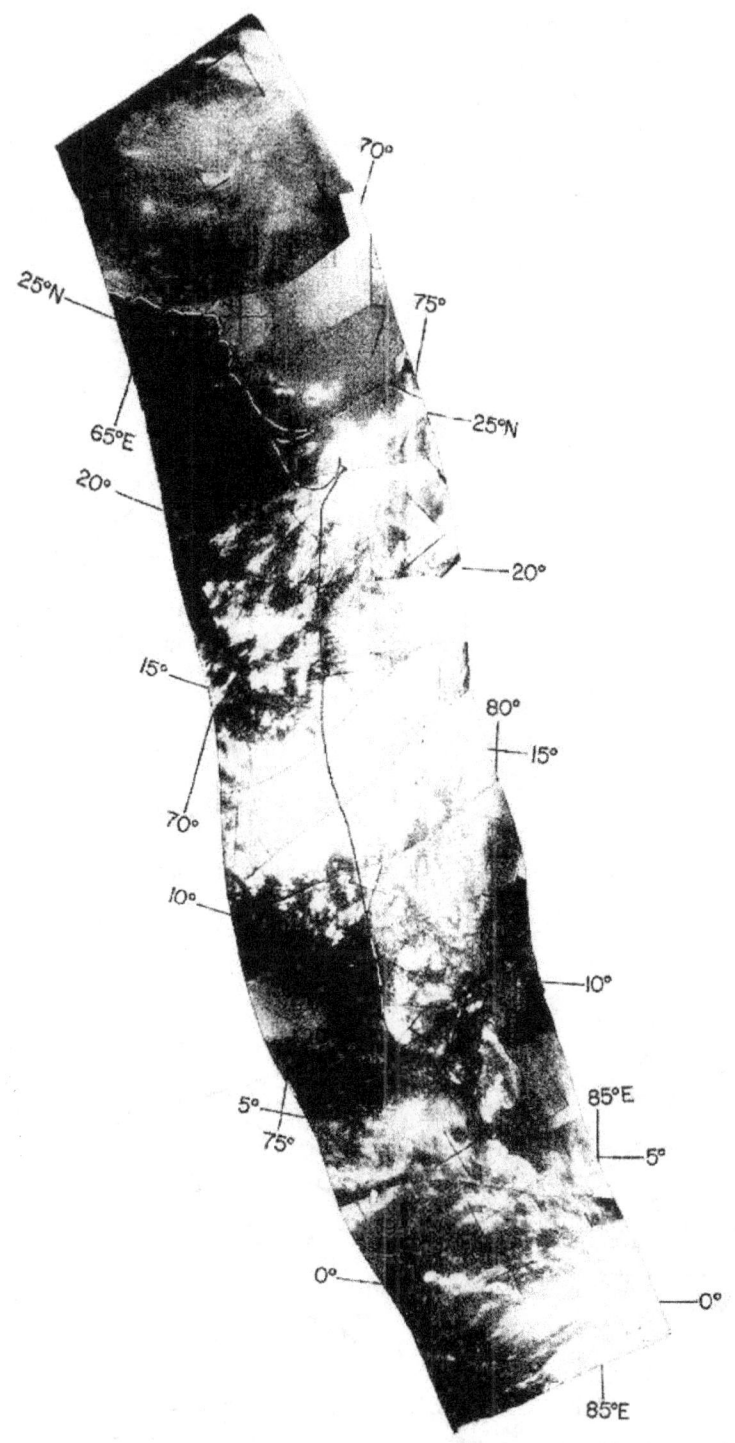

Figure 8-50. TIROS VI cloud photos taken at 0900Z 4 December 1962 when an upper trough and a lower disturbance were superposed.

Frank (1969) showed that as westward-moving cloud systems in the North Atlantic moved beneath southwest flow east of an upper-tropospheric trough, convection rapidly increased, and decreased as they moved west of the upper trough. According to Simpson (1970b), these changes stem from initial stimulation and then constraint of the upper-level outflow as the lower system moves beneath the upper trough (see Figure 8-53, opposite).

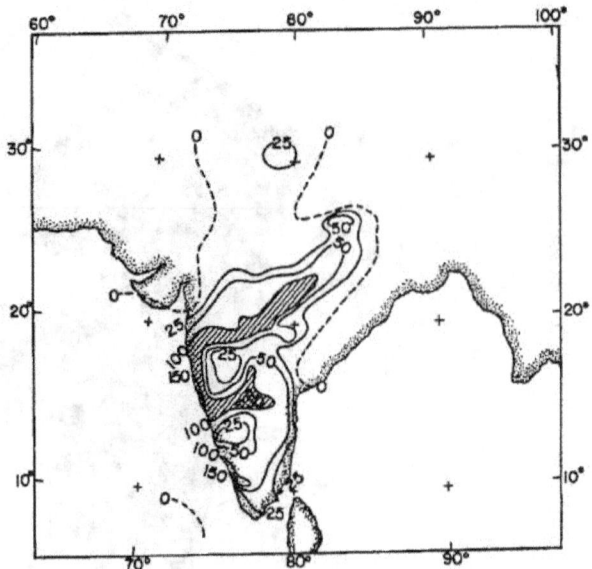

Figure 8-51. Total rainfall (mm) over India from 1 through 6 December 1962.

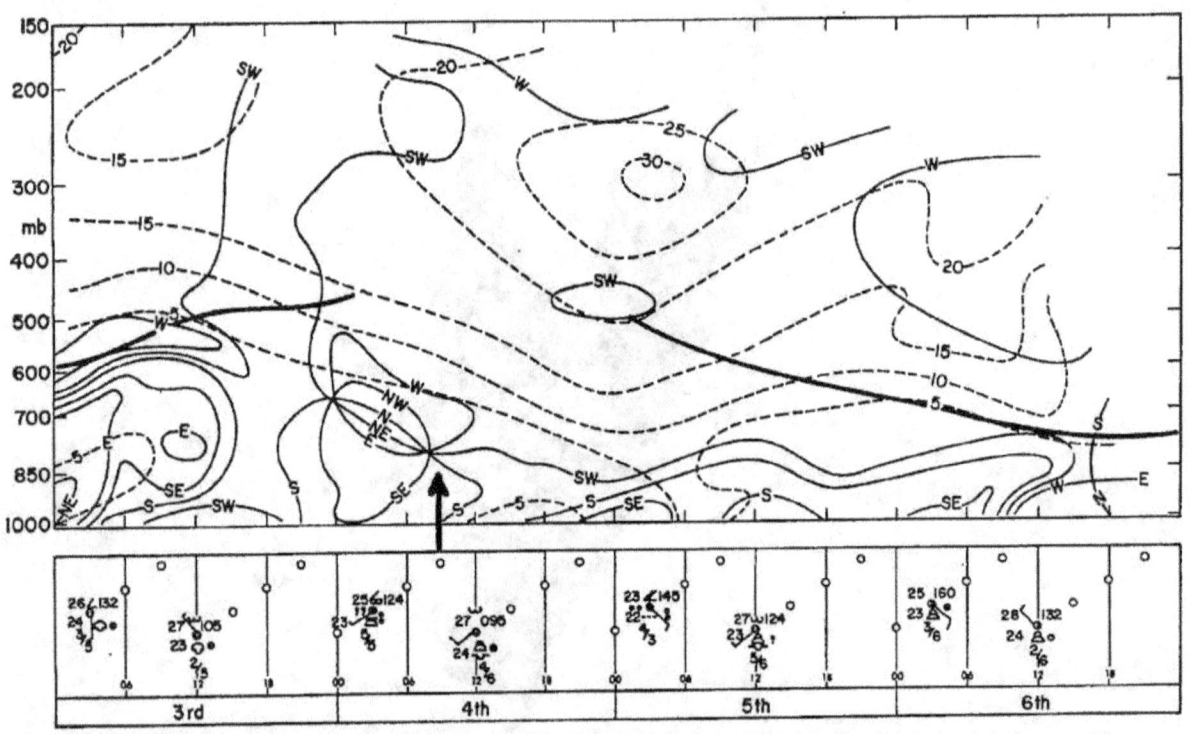

Figure 8-52. Time cross-section for Bombay (19°N, 73° E) from 3 to 6 December 1962. Solid lines are isogons; dashed lines: isotachs (ms⁻¹); thick lines: upper limit of RH more than 50 percent. The arrow marks the time of the satellite photo in Figure 8-50.

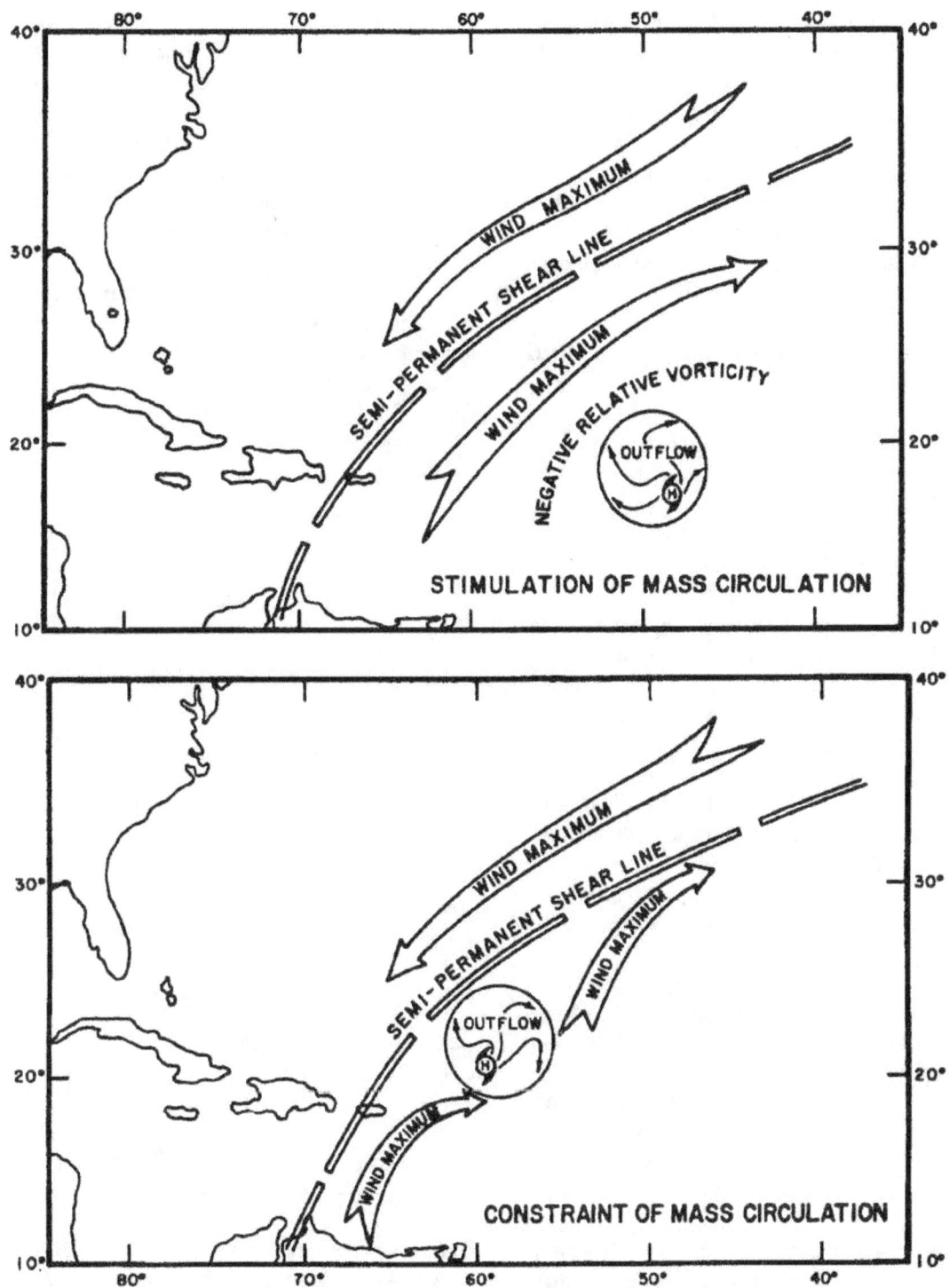

Figure 8-53. Role of the Atlantic upper-tropospheric trough in the stimulation or constraint of mass circulation in a tropical cyclone. As the cyclone approaches the wind maximum south of the trough (anticyclonic-shear area), mass circulation in the cyclone is stimulated. As the cyclone passes across the upper-trough axis and moves beneath cyclonic shear, mass circulation is constrained (after Simpson, 1970b).

Since superposition is unseasonable, forecasters should be on the alert for the unseasonable disturbance. In the first example in Figure 8-50, that would be the tropical vortex; in the second, both the monsoon depression and the upper westerly troughs. Predicting that two approaching disturbances will superpose is not too hard, but anticipating when they will interact and for how long can be difficult. Frequent satellite pictures that show deepening clouds help greatly.

8.4.8 The Near-Equatorial Tradewind Convergence (NETWC).

8.4.8.1 Convergence Over the Oceans. Over the southeast Pacific and the South Atlantic, the southeast tradewinds cross the equator and converge with the northeast tradewinds between 5 and 10° N, where a cloud band (the NETWC) is found.

A similar convergence occasionally appears over the western Indian Ocean during transition seasons between the monsoons (see 6.3.6.4). In previous chapters, mean charts of gradient-level winds, cloudiness, and OLR all confirm that the most persistent disturbed weather in the tropics occurs along the near-equatorial tradewind convergence of the North Atlantic and the central and eastern North Pacific, where ships report rain more than 30 percent of the time; elsewhere over the tropical oceans, 20 percent is rarely exceeded (Crutcher and Davis, 1969).

Not only are there more rain days under the NETWC, but the rainfall per rain day is less than elsewhere. Two coral islands, Washington (5° N, 160° W) and Kwajalein (9° N, 168° E), have the same average annual rainfall (2,592 and 2,630 mm) but ships report twice as many rain days at the former, influenced by the NETWC, than at the latter, where moving disturbances usually cause the rain.

The NETWC never moves east-west and only slowly north-south, probably because it overlies a warm eastward-moving ocean current. OLR measurements show that NETWC clouds are not as deep as those over the equatorial continents and Indonesia (Figure 6-22). This was confirmed by research Aircraft flying over the cloud at 20,000 feet (6 km), as well as by rare satellite-detected lightning discharges (Figure 10-2) (Orville and Henderson, 1986) and the absence of cirrus in Nimbus 7 satellite pictures (Stowe et al., 1989). The cloud tops may well be overlain by remnants of the tradewind inversion.

Twelve meteorological/oceanographic research flights shuttled between Honolulu and Tahiti from 29 November 1977 to 5 January 1978 (Ramage et al., 1981) (Figure 8-54). They revealed large intensity changes in the central Pacific NETWC, apparently stemming from changes in the South Pacific and North Pacific tradewinds.

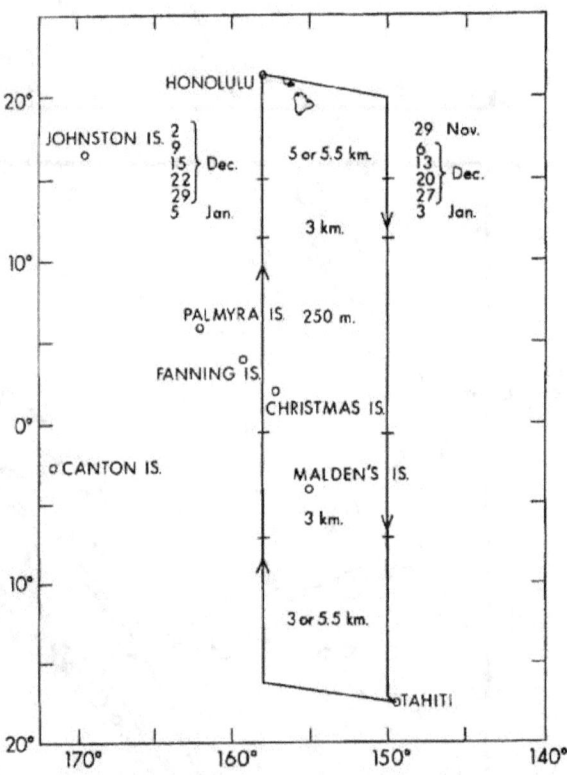

Figure 8-54. Flight tracks and altitudes flown by NOAA P-3 research aircraft, November 1977-January 1978 (Ramage et al., 1981).

Clouds in satellite pictures of the NETWC along 150° W; 158° W varied equally. For the period encompassing the shuttle flights (38 days), intensities were the same along 150 and 158° W (476 NM (882 km) apart) on only 15 days and persisted from one day to the next on 15 days at 150° W and on 11 days at 158° W.

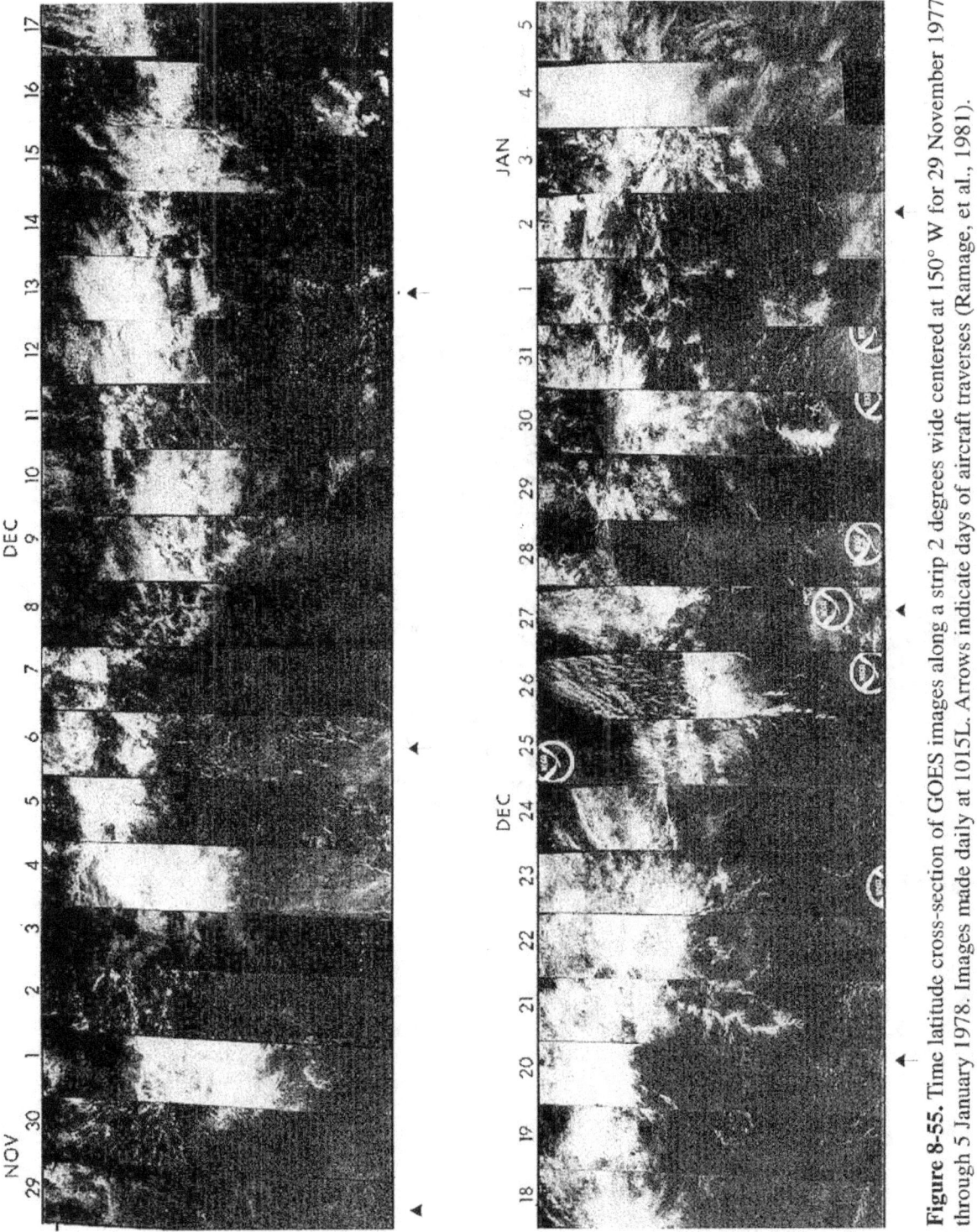

Figure 8-55. Time latitude cross-section of GOES images along a strip 2 degrees wide centered at 150° W for 29 November 1977 through 5 January 1978. Images made daily at 1015L. Arrows indicate days of aircraft traverses (Ramage, et al., 1981).

235

In the NETWC region the aircraft flew at an altitude of 820 feet (250 meters). The northern and southern edges of the convergence zone were sharply defined in the wind and moisture fields no matter how severe the weather.

When the NETWC was active, it comprised lines of convergence and rain cloud, separated by relatively dry strips in which the air was sinking. There were no thunderstorms.

The schematic cross section of a vigorous NETWC (Figure 8-56) is based on shuttle flights and on subsequent 20,000 feet (6 km) flights over the NETWC along 110° W. The easterly component of the wind goes into the paper. Air follows helical trajectories with east-west axes. Consequently, air parcels are kept within the cloud long enough for droplets to grow to raindrops and fall out. The lines of cloud spread out just beneath the raised tradewind inversion; the uniform surface thus presented to weather satellites often hides the lines beneath.

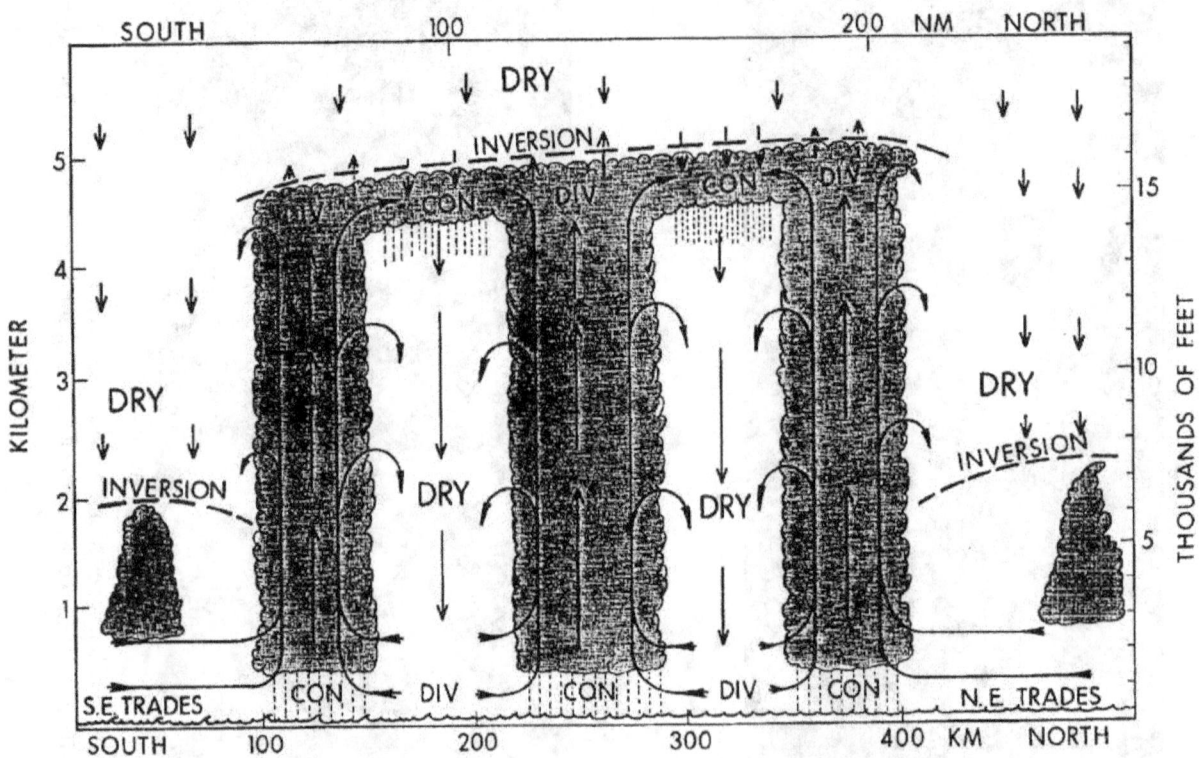

Figure 8-56. Schematic meridional cross-section of a vigorous near-equatorial tradewind convergence.

The NETWC is a mechanical phenomenon, resembling in some ways the orographically-lifted cloud lines that account for most of the rain on Mt. Waialeale (6.3.2). The rapid intensity changes seem to arise from mutually independent velocity changes (some possibly caused by surges) in the NH and SH tradewinds, and from complex interactions with the upper troposphere. Equatorward-moving shear lines might sometimes be responsible, but small-scale local fluctuations are probably more common. No evidence of westward-moving easterly waves has been found.

A trough in the upper-tropospheric west-erlies, often preceded by a wide band of middle and high clouds (see 8.4.6), may interact with the NETWC (Giambelluca, 1986). In the Pacific, as the trough ap-proaches from the west, it induces the tradewinds to become more easterly and weaken. Convergence into the NETWC then diminishes and clouds break up. The middle- and upper-tropospheric rising mo-tion associated with the cloud band just east of the upper trough may either be decoupled from the NETWC or induce deep convection and thunderstorms in it. Once the upper trough has passed, the tradewinds freshen and shift from east back to northeast, and the NETWC is re-established.

McGurk and Ulsh (1990) averaged 35 cases of interaction between a trough in the upper-tropospheric westerlies and the NETWC during the winter of 1983-84 over the eastern North Pacific. They used vapor images from GOES to estimate moisture distribution. Before the trough extended far enough south for interaction to occur, NETWC moisture (cloudiness) diminished to the east. At the interaction, north of the NETWC, moisture greatly increased east of the trough and diminished west of the trough. The response was similar, but much less in the NETWC. By comparison, when upper-tropospheric troughs were absent, the NETWC was generally moister and more convectively unstable. During the period studied, troughs and NETWC were interacting 80 percent of the time over the eastern North Pacific.

As shown in Figure 8-57, an upper-tropospheric trough approaching Hawaii from the west at 0000Z on 2 December 1985 was preceded by a vigorous STJ, clearly delineated in the satellite image by a broad band of middle and high clouds. A well-developed NETWC lying along about 9° N was somewhat weakened just ahead of the upper trough. Conforming to the McGurk and Ulsh averages, the line linking deepest cloud to the east and least cloud to the west of the trough lay about 5° north of the NETWC.

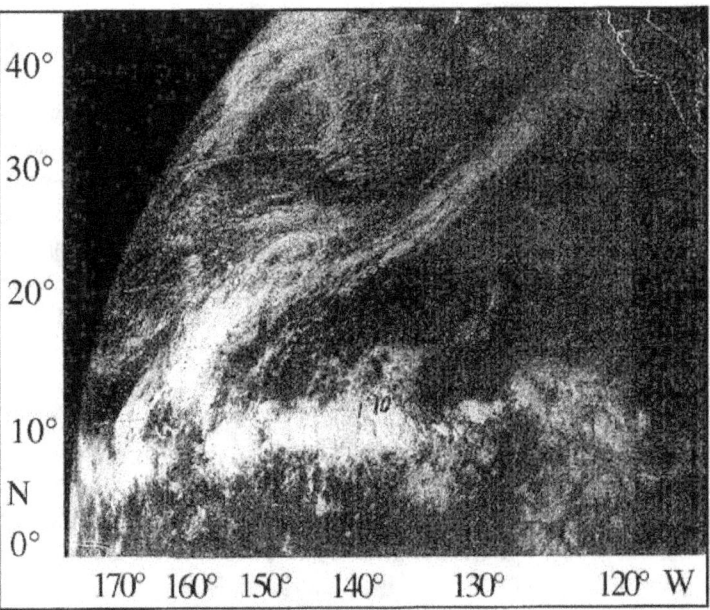

Figure 8-57. GOES visible image (0000Z 2 December 1985) of an upper-tropospheric trough interacting with a near-equatorial tradewind convergence.

Over Southeast Asia during winter, the northeast monsoon extends to the equator (Figure 4-9). Then, as with the NETWC, an approaching trough in the upper-tropospheric westerlies weakens the monsoon and improves equatorial weather (Figures 8-43 and 8-45).

Except when a trough in the upper-tropospheric westerlies is involved, flying weather in the NETWC is not severe. Turbulence is slight or moderate and jet aircraft easily top the clouds. At the surface, prolonged rain under one of the cloud bands might significantly lower visibility.

8.4.8.2 Convergence Over Northern South America.
Over South America, surface pressure decreases westward to climatological heat lows over western Colombia, the Gulf of Panama, and over Bolivia. In response, easterlies blow down the gradient, and the Atlantic NETWC extends as far west as the Andes (Walters et al., 1989). Intensified by orography, weather is usually worse and more variable along the continental NETWC than out to sea, and thunderstorms are common (Figure 10-2). As over the oceans, the convergence is farthest north in the

NH autumn and farthest south in the NH spring and oriented roughly east-west. In March/April, it enters the Brazilian coast south of the Amazon mouths, moves rapidly north in May and June, and reaches 10° N in August/September. Then follows a rapid return to south of the Amazon by December. Western Amazonia and the eastern foothills of the Andes, within 5° of the equator, experience little annual variation in rainfall. Farther south and farther north, rains are heaviest in the summer. As over the oceans, weather some distance from the NETWC is generally fair, with the tradewind inversion limiting convection. When the inversion rises or weakens, popcorn cumulus develops during the afternoon.

Extensive field programs in Venezuela during the wet seasons of June to October of 1969 and 1972 (Riehl, 1979b) and over the Amazon basin during the dry season (July-August) of 1985 (Garstang et al., 1988) and during the wet season (April-May) of 1987 (Greco et al., 1990) included research aircraft flights and special aerological soundings. The Venezuelan programs are extensively documented; papers on the Amazon projects are still appearing. The regions adjoin, and although the topographies differ, disturbances may affect both at the same time, producing more rain near the NETWC. What follows summarizes many research findings.

In the NH summer, when the NETWC often lies over Venezuela, surface westerlies sometimes appear along the north coast; they may be as deep as 7,000 to 10,000 feet (2 to 3 km), suggesting development of a weak monsoon trough to the north. Rarely, disturbances that resemble monsoon depressions form in the trough and cause heavy rain to fall on the eastern slopes of the Andes (Riehl, 1979b).

Farther east, weak and shallow vortexes have been detected by research aircraft. A surge from the SH may cause the westerlies. When they are absent, Venezuelan wet-season rainfall is less than normal, especially along east-west segments of the coast, where the tradewinds diverge in response to stress-differential (see Figure 2-8).

Over South America north of 10° S and east of the Andes, easterly winds prevail below 300 mb throughout the year (Chu, 1985), while an easterly jet has been observed near 10,000 feet (3 km) over Venezuela. Squall lines, resembling those of West Africa, develop in this environment or even over the Atlantic, and move westward (Fernandez, 1982). The associated cloud clusters are oriented roughly north-south and account for most of the rainfall. During their passage, they disrupt the NETWC.

Surges from either hemisphere may generate squall lines. A favored source region is just off the east coast between 7° N and 5° S, where land breezes and tradewinds converge at night and cause vigorous convection. By late afternoon, the squall line has moved west and intensified over the coastal mountains. It continues inland at speeds ranging from 5 knots (weak) to 32 knots (strong). Squall lines may also develop over land. The associated cloud clusters undergo Diurnal variations as they move. Extensive cloud clusters tend to be most active at night; when convection is scattered, it is most common during the afternoon and evening.

Annual variation in rainfall is controlled by the meridional movement of the NETWC and by the frequency of squall lines. There are more of these in the wet season than in the dry season. Rainfall may vary considerably from year to year, in response to variations in NETWC latitude and number of squall lines. Northeast Brazil is especially prone to droughts. Only when the NETWC moves anomalously south of its normal southernmost latitude is northeast Brazil assured of adequate rain (Hastenrath, 1985).

Forecasts should incorporate westward movement of weather systems, allowing for large ranges in speed and duration. Fair tradewind weather and scattered convection can occur at any time of the year, but are more likely during the dry season.

Conversely, although squall lines can occur year-round, they are more common in the wet season, and give most rain near the NETWC. Forecasters may get a jump on squall-line development by detecting surge generation from middle latitudes.

8.4.9 Midsummer Dry Spell. Over Southeast Asia, this dry spell shows up in long-term average pentad rainfall (Figure 6-38) and can be followed synoptically. Figure 8-58a, from Sadler et al. (1968), shows a situation in which a ridge line had moved

northward over southeast Asia by 15 July 1967; satellite pictures revealed decreasing convection. As shown in the figure, The ridge continued northward on 19 July. Figures 8-58b-d are 700-mb analyses for 11, 15, and 19 July 1867.

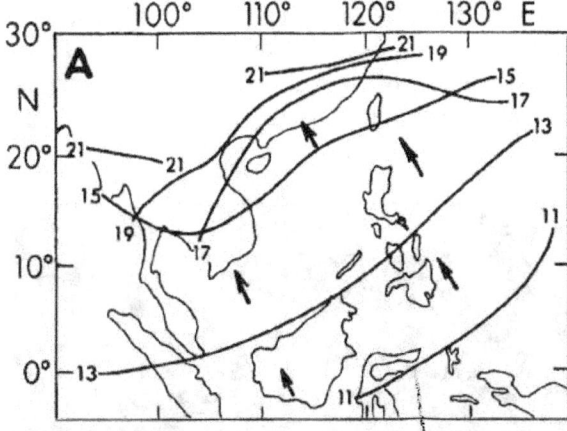

Figure 8-58a. Positions of the 700-mb ridge on alternate days from 11 to 21 July 1967 (adapted from Sadler et al., 1968).

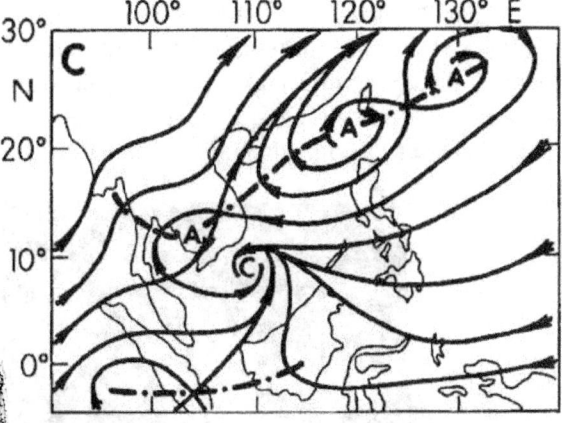

Figure 8-58c. 700-mb analysis, 15 July 1967 (adapted from Sadler et al., 1968).

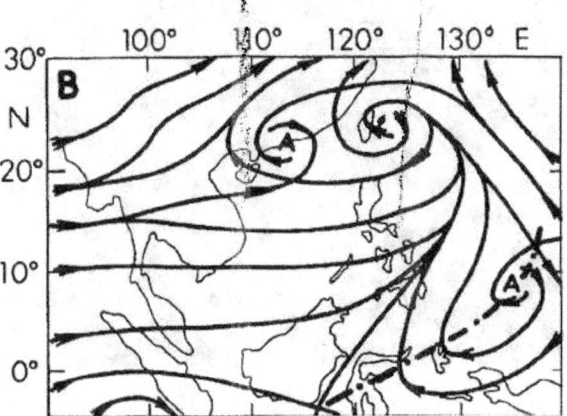

Figure 8-58b. 700-mb analysis for 11 July 1967 (adapted from Sadler et al., 1968).

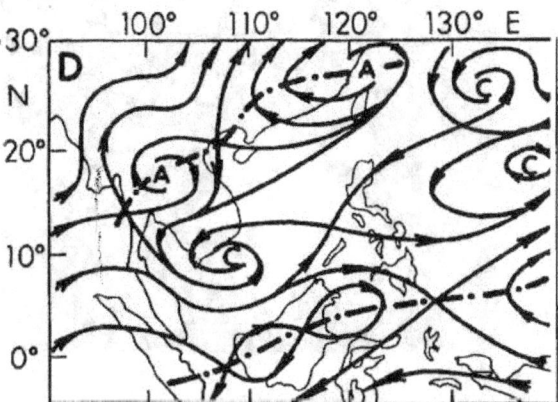

Figure 8-58d. 700-mb analysis, 19 July 1967 (adapted from Sadler et al., 1968).surge generation from middle latitudes. Southern hemisphere winter surges ahead of vigorous cold fronts may be the easiest to anticipate.

Cheng (1978), in a detailed study of the dry spell used 1-day rainfall averages and concluded the the spell usually comprises three shorter periods. Of these, 92 percent were associated with ridges and 8 percent occurred in the region of divergence west of a typhoon.

On 11 July 1985, the weather satellite photo in Figure 8-59 showed a distribution typical of the mid-summer dry spell. A ridge of high pressure, accompanied by fine weather, extended along about 20° N from the central Pacific to Burma. It persisted from 10 to 18

July across South China and the northern South China Sea.

Over India, "breaks" in the summer monsoon rains occur when the monsoon trough, usually lying along the Ganges Valley, moves northward to the Himalayas and is replaced by a ridge (Rao, 1976). At times, this shift can extend as far east as the Philippines. Although a break is more likely in the second half of August (Figure 6-39), it can occur at any time in the summer and may be a manifestation of the 30- to 60-day oscillation (see 12.4.2).

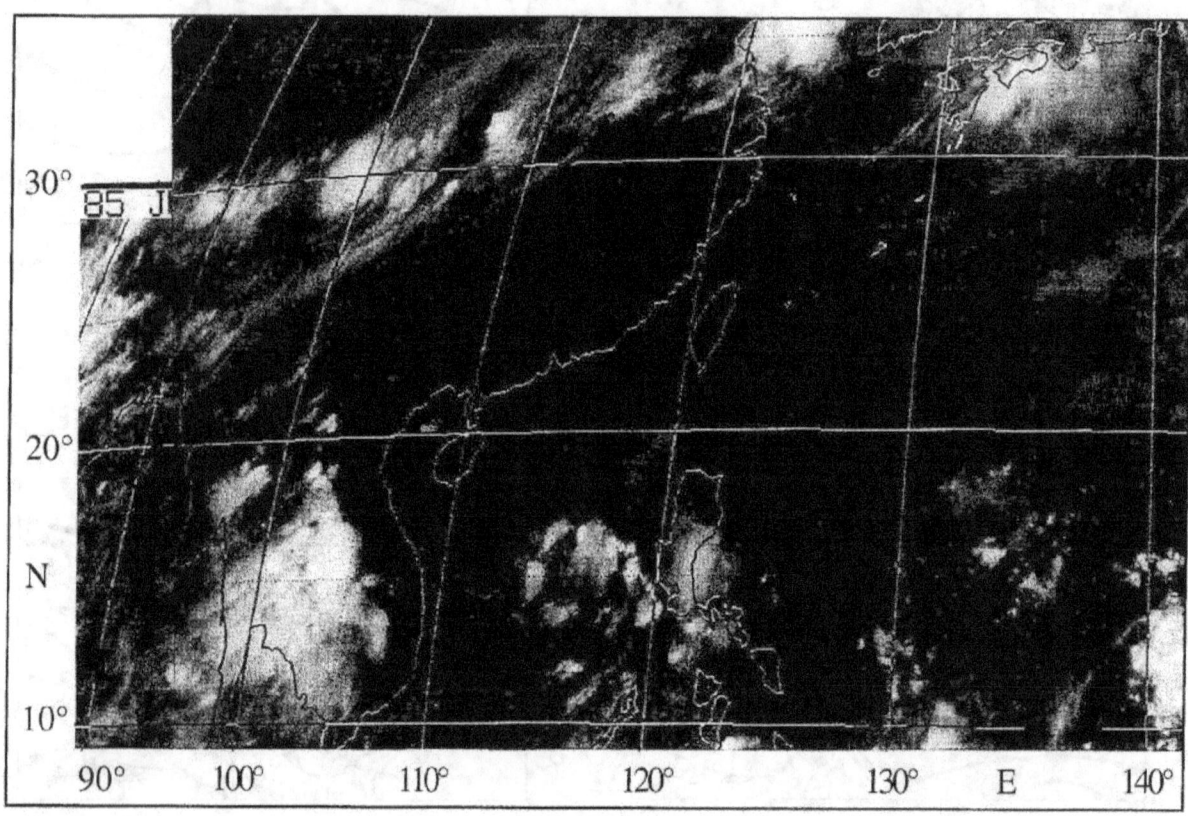

Figure 8-59. GMS IR image (1600Z 11 July 1985) showing a midsummer dry spell over South China and the South China Sea.

Chapter 9

TROPICAL CYCLONES

9.1 GENERAL

The general term "tropical cyclone" is subject to various interpretations. In this report, the term is used in its broadest sense: that is, as a "non-frontal, synoptic-scale cyclone, developing over tropical or subtropical waters and having a well-organized circulation." This definition implies nothing about wind speed or weather; however, the more destructive tropical cyclones are warm-cored and have strongest winds near the surface that decrease with height.

Intense tropical cyclones (i.e., hurricanes or typhoons) are the tropics' most impressive phenomenon. The destructive potential of their high winds, rain, and storm tides is well known; tropical meteorologists must be thoroughly versed in them. Most research in tropical meteorology concentrates on tropical cyclones. There are many references to this subject, but only a few will be mentioned here. Useful general works include those by Riehl (1979a), the World Meteorological Organization (1979), Anthes (1982), and Elsberry (1987). Satellite observations of tropical cyclones and their analysis were extensively discussed by Dvorak (1984) and by Dvorak and Smigielski, (1990). In this chapter, the essential global characteristics of tropical cyclones are summarized with sections on structure, classification, climatology, formation, dissipation, movement, and forecasting.

9.2 STRUCTURE OF MATURE TROPICAL CYCLONES

9.2.1 General. Even though tropical cyclone forecasting is highly centralized, meteorologists must be familiar with the major features of intense tropical cyclones so as to properly interpret and apply these forecasts to terminal and local weather conditions. Detailed observations, especially those obtained by reconnaissance aircraft, have enabled the structure of tropical cyclones to be described. Satellite pictures have also contributed significantly. This section uses an excellent survey paper by Miller (1967) in discussing the characteristics of the wind, temperature, and cloud distributions in tropical cyclones. Wherever they occur, tropical cyclones have similar features.

9.2.2 Winds. Low-level storm circulation comprises three distinct areas:

• In the outer portion, extending from the storm periphery inward to the edge of the zone of maximum winds, wind increases toward the center.

• The annulus of maximum winds surrounding the eye is the most outstanding feature of the mature tropical cyclone. It is about 5 to 10 NM (10 to 20 km) wide and coincides with the wall cloud, site of the most vigorous convection and the heaviest rain in the storm.

• The eye is the innermost part of the storm. Here the wind weakens rapidly toward the eye center. As determined by the radius of the eye-wall cloud, the eye can be as little as 3 NM (6 km) to as much as 100 NM (185 km) across.

Izawa (1964) prepared composite models of the circulation in Pacific typhoons, based on 14 storms approaching Japan. His composite vertical cross section of the mean tangential speeds is shown in Figure 9-1. Between the surface and 3,000 feet (1 km), the wind increases considerably, but between 3,000 and 20,000 feet (1 and 6 km) there is little change. Izawa attributed the lower surface winds to friction since most of his data was collected from coastal and large island stations.

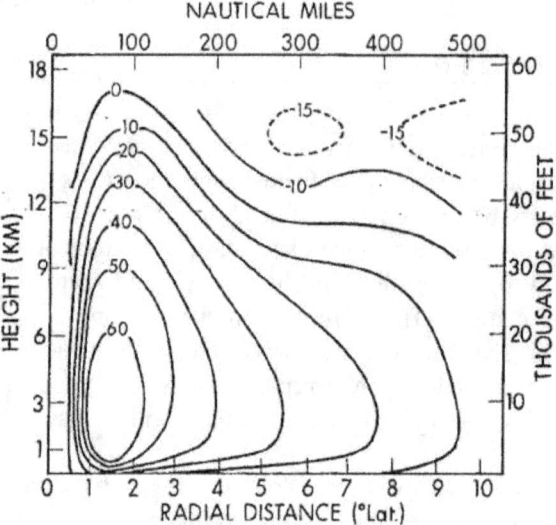

Figure 9-1. Vertical radial cross section of the mean tangential velocity (knots) in 14 Pacific typhoons. Positive values, cyclonic circulation; negative values, anticyclonic circulation (after Izawa, 1964).

Over the ocean, winds increase less with height near the surface. In general, western Pacific typhoons are larger than North Atlantic hurricanes. Tropical cyclone advisories for the North Atlantic and North Pacific include the observed and forecast radii of 50-knot sustained winds in addition to the maximum sustained winds.

The circulation of intense tropical cyclones extends upward to around 46,000 to 49,000 feet (14 to 15 km), close to the tropical tropopause. Since the cyclones are warm-cored, cyclonic circulation weakens with height, as shown in Figure 9-1. But up to about 20,000 feet (6 km), wind shear in the vertical is small. The circulation may be divided into three layers.

The inflow layer extends from the surface to about 10,000 feet (3 km). It contains a pronounced component of motion toward the storm center. This inflow is largely confined to the planetary boundary layer, below 3,000 feet (1 km). However, tropical cyclone circulation composites constructed by Gray (1978) show inflow extending up to 23,000 feet (7 km).

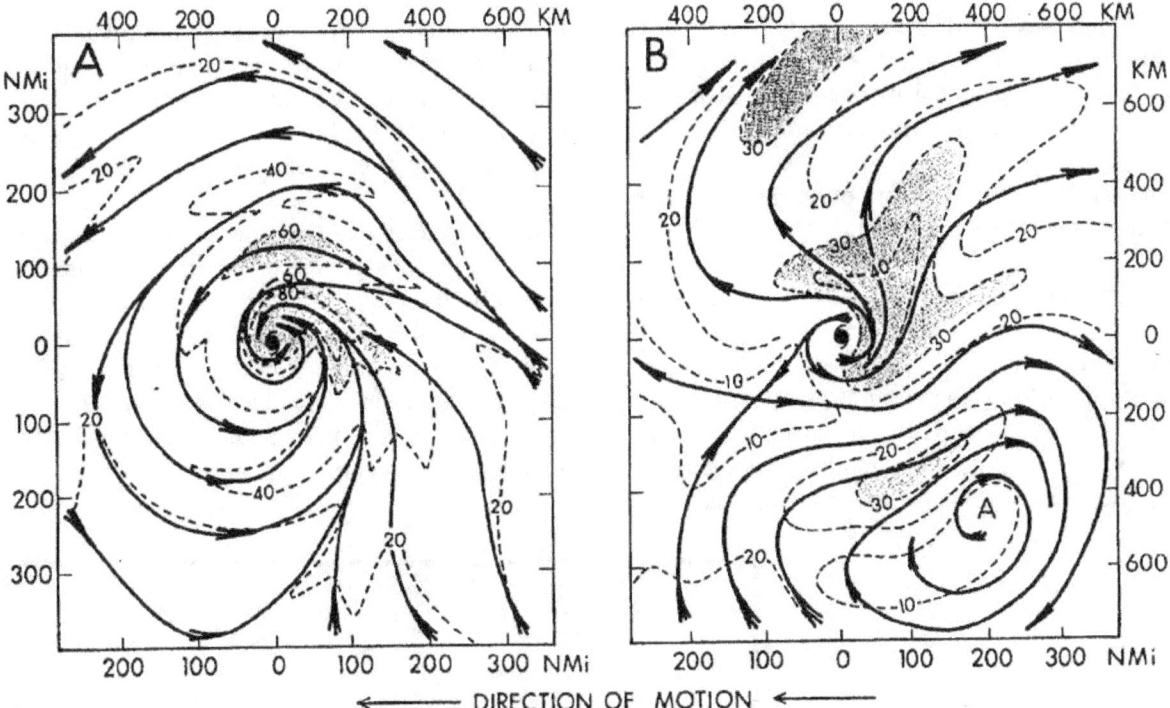

Figure 9-2. Kinematic analyses of (A) the lower-tropospheric and (B) the upper-tropospheric circulation in Hurricane Donna on 10 September 1960. Areas of speed maxima are shaded (after B.I. Miller, 1967).

In the middle layer from about 10,000 to 25,000 feet (3 to 7.6 km), the flow is mostly tangential, although Gray would disagree.

The outflow layer extends from 25,000 feet (7.6 km) to the top of the storm with maximum outflow in mature storms near 39,000 feet (12 km). The characteristic horizontal circulation patterns in the inflow and outflow layers are shown in Figure 9-2. Low-level inflow is most pronounced in the rear semicircle. Typically, winds are strongest to the right of the direction of movement and in the eye wall.

In the upper troposphere, outflow is cyclonic. This circulation is much smaller than near the surface and is surrounded by anticyclonic flow (corresponding to the negative values of the tangential winds in Figure 9-1). The strongest winds and strongest divergence extend outward in the right semicircle of the storm. Such a pattern is often reflected in the cirrus streamers shown in satellite pictures. A storm's outflow pattern depends on the environmental winds and on the dynamics of the storm itself (see 9.5.5).

Very few measurements have been made above 20,000 feet (6 km) in the central regions of intense tropical cyclones; as a result, Figures 9-1 and 9-2B may not be truly representative of the upper tropospheric circulation.

On 17 September 1990 the NASA DC-8 research Aircraft flew at 41,000 feet (12.5 km) across Super Typhoon Flo. The center was located near 24° N, 129° E, and was moving north at 6 to 8 knots. A dropwindsonde in the eye recorded a surface pressure of 891 mb; maximum surface winds were estimated to be 140 knots. Maximum flight-level winds on four crossings of the eyewall averaged 96 knots, a much smaller decrease with height than suggested by Figures 9-1 and 9-2. Even though pressure in the eye was about 2 mb higher than in the eyewall, the circulation at 41,000 feet was strongly cyclonic, and flowed away from the center only beyond a radius of 160 to 270 NM (300 to 500 km).

243

Fett (1964) was one of the first to study hurricanes using manned spacecraft and early satellite photographs. He identified two features found in many cyclones at some time during their life-cycle:

• A relatively clear annular zone of subsidence around the rim of the storm's high-cloud shield.

• An outer convective band beyond this annular zone.

Later, Fujita et al. (1967) proposed a model (Figure 9-3) of the outflow pattern from storms that have both inner and outer rainbands or only inner rainbands. To simplify the analysis, the rainbands are shown as concentric circles rather than the spirals to which they actually conform. They deduced that vertical transport of momentum by outer rainbands alters the outflow wind pattern, resulting in formation of an upper-level shear-line and a cloud-free annulus which separates the inner cirrus shield from the outer rainbands.

In summary, the three-dimensional wind structure of tropical cyclones comprises air flowing into the cyclone in the lower layers, rising primarily in the eye-wall cloud and the other rainbands, and finally flowing outward from the cyclone top and sinking some distance away. Forced sinking inside the eye, by warming the air through adiabatic compression, contributes to low surface pressure.

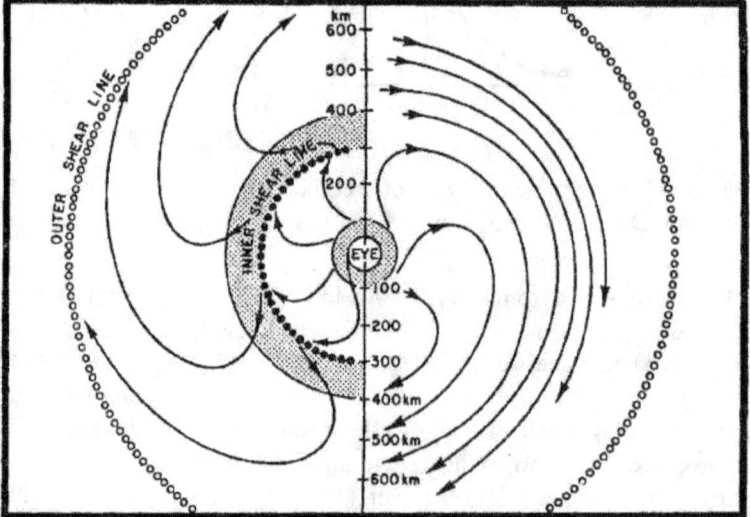

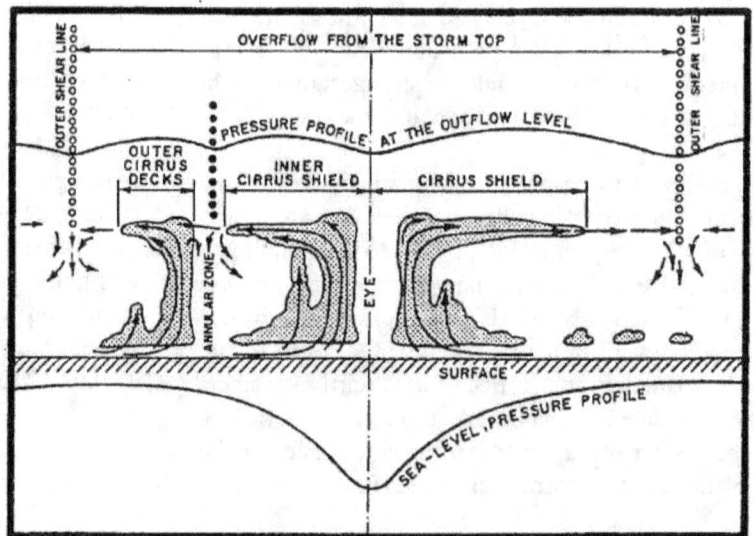

Figure 9-3. Fujita typhoon models. The figure on top shows upper-tropospheric outflow winds; the bottom figure is the vertical cross section. The pressure along both the outer and inner shear lines is a little low, creating convergent outflow winds that subside somewhat (after Fujita et al., 1967).

9.2.3. Temperatures. Tropical cyclones of at least tropical storm intensity are warm-core direct atmospheric circulations in which warm air rises and cold air sinks (except in the eye, where warm air sinks). They convert heat energy to potential energy and potential energy to kinetic energy. Their primary energy sources are latent heat of condensation released in the eye wall and in the spiral rainbands, and sensible heat supplied from the ocean surface.

As surface winds spiral in toward the wall cloud of a tropical cyclone, speed increases along the trajectory. Evaporation from the ocean, and the rate at which latent heat is added to the air, also increase. Along the trajectory, pressure decreases. The resulting adiabatic cooling is counteracted by sensible heat added from the ocean. In the annulus surrounding the eye, the pressure gradient is very large and the winds are strong. In this zone, the air receives most latent and sensible heat from the ocean. This causes the wall cloud to be much warmer than the peripheral air, and so maintains a direct, energy-producing circulation.

Figure 9-4 shows the temperature anomalies measured in Hurricane Cleo on 18 August 1958 (LaSeur and Hawkins, 1963). The upper troposphere warmed most

with temperatures 10° C or more above normal. The largest horizontal temperature gradients occur in the mid-troposphere and are concentrated in a narrow band extending across the wall-cloud.

The record is shared by Super Typhoons Rita on 23 October 1978 (east of Guam) and Vanessa on 26 October 1984 (between Guam and the Philippines). At 700 mb, reconnaissance aircraft measured 31° C in the eyes, for an anomaly of 23° C. Gradients are small within the eye, especially at lower levels. Outside the eye-wall, temperatures in the lower troposphere are slightly below normal, especially in the rear quadrant of the cyclone, where the surface waters have been cooled by evaporation and upwelling as the eye passed. The ultimate source of both latent heat (evaporation) and sensible heat is the warm sea surface (see 9.5.1).

Riehl (1954) has compared tropical cyclones to simple but very inefficient heat engines with only about 3 percent of the total released latent heat being converted into kinetic energy. Much of the rest of the heat is converted into potential energy and exported through the outflow layer.

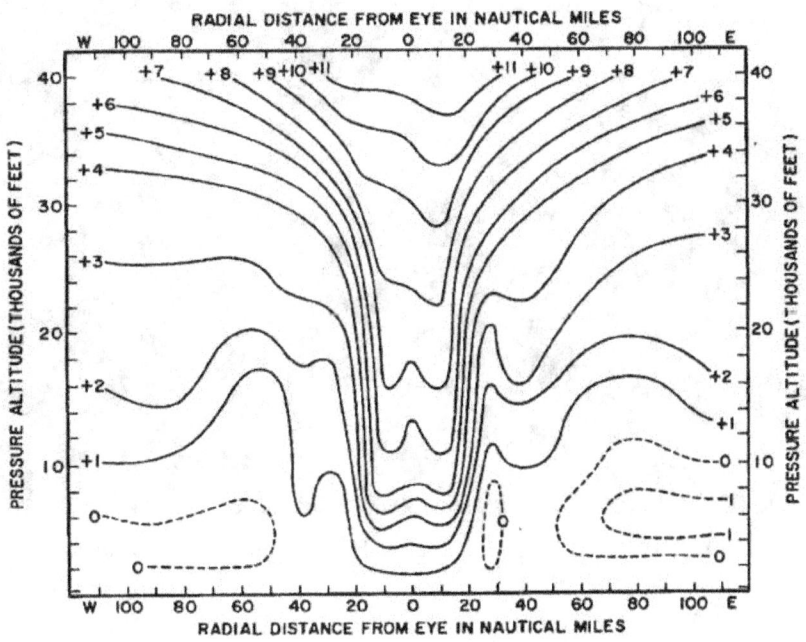

Figure 9-4. Vertical cross-section of temperature anomalies (° C) relative to the mean tropical atmosphere, Hurricane Cleo, 18 August 1958 (after LaSeur and Hawkins, 1963).

9.2.4 Clouds. The major convective cloud systems ("rainbands") in tropical cyclones lie along spirals shown by radar scans and satellite pictures. Upward motion is concentrated in the rainbands and especially in the wall-cloud, where updrafts of 10 to 26 knots have been measured. Vertical transport of heat and conversion of potential to kinetic energy are concentrated in the rainbands.

Figure 9-5 reproduces the Kadena (26° N, 128° E) radarscope as the eye of Typhoon Nelson lay about 85 NM (160 km) southeast of the station, moving northeast at 12 knots. At Kadena, pressure fell to 987 mb, maximum sustained winds reached 38 knots, and rainfall totaled 8.35 inches (212 mm). Note evidence for an inner eye, and how attenuation of the radar echo renders the southeastern wall cloud almost invisible. The cirrus shield near the cyclone center often prevents the convective banded structure from being seen in satellite pictures at visible wavelengths. However, computer techniques can enhance the brighter areas caused by cumulonimbus penetrating the cirrus shield and thus help to locate the rainbands. Microwave imagers can also detect the spiral rainbands beneath a cirrus overcast. High-resolution infrared sensors give more detail on the convective activity. The more intense the storm, the higher the convective tops. It is not uncommon for them to exceed 49,000 feet (15 km), especially in the eyewall.

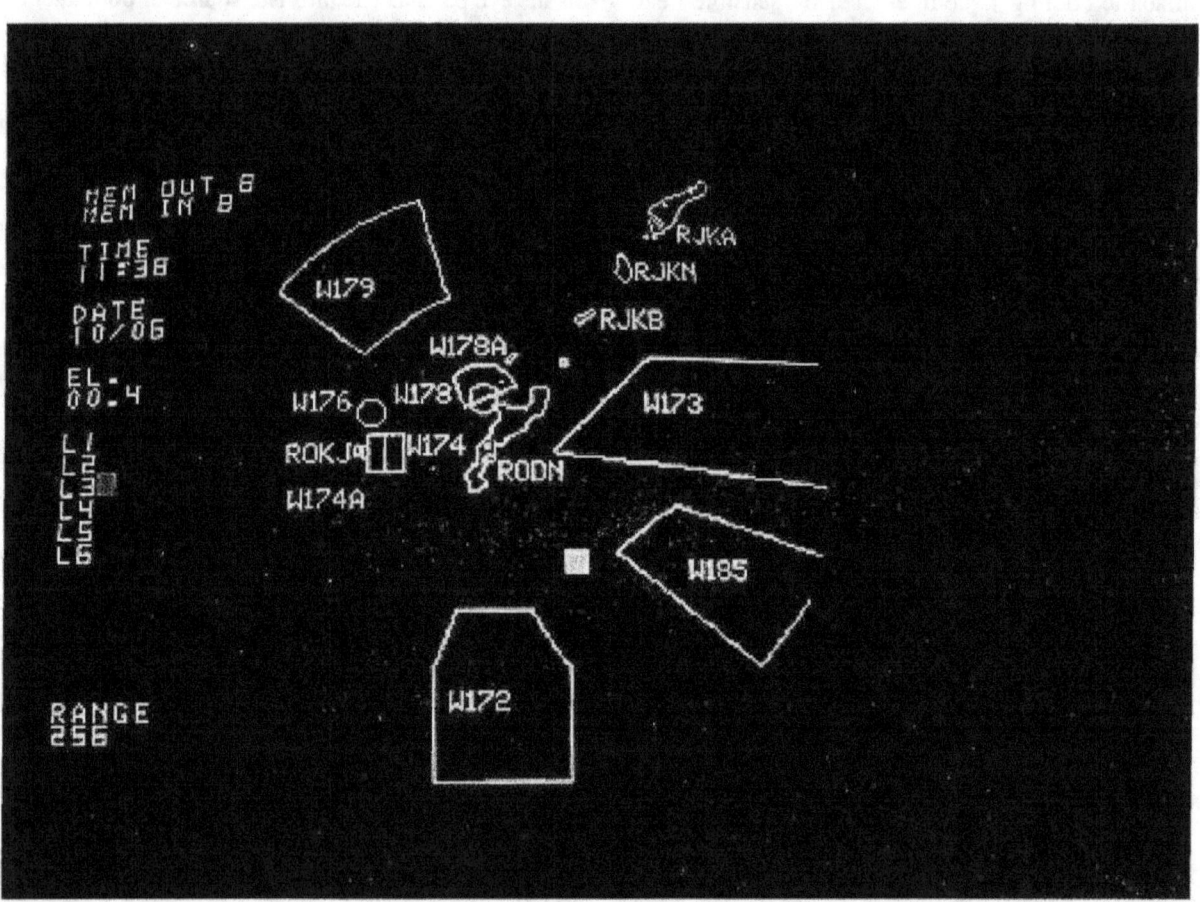

Figure 9-5. Radarscope photo of Typhoon Nelson, centered 157 km southeast of Kadena (RODN) at 1138Z 6 October 1988.

At the time the Figure 9-6a photo was taken, the central surface pressure was 891 mb and the maximum estimated surface wind was 140 knots. Emitting temperature of the uniform mass of cirrostratus surrounding the eye (white) ranged between -70.2 and -75.2° C. Within it, the temperature of protruding cumulonimbus tops (gray) ranged between -76.2 and -80.2° C. As shown in Figure 9-6b, the inner surface of the eyewall ranged over eight enhancement segments (1-8) from the sea surface to -75.2° C.

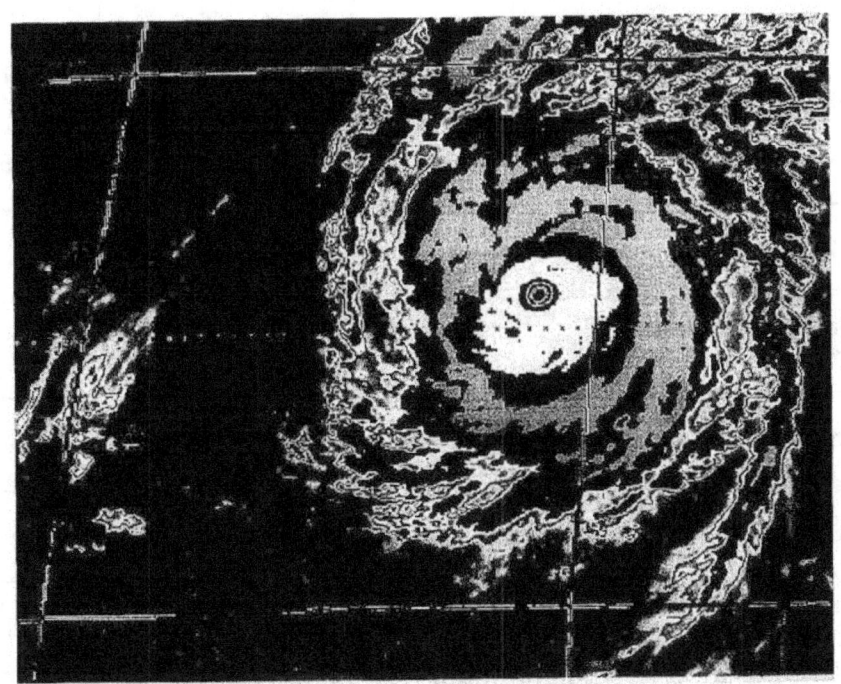

Figure 9-6a. Enhanced GMS IR image of Super Typhoon Flo, centered near 26° N, 129° E at 0538Z 17 September 1990 (from NOAA/NESDIS, 1983).

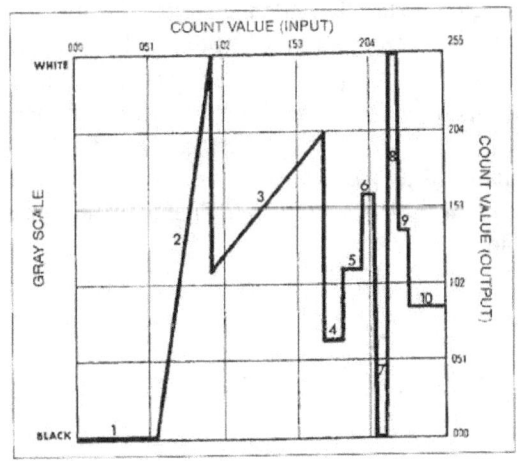

Figure 9-6b. The enhancement curve used to shade-code the image in Figure 9-6a (from NOAA/NESDIS, 1983). The table gives temperature data and commentary for each enhancement segment.

Segment Number	° C Temperature	Comments
1	56.8 to 28.3	No Significant Data
2	27.8 to 9.3	Low Clouds
3	8.8 to -30.2	Cirrus Outflow Pattern
4	-30.7 to -41.2	Dark Gray
5	-42.2 to -53.2	Medium Gray
6	-54.2 to -63.2	Light Gray
7	-64.2 to -69.2	Black
8	-70.2 to -75.2	White
9	-76.2 to -80.2	Top Medium
10	-81.2 to -110.2	Top Dark

9.3 CLASSIFICATION AND DEFINITION OF TROPICAL DISTURBANCES

TABLE 9-1. Areas of occurrence of intense tropical cyclones and regional terminology (obtained from World Meteorological Organization, 1979). In some regions, "Tropical Storm" is subdivided into "Tropical Storm" (34 to 47 knots), and "Severe Tropical Storm" (48 to 63 knots). Extreme surface wind gusts in tropical cyclones may be 30 to 50 percent higher than the reported sustained surface wind.

REGION	RANGE OF MAXIMUM WINDS (KTS)	
	34—63	64—165
Western North Pacific	Tropical Storm	Typhoon
Bay of Bengal and Arabian Sea	Cyclone	Severe Cyclone
South Indian Ocean	Tropical Depression	Tropical Cyclone
South Pacific	Tropical Depression	Tropical Cyclone
North Atlantic and Eastern North Pacific	Tropical Storm	Hurricane

Tropical cyclones form over all the tropical oceans except for the South Atlantic and the South Pacific east of about 130° W. Regional differences in terminology are listed in Table 9-1, above.

In addition to the Table 9-1 classifications, the United States meteorological services define a "tropical depression" as a weak tropical cyclone with a surface circulation incorporating one or more closed isobars, and highest sustained winds (averaged over one-minute or longer periods) of less than 34 knots.

When sustained winds reach or exceed 130 knots, U.S. meteorologists use the terms "super hurricane" or "super typhoon." They define "tropical disturbance" as a discrete system of apparently organized convection, generally 75 to 250 NM (150 to 500 km) across, originating in the tropics or subtropics, having a non-frontal migratory character and having maintained its identity for a day or more. Tropical disturbances may subsequently intensify into tropical cyclones.

9.4 GLOBAL CLIMATOLOGY OF TROPICAL CYCLONES

Tropical meteorologists should realize that historical records of tropical cyclone intensity are incomplete. Over oceans lacking aircraft reconnaissance, the strongest surface winds were seldom encountered by ships or island stations. Even today, storm intensity estimates based on satellite pictures occasionally differ from aircraft reconnaissance observations by about 20 knots. These limitations must be kept in mind when the frequencies of various classes of tropical cyclones are compared. Before weather satellites, many smaller tropical cyclones were missed entirely. Especially in the eastern North Pacific and the South

Indian Ocean, tropical cyclones are observed to be much more numerous with the advent of weather satellite observations. This section presents the global climatology of tropical cyclones. An excellent survey paper by Gray (1978) provided most of the data.

In Figure 9-7, the average number of tropical cyclones per month relative to both the calendar and solar year is presented. Note that the maximum frequency in the SH occurs earlier in the solar year (January) than in the NH (August).

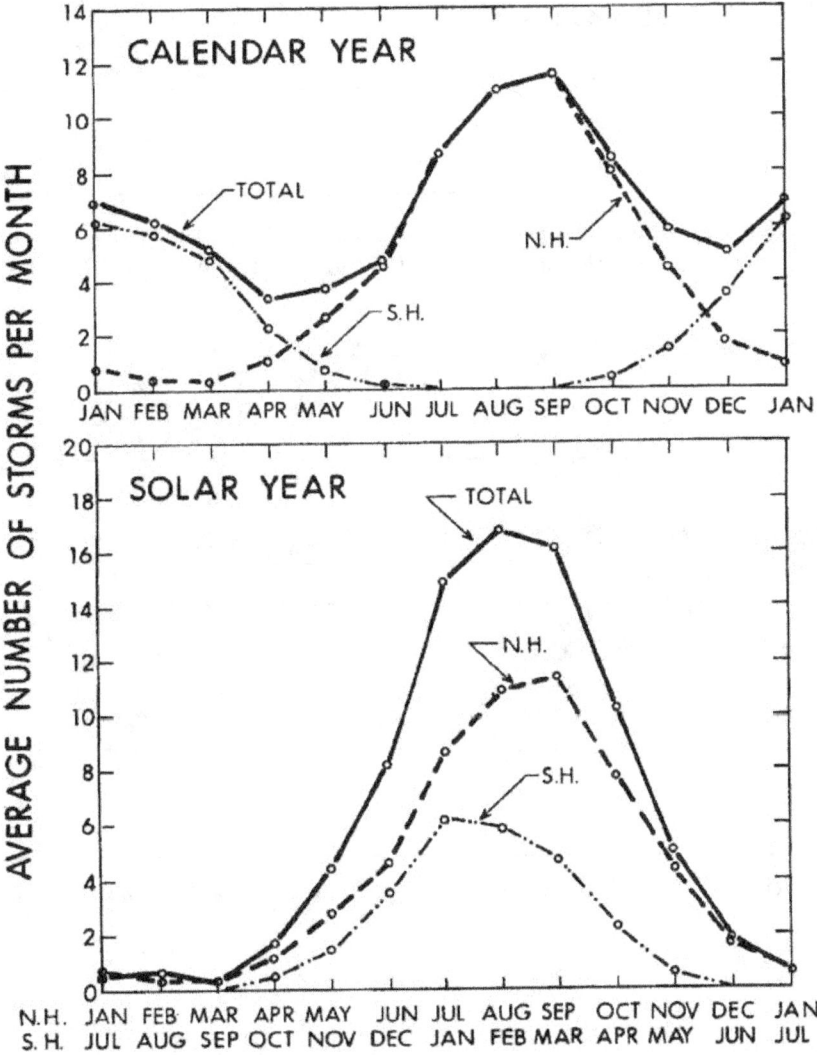

Figure 9-7. Average monthly number of tropical cyclones of tropical storm or greater intensity, relative to calendar and solar years for the northern and southern hemispheres and the globe (from Gray, 1978).

Figure 9-8 gives the average annual frequency of tropical cyclones in each generating area. An annual average of only 80 tropical cyclones develops worldwide (Gray, 1985).

Figure 9-9 shows 3 years of representative tracks. Nearly half the storms form in the North Pacific, about one-third in the Indian Ocean and Australian waters, and only about 11 percent in the North Atlantic.

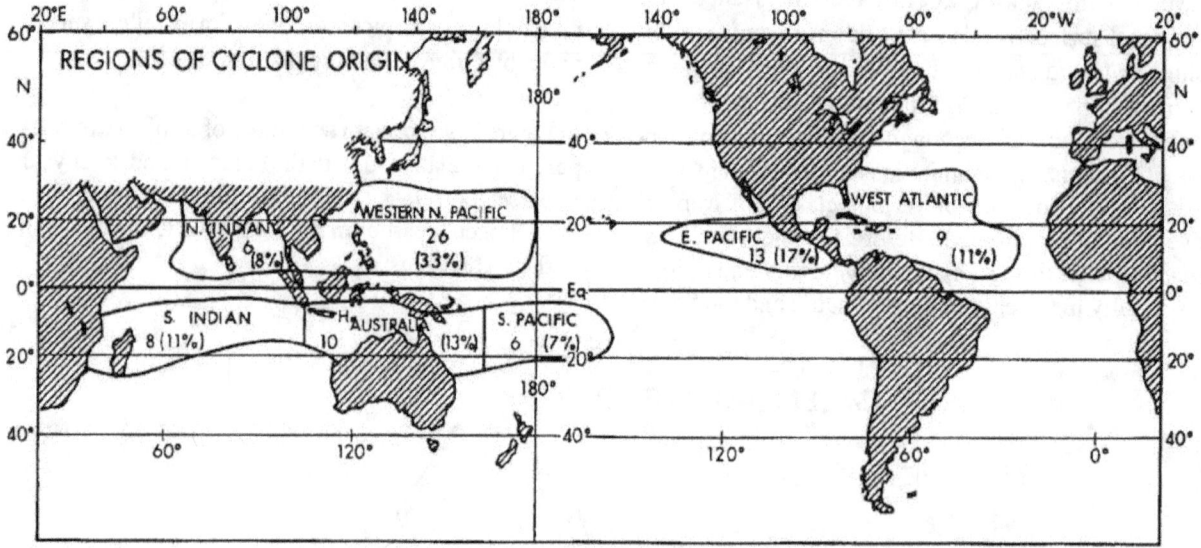

Figure 9-8. Average annual number (and percentage of global total) of tropical cyclones that reach tropical storm or greater intensity in each development area for 1958 to 1977 (from Gray, 1978).

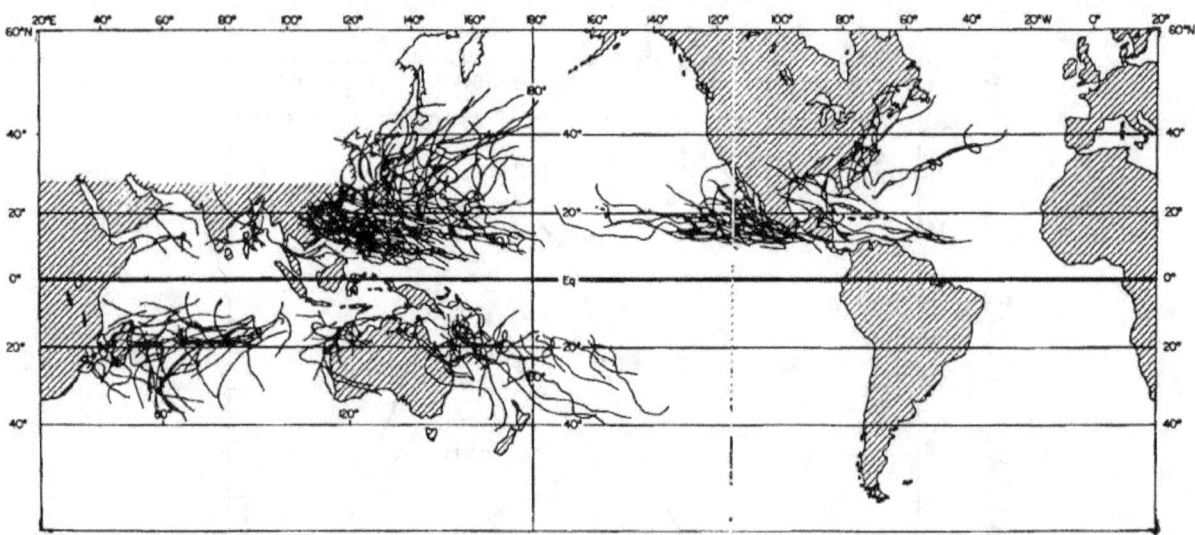

Figure 9-9. The tracks of tropical cyclones for a 3-year period (Gray, 1978).

TABLE 9-2. Average monthly frequency of tropical cyclones of at least tropical storm intensity for each major basin (storm-development area). The data is from Crutcher and Quayle (1974), except for the eastern North Pacific, for which 22 years with satellite observations (1967-1988) have been used; in this area, in earlier years not all tropical cyclones were picked up by ship reports (Sadler, 1964). Cyclones are ascribed to the month in which they began; the monthly values add up to the annual value. For the other areas, records comprise at least 70 years. Minus signs (-) indicate less than .05. Monthly values cannot be combined because single storms overlapping 2 months were counted once in each month and once in the "YR" column.

BASIN AND STAGE	JAN	FEB	MAR	APR	MAY	JUN	JUL	AUG	SEP	OCT	NOV	DEC	YR
North Atlantic													
Tropical Storms	-	-	-	-	0.1	0.4	0.3	1	1.5	1.2	0.4	-	4.2
Hurricanes	-	-	-	-	-	0.3	0.4	1.5	2.7	1.3	0.3	-	5.2
Total	-	-	-	-	0.2	0.7	0.8	2.5	4.3	2.5	0.7	0.1	9.4
Eastern North Pacific													
Tropical Storms	-	-	-	-	-	0.9	2.3	1.6	1.2	1	0.3	-	7.4
Hurricanes	-	-	-	-	0.3	1	1.7	2.4	1.8	0.9	-	-	8.2
Total	-	-	-	-	0.5	1.9	4	4	3	1.9	0.3	-	15.6
Western North Pacific													
Tropical Storms	0.2	0.3	0.3	0.2	0.4	0.5	1.2	1.8	1.5	1	0.8	0.6	7.5
Typhoons	0.3	0.2	0.2	0.7	0.9	1.2	2.7	4	4.1	3.3	2.1	0.7	17.8
Total	0.4	0.4	0.5	0.9	1.3	1.8	3.9	5.8	5.6	4.3	2.9	1.3	25.3
Southwest Pacific and Australian Area													
Tropical Storms	2.7	2.8	2.4	1.3	0.3	0.2	-	-	-	0.1	0.4	1.5	10.9
Hurricanes	0.7	1.1	1.3	0.3	-	-	0.1	0.1	-	-	0.3	0.5	3.8
Total	3.4	4.1	3.7	1.7	0.3	0.2	0.1	0.1	-	0.1	0.7	2	14.8
Southwest Indian Ocean													
Tropical Storms	2	2.2	1.7	0.6	0.2	-	-	-	-	0.3	0.3	0.8	7.4
Hurricanes	1.3	1.1	0.8	0.4	-	-	-	-	-	-	-	0.5	3.8
Total	3.2	3.3	2.5	1.1	0.2	-	-	-	-	0.3	0.4	1.4	11.2
North Indian Ocean													
Tropical Storms	0.1	-	-	0.1	0.3	0.5	0.5	0.4	0.4	0.6	0.5	0.3	3.5
Cyclones (Winds 50 knots or more)	-	-	-	0.1	0.5	0.2	0.1	-	0.1	0.4	0.6	0.2	2.2
Total	0.1	-	0.1	0.3	0.7	0.7	0.6	0.4	0.5	1	1.1	0.5	5.7

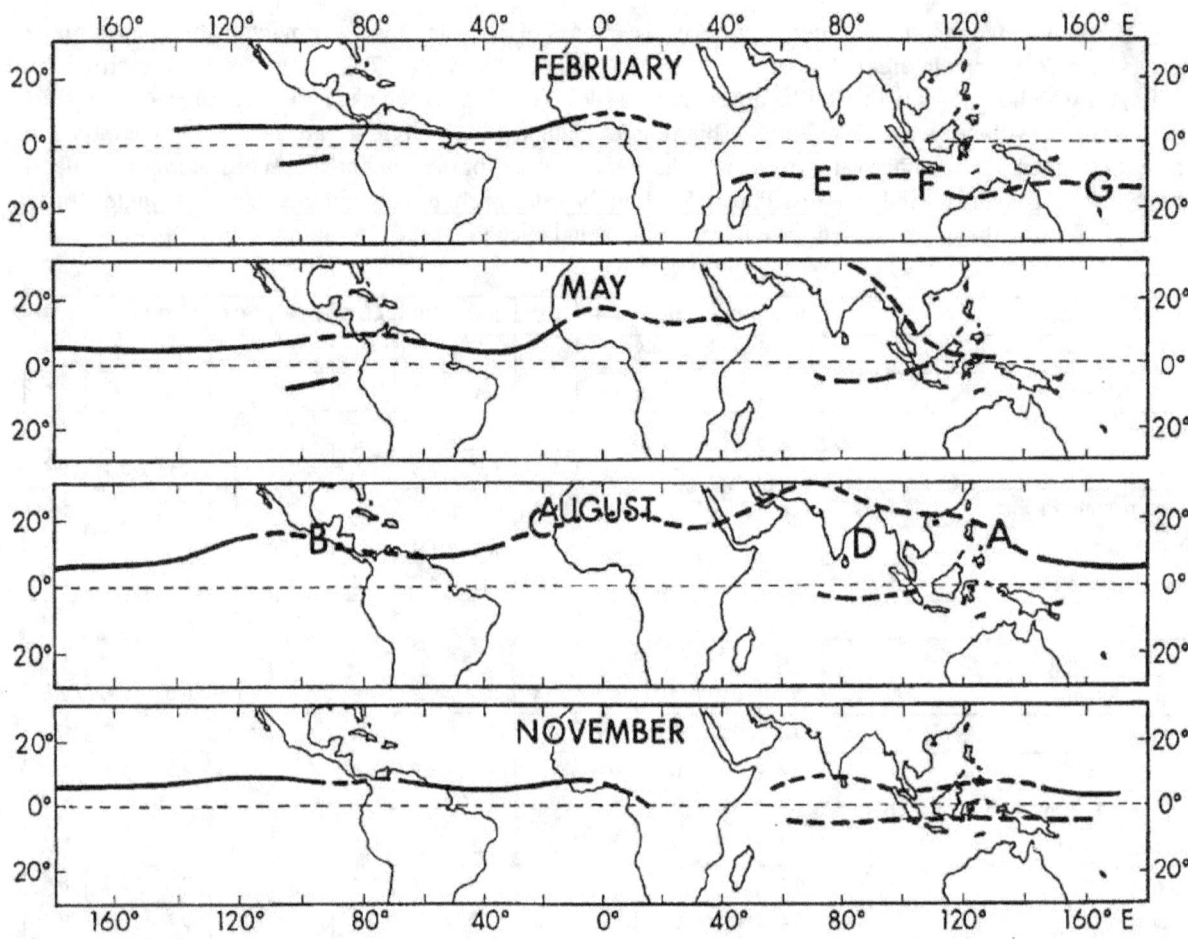

Figure 9-10. Locations of near-equatorial tradewind convergence (solid lines) and the monsoon troughs (dashed lines) at the gradient level over land and at sea level over the oceans during February, May, August, and November (adapted from Atkinson and Sadler, 1970).

In many areas, the climatology of tropical cyclone occurrence goes back a hundred years or so. However, if significant differences are apparent between earlier cyclone statistics and those based on more recent data, only the latter is used. Any long-term trends in tropical cyclone frequency defy identification in most areas because of long-term changes in data and methods of tracking.

Tropical cyclones originate in surface troughs, characterized by westerly winds on their equatorward sides and easterly winds on their poleward sides. Figure 9-10 shows the average locations of the monsoon troughs and near-equatorial tradewind

convergence for February, May, August and November. Following Sadler (1967), the areas of vortex origin are labeled with capital letters; the characteristics of each are briefly discussed below.

• **Western North Pacific (Area A).** This is the most active region. July through November accounts for 77 percent of the cyclones, but they can occur in any month, and they can be as intense in January or February as at the height of the typhoon season. About 70 percent of tropical storms intensify into typhoons. Tropical depressions have a similar distribution to tropical storms and typhoons. More than 80 percent of tropical depressions further intensify into tropical storms or typhoons. From September to December, and often in El Niño years, the monsoon trough may extend east to the date line or beyond, causing cyclones to form in the Marshall Islands area or even to the south of Hawaii.

• *South China Sea (Area A)*. About 15 percent of western North Pacific storms develop in the South China Sea. The season extends from May to December, with most activity in July through September. A minimum during June coincides with the ridge of high pressure that is moving north to give the early July dry spell over south China (see Figures 6-38 and 8-58).

• *Eastern North Pacific (Area B)*. This area ranks second only to the western North Pacific in the number of tropical cyclones. Most storms form from June through October. More than half occur in August and September when the monsoon trough is farthest north; 53 percent of the tropical storms become hurricanes. Once formed, most cyclones move westward or northwestward into a region of cooler sea surface and strong vertical wind shear, as shown by Figure 9-11 (Sadler, 1964). Consequently, most dissipate before they reach populated areas (see 9.6). A rare vortex travelling westward south of Hawaii can intensify into a hurricane and affect the central Pacific, as did Hurricane Sarah in 1967. Sarah passed Johnston

Island (17° N, 170° W) and moved directly over Wake Island (19° N, °167 E), causing considerable damage. The monsoon trough usually extends westward to about 120° W; however, during active periods it can reach to near the date line.

• *North Atlantic (Area C)*. In this region, about 80 percent of the tropical storms and hurricanes occur in August to October; 55 percent of tropical storms reach hurricane intensity.

Over the eastern Atlantic, Area C resembles Area B. Both are narrow warm-water zones between cooler water to the north and south (see Figure 5-1). Maximum cloudiness also occurs in the westerlies south of the monsoon trough (see Figure 6-16). As in Area B, the monsoon trough merges in the west into a tradewind convergence; the African vortexes usually decay as they move into the unfavorable environment of the tradewinds. However, several vortexes intensify into storms and hurricanes in the mid-Atlantic before leaving the trough, while others may persist as low-level vortexes (not necessarily at the surface) all the way to the Caribbean before dying, or rarely intensifying. These observations are illustrated by the life histories of cyclones emerging from Africa during August 1963 (Figure 9-12, Aspliden et al., 1965-1967).

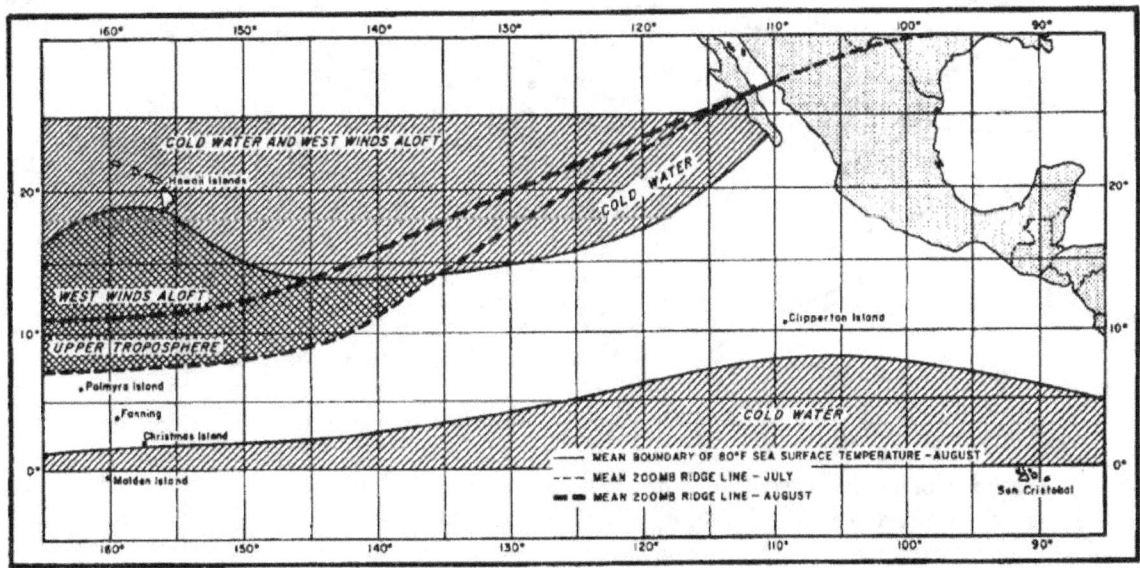

Figure 9-11. Mean position of the 200-mb ridge in July and August and mean August 26.7° C sea-surface isotherm in the eastern North Pacific (Sadler, 1964).

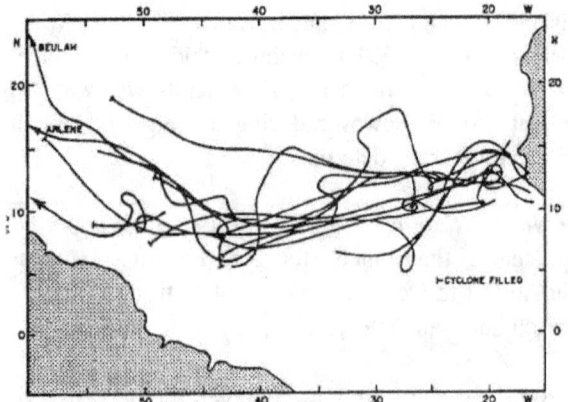

Figure 9-12. Tracks of surface cyclones over the tropical Atlantic during August 1963 (after Aspliden et al., 1965-1967).

Of the 11 cyclones tracked during August 1963, only two developed into hurricanes. Similarly, during Phase III of GATE (between 20 August and 23 September 1974), ten African vortexes moved over the Atlantic; six dissipated before reaching 45° W. Of the remaining four, two became hurricanes (Sadler and Oda, 1979).

During Phases I and II of GATE, between 26 June and 19 August 1974, no West African vortexes reached 45° W. Westward moving vortexes were confined to Phase III; Sadler and Oda ascribed this fact to northward movement of the trough over the ocean and southward movement of the trough over West Africa, resulting in an east-west alignment. This helped prolong the lives of West African vortexes, and resembles patterns in other years. Vortexes that leave West Africa between 20 August and 10 September have the best chance of crossing the Atlantic.

Sadler (1967) developed the schematic model associated with low-level cyclones in the North Atlantic shown in Figure 9-13. Visualize the model as either a synoptic picture of a chain of cyclones or as the life history of one cyclone. In the model, the initial vortex fails to develop beyond tropical depression intensity (A and B). The circulation remains a vortex in the easterlies (C) to beyond mid-Atlantic before decaying further to an open wave (D). Associated cloud masses move westward with the circulation; the dominant system south of the trough may fluctuate greatly from day to day.

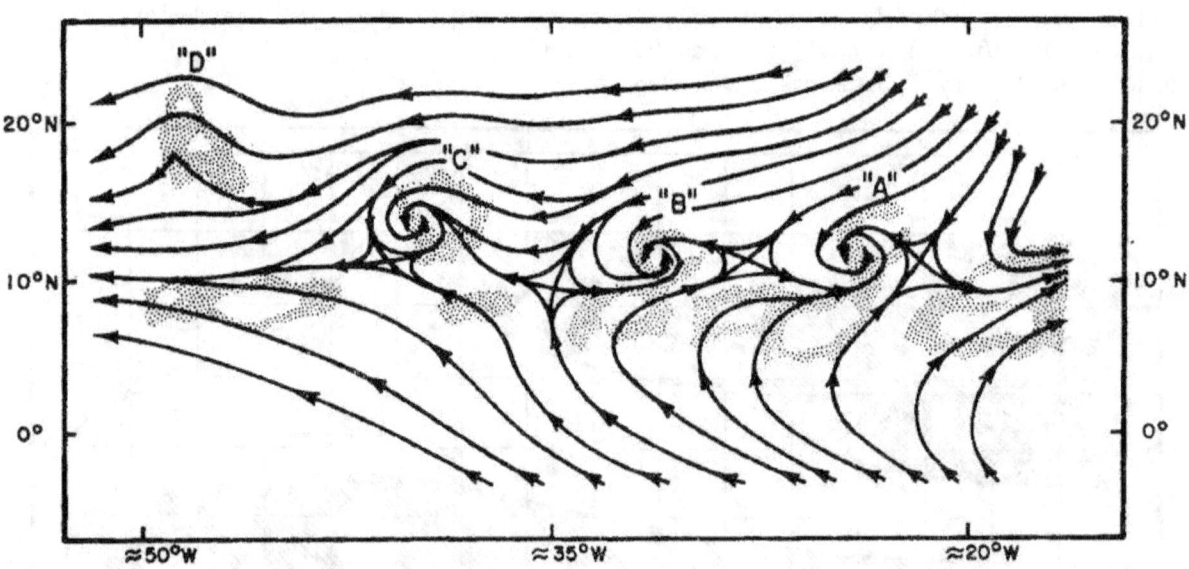

Figure 9-13. Model of low-level cyclones in the tropical North Atlantic depicting either a chain of cyclones or the life history of one cyclone. Major satellite-observed cloud areas are stippled; the clear areas within the stippling represent deep, more convective cloud groups (after Sadler, 1967).

• *North Indian Ocean (Area D).* Over both the Arabian Sea and the Bay of Bengal, most tropical storms are confined to early summer and autumn, when the surface trough lies near 10° N (Figure 6-33). Between June and September, the heat/monsoon trough, which has developed discontinuously over Pakistan and the Ganges Valley, ensures that no tropical storms can form over the Arabian Sea. Only when the monsoon trough extends over the extreme north of the Bay of Bengal can development occur there (see 8.3.2); because the environment is baroclinic, it is usually restricted to monsoon depressions.

• *Southwest Pacific and Australia (Areas F and G).* In these areas and over the southwest Indian Ocean, a third or less of all tropical storms intensifies into hurricanes, a significantly lower proportion than over the North Pacific and the North Atlantic. Poor observing networks here could be to blame for unreliably low intensity estimates.

The median recurvature latitude for tropical cyclones is 15°, compared to 24° north of the equator. In general, tropical cyclones lying equatorward of the upper tropospheric subtropical ridge move toward the west and poleward. As they cross the ridge, the zonal component of their movement shifts to the east. Over the western North Pacific the 200 mb subtropical ridge moves 1,000 NM (2,000 km) poleward between winter and summer, but over the southwest Pacific/Australian area, the ridge remains between 10° and 15° S throughout the year; tropical cyclones recurve soon after development. The season is almost as prolonged as in the western North Pacific. This may not be a coincidence; early or late season cyclones in one region could trigger late or early season cyclones across the equator in the opposite region, and so extend the tropical cyclone season in both regions. Keen (1982) identified 22 cross-equatorial pairs over the Pacific in about 10 years (Figure 9-14).

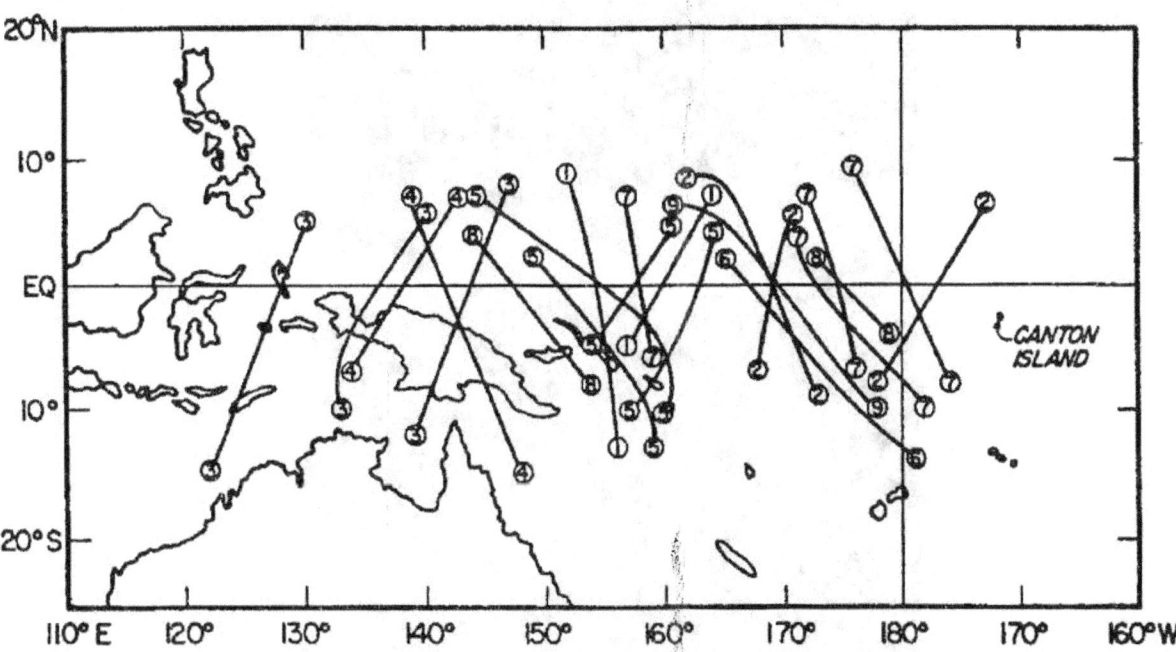

Figure 9-14. Twenty-two cross-equatorial named tropical cyclone pairs over the Pacific Ocean, September 1971 through January 1980. Circled numbers indicate points of origin and season of occurrence (1 = 1971-72 and 9 = 1979-80) (from Keen, 1982).

The satellite image in Figure 9-15 shows a tropical cyclone pair over the West Pacific. The northern cyclone became Super Typhoon Russ, which later passed south of Guam (frontispiece). The "twin" tropical cyclone (Joy) in the Coral Sea subsequently moved westward, then south, finally going ashore near Townsville (19° S, 147° E) on 27 December.

Surface westerlies and bad weather coincide along the equator, where cirrus plumes reveal strong upper tropospheric easterlies, evidence of an active Walker Circulation (see Figure 2-7). Twin development is usually preceded by a westerly wind surge along the equator (8.4.4). A succession of such events, at intervals of a few weeks, may generate El Niño (6.3.5).

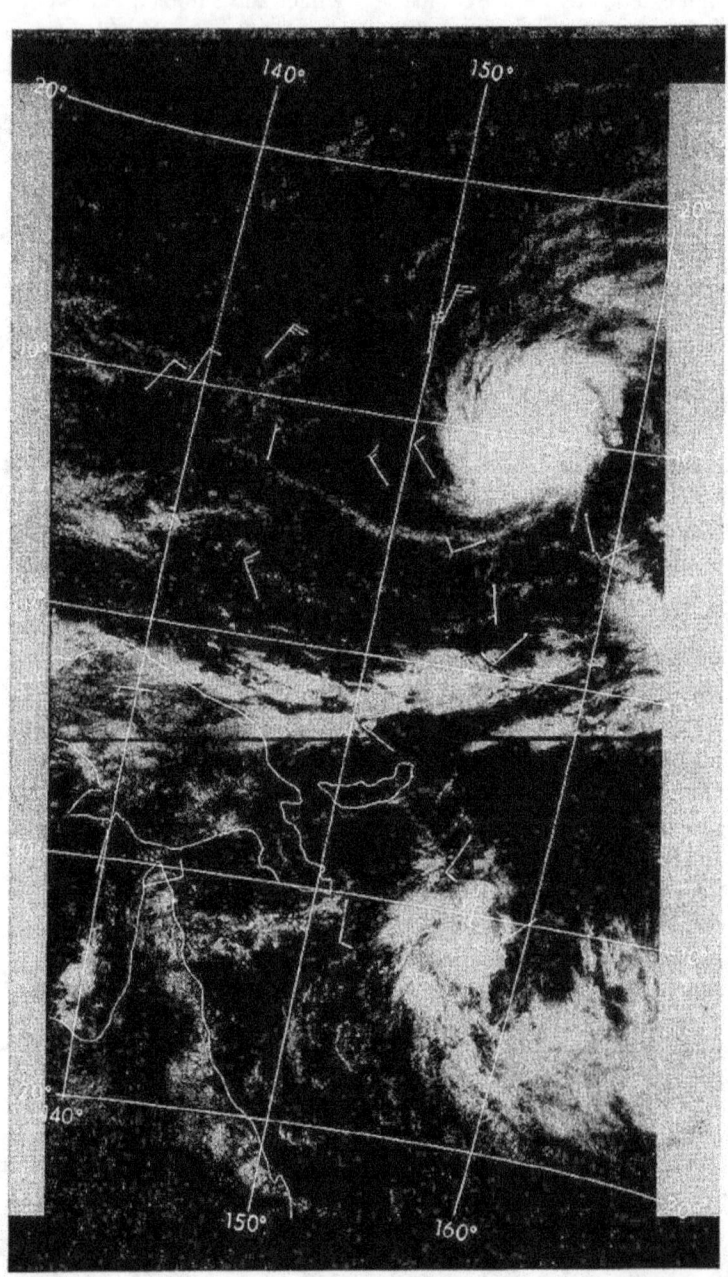

Figure 9-15. DMSP visible image of twin cyclone development at 2330Z, 17 December 1990. Observed surface winds are plotted.

• *Southwest Indian Ocean (Areas E and F).* This region incorporates the ocean from the coast of Africa to 110° E. Seventy-three percent of the storms occur from January through March; 50 percent reach hurricane intensity. As over the western and southwestern Pacific, 14 twin developments occurred north and south of the equator in the Indian Ocean in 11 years (nine in November-December). Figure 9-16 shows the tracks (Mukerjee and Padmanabham, 1977).

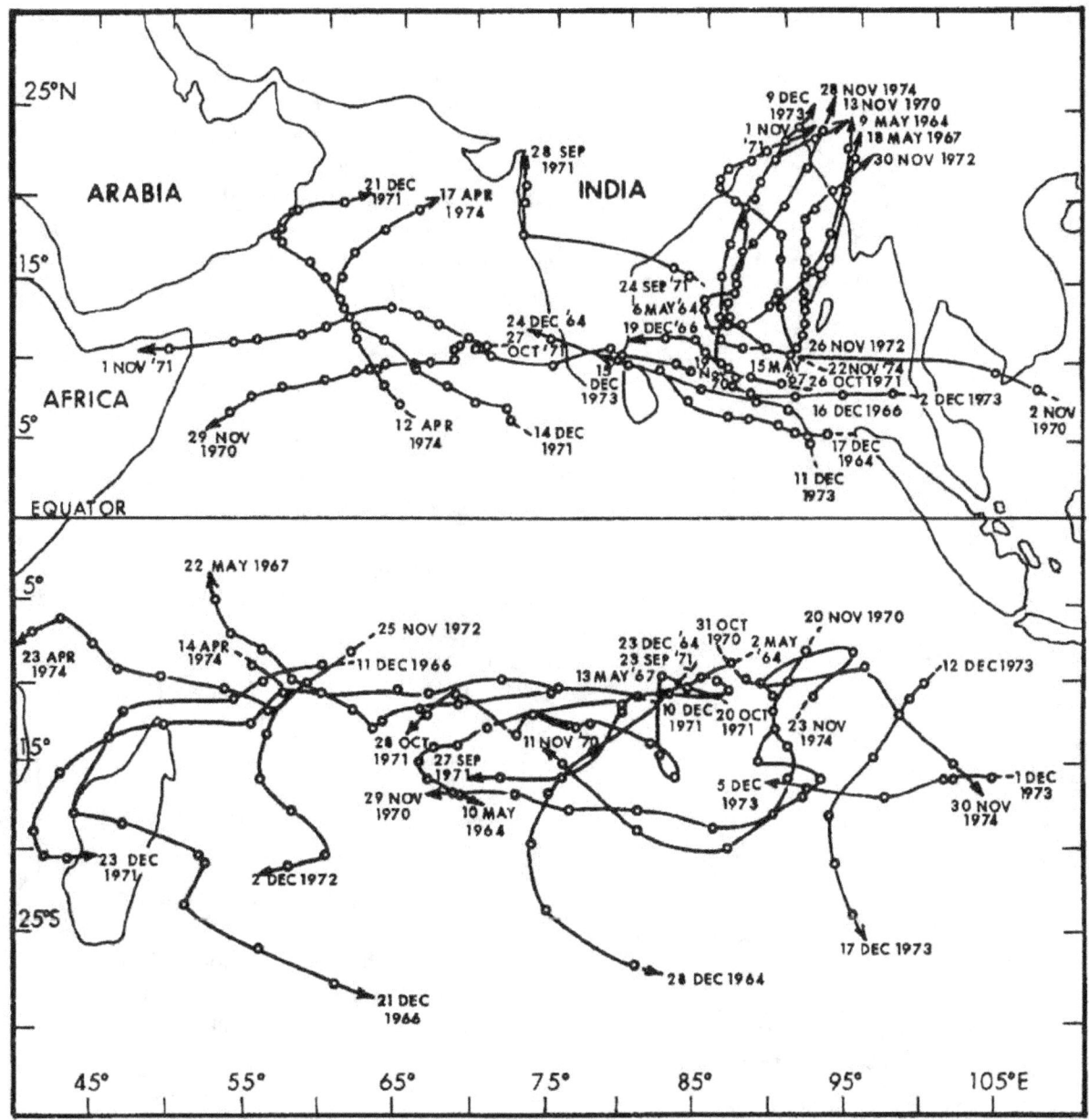

Figure 9-16. Tracks of pairs of tropical cyclones over the Indian Ocean, 1964 to 1974 (Mukerjee and Padmanabham, 1977).

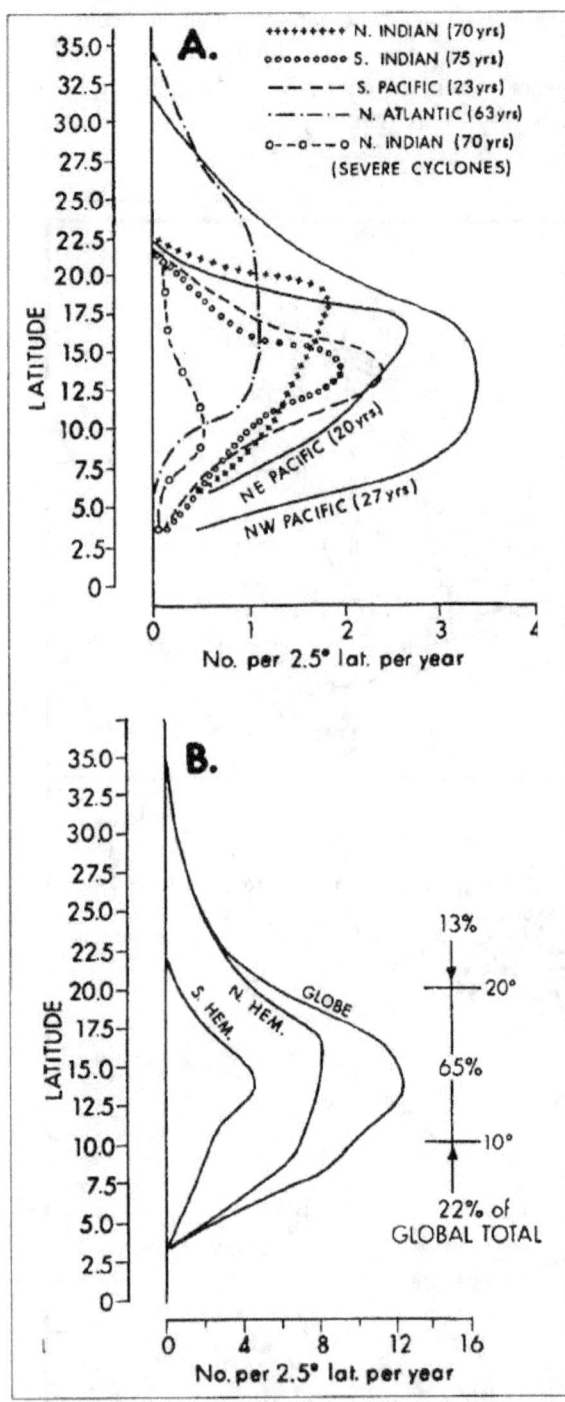

The latitude at which initial disturbances that later became tropical storms were first detected is shown for each development region in Figure 9-17A, at left. The regions differ significantly, especially in the NH. In the North Atlantic, proportionately more storms develop above 20° N than in the other regions. In the western North Pacific, storms develop over a broad meridional range because of relatively large seasonal and shorter period movements of the monsoon trough and because of developments initiated by upper-tropospheric cyclones. In the SH, the monsoon trough moves less, with most storms developing within 250 NM (500 km) of the mean trough position near 13° S. Over the Bay of Bengal, if monsoon depressions are not considered, the peak development latitude shifts south to 10° N.

The combined graphs for NH, SH, and globe are shown in Figure 9-17B. Overall, 65 percent of the disturbances that became tropical storms were first detected between 10 and 20 degrees latitude, 13 percent poleward of 20 degrees, and 22 percent equatorward of 10 degrees.

The number of tropical cyclones in any month or year undergoes a large interannual variability the average is seven for August in the western North Pacific, but since 1940 the range has been from two in 1947 and 1977 to 18 in 1950. Figure 9-18 shows the frequency

N. ATLANTIC	113251 3112		
E. N. PACIFIC	1 221232 222 1		
W.N. PACIFIC		112 11211 12111 21	1
S.W. PAC., AUSTR.	3 32321 221 1		
S. W. INDIAN	1151321321		
N. INDIAN	141041		

4 6 8 10 12 14 16 18 20 22 24 26 28 30 32 34 36 38
ANNUAL FREQUENCY

Figure 9-18. Frequency distribution of the annual number of tropical cyclones of tropical storm intensity or greater in various development areas (Gray, 1978).

Figure 9-17. (A) Latitude at which initial disturbances that later became tropical storms were first detected in each development area. Average number of years of data are given in parentheses. (B) Combined data for the NH, SH, and globe (adapted from Gray, 1968).

The global climatology presented in this report can be supplemented by regional publications that incorporate much more detailed climatologies. These publications include tropical cyclone tracks for each month (or, in some areas, for semimonthly or 10-day intervals), frequency distributions and means of tropical cyclone movement for various latitude-longitude rectangles, probabilities of tropical cyclones affecting particular locations based on their present position, and many other detailed statistics. The following annual reports give tropical cyclone tracks and intensities and movement forecast errors, as well as analyses and satellite pictures:

• *Annual Tropical Cyclone Report,* Joint Typhoon Warning Center, Guam: west and South Pacific and Indian Ocean.

• *Tropical Cyclone Summaries,* Royal Observatory, Hong Kong: North Pacific west of date line.

• *Monthly Weather Review,* North Atlantic and eastern North Pacific.

• *Mausam,* India Meteorological Department: Bay of Bengal, Arabian Sea and North Indian Ocean.

• *Australian Meteorological Magazine,* southwest Pacific and Australian area.

• *Technical Report - C5,* Mauritius Meteorological Service: southwest Indian Ocean.

9.5 TROPICAL CYCLONE FORMATION

Several conditions must be met before tropical cyclones can form.

9.5.1 Adequate Source of Surface Energy. This generally limits the area to seas or oceans with surface temperatures of 79° F (26° C) or higher. According to Gray (1978), these are also the regions in which the warm water is at least 200 feet (60 meters) deep and where mixing by the winds will not lower the surface temperature too much. The 26° minimum requirement may not necessarily be valid. Tropical-type storms have formed in the Baltic Sea (Bergeron, 1954) and the Mediterranean (Meyencon, 1983), and even over the Arctic Ocean (Emmanuel and Rotunno, 1989; Businger, 1991), where the sea-surface temperature was certainly less than 26° C. In 1980, Atlantic Hurricanes Ivan and Karl developed where SSTs were 73° F (23° C) and 68° F (20° C), respectively.

A small horizontal SST gradient, most often found where the temperature is high, is needed. Then, a surface air parcel spiralling inward from the cyclone periphery toward the eye will replace heat lost by adiabatic expansion by heat gained from the ocean, and will reach the eye wall much more buoyant than the environmental air surrounding the cyclone. If a surface air parcel undergoes a 60 mb pressure drop as it moves along the pressure gradient toward the eye, it should cool adiabatically by about 9° F (5° C), but it is observed to maintain a nearly constant temperature. The only way this can happen is for the ocean to warm the air. Conversely, the low heat capacity of a land surface prevents it from heating the storm; by inhibiting buoyancy in the center of the storm, it either prevents development or causes a developed storm to dissipate.

9.5.2 Wind Shear. Weak shear in the vertical of the tropospheric wind allows the heat released by condensation to concentrate in a vertical column, enhancing the surface pressure fall. Figure 9-19 shows where shear is favorable. Above the tradewinds, winds must be easterly or weak westerly. During summer, surface westerlies of the Asian monsoon may extend across the Philippines and into the Caroline Islands, causing weak wind shear and favoring cyclone formation in Micronesia. A similar effect is produced over the western Atlantic. The strong summer shear over the Arabian Sea and the Bay of Bengal partially accounts for the absence of tropical storms there. The significant summer shear over the South China Sea is sometimes reduced when an upper-tropospheric trough extends westward from the North Pacific across the Philippines and therefore favors tropical storm development (Sadler et al., 1968). In contrast, the weak shear throughout the year near 10° N over the western North Pacific allows development in any month, while during summer this region encompasses the largest area of weak climatological shear (from 5° to 30° N). In the western North Atlantic, mean wind shear less than 10 knots extends to 33° N; proportionately more cyclones develop in this region at higher latitudes than anywhere else (see Figure 9-17).

9.5.3 Preexisting Disturbance. In general, active cyclone formation is confined to regions of weak cyclonic shear in a summertime surface trough (Gray, 1968). Local concentrations of vorticity in the trough can help triggering mechanisms start tropical cyclogenesis. Besides strong wind shear, absence of a surface trough over the Arabian Sea also contributes to the absence of summer tropical cyclones there. A similar, but less marked, effect is present over the Bay of Bengal and the South China Sea. In all three areas, tropical cyclones are more likely to form in the transition seasons when surface troughs are present and wind shear is least. Surface troughs are absent throughout the year over the southeast Pacific east of about 130° W, and over the South Atlantic.

9.5.4 Earth's Rotation. Almost all cyclones form poleward of 5° latitude, since a strong rotation can be generated only where the Coriolis force exceeds a certain minimum value. Disturbances near the equator rarely develop into tropical storms. However, Typhoon Kate (1970) grew to typhoon intensity at 4.5° N and moved east to west along 5° N before recurving northward (Holliday and Thompson, 1986). Kate was very small compared to other typhoons with similar central pressures.

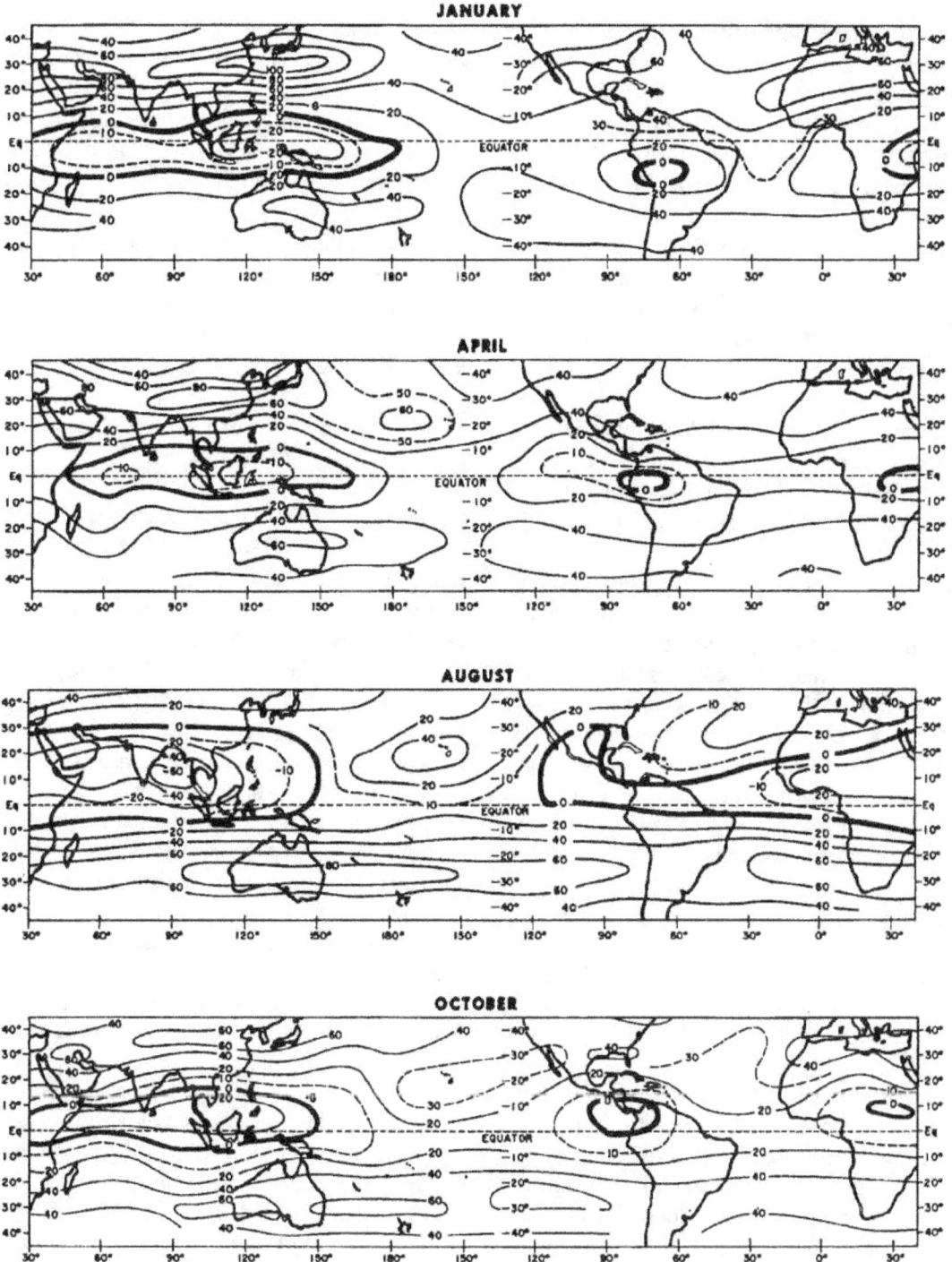

Figure 9-19. Average zonal wind-shear (knots) between 850 mb and 200 mb over the tropics for January, April, July, and October (after Gray, 1968).

261

9.5.5 Upper-Tropospheric Outflow. All the preconditions for tropical cyclone development so far listed prevail together over vast ocean areas for weeks or even months at a time. Yet, in an average year, only 80 disturbances intensify to tropical storms or beyond. In numerical models, the preconditions lead to hurricanes only if the model starts out with a surface cyclone of about 20 knots or more (Gray, 1978), more than double what has usually been observed. Even then, no model duplicates occasionally observed surface pressure falls of more than 50 mb in 24 hours. Something is missing.

The notion that the upper troposphere holds the key to rare and often rapid intensifications has become more generally accepted (see, for example, Riehl, 1979a). In most generating regions, climatologically divergent upper-tropospheric easterlies overlie and extend equatorward of the monsoon trough (Figures 4-13 to 4-16). Poleward of the generating region, troughs in the upper-tropospheric westerlies or cold lows in the TUTT may also cause divergence far enough east and southeast of the strong peripheral winds to ensure weak wind shear in the vertical (see Figure 8-21). These divergent outflows to north and south, acting on a surface disturbance, may cause the surface pressure to fall enough for a tropical cyclone to form (see Sadler, 1976a; Elsberry, 1987).

As a tropical cyclone develops, the surface pressure fall in the eye is usually punctuated by short periods of no change or of slight rises (see Sadler, 1976b). This suggests that the circulation responds to a succession of intensification impulses.

Tropical meteorologists, used to near-simultaneous divergence/convergence compensations in the atmospheric column leading to insignificant surface pressure changes, have trouble imagining an unbalanced upper divergence sudden and intense enough to cause the observed surface pressure falls. Perhaps the divergence stems from ephemeral small-scale jet "streaks" that escape discovery in jet aircraft winds or in cloud motion vectors derived from satellite images. Reconnaissance and research aircraft have only rarely probed the upper troposphere near potential tropical storms. In the U.S. Navy typhoon field project of 1990 (Elsberry, 1989), 1-second

measurements made on the NASA DC-8 flying at 190 mb over Typhoon Flo may give some clues.

To summarize, there is general agreement on conditions necessary for tropical cyclone development. These are:

• A warm, small-temperature-gradient ocean more than 5° from the equator

• A surface trough overlain by weak shear through the troposphere

• An essential condition of intense divergence (outflow channel) in the high troposphere.

For example, if large-scale atmospheric circulation anomalies at various levels combine to produce strong wind shear above a development area usually characterized by weak shear, cyclone development will be prevented. The converse would favor development where cyclones are normally rare.

There may be regional differences. In the North Atlantic, African vortexes in the monsoon trough have been identified as initiating disturbances. The TUTT in the North Pacific and North Atlantic is a source of cyclones, especially those forming north of 20° N. The sequence illustrated by Figure 8-21 is usually prolonged. Divergence east of a cold low, leading to convection, may eventually reduce surface pressure enough to produce a depression in the tradewinds. If upper-level conditions remain favorable, this circulation may become a tropical storm, and rarely, a typhoon or hurricane.

It was previously thought that easterly waves grew into tropical cyclones. Over the Atlantic from 1960 to 1964, only 6 of the 43 storms or hurricanes (14 percent) were attributed to easterly waves. From 1965 to 1969, with almost continuous satellite coverage and more aircraft reports, not one of the storms was associated with easterly waves. It should be noted that in that period the term "easterly wave" was replaced by the more general term "tropical wave," defined in Chapter 8. Even so, tropical waves were said to account for only 4 of the 28 storms occurring between 1967 and 1969.

Lack of conventional data over the central tropical Atlantic still handicaps research into tropical cyclone origins. Aerial reconnaissance starts only after a disturbance has moved west of 55° W. Annual summaries published in the *Monthly Weather Review* suggest that mid- and upper-tropospheric circulation features and their interaction with lower levels may be more important over the Atlantic than elsewhere.

Early in the hurricane season, mid-latitude influences are important. For example, development of two hurricanes and one tropical storm in June 1968 was related to baroclinic processes in the mid-troposphere. Carlson (1969a,b) has shown that during the heart of the hurricane season (mid-July to mid-September), most Atlantic tropical storms result from intensification of low-level cyclonic vortexes that originate over Africa (8.3.3). They generally intensify over the warmer waters of the western North Atlantic. One purpose of GATE in 1974 was to observe the evolution of African vortexes into tropical cyclones, but this was thwarted by too little tropical cyclone activity.

Early and late in the Atlantic season, cyclones tend to form in the southwestern Caribbean when the monsoon trough extends eastward from the eastern Pacific. The role of cold lows in the subtropics has already been mentioned. In the western North Pacific, 85 to 90 percent of tropical cyclones form in the surface monsoon trough; 10 to 15 percent develop farther north as the TUTT acts on the surface tradewinds (Figure 8-21). In the other tropical cyclone regions most developments take place when vortexes in the surface monsoon trough intensify.

On 11 July 1989, Typhoon Gordon suddenly developed from convection located near the center of a cold low within the TUTT (Joint Typhoon Warning Center, 1990). Within 30 hours, the central pressure fell 70 mb. As mentioned earlier, isolated, short-lived deep convection often develops near the centers of cold lows, but a typhoon? Upper-tropospheric convergence into the cold low would seem to preclude this possibility, making this very rare event a mystery.

9.6 TROPICAL CYCLONE DISSIPATION

When conditions that favor development or persistence disappear, hurricanes or typhoons dissipate.

9.6.1 Surface Energy Source Removed. When the surface supply of sensible heat to the core is cut off, the buoyancy of the system is destroyed (Bergeron, 1954; Miller, 1964). Since only a uniformly warm ocean surface can supply the necessary heat, the circulation weakens rapidly when it moves over land. The isothermal inflow at the surface stops, the air is now cooled by adiabatic expansion, and the horizontal buoyancy gradient disappears. The same effect is produced when relatively cold or dry surface air penetrates the core.

Forecasters should remember that cold air reaches the eye wall from the west, or equatorward of the center. This happens because air moving in from north is picked up by the strong winds spiralling around the eye; it does not penetrate the eye until it has first passed to the west or south. If air swirls around the center before reaching the wall, the outward-directed temperature gradient (and the cyclone intensity) may temporally increase before the system starts to fill (Brand and Guard, 1979). After recurvature, a tropical cyclone may move too fast for the cold air to catch up; hurricane winds then persist into high latitudes. The worst winds ever experienced in New Zealand were caused by a tropical cyclone that moved south-southeast at 22 knots from the vicinity of New Caledonia in April 1968 (Hill, 1970).

The stress of cyclonic winds transports ocean-surface water outward from a tropical cyclone center (5.2.2).

Ensuing upwelling of a stratified ocean (more likely above 20° latitude) may lower the sea-surface temperature by as much as 9° F (5° C) (Leipper, 1967; confirmed by many more recent measurements), but with a lag too great for the energetics of a moving cyclone to be affected. However, should the cyclone be stationary, it would weaken as its buoyancy decreased. For example, in December 1989, Typhoon Jack remained stationary for 48 hours. In the first 36 hours, maximum winds decreased from 120 to 30 knots.

9.6.2 Excessive Wind Shear. If a tropical cyclone moves beneath a strongly shearing layer, the low-level center may become detached from the rest of the circulation, be weakened by subsidence, and quickly dissipate. This sequence usually shows clearly in satellite images. Dissipation by shearing often occurs in the eastern North Pacific (Figure 9-11) and in the eastern North Atlantic, as tropical cyclones move west or northwest into the tradewinds and beneath upper-tropospheric westerlies.

9.6.3 Upper Tropospheric Convergence. As a tropical cyclone approaches an upper-tropospheric trough, it generally recurves poleward beneath upper tropospheric divergence east of the trough, and remains strong. It may even intensify as convection is enhanced beneath outflow and rising motion associated with acceleration of an upper tropospheric jet stream (Rodgers et al., 1991). Occasionally, if the upper trough is moving rapidly across the tropical cyclone track, it may pass over the cyclone. Upper-tropospheric convergence then fills the cyclone (see Figure 8-53).

9.7 TROPICAL CYCLONE MOVEMENT

Various climatological publications provide generalized mean tracks of tropical cyclones by month or season. These tracks are drawn along axes of maximum cyclone frequency. They are useful for some purposes; however, they hide the characteristically large variability of tropical cyclone tracks that can be revealed by charts of the tracks of individual cyclones compiled from many years of record (see Figure 9-9). Recent tropical cyclone tracks, based on aircraft reconnaissance and satellite pictures, are more accurate than tracks for earlier years, especially where other data is sparse.

The annual meridional movement of the subtropical ridge lines controls the mean latitude of recurvature of tropical cyclones. Data for the western North Pacific is shown in Figure 9-20. January and February storms generally dissipate before they can recurve. In the western North Atlantic, mean recurvature positions determined as a function of both latitude and longitude are shown in Figure 9-21 (Colòn, 1953).

The environment in which a cyclone is embedded largely controls its track.

The environment affects tropical cyclone motion. The influence of the environmental wind field increases as a cyclone recurves and moves poleward. South of the subtropical ridge and within the tropics, the cyclone dominates the circulation for a considerable distance from the center and throughout the troposphere. The warm core of the cyclone transports large amounts of latent heat of condensation to the middle and upper troposphere, inflating isobaric surfaces and developing a ridge at these levels northeast of the cyclone center. This helps maintain westward movement. If a cyclone recurves poleward where the ridge is weak or broken, its depth decreases from the

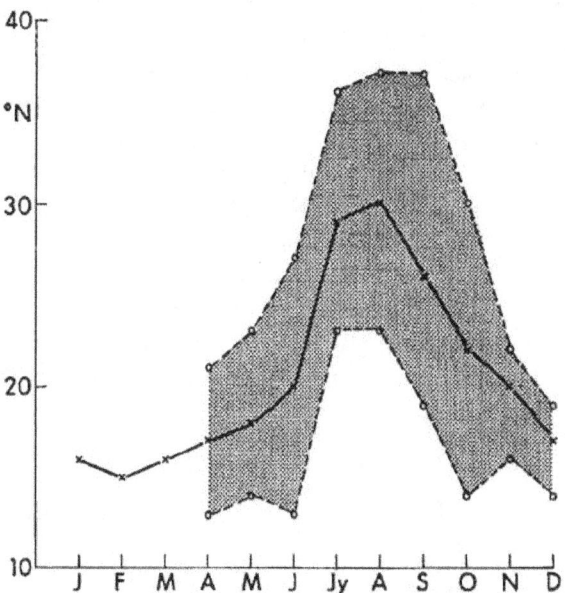

Figure 9-20. Mean recurvature latitudes and ranges for western North Pacific tropical cyclones, 1965-1982 (from Guard, 1983).

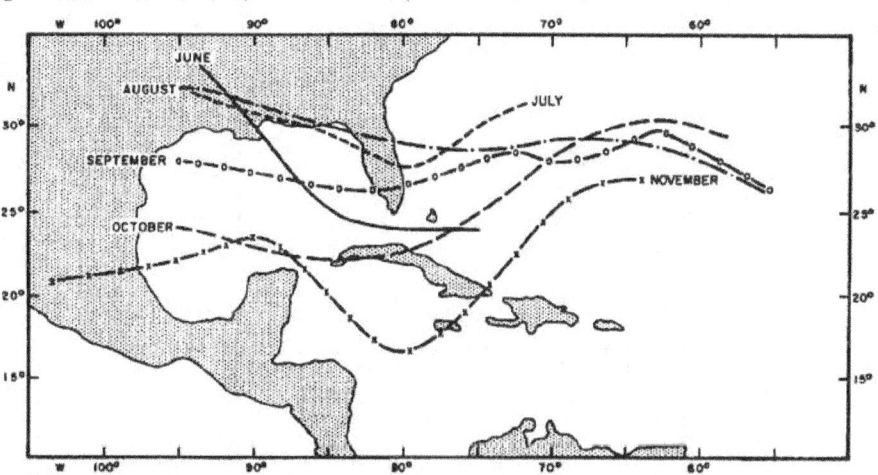

Figure 9-21. Mean monthly recurvature positions for tropical storms and hurricanes in the North Atlantic (after Colòn, 1953).

top down. Then, as it moves beneath the upper westerlies and starts travelling eastward, the closed circulation is confined to the lower troposphere. Thus, the influence of the cyclone on its surroundings decreases after recurvature and the large-scale environment is then largely responsible for steering the storm.

265

9.8 TROPICAL CYCLONE FORECASTING

9.8.1. General. Forecasting tropical cyclone movement is highly specialized and can best be accomplished in central locations by experienced forecasters using all available data. The National Hurricane Center in Miami, Florida, prepares forecasts for all tropical cyclones of depression intensity or greater in the North Atlantic, Caribbean, Gulf of Mexico and eastern Pacific. In the western North Pacific, tropical cyclone forecasting is centralized at the Joint Typhoon Warning Center (JTWC) on Guam (13° N, 145° E), which issues warnings for the Indian Ocean and the southwest Pacific, as well.

Other weather centers also issue warnings; in some regions the warning areas overlap; differing forecasts for the same storm may confuse users. The techniques used at the various centers change from time to time in response to research results and to changes in methods of observing and tracking the storms. These techniques will be discussed here briefly, since field forecasters should know the capabilities and limitations of tropical cyclone forecasts. Figure 9-22 provides 1989 error statistics for selected techniques in the western North Pacific.

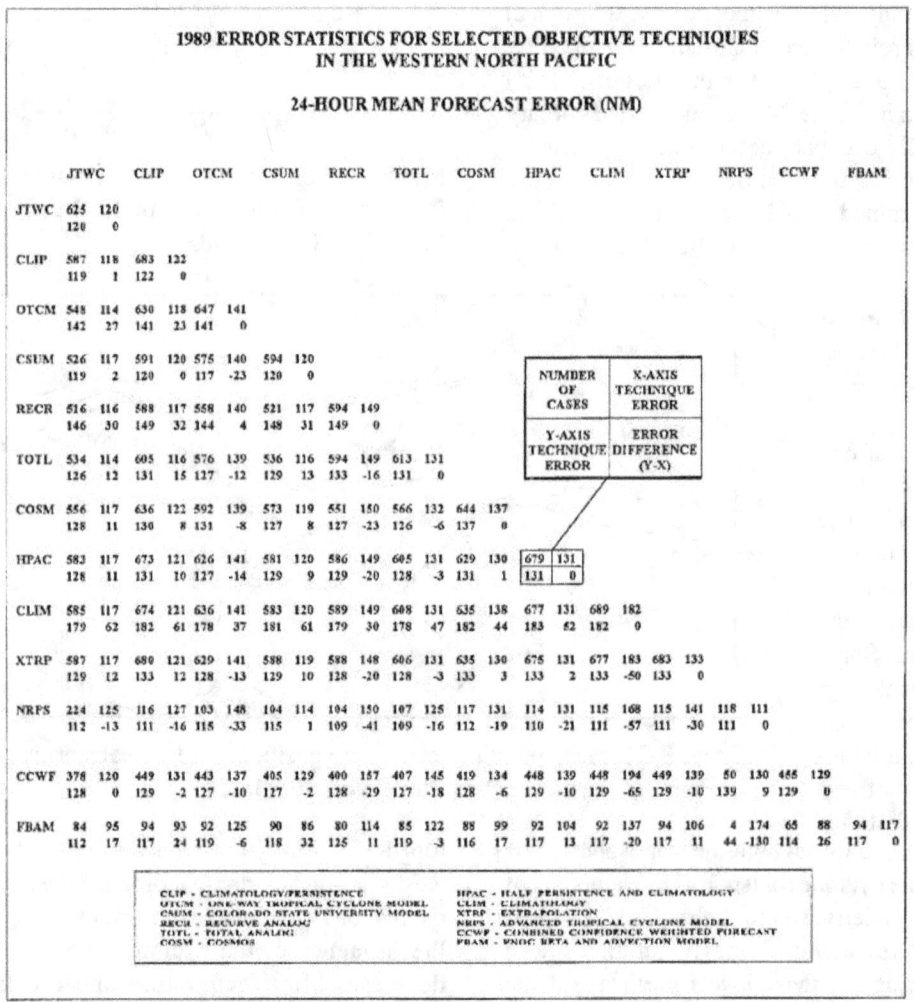

Figure 9-22. 1989 error statistics for selected objective tropical cyclone forecasting techniques in the western North Pacific.

The various centers compare and evaluate old and new techniques in real time, but until now, no one system has proved to be best under all circumstances. An official forecast is always a subjective decision; the techniques help, and establish a kind of probability envelope, but the forecaster has the last word. Results support this procedure; rarely, over the course of a season, does any one technique better the official forecast. Detailed comparisons made at JTWC are illustrated in Figure 9-22 on the preceding page. In the figure, the "X-axis" refers to techniques listed horizontally, and "Y-axis" to techniques listed vertically. In each box, only forecasts made at the same times for the same tropical cyclones are tallied; thus the comparisons are homogeneous.

9.8.2 Positioning.
Satellite imagery has revolutionized cyclone reconnaissance. At least two-thirds of all cyclone-center fixes are based on satellite pictures. Ending aircraft reconnaissance in the western North Pacific has made weather satellite observations even more important. The geosynchronous satellite makes it possible to continuously monitor cyclone positions and cloud motions. Although the cyclone center position as indicated by the satellite may differ from the true position, it is sufficiently accurate to support operations. Just what constitutes the cyclone center (the geometrical center of the eye) as defined by the wall cloud (satellite), the minimum pressure (aircraft reconnaissance), the maximum temperature, or the circulation center is debated. Differences can amount to as much as 30 NM (55 km). Because of this and occasional mapping displacement by the satellite, there will always be some disagreement on precise center location, particularly in poorly-defined cases.

9.8.3 Forecasting Techniques.

9.8.3.1 Movement. Most tropical cyclone forecasting centers combine subjective and objective techniques to forecast movement. The subjective techniques depend on the experience of individual forecasters and rules of thumb. Objective techniques include persistence forecasts based on extrapolation of past positions, average tropical cyclone movements in various areas based on climatology, and dynamic or statistical forecast methods. Those based on satellite-observed cloud patterns (e.g., Lajoie, 1981) have proved hard to apply and are not used much.

Dynamic techniques apply the same numerical prediction methods that are a feature of mid-latitude forecasting. Thus, cyclone motion is automatically included in the predicted pressure or pressure-height fields. Since the dynamic model does not depend on statistical relationships between the current cyclone and historical cyclones, the model's reliability should be independent of the details of the cyclone's past behavior. The cyclone track is thus explicitly predicted. Some dynamic models incorporate sophisticated procedures such as nested grids and two-way boundary interactions with larger-scale model output. Insufficient data prevents the synoptic field from being analyzed accurately enough to meet the initial field requirements of numerical models. In the North Atlantic, measurements by reconnaissance aircraft have been used to fill in gaps. Perhaps new sensors being developed for future satellites will lead to improvements in analyses.

Statistical techniques have been applied to movement forecasting. They use multiple-discriminant-analysis methods to select statistical predictors from grid-point fields of surface pressure, upper-air pressure-heights, winds, thickness values, etc. Persistence, incorporating past positions of the storm, is often included. Performance of statistical techniques is also sensitive to input data and will generally be best where there is an adequate network of surface and upper-air observations.

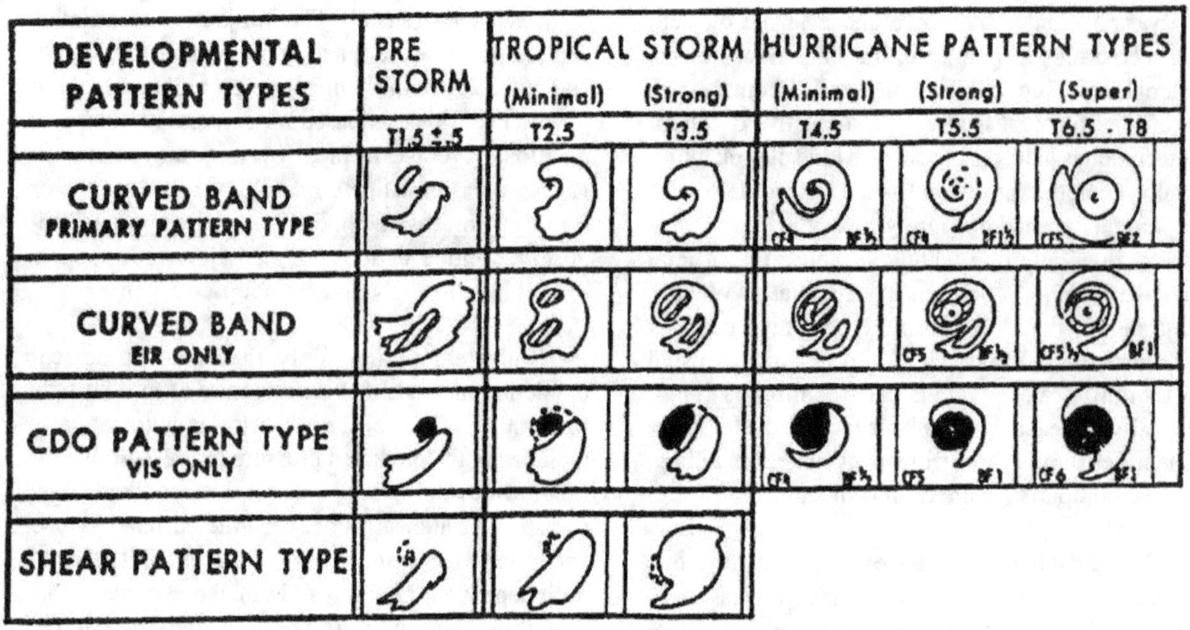

DEVELOPMENTAL PATTERN TYPES	PRE STORM	TROPICAL STORM		HURRICANE PATTERN TYPES		
		(Minimal)	(Strong)	(Minimal)	(Strong)	(Super)
	T1.5 ±.5	T2.5	T3.5	T4.5	T5.5	T6.5 - T8
CURVED BAND PRIMARY PATTERN TYPE						
CURVED BAND EIR ONLY						
CDO PATTERN TYPE VIS ONLY						
SHEAR PATTERN TYPE						

Figure 9-23. Developmental cloud pattern types used in intensity analysis of satellite images. Pattern changes from left to right typify 24-hourly changes. EIR denotes "enhanced infrared" (Dvorak, 1984).

9.8.3.2 Intensity. Intensity forecasting relies heavily on interpreting satellite pictures (Dvorak, 1975, 1984); see Figure 9-23. Enhanced infrared and digital infrared satellite data now help determine intensity. The techniques all use cloud features and rules based on a statistical model of tropical cyclone development to determine the current (and estimate the 24-hour) intensity of a tropical cyclone. The visible and infrared techniques differ mainly in the cloud features that are used in the analysis and how they are measured.

Recently, a panel of tropical cyclone meteorologists differed on what processes controlled intensification (Elsberry et al., 1992). Some favored convection, while others emphasized SST or changes in the upper troposphere. All agreed that improved forecasts require more observations, especially from jet aircraft.

Aircraft to Satellite data Relay (ASDAR) on commercial jets (3.1.4), might help, but prospects are discouraging.

9.8.4 Position Forecast Accuracy. The annual reports listed earlier tabulate errors in tropical cyclone position forecasts (see also Figure 9-22). For each forecast, the error is the distance between the forecast position and the observed position as determined by the "best track" based on careful post-analysis. Forecasts of tropical cyclone intensity or peripheral winds are not usually evaluated. At JTWC, Guam (Annual Tropical Cyclone Reports), NHC, Miami (De Maria et al., 1990) and in the Australian Bureau of Meteorology (Barnes, 1990) forecast models combining statistics and dynamics gave the smallest position errors.

Figure 9-24 and Table 9-3 present and analyze position errors in 24-, 48-, and 72-hour forecasts made for western North Pacific tropical cyclones at JTWC, Guam, and for North Atlantic tropical cyclones made at NHC, Miami, from 1970 through 1989. Average errors over the 20-year period differ little. Beyond that, significant differences may stem from the mean latitude of North Atlantic tropical cyclones being 7 degrees farther north (27.7° N) than the mean latitude of western North Pacific tropical cyclones (20.4° N) (Pike and Neuman, 1987).

For the North Atlantic, the modest improvement over time (time trend Correlation negative) is to be expected from better mid-latitude numerical weather predictions, particularly for more than 1 day in advance. No corresponding effect is discernible in the western North Pacific where tropical cyclones stay deep within the tropics for most of their lives. The latitude effect is also seen in the standard deviations of forecast errors, which are more than twice as large in the North Atlantic.

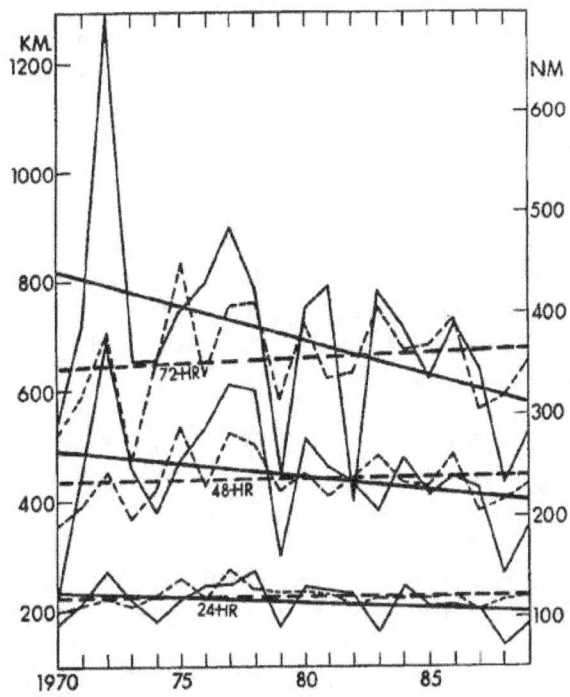

Figure 9-24. Yearly averages of tropical cyclone position-forecast errors compared to best tracks, 1970 through 1989 for NHC, Miami (solid lines) and JTWC, Guam (dashed lines). Heavy lines show linear trends.

TABLE 9-3. Average errors (NM) and trends in tropical cyclone position forecasts made at JTWC, Guam (13° N, 145° E) and NHC, Miami (26° N, 80° W) for 1970 through 1989.

ERRORS AND TRENDS	24-Hour		48-Hour		72-Hour	
	Miami	Guam	Miami	Guam	Miami	Guam
Average Error	113	119	239	235	374	353
Standard Deviation	20	10	61	26	103	47
Time Trend Correlation	-0.30	-0.01	-0.25	0.08	-0.38	0.13
Trend (NM yr)	-1.03	-0.03	-2.64	0.32	-6.85	0.27
Miami/Guam Correlation	0.33	-	0.55	-	0.46	-

Average position errors are exactly correlated with forecast periods. This means that a statistical forecast that can be issued as soon as the observations are received has a time advantage (equivalent to a 15-25 NM/30-50 km accuracy advantage) over a forecast for the same valid time, but made later by numerical models. Said another way, shortening the time to react to warnings by 2 hours in the western North Pacific would "improve" forecasts much more than they have been over the past 20 years. Warning users should be made aware of this.

The best standard to use in evaluating forecast accuracy is persistence; i.e., extrapolation from past positions. When cyclones move along climatological tracks, both forecast and persistence errors tend to be low. Errors increase when cyclones move erratically, especially during recurvature. Not surprisingly, improvement over persistence increases with increasing forecast interval. For 12-hour forecasts, extrapolations from positions 6 to 12 hours before are hard to beat. Ramage (1980) found that more than half the forecast error variance could be explained by the errors in persistence forecasts. He also found that errors in the western Pacific and the Atlantic persistence forecasts showed the same trends: Improvement to 1971 or 1972, then deterioration to the levels of the mid-60s. The moderately significant correla- tions between Guam and Miami forecast errors in Table 9-3 suggest that this relationship has persisted into the 1980s and that the variability in tropical cyclone tracks may respond to hemispheric causes.

Movement forecasting is more difficult in some tropical cyclone basins than in others. Pike and Neumann (1987) used the same persistence/ climatology model to make 24-hour movement forecasts for every basin for periods of 33 years or more. The errors, shown in Table 9-4, measure the forecast difficulty, and are generally related to mean storm latitude. The southwest Pacific/Australia region is the exception, possibly because storms there recurve at lower latitudes than elsewhere (see 9.4).

TABLE 9-4. Average errors in 24-hour persistence/climatology movement forecasts of tropical cyclone centers for tropical cyclone basins (Pike and Neumann, 1987).

Basin	Mean Storm Latitude	Error	
		NM	Km
Southwest Pacific/Australia	20.1° S	130	241
North Atlantic	27.6° N	113	210
Western North Pacific	20.4° N	99	184
Southwest Indian	18.4° S	87	161
Eastern North Pacific	17.9° N	78	144
North Indian	15.7° N	63	117

Average tropical cyclone position forecast errors can be applied to daily forecast positions for various time intervals to give probability density distribution. The distribution of position errors around the forecast positions is closely approximated by "circular-normal," or in some cases "elliptical-normal," distributions (U.S. Navy Weather Research Facility, 1963). In the circular-normal distribution, the probabilities of observed storm positions lying within various distances from the forecast positions can be represented by a family of concentric circles, whose radius kS_v is related to the probability level desired. S_v is the standard vector deviation of the observed positions about the forecast positions and the multiplier (k) relates to the various probability levels, (P), as shown below.

P (%) = 30   50   70   90   99

k = .60 .83 1.10 1.52 2.15

For example, there is a 70-percent probability that the observed position of the storm would be within a radius of $1.10S_v$ of the forecast position.

A study of statistical forecast-error data for Atlantic storms showed that the S_v values were nearly equal to the average forecast error. Therefore, as a first approximation, representative average forecast errors obtained from Table 9-3 can be used for S_v to establish probability levels in tropical cyclone forecast positions. For western North Pacific tropical cyclones, the 24-hour forecast error is 119 NM (221 km); S_v equals 119 x 0.83 = 99 NM, which is the 50-percent probability radius. Figure 9-25 illustrates the procedure.

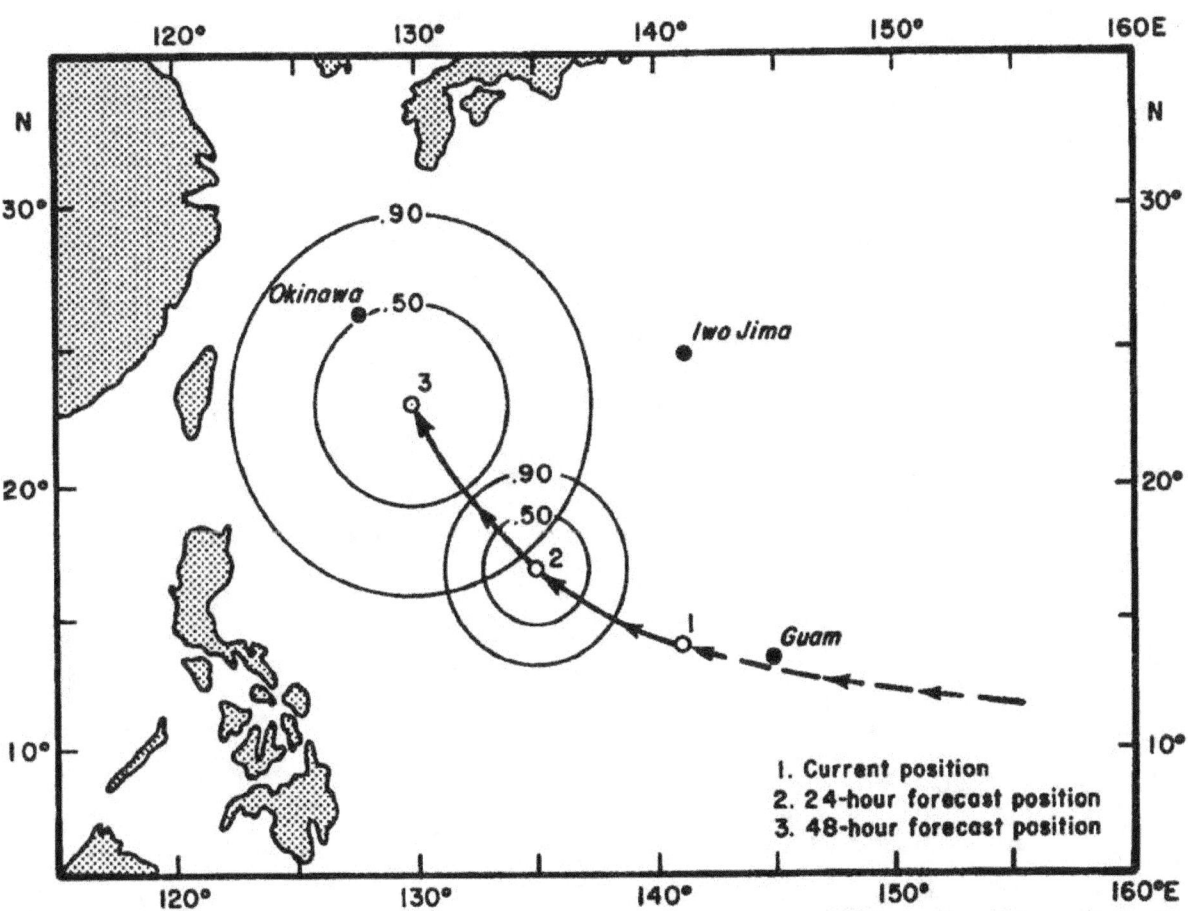

Figure 9-25. How to use average root-mean-square vector errors to establish areal confidence limits for typhoon forecasts.

9.8.5 Summary. Despite the disproportionate effort made to observe and understand tropical cyclones (as mentioned in Chapter 1), the results shown in Figure 9-24 are discouraging, even though movement forecasts at JTWC were improved when reconnaissance delineated the mid-tropospheric subtropical ridge (Shoemaker et al., 1990). Incorporating dropwindsonde data from reconnaissance made on the periphery of hurricane Debby (1982) improved 12-hour movement forecasts made by a numerical model (Lord and Franklin, 1987). In a recent review, Chan and Holland (1989) were pessimistic. They concluded that satellites and aircraft reconnaissance have not helped improve intensity and position forecasts much.

Chapter 10

SEVERE WEATHER IN THE TROPICS

10.1 GENERAL

The tropics experience a wide variety of weather; from the persistent, almost cloudless skies of many subtropical lands during the dry season, to the violent winds and rain of intense tropical cyclones. The term "severe weather" with its range of meanings, is often defined in relation to its effects on human activities.

In this chapter, the following phenomena, often classified as severe weather, are discussed: thunderstorms, tornadoes, waterspouts, hail and related gusty surface winds, dust storms, extreme winds associated with tropical cyclones, turbulence, icing, and ocean waves. Heavy rains, also severe weather, have already been treated in Chapter 6.

10.2 THUNDERSTORMS

10.2.1 Thunderstorm Frequency. Thunderstorms are more common in the tropics than in higher latitudes. Figure 10-1 shows the mean annual number of thunderstorm days over the Earth (World Meteorological Organization, 1956b). WMO defines "thunderstorm day" as an observational day during which thunder is heard at a station. Tropical continents receive the most thunderstorms, with centers of over 140 days annually in South America, Africa, and southeast Asia. This distribution is confirmed by lightning observations made from DMSP satellites (Orville and Henderson, 1986) and plotted in Figure 10-2.

There are probably more thunderstorms per thunderstorm day over the continents than over the oceans. Note (in Figure 10-2) how few lightning flashes were recorded over the near-equatorial tradewind convergence in the central and eastern Pacific and the western Atlantic (see 8.4.8.1).

Over tropical oceans, almost all thunderstorms are associated with synoptic disturbances, while over land, air-mass thunderstorms due to convective heating and orographic lifting are common.

MEAN ANNUAL NUMBER OF THUNDERSTORM DAYS

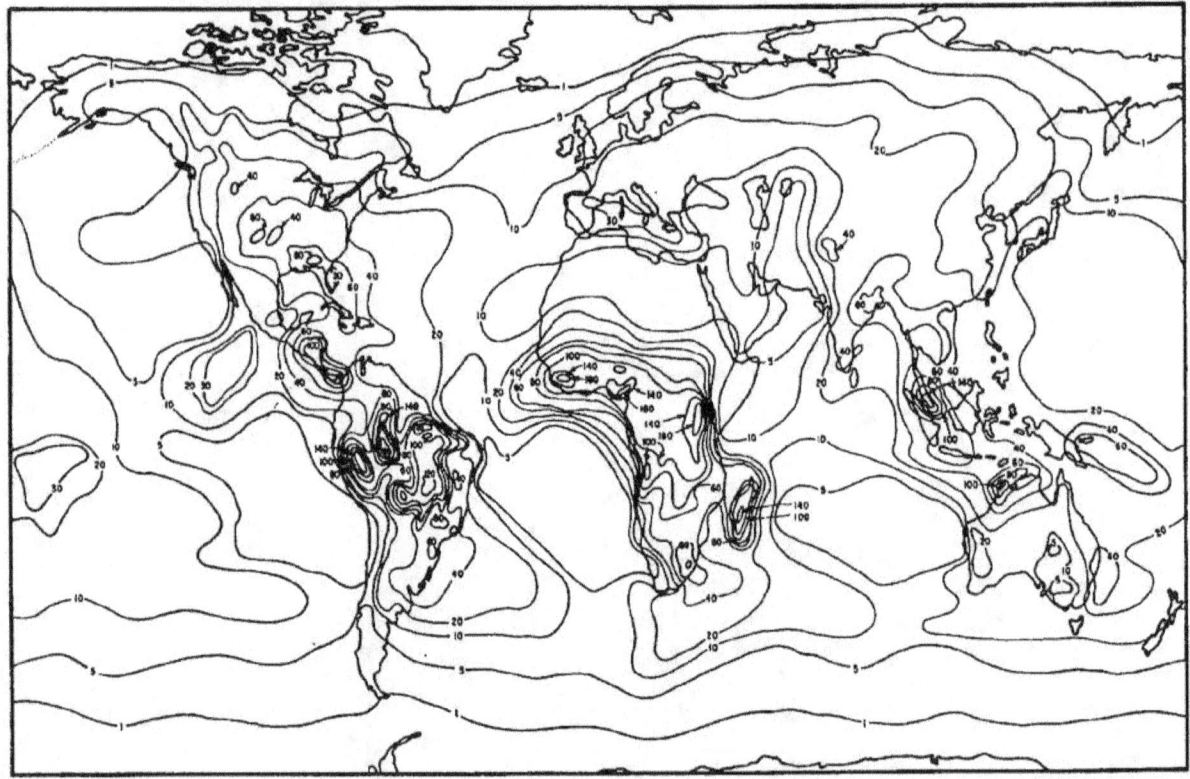

Figure 10-1. Mean annual number of thunderstorm days over the globe (World Meteorological Organization, 1956b).

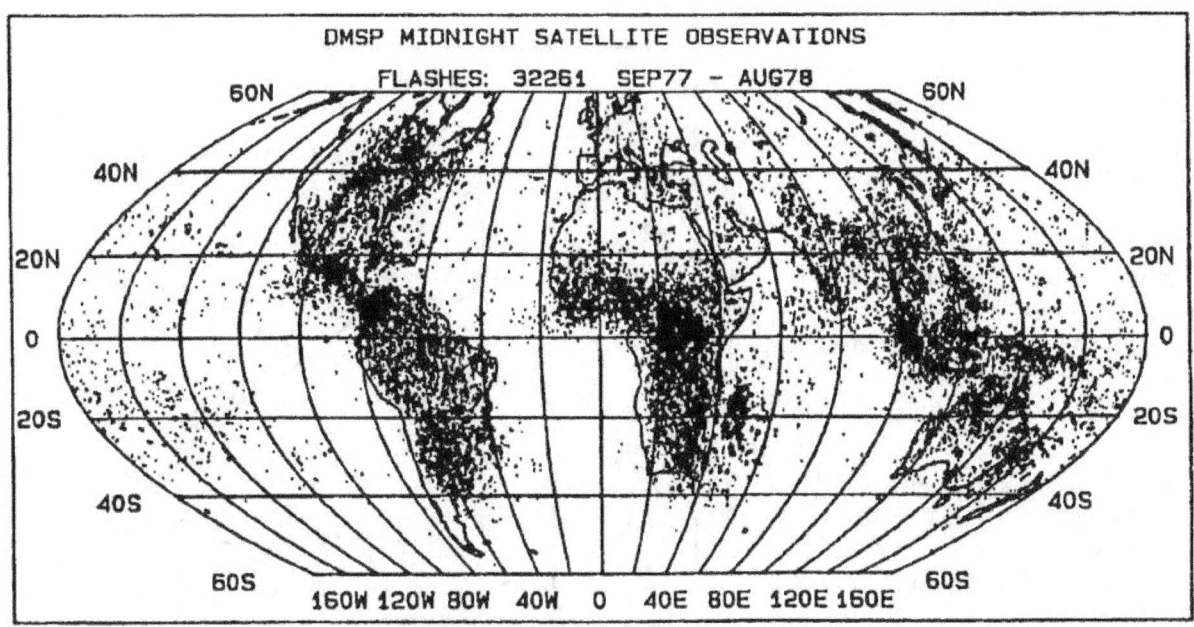

Figure 10-2. One year of midnight lightning locations detected on DMSP images for 365 consecutive days from September 1977 to August 1978 (Orville and Henderson, 1986).

In the equatorial belt, most thunderstorms occur over land (Figure 10-3). 82 percent of observed thunderstorms are confined to South America, Africa and Indonesia, while the oceans account for only 18 percent. Figures 10-1, 10-2, and 10-3 give only the broad-scale distribution; more detail, such as monthly frequencies, diurnal variations, etc., can be found in climatological publications.

A word of caution: Of all commonly reported weather elements (such as temperature, humidity, rainfall, and pressure), thunderstorm reports are often the least reliable. For example, stations making observations only during daylight invariably record fewer thunderstorms than comparable first-order stations operating 24 hours a day.

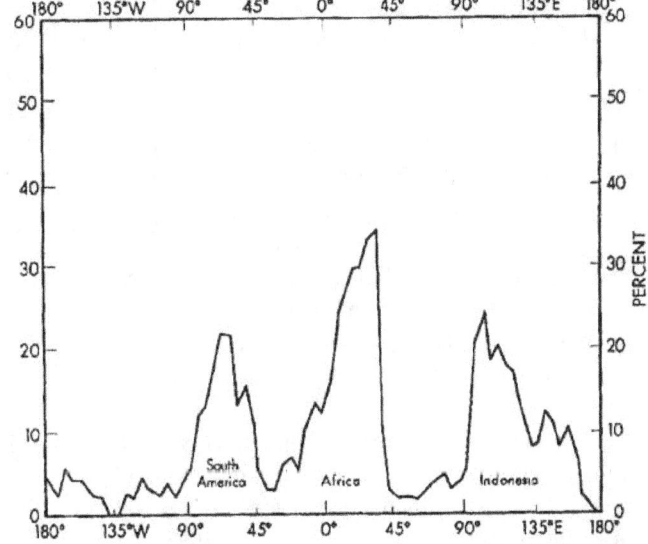

Figure 10-3. Mean annual percentage of days with thunder by 5-degree longitude intervals for the equatorial belt between 10° N and 10° S (adapted from Ramage, 1968).

275

The WMO definition of a "thunderstorm day" was not always generally accepted. Reliable thunderstorm statistics are hard to obtain; the following examples illustrate the problem:

• Figure 10-4 (Atkinson, 1967a) gives the mean annual number of thunderstorm days over southeast Asia based on a careful study of all available data. In some areas, the annual thunderstorm frequency is more than double that shown in Figure 10-1.

• Figure 10-5 shows the mean annual number of thunderstorm days over India from 1931 to 1960 (Alvi and Punjabi, 1966). Analysis in this figure is based primarily on thunderstorm statistics from first-order 24-hour observing stations. Compare the isobronts (lines connecting equal average numbers of days with thunderstorms) in this figure with the reported annual means for many other stations also plotted on the map.

• In Honolulu (21° N, 158° W), during the winter of 1970-71, the first-order airport station reported thunder or lightning on 32 days. A sferic detector, with a range of 15 NM (30 km) and operated at the University of Hawaii, 6 NM (11 km) from the airport, was activated on 69 days (Sherretz et al., 1971).

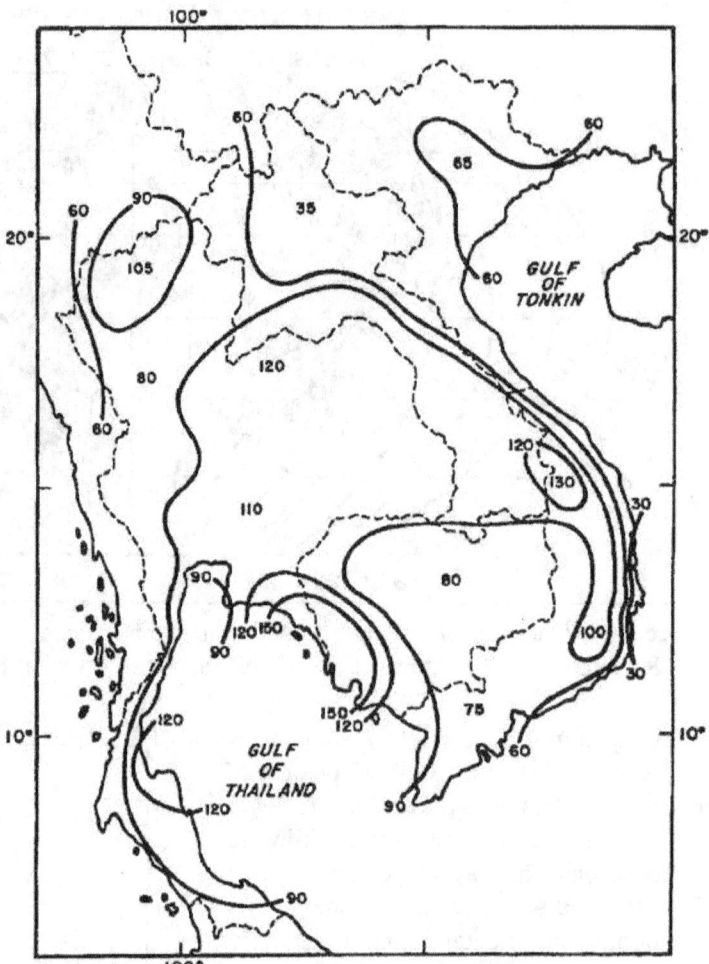

Figure 10-4. Mean annual number of thunderstorm days over Southeast Asia (after Atkinson, 1967a).

Ground-based lightning detection, although restricted to a few locations, is both accurate and reliable, and continuous lightning detection from GOES is planned for the mid-1990's (see 3.2.1). But until then, the examples above emphasize the great care needed to correctly interpret thunderstorm statistics derived from current observing techniques. Most areas experience large year-to-year variations in annual and monthly numbers of thunderstorm days. A detailed study (Atkinson, 1967a) of thunderstorm variability over 12 years at 28 stations in Thailand, illustrates this. To increase the sample size, the individual monthly numbers of thunderstorm days were combined for station-months which had the same mean number of thunderstorm days and standard deviations, and frequency distributions determined for these grouped samples. Only station-months with a mean of 3 or more thunderstorm days were considered. Figure 10-6a (on page 10-6) shows the resulting frequency distribution for selected categories of monthly means.

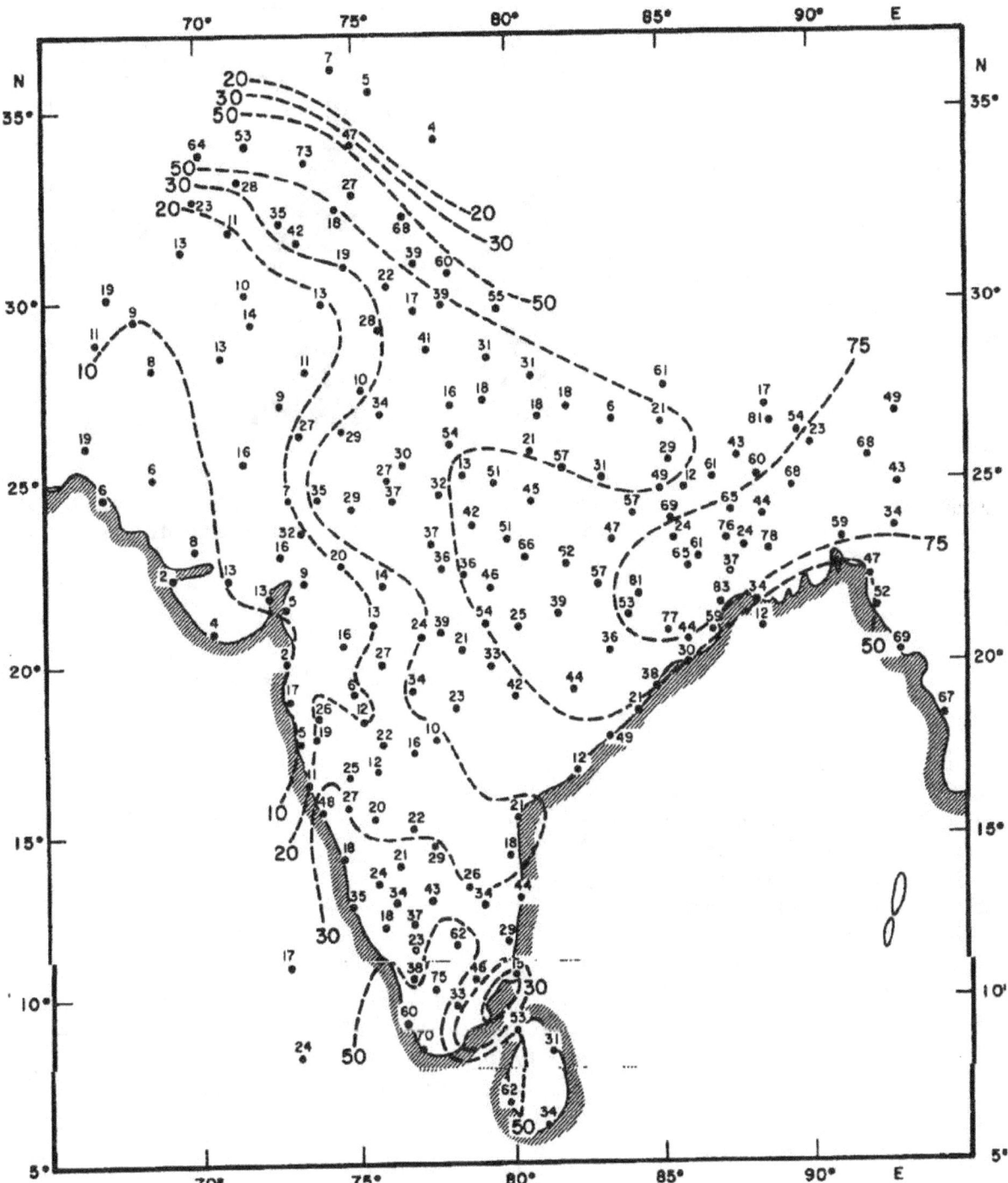

Figure 10-5. Mean annual number of thunderstorm days over India during the period 1931-1960. Individual station means are plotted (after Alvi and Punjabi, 1966).

277

In Figure 10-6a, a bar graph showing a monthly mean of 3.6 days includes the individual yearly values for station-months with a mean of between 3.0 and 3.9 days. Since the resulting distributions are approximately normal, standard deviations were computed for each 1-day mean increment.

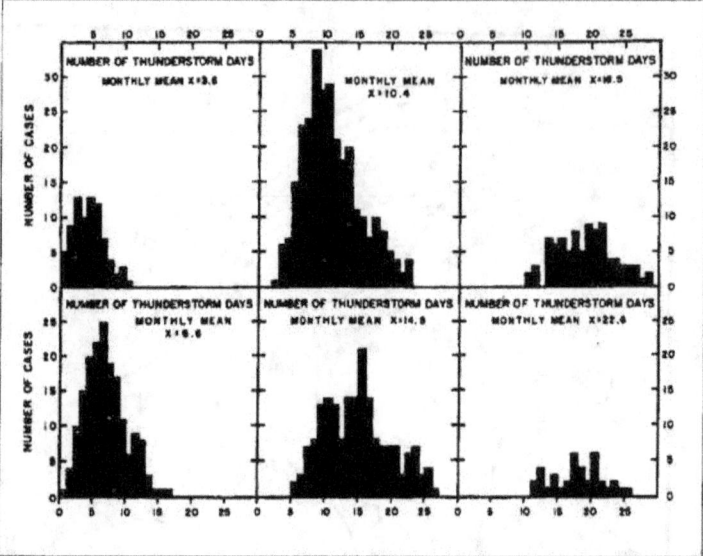

Figure 10-6a. Frequency distributions for selected categories of monthly means of the number of thunderstorm days for stations in Thailand.

In Figure 10-6b, the standard deviations computed from 1-day mean increments in Figure 10-6a are plotted against the mean monthly number of thunderstorm days; a regression line is fitted by eye to these points. The means and standard deviations increase together up to a mean of about 16 days, above which the standard deviation decreases for further increases in the mean.

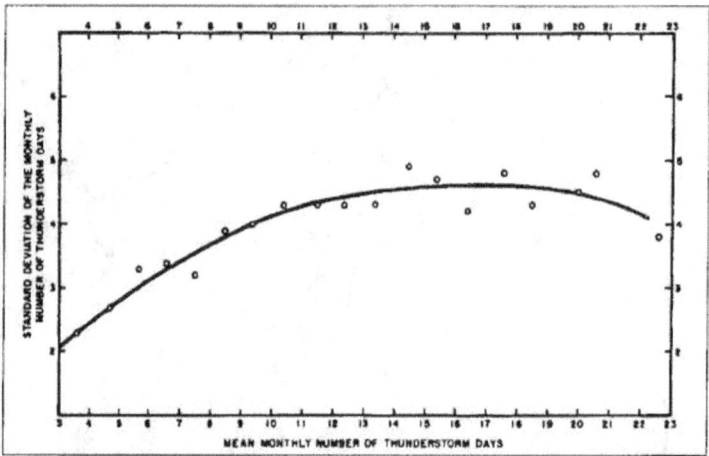

Figure 10-6b. Standard deviations for selected categories of monthly means and line of best fit in the number of thunderstorm days for stations in Thailand.

To determine the percentage range in thunderstorm days that can be expected in any month, cumulative frequency distributions of the yearly values for station-months with similar means (such as those in Figure 10-6a) were plotted. The initial family of curves was smoothed slightly and used to interpolate the percentage values for monthly means of 3 to 24 thunderstorm days. This derived family of curves shown in Figure 10-6c can serve as a model for estimating the year-to-year variation in thunderstorm days for any month, given the mean monthly value.

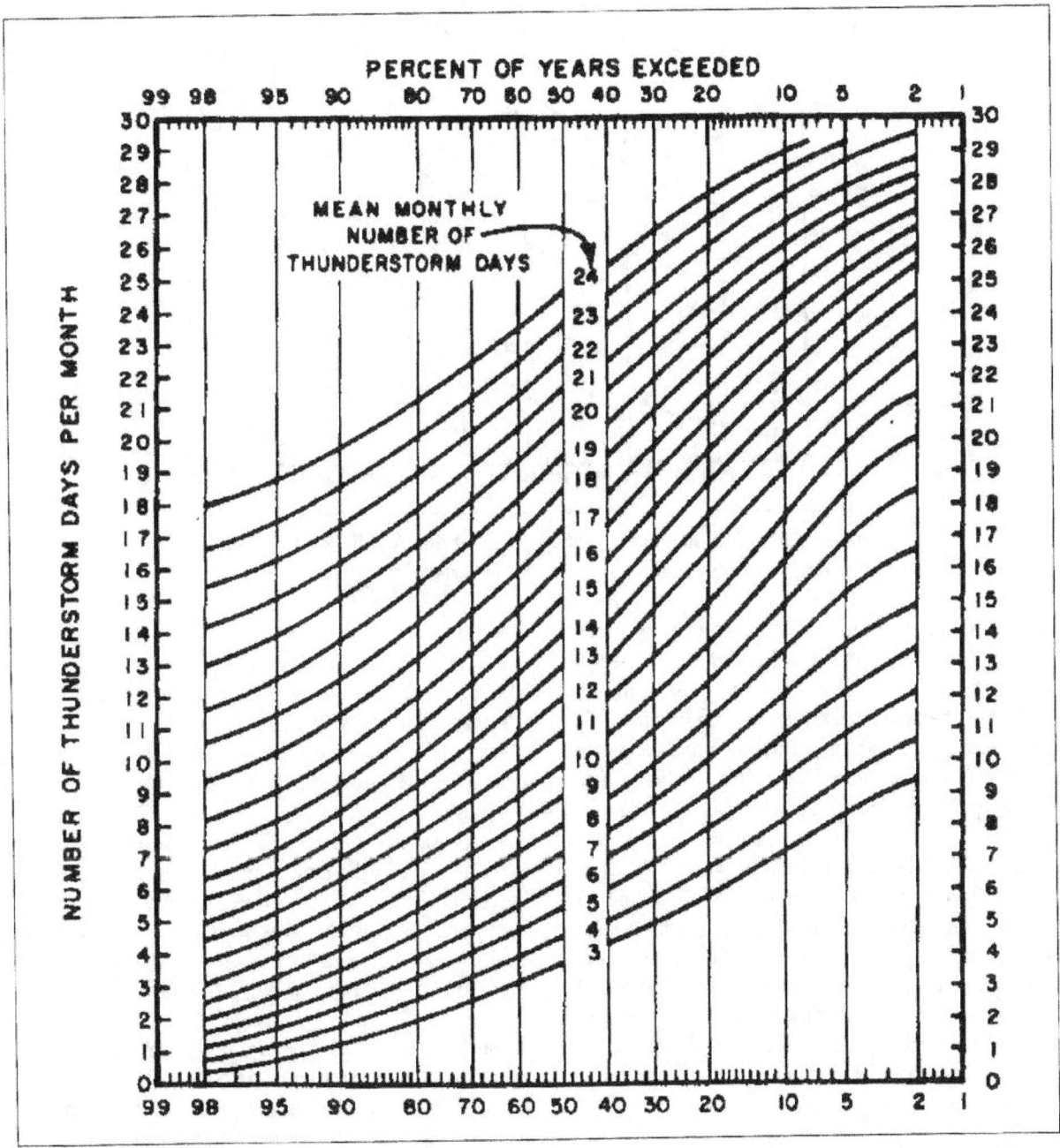

Figure 10-6c. Cumulative frequency distributions for the monthly number of thunderstorm days according to the mean values for stations in Thailand.

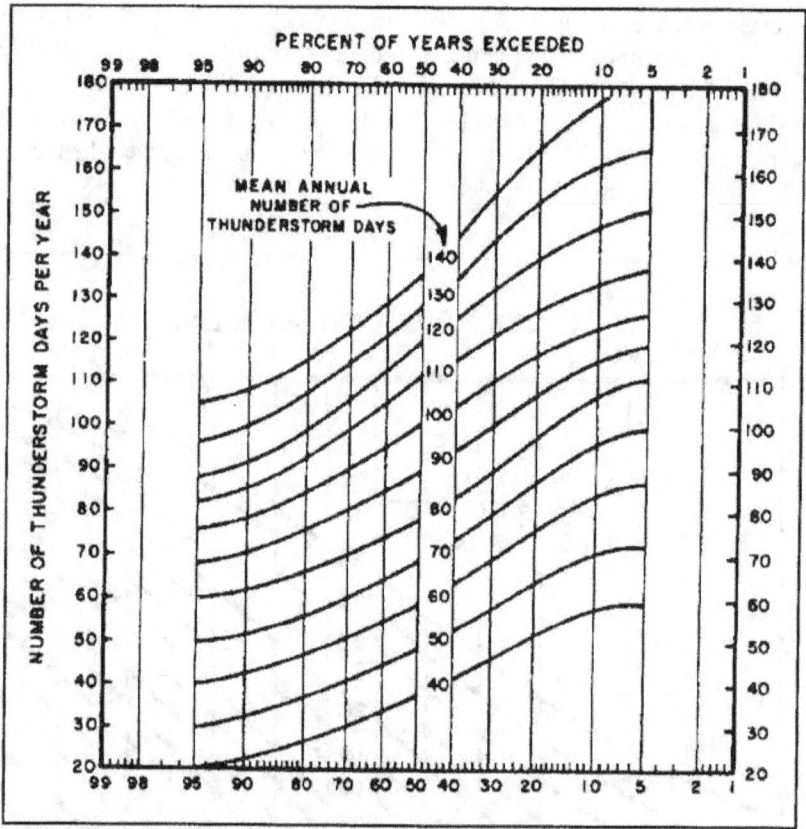

Figure 10-6d. Cumulative frequency distributions for the annual number of thunderstorm days according to the mean values for stations in Thailand.

Climatological records usually contain only mean monthly or mean annual numbers of thunderstorm days. If the frequencies are assumed to be normally distributed, then we can use the curves in Figure 10-6c to estimate the probability of thunderstorm days for a month; the curves in Figure 10-6d can be used to estimate probabilities for a year.

Use the figures to estimate monthly probabilities as follows: If you seek the probability of observing 15 thunderstorm days in a month having a long-term mean of 9 days, move up the curve in Figure 10-6c representing this mean until it intersects the horizontal line corresponding to 15 days; read the percentage of occurrence from the top scale (in this case, 10 percent). Therefore, you could expect 15 or more thunderstorm days in this month in 1 year out of 10.

Suppose you need to know the 90-percent range (from the 95th to the 5th percentile) of thunderstorm days

for a station with a monthly mean of 16 thunderstorm days. Entering the curve representing this mean, read the number of days at the 95th and 5th percentiles (in this case 9 and 24 days). This tells you that in 90 percent of the years this station can expect between 9 and 24 thunderstorm days in this month.

Use the same procedure (in Figure 10-6d) to relate the variation in the annual number of thunderstorm days to the mean annual value. Because of the small sample size, which gives much larger errors of estimate at the extreme percentiles, the curves are limited to the 90-percent range of values.

Although the models for Figures 10-6a, b, c, and d were developed for stations in Thailand, they might also be used for other tropical lands where similar year-to-year variations in the annual and monthly numbers of thunderstorm days occur. Diurnal variation was discussed in Chapter 7.

10.2.2 Thunderstorm Duration. Rain falls from individual thunderstorm cells for an hour or less; however, continuous thunderstorms last for many hours (6.3.1.2). Climatological summaries are usually based on hourly- or 3-hourly synoptic observations and are inadequate to determine thunderstorm durations; local and special observations are needed.

Figure 10-7a (right) shows the durations of all thunderstorms at Mactan Air Base (10° N, 124° E) in the Philippines for June through August 1965, based on WBAN forms for surface observations (Atkinson, 1967b).

Figure 10-7b (below) shows the frequency distribution of thunderstorm durations at Mactan for the rainy seasons (June-November 1965 and May-November 1966) based on data similar to that used to prepare Figure 10-7a. Each period of thunderstorm activity is counted as a separate occurrence regardless of the length of the intervening interval.

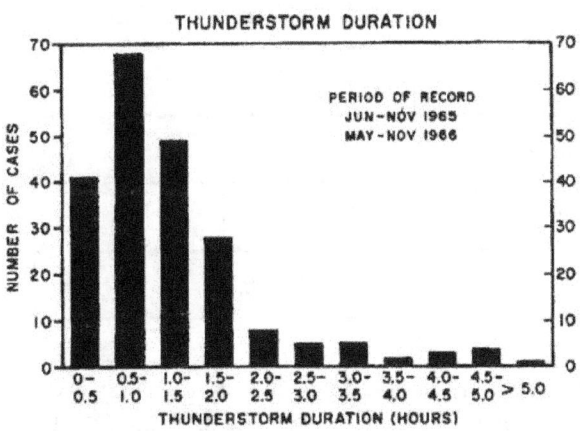

Figure 10-7b. Frequency distribution of thunderstorm durations considering all separate occurrences at Mactan Air Base, Philippines.

Figure 10-7c combines successive periods of thunderstorm activity separated by less than 2 hours into one occurrence. Of course, this produces more long-duration storms. At Mactan, thunderstorms commonly last from 30 minutes to 1 hour, while very few last more than 2 hours. Mactan, on a small island 10 km east of Cebu City, experiences mostly nocturnal thunderstorms, as shown in Figure 10-7a; they are probably generated from convergence between a land breeze and the prevailing offshore wind.

Representations similar to Figure 10-7a, b, and c are easy to prepare and can help to acquaint newly-assigned forecasters with the thunderstorm activity patterns at any station.

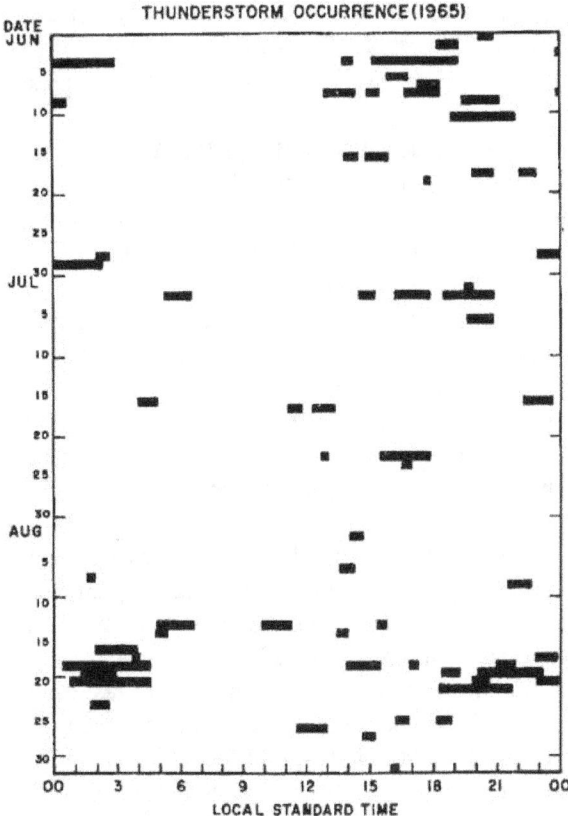

Figure 10-7a. Thunderstorm occurrence, Jun-Aug 1965 at Mactan Air Base, Philippines.

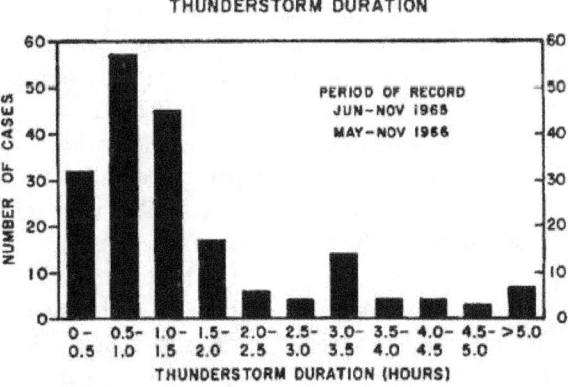

Figure 10-7c. Frequency distribution of thunderstorm duration derived by considering successive periods of thunderstorm activity separated by less than 2 hours as one occurrence at Mactan Air Base, Philippines.

281

The average thunderstorm duration at any station can be estimated by dividing the mean number of hourly observations of thunderstorms in any period (i.e., month, season, year) by the mean number of thunderstorm days in that period (Table 10-1). Durations range from 0.9 hours at Chiangmai and Songkla to 2.1 hours at Bangkok. The overall average of 1.5 hours is probably typical of most of the tropics.

10.2.3 Thunderstorm Heights. As in higher latitudes, the heights of tropical thunderstorm tops vary considerably; most range between 30,000 and 50,000 feet (9 and 15 km). In general, tops are higher over the continents than over the oceans (see Figure 6-22), probably because of surface heating and orographic lifting. Except in intense tropical cyclones, where cloud tops in the eye wall and feeder bands may range between 50,000 and 60,000 feet (15 and 18 km), thunderstorm tops over the tropical oceans lie between 30,000 and 40,000 feet (9 and 12 km).

Deshpande (1964) analyzed cumulonimbus tops over India during the summer monsoon (June-September) of 1957 through 1962 using 322 reports made from jet aircraft flying near the tops. The observations were fairly evenly distributed over India. The resulting frequency distributions showed that most of the tops lie between 35,000 and 50,000 feet (10.7 and 15.2 km) with a median height between 40,000 and 45,000 feet (12.2 and 13.7 km). 60,000 feet (18.3 km) marked the upper limit. There was no significant regional variation.

Five years of radar data was used to determine the height of cumulonimbus tops over the plateau and plains of central India (where local orographic effects are slight) (Thomas and Raghavendra, 1977). During the pre-monsoon months of March to May, when a showers regime prevails, 12 percent of all tops exceeded 40,000 feet (12 km). During the summer monsoon months of June to September, the figure was only 6.2 percent. This is considerably lower than what Deshpande found and emphasizes the difference between a continuous series of measurements such as these and the spot observations Deshpande used.

TABLE 10-1. Thunderstorm duration, shown by the ratio of the mean annual number of hourly observations reporting thunderstorms to the mean annual number of thunderstorm days for stations in southeast Asia.

LOCATION	LONGITUDE, LATITUDE	DURATION
THAILAND		
Chiangmai	19° N, 99° E	0.9
Korat	15° N, 102° E	1.5
Bangkok	14° N, 101° E	2.1
Songkia	7° N, 107° E	0.9
VIETNAM		
Pleiku	14° N, 108° E	1.6
Saigon	11° N, 107° E	1.5
Soctrang	10° N, 106° E	1.6
PHILIPPINES		
Clark	15° N, 121° E	1.5
Cebu	10° N, 124° E	1.7

Puniah (1973) analyzed about 6 years of mainly over-ocean flights east and southeast of Calcutta (23° N, 88° E); 86 percent of the tops ranged from 32,000 to 42,000 feet (9.8 and 12.8 km), and were very rarely higher. Overall, tops were about 6,500 feet (2 km) lower than the summer monsoon tops reported by Deshpande (1964).

In the tropics, cumulonimbus tops are about the same height as cirrus (see 6.2.6.6) but they must reach 30,000 feet (9 km) before developing into thunderstorms.

10.3 TORNADOES AND WATERSPOUTS

A tornado is a violently rotating column of air, pendant from a cumulonimbus, that touches the ground. The sense of rotation is usually cyclonic. On a local scale, it is the most destructive of all atmospheric phenomena; wind speeds are estimated at from 90 to 270 knots in its vortex, several hundred meters in diameter. All continents have tornadoes, but they are most common in the United States and Australia. Although they can occur throughout the year and at any time of the day, they are more frequent in spring and during the middle and late afternoon. A tornado over water (a waterspout) is much less violent. Waterspouts develop more often in the tropics and subtropics (Huschke, 1959). A "funnel cloud" is a tornado that does not touch the ground.

Tornadoes and their accompanying death and destruction are well-known to mid-latitude residents; detailed statistics have been compiled. Tornadoes are rare in the tropics for which few or no statistics are available. Perhaps their reported frequency will increase as observing networks expand and communications improve.

The India Meteorological Department recorded 42 tornadoes between 1951 and 1980 (Singh, 1981). Thirty-three occurred in March through May; as Figure 10-8 shows, 27 were confined to West Bengal and Bangladesh. Development there was helped by low-level moisture from the Bay of Bengal and the lakes and rivers of Bangladesh. The tornadoes were associated with scattered thunderstorms that generally formed east of troughs in the upper-tropospheric westerlies. Usually, upper-tropospheric and middle-to lower- tropospheric wind maxima were present, with strong wind shear between the lower maximum and the surface. These preconditions resemble those of the Gulf Coast region of the United States.

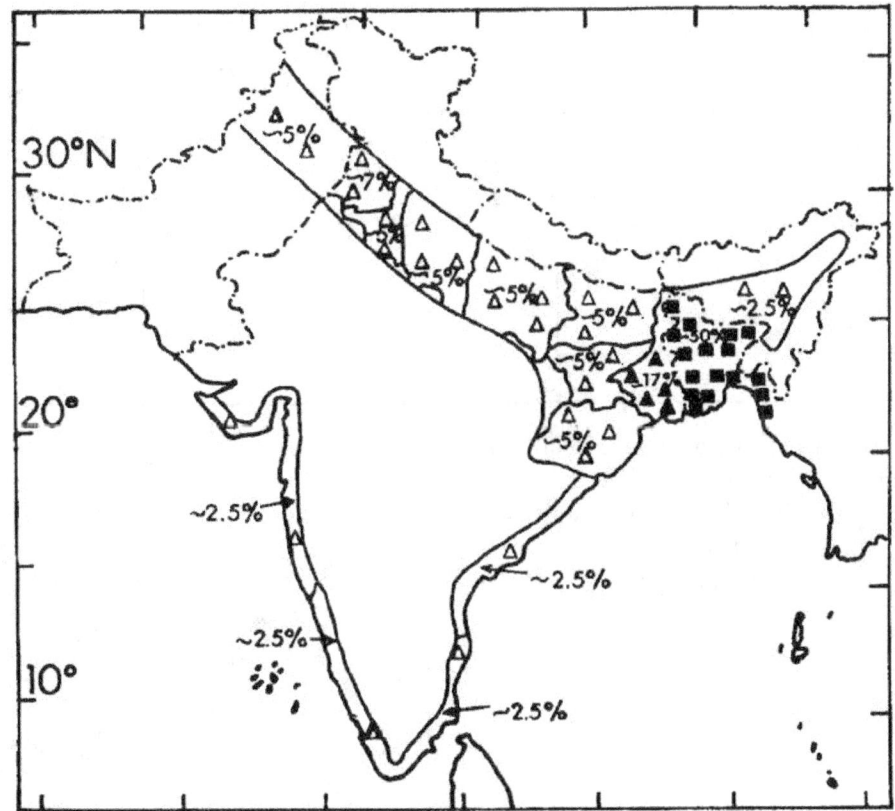

Figure 10-8. Percentage frequency of occurrence of tornadoes over India, 1951 to 1980; total number, 42 (Singh, 1981).

Tornadoes, waterspouts and funnels are fairly common during the winter in Hawaii (Schroeder, 1976). The clustering of observations around airports (Figure 10-9) strongly suggests many more unreported occurrences. Funnels are most likely just east of a deep trough in the upper-tropospheric westerlies, when the trade winds and the tradewind inversion are absent. Heavy rain often falls.

Prior to 1960, tornado data for southeast Asia was generally lacking. Since then, funnel clouds have been reported in the region, most often in the Mekong Delta of Vietnam (Sands, 1969). Waterspouts have been observed near Hong Kong (22° N, 114° E) and on Guam (13° N, 145° E).

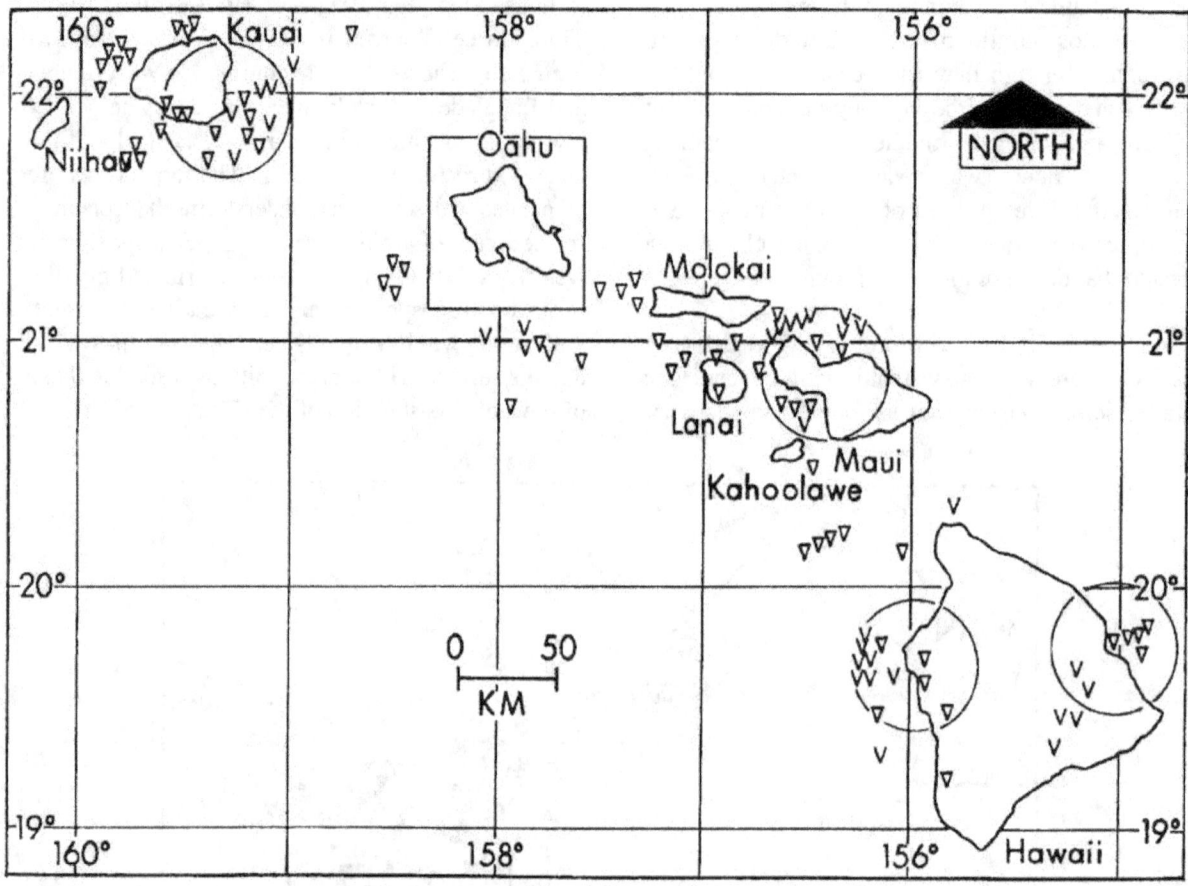

Figure 10-9. Locations of reported funnels aloft (V) and waterspouts/tornadoes (▽) for the Hawaiian Islands, 1961-1974. 30-km radius circles are drawn around the major airports. The island of Oahu is enlarged at right (Schroeder, 1976).

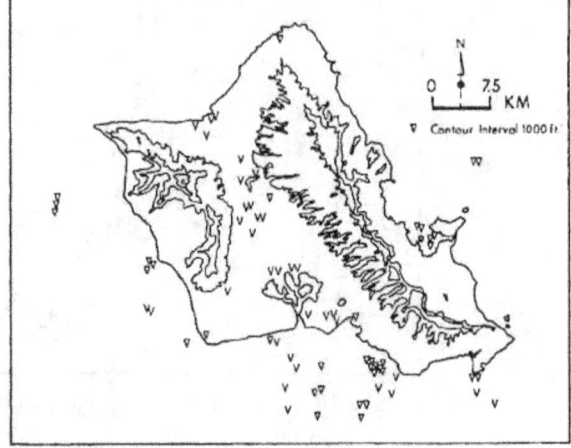

Gerrish (1967) prepared a detailed climatology of tornadoes, waterspouts, and funnel clouds over south Florida within 75 NM (140 km) of Miami (26° N, 80° W) for 1957 through 1966. As in Hawaii, many occurrences are missed by regular reporting stations. Gerrish therefore added to the reports observations from small airfields, park rangers, news media, and other unofficial observers. He found that in the 10-year period studied, at least 56 tornadoes and 218 waterspouts occurred on 47 and 139 days, respectively. In addition, on 224 days, 314 funnel clouds were observed in the absence of tornadoes and waterspouts.

Figure 10-10 (below) shows the results of Gerrish's study and Figure 10-11 (next page) locates the events. Most occurred during the warm season (May through October). Tornadoes are equally likely through the summer, while waterspouts and funnel clouds have a sharper summer maximum. Diurnal variations differ. Tornadoes are most common in late afternoon; funnel clouds, in early afternoon. Waterspouts favor the morning with a slight secondary maximum in late afternoon. In Figure 10-11, all phenomena are concentrated near the coast; once again, this distribution is probably biased by population density.

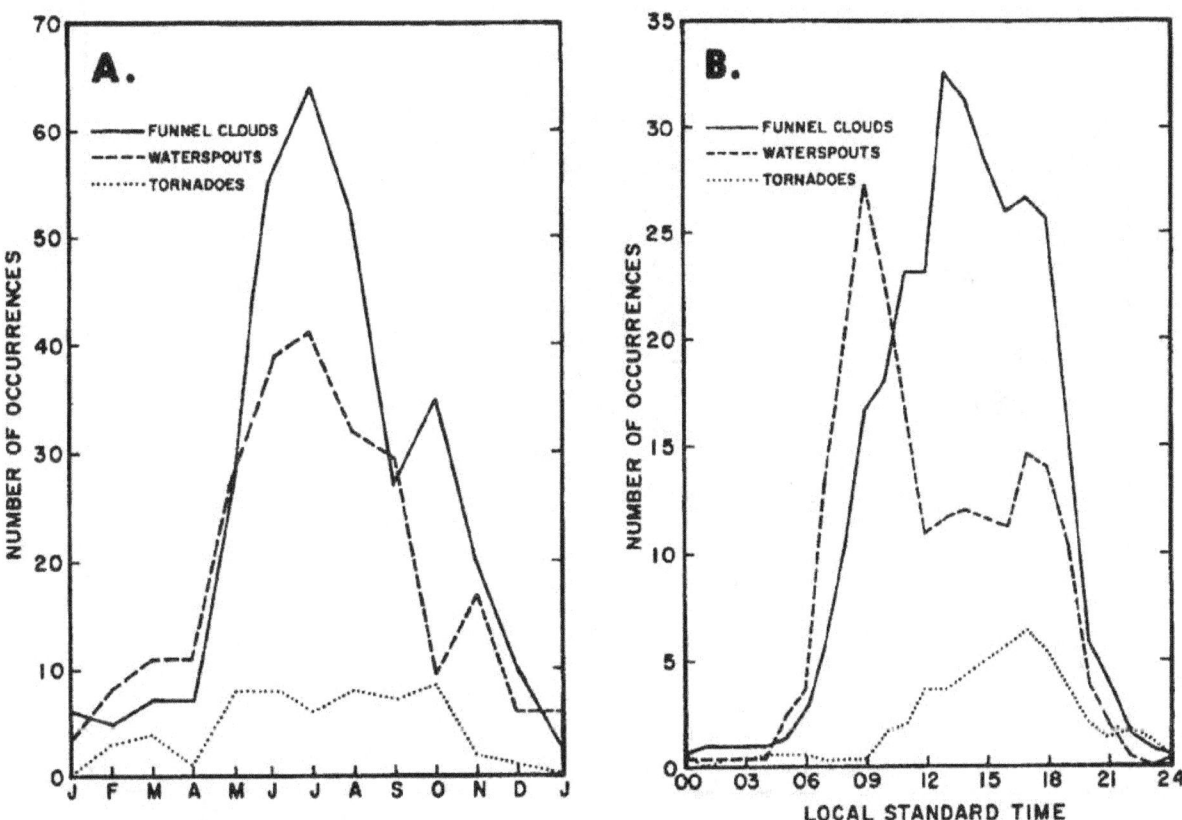

Figure 10-10. Occurrence of funnel clouds, waterspouts, and tornadoes within 140 km of Miami, Florida, 1957 to 1966; (A) monthly, (B) Diurnal variation (hourly values smoothed by a 3-hourly running mean) (adapted from Gerrish, 1967).

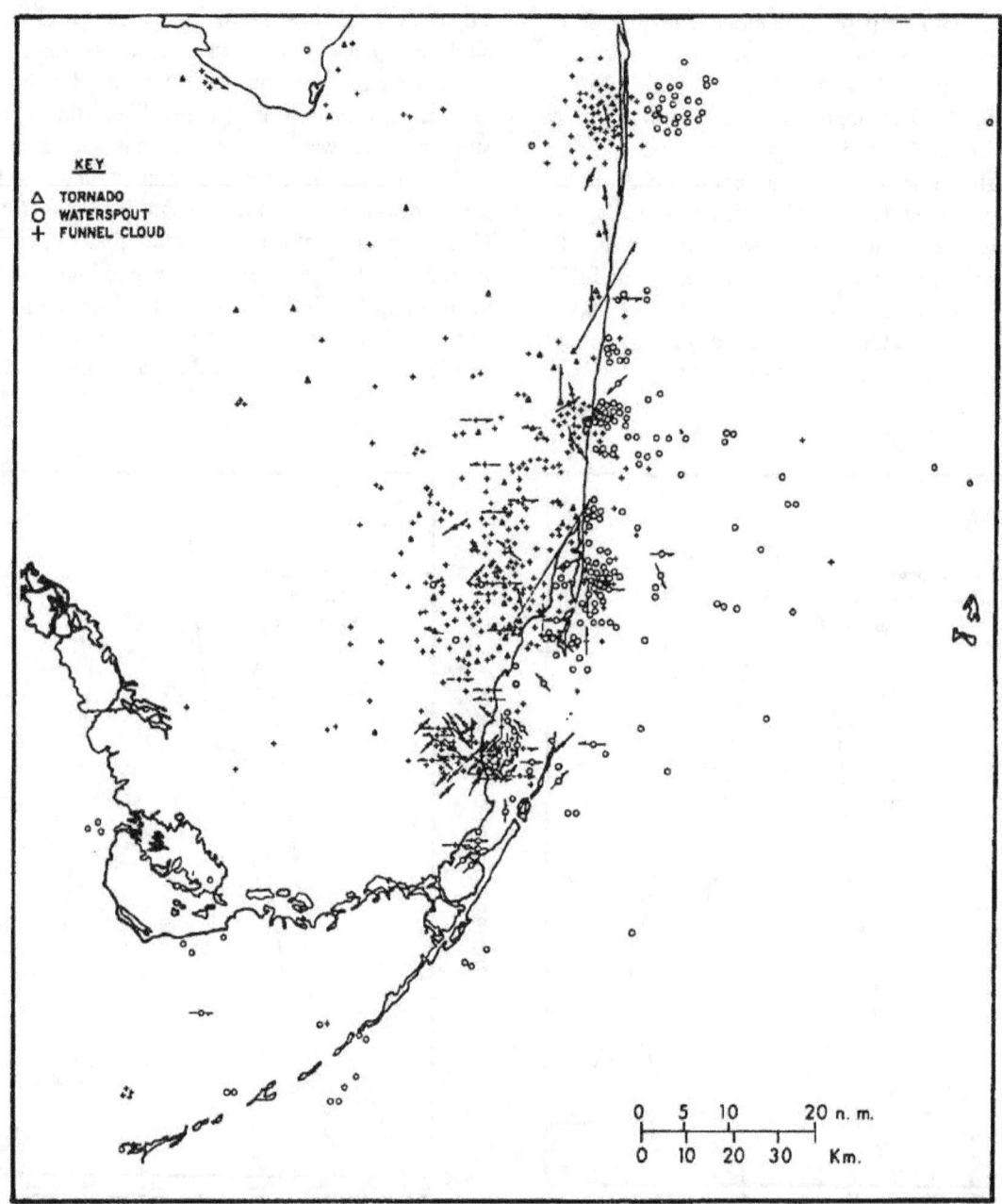

Figure 10-11. Distribution of tornadoes, waterspouts, and funnel clouds in the south Florida area, 1957 to 1966. Arrows show reported direction of movement (after Gerrish, 1967).

Florida tornadoes, particularly those in south Florida, are generally neither as severe nor as destructive of life and property as those in the central United States. Gerrish (1967) found that tornadoes are more often accompanied by lightning than are waterspouts. He cautioned, however, that waterspouts are potentially

dangerous to light aircraft and boats. During the cool season (November through April), almost all tornadoes and many waterspouts are associated with fronts or squall lines. But during summer, they usually accompany surface ridges or other "benign" tradewind patterns with no upper-level jet stream; this makes warm season forecasting extremely difficult. Tornadoes generally accompany significant radar echoes; however, waterspouts occasionally occur with clouds having small or no radar echoes. Relating tornado occurrences to temperature and moisture aloft, Gerrish found that less instability and moisture are prerequisite to tornadoes in south Florida than along the Gulf Coast. Also, tornadoes, waterspouts and funnel clouds are unlikely when the tradewind inversion is lower and stronger than normal.

During the Lower Florida Keys Waterspout Project of May to September 1969 (Golden, 1973), 397 waterspouts and funnel clouds were observed within 50 NM (90 km) of Key West (25° N, 82° W), many more than recorded by Gerrish and shown in Figure 10-10. 95 waterspouts were documented by aircraft during the project. They generally formed in lines of growing cumulus congestus, and lasted from 1 to 22 minutes. In the more intense of these, surface tangential winds exceeded 140 knots (Golden, 1974). In September 1974, instrumented aircraft flying above 1,300 feet (400 meters) penetrated 16 lower Florida Keys waterspouts (Leverson et al., 1977), seven of them more than once. Never did maximum vertical velocities exceed 20 knots, nor did tangential velocities exceed 60 knots.

10. 3. 1 Hurricane-generated tornadoes.

Novlan and Gray (1974) and Gentry (1983) developed a climatology of hurricane-spawned tornadoes over the United States. For the period 1948 through 1981 (excluding Hurricane Beulah in 1967 which had 141 tornadoes), tornado-producing hurricanes spawned an average of 10 each. The geographical distribution shown in Figure 10-12 emphasizes the contributions of a few cyclones. These reports agree with others (Smith, 1965; Pearson and Sadowski, 1965; and Hill et al., 1966) in locating tornadoes generally in the right front quadrant of the hurricane, within 150 NM (300 km) of the center, as shown in Figure 10-13, next page.

The area of greatest tornado concentration lies outside the general area of hurricane-force winds, the mean radius of which is only about 35 NM (70 km) during tornado occurrence. After first eliminating tornadoes associated with the extratropical stage of the tropical cyclones, Hill et al. (1966) concluded that 94 percent occurred between azimuths of 10 and 120°. Most of these tornadoes were 60 to 240 NM (110 to 440 km) from the cyclone center. Hurricane tornadoes accompanied thunderstorm cells in both the inner and outer rainbands (as well as in scattered outer convective cells); the stronger cells of the outer rainbands of the storm are most likely to spawn tornadoes.

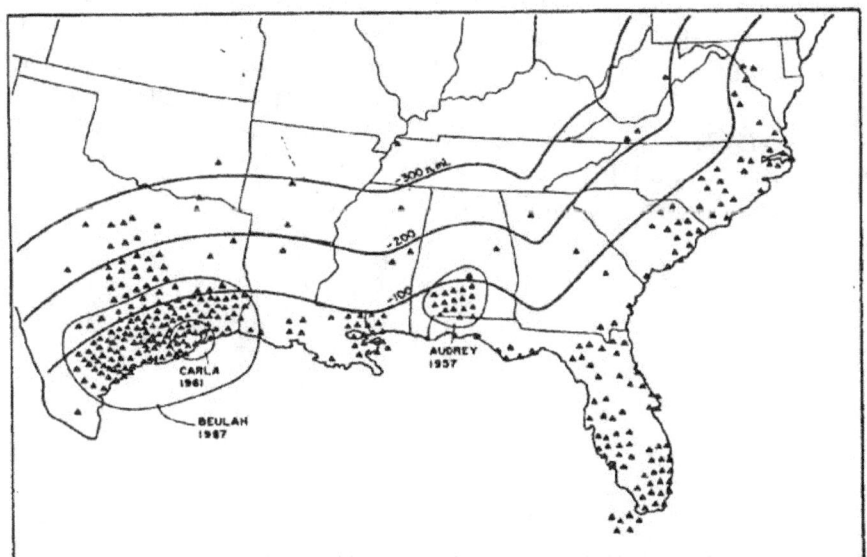

Figure 10-12. Distribution of hurricane tornadoes, 1948 to 1972 (Novlan and Gray, 1974).

287

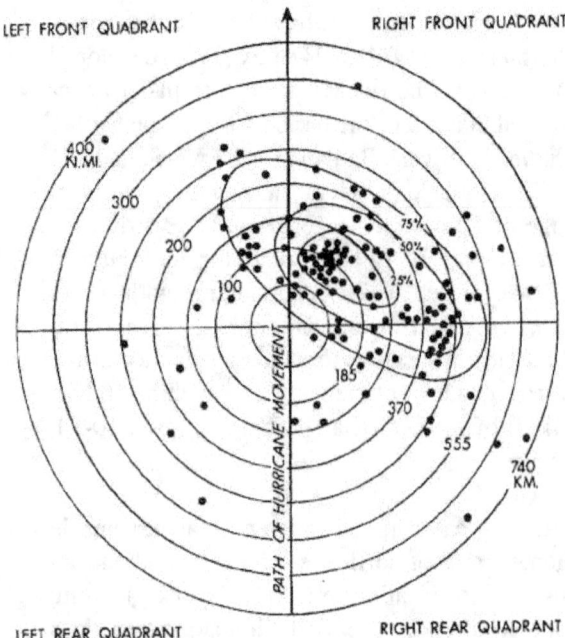

LEFT FRONT QUADRANT   RIGHT FRONT QUADRANT

LEFT REAR QUADRANT   RIGHT REAR QUADRANT

Figure 10-13. Hurricane tornadoes over the U.S., 1955-1964, located with reference to the center and direction of movement of the hurricanes. Isopleths enclose percentages of the total number of cases (adapted from Pearson and Sadowski, 1965).

Smith (1965) found that 82 percent of hurricane tornadoes from 1955 to 1962 occurred between 0300 and 0400L and between 1500 and 1600L. For Hurricane Isbell in 1964, Pearson and Sadowski (1965) found a pronounced peak between 1500 and 1800L; however, little diurnal variation was evident for the remaining cases in 1964.

During Hurricane Beulah (1967), 80 percent of the tornadoes occurred between 0300 and 1500L, with a pronounced peak (32 percent) between 0900 and 1200L (Orton, 1970).

Hurricane tornadoes appear to favor daylight, but this may be only because they are easier to observe then. Gentry pointed out that all hurricanes crossing the coast of the United States between Brownsville, Texas, and Long Island, New York, have associated tornadoes. Convection is usually most intense in the right front quadrant of the hurricane. As Figure 10-13 shows, tornadoes usually develop in this quadrant in air that has moved over land for less than 25 NM (50 km). The frictional slowing of the surface wind then creates a large vertical shear strongly favoring development.

Forecasters could use Novlan and Gray's (1974) advice:

> "When a hurricane moves inland and starts to decay, persons located particularly in the right front section of the storm should be observant for surface winds of 13 - 17 knots while overhead the low level clouds appear to be moving with much greater velocities. These conditions signify tornado potential in the regions where cumulus convection is occurring."

Compared to the United States, hazards in other regions are likely to be fewer. Over Japan between 1950 and 1971, tornado-producing typhoons averaged only 2.3 tornadoes each (Fujita et al., 1972). Over the oceans, where frictionally induced increase in the shear of wind in the vertical is very small, waterspouts have never been reported in association with tropical cyclones. In Figure 10-14, next page, Novlan and Gray (1974) provide a checklist for hurricane-tornado forecasting.

TORNADOES LIKELY	TORNADOES UNLIKELY
Intense hurricanes, or those tropical cyclones increasing in intensity just before landfall.	Weak hurricanes or filling tendency just before landfall.
After hitting land, hurricanes fill at rates greater than 30 mb 12 h^{-1}.	After hitting land, hurricanes fill at less than 10 mb 12 h^{-1}.
Once on shore, the center rapidly cools and becomes 6° C colder than 100 NM (185 km) from the storm center.	Only small central storm cooling.
Vertical wind-shear profiles surface to 5,000 feet (1.5 km) of 40 knots or more; surface winds only 15-20 knots.	Surface to 850-mb wind shears less than 40 knots.
RECOMMENDED TORNADO WATCH AREAS	
General area: Where surface pressure is between 1004-1012 mb.	Specific area: Along strong radar-observed rainbands, particularly the outer rainbands.
Specific area: Areas of vertical wind shear greater than 40 knots from surface to 850 mb; surface winds 15-20 knots.	Specific area: Within the "preferred sector" 60-250 NM (110-460 km) from the center of the hurricane and at an azimuthal range of 0 to 120° with respect to *true* north.
BEGIN **TORNADO WATCH**	**END** **TORNADO WATCH/WARNING**
When the center is 100-150 NM (180-280 km) off shore and the first rainbands start to come on shore.	When the rainbands begin to break up and the vertical wind shear (sfc-850 mb) falls below 40 knots.
NOTE: Significant dry air intrusions in the right rear quadrant indicate a potential for tornado "family" outbreaks.	

Figure 10-14. A Hurricane-tornado forecasting checklist (Novlan and Gray, 1974).

10.4 HAIL

"Hail" is defined as precipitation in the form of balls or irregular lumps of ice produced by convective clouds, usually cumulonimbus (Huschke, 1959). By WMO convention, hail is 5 mm or more wide; smaller ice particles are termed "ice pellets." Large hail is more than 2 cm across. The following short summary of hail in the mid-latitudes, from Cole and Donaldson (1968), probably applies to the tropics as well.

• Large hail is found in, alongside, and under well-developed thunderstorms whose tops may extend above 50,000 feet (15 km).

• Small hail or ice pellets may occur in any thunderstorm and even in some large cumulus.

• Hail reaching the ground is most common over mid-latitude mountains and adjacent plateaus, where some places may experience five to ten hailstorms a year.

• Thunderstorm cells are about 4 to 8 NM (8 to 16 km) wide; the instantaneous width of the hailing area at the ground ranges from 1 to 3 NM (2 to 5 km).

• Although hail can form only at altitudes above the freezing level, it may be encountered in flights from the surface to very high levels.

The chance of hail reaching the ground increases with the height of the radar-echo tops of the associated thunderstorm. For example, half the thunderstorms in New England with radar-echo tops above 50,000 feet (15 km) produced hail at the ground. In Texas, hailstorm echo tops occasionally exceed 60,000 feet

(18 km). Hail melts much less before reaching the ground in the United States than in the tropics. Thus, it is not surprising that hail measured at the ground in the United States is twice as likely to be larger than 1 inch (25 mm) than over India.

Over the tropics, hail is fairly rare compared to higher latitudes. This is because the freezing level is usually above 15,000 feet (4.5 km) and thunderstorm updrafts are seldom strong enough to support hail growth. However, hail has been observed; meteorologists must consider the possibility of at least small hail in all severe tropical thunderstorms.

In Hawaii, hail falls five to ten times a year. It is almost always quite small and affects a small area. But on 30 January 1985, as a cold front passed through, thunderstorms dropped hail over much of the northeastern and southeastern parts of the island of Hawaii. Takahashi (1987) analysed 107 hailstones with an average width of 0.71 inches (18 mm). Since records were begun, south Florida has experienced three storms giving large hail. On 29 March 1963, a severe thunderstorm produced hailstones up to 3 inches (75 mm) wide in Miami as the axis of a cold upper low crossed the city (Neumann, 1965).

Frisby (1966) and Frisby and Salmon (1967) made the most complete survey of hail in the tropics (between the Tropics of Cancer and Capricorn) known. They gathered data for many years from the literature and through extensive correspondence with individuals and meteorological services. Since hail is so rare and localized, routine climatological summaries often miss most of the events.

There is enough data for Africa, India, and Australia to allow isopleths of hail occurrence to be drawn (Figure 10-15). Although hail is rare at African coastal stations, most inland stations, including those near sea level, have occasionally reported hail. Hail is commonest over the highlands of East Africa where in a few places the mean frequency exceeds 5 days a year. In the Kenya highlands, hailstones more than 1 inch (25 mm) across are not uncommon, although here and elsewhere in the tropics hailstones are generally less than 0.4 inches (10 mm) wide.

In India, hail is most common in the north and along the southwest coast. In tropical Australia, hail is primarily confined to Queensland and is rarer than over parts of Africa and India.

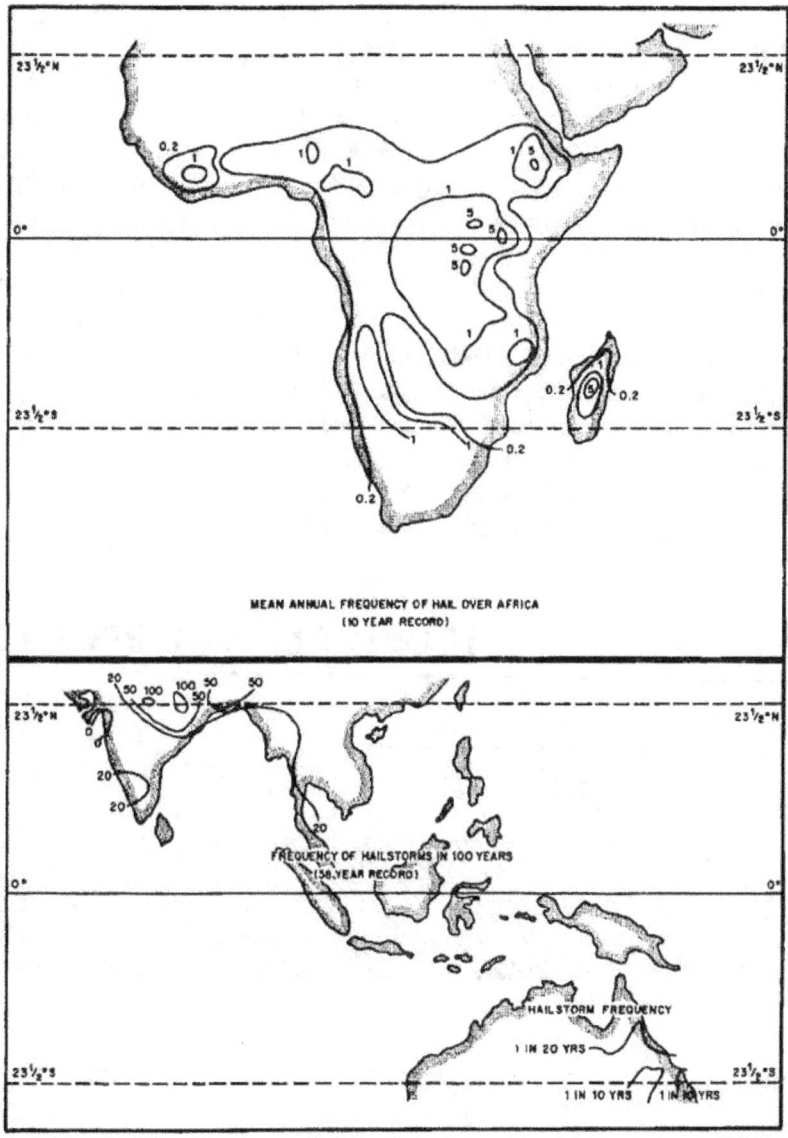

Figure 10-15. Frequency of hail occurrences at the surface in parts of the tropics (adapted from Frisby and Sansom, 1967).

291

Frisby (1966) analyzed a scatter diagram of annual hail frequency at tropical stations versus station elevation. She found that on average most stations below 4,000 feet (1,200 meters) experience fewer than 2 hail days a year, while at higher stations, hail falls on 3 to 8 days a year. There was considerable scatter of points about the line of best fit, suggesting that other factors besides elevation play roles in hail occurrence.

Frisby also prepared monthly hailstorm distribution graphs for stations with reliable data and then composited the graphs for areas that appeared to be part of similar hail regimes. Three main tropical hail regimes were identified: Zone 1 (10° N to 23.5° N), Zone 2 (10° N to 10° S), and Zone 3 (10° S to 23.5° S).

Figure 10-16 indicates the months of highest hail frequency in each zone. In Zones 1 and 3, hail principally falls in spring, the transition season preceding the summer rains. This is when increased insolation, combined with incursions of cool air aloft associated with upper westerly troughs, create maximum instability.

In the NH (Zone 1), the months are usually February to May and in the SH (Zone 3), September to December. In the equatorial region (Zone 2), hail appears to have two seasons; in some places it occurs throughout the year.

TROPICAL HAIL REGIMES

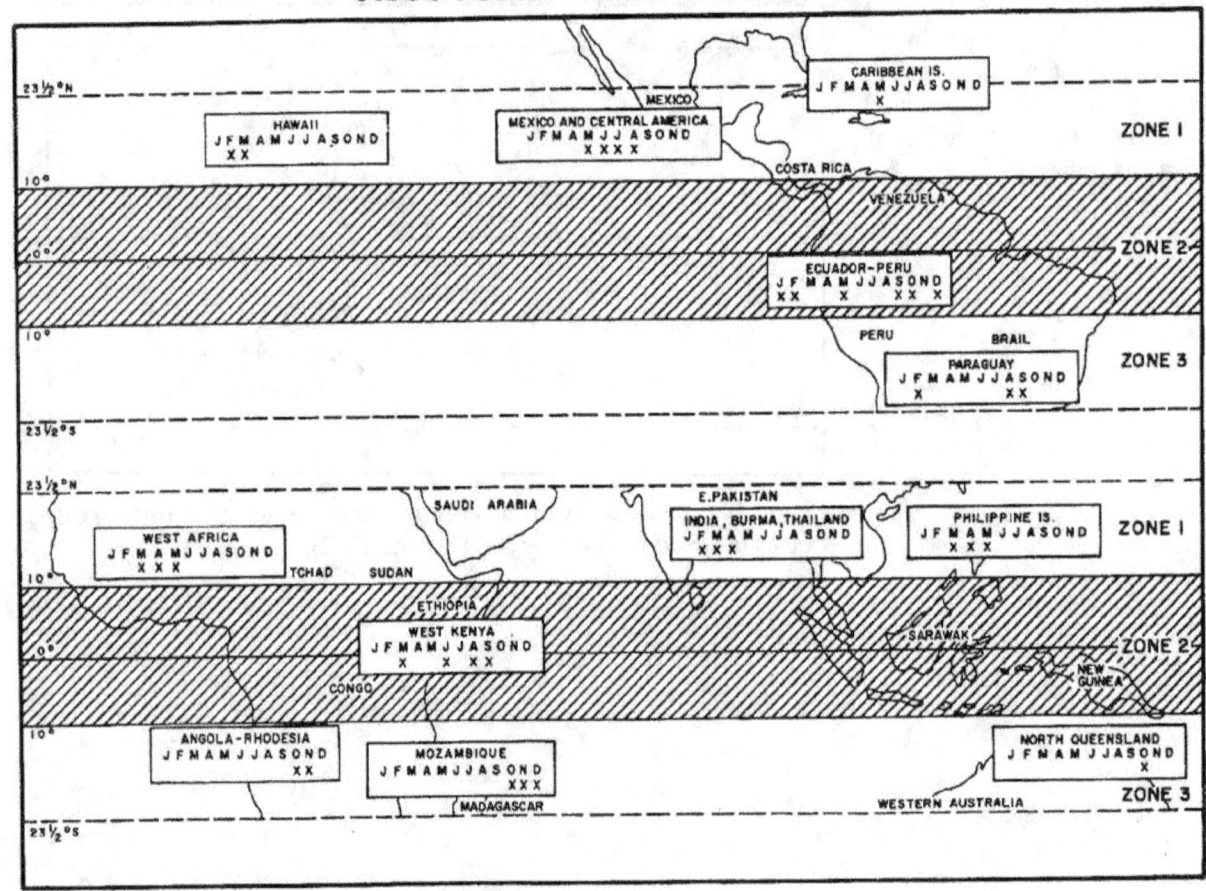

Figure 10-16. Hail regimes for three tropical zones. The X's indicate months in which hail at the surface is most likely (adapted from Frisby, 1966).

10.5 EXTREME WINDS

10.5.1 General. In the tropics, extreme winds are usually caused by thunderstorm downrush or by tropical cyclones. In the subtropics strong winds may also accompany deep extratropical cyclones. Channeling by terrain may further accentuate strong winds. A vigorous surge in the winter (northeast) monsoon can cause 50-knot winds through the Taiwan Strait and passes of the Annam mountains in southeast Asia. Figure 8-36 showed how passes across Central America funnel the trade winds. Strong low-level winds often blow over the Horn of Africa during summer; these thermally driven winds can exceed 100 knots in the lowest 6,500 feet (2 km) (Findlater, 1977).

"Highest sustained wind speed" should be distinguished from "peak gust." The *sustained speed* is the mean speed over some short averaging period (1 minutes, 5 minutes, or 10 minutes) while the *peak gust* is the highest value recorded by an anemograph. Peak gusts usually last less than 20 seconds. Compared to 1-minute sustained winds, the peak gusts may be 20 to 50 percent higher (the gust factor), depending on topography. Anemograph type, height and exposure, and other circumstances should be determined for each location. Inland, a 50-percent gust factor is typical, while along the coast or over the ocean (where friction is less), 20 percent is more likely. For example, as Typhoon Russ approached and passed to the south of Guam (Frontispiece), near the center of the island, at Agana Naval Air Station (13° N, 145° E) gusts were highly correlated with sustained surface wind speed (r = 0.96) and were 37 percent stronger. In the following discussion, peak gusts are used since they are easier to determine from anemograms than sustained winds. If desired, the sustained winds associated with peak gusts can be estimated by applying the appropriate gust factors.

10.5.2 Thunderstorm Winds. Almost all tropical thunderstorms (except continuous thunderstorms) produce downrush winds; however, speeds depend on geography, season, time of day, and the character of individual thunderstorm cells. The strongest winds are seldom measured by the surface station network. Even so, enough thunderstorm wind data has been recorded on anemograms to allow a fairly good description.

In India, strong downrush winds produced by thunderstorms are called "squalls," which are defined as sudden increases in wind of at least three intervals on the Beaufort scale, reaching a speed of 24 knots or more and lasting at least 1 minute. A "severe squall" reaches 43 knots or more.

Sharma (1966) used anemograms for 1954 to 1963 to study squalls at Nagpur (21° N, 79° E) in central India (Figure 10-17, next page).

Squalls were recorded in every month except November. They were most common from March through July, and reached an average of 12 in June. Highest gusts of 74 knots occurred in both March and April. With the squalls, pressures generally rose and temperatures fell suddenly (see Figure 10-17e, "Temp Changes"). Pressure changes ranged from -1.3 mb to +4.1 mb; temperature changes, from +11.2° F to -27.4° F (+6.2° C to -15.2° C). Squall frequency averaged 36 over the year and ranged from 16 in 1954 to 51 in 1962. Fully 80 percent of all the squalls and all of the severe squalls occurred with thunderstorms; the rest occurred with heavy showers. Over 80 percent of the squalls occurred between 1200 and 2100L. The frequency distribution of squall duration is shown in Table 10-2.

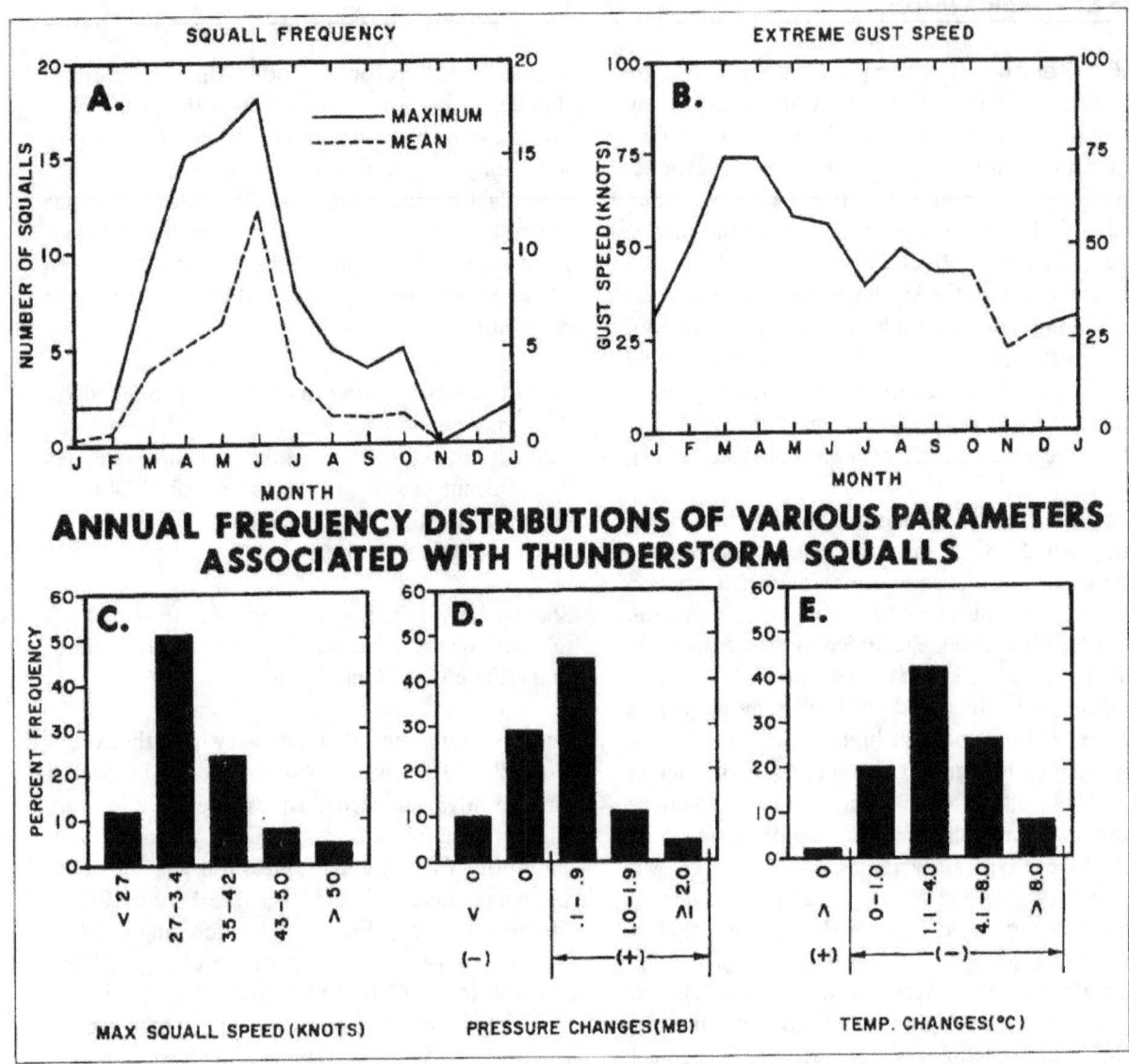

Figure 10-17. Climatology of thunderstorm squalls and related phenomena at Nagpur (21° N, 79° E), India: (A) mean and maximum squall frequency, (B) monthly extreme gust speeds with squalls, and frequency of distribution of (C) maximum squall winds, (D) pressure changes and (E) temperature changes associated with squalls (adapted from Sharma, 1966).

TABLE 10-2. Frequency distribution of squall duration at Nagpur.

Duration (min.)	1-5	6-15	16-30	>30
Percent Frequency	54	45	9	2

Leeward of mountain ranges, monsoon winds are dried by descent and convectively destabilized; this favors a showers regime. Afternoon thunderstorms may then develop and cause squalls. At Madras (13° N, 80° E) on the east coast of peninsular India, an average of one to two squalls occur each year; 74 percent during summer. The squalls are not severe; 65 to 70 percent range between 27 and 41 knots. Gusts over 54 knots are extremely rare (Subba Reddy, 1985). Squally lee-side thunderstorms have also been observed over eastern north Sumatra when westerlies

Apart from tropical cyclones affecting stations in the south coast and peninsular region, almost all extreme winds were caused by thunderstorms. Peak gusts ranged from 40 to 73 knots. The strongest among these were generally confined to the mountains of north Thailand or adjacent to the Gulf of Thailand. In the interior, the highest gust in any year tends to occur in April and May, while along the coast of southern Thailand the peaks occur from July through November. For extreme wind analysis of these data, see 10.5.5.

TABLE 10-3. Peak gusts seasonal averages and absolutes at Cape Kennedy.

	Winter	Spring	Summer	Fall
Average Peak Gust (knots)	20	19	17	17
Absolute Peak Gust (knots)	34	53	48	46

prevail; in general, they are rare in the dry subsiding air and may depend on moisture brought inland by a sea breeze.

Carter and Schuknecht (1969) analysed wind gusts occurring within 1 1/2 hours of thunderstorms at Cape Kennedy (29° N, 81° W), Florida, for 1957 to 1967. The seasonal averages and the absolute values are shown below (November is included in the "Winter" column).

Thunderstorm gusts over 40 knots are rare at Cape Kennedy, which seems to typify many tropical and subtropical coasts where thunderstorm gusts are generally less than those observed inland.

Atkinson (1966) used data for 1956 to 1966 in studying extreme winds at 13 stations in Thailand.

Over West Africa, squall lines are most intense (and accompanying winds strongest) at the end of a long dry season when lapse rates are most unstable. For the same reason, the second of two closely-spaced squall lines is less intense than the first, and squalls are more vigorous in the east than in the west of the region. There does not appear to be much diurnal variation. Squall lines are rare over southern Africa and may have occurred during the SH summer in Zambia and Angola (Leroux, 1983).

In the United States, major weather-related passenger aircraft disasters near airfields led Fujita and Caracena (1977) to ascribe the cause to severe downdrafts (microbursts) generated by nearby thunderstorms. As a microburst descends, the ground diverts it horizontally.

When aircraft taking off or on final approach to landing encounter bursts of wind in the direction of flight, the headwind component and lift are reduced: The aircraft can stall and crash. Wilson et al. (1984), using doppler radar measurements, proposed the model shown in Figure 10-18 and recommended installation of Doppler radars at airports to detect microbursts and warn pilots. Compared to the United States, the weaker thunderstorms of the tropics are less likely to cause severe microbursts. However, three aircraft have crashed in the tropics: One at Kano (12° N, 9° E), one at Doha (25° N, 52° E), and another at Pago Pago (14° S, 171° W) (Fujita, 1985). A thunderstorm near or over an airport is dangerous to approaching and departing aircraft. Caracena et al. (1990) explained microbursts and illustrated them with color photographs that included time-lapse sequences.

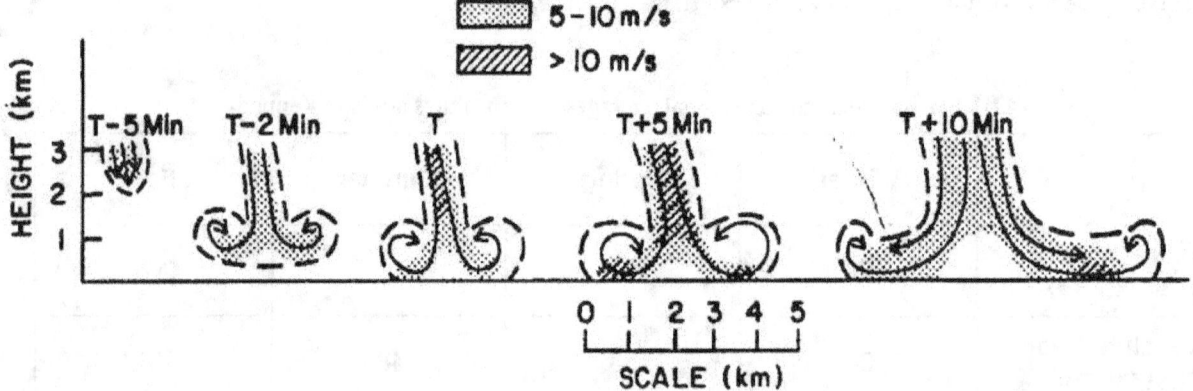

Figure 10-18. Vertical cross section of the evolution of the microburst wind field. T is the time of initial divergence at the surface. Shading denotes the vector wind speed (from Wilson et al., 1984).

10.5.3 Duststorms.

10.5.3.1 General. Any gust of wind raises dust in the desert. The conditions necessary for a duststorm are simple: A dry surface, covered to some depth with dust or fine sand, and a wind strong enough to lift these particles into the air. Great duststorms usually develop in the cool season as strong winds in polar outbreaks sweep across the desert. Friction and convection raise clouds of dust that fill the air beneath the usual subsidence inversion; dust often obscures the sun and reduces visibility to fog level. Winter dust storms have been well observed and described in the Sahara. Figure 8-37 illustrates a duststorm-generating cold front moving across North Africa. Intense mesoscale duststorms, or "haboobs," summer phenomena, commonly occur around and to the south of Khartoum (16° N, 33° E).

Forecasters should remember that major winter duststorms accompany strong cold outbreaks when a branch of the Hadley cell accelerates and a surge develops. The duststorm edge may advance much faster than the following winds would indicate; poor visibility and the low-level jet endanger flying. In summer, the same weather conditions that produce severe thunderstorms or tornadoes may also cause duststorms; surges may sometimes be responsible. Besides the meteorological prerequisites, the ground surface must be favorable; i.e., very dry, with ample dust or fine sand.

Duststorms and dust devils not only reduce visibility, but are also accompanied by turbulence that may endanger light aircraft or helicopters. Dust in the air warns of turbulence, but no dust aloft does not guarantee no turbulence; it means only that the surface is dust-free.

10.5.3.2 Winter Duststorms. A vigorous cold front moving rapidly southeastward across North Africa and southwest Asia heralds onset of the "harmattan" and possibly a severe duststorm. North of the subtropical ridge, increasing southwesterlies ahead of the front raise the dust first. Scattered showers at the front may locally lay the dust, but in general, strong gusty winds at the front and strong northwesterlies in the cold air maintain the duststorm. South of the subtropical ridge a surge often develops (8.4.4). Freshening northeast winds then begin to blow dust well ahead of where the front might reasonably have been expected to be. Surface winds in the surges often exceed 30 knots (Kalu, 1979). Dust crossing the Sahel and reaching the Gulf of Guinea may have been raised in the region south of the Tibesti massif in northern Chad, and carried in a plume toward the southwest. Jalu and Dettwiller (1965) described an exceptionally severe duststorm that reached 10° N on 27 March 1963 and raised thick dust over an area of 87,000 NM² (300,000 km²) in the central Sahara. Behind the rather sharply defined edge of the advancing dust cloud, surface temperatures were more than 10° C lower than in the clear air ahead. Jalu and Dettwiller ascribed the difference to interception of insolation by the thick dust. Their conclusion that the dust served to accentuate the cold front along its leading edge is not necessarily valid. If events had likely followed the sequence described in 8.4.4, the front would have long since disappeared as surface northeast winds increased well ahead of the surface air-mass discontinuity. Then, the dust raised by these winds would create its own accompanying "front." The large daytime temperature gradient across the leading edge of the dust cloud would have produced a strong low-level jet that would raise more dust. It is probable that, once a threshold of radiational opacity is crossed, local feedback may well intensify a duststorm. Subsidence following the cold surge usually restricts dust to the lowest 10,000 feet (3 km).

About two or three times a month, cold fronts moving down the Persian Gulf are followed by strong north or northwest winds referred to as the "Shamal." Raised dust may reduce visibility for 24 to 36 hours.

Figure 10-19 shows duststorms accompanying a cold outbreak that crossed Saudi Arabia at 25 knots on 31 January 1991. Strong winds, both ahead of and behind the front, raised the most dust. Once or twice each winter, an intense Shamal accompanying a cold surge lasts 3 to 5 days (Walters and Sjoberg, 1988). Winds may reach 50 knots and duststorms seriously reduce visibility. Figure 10-20 (next page) shows a Shamal duststorm moving south across northeastern Arabia and the Persian Gulf. Infrared imagery clearly distinguishes the dust from the warm Gulf waters, but this is much more difficult over the desert. In March, a low moving east across central Arabia is preceded by a hot, dry southerly wind (the "Aziab") that raises dust over southeastern Arabia.

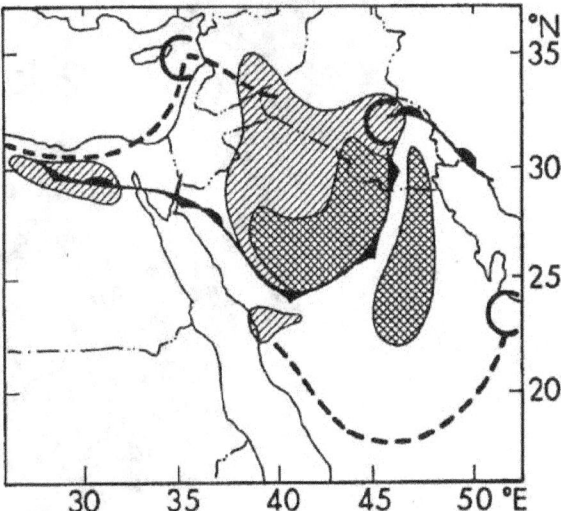

Figure 10-19. Schematic for 1200Z 31 January 1991 showing duststorms accompanying a cold front crossing Saudi Arabia. The cross-hatched areas have visibilities of 0-3 miles; the hatched areas, 3-6 miles. Dashed lines show the frontal position at 1200Z 30 January and 1200Z 1 February.

Figure 10-20. DMSP IR image for 19 October 1984 of a duststorm generated by strong northwest winds (Shamal) over the Persian Gulf and northeast Arabia. The dust edge can be seen south of the clouds of a tropical jet stream. The northern edge of the duststorm is hard to detect.

10.5.3.3 Summer Duststorms. Table 10-4 shows that southward-moving winter duststorms affect Khartoum, but that summer duststorms, which generally move in from the south, are most common. According to Sutton (1925) summer duststorms are convective phenomena; one in three is followed by a thunderstorm and two in three by rain. The storm fronts are usually between 10 and 15 NM (20 and 30 km) long and affect a locality for about 3 hours. These "haboobs" are generally of mesoscale size, most often occur between 1600 and 1800L, and can seldom be traced between neighboring stations. They originate in complex interactions between synoptic disturbances, probably surges (see Figure 8-39), and local orographic and heating gradients.

TABLE 10-4. Frequency of duststorms at Khartoum (16° N, 33° E) based on 8 years of observations (from Sutton, 1925). NOTE: Direction of approach could not be determined for all storms.

	Jan	Feb	Mar	Apr	May	Jun	Jul	Aug	Sep	Oct	Nov	Dec
Total	5	7	11	13	21	48	37	23	25	5	0	1
From NE-SW	5	7	8	4	3	5	1	1	1	2	0	1
From SE-SW	0	0	0	4	12	26	16	11	15	2	0	0

Rain accompanying the storm lays the dust and reduces the chances of a rapid sequence of storms. Sutton thought that the frequency decrease between June and August at Khartoum might be partially due to sprouting of a dust-inhibiting ground cover. Similar duststorms are common during summer near the borders of other deserts (Powell and Pedgley, 1969).

In summer, the Persian Gulf lies north of the heat trough, and the shamal prevails (Walters and Sjoberg, 1988). When it freshens, it raises dust over the desert. Disturbances similar to those responsible for haboobs can lower visibilities to near zero, as winds exceed 30 knots.

At Agra (27° N, 78° E), 80 percent of all duststorms ("Andhi") occur in May and June (1 day in 4) prior to summer monsoon rain. The storms are also mesoscale phenomena; 64 percent occur between 1500 and 2000L and only 25 percent between 0400 and 1100L. In spring, they may be triggered by disturbances moving in from the west (Sreenivasaiah and Sur, 1939).

Joseph et al. (1980) made case studies of Andhi at Delhi (29° N, 77° E), which averages one storm in April and three each in May and June. Maximum gusts ranged from 25 to 54 knots and minimum visibility from 360 to 2,950 feet (110 to 900 meters). Often no rain fell at Delhi; radar showed that the squall line responsible for the dust could be as far as 15 NM (30 km) from the originating cumulonimbus.

Over the deserts, subsidence in winter limits the depth to which dust can penetrate to 10,000 feet (3 km). In summer, vigorous convection spreads dust upward to between 13,000 feet (3.9 km) and 16,500 feet (5.1 km). The very fine dust above 10,000 feet (3 km) does not dissipate between duststorms, but persists through the summer, seriously reducing visibility and the range of electrooptical devices. A satellite image on 25 June 1985 (Figure 10-21) reveals widespread atmospheric dust with no obvious source region. The swirl of dust over southeastern Arabia has probably been caused by a heat low acting for some time on a reservoir of atmospheric dust.

Figure 10-21. DMSP IR image for 25 June 1985, showing widespread dust covering Arabia and the neighboring seas.

High-latitude meteorologists have trouble telling snow from cloud in satellite images. Distinguishing a dust cloud from the underlying surface may be equally hard. Infrared images don't help much when the surface temperature is close to the dust cloud temperature (there is a similar problem with stratus or fog over cold upwelled water). In winter, over the sea, (which is much warmer than the air), dust can be easily recognized in both IR and visible images. Over the relatively cold desert, however, the temperature difference (and consequently the contrast in the IR image) is much less—see Figure 10-20. In summer, the situation is reversed. Over the relatively cool sea, visible images show dust clouds well, but the small temperature difference between sea and dust blurs the IR image. Over hot land, IR picks up the temperature contrast with the dust cloud, but as usual, the visible image seldom distinguishes dust cloud from desert.

To better reveal dust, Lee (1989) suggested merging the IR image over land with the visible image over the sea. Figure 10-22 shows IR (A) and visible (B) images of the Red Sea and its neighborhood taken on 25 August 1989. A duststorm over the eastern Nubian desert is clearly differentiated from clouds by IR, but only the visible image shows the dust extending over the Red Sea. Figure 10-22C combines (A) and (B) and clearly defines the limits of the duststorm.

10.5.3.4 Dust Devils. These are summer convective phenomena; they are not associated with convective cloud. Hess and Spillane (1990) studied pilot reports of dust devils over Australia and combined their findings with earlier results. They found that up to four dust devils could exist at once over an area of 3.5 square nautical miles (12 km^2).

In the Persian Gulf region (Walters and Sjoberg, 1988), the strongest dust devils occur near the coast just inland of the edge of the sea breeze. Here, environmental surface winds are light and horizontal wind shear strong. The sandy coastal zones of Somalia, Yemen, and Djibouti often experience dust devils (Vojtesak et al., 1990).

Dust devil conditions and characteristics should be similar over other tropical deserts. Conditions favoring dust devils are:

• Relatively level, fairly smooth terrain.

• Intense surface heating, with lapse rates near the surface much greater than dry adiabatic.
• Winds near the surface relatively light and variable; some shear in the vertical.

• Strong downdrafts producing horizontal wind shear and vorticity; characteristics range widely.

Dust devil characteristics are:

• Height: Up to 6,500 feet (2 km)

• Diameter: Less than a foot (0.3 meters) to more than 330 feet (100 meters).

• Duration: A few seconds to an hour.

• Peak winds: Up to 40 knots.

10.5.4 Tropical Cyclone Winds. The strongest winds at the earth's surface (excluding tornado winds) are caused by intense tropical cyclones. Super Hurricane Camille, which hit the Gulf Coast of the United States in August 1969, had a peak recorded gust of 175 knots. Storms are rarely this intense in the North Atlantic, but super typhoons are well-known to residents of coasts and islands in the western North Pacific. Except for the specially-designed Dines pressure-tube anemograph at the Royal Observatory, Hong Kong, anemometers often break or are blown away when wind gusts exceed 100 knots. Maximum winds can be roughly estimated from the resulting storm damage. A better method uses relationships between the minimum sea-level pressure in a storm's eye (obtained from surface or dropsonde measurements) and the strongest winds (maximum sustained 1-minute averages) in the tropical cyclone. From anemograms at coastal and island stations in

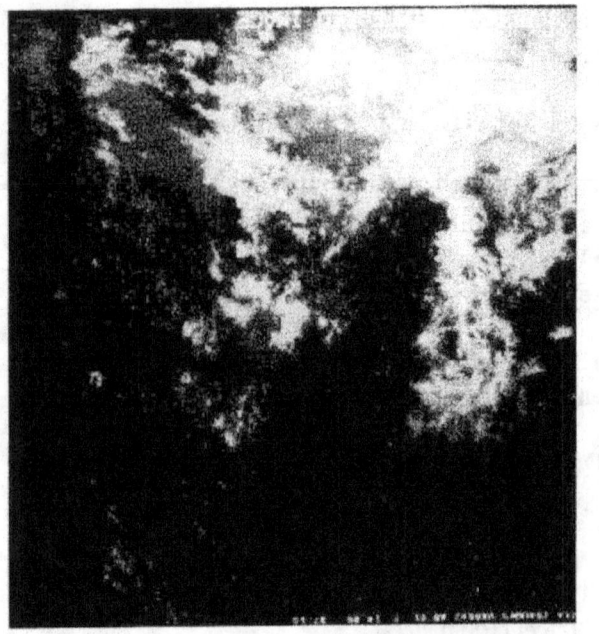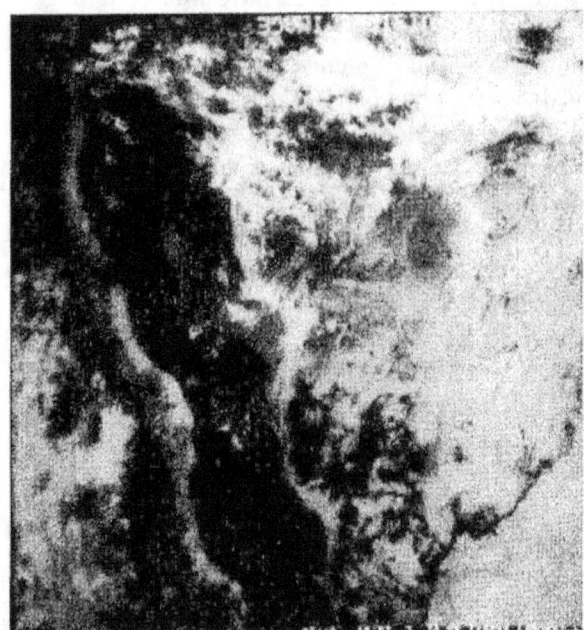

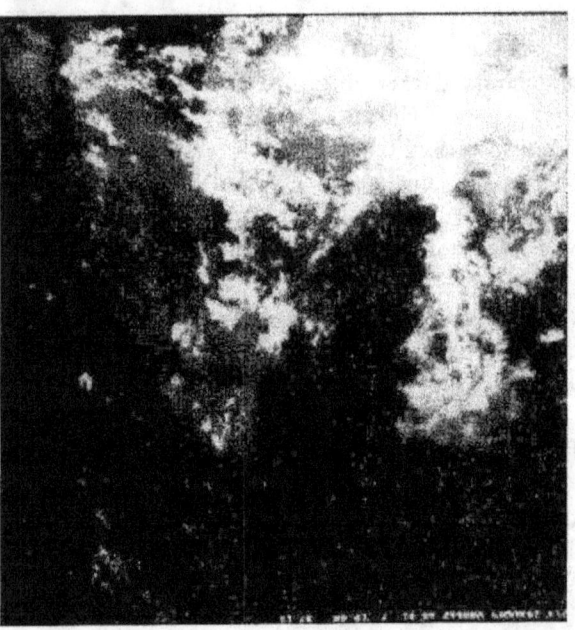

Figure 10-22. DMSP images for 25 August 1989, showing a duststorm extending from the Nubian Desert over the Red Sea. (A) top left, IR image; (B) top right, visible image; (C) above, compositeimage (IR over land, visible over the sea) (Brooks, 1990).

the western Pacific, Atkinson and Holliday (1977) selected 76 cases in which it was very likely maximum winds were experienced during cyclone passage. They determined that

$$V_m = 6.7 (1010 - Pe)^{0.644}$$

where V_m is the maximum sustained surface wind in knots and Pe is the lowest surface sea-level pressure (mb) in the storm center-see Figure 10-23.

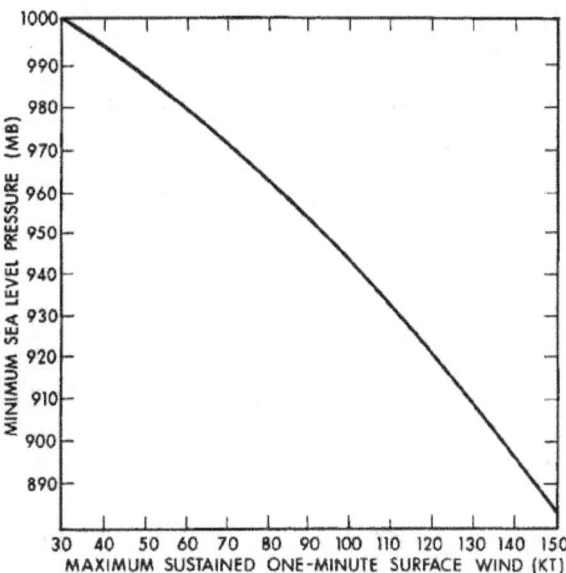

Figure 10-23. Curve of best fit between maximum sustained surface wind speeds and minimum sea-level pressures for 76 western Pacific tropical cyclones (from Atkinson and Holliday, 1977).

The lowest pressure ever recorded (870 mb in Super Typhoon Tip on 12 October 1979) corresponds to a maximum sustained surface wind of 162 knots. Eye pressures are measured on reconnaissance flights. As there are now far fewer such flights, this technique can seldom be used.

Coastal and island stations suffer the full fury of the maximum winds in tropical cyclones, which quickly weaken as they move inland; surface friction increases and the oceanic source of heat and buoyancy is cut off. Topography plays a large part in locally determining how strong the winds will be. Taniguchi (1967) showed that typhoons passing over northern Luzon in the Philippines seldom cause damaging winds at Clark Air Base (15° N, 121° E) because of the protection by surrounding mountains. Since 1945, the highest recorded gust at Clark was 83 knots; gusts exceeding 50 knots were rarely observed. In November 1957, the center of Typhoon Kit (which on the previous day had a central pressure of 915 mb) passed within about 24 NM (45 km); surface winds at Clark never exceeded 10 knots, even though there was extensive damage on the periphery of the base. Similarly, stations on the west coast of Taiwan experience much lighter winds from tropical cyclones than stations on the east coast.

Strong winds damage structures in two ways: through *static* effects, in which the force exerted is proportional to the square of the wind speed, and through *dynamic* effects, in which unstable oscillations may be set up by the wind. Old buildings with thick walls are susceptible to static effects; forecasters should remember that the static force of a 100-knot gust is double that of a 70-knot gust. Tall, flexible steel-framed buildings respond to dynamic effects. These can be reduced by structural damping, but since this is hard to extrapolate from models, an instrumented eight-story steel-framed building on an exposed site in Hong Kong is being used to directly monitor the effects of typhoon winds (Mackey, 1969).

The National Weather Service uses the Saffir/Simpson Damage Potential Scale (Table 10-5) to alert public safety officials when a hurricane is within 72 hours of landfall (Simpson and Riehl, 1981).

TABLE 10-5. Saffir/Simpson Damage Potential Scale ranges.

Scale Number	Central Pressure (mb)	Winds (knots)	Surf (feet)	Damage
1	980 or more	64-82	4-5	Minimal
2	965-979	83-96	6-8	Moderate
3	945-964	97-113	9-12	Extensive
4	920-944	114-135	13-18	Extreme
5	less than 920	>135	>18	Catastrophic

10.5.5. Extreme Wind Analysis. Extreme-value statistical distributions are often used to estimate the probability of various meteorological extremes; e.g., wind speeds, temperatures, and rainfall (see 6.3.13). These distributions are usually applied to series of annual extremes (i.e., the highest value of each year over a period of years) to determine realistic design values for structures such as dams, and for air conditioning or heating needs. Because of its simplicity and wide applicability, the Gumbel double-exponential distribution (1958) already described and applied in 6.3.13, is the most frequently used model. Figure 10-24 shows double-exponential distribution fitted to the annual peak gusts at Kadena Air Base (26° N, 128° E), Okinawa, for 1945 to 1988. Once every 100 years, Kadena can expect a peak gust of 136 knots, while the 2-year return period or median value is 76 knots.

In Figure 10-25, peak gust values for 2-year and 100-year return periods are shown for some tropical stations, based on data from a variety of sources. Values for other return periods can be determined by

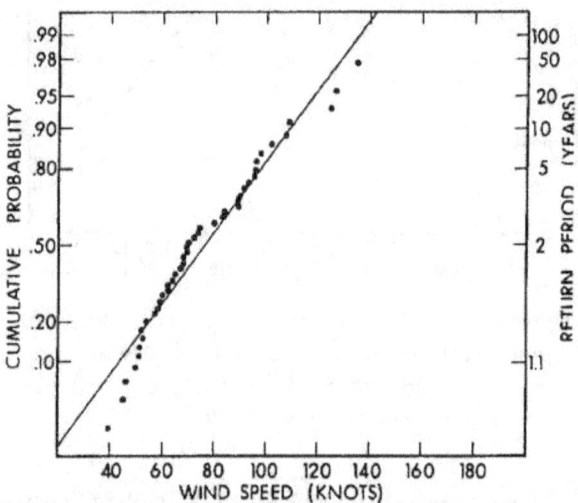

Figure 10-24. Double-exponential distribution (full line) fitted to the annual peak gusts at Kadena Air Base (26° N, 128° E), Okinawa, 1945-1988.

plotting the 2-year and 100-year return period values in the nomogram in Figure 10-26 (on page 10-34), drawing a straight line connecting the points, and

304

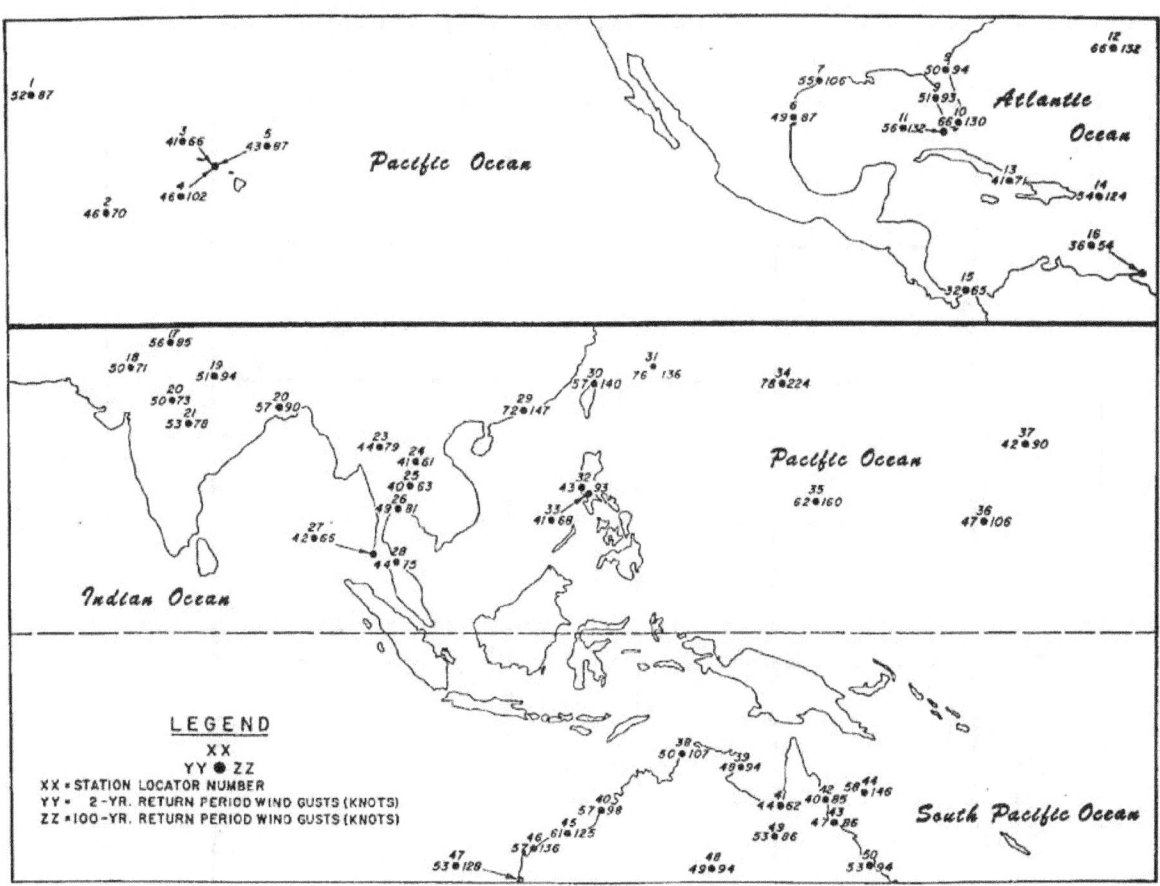

Figure 10-25. Expected extreme wind gusts (knots) for 2-year and 100-year return periods for selected tropical stations.

Stations with the greatest extremes often experience tropical cyclones; even there, however, station location and topography affect the results. For example, the 100-year return period wind at Guantanamo Bay (20° N, 75° W) is only 71 knots, compared to 124 knots at San Juan (18° N, 66° W). Guantanamo is sheltered by nearby mountains and by the island of Hispaniola to the east, whereas San Juan is exposed to the full force of Atlantic hurricanes. Coastal stations in Australia and the southeastern United States have about the same expected extremes, but the extremes in the western North Pacific are

significantly greater. Extremes are much less at inland stations where thunderstorms are chiefly responsible; values range from 35 to 55 knots for the expected 2-year extremes and from 60 to 90 knots for the 100-year extremes.

Okulaja (1968) analyzed extreme winds at Lagos (6° N, 3° E), Nigeria, where thunderstorms are the primary cause. Based on a 14-year period of record, the 2-year and 100-year expected extreme gusts for Lagos are 44 and 67 knots, respectively.

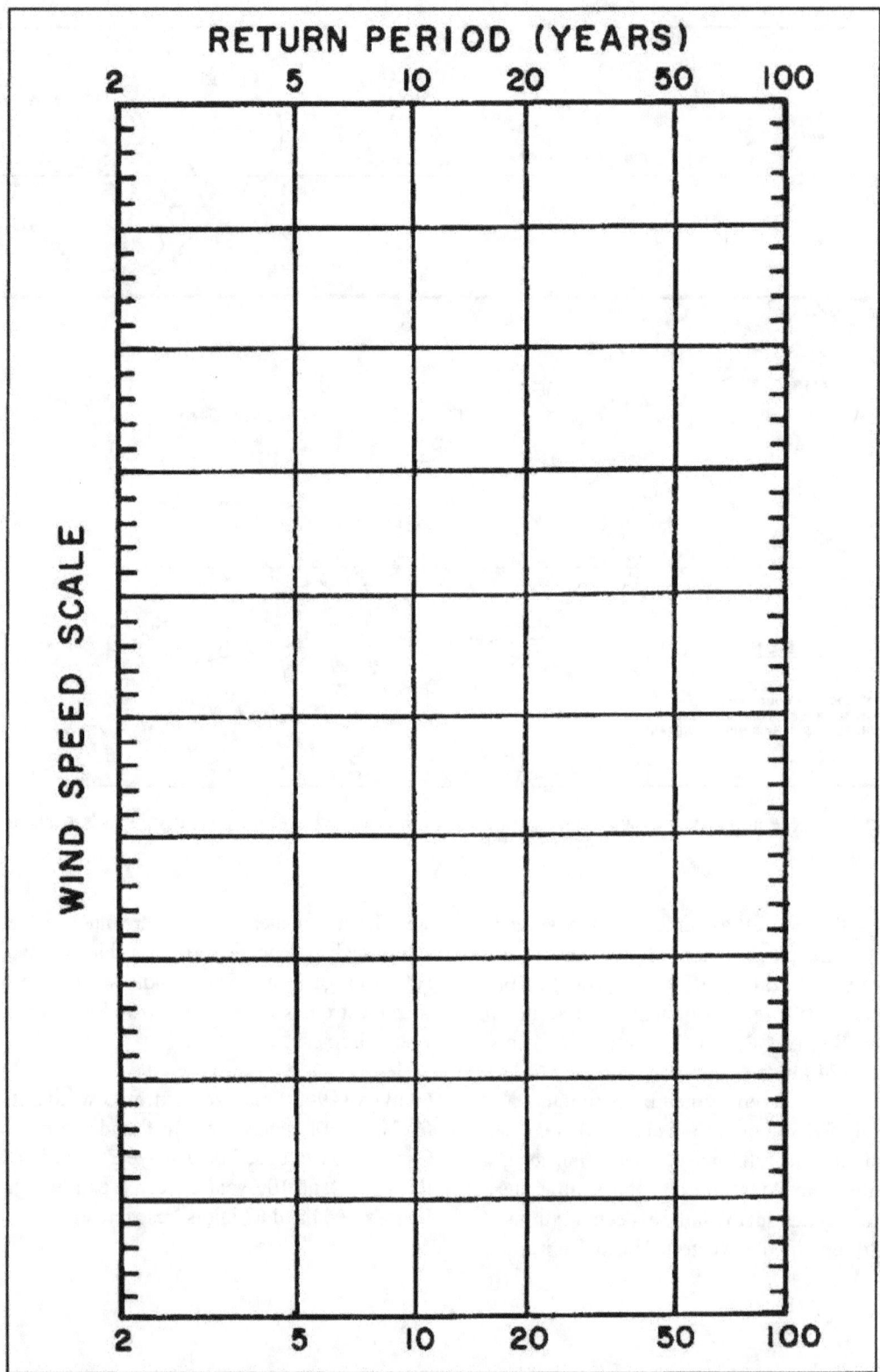

Figure 10-26. Nomogram for determining expected extreme wind gusts for various return periods based on the 2-year and 100-year return-period values (based on the Gumbel double-exponential distribution).

10.6 TURBULENCE

10.6.1 General. Turbulence affecting aircraft operations is a major problem in higher latitudes. In Figure 10-27, Saxton (1966) gives the relative frequency of significant turbulence at various altitudes for mid-latitude land areas. This schematic applies as well to the tropics, even though turbulence is much rarer there.

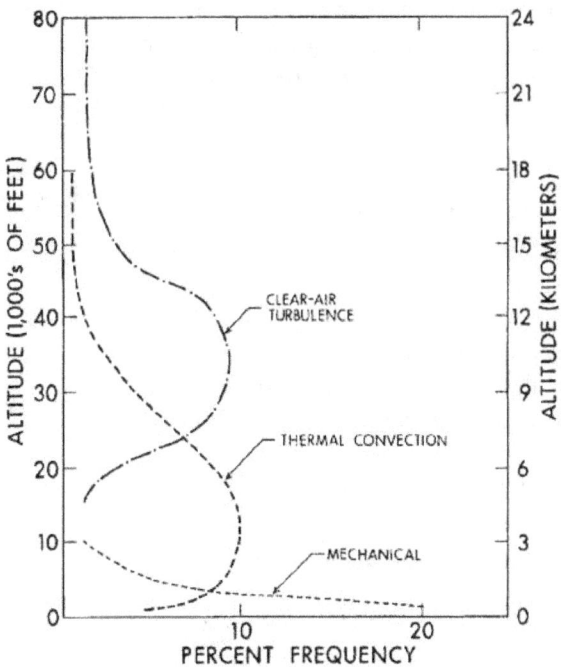

Figure 10-27. Relative frequency of various types of turbulence versus altitude (after Saxton, 1966).

10.6.2 Mechanical Turbulence. Mechanical turbulence, a function of surface wind speed and terrain roughness, dominates the lowest few thousand feet of the atmosphere. Because of generally light winds in the tropics, low-level mechanical turbulence is less common there than in higher latitudes, although it occurs leeward of mountain ranges when surface winds reach about 30 knots. Daytime insolation of land surfaces in the tropics characteristically causes "bumpiness" up to 10,000 to 20,000 feet (3 to 6 km), especially when an LLJ is present.

10.6.3 Convective Turbulence. Convective turbulence, which is generally associated with cumuliform clouds, dominates the middle troposphere and extends as high as 18 km around severe thunderstorms.

In the tropics, convective turbulence over land is likely in association with thunderstorms and is more severe with more intense thunderstorms. It also depends on the *stage* of the thunderstorm; it is most severe during the formative and mature stages. In 1964-65, India's Meteorological Department (1968) had pilots observe turbulence in or near thunderstorms or towering cumulus. Of the 182 reports, 55 percent were of light turbulence, 44 percent of moderate turbulence, and 1 percent of severe turbulence.

Tropical thunderstorms pose little threat to aircraft equipped with weather radar or guided by ground-based radar, since the more intense cells can usually be circumnavigated. Thunderstroms are most hazardous during landing and takeoff, when microbursts may destroy lift and cause an aircraft to stall (see 10.5.2). At jet aircraft flight levels, tropical cyclones pose few problems. In the western Pacific, commercial airlines often fly over the eyes of hurricanes and rarely experience more than moderate turbulence.

10.6.4 Clear Air Turbulence (CAT). Clear air turbulence (CAT) dominates the upper troposphere and stratosphere; although it is infrequent between 50,000 and 75,000 feet (15 and 25 km), no upper limit is known.

In association with mountain waves (caused by strong flow perpendicular to mountain ranges) or with strong wind shear in the vertical near jet streams, CAT is rarer in the tropics than in middle latitudes, perhaps because favorable synoptic situations are also rarer. A survey paper by Saxton (1966) that includes frequency distribution with altitude (Figure 10-27), is the basis for the following information. CAT is

307

normally found in thin horizontal sheets, about 2,000 feet (600 meters) thick, 10 to 20 NM (20 to 40 km) wide and 50 to 100 NM (100 to 200 km) long, oriented along the wind flow. Its dimensions suggest that CAT is related to the mesoscale structure of the atmosphere and is mainly caused by breaking (unstable) gravity waves. In the free atmosphere, gravity waves often occur in stable layers with strong wind-shear in the vertical. Occasionally, in the presence of sufficient moisture, cloud patterns reveal gravity waves 3,000 feet (1,000 meters) or more in length. However, these long waves are not particularly turbulent for subsonic aircraft; they are more properly classified as "undulance" and are characterized by laminar flow. Under certain conditions of density stratification and wind shear, waves shorter than some critical value amplify exponentially and break down the laminar flow into turbulent eddies. Likewise, for given wave length and density stratification, increasing wind shear in the vertical may make the flow unstable. In most empirical studies, strong vertical wind shear is most consistently correlated with CAT.

Aircraft observations of CAT comprise three categories: pilot reports; those obtained using the aircraft as a sensor; and those obtained from a sensor mounted on an aircraft. These data have provided considerable insight into the relative climatological frequencies of CAT intensity over many regions and altitudes. Many investigators have tried to find numerical values of meteorological elements or combinations of elements which correlate well enough with CAT to be used in forecasting occurrence and strength, but progress has been painfully slow. The reasons, according to Saxton, are these:

• Most CAT reports are subjective; only positive encounters are usually reported.

• Meteorological data available to operational forecasters lacks the detail (horizontally, vertically and temporally) for computing the desired mesoscale variables where turbulence occurs.

Even where CAT and observed meteorological elements are significantly related, inability to forecast these elements restricts application of the relationships.

CAT forecasts combine objective and subjective techniques. Forecasts will become more automated as numerical models improve. Saxton's concern that the CAT problem can never be addressed through forecasting because of the limitations imposed by the spacing of upper-air stations and the long intervals between soundings may be partly removed if wind and temperature profilers become common (see 3.2.2).

The International Civil Aviation Organization (ICAO) sponsored a world-wide data collection program to determine the relative frequency of CAT. During the four 5-day collection periods, one for each season in 1964-65, pilots were asked to make turbulence reports on all flights at and above 20,000 feet (6 km). They used the following scale:

• Very Light CAT — Perceptible

• Light CAT — Slight discomfort

• Moderate CAT — Difficulty in walking

• Severe CAT — Loose objects dislodged

• Extreme CAT — aircraft violently tossed about

Turbulence in cirrus or cirrostratus, not connected to convection, was also classified as CAT. The data, collected and analysed by various meteorological centers, is summarized for the tropics in Table 10-6.

As shown, in the table, CAT is not a serious problem in the tropics. Extreme turbulence was not reported and severe turbulence was very rare. These cases, and many of the moderate cases, occurred near cool season upper-tropospheric troughs extending from mid-latitudes into the tropics. Mountain-induced waves in central Africa were responsible for some moderate turbulence.

TABLE 10-6. Percent frequency of various categories of CAT by 5-degree latitude-longitude rectangles in the tropics, for flights above 20,000 feet (6 km) (for the 1964-65 ICAO collection periods).

Area	Turbulence Category			
	None	Light	Moderate	Severe
Caribbean	86.2	9.4	4.4	0
Mexico and Central America	92.6	4.9	2.5	0
South America and South Atlantic	92.1	5.8	1.9	0
North Pacific 0°-15° N	94.2	5.3	0.5	0
North Pacific 15°-30° N	89.8	6.2	3.9	0.1
Africa South of 15° N	94.8	3.7	1.4	0.1
Australia, South Pacific	92.5	5.3	2	0.2
Middle East and Southeast Asia 15° S-0°	96	2.7	1.3	0
Middle East and Southeast Asia 0°-15° N	93.1	3.4	3.1	0.4
Middle East and Southeast Asia 15° N-30° N	92.3	5.6	2.1	0

A later study of CAT on air routes to and from India for 1966 through 1969 (Gupta et al., 1972) found moderate and sometimes severe CAT above 25° N over India associated with the subtropical jet stream and troughs in it. Only light or moderate CAT occurred south of 25° N. Severe CAT was confined to winter. As already mentioned, CAT was most often associated with strong wind shear in the vertical. Severe turbulence occurred only with shears of greater than 4 knots 1,000 feet[1] (13.5 knots km[1]), with wind increasing with height.

Between July and October 1963, a specially-instrumented jet aircraft (Project TOPCAT) was used to investigate CAT in the vicinity of the winter subtropical jet stream over Australia (Spillane, 1968). Near the jet stream core (about 200 mb), all intensities of CAT were five times more common than for all flights over Australia near 20,000 feet (6 km) altitude.

Severe CAT is extremely rare in the tropics; it is almost impossible to forecast accurately. In satellite images, cloud bands transverse to the flow generally indicate moderate to severe turbulence. During the cool season, near the sharp upper-tropospheric troughs that extend deep into the tropics, light and occasionally moderate CAT may be experienced. It may also occur in strongly shearing layers near the tropical tropopause (about 52,000 feet (16 km)). For example, during the Line Islands Experiment in March and April 1967, radiosonde observations between 6° N and 1° N revealed extremely strong wind shears exceeding 12 knots 1,000 feet[1] near the 16-km level (Madden and Zipser, 1970) who thought it likely that they would be accompanied by considerable turbulence.

10.7 ICING

Aircraft icing is of minor concern in the tropics. Few large aircraft fly at the altitudes of greatest icing potential and visual or radar avoidance of cumulus congestus or cumulonimbus is common practice. Modern aircraft equipment also reduces or eliminates certain icing hazards. The physical factors affecting aircraft icing are the same in the tropics as in higher latitudes.

Icing is possible if the aircraft surface is colder than 0° C and if supercooled water droplets are present. Even then, the amount and rate of icing depends on the relationships among meteorological and aerodynamic factors such as temperature, liquid-water content of the air, droplet size, collection efficiency, and aerodynamic heating.

In the tropics, the freezing level lies at about 15,000 feet (4.5 km), with little synoptic, annual, or diurnal variation.

In stratiform clouds, icing is most likely between about 17,000 and 22,000 feet (5 and 7 km). In the tropics, middle- or upper-level cyclones or dissipating convective systems may produce widespread altostratus or nimbostratus at these levels; higher up, icing is much less likely.

In cumuliform clouds, the zone of probable icing is smaller horizontally but larger vertically than in stratiform clouds. The cloud's stage of development controls the icing risk. icing intensities may range from a trace in small supercooled altocumulus to moderate to severe in cumulus congestus or cumulonimbus. In cumuliform clouds, icing is usually clear or mixed.

Aircraft icing is most likely with temperatures between 0° C and -20° C; but it has been reported at temperatures below -40° C in the upper parts of thunderstorms.

10.8 OCEAN WAVES

10.8.1 Storm Surges. A storm surge is an abnormal rise of the sea along a shore; it results primarily from the winds of a storm. It is measured as the difference between the actual elevation of the sea surface and the elevation expected in the absence of a storm (i.e., the predicted astronomical tide).

Along low coasts, the storm surge is usually the chief destructive agent of tropical cyclones; with intense cyclones, the sea may penetrate 10 to 20 NM (20 to 40 km) inland. Since the surge generated at the coast must move inland as a gravity wave, the peak rise at the end of a long channel may lag the peak storm winds by many hours. Interaction with the tide may also shift the time of severest flooding several hours away from the time of greatest storm intensity. Besides the winds and the tides, the extent of coastal flooding depends also on the pressure deficiency in the storm compared with the storm's environment, the size and travel speed of the cyclone, the bottom topography near the cyclone's landfall and other factors. Most important in determining the maximum storm-surge height is the maximum wind speed in the cyclone as it heads inland. Since this in turn is related to the cyclone's minimum sea-level pressure, Conner et al. (1957) related the maximum storm-surge height (H) on the open coast to the minimum sea-level pressure P_o, (in mb) for 30 hurricanes entering the United States from the Gulf of Mexico prior to 1956. In their linear regression equation: H (in feet) = 0.154 (1019 - P_o). The resulting correlation coefficient was 0.68. The equation worked well with ten Atlantic hurricanes that moved inland; but when hurricanes skirted the coast, the equation overestimated the maximum surge height.

Hoover (1957) made a similar study of Atlantic and Gulf Coast hurricanes. His results are shown in Figure 10-28 for 18 Atlantic coast hurricanes ($r = 0.86$) and 24 Gulf of Mexico hurricanes ($r = 0.81$). Conner's regression line for Gulf Coast hurricanes is also shown in Figure 10-28. According to Hoover, his regressions tend to underestimate the probable maximum storm-

surge heights since only observed storm surge data were used in their derivation.

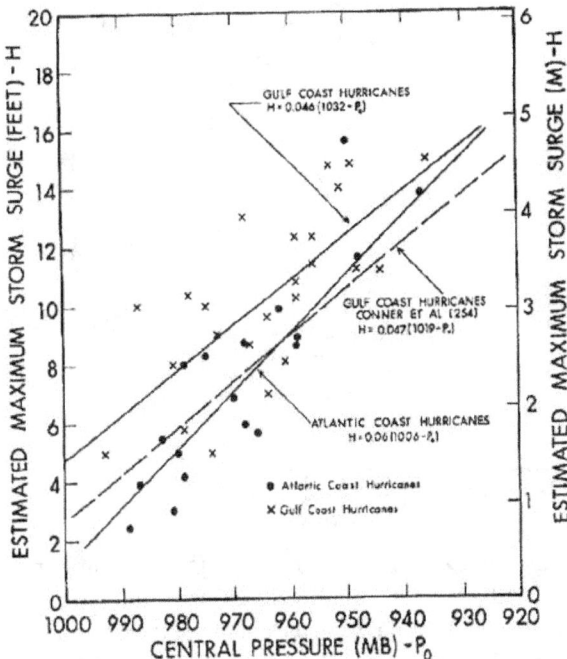

Figure 10-28. Regression of maximum storm-surge height on central pressure for Atlantic and Gulf Coast hurricanes (adapted from Hoover, 1957 and Conner et al., 1957).

Four typical storm-tide profiles are shown in Figure 10-29, next page. In each case, the cyclone track was almost perpendicular to the coast. The storm surge was highest just to the right of the cyclone's center (near the location of maximum winds). From here, the surge height fell off rapidly with distance.

Hoover's regression equation can be used to compute the maximum storm-surge height accompanying Super Hurricane Camille in August 1969. Just before landfall, the hurricane's central pressure was 905 mb, which gives a predicted peak storm-surge of 19.2 feet (5.85 meters). The observed peak was 26.4 feet (7.5 meters). Hoover had no cyclones of Camille's intensity in his sample.

311

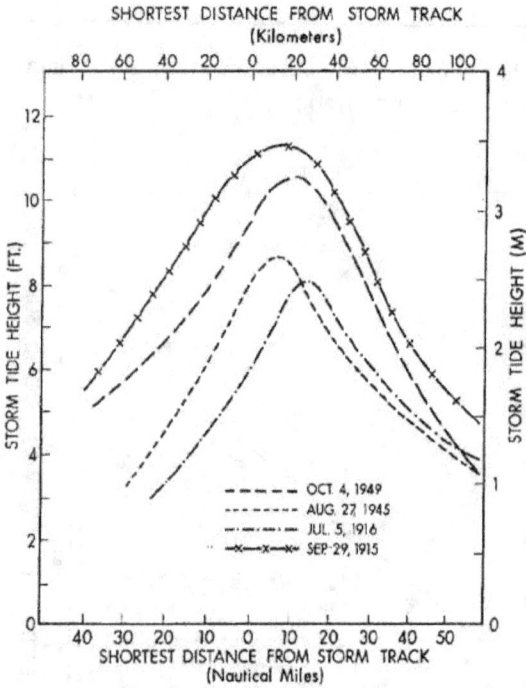

SHORTEST DISTANCE FROM STORM TRACK
(Kilometers)

OCT. 4, 1949
AUG. 27, 1945
JUL. 5, 1916
SEP. 29, 1915

SHORTEST DISTANCE FROM STORM TRACK
(Nautical Miles)

Figure 10-29. Storm-tide height profiler for four hurricanes that entered the U.S. Gulf coastline west of Tallahassee, Florida, from the south-southeast, south, and south-southwest, (adapted from Hoover, 1957).

Jelesnianski (1966, 1967) developed numerical models to compute storm-surge heights associated with tropical cyclones. Besides the elements identified by earlier investigators, he included latitude, wind stress, and the depth contours of the basin. His "standard cyclone" was centered at 30° N and had a maximum wind of 87 knots. His standard basin had a linear slope typical of the continental shelf. He computed storm-surge height profiles for various coast-crossing angles and radii of maximum winds. A detailed description of these models or of the many assumptions used in their derivation, however, is beyond the scope of this report.

Sugg (1969) based a mean storm-surge profile on data from 19 great Atlantic hurricanes (with central pressures of 950 mb or less) making landfall in the United States south of 35° N from 1900 to 1969. In Figure 10-30 (next page) 84 maximum storm-surge

height measurements, made at various distances from hurricane centers as they crossed the coast, are plotted. The dashed curves enclose the estimated range of storm-surge heights. Data points falling outside these curves reflect local influences of bays, rivers, inland waterways, etc. The highest storm-surge (24.6 feet/ 7.5 meters), near Pass Christian (30° N, 89° W), Mississippi, was caused by Super Hurricane Camille in August 1969. The computed storm-surge profile from one of Jelesnianski's (1974) models is included in Figure 10-30 for comparison. There is good agreement, although the maximum heights with respect to the storm center differ somewhat.

Some extreme events defy explanation. At 2215L on 21 October 1972, Hurricane Bebe drove a solitary wave across the coral Atoll of Funafuti (9° S, 179° E) to a depth of 13 feet (4 meters); it created a new island along the eastern shore, 20 NM (37 km) long, averaging 121 feet (37 meters) wide and rising 11.5 feet (3.5 meters) above the sea. Coral rubble that had previously accumulated up to 40 feet (12 meters) beneath the sea surface on the open ocean side of the Atoll had been torn out and piled up by the wave. The total area of the Atoll was increased by one-third (Maragos et al., 1973). Although Bebe passed across other islands in the Ellice chain before hitting Viti Levu, Fiji, the experience of Funafuti was not repeated. Since neither orographic nor bathymetric features could account for the event, one is left with a coincidence: Bebe's eye was moving in a tight loop near the island when the wave struck.

It's obvious that the success of any storm-surge model depends critically on accurate forecasts of storm movement, intensity, and wind profile.

10.8.2 Tsunamis are not really "tidal waves," but rather ocean waves produced by submarine earthquakes, landslides, or volcanic eruptions. These waves may have enough energy to cross and sometimes recross entire oceans as gravity waves. Because of their long wave length, tsunamis cannot be detected from a ship in mid-ocean, but their enormous dimensions become obvious at coasts.

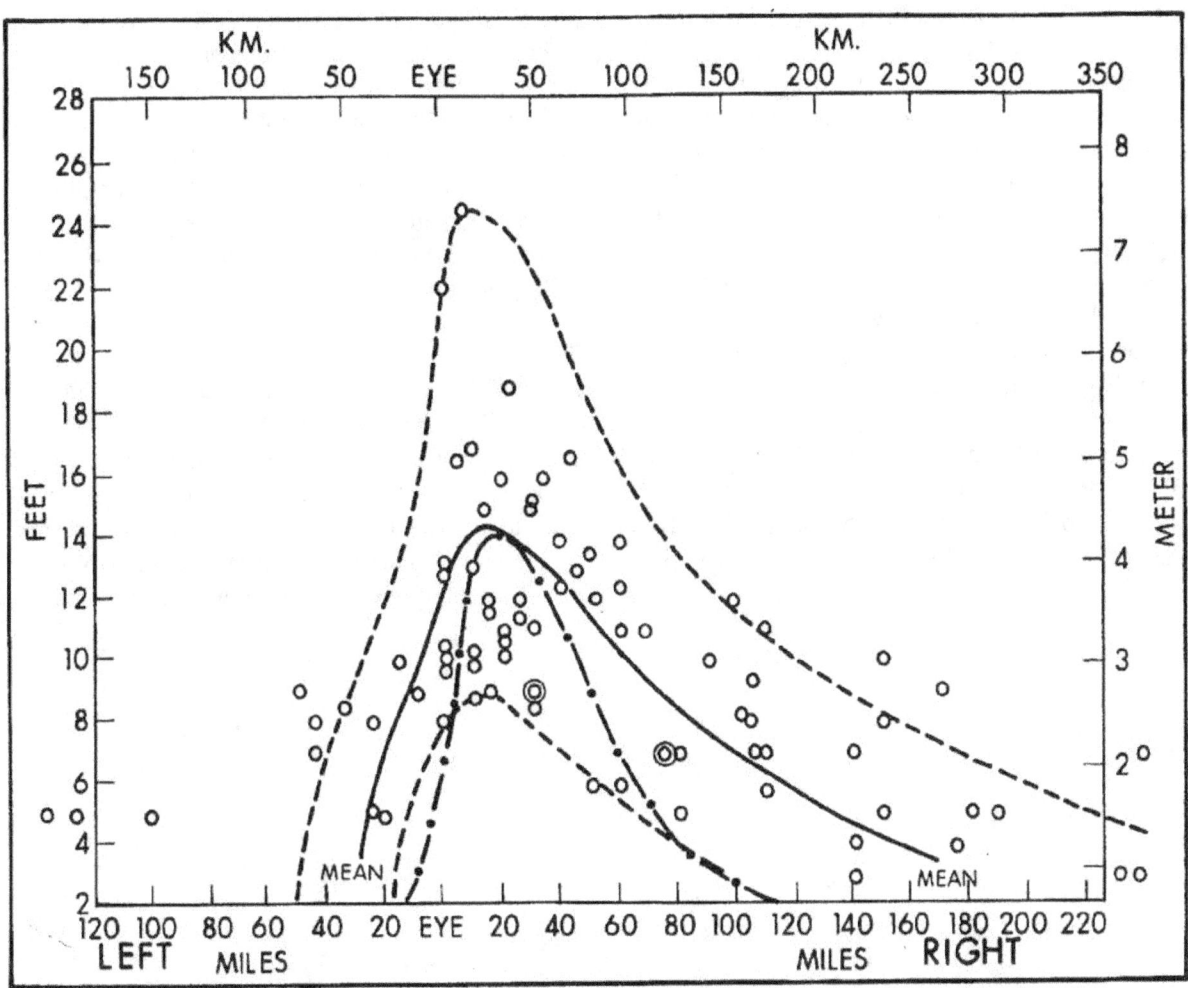

Figure 10-30. Storm-surge heights on the U.S. coasts south of 35° N for great hurricanes (central pressures 950 mb or less). The solid line gives the mean value and the dashed lines the limiting values. Double circles denotes two occurrences at that point (adapted from Sugg, 1969). The dashed-dot line is the computed storm surge profile of a standard storm (see text) derived from Jelesnianski (1974). Note higher surges on the right-hand (onshore winds) sides of the storms.

Tsunamis steepen and increase in height as they begin to move over shallow water, inundating low-lying areas; where local submarine topography is extremely steep, they may break and cause devastation. Meteorologists may become involved in receiving and disseminating tsunami warnings since the international meteorological telecommunication system has been given this task. The following description of tsunamis and the tsunami warning system is summarized from Argulis (1970). The most popular theory holds that vertical displacement of water by underwater earthquakes generates tsunamis; however, only a small fraction of ocean-bottom earthquakes cause tsunamis. Earthquakes of magnitude 6.5 or more are usually necessary before a tsunami is likely. Other factors affecting tsunami generation and magnitude are water depth, earthquake depth, and efficiency of energy transfer. Tsunamis propagate radially from their source with wave lengths up to 400 NM (750 km) and wave periods of several minutes to an hour. Their speed (S) through water is given approximately by:

$$S = (gh)^{0.5}$$

where *g* is the acceleration of gravity (32 feet s $^{-2}$ or 9.8 m s^{-2}) and *h* is the depth of water through which the wave is travelling. For example, in water 18,000 feet (5,500 meters) deep the tsunami travels at about 450 knots; in 900 feet (274 meters) the speed is 100 knots; and in 60 feet (18 meters) it is 25 knots. This physical relationship enables arrival time to be predicted if ocean depth and time of the tsunami-generating shock are known. As the tsunami begins to cross the rising bathymetry near a coast, friction becomes an active force, causing the wave to slow and its height to increase.

The variability of tsunami runup at different locations can be extreme. The 1960 tsunami generated by an earthquake off Chile caused a 35- foot (10.7-meter) runup at Hilo (20° N, 155° W) and only 9 feet (2.7 meters) at Papaikou 7 kilometers away. Runup is controlled by change in coastal slope, angle between the direction the wave is propagating and the coastline, coastal shape, and underwater terrain features. Also, although a shoreline may not be in the direct path of an advancing tsunami, bathymetry can cause a deflection. Thus, all major tsunamis should be considered potentially dangerous, even though they may not cause uniform damage on every coast they strike. Potential danger zones are less than 50 to 100 feet (15 to 30 meters) above sea level and within a mile (one to two km) of the shore.

The National Ocean Survey, NOAA, which is responsible for issuing tsunami warnings, operates a Seismic Sea Wave Warning System (SSWWS) that employs a worldwide network of seismological stations to detect and locate earthquakes. They also operate a network of tide-gauges to record passing tsunami waves and a communication system to speed transmission of information and warnings to various countries.

10.9 VOLCANIC ASH

Explosive volcanic eruptions may inject vast amounts of fine ash into the atmosphere. Jet aircraft engines that ingest ash-laden air may be severely damaged. In 1982, two separate eruptions of volcanic ash from Mount Galung in west Java enveloped two night-flying jet aircraft and almost caused them to crash. The eruptions had started only a short time before and no warnings had been issued (Smith, 1983).

Such coincidences may be hard to avoid, but once an eruption has been reported, an ash plume can be readily detected in satellite images and its future movement anticipated. Forecasters should use IR images in distinguishing tropospheric ash, which generally falls out soon after the eruption ends, from persistent stratospheric dust injected by major eruptions such as El Chichon (17° N, 93° W) in April 1982, and Pinatubo (15° N, 120° E) in June 1991

Chapter 11

TROPICAL ANALYSIS

11.1 GENERAL

Godske et al. (1957) gives the aim of weather map analysis to be: "After careful consideration of their representativeness and reliability, *all available meteorological data* must be fitted into the most probable system of ideal and modified three dimensional tropospheric models." [italics added]

A skilled analyst's reasoning can seldom be quantified; knowledge is hard to transfer. Analysis, carried out as an integral part of forecasting and operational research, is important, even though immeasurable. During analysis, the meteorologist mentally stores impressions of data that consciously or subconsciously he taps when preparing the forecast. Equally significant and subtle impressions cannot stem from contemplating machine-analyzed processed data.

Post-analysis may lead to understanding of past errors and decrease the chance of similar errors in the future. No better way has yet been devised than analysis-forecasting, closely interacting with operational research, for rapidly identifying problems, developing, testing and modifying hypotheses and compiling useful statistics. Ideally, the analyst-forecaster and the operational researcher should be one.

Analysis techniques for the mid-latitudes and the tropics differ greatly, largely because in the tropics simplified pressure/wind relationships such as the geostrophic wind are of questionable validity. In the mid-latitudes, pressure (height) analyses combined with the Norwegian air mass/frontal models can describe most synoptic-scale weather disturbances in the lower troposphere. In the tropics, except in cyclones, pressure (height) and weather are not too well linked. In mid-latitudes, most cyclones develop and intensify in response to baroclinic instability associated with large horizontal temperature gradients. In the tropics, mid-tropospheric cyclones

develop in a baroclinic environment (wind shearing in the vertical) while tropical cyclones develop in a barotropic environment (wind shearing in the horizontal). Modelers often disagree on which is more important to development (Lighthill and Pearce, 1981). In the tropics, most synoptic-scale systems are weak; diurnal variations, local topographic effects and cumulus convection play larger roles in determining weather than in mid-latitudes, and assume greater importance in daily analysis and short-range forecasting. For this reason, sub-synoptic analysis is essential in the tropics.

In this chapter, tropical analysis techniques are reviewed and evaluated. Very little of this material is new or original: subjective analysis techniques have not changed much since the classic texts of Palmer et al., (1955), Riehl (1954), and Saucier (1955). Improvements can generally be attributed to more conventional data (especially upper-air observations), more and better aircraft reports, and cloud motion vectors derived from sequences of geostationary satellite pictures. In addition, cloud patterns revealed by satellites have been identified with tropical synoptic models (Chapters 8 and 9), and so can greatly increase analytical precision in regions lacking any other data. As a result, meteorologists have come to rely on satellite data much more and should become experts in their interpretation and application.

Four major topics are covered. First, data collection and evaluation show how to interpret the various data available to tropical analysis. Second, manual analysis techniques, especially those using streamlines and isotachs (kinematics) are summarized. Third, automated analysis techniques and products and the outlook in this field are surveyed. Fourth, auxiliary aids to tropical analysis are described. The role of weather-radar data in tropical analysis and forecasting is covered in Chapter 12.

11.2 DATA COLLECTION AND EVALUATION

11.2.1 General. As was shown in Chapter 3, a wide variety of data is available for tropical analysis. Nevertheless, data collection and evaluation are more difficult in the tropics than in higher latitudes due to fewer stations, poorer communications, and local effects on the observations. Tropical weather units must have aggressive data-collection and quality-control programs aimed at acquiring all observations and making them available to the analysts. This demands intimate knowledge of station networks, including the type and frequency of observations for each station, the communication schedules for weather-data transmissions, meteorological codes, and special data sources such as aircraft reconnaissance observations and drifting data buoys.

Information on networks, observational and broadcast schedules, and codes can be found in Air Weather Service pamphlets and regulations, Federal Meteorological Handbooks (prepared by the U.S. Departments of Commerce and Defense), and in various WMO publications (World Meteorological Organization, 1986). In addition, weather units in the tropics may wish to use automated data surveys made periodically by the Air Force Global Weather Central (AFGWC) and other centralized units to compare their local data receipt with those of the centrals. In this way, deficiencies in the station's program can be revealed.

Paradoxically, in recent years some local weather units have received more centralized products than they could effectively use. Thus, they must be aware not only of all data sources, but must also select only the products that contribute to their daily analysis and forecasting. Ideally, station chiefs at each station monitor routine conventional observations and ensure that all available data required is collected and plotted; they also survey and evaluate centralized products and determine their usefulness.

Tropical meteorologists, even more than their mid-latitude counterparts, must know how the observations are made, their accuracy, and their representativeness in order to analyze distributions and integrate various types of data into one composite analysis (e.g., upper-wind data from radar and visually tracked balloons, aircraft reports, and satellite cloud-motion vectors).

As an integral part of synoptic analysis, data is evaluated as the analysis proceeds. This is best achieved by comparing observations to neighboring observations for the same time, to observations above and below the analysis level, and to earlier observations at the same station. In other words, the analyst evaluates data *horizontally, vertically,* and *chronologically.* Observations can be checked against climatology to identify gross errors; this is especially useful for isolated stations. Synoptic analysis constructs, from spot observations, a graphical picture of the entire field of a meteorological variable over a given region for a given moment in time. Since analysis extrapolates outward from points of observation, synoptic reports should ideally be error-free and representative on the analysis scale; that is, data used in a synoptic-scale analysis should be free from purely local effects. However, in the tropics, local effects often overshadow synoptic changes, especially near the surface. Ignoring radiational heating and cooling, topography, local winds, and convection at or near the station at observation time can lead to serious errors in data interpretation and analysis. On the other hand, local effects reflected by the data may be entirely appropriate for a mesoscale or microscale analysis. Tropical forecasters must be able to separate local from synoptic influences. A useful rule of thumb holds that local effects are least around local noon and midnight—transition times between daytime and nighttime circulations. The analyst first assumes that observations are correct, even when they apparently contradict preconceived ideas. They are rejected only on the basis of careful physical reasoning.

11.2.2 Surface Observations.

11.2.2.1 Pressure. Various factors cause surface pressure reports to be unrepresentative. Mountain chains produce large wave-like distortions in the surrounding surface pressure patterns. For example, the mountains of the Philippines and Central America cause lee-side troughs in strong tradewinds (see Figure 8-36). The Western Ghats of India act the same way on the southwest monsoon. Much smaller topographic features also produce effects comparable to the pressure changes accompanying weather disturbances. On the island of Hawaii, pressure on the windward side can be as much as 2 mb higher and on the leeward side, as much as 4 mb lower than it would be without the island. A venturi effect in channels between large islands can reduce the pressure by 1 mb or more. Heavy thunderstorms can cause brief local pressure rises of about 0.5 mb.

Instrument errors can make surface pressure measurements unrepresentative. In many regions, barometers are not calibrated. This problem is magnified in the tropics, where stations are usually hundreds of miles apart; pressure reports can seldom be checked against nearby measurements. Pressure gradients are very small, especially near the equator; errors of 1 or 2 mb can lead to significant errors in locating pressure centers. Passing synoptic disturbances may cause the pressure to vary less than with the semidiurnal pressure wave.

Ship pressures are often wrong. The National Meteorological Center (NMC) (U.S. Weather Bureau, 1963) assumes a standard deviation of the errors of 1 to 2 mb for merchant-ship pressures and 1 mb for weather-ship pressures, large enough to handicap tropical pressure analysis.

Over tropical continents, reduction of pressure to sea-level may introduce another error. NMC assumes that the sea-level pressure will have a standard deviation of error of 0.5 mb for each thousand feet (1.6 mb for each kilometer) of station elevation. This error, when combined with the other errors, renders sea-level pressures from high-level tropical stations virtually worthless for synoptic analysis. Sea-level pressure analyses are of little use, especially equatorward of 20 degrees. This will be discussed further in 11.3.

Local pressure changes are undoubtedly more accurate than the absolute values of pressure. In mid-latitudes, 3-hourly pressure tendencies can indicate motion and development of synoptic systems. In the tropics, the large diurnal pressure variation (comparable to the small changes stemming from synoptic influences) renders 3-hourly changes useless. However, for a station on a small island, the average oceanic diurnal pressure variation (Figure 7-4) can be subtracted from the observations, and so allow most of the synoptic signal to be revealed. Because of local heating and cooling cycles, individual average diurnal pressure variation curves need to be determined for each land station, but it is worth doing. At Hong Kong (22° N, 114° E), for example, the pressure, measured every hour, is "adjusted" in this way. It could be further adjusted to take account of the different variations associated with clear and cloudy skies (see Figure 7-5). Of course the adjusted values are not included in synoptic reports. 24-hour pressure tendencies eliminate the diurnal variation, but the coarse time resolution hides the effects of small-scale systems. Their use is discussed in 11.5.5.

11.2.2.2 Temperature and Dew Point. In addition to the normal instrumentation and exposure errors, unrepresentative temperatures in the tropics are often caused by local convection. This is due primarily to evaporative cooling by rain of the air beneath clouds and to the downrush of cold air from showers and thunderstorms. These downdrafts can lower the surface temperature well below the representative surface wet-bulb temperature. For example, a record low temperature of 70° F (21° C) at Canton Island (2° S, 172° W) occurred in heavy rain (Figure 11-1, next page). Cold downdrafts may well have caused record cold at many other tropical stations.

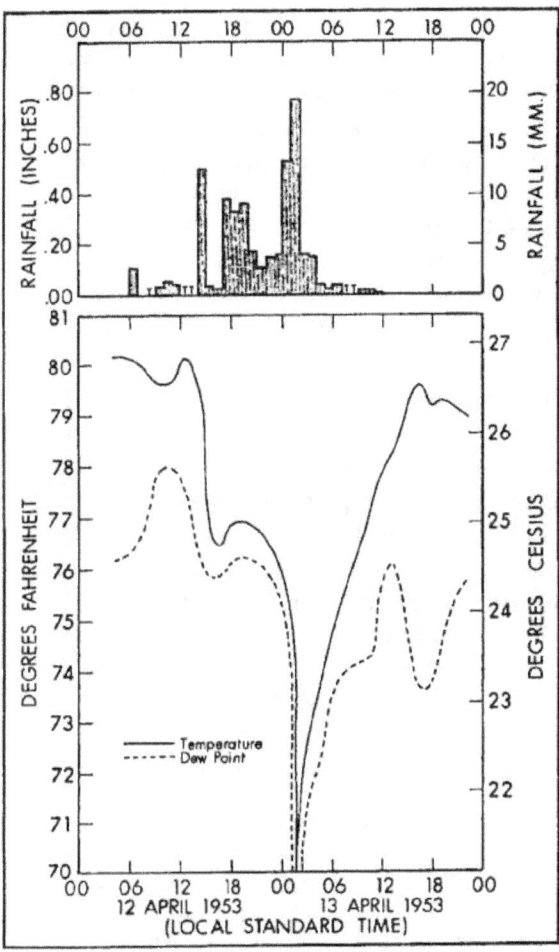

Figure 11-1. Hourly temperature, dew point, and rainfall for Canton Island for 12 to 13 April, 1953 (after Palmer et al., 1955).

Besides rapid local temperature fluctuations, convection and land-sea breezes cause large horizontal temperature gradients at the surface. Over homogeneous surfaces in higher latitudes such gradients are often associated with fronts but in the tropics they do not separate different air masses. Over tropical continents, particularly in summer, diurnal temperature variations dominate and are generally much larger than temperature changes due to synoptic influences.

The representativeness of the surface dew point depends on the same local influences that affect temperature, but the dew point is not so responsive (see Figure 7-3). In humid regions, dew point varies

diurnally in phase with the temperature, with an afternoon maximum and a morning minimum. In arid regions, and to the lee of mountains, the cycle is reversed. Diurnal dew-point variations may often overshadow synoptically induced dew-point changes. Over tropical continents in winter, the dew point changes as a front passes. Widespread subsidence and suppressed convection in the tradewinds may be reflected in below-normal dew points (see 5.3.5). Over some tropical continents, such as parts of Africa, dew points, in conjunction with low-level kinematic analyses, can be used to delineate boundaries between moist and dry air masses.

Twenty-four-hour temperature changes are not well correlated with synoptic influences in the tropics. Sometimes 24-hour dew-point changes are useful, since cool-season fronts penetrating the tropics may lower the dew point more than the temperature, especially if the front is accompanied by little cloud; 24-hour dew-point change charts are useful during these periods. Overall, dew points are more representative and conservative than temperatures.

11.2.2.3 Wind. If measured properly, surface winds are representative of synoptic influences over the open ocean. But even tiny, flat, isolated islands slow the wind. At Willis Island (16° N, 150° S) (Figure 7-2), wind measured at 12 feet (3.7 meters) elevation on the reef averaged the same as wind measured at 33 feet (10 meters) on the island. A typical vertical profile would require the latter to be 20-percent stronger. Over tropical continents and mountainous islands, winds do not usually reflect synoptic influences. Pronounced land-sea breeze regimes affect coasts; mountain-valley winds affect the interior. At inland stations on flat terrain, wind directions may reflect synoptic influences around midday, when low-level mixing is taking place and when any diurnal cycle is in transition. Differences in instrumentation, surrounding vegetation, and anemometer heights and exposure may cause spurious differences in wind velocities between neighboring stations. As with temperatures and dew points, surface winds are affected by nearby convection. In regions not affected by tropical cyclones, the strongest winds are usually caused by downrush from heavy showers or thunderstorms.

The representativeness of surface wind reports depends on the observing procedure and the limitations of various meteorological codes. In surface synoptic reports, wind speed represents the sustained speed over 1 minute for U.S. stations and over 10 minutes for other countries; some U.S. bases overseas use a 2-minute average. Maximum gusts lasting for less than 1 minute are not reported. In aviation weather reports, the wind observation coded in METAR represents the mean velocity over the 10 minutes preceding the observation. "Maximum wind speed" is included in the report if, during this 10 minutes, it exceeds the mean speed by at least 10 knots. The averaging period may vary among stations. For example, stations with anemographs may give 10-minute averages, while reports from other stations may be based on shorter periods. These and other observational differences require that surface wind observations for all land stations be carefully evaluated. Only those reports that appear to reflect synoptic influences should be used in pressure and kinematic analyses.

Surface wind reports from ships are probably the most representative and useful for synoptic analysis. Verploegh (1967) has estimated the standard error of ship wind-speed observations to be 0.6I (knots), where I is the Beaufort interval reported by the observer. Table 11-1 converts Beaufort number to knots at a height of 33 feet (10 meters) above the surface. For example, the estimated standard error of a wind speed of 20 knots would be 3 knots (0.6 times 5). Over the western Pacific between 10° N and 10° S, Morrissey et al. (1988) estimated the error to be 5 knots. The direction error is estimated to be +/-10 when wind is reported in tens of degrees. Besides the standard error of estimating winds, wind measurements depend on anemometer height and exposure. Bunting (1968) found a large range in the distribution of anemometer heights for non-Navy ships. The average height was about 66 feet (20 meters) with a range from 26 to 121 feet (8 to 37 meters).

TABLE 11-1. Relationship of Beaufort wind-scale number and wind speed (knots) as included in synoptic reports (from World Meteorological Organization, 1966).

Beaufort Number	Mean Speed at 10 Meters Above Surface	Beaufort Number	Mean Speed at 10 Meters Above Surface
0	<1	7	28-33
1	1-3	8	34-40
2	4-6	9	41-47
3	7-10	10	48-55
4	11-16	11	56-63
5	17-21	12	>63
6	22-27		

TABLE 11-2. Ratio of the wind speed at various levels to the wind speed at 66 feet (20 meters) elevation, based on the one-seventh power law.

Height (meters)	10	15	20	25	30	35	45	50
Height (feet)	33	49	66	82	98	115	148	164
Ratio	*0.91*	*0.96*	*1*	*1.03*	*1.06*	*1.08*	*1.1*	*1.14*

The one-seventh power law gives a good estimate of the variation of wind speed with height near the surface:

$$V/V_o = (Z/Z_o)^{1/7}$$

Where V is the wind at some level Z and V_o is the wind at some reference level Z. Table 11-2 shows the ratio of the wind speed at various levels to the wind speed at 66 feet (20 meters). For the usual range of anemometer heights, wind-speed differences of up to +/-10 percent from the winds measured at 66 feet (20 meters) could be expected. Ships do not include anemometer heights in synoptic reports. More ships are being equipped with anemometers; while only 20 percent had them in 1962, 75 percent were equipped in 1986. On average, winds measured by anemometers exceed winds estimated from sea-surface conditions by about 2 knots (Ramage, 1987). Synoptic reports now indicate whether an anemometer is being used, so that analysts can make rough adjustments. Indirect wind indicators can be used. For example, wind "packs" cloud against windward ranges, while shear lines parallel the flow.

11.2.2.4 Cloudiness, Rain, and Visibility. Once more, local effects and large diurnal variations make cloud and rain observations taken from tropical continental or large island stations unrepresentative of the general conditions over surrounding areas. Observations from ships or atolls are more representative, but usually too scattered to allow cloudiness boundaries to be determined. These difficulties are obvious to anyone who has tried to use surface observations in constructing nephanalyses in the tropics. Satellites have removed most of the problems, but to provide accurate local-area or point forecasts, tropical meteorologists must be thoroughly familiar with local effects on cloudiness and rain and their diurnal variations; detailed study of current and past satellite imagery is essential. Satellite pictures are available from the National Environmental Satellite, Data and Information Service (NESDIS), as polar hemispheric visual and infrared mosaics. Geostationary satellite pictures are also available. The Cooperative Institute for Research in the Environmental Sciences (CIRES) of the University of Colorado archives DMSP records. Station files of GOES, GMS, METEOSAT, and ESA pictures can also be consulted.

During daylight, reports of cloud types and amounts are limited by the experience and training of the observer and by the meteorological codes that do not permit a complete description of tropical clouds. At night, especially when there is no moon, cloud reports are much more uncertain. For example, cirrus is usually present over most of the tropics but since it cannot often be seen at night, it is not reported. Baer (1956) found that at Tucson (32° N, 111° W), Arizona, average cirrus cloudiness on moonless nights was about half that on nights with a full moon. This leads to biases in mean cloudiness and in the difference between mean daytime and nighttime amounts. Observations and statistics of low cloudiness are generally more reliable than those of middle and high cloudiness and more likely to show true diurnal variations.

The 6- and 24-hour rainfalls should be plotted on surface charts to aid in identifying disturbed weather. Even though the convective nature of tropical rain causes the amounts to cover a wide range, the knowledge is still useful and should supplement successive satellite pictures in determining the movements and intensity changes of rain areas. It is important to differentiate showers and steady rain. In the tropics, showers predominate, but steady rain is more common than once believed. Widespread steady rain may fall from nimbostratus east of upper-tropospheric cyclones or in the later stages of dying tropical cyclones (Zipser, 1977). It is very heavy beneath continuous thunderstorms (see 6.3.1.2).

Movement and varying intensities of showers or thunderstorms can cause visibility at tropical stations to fluctuate rapidly. For example, as a heavy shower passes, visibility can go from "unlimited" to less than 1,500 feet (400 meters) and back to unlimited in 15 to 30 minutes or less. On the other hand, poor visibility associated with fog and low stratus can persist for days when warm, moist air moves over cool water or up a mountain slope. Fog may develop in mountain valleys in response to radiational cooling, but usually dissipates during the day. The synoptic situation, the topography of the stations reporting poor visibilities, and the sequence of weather must all be considered in determining the representativeness of visibility reports.

The World Meteorological Organization has yet to devise a procedure for reporting visibility that adequately meets the needs of aviation. In WMO code, the minimum visibility in the horizon circle surrounding the observer is reported. But for aviation weather reports, many countries use the visibility index, which is the greatest visibility attained or surpassed through half of the horizon circle, not necessarily in contiguous elements. When a shower or a thunderstorm restricts visibility, the minimum visibility can differ greatly from the visibility index (sometimes called the "prevailing visibility"). Therefore, to evaluate visibility reports, the forecaster must know which coding procedures are being used.

11.2.3 Upper-Air Observations. Direct observations are made by sensors travelling through or suspended in the atmosphere. Rawinsonde, pilot-balloon, and aircraft measurements are examples. Indirect observations include estimates of winds at various levels derived from profilers, single or successive satellite pictures and vertical temperature soundings made from satellite or ground-based radiometers.

11.2.3.1 Rawinsonde and Pilot-Balloon. The rawinsonde is an instrument package carried aloft by a balloon. During the flight, it measures and transmits temperature, pressure and humidity data, while the ground tracking of the balloon by radio-direction-finding equipment or by radar determines wind velocity. The balloons generally reach 65,000 to 100,000 feet (20 to 30 km). Careful field trials (Hardin, 1955; Meteorology Working Group, 1965) have determined the following root-mean-square (RMS) accuracies of the measurements:

Temperature:	*0.5 to 1.0 C*
Pressure:	*2 mb (surface to 50 mb), 1 mb (above 50 mb)*
Humidity:	*5 percent (temperature above 0° C), 10 to 20 percent (temperature below 0° C).*

Such accuracies are unlikely to be realized in routine measurements. Several varieties of radiosonde are used in the tropics. Some have been so poorly calibrated that data from them cannot be analytically meshed with neighboring data. The European Center for Medium Range Weather Forecasts (ECMWF) identifies the offending stations.

The square of the error in a calculated constant-pressure surface height equals the sum of the squares of the height errors due to errors in the measurements of temperature and pressure. Table 11-3 (next page) shows the distribution of height errors resulting from an RMS temperature error of 1.0° C and an RMS pressure error of 2.0 mb. The temperature error largely determines the height error.

TABLE 11-3. Root-mean-square (RMS) pressure-height errors (meters) arising in radiosonde observations with an RMS temperature error of 1.0° C and an RMS pressure error of 2.0 mb.

Pressure Level (mb)	Due to Temperature Error	Due to Pressure Error	Total Error
1,000	0	0	0
700	10	1	10
500	20	3	20
300	35	7	36
200	47	11	48
100	67	22	71
50	87	9*	87

*Pressure-height error reverses sign above the tropopause, which is usually near the 100-mb level in the tropics.

The RMS temperature and pressure-height errors are enough to limit the usefulness of temperature and pressure-height analyses in the tropics. A further handicap is the questionable validity of the geostrophic and thermal wind relationships in low latitudes. However, 24-hour pressure-height changes at an individual station can be useful in determining trough and ridge passages and intensity changes.

Radiosonde relative humidity measurements are not very accurate. At best, the RMS error is about 5 percent at temperatures above 0° C. This accuracy is possible only if the humidity element has not already been exposed to high humidities in passing through clouds or rain. In disturbed weather in the tropics, RMS errors of 10 percent for temperatures above 0° C and 10 to 20 percent for temperatures below 0° C may be more likely. These errors correspond to a dew-point error of about 2 to 3° C in the lower troposphere, and an error of 3 to 6° C in the upper troposphere. Most humidity elements cease to function properly at temperatures below -40° C, which corresponds to a pressure of about 250 mb in the tropics.

The humidity element can seldom measure the extreme dryness of the air subsiding above the tradewind inversion. This prevents forecasters from inferring where the air came from.

The winds in radiosonde and pilot-balloon reports represent vector winds averaged through atmospheric layers of varying thickness. Between the surface and 23,000 feet (7 km), the average is calculated over a 2-minute interval with a 1-minute overlap. Thus, wind data for the 3rd minute is determined from the balloon positions projected on a horizontal surface at the 2d and 4th minutes. From 7 to 14 minutes, 2-minute non-overlapping intervals are used, and above 46,000 feet (14 km), 4-minute intervals with 2-minute overlaps are used. Since 30-gram pilot balloons rise at 600 to 650 feet (180 to 200 meters) min⁻¹ and 100-gram pilot balloons and radiosonde balloons rise at 900 to 1,100 feet (280 to 340 meters) min⁻¹, wind observations below 46,000 feet (14 km) represent the average vector wind through layers about 1,300 to 2300 feet (400 to 700 meters) thick.

TABLE 11-4. Root-mean-square errors (knots) in rawinsonde wind speeds according to altitude and the magnitude of the mean-wind vector. (Meteorology Working Group, 1965).

Altitude Range; zero to:		Mean Wind Vector (knots)		
Feet	Km	<30	30-60	60-90
10,000	3	2	5	10
20,000	6	3	7	15
39,000	12	4	14	30
59,000	18	6	21	45

The accuracy of radiosonde wind measurements depends on altitude and wind speed. Direction is generally accurate to within 5 degrees; however, since wind direction is often reported or plotted to the nearest 10 degrees, a deviation of +/-10 degrees from the plotted direction should be assumed for streamline analysis. When the speed is less than 5 knots, direction should be assumed to be variable and more leeway given in analysis. The probable RMS errors are related to altitude in Table 11-4.

In many parts of the tropics, the zonal-wind component reverses with height in the troposphere. This results in a small angle being maintained between the local zenith and a line from the tracking radar to the balloon. In these cases, fairly accurate measurements may be expected, provided the tracking system can cope with rapid changes in elevation. For most of the tropics, the mean wind vector is less than 30 knots, so that errors of only about 4 knots would be expected. Beneath a strong subtropical jet stream, the mean wind vector through the troposphere could exceed 30 knots. For practical purposes, an accuracy of +/-10° in direction and +/-10 percent in speed can be used.

Since pilot balloons must be tracked visually, the height achieved by the observations is limited by darkness (unless the balloon trails a light) and rain or clouds between theodolite and balloon. Even pibals that reach only 1,500 to 3,000 feet (500 to 1,000 meters) before entering cloud are very useful to gradient-level wind analysis, especially when the surface winds are greatly influenced by local effects.

11.2.3.2 Aircraft. Aircraft reports supplement the inadequate network of ground-based upper-air stations and enhance tropical analysis. On jet aircraft, winds are measured with doppler radar or inertial navigation equipment. The estimated RMS temperature errors are slightly larger, while the wind and height errors are about the same as for radiosondes. Aircraft reports include information on cloudiness, icing, turbulence, and other significant weather. Elevation differences between aircraft and clouds generally determine how accurately cloud base and top heights are estimated. A height difference of 10,000 feet (3,000 meters) means that the estimates will be within 2,000 feet (600 meters). Cloudiness estimates (scattered, broken, or overcast) are generally reliable. Cirrus is seldom reported but may be present. Aircraft cloud reports supplement satellite photographs and surface observations: AIREP extent and frequency will increase as more ASDAR systems are installed (3.1.4).

11.2.3.3 Weather Reconnaissance. As pointed out in Chapter 1, weather reconnaissance flights are almost entirely confined to tropical cyclones. Even these flights were seriously curtailed when tropical cyclone reconnaissance in the western North Pacific was suspended. Reconnaissance is limited to North Atlantic tropical cyclones. Data is processed on board the aircraft and automatically transmitted to NHC, Miami. RMS errors for aircraft reconnaissance observations are 0.5° C for temperature and 3-5 knots for doppler wind observations. In heavy rain, wind speed may be underestimated.

In addition to flight-level observations, reconnaissance aircraft probe the atmosphere beneath the aircraft with dropwindsondes; these are similar to radiosondes, but are lowered to the surface by parachute. They measure temperature, pressure, humidity, and winds. Performance is about the same as for radiosondes. Once again, measurement of relative humidity is the problem, both in the aircraft and on the sondes.

11.2.3.4 Satellite-Derived Wind and Temperature. Geostationary satellite data is used in calculating winds. Central processing offices use a pattern-matching technique to calculate low-cloud and high-cloud motions (see Figure 3-7). These vectors portray the true winds at cloud level fairly accurately (Hubert and Whitney, 1971). At many stations receiving sequences of geostationary satellite pictures, cloud motion vectors can be readily determined. Thus, the information normally available can be enhanced where other data is scanty or where significant weather is occurring. In rain-free areas over the ocean, surface wind speeds are estimated from microwave images (SSM/I) on DMSP.

Atmospheric temperature and moisture profiles can be calculated from radiances measured from satellites. The TIROS Operational Vertical Sounder (TOVS) has for several years provided synoptic data that has been incorporated into numerical weather prediction models and has somewhat improved the forecasts (Menzel and Chedin, 1990). But the vertical resolution of the profiles remains inadequate to replace or supplement radiosonde measurements in the tropics. Better ground-truth data from superior radiosonde measurements and new radiometers with better horizontal resolution should improve matters in the 1990s.

11.3 MANUAL ANALYSIS TECHNIQUES

11.3.1 General. This report has already demonstrated that for both synoptic and climatological understanding, the tropical wind field contains more detailed information than the pressure or pressure-height fields. What follows focuses on analyzing the winds. Best results are likely if the analyst also sketches in isobars or height contours and makes sure that general pressure-wind relationships, true everywhere, are satisfied. For example, winds should flow cyclonically around a low and anticyclonically around a high; they should curve cyclonically through a trough and anticyclonically across a ridge. Except near the equator, and in upper levels over tropical cyclones, winds should blow across the isobars toward lower pressure with the angle being greatest in the surface layers. On the equator, flow should be perpendicular to the isobars toward lower pressure. In other words, wind and pressure analyses should always be mutually consistent. Although some automated tropical analysis programs have been developed, manual analysis is still widely used in the tropics. The techniques have changed little since they were developed by Riehl, Palmer Saucier, and others during the 1940s and 1950s. The most significant change has been the integration of satellite data into the analyses. The principles of tropical analysis and the examples presented here provide only a foundation for skill. To be an accomplished analyst demands considerable practice in working with sequences of real data. To this end, the USAF conducts tropical meteorology courses and training programs.

11.3.2 Instruction in Kinematic Analysis.

11.3.2.1 General. All analysts should follow certain procedures. Kinematic analysis constructs a continuous representation of the wind field from observations of two-dimensional horizontal wind vectors for each analysis surface. Since wind is a vector, with both direction and speed, it can be represented by a stream function field, in which the field is non-divergent and the distance between streamlines is inversely proportional to the wind speed

(Byers, 1974). It is evident that isobars are stream function lines for gradient-equilibrium flow on a horizontal surface. Stream function analysis is sometimes incorporated on maps disseminated by weather centrals; although it gives a better depiction than isobars in low latitudes, it is a rather crude representation of the flow (Figure 11-10). The preferred analysis method requires two sets of lines (streamlines and isotachs) to represent the field. Streamlines are everywhere tangential to the instantaneous wind direction, while isotachs link points with the same speed. If the lines are sufficiently close, wind direction can be determined from the streamlines and wind speed from the isotachs at any point on the chart. Streamlines should not be confused with trajectories. A trajectory links the points successively occupied by a moving air particle, and coincides with a streamline only in steady-state flow, or on climatological charts (Figures 4-9 to 4-16). In these relatively simple patterns, several anticyclones are embedded in the subtropical ridge; the low-level trough contains a series of cyclones with easterly tradewinds equatorward of the subtropical ridge. Circulation features are defined below.

• *Asymptotes.* These are streamlines away from which neighboring streamlines *diverge* (positive asymptotes) or toward which they *converge* (negative asymptotes). Asymptotes may or may not represent lines of true horizontal mass divergence or convergence, depending on the distribution of wind speed in the area. Therefore, it is better to use the terms "asymptotes of difluence or confluence" rather than "asymptotes of divergence or convergence."

• *Waves.* These perturbations in the streamlines are analogous to the wave-like arrangement of troughs and ridges in isobars. Waves that do not extend across the entire width of the current in which they are embedded are called "damped waves." In this case, the streamlines on one or both sides of the current have smaller amplitudes than those in which the wave

• *Singular Points.* These are points into which more than one streamline can be drawn, or about which streamlines form a closed curve. The wind is calm at singular points, and the speeds immediately adjacent to a singular point are always relatively light. There are three classes of singular points: *cusps, vortexes,* and *neutral points.*

• *Cusps.* These are intermediate patterns in the transition between a wave and a vortex. Cusps are relatively unimportant in routine synoptic analysis since they exist only briefly in any one horizontal plane; therefore, there is rarely enough data to determine their presence. Figure 11-2 illustrates cyclonic and anticyclonic cusps in an east-west current.

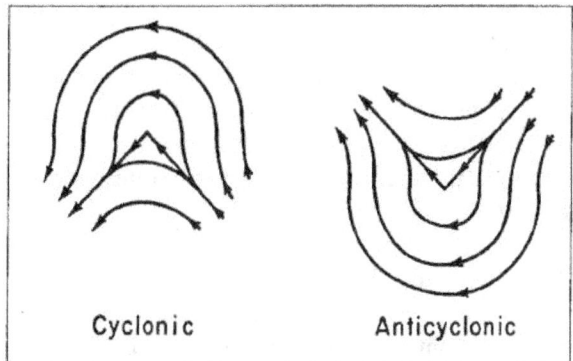

Figure 11-2. Example of cyclonic and anticyclonic cusps in an east-to-west current (after Palmer et al., 1955).

• *Vortexes.* These are cyclonic or anticyclonic circulation centers and anticyclonic and cyclonic outdrafts and indrafts. Figure 11-3 illustrates the six basic types of vortex possible in streamline analysis plus pure outdrafts and pure indrafts (the latter two may be occasionally observed on or near the equator). At low levels, anticyclonic outdrafts and cyclonic indrafts are common and coincide with highs and lows in the pressure field. At upper levels, any of the combinations may occur; cyclonic outdrafts are common in the upper troposphere over intense tropical

cyclones. Data is usually too sparse at upper levels to determine the outdraft or indraft characteristics of vortexes; when in doubt, favor cyclonic indrafts and anticyclonic outdrafts. Above newly-developed intense convection, particularly in the deep tropics, pure outdrafts have been observed in the upper troposphere. After a time, those away from the equator change into anticyclonic outdrafts.

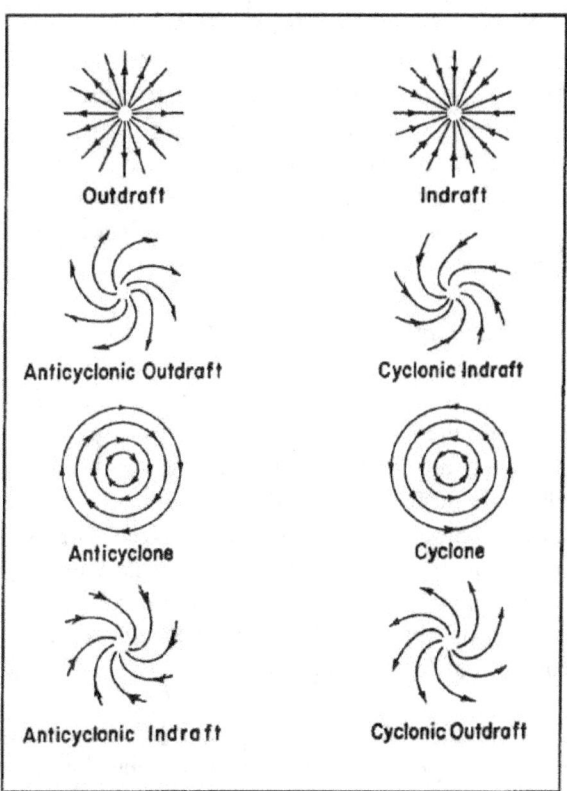

Figure 11-3. Models of pure outdrafts, pure indrafts, and six types of vortexes possible in streamline analysis (after Palmer, et al., 1955).

• *Neutral Points.* At these points, two asymptotes, one of directional difluence and one of directional confluence, appear to intersect. They correspond to cols in the pressure field, representing a "saddle" between two areas of anticyclonic flow (high pressure) and two areas of cyclonic flow (low pressure).

11.3.2.2 Streamline Analysis. Wind direction can be analyzed by first constructing an isogon field, as in Figure 8-52. An isogon joins points in a plane with the same wind direction. Even in research, isogon analysis is being supplanted by numerical objective analysis programs and is no longer used in the field. Streamlines are drawn parallel to the observed wind directions. The end result is a set of streamlines everywhere tangential to the wind arrows. To illustrate the procedure, Figure 11-4 (next page) shows the basic wind data and Figure 11-5 shows the streamlines. An isotach analysis is also included in this figure. In streamline analysis, interpolations among the winds are carried out by eye and the streamlines are drawn directly using a trial and error approach. The accuracy achieved depends significantly on the skill of the analyst.

To do a streamline analysis, follow these steps:

(1) Before starting the analysis, tentatively locate and mark the dominant features of the chart by using the plotted data, including the pressure or pressure-height fields, continuity (previous analyses), and the monthly wind climatology of the level being analyzed. These include features such as the subtropical ridge lines and monsoon troughs that generally lie east-west and are axes of maximum streamline curvature, asymptotes of confluence, fronts, etc.

(2) Tentatively locate and mark anticyclonic and cyclonic vortexes along the ridge and trough lines, respectively, by using the plotted data, isobaric charts, continuity and satellite pictures.

(3) Tentatively locate and mark neutral points associated with the vortexes.

(4) Mark center positions of tropical depressions and tropical storms or typhoons (hurricanes), according to information from tropical cyclone advisories or warnings.

(5) After completing Steps 1-4, and before drawing the first streamline, visualize how the finished analysis will appear.

(6) Now sketch streamlines around the dominant features, such as the NH and SH subtropical ridges and associated anticyclones and neutral points. Next, turn to any large areas of undisturbed flow, such as the tradewinds or the southwest monsoon. Tropical cyclones, with centers fixed by aircraft reconnaissance or satellite, can then be located. Finally, the analysis can be completed in equatorial regions; the analyst will depend mainly on satellite pictures to locate significant features.

(7) On low-level maps, work outward from anticyclonic centers and inward toward cyclonic centers to ensure a more rapid and elegant analysis. This guarantees that divergent flow around anticyclones and convergent flow around cyclones are readily apparent and accurately depicted.

(8) Do not draw a streamline through every report; this clutters the map and prevents proper spacing of the streamlines. Analysts tend to distort streamlines to make them run through data points; this may cause sharp bends or departures from the requirements of tangency.

(9) Keep streamline-free areas around neutral points relatively small. For example, on the surface-gradient-level analysis, this should be encompassed by the 5-knot isotach and on upper-level charts by the 10-knot isotach. To accomplish this, draw the neutral-point intersections first and then the remaining streamlines associated with the neutral points while bearing in mind the reported wind speeds.

(10) Avoid distorting the analysis over too large an area on the basis of one report that lacks substantiating data. It is sometimes easy to draw a wave in the tradewinds on the basis of one report. Maintaining continuity reduces this risk. Also, before introducing a cyclonic vortex into a broad zonal current, the analyst should consult satellite pictures for evidence of accompanying disturbed weather.

(11) Ensure that streamlines approach each other at very small angles along asymptotes; avoid "Y-shaped" mergers.

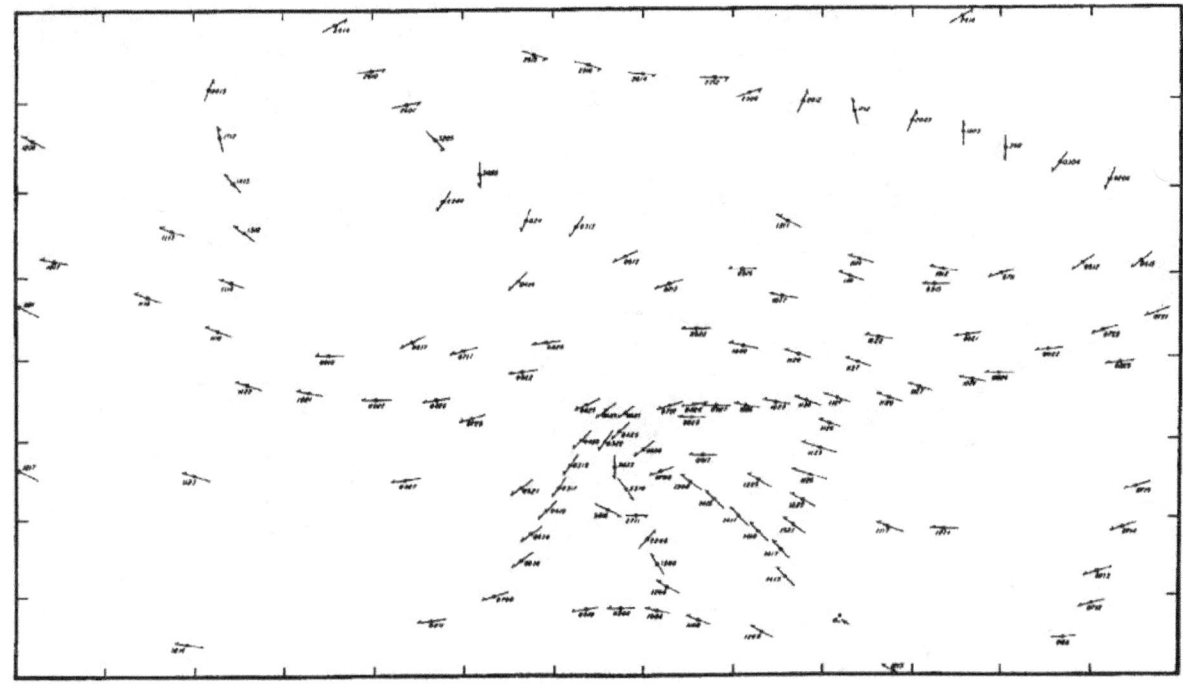

Figure 11-4. Wind data for kinematic analysis shown in Figure 11-5 (wind speeds in knots) (Dept. of Weather Training, Chanute AFB, IL).

(12) Remember that streamlines usually *cross* isobars. Despite their recognition that in the tropics, especially at the surface, geostrophic and actual winds differ, some analysts unconsciously try to draw streamlines *parallel to* isobars.

(13) Beware of concluding that light wind (less than 10 knots) is evidence of a neighboring singularity; it may reflect a singularity above or below the analysis level.

(14) Never forget the requirement that streamlines parallel wind directions to avoid the most common error in streamline analysis. However, in regions of very light winds, some leeway is permitted to achieve a reasonable pattern.

(15) Streamlines should not be crowded, but there should be enough to permit easy interpolation of wind direction at any point.

(16) Having finished the preliminary streamline analysis, analyze the speed field following the suggestions in the next subsection.

11.3.2.3 Isotach Analysis. After the preliminary streamline analysis is completed, the wind-speed field is analyzed. The analyst has already considered wind speeds during the streamline analysis to help place singular points, ridges, troughs, etc. Isotach patterns resemble those of other scalar fields such as pressure or temperature; there are centers of maximum and minimum values and saddles or cols. The following relationships between isotach and streamline patterns can be applied during isotach analysis. Many of them are illustrated by the kinematic analyses of Figures 4-9 to 4-16, and 11-5.

(1) The major axes of elongated speed-maxima roughly parallel the streamlines, particularly in the major zonal currents. There is usually an elongated speed-maximum near the center of each streamline current. In very broad currents, two or more elongated speed maxima may lie side by side.

(2) Isotachs on either side of a speed maximum also roughly parallel the streamlines. The isotachs are closer together here than along the axis of the maximum.

329

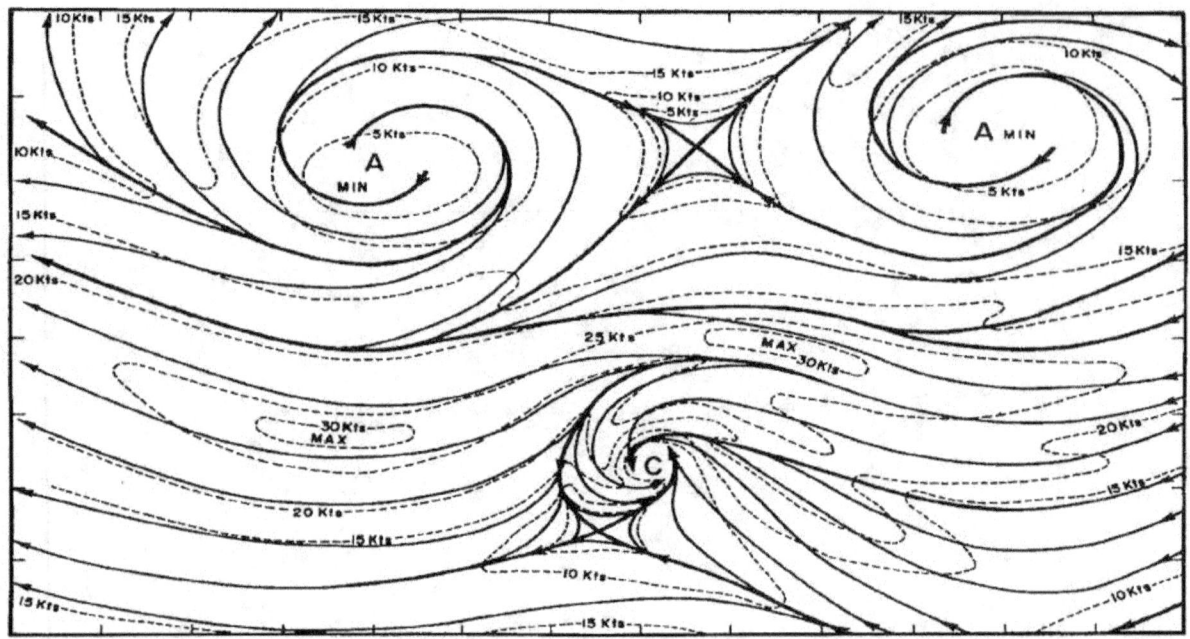

Figure 11-5. Kinematic analysis based on wind data in Figure 11-4 (Dept. of Weather Training, Chanute AFB, IL).

(3) At all singular points, the speed is zero. All other speed minima must have values greater than zero.

(4) Winds are generally light where streamlines curve sharply, and where there is a singular point just above or below the level being analyzed.

(5) Close to neutral points, isotachs should approximate ellipses. Farther from a neutral point, the isotach ellipse tends to resemble a four-pointed star with the points lying on the asymptotes that cross at the neutral point.

Finally, make any mutual adjustments in the streamline and isotach analyses to ensure a reasonable and consistent analysis that logically follows the analyses made for the previous synoptic time. The finished patterns should be generally smooth and elegant, difficult to define but easy to recognize.

11.3.2.4 Frontal Analysis. Satellites have shown that fronts extend into the tropics during the cool season. Even though an air mass discontinuity may no longer exist, a band of cloud can persist along a shear line extension of a cold front (see Figures 8-29 and 8-30).

The tropical meteorologist should be familiar with surface-layer wind flow associated with the two types of front shown in Figure 11-6: A, in which both warm and cold air are rising, and B, in which warm air rises and cold air sinks. Since, on the synoptic scale, air cannot be exchanged across the front, the front must move at the same speed as the component of the wind perpendicular to it. It follows then, that an asymptote can only coincide with a stationary front. Finally, cyclonic vorticity is concentrated in the frontal zone.

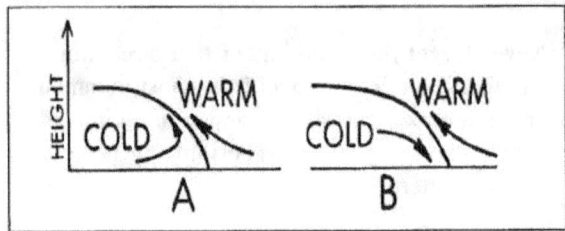

Figure 11-6. Schematic cross section of two types of front. (A) cold and warm air both rising. (B) cold air sinking, warm air rising.

These specifications are used in constructing the NH frontal wind models shown in Figures 11-7 and 11-8. The frontal segment is short enough to allow the assumption that winds are constant parallel to the front.

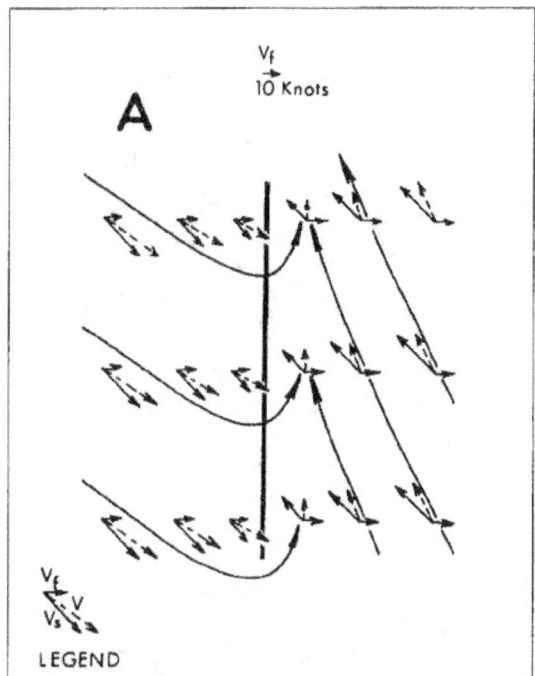 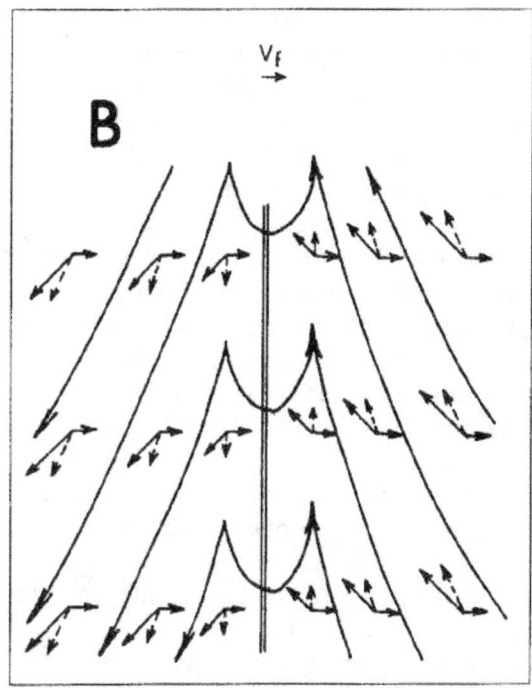

Figure 11-7. Schematic kinematic analysis of surface winds around a moving front (double line). A: Cold and warm air rising. B: Cold air sinking, warm air rising. V_s = wind flow vector relative to the front; this changes by 180° through the front. V_f = frontal movement vector. V= resultant of V_s and V_f = actual wind (Ramage, 1957).

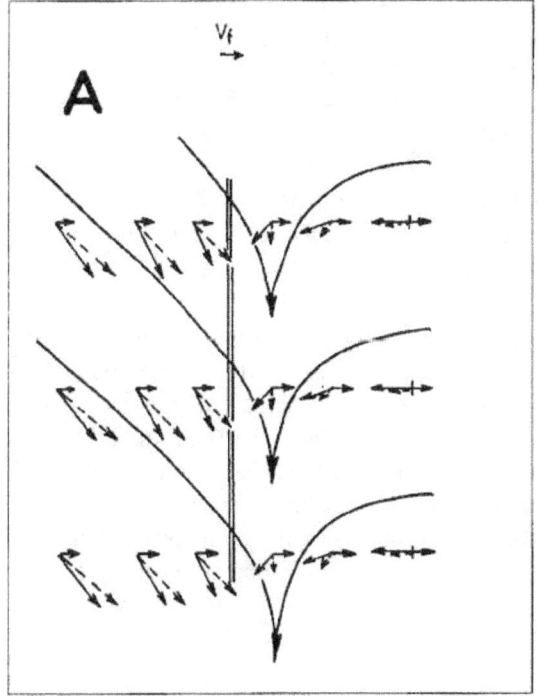 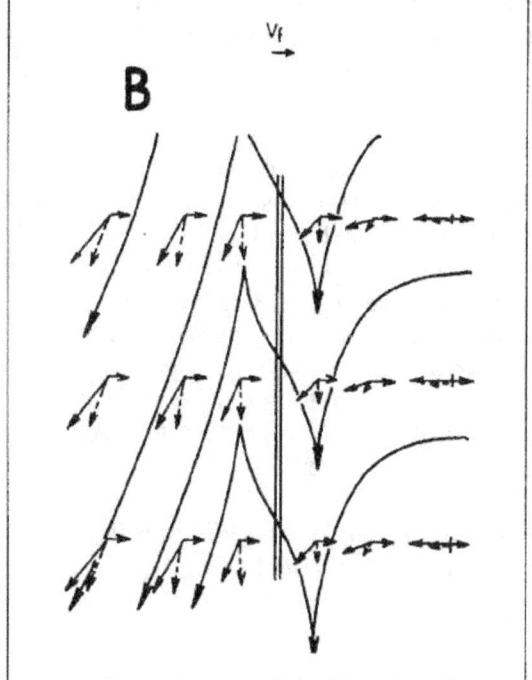

Figure 11-8. Same as Figure 11-7, except that V_s does not change direction through the front.

In Figure 11-7 (opposite), if the front is oriented SW-NE, the circulation is typical of that around a cold front *north* of the subtropical ridge, where northwesterlies replace southwesterlies as the front passes (see Figure 8-29). In the type A front, although convergence is greatest at the front, the asymptote of confluence is displaced ahead of the front and may not necessarily coincide with a cloud line. With the type B front, an asymptote of difluence appears behind the front. Successive analyses showing one or more asymptotes approaching the front indicate that the front is moving slower.

In Figure 11-8, with the front oriented WSW-ENE, the circulation is typical of that around a cold front *south* of the subtropical ridge, moving ahead of a surge in the tradewinds or in the winter monsoon. The asymptotes appear as in Figure 11-7, but most importantly, there need be no change in wind direction as the front passes— only an increase in speed. The tropical analyst should remember that a streamline represents the instantaneous field of wind direction, and not the trajectory of an air parcel. Thus, all moving fronts are crossed by streamlines; as long as the wind component perpendicular to the front equals the frontal speed, air-mass separation is ensured.

Figure 11-9 shows a synoptic example of isobaric and kinematic analyses around a frontal system. Note the pattern surrounding the wave cyclone. Frontal models underline how important the isotach field is in determining where convergence or divergence is concentrated (see also Figure 6-16b). Without it, the asymptotes would tend to overpower interpretation, and even force the analyst to distort the streamline pattern so as to make a confluence asymptote coincide with the most severe weather.

At many stations, middle-latitude contour-height or pressure analyses are merged with tropical kinematic

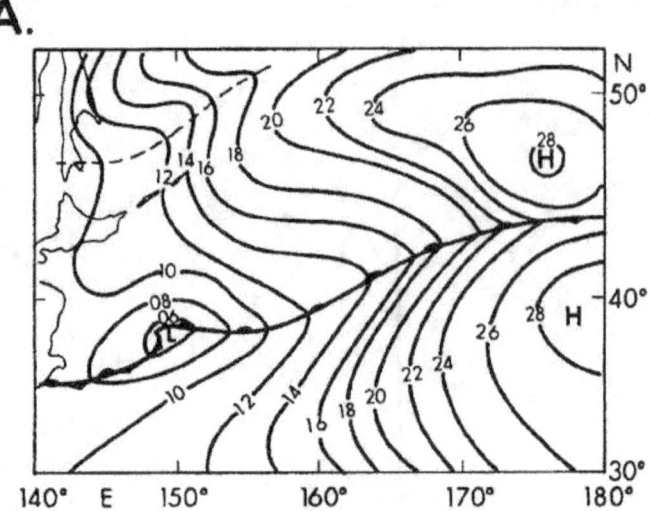

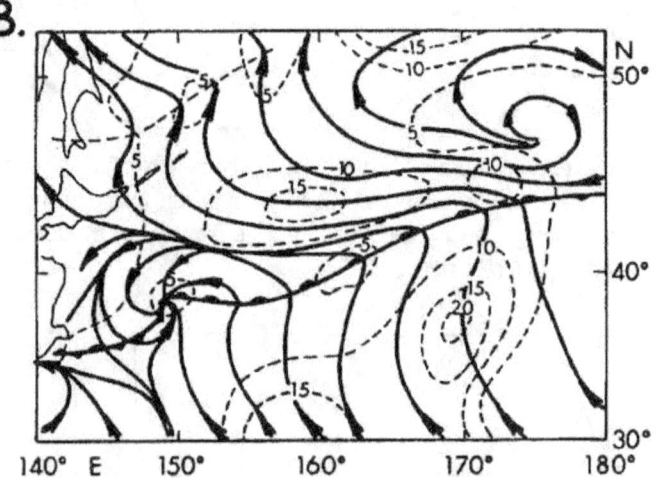

Figure 11-9. Surface chart for 1200Z 22 July 1956. (A) Isobaric analysis (mb). (B) Kinematic analysis (isotachs labeled in knots.)

analyses. For levels above the surface, this poses no problem, as long as the analyst remembers that streamline spacing is not related to wind speed. However, merging surface charts produces a very messy result, since the cross-isobaric component of the wind ensures that isobars and streamlines cannot coincide. Preferably, kinematic analysis should be extended as far poleward as operations require.

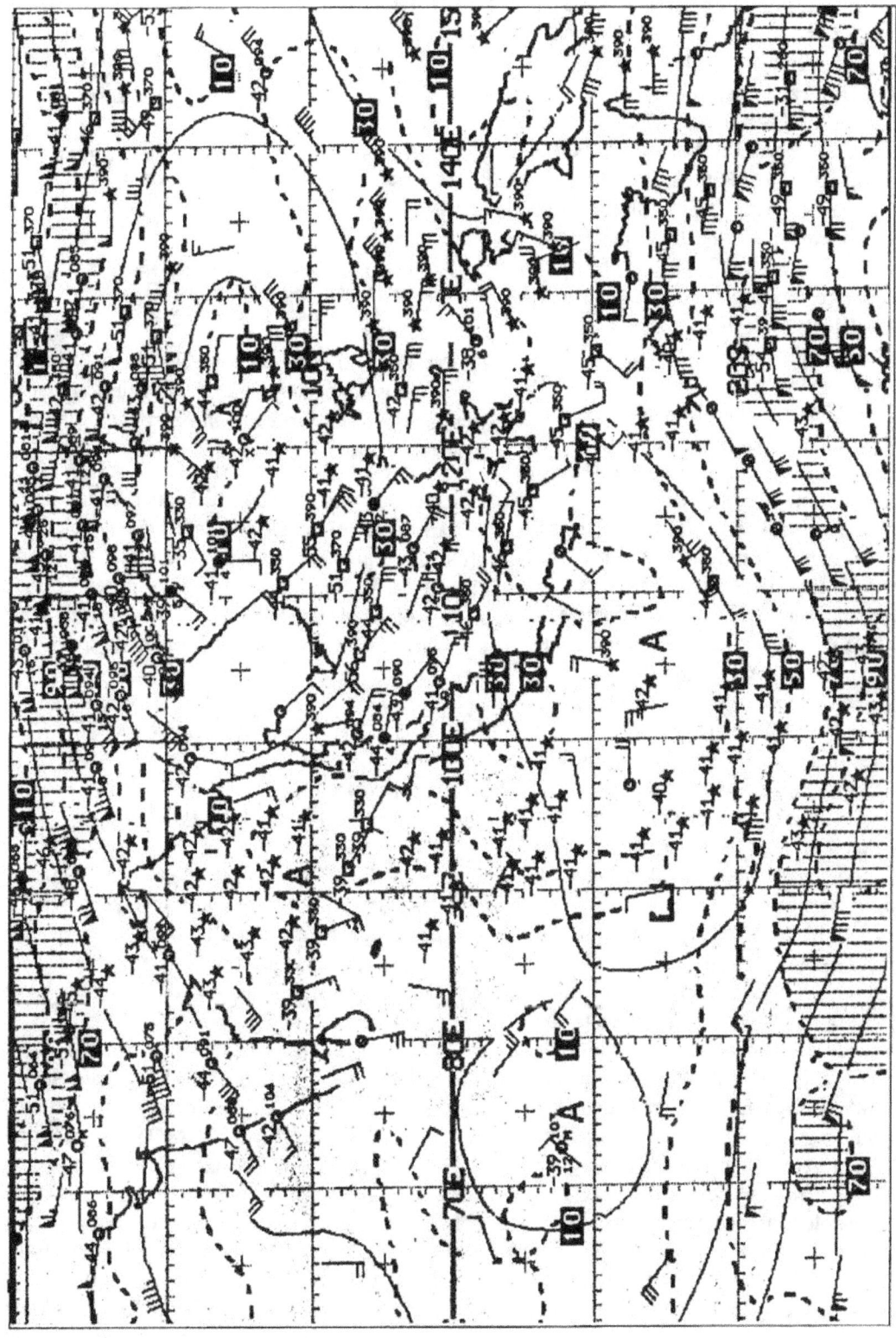

Figure 11-10. Chart for 250-mb winds (knots) and temperatures (° C), for 1200Z 26 November 1985. Stream function and isotachs shown. Areas with speeds greater than 70 knots are hatched.

(AIREPs) have become more important to tropical analysis because of more flights, higher accuracy, and improved communications. Sadler (1965) early demonstrated the usefulness of AIREPs when he prepared global tropical streamline analyses for various levels on 10 December 1963. By using AIREPs, off-time reports, and PIBALs, he more than doubled the information available from RAWINs alone. Since then, reports from 250 mb (jet aircraft) have continued to increase, but they have decreased for the lower levels. In a typical 250-mb operational chart for the global tropics (part of which is shown in Figure 11-10), AIREPs total 97, compared to 135 rawin and 16 cloud motion vector reports.

Aircraft wind measurements made with doppler radar and inertial navigation systems are as accurate as radiosonde observations, but, when many AIREPs are plotted on constant-pressure-level charts, inconsistencies are evident, especially in reported wind speeds. Much of this stems from height and time differences between the AIREPs and the charts. On upper- tropospheric charts the wind shear between the 300- and 200-mb levels, as determined from rawinsonde observations, can be applied to AIREPs to adjust the reported speeds to the analysis level. (If the maximum wind level lies between 300 and 200 mb, then the shear between 300 mb and the level of maximum wind may more appropriately be used, depending on the analysis level). Since wind direction does not vary much through the layer between 300 and 200 mb, wind shear per thousand feet can be estimated by dividing the difference between 300 and 200 mb wind speeds by nine. This shear, interpolated between radiosonde stations, can be applied to AIREP data (Figure 11-11). In this example, the adjusted AIREP wind speed of 74 knots, compared to the 200-mb wind speed of 80 knots at the rawinsonde station to the west, indicates that the jet-stream core is probably north of the AIREP location. Based only on the two rawinsonde reports, the jet-stream core could easily have been analyzed as being on or south of the AIREP position. Where there are few rawinsondes, climatological shear values may be used in adjusting AIREP data to the analysis level. The wind speeds at the analysis level, interpolated from AIREPs, can be plotted in brackets to facilitate isotach analysis.

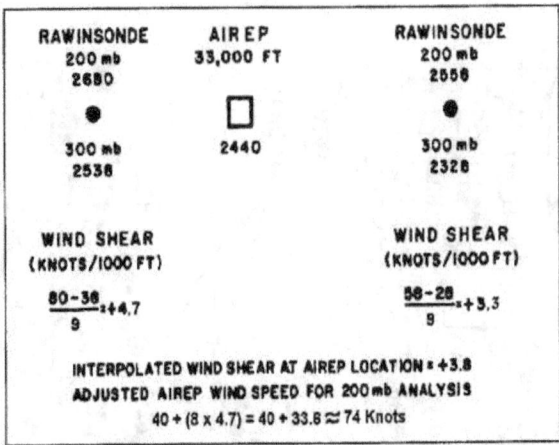

Figure 11-11. Illustrating how AIREP wind speeds can be adjusted to analysis level using the computed wind-shear in the vertical from surrounding rawinsonde stations.

It is not so easy to adjust AIREP data to compensate for variations with time. Apparent inconsistencies can often be removed by making the effort. Lenhard (1967) used a number of previous observational studies of wind variability to derive the following equations relating wind variability to wind speed and time or distance:

Time Variability $\quad S_t = (2.9 + 0.1W)t^{0.5}$

Horizontal Space
Variability $\qquad S_d = 4(2.9 + 0.1W)(1.85t)^{0.5}$

S_t or S_d is the RMS change in wind with time or distance (knots), W is the resultant wind speed (knots), t is the time in hours, and d is distance in nautical miles. Thus, the change in wind over a distance of 8.6 NM (16 km) is equivalent to the change in wind at a point during an hour.

Figure 11-12, next page, gives the graphical solution to these equations for periods of up to 6 hours (or horizontal distances of up to 52 NM/96 km), and wind speeds up to 120 knots. For example, for $t = 4$ hours and $W = 60$ knots, the resulting RMS wind variability is 18 knots. While it is not possible to make objective corrections to account for wind variability with time, the RMS variability values can roughly help determine how much the AIREP wind speeds should be smoothed (after they have been adjusted for height differences).

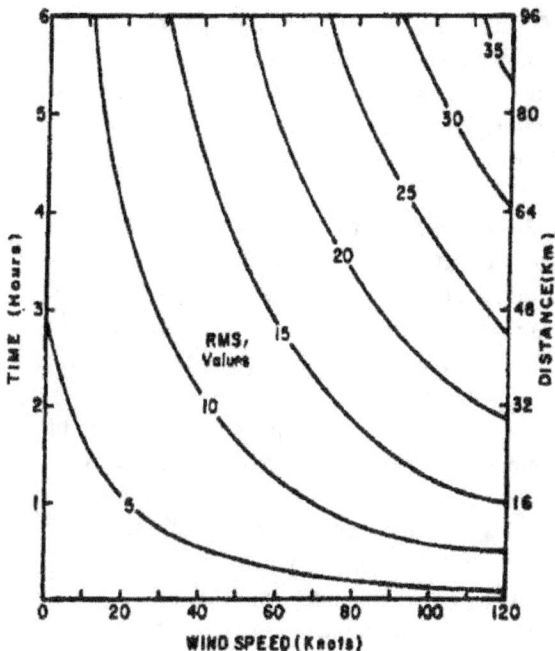

Figure 11-12. Root-mean-square variability (knots) of upper-air wind speeds as a function of wind speed and time or distance (adapted from Lenhard, 1967).

In some cases, locations of off-time AIREPs can be adjusted with respect to moving weather features such as tropical cyclones or troughs in the westerlies. AIREPs made before synoptic map time are displaced in the direction the system is moving and AIREPs made after map time are displaced in the opposite direction. The displacement distance for an AIREP equals the distance traveled by the system between map time and AIREP time.

Figure 11-13 illustrates the location adjustment for two AIREPs near a rapidly moving mid-latitude trough. Before adjustment, the AIREPs seem to show a ridge; after adjustment, they pinpoint the trough

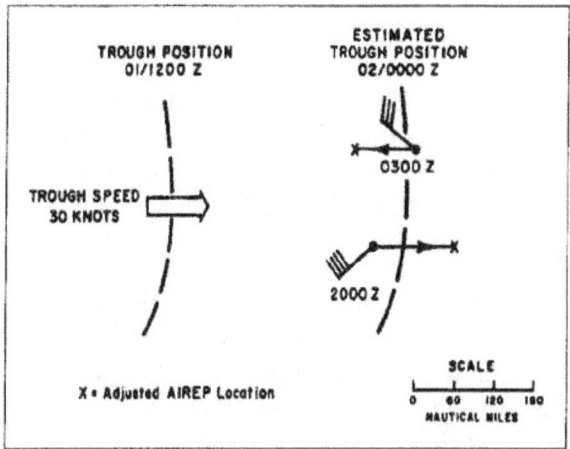

Figure 11-13. Illustrating how AIREP positions can be adjusted according to the movement of synoptic-scale features.

position. Aircraft reconnaissance reports around tropical cyclones can be similarly adjusted. AIREP positions need be adjusted only near sharply defined circulation features; therefore, on any one map only a few adjustments are made. Further examples of the use of asynoptic data in analysis are given in *Use of Asynoptic Data in Analysis and Forecasting*, AWS TR-225 (1960).

Once the AIREP wind data is adjusted for wind shear in the vertical and location (when appropriate) and the normal variability of wind with time is taken into account, more than 80-90 percent of AIREP data is found to fit well with conventional and satellite-derived wind data. Where AIREPs are plentiful, reports that are obviously wrong usually stand out and can be removed. Isolated AIREPs, where other wind data is sparse, can be compared to climatology and satellite pictures to identify those that appear to be badly wrong.

11.3.4 Use of Climatology.

11.3.4.1 General. It is virtually impossible to produce consistently good kinematic analyses without carefully considering the wind climatology, a most important analysis aid if synoptic data is scanty. In addition, short-tour policies for most tropical areas mandate that each weather unit maintain comprehensive climatological files to provide newly-arrived forecasters with "instant experience." Kinematic analyses of resultant wind fields are indispensable (see Figures 4-9 through 4-16). The resultant wind is the vector, averaged over a period of years, of all winds at a particular level at one station. If possible, at least 5 years should be averaged. The climatological charts should depict resultant winds and (if possible) "steadiness" (but see Figure 11-15 on Page 11-22). "Steadiness" is the ratio of the resultant wind speed to the mean scalar wind speed, expressed in percent. Although not all the standard pressure levels are routinely analyzed, the wind climatology for these levels (surface/gradient, 850 mb, 700 mb, 500 mb, 300 mb, and 200 mb) should be available for training and operational planning. The following sections describe how to use resultant-wind climatology in analysis, particularly where synoptic data is scarce.

11.3.4.2 Streamlines. From the resultant streamlines, one can determine the mean locations of major circulation features; e.g., subtropical ridges, low-level monsoon troughs, heat lows, near-equatorial tradewind convergence, and upper-tropospheric troughs. On the resultant charts, lines should be drawn through the centers, neutral points, and points of maximum curvature of each system. The slope in the vertical of major circulation systems, as shown by resultant wind charts and mean meridional cross sections, may be applied to synoptic analysis. Composite monthly charts superimposing the locations of major circulation features of the various levels (see Figure 11-14), and including jet-stream locations, aid in analysis and training.

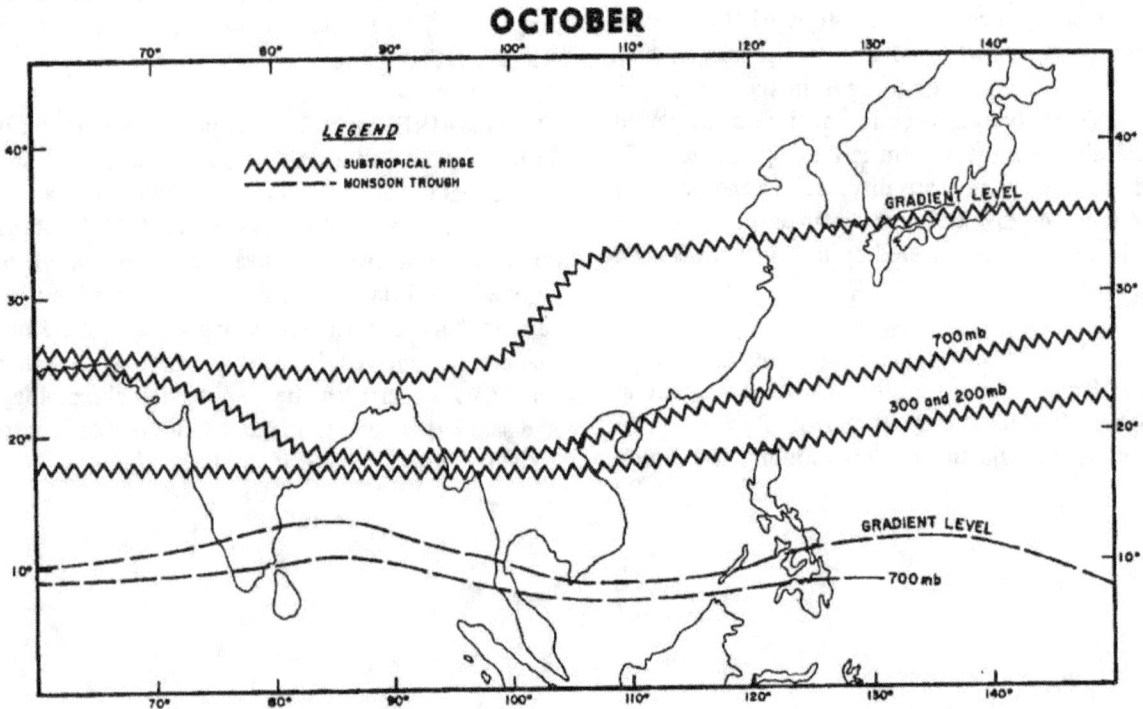

Figure 11-14. Chart of the mean positions of major circulation features over the western North Pacific and southern Asia for October.

Since climatological systems are usually present on synoptic charts, wind climatology will alert the analyst as to what to expect. In most cases, the systems will be better defined on the synoptic charts than on the mean charts because climatology smooths out short-period variations. For example, over the North Pacific and North Atlantic during the warm season, resultant winds turn cyclonically through the upper-tropospheric troughs by 90 degrees or less. On synoptic charts, wind shifts of 180° across the troughs are common.

Resultant wind charts help identify significant circulation anomalies that often accompany significant departures from the normal cloud and weather patterns for that month (see also Chapter 12). They also help determine the boundaries of local-area charts to be analyzed by forecasting units. The analysis area should encompass the mean position and normal variations of the major features that affect local-area weather. Stations with a large annual variation in circulation patterns, especially in monsoon regions, may shift their analysis areas so as to adequately track tropical systems during summer and mid-latitude systems during winter.

11.3.4.3 Isotachs. Resultant wind isotachs give a pretty good idea of wind variability. Figure 11-15 relates monthly mean resultant-wind speeds and steadiness at the gradient level, based on monthly data from 23 stations well distributed throughout the tropics. Similar relationships are found for higher levels; however, the lines of best fit may differ. Thus, for practical purposes, isotachs of the resultant wind could stand in for steadiness.

Figure 11-16 relates resultant gradient-level wind speed and the percent of time the wind blows from ±45 degrees of the resultant direction, computed for the same station months used in Figure 11-15. Wiederanders (1961) related modal wind direction (wind direction most often observed) to resultant wind direction and steadiness at six Pacific area stations. His sample consisted of 360 cases (12 months x 5 levels x 6 stations) (Table 11-5). The results imply that resultant directions are close to modal directions when steadiness exceeds 30 percent or when resultant speeds (at gradient level) exceed 10 knots.

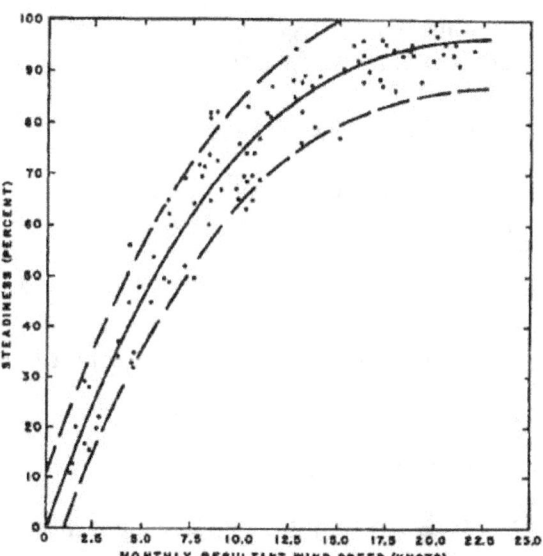

Figure 11-15. Relationship between the monthly mean resultant wind speed at the gradient level and the "steadiness." Dashed lines enclose departures of less than ±10% from the line of best fit (Atkinson and Sadler, 1970).

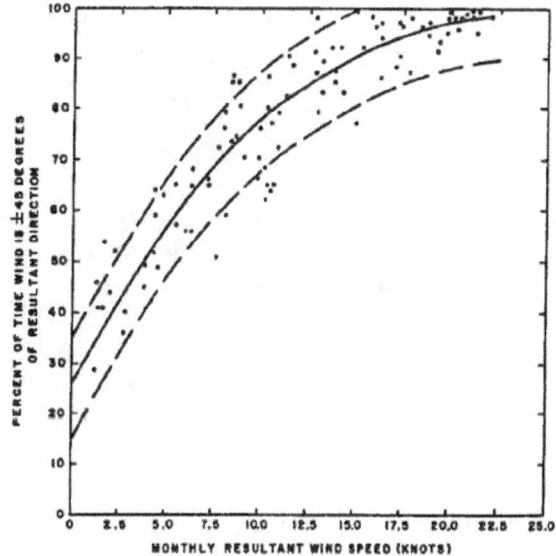

Figure 11-16. Relationship between the monthly mean resultant wind speed at the gradient level and the percent of time that the wind is within ±45 percent of the resultant direction. Dashed lines enclose departures of less than ±10 percent from the line of best fit (Atkinson and Sadler, 1970).

TABLE 11-5. Percentage frequency of modal wind directions differing by less than one compass point from resultant wind directions (after Wiederanders, 1961).

Steadiness (percent)	0.19	20-29	30-49	50-69	70-100
Percentage Frequency	25	75	75	90	100

Relationships between resultant-wind speed and wind variability help in synoptic analysis and forecasting. Where resultant winds are relatively light, wind direction fluctuates greatly from day-to-day. The type of circulation (trough or ridge) associated with a resultant wind-speed minimum indicates the type of synoptic circulation (cyclone or anticyclone) to be expected. Conversely, circulation centers are seldom found in regions with strong resultant winds. Altitude determines what is meant by "light" and "strong." At the gradient level, relatively light winds may be less than 10 knots; relatively strong winds, more than 15 knots. At 200 mb, the criteria are more likely to be 20 knots and 40 knots, respectively.

Charts of resultant winds can be used in checking for errors in synoptic reports. For example, a single report of a west wind in an area normally having moderate to strong easterlies should be carefully examined for plotting or transmission errors and compared to surrounding reports and satellite pictures before it is accepted. Occasionally, such errors cause vortexes to be incorrectly analyzed in the tradewinds and even carried on subsequent analyses on the basis of "continuity." In another example, a report of a 20-knot wind in the lower troposphere where the resultant wind speed is less than 5 knots might be a clue to storm formation, but one should look at the latest satellite picture for supporting evidence of a vortex.

11.3.5. Use of Satellite Data. Chapter 4 of *Application of Meteorological Satellite Data in Analysis and Forecasting*, AWS TR-212 (1969) (ESSA NESC TR 51), discusses the use of satellite data in tropical analysis and incorporates satellite images. Since the report is readily available, details will not be presented here. Other publications include Weldon (1979), Dvorak (1984), Fett and Bohan (1986), and Dvorak and Smigielski (1990). Weather satellite images have illustrated the earlier chapters of this report and have formed the basis for the cloudiness and rainfall climatologies of Chapter 6. In summary, the following useful information can be obtained by applying the techniques covered in AWS TR 212:

• Estimates of upper-tropospheric winds from cumulonimbus plumes and cirrus shields.

• Estimates of lower- and upper-tropospheric winds from cloud-motion measurements from consecutive geosynchronous satellite pictures.

• Estimates of maximum wind speeds in tropical cyclones.

• Location of major trough and ridge systems in the mid-latitude part of the analysis area.

• Location of subtropical jet stream.

• Estimate of low-level wind direction from orientation of cumulus lines.

• Location of frontal zones.

• Location of cyclonic vortexes in the upper and lower troposphere.

• Location of low-level ridges in the subtropics.

• **Location of large-amplitude upper troughs that extend into the tropics.**

• Estimates of surface wind direction and/or speed from sea breezes, topographic effects, sunglints, anomalous cloud lines in fog and stratus, etc. Many diagrams in Chapter 7 show diurnal variation curves crossing zero near local noon and midnight. Thus, sequences of geostationary satellite images for these times give the forecaster the least distorted view of synoptic and mesoscale weather systems, and help determine tracks and intensities.

Weather satellite images have become essential to tropical analysis, but it is worth remembering that they have not solved all the forecasting problems, as many expected they would back in 1960. Well-defined weather systems, such as tropical cyclones, troughs in the upper-tropospheric westerlies, near-equatorial tradewind convergence, fronts and shear lines are easy to recognize and track from one day to the next. Analysts can usually track squall lines and most cloud clusters from hour to hour on geostationary satellite images, but often have trouble over the longer intervals between passes of polar-orbiting satellites. Most of the time, cloud signatures cannot be linked to circulation features; models based on the images apply to a small fraction of the "organized" cloud systems seen by the satellite. Satellites tell us what clouds are there, but most of the time they don't tell us why they're there or help us forecast development, dissipation or motion.

From the satellites we have learned that a typical tropical cloud system is much more variable than we had expected, and that merely extrapolating movement rarely ensures a good forecast. Perhaps exhaustive studies of surges will lead to new models and to explaining some of what is now mysterious. This will come only if satellite images are closely linked to the other important variables of winds, lapse rates and moisture, and satisfy three-dimensional continuity.

11.3.6 Post-Analysis Programs. LaSeur (1960) recommended an active post-analysis program that calls for reanalysis of the most recent charts after the immediate pressure of issuing forecasts is past. The analyses are revised in the light of late data and subsequent synoptic developments. In addition, the analyses and satellite data are carefully compared to ensure consistency. Post-analysis should compare charts at various levels for vertical consistency and

prepare chronological continuity charts showing successive locations of major circulation features. Depending on available staffing, schedules, and other demands, post-analysis can be carried out either once or twice daily. Aggressive post-analysis ensures that the best possible continuity is established, to the benefit of the next analysis cycle.

11.3.7 Examples of Kinematic Analyses. To illustrate analysis principles, a set of kinematic analyses for the western North Pacific/Asia region is shown in Figures 11-17 through 11-19.

Figure 11-17 is the composite lower-tropospheric chart in which the 850-mb flow (the arrows and the first set of winds in parentheses directly under the station) is analyzed. To facilitate the analysis, the 3,000-foot/1-km) winds (shown as the second set of winds in parentheses) and ship winds (shown in wind-barb form) are included. Scalloped lines enclose areas of significant cloudiness (multi-layered broken to overcast cloud and/or areas of active convection). Although used in the analysis, surface synoptic data has been omitted here to avoid cluttering. Wind reports considered to be not representative or wrong are marked with "X"s. The major synoptic features are Tropical Storm Dot, just southwest of Japan, the monsoon trough lying generally east-west near 20° N, and broad westerlies between about 5 and 20° N. Tropical Storm Dot interrupts the monsoon trough circulation with westerlies between about 120 and 130° E and 5 and 30° N. This situation is common during the typhoon season when typhoons or tropical storms may move across the Japan/Okinawa region. Over southeast Asia, the strong southwest monsoon is accompanied by cloudiness and convection. Note the sharp break in cloudiness along the east coast of Vietnam caused by the downslope flow over the Annam Mountains.

In the 500-mb analysis (Figure 11-18) the low-latitude ridge and the monsoon trough along 20° N are well defined. Anticyclones cover China while the cyclonic circulation of Tropical Storm Dot is clearly evident southwest of Japan. Note the generally light winds at 500 mb; only near the tropical storm do winds exceed 20 knots.

Figure 11-17. 850-mb kinematic analysis and nephanalysis for the western North Pacific and Asia for 0000Z 26 July 1967. Wind reports that are probably wrong or not representative on this analysis scale, are indicated by "X"'s. Wind speeds in knots (from analyses made at the University of Hawaii).

Figure 11-18. 500-mb kinematic analysis and nephanalysis for the western North Pacific and Asia for 0000Z 26 July 1967. Wind reports that are probably wrong or not representative on this analysis scale, are indicated by "X"s. Wind speeds in knots (from analyses made at the University of Hawaii).

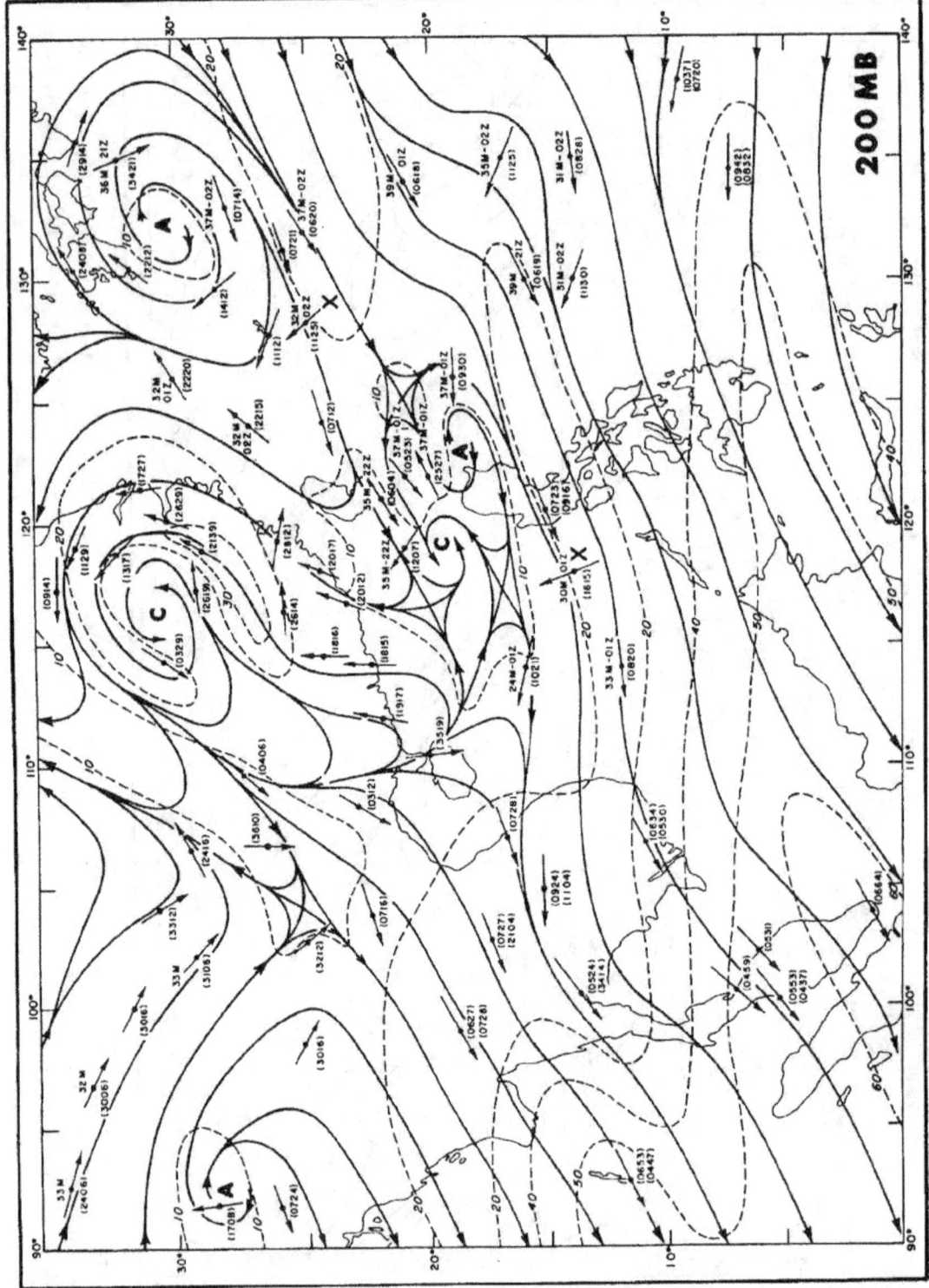

Figure 11-19. 200-mb kinematic analysis and nephanalysis for the western North Pacific and Asia for 0000Z 26 July 1967. Wind reports that are probably wrong or not representative on this analysis scale, are indicated by "X"'s. Wind speeds in knots (from

342

The 200-mb analysis is shown in Figure 11-19. The circulation features are simpler than at the lower levels. Easterlies dominate most of the region south of 25° N. The cyclonic circulation of Tropical Storm Dot has given way to an anticyclone east of the low-level center. Over eastern China an intense cyclone overlies generally anticyclonic circulations at 850 and 500 mb. East of the upper cyclone, Figure 11-17 showed extensive convection (see Figure 8-18). After developing in the upper-tropospheric trough over the western North Pacific, this cyclone drifted westward over China.

This analysis set illustrates how most of the synoptic-scale cloud features are related to circulation features at various levels, as well as to terrain influences. As over much of the tropics, the lower- and upper-tropospheric circulations in this area and season are linked more closely to the cloud and weather patterns than are the 500-mb circulations. These associations may not be so obvious in those parts of the tropics where data is scarce.

11.3.8 Recommended Analysis Levels. In general, the levels best suited to relating circulation features to weather patterns in the tropics are a near-surface level which is relatively free of frictional effects and a level in the upper troposphere. There are exceptions, particularly the mid-tropospheric cyclones that have already been discussed in Chapter 8. In most tropical areas, the gradient-level over land, depicting the friction-free flow at about 3,000 feet (900 meters) elevation, combined with the surface level over the sea, and the 200- or 250-mb level are suitable choices for the basic charts to be analyzed. Fortunately for tropical meteorologists, most observations are made at these levels. The data available and its accuracy has been discussed in earlier sections.

In view of the wide variety of wind-data sources, it may be more appropriate to consider making "layer" analyses rather than analyses at fixed levels. Such charts can give the most information to analysts. As shown in Figure 11-20, a near-surface layer depiction could include the regular surface observation (including pressure), the 3,000-foot/900-meter wind

(indicated by an arrow), and the 850-mb wind velocity, indicated numerically. Adding the 850-mb wind greatly aids subjective interpolation of the low-level wind-speed profile when making an isotach analysis. It also helps streamline analysis when the wind direction changes significantly with height.

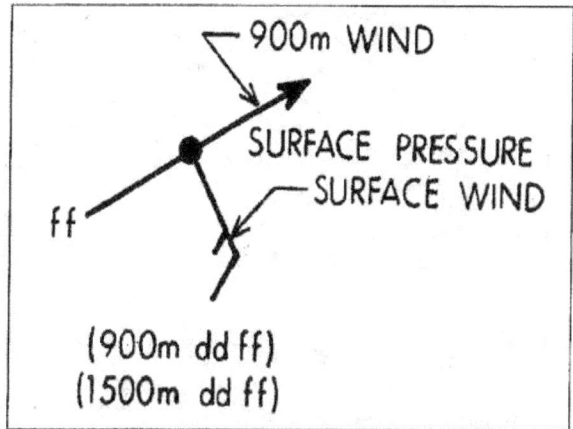

Figure 11-20. Recommended plotting model for composite lower-tropospheric wind charts.

Growing interest in climate change and air-sea interaction is resulting in more surface observations over the tropical oceans from ships, buoys (both anchored and drifting), and small islands, including remote weather stations (see Chapter 3). Thus, in the near-surface layer, emphasis should shift toward observations made at the surface, supplemented by gradient-level winds. Since low cloud motion vectors calculated from successive geostationary satellite pictures are far more numerous than winds from all other sources over the tropical oceans, how might they better contribute to analysis? Sadler and Kilonsky (1985) derived climatological shears between ship winds and low-cloud motion vectors for the eastern tropical Pacific; Figure 11-21 shows shears exceeding 10 knots along the equator. They then successfully applied the climatological shear to cloud-motion vectors averaged over a single month to obtain corresponding surface winds. Lander (1986) has found support for the hypothesis that the shear equals the thermal wind, which in turn is determined by the pattern of sea-surface temperature (SST). Thanks to satellites, the SST can now be accurately estimated at weekly intervals. Research at the University of

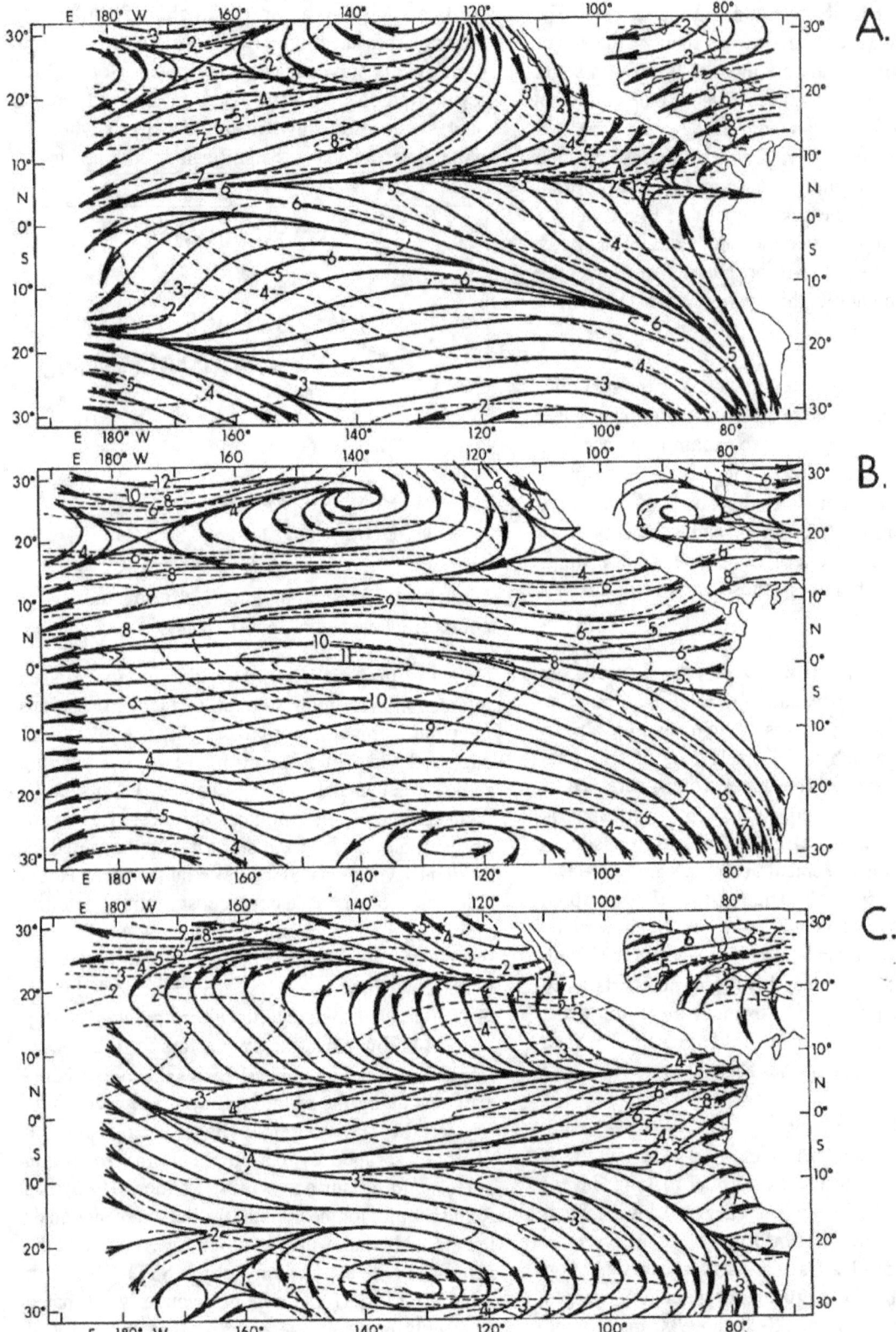

Figure 11-21. January long-term mean (A) ship winds and (B) satellite cloud motion vectors; (C) wind shear in the vertical between the two data sets (m s^{-1}) (Sadler and Kilonsky, 1985).

Hawaii raises the possibility that central offices, which now distribute low cloud motion vectors, could before long also derive and distribute useful surface wind estimates.

The DMSP microwave imager (SSM/I) senses surface wind speed over the ocean in rain-free areas (see 3.1.5). For the period July 1987 through June 1988, Atlas et al. (1991) first made kinematic analyses of winds observed from ships and buoys. They then derived wind directions from the analyses, assigned them to speeds estimated by SSM/I, and combined the two datasets. Increasing the number of observations in this way should be operationally feasible.

For the upper-tropospheric chart, RAWIN data at two or three pressure levels (e.g., 300, 250 and 200 mb) may be presented in numerical form with arrows depicting wind directions at the analysis level. To allow adjustments, time and height must be indicated for each AIREP. Level-of-maximum-wind reports may also be plotted.

The inclusion of temperature and height or D-value data should be optional, depending on the season and location of the reports. In the subtropics during the cool season, pressure-height/temperature data used in conjunction with the geostrophic and thermal-wind relationships can improve wind analysis. Avoid plotting information that has no value to diagnosis or forecasting. Figure 11-22 recommends plotting models for rawins (pilot balloons) and aircraft wind reports on upper-air charts. Offtime data should be plotted in red and late data (that added to the chart after the preliminary or operational analysis has been made) in green.

Although time, height, and distance adjustments can be made to jet-aircraft winds, it may happen that stations heavily dependent on AIREPS would prefer to analyze the 250-mb level rather than 200 mb (because most jets fly closer to 250 mb) and reduce the height adjustments needed. If 250 mb is chosen

as the analysis level, rawins may not always include 250-mb data. In this case, an average of the 300- and 200-mb winds will be an adequate estimate of the 250-mb wind, because in the tropical troposphere, the strongest winds are usually above or near 200 mb.

Besides the basic charts described above, a unit's operational requirements must determine what other levels should be analyzed. For example, forecasts for non-jt aircraft operating between 10,000 and 20,000 feet (3 and 6 km) may require 700- or 500-mb analyses if adequate wind analyses or forecasts are not available from weather centrals. Over South Asia during the southwest monsoon, 700- or 500-mb analyses help identify and track ridges and mid-tropospheric cyclones. During the northeast monsoon, 500-mb analyses do the same with mid-tropospheric short waves (vorticity centers) that may trigger surface cyclogenesis and monsoon surges. In summer over northwest Africa, weather-causing cyclones are often best-defined near 700 mb. Over the plateau of eastern and Southern Africa, the 700-mb level is relatively free of the terrain effects that influence lower-level winds. Although many factors determine choice of analysis levels, it is well to remember (Gabites, 1963) that the forecaster should concentrate on only two or three levels and not disperse his energy in sketchily treating too many.

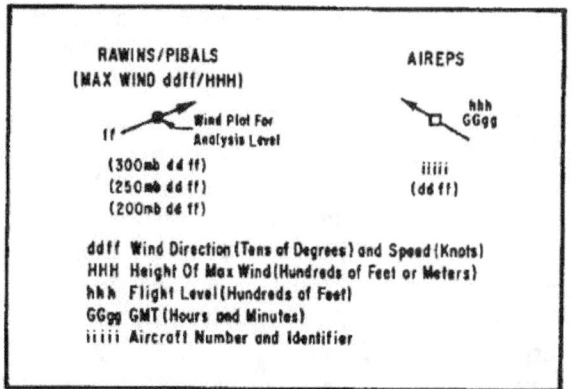

Figure 11-22. Recommended plotting models for composite upper-tropospheric wind charts. Barbs can be substituted for numerals to indicate wind speed.

11.3.9 Frequency and Scale of Analysis.
Compared to higher latitudes, most tropical weather systems move slowly and usually develop slowly. It often takes days for a tropical cyclone to grow from an initial disturbance, whereas extratropical cyclones can deepen explosively in 12 to 24 hours. In higher latitudes, 3- or 6-hourly analyses of surface pressure and frontal systems can benefit short-range local-area forecasts. In the tropics, however, surface pressure analyses intermediate between the times of upper-air soundings (0000 and 1200Z) are likely to be of little use and often not worth the effort. Even analyses at 6-hourly intervals may not be frequent enough to pick up squall-line development or intensity change. Therefore, analyses made at 12-hour intervals usually suffice for operational and forecasting purposes. It is better to make high-quality analyses twice a day than to make less accurate and less detailed analyses four times a day. In fact, Riehl (1966) suggested that in some parts of the tropics during the dry (or "good") weather season, once-daily analyses would satisfy local forecasting needs. Military operations probably require at least twice-daily analyses. The chosen times may depart from 0000 and 1200Z. Over Africa, more lower-tropospheric wind data is available at 0600Z than at 0000Z, while in India the favored hour is 0900Z.

The map scale for analysis should depend on many factors—the spacing of reporting stations in the densest part of the network, the total area covered by the analysis, practical chart-size limits, etc. For tropical analysis, a mercator projection is best, regardless of the scale. The scale should be the same for all levels and cross sections, to allow overlaying on light tables. A 1:10 million scale may be appropriate for regional analyses, while a 1:20 million or 1:40 million scale might be better for hemispheric or global analyses. It may not always be necessary to plot all surface observations. Instead, only stations that make upper-wind measurements and other stations selected to provide good coverage suffice. Because of their representative winds, all ship reports should be plotted. Map scales of 1:2.5 million (or even larger) may be suited to briefing or display or for plotting radar scans or reconnaissance reports.

11.3.10 Operational Procedures. Kinematic analysis of the wind field at various levels is most commonly practiced in the tropics. Some tropical stations, whose area of responsibility extends into mid-latitudes, combine sea-level pressure or pressure-height analyses in middle latitudes with kinematic analyses in the tropics. The two lines cannot be easily merged, especially in the surface layer. The area covered by each type of analysis may depend on season and synoptic situation. Stations whose primary interest is the tropics should produce high-quality kinematic analyses, while relying on facsimile charts from mid-latitude weather centers for pressure (pressure-height) and frontal analyses. Synoptic mid-latitude analyses are usually available by the time all the tropical data has been received and plotted. Despite projection differences, frontal positions and selected isobars (contours) can be quickly transferred to tropical mercator-projection charts. Satellite images can help verify frontal positions, especially their extensions into the tropics. Preferably, the analyst can use this information in applying frontal wind models (11.3.2.4) and so extend his kinematic analysis into middle latitudes. Once credible analyses are being received from one or more mid-latitude centers, a tropical station may greatly reduce the number of mid-latitude reports plotted on its charts.

The following charts should be displayed for the analyst: monthly climatological charts for all levels analyzed, and large-scale terrain maps showing station locations. By studying these, the analyst can become familiar with the local topographic effects that influence the representativeness of weather reports and often control short-period weather changes. The terrain contours on weather plotting charts are not sufficiently detailed. All analyses should be made on a light table with the previous charts underlain to ensure continuity. Too often, the analyst ignores continuity or climatology, ensuring poor work. Acetate sheets of proper size and marked at selected latitude-longitude intersections are essential to good analysis. There should be enough acetates for each set of isopleths—isobars/contours, streamlines, isotachs—at each analysis level.

346

11.3.10.1 Surface/Gradient-Level Charts. These charts contain the most information and demand more analytical effort than higher-level charts. Step-by-step:

(1) Underlay plotted chart with the last two 12-hourly analyses and overlay acetate for pressure analysis.

(2) Transfer sea-level isobars and fronts from centrally-prepared mid-latitude analysis. Rely heavily on satellite pictures and continuity to locate fronts.

(3) Use later conventional data, ship/aircraft reports, and satellite pictures in adjusting frontal positions.

(4) From weather satellite pictures, prepare a rough nephanalysis of the major cloud systems and transpose it on to the plotted map in a suitable color. Outline only those cloud systems that are hundreds of nautical miles wide and that appear to contain active convection. Apparent centers of vortical cloud systems should be especially noted. Abbreviated remarks can further describe the cloud systems; e.g., "isold," "sct," "bkn," or "many Cbs," "multi-lyrd clds," "jet-stream cirrus shield."

(5) Complete the pressure analysis of the chart for the tropics using a 2-mb interval in the tradewinds and a 1-mb interval near the equator, where gradients are small. Adjust the analysis to fit the plotted gradient-level winds, especially more than 5 to 10° from the equator.

(6) Trace the isobaric analysis (in light pencil) and the frontal analysis (in the appropriate color or symbols) from the acetate to the plotted map and place a clean acetate over the plotted chart for the streamline analysis. (This step is optional, as the isobaric/frontal analysis could be left on its acetate and traced later when the kinematic analysis is finished.)

(7) Make a surface streamline analysis by considering and evaluating the plotted data, the pressure field, continuity, climatology, and the latest satellite data. Apply frontal kinematic models.

(8) Place a clean acetate over the streamline analysis and analyze the surface wind speeds. Isotachs should be drawn for 5- or 10-knot intervals, except near tropical cyclones, where greater intervals are appropriate. Considerable smoothing may be needed to ensure a reasonable-looking analysis.

(9) Plot data received since starting the analysis and adjust the analyses as necessary.

(10) Trace the analyzed fields onto the plotted chart. The isobars should be drawn lightly with a lead pencil, allowing the pressure patterns to be seen but not obtrude. Fronts should be traced in the appropriate colors or by a lead pencil with the appropriate symbols. Streamlines should be traced with a purple pencil or another selected color, with enough arrowheads to identify flow direction at any point. Isotachs should be traced as dashed lead-pencil lines, and be labeled for wind speeds. It is recommended that isotach minima be shaded light green and isotach maxima shaded light yellow. Cyclonic centers should be labeled "C" and anticyclonic centers "A." The appropriate symbol locates tropical cyclone centers. Beneath, the maximum surface wind speed, the storm name, and the advisory or warning number should be entered.

(11) If the chart is to be used for briefing, the appropriate colored symbols should be entered at stations reporting rain or lowered visibility.

11.3.10.2 Upper-Level Charts. Analyzing upper-air charts is more straightforward, since what is needed is an accurate kinematic analysis rather than the kinematic, pressure, and weather analyses performed on the surface/gradient-level charts. Step-by-step:

(1) Underlay the plotted chart with the last two 12-hourly analyses and overlay an acetate sheet for streamline analysis.

(2) If available, transpose selected height contours from a centrally-prepared mid-latitude analysis. In the mid-latitudes and subtropics, where westerlies

generally prevail, height contours usually parallel the flow and can be used as streamlines. The west-east contour that borders the zonal westerlies on their equatorward side should first be drawn in, and the contours to poleward should then be added. Height contours from the 300- or 200-mb analyses can be applied to a 250-mb streamline analysis. Lack of a centrally-prepared pressure-height analysis requires that a streamline analysis be made of the subtropical and middle latitude winds.

(3) Equatorward of the height contours transposed to the acetate in the previous step, consider and evaluate the plotted data, continuity, climatology, and satellite data in making a streamline analysis.

(4) Place a clean acetate sheet over the streamline analysis and analyze the wind speeds. For an upper-tropospheric chart, isotachs should be drawn at 10-knot intervals up to 20 knots and beyond that at 20-knot intervals.

(5) Trace analyzed fields on the plotted chart with a soft lead pencil; streamlines in solid lines with enough arrowheads to define the flow direction at any point, and isotachs in dashed lines, labeled for speed. Isotach minima should be shaded in light green and isotach maxima in light yellow. Jet streams are located by heavy lines with arrowheads drawn parallel to the streamlines along the axes of isotach maxima greater than 60 knots. Cyclonic and anticyclonic centers are labeled "C" and "A," respectively. Tropical cyclone centers should be shown in the same way as on the surface/gradient level chart.

For the most part, Automated Weather Distribution Systems (AWDS) have replaced manual analysis with machine analysis (11.4.4), but analysts trained to use the procedures described above are still needed to reanalyze machine products, incorporate data that had missed the collection cut-off, or analyze from scratch whenever the AWDS communications or processing components break down.

11.4 AUTOMATED ANALYSIS TECHNIQUES

11.4.1 General. Global weather prediction and climate prediction have begun to emphasize surface heat-exchange in the tropics. Since the most important component of this is the surface wind, numerical analysis of tropical winds has become more than peripherally significant to global analysis centers.

Weather centrals are undertaking two types of automated kinematic analysis. They roughly correspond to direct subjective analysis of the wind field as carried out in the tropics and to subjective extrapolation of pressure analysis from higher latitudes into the tropics.

11.4.2 Objective Wind Analysis. As an example, since September 1983, the Australian Bureau of Meteorology has analyzed tropospheric winds between 40° N and 40° S and 70° E and 180° E (Davidson and McAveney, 1981). The analysis uses an optimum interpolation scheme (Bedient and Vederman, 1964) in which the mean square interpolation error is minimized. The analysis sequence starts with climatology as a first-guess field. The zonal and meridional components of the wind are first analyzed and then combined (Figure 11-23). The objective analyses agreed well with subjective analyses made during Winter MONEX. Compared to rawin measurements, the near-surface chart had an RMS error of 6.8 knots and the 200-mb chart an RMS error of 8.9 knots. Divergence fields calculated from the winds matched well with cloudy and clear areas in satellite pictures. Provided preliminary error-checking equals that of a human analyst, objective wind analyses are probably as good as subjective analyses.

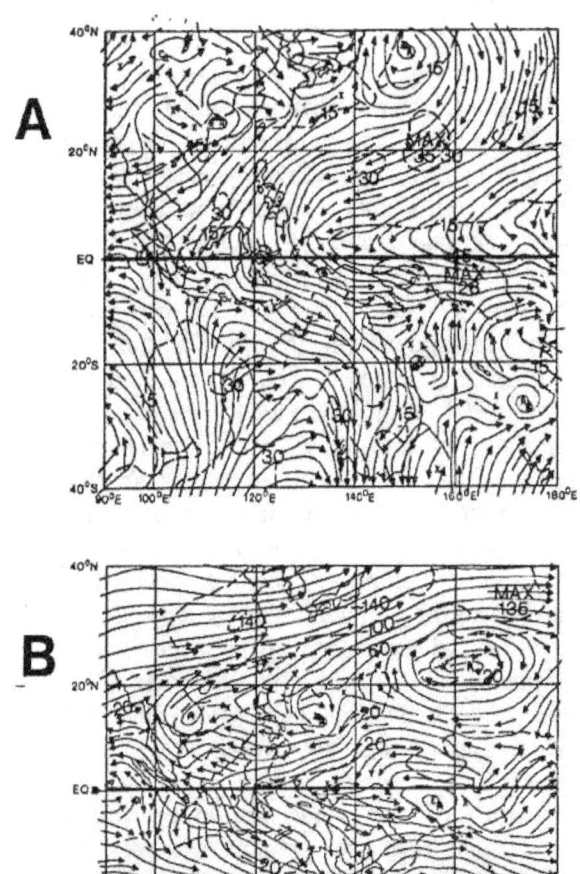

Figure 11-23. Objective kinematic analyses for 0000Z 2 December 1978. Wind speed in knots (A) at 950 mb and (B) at 200 mb (Davidson and McAvaney, 1981).

11.4.3. Numerical Weather Prediction Analysis. According to Trenberth and Olson (1988), at both NMC and ECMWF,

"...analyses are produced using a four-dimensional data assimilation system that uses a set of first guess fields as the base for integrating observations to produce the analysis. The first guess analysis at both centers is either a 6- or 12-hour forecast from a Numerical Weather Prediction (NWP) model using a previous analysis as its initial condition. The first guess carries forward information from the previous analyses. It is dependent upon the veracity of the NWP model and it can be biased. The subsequent procedure itself is based upon a scheme that makes use of the statistical mean errors expected in both the first guess and in the observations. The analysis is initialized using nonlinear normal mode initialization, a procedure designed to ensure that the resulting fields are dynamically consistent with each other while appropriately emphasizing relatively slow-moving, meteorologically significant components and damping spurious gravity waves."

Although the centers use similar procedures, Trenberth and Olson found that, in the tropics, RMS differences in the east-west and north-south components of the analyzed winds often exceeded 10 knots above 500 mb; the computed divergent wind fields, and hence associated vertical motions, significantly disagreed (see also Lambert, 1989). A new spectral statistical interpolation analysis introduced at NMC on 25 June 1991 eliminates the need for initialization, produces more balanced analysis increments, and readily assimilates new, non-conventional observations.

The WMO Working Group on Numerical Experimentation (1987) is continuing to compare operational tropical (40° N to 40° S) analyses made at various numerical weather centers. For May 1987, between 90° E and the dateline, the ECMWF and the Australian Bureau of Meteorology showed RMS differences of about 8 knots at 850 mb and 14 to 16 knots at 200 mb; these were about the same as differences between the analyses and points at which observations were made. Studies underway will include analyses from the United Kingdom Meteorological Office in the comparisons.

Errors of this size between analysis and observations (on which the analyses were based) mean that tropical forecast offices cannot depend on centralized numerical analyses. They must continue to make their own careful kinematic analyses, preferring objective wind analyses as back-up.

11.4.4 Computer-Aided Analysis. The Automated Weather Distribution System (AWDS) now being installed in AWS forecast offices obtains grid-point data from GWC and analyses them on a screen. New data can be added, the field reanalyzed by the operator, and a hard copy made. Zooming allows the scale of the screen map to be changed. Various fields can be overlaid, including satellite images.

Auxiliary charts, diagrams, and other aids support tropical analysis and diagnosis. Their types and numbers differ among operational weather units, which should periodically check to see if the ones they use are still needed.

Chapter 11

11.5 AUXILIARY AIDS

11.5.1 Time Cross Sections. These have been used in both operations and research. They show the distribution of radiosonde and/or wind measurements above a station over a period of time. In the Johnston Island (17° N, 170° W) vertical cross section shown in Figure 11-24 (next page), wind directions are indicated by arrows (north toward the top of the page) and speeds by numerals. The 6-hourly surface synoptic observations are also plotted. In the tropics, disturbances usually move from east to west, that is, from right to left on a chart. In the time sections, time increases from left to right; prior conditions are shown to the left of subsequent conditions. Radiosonde data may also be plotted on time sections. Dew points, or the depressions of dew points below temperatures, including those at significant levels, help monitor changes in moisture distribution, especially within the near-surface moist layer. Time sections help maintain continuity in the plotted elements during the periods between synoptic charts and in the layers between analysis levels. Time sections make identifying wrong or non-representative reports easier. For example, the circled wind report in Figure 11-24 suggests an abrupt change with height from easterly to southerly winds and back to easterly near 5,000 feet (1.5 km). Transmission is probably to blame; the actual wind direction is likely to be 70° rather than 170°. Had this wind been plotted as 170° and so analyzed on a synoptic chart, a serious error would have resulted. Time sections indicate the height and thickness of transition layers between easterlies and westerlies. Synoptic kinematic analysis in such transition layers is often hard because winds are light and variable.

Time sections should be maintained for the home station, for other stations for which forecasts are issued, and for nearby upstream stations. Since hindsight is an exact science, time-section post-analysis often helps research. In operations, however, time sections are analyzed stepwise as data becomes available. In general, extrapolating trends doesn't help forecasting much. Possible types of analysis follow:

11.5.1.1 Wind Direction. Wind direction can be analyzed in two ways: by drawing isogons (see Figure

8-52) or, more commonly, by drawing pseudo-streamlines parallel to the wind directions as plotted on the section (see Figure 8-22). On the section, singular points are found at the intersections of the sloping axes of moving singularities in the horizontal wind field with the plane of the time section. Associating singular points on the horizontal and vertical charts helps the meteorologist to better understand the three-dimensional structures of the parent circulations.

11.5.1.2 Wind Speed. At some stations, cloud and weather changes are associated more with speed changes than with direction changes. Therefore, isotach analysis on time sections may help reveal this weather-modifying mechanism. As on horizontal wind charts, isotachs are drawn at 5, 10, or 20 knot intervals. Some smoothing is necessary. Isotach analysis on sections helps detect wind-speed maxima and minima, and over time determine the usual levels at which the extremes occur.

11.5.1.3 Moisture. Moisture can be analyzed in several ways: dew points, dew-point spread, or departures of dew point from normal. In dew-point analysis, a 4 or 5° C interval may be sufficient; for dew-point spread or departure from normal, a 2° C interval should be used. The boundary of the moist layer can also be drawn. LaSeur (1960) recommended using the height of the 5 g kg^{-1} moisture surface during the rainy season and the 3 g kg^{-1} surface during the dry season to indicate the depth of the moist layer.

11.5.1.4. Temperature/Height. Isotherms or contours (or their departures from normal) could be analyzed on time sections, but for many tropical stations they would provide little useful information. The 24-hour pressure-height changes are useful. When they are plotted in the middle of each 24-hour period for which the difference is determined, they are controlled by the intensity and movement of circulation systems. For example, a pressure-height rise before a trough passes indicates weakening; equal and opposite pressure-height changes ahead of and behind a trough indicate no movement or intensity changes; the zero isopleth should then coincide with the wind shift line.

351

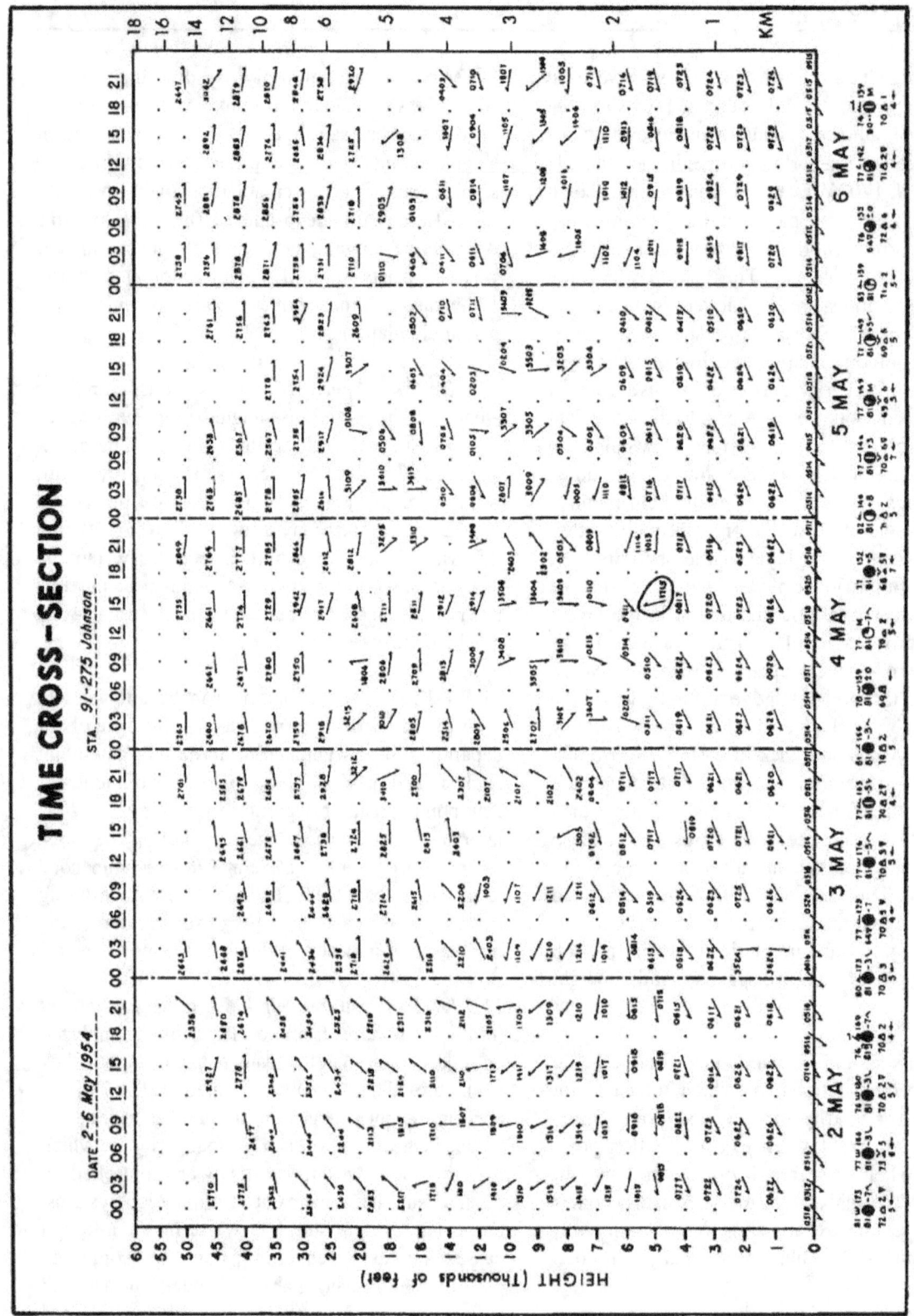

Figure 11-24. Vertical time cross-section of winds (knots) above Johnston Island (17°N, 170° W) on 2-6 May 1954. Circled report is probably wrong (Dept of Weather Training, Chanute AFB, IL)

11.5.2 Space Cross-Sections. These give a synoptic picture in the vertical of the distribution of various elements at a number of stations along a particular route, spaced along the abscissa according to distance. Analyses resemble those for time sections; time-section forms can usually be modified for use as space sections. In some cases, daily space cross-sections may be prepared for a route instead of a time cross section for each station along the route. Space cross-sections may also be used in route briefing; time cross-sections could not.

The Automated Weather Distribution System (AWDS) can plot, analyze and store the scalar variables depicted on time and space cross-sections. New soundings can be added and old soundings removed, the analyses kept up-to-date, and hard copies made.

11.5.3 Thermodynamic Diagrams. These are less useful to tropical analysis than they are to frontal, air-mass, cloud, and stability analyses in higher latitudes. Nevertheless, they are of some help in tracking the vertical distribution and changes of atmospheric moisture.

For AF weather units, the standard thermodynamic diagram is the Skew T Log P chart which is printed in several different versions. Stability indices may not help tropical forecasts, and they should be computed only where and when local forecast studies have confirmed their value. Generally, Skew T diagrams are more useful over tropical continents than over tropical oceans. This is because over land the diurnal convective cycle and day-to-day changes in low-level moisture are larger. Even then, weather systems control lapse rates. During the wet season, the air is more stable in unsettled rainy weather and less stable in fair weather with isolated thunderstorms (see 6.3.1). Thus, a Skew T diagram contains very little predictive information.

Vertical dew-point profiles are useful. Daily values should be compared to the mean dew-point profile for the same calendar month, plotted on the same Skew T diagram. Alternatively, plot the monthly mean dew-point profiles on acetates which can be overlain on the daily profiles for comparison. Mean monthly dew points for standard pressure levels are available in monthly climatic data summaries (NOAA/ NESDIS, 1948) or can be obtained by AF weather units from USAFETAC.

11.5.4 Checkerboard Diagrams. Checkerboard diagrams, on which hourly weather observations are plotted, have been used extensively by AF weather units. A sample is shown in Figure 11-25, next page. (This checkerboard has been truncated; in standard checkerboards, 24 boxes in each row allow a day of hourly observations to be plotted.) The same form can also be used to depict hourly or 3-hourly observations for many stations on a given day. Then the stations should be ordered geographically. For example, stations along an air route should be arranged sequentially from one end of the route to the other. This readily reveals the current weather distribution. Checkerboards also make computation of 24-hour changes easy. Various weather phenomena should be emphasized on checkerboards by color shading or symbols.

Checkerboards provide a statistical aid to analysis and forecasting. For example, during the rainy season, the diurnal cycle of convection can be deduced from the hourly observations made over the previous weeks. The relative importance of local and synoptic effects on winds and convection can be estimated by comparing hourly observations with corresponding synoptic analyses. Typical diurnal cycles of fog and stratus can also be established. For example, when fog occurred at Don Muang (14° N, 101° E) during January 1968 (Figure 11-25), it generally set in between 0500 and 0600L and broke up between 0800 and 1000L. Thus, when fog was expected during this period, it could have been confidently forecast to occur between 0600 and 0900L.

11.5.5 Pressure-Change Charts. Twenty-four-hour pressure-change (isallobaric) charts have been used by various tropical weather units for many years. The charts can be prepared by subtracting yesterday's pressure from today's at fixed reporting stations. Where stations are few, pressures can be interpolated from the isobaric analyses at 5° grid points and the interpolated values used to compute the 24-hour changes. In the analysis, negative isallobars are drawn in red, positive isallobars in blue, and zero isallobars in purple. One-mb spacing is appropriate.

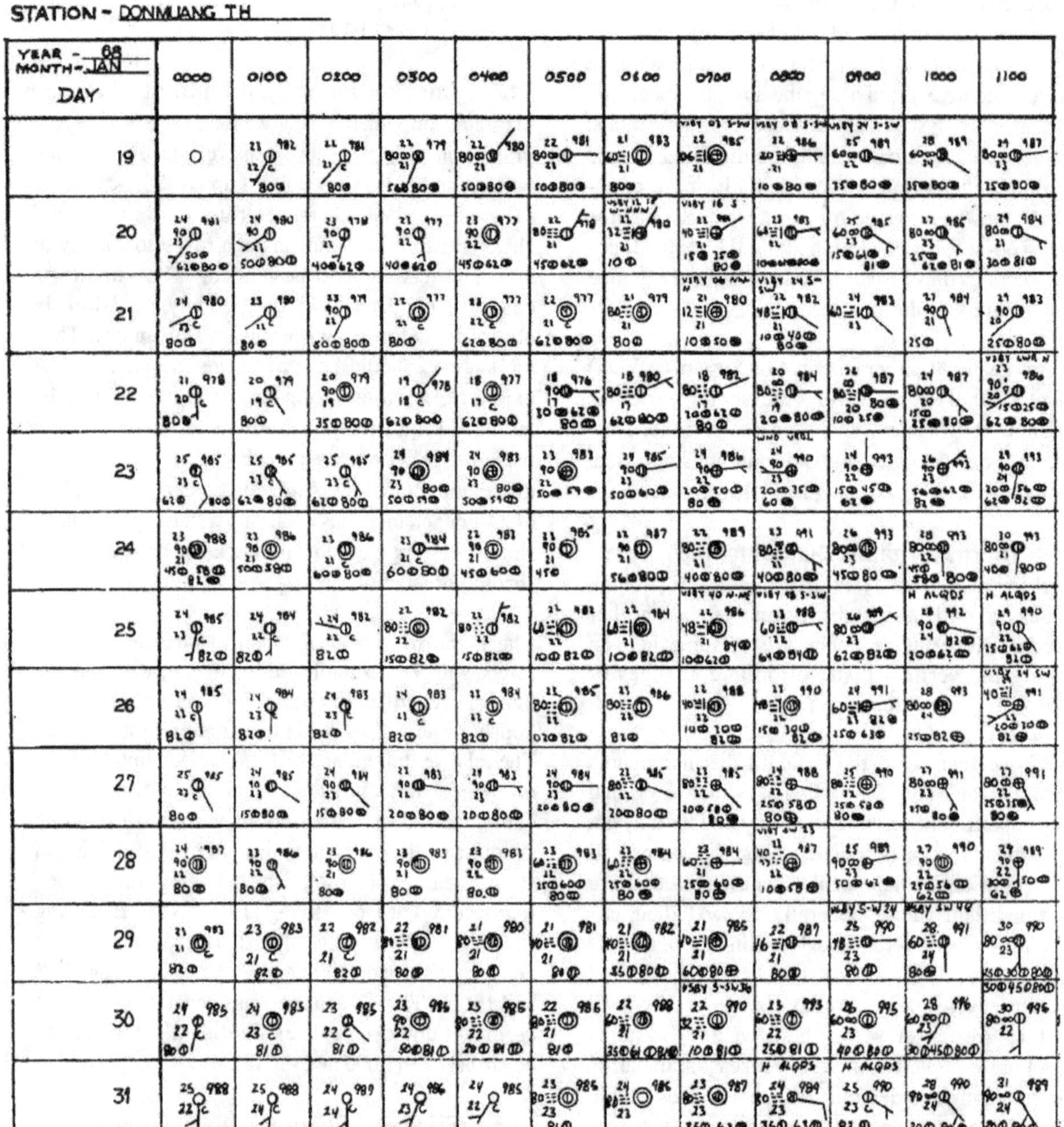

STATION - DONMUANG TH

Figure 11-25. Checkerboard plot of hourly surface observations at Don Muang (14°N, 101°E), Thailand, 19 to 31 January 1968 (Dept of Weather Training, Chanute AFB, IL).

According to LaSeur (1960), 24-hour pressure tendencies have little synoptic value from the equator to about 10 degrees on either side, where the value increases. Dunn (1940) used 24-hour isallobaric charts in studying hurricane development over the North Atlantic. They have since been used elsewhere in the tropics to monitor potential tropical cyclone formation. In particular, areas with 24-hour pressure falls of 2 or 3 mb or more were closely watched. Nowadays, isallobaric charts are used less because developing cyclones can usually be detected in satellite pictures before they cause pressure falls in the sparse station network. If the very large area changes in surface pressure discussed in 4.2.2 are monitored on charts of 24-hour pressure tendencies, forecasters can distinguish between synoptic and regional causes for the changes; they may also be helpful in detecting surges.

Twenty-four-hour isallobaric patterns are much harder to relate to minor tropical weather disturbances. Observational errors, local convective and radiational effects on surface pressure, scarce data in many areas, and widespread synchronous pressure falls (see Figure 4-7) complicate the problem. Therefore, each unit must decide whether plotting 24-hour pressure change charts is worth the effort.

11.5.6 Wind-Shear Charts. Charts of the total tropospheric wind shear help in forecasting tropical cyclone development. As discussed in Chapter 9 and further elaborated by Gray (1978), small wind shear in the vertical is one of the necessary conditions for tropical cyclone development. Since the shear can change considerably from day-to-day, the analyzed shear field should be frequently examined to determine which of many minor tropical disturbances shown in satellite pictures are most likely to develop. In the past, scarce upper-air data, and the time required to plot the information, stopped routine preparation of wind-shear charts. Today, more data and automated objective wind analysis programs, as well as office personal computers, make wind-shear charts easy to prepare. For example, the National Hurricane Center prepares integrated wind velocity charts for the lower- (1,000 to 600 mb) and upper- (600 to 200 mb) tropospheric layers and a chart of wind shear between the two layers (Simpson, 1970a). Stations making kinematic analyses of the lower and upper troposphere on the same map scale can superimpose them to estimate areas of light vertical wind shear, say less than 10-20 knots.

11.5.7 Rainfall Analyses. Separate analyses of 6- or 24-hour rainfall may help determine the presence, intensity, and movement of tropical disturbances. The more observations, the better the charts. As mentioned in 6.3.1.2, the mesoscale nature of rainfall ensures large point-to-point differences in amounts. Therefore, only generalized analyses should be based on rainfall reports from synoptic stations; e.g., outlining areas receiving measurable rain. The maximum amounts measured may give a rough idea of how intense a disturbance is. Whenever available, satellite data and weather radar reports should be integrated into the rainfall analysis. For local-area rainfall analysis, radar is the best tool (see 12.2.6).

Rainfall can be analyzed on acetates overlain on surface synoptic charts, or the rainfall reports can be plotted on separate maps and then analyzed. AWDS analyses are also possible. Stations that did not include rainfall in their reports should be indicated. In India, rainfall accompanying monsoon depressions is plotted and carefully analyzed. Rainfall charts can contribute to forecast studies when combined with other data. For example, they can help determine whether disturbances move into an area or develop and dissipate without moving much.

Chapter 12

TROPICAL FORECASTING

12.1 GENERAL

Palmer (1951a) surveyed the history of the various approaches to tropical analysis and forecasting. He described three methods (or "schools of thought") which have been applied to the tropics: the *climatological* method, the *air-mass* method, and the *perturbation* method.

• The climatological method was based on the conviction that day-to-day tropical weather differs little from the monthly and annual means. Therefore, the best guide to forecasting would be a detailed knowledge of the mean values of the elements being forecast.

• In the air-mass method, the techniques of air-mass and frontal analysis developed in higher latitudes were applied to the tropics. The "equatorial front" was thought to resemble the polar front; tropical storms resulted when waves on it occluded.

• The perturbation method evolved from tropical meteorology research conducted during and after World War II at the Institute of Tropical Meteorology in Puerto Rico and the University of Chicago. In this concept, the basic currents in the tropics were considered to be generally zonal, but subject to wave-like and sometimes unstable perturbations accompanied by characteristic weather and pressure patterns that could be identified on synoptic charts.

Subsequently, tropical meteorology research has tended to confirm the basic concepts of the perturbation school. More synoptic observations and weather satellite data have shown the tropics to be dominated by a variety of circulation and weather systems (or perturbations) and that significant day-to-day weather changes are common, especially during the rainy seasons. Our ability to forecast these changes, however, has made painstakingly slow progress due to our incomplete knowledge of tropical dynamics and still inadequate data. Riehl's opinions

on tropical weather prediction made in a 1966 planning report to the World Weather Watch are still relevant:

"No generally valid methods to predict synoptic disturbances on the 24- to 48-hour time scale, based on analogue, physical, or statistical techniques, have been perfected. The variety of synoptic-scale disturbances in different parts of the tropics is great. Because of severe deficiencies in networks it has not yet been possible to obtain their complete description and to analyze their mechanics and energetics. Models of certain simple types of disturbances do exist, but their practical use is restricted by the limited number of days when the models apply. Hence, no powerful dynamic tools comparable to those outside the tropics are available for prediction."

Riehl went on to say that the primary tools available to tropical meteorologists are climatological and kinematic. The kinematic method relies heavily on extrapolation, occasionally tempered by qualitative consideration of interactions between tropical and mid-latitude circulations, and between lower- and upper-tropospheric circulations within the tropics. The kinematic techniques are applied mainly to wind-circulation features to predict tropical weather patterns. Their application to upper-air temperature and height fields has not helped prediction.

Alaka assessed tropical prediction in a 1964 WMO Technical Note:

"Short-range weather forecasting consists essentially of predicting the development and movement of weather-producing disturbances in the tropics. Before the future state of the atmosphere can be accurately predicted, its present state must be known. Therefore, the problems of forecasting in the tropics stem largely from inadequate analysis and deficiencies in our knowledge of the structure and properties of tropical disturbances. Improvement in forecasting is contingent on the removal of those deficiencies."

Conditionally unstable most of the time, the tropical troposphere is often in a delicate balance. Small changes in stability, even those too small to measure, can result in large weather changes.

Prior to weather satellites, tropical weather forecasts seldom bettered climatology or persistence. Most practitioners believed that the usually unexpected weather changes resulted from movement of previously unobserved weather. They also believed that a sequence of satellite pictures would enable them to identify and track disturbances. Then, by simply extrapolating the movement, they could greatly improve weather forecasts. This did not happen. Although satellite pictures filled observation gaps, they also showed that in the tropics, and especially between 10° N and 10° S, disturbances not only moved erratically, but changed intensity rapidly.

By 1992, many of the data shortages bemoaned by Alaka have been removed, but even with the most investigated of all tropical phenomena, the tropical cyclone, progress has been slow. As Figure 9-24 showed, the only improvement in forecasting tropical cyclone movement over the past 20 years has probably arisen from improvements in mid-latitude forecasts. More data, then, has demonstrated the complexity of tropical weather forecasting.

The introduction of global dynamic models has improved the prospect of a useful prediction model, but their performance in the tropics has barely bettered persistence, and until now (Section 12.3) their forecasts have never been compared to an optimal combination of persistence and climatology. In recent years, conflicts in southeast Asia, southwest Asia,

Africa, and Central America and field programs such as IIOE, BOMEX, ATEX, GATE, MONEY and TAME have stimulated research in tropical meteorology.

To develop forecasting techniques for southeast Asia, the Air Force Cambridge Research Laboratories (AFCRL) (now Phillips Laboratory) and the Naval Weather Research Facility, or NWRF (now the Naval Oceanographic and Atmospheric Research Laboratory, or NOARL) set up concentrated applied research programs, both in-house and at the University of Hawaii. As a consequence, daily forecasts were improved, largely because all available types of data (conventional, satellite, radar, aircraft reports, and climatology) were integrated into the research. The same approach might lead to better forecasts in other parts of the tropics.

In the past, most operational effort in tropical meteorology was devoted to *analysis*. Forecasting was given little emphasis, but nowadays, automated data-handling programs are freeing time for forecasting.

Forecasting suggestions made in earlier chapters will not be repeated here. This chapter focuses more on the *philosophy* of tropical forecasting than on proven forecasting techniques. The techniques that are included illustrate the variety of approaches possible. Experience confirms that without satellite data, accurate tropical analysis and forecasting would be almost impossible. Likewise, very-short-range local area and terminal forecasts depend on satellite pictures. We have also learned that simple-minded extrapolation from a series of satellite pictures (especially when intensity changes cannot be foreseen) can make forecasts worse than they would have been without the pictures. This chapter assumes that satellite and radar data are available at first-order tropical weather units. Although short-range forecast techniques are emphasized, the chapter includes sections on medium-and long-range forecasting and numerical weather prediction.

12.2 SHORT-RANGE FORECASTING TECHNIQUES

12.2.1 General. For the purpose of this report, short-range forecasts are defined as those made for periods or valid times up to 36 hours in advance. Within this period, forecasts for 0 to 12 hours ahead have been termed "nowcasts" (Browning, 1981). This subdivision may have some value in higher latitudes, but it is largely irrelevant in the tropics. Short-range forecasts based on careful analysis and extrapolation of existing circulation and weather patterns modified for diurnal and topographical effects, show a little skill. The shorter the forecast period and the stronger the diurnal variation, the better the forecast. A number of short-range forecast tools are discussed: circulation and cloud prognostic charts, climatological aids, local forecast studies, stability indices, and radar. The most successful tropical forecasters use a systematic approach and intelligently integrate data from many sources into their routines.

12.2.2 Circulation and Cloud Prognostic Charts. The starting point for short-range weather forecasts, especially those in the 12- to 36-hour time period, should be circulation and cloud prognostic charts based on recent synoptic analyses and satellite pictures. These charts should be on the same map projections and scales as the analyses.

12.2.2.1 Circulation Prognostic Charts. In most tropical areas, the basic chart should be for the surface/gradient level. The forecast positions of mid-latitude circulation and cloud features are generally available in transmissions from higher latitude weather centrals. The tropical parts of the charts feature extrapolation of systems shown in the most recent kinematic analyses. Just as isotach fields are harder to analyze than streamlines, predicting them is also hard. Winds increase suddenly as surges develop. This means that forecasting the movement and evolution of weather systems associated with features in the wind-speed field (e.g., regions of speed convergence or strong horizontal wind shear) will be more difficult than forecasting those weather systems associated with changes in wind direction (e.g., troughs or cyclones). However, the attempt should be made, because divergence and convergence (and hence vertical motion and weather) are more commonly determined by speed gradients than by streamline patterns.

12.2.2.2 Cloud Prognostic Charts. These should be based on climatology and recent nephanalyses prepared from satellite and other sources. Chapter 11 recommended that nephanalyses outlining convective regions be entered directly on the gradient/surface analyses. In this way, successive charts can be used in extrapolating moving circulation and cloud systems. Video sequences of geostationary satellite images depict movement and change in weather systems even more graphically.

Forecasts should be made of changes in orographically influenced cloud systems as they move or as the circulation changes. Climatology, including diurnal variation of convection (especially over land) and relationships between circulation and cloud patterns (e.g., cloudiness maxima equatorward of the monsoon trough and along asymptotes of convergence) can be incorporated into the cloud prognostic charts. However, the forecaster should never forget that even along the most persistent tropical system (the near-equatorial tradewind convergence), cloudiness often fails to persist (Figure 8-55). Satellite studies by Simpson et al. (1968) revealed cloud clusters moving westward across the North Atlantic tradewinds during the tropical storm season. Regardless of the circulation features associated with the clusters, direct extrapolation of the clusters themselves may help in forecasting changes in cloudiness and rainfall along their tracks. Remember, though, that cloud clusters often appear or disappear without any detectable reason because lack of observations prevent the accompanying circulation changes from being monitored, and that in the statistical study of cloud clusters by Williams and Gray (1973), 27 percent of the clusters disappeared from one day to the next (see 6.2.5). In using cloud prognostic charts there is a tendency to over-forecast the extent of active convection, but this may sometimes be desirable to compensate for errors in forecasting movement and development of the cloud system.

12.2.3 Climatological Aids.

12.2.3.1 General. A thorough knowledge of climatology has been stressed throughout this manual. Although tropical meteorologists agree, many tropical stations lack climatological information in the formats best suited to analysis and forecasting. The importance of displaying monthly climatological resultant-wind maps for all levels analyzed has already been emphasized. At many stations, excellent climatological displays help in forecasting and acquaint aircrews and other users with the local climatology. Formats are many; a sample, based on data for Saigon/Ho Chi Minh City is shown in Figure 12-1.

12.2.3.2 Conditional Climatology Summaries. These summaries give the probabilities of various categories of some meteorological elements occurring at a range of future times based on the values at the time the forecast is made (the initial time). For example, having Category X_i at the initial time, a summary gives the probabilities of Categories X_1, X_2, X_3, ... being observed Y hours later. Where marginal weather is uncommon, conditional climatologies (CCs) do not provide much more information than climatology.

Although most commonly applied to ceiling and visibility, conditional climatologies can be constructed for any categorizable meteorological element, including rain occurrences or amounts, cloudiness, and thunderstorms. Where convective rainfall dominates, CCs of ceiling heights help forecasting to some extent, but CCs of visibility are of limited value. Samples of CC ceiling and visibility summaries for a station with a low cloud and fog regime (Danang, 16° N, 108° E) for December, and a convective regime (Saigon, 11° N, 107° E) for June are shown in Figure 12-2.

Wind-stratified CCs have been widely used by AF weather units in the extratropics and have been prepared for many tropical stations. Where wind direction and various levels of marginal weather are highly correlated, wind-stratified CCs can aid forecasting significantly. For example, at Clark Air Base (15° N, 121° E) when low-level winds shift to the southwest, weather is likely to deteriorate. The wind-stratified CCs were designed to account for synoptic and local effects. In many tropical areas, large day-to-day weather changes are often unaccompanied by changes in wind direction. Also, winds over tropical continents tend to be lighter and more variable than in higher latitudes, while wind directions can be changed by nearby convection. These reasons limit the usefulness of wind-stratified CCs in the tropics.

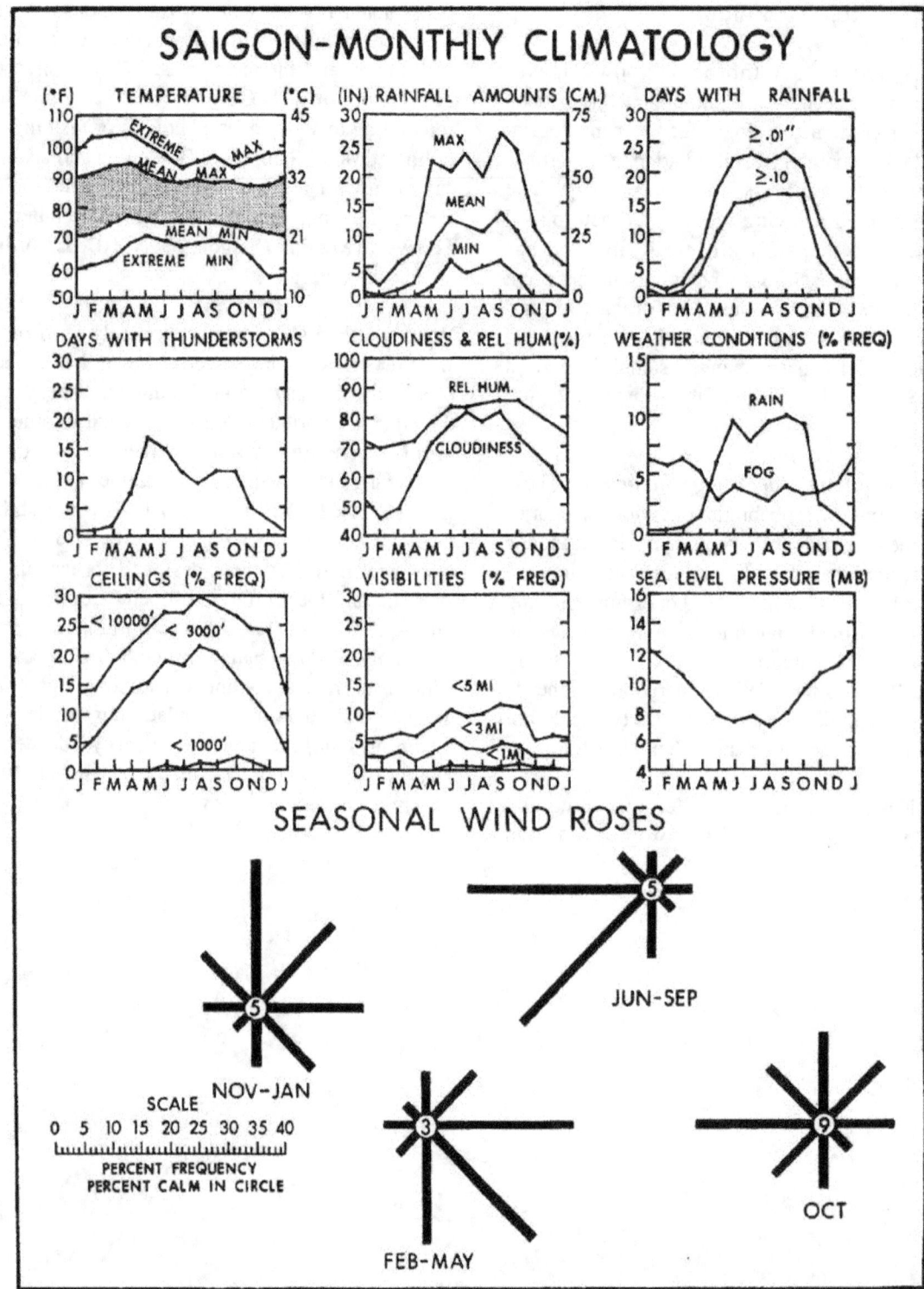

Figure 12-1. Monthly climatologies of weather variables at Saigon/Ho Chi Minh City (11° N, 107° E).

STATION 48355 DA NANG VIETNAM APT   FOR 1957-65   DECEMBER   HOUR 07 LST

CEILING

CEILING CATEGORY INITIAL	LATER	1	2	3	4	5	6	12	24
A		Not observed							
B		Not observed							
C	A				40			20	
	B								
	C	80	60	40			20		
	D	20			20				
	E	20	40	40	40	80	80	60	100
	F			20		20		20	
	CASES	5	5	5	5	5	5	5	
D	A								
	B	10	9	9					
	C				18	10	11		
	D	50	55	36	27	20	22		
	E	40	18	36	36	70	67	89	82
	F		18	18	18			11	18
	CASES	10	11	11	11	10	9	9	11
E	A								
	B								
	C			2	1	2	3	3	2
	D	2	2	4	4	5	2	2	9
	E	83	74	64	62	55	55	59	59
	F	16	23	30	33	38	40	37	30
	CASES	128	124	116	126	121	114	120	115
F	A								
	B								
	C		3	2					3
	D	2						3	1
	E	15	14	20	17	26	14	33	51
	F	83	84	79	83	74	86	64	45
	CASES	81	74	66	69	70	64	75	71

INITIAL CATEGORY	ALL.	A	B	C	D	E	F
PERCENTAGE	100.0			2.1	4.6	57.0	36.3
total obs	237			5	11	135	86

VISIBILITY

VISIBILITY CATEGORY INITIAL	LATER	1	2	3	4	5	6	12	24
J		Not observed							
K		Not observed							
L	J								
	K							13	
	L	43	43	29	29	..	13		25
	M			14	14			13	
	N	14				13	13		
	O	43	57	57	57	63	63	88	75
	CASES	7	7	7	7	8	8	8	8
M	J								
	K								
	L	17	17			17	17		
	M	50	17	17					
	N	17	33	33	33	33	50	17	
	O	17	33	50	50	50	50	83	100
	CASES	6	6	6	6	6	6	6	6
N	J	6							
	K								
	L	6	19	19	13	6	6		
	M	13	13	13	6			7	
	N	56	19	19	6	6	6	7	
	O	19	50	69	81	88	87	100	
	CASES	16	16	16	16	16	16	15	15
O	J								
	K								
	L	1		.1	0	2	2	3	3
	M	1	1		1	1		3	3
	N	3	5	4	4	3	4	4	7
	O	96	93	94	95	94	94	92	87
	CASES	229	227	215	228	229	223	226	217

	ALL	J	K	L	M	N	O
	100.0			3.1	2.3	6.1	88.5
	261			8	6	16	231

STATION 48900 SAIGON VIETNAM/TAN SON NHUT APT   FOR 1957-65   JUNE   HOUR 13 LST

CEILING

CEILING CATEGORY INITIAL	LATER	1	2	3	4	5	6	12	24
A		Not observed							
B		Not observed							
C	A								
	B								
	C								
	D								
	E								
	F	100	100	100	100	100	100		
	CASES	1	1	1	1	1	1		
D	A								
	B								
	C	50				50	50		
	D	50	50	50					
	E					50	50		
	F		50	50	100	50		50	100
	CASES	2	2	2	2	2	2	2	1
E	A								
	B								
	C								
	D	2	2	7	2		4		2
	E	69	48	13	6	7	7	10	46
	F	29	50	80	92	93	89	90	48
	CASES	55	44	45	51	57	57	59	48
F	A								
	B				1				
	C		3						
	D			1	5	2	2		2
	E	11	8	1	2	8	6	11	34
	F	89	90	98	93	88	90	89	64
	CASES	82	79	80	82	84	82	80	61

INITIAL CATEGORY	ALL	A	B	C	D	E	F
PERCENTAGE	100.0			.6	1.2	42.5	55.7
TOTAL OBS	167			1	2	71	93

VISIBILITY

VISIBILITY CATEGORY INITIAL	LATER	1	2	3	4	5	6	12	24
J		Not observed							
K		Not observed							
L	J								
	K								
	L								
	M	25							
	N	25							
	O	50	100	100	100	100	100	100	100
	CASES	4	4	4	4	4	4	4	3
M	J								
	K								
	L								
	M								
	N	25					25		25
	O	75	100	100	100	100	75	100	75
	CASES	4	4	4	4	4	4	4	4
N	J								
	K								
	L	8		8	8	8	8		
	M		8	8	8				
	N	38	8	8	8		15		
	O	54	77	77	75	92	85	100	100
	CASES	13	13	13	12	13	13	13	13
O	J			0	2	1	0		0
	K	0	0	1	1	0	1		
	L	1	3	3	3	3	4		2
	M	0	2	3	2	1	2		
	N	3	3	4	6	7	9	3	5
	O	95	90	87	88	88	85	97	91
	CASES	229	230	231	231	229	227	225	224

	ALL	J	K	L	M	N	O
	100.0			1.6	1.6	5.1	91.4
	255			4	4	13	234

CATEGORY	CEILING CRITERIA FEET	METERS	CATEGORY	VISIBILITY CRITERIA MILES	KILOMETER
A	< 200	< 60	J	< 1/2	< 0.8
B	≥ 200 but < 500	≥ 60 but < 150	K	≥ 1/2 but < 1	≥ 0.8 but < 1.6
C	≥ 500 but < 1000	≥ 150 but < 300	L	≥ 1 but < 2	≥ 1.6 but < 3.2
D	≥ 1000 but < 1500	≥ 300 but < 450	M	≥ 2 but < 3	≥ 3.2 but < 4.8
E	≥ 1500 but < 5000	≥ 450 but < 1500	N	≥ 3 but < 5	≥ 4.8 but < 8
F	≥ 5000 or none	≥ 1500 or none	O	≥ 5	≥ 8

Figure 12-2. Conditional climatology summaries for Saigon (11° N, 107° E) and Danang (16° N, 108° E).

12.2.3.3 Diurnal Pressure and Temperature Data. Since, in the tropics, diurnal variations of temperature and pressure are normally much larger than interdiurnal changes, their mean hourly values for month or season can help forecasting. Charts (such as those shown in Figures 7-4 and 7-5) or tables can be readily prepared. In many cases, mean seasonal charts (e.g., for wet and dry seasons) can be useful. Short-range forecasts of altimeter settings can be based on these tables, using the following formula:

Forecast altimeter setting **equals** *current altimeter setting*
plus
departure from the mean at forecast hour
minus
departure from the mean at current hour

Of course, adjustments should be made whenever well-defined pressure systems are forecast to affect the station. Diurnal temperature variation can be treated similarly (Figure 7-3). Since cloudiness and rain influence the magnitude of the diurnal cycles of temperature and pressure, high accuracy would require separate diurnal temperature tables for rainy or cloudy days, and for fair days. When the pressure tables are combined with tables of diurnal temperature variation, forecasts of density altitude can be prepared. Guidance is given in AWS/TR-79/005.

12.2.3.4 Variation of Weather. Monthly tables or graphs showing percent frequency by hour of various weather elements should be prepared for the local and representative stations of interest (see Figures 7-31 and 7-33). These presentations may include selected ceiling and visibility categories, rainfall, thunderstorms, fog, wind velocity, and other categories. Knowledge of this climatology keeps forecasts within reasonable limits and indicates most likely occurrence times. A sample climatological display for Saigon/Ho Chi Minh City is shown in Figure 12-3, below. Note the double maximum in the frequency of visibility values below certain limits. Comparing the visibility, rainfall, and fog frequency curves reveals that fog usually reduces morning visibility, while showers or thunderstorms reduce visibility in the afternoon and evening. If fog is forecast, it is most likely between 0400 and 0900L; visibility quickly improves after 0800L. Ho Chi Minh City's ceiling-frequency curves show that ceilings below 1,000 feet (300 meters) are most likely with early morning fog, while ceilings from 1,000 to 3000 feet (300 to 900 meters) are common around midday due to convective clouds. Note that although ceilings between 3,000 and 10,000 feet (900 and 3,000 meters) are rare during the day, they occur occasionally at night, probably after showers and thunderstorms have dissipated.

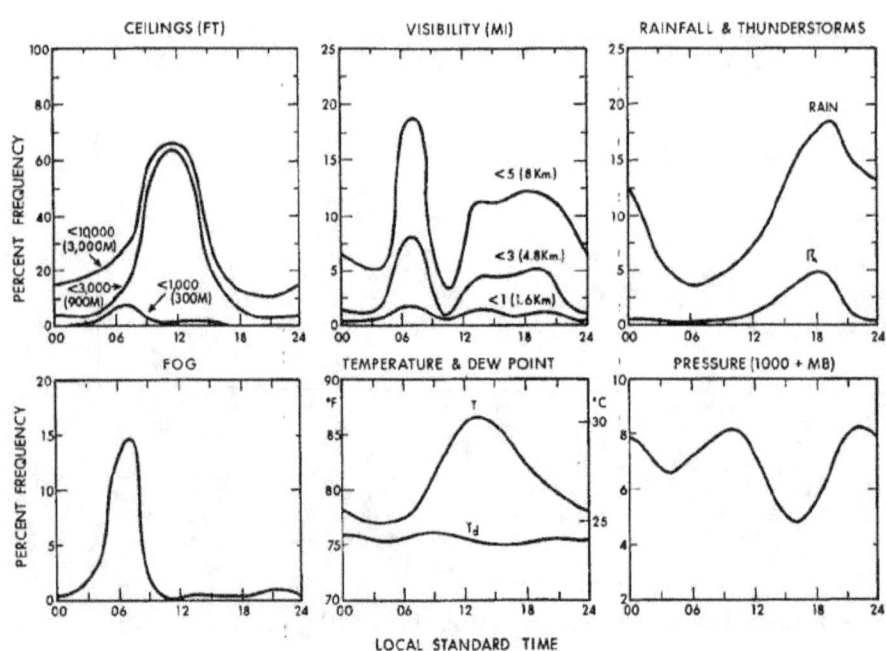

Figure 12-3. Climatological display for Saigon (11° N, 107° E) in August.

12.2.3.5 Sources of Climatological Data. If the data you need is not available locally, it can usually be obtained from USAFETAC. Most station climatological data is included in USAFETAC's "Surface Observation Climatic Summaries" (SOCS), formerly known as the "Revised Uniform Summary of Surface Weather Observations," or "RUSSWO." Much of this information appears in publications such as USAFETAC's "Station Climatic Summary" series, published in one volume per continent and updated about every 3 years. A similar source is the Navy's extensive "Surface Meteorological Observation Summaries" (SMOS). Summarized data for oceanic areas can be found in the USN Marine Climatic Atlas series. Other, older sources such as the Weather and Climate or Marine Climate sections of "National Intelligence Summaries" can be helpful. The *Joint US Navy/US Air Force Climatic Study of the Upper Atmosphere*, published in 12 volumes (one per month), gives climatological standard-level winds, temperatures, dew points, and certain other specialized summaries for the entire planet.

Atlases and climatological publications from various national meteorological services contain data useful for forecasting. Bibliographies from NOAA/EDIS and USAFETAC can be used to locate other sources. The Air Weather Service Technical Library can prepare, on request, tailored subject bibliographies. USAFETAC'S narrative regional, electrooptical, and refractivity climatological studies, if completed for the area of interest, provide a relatively detailed climatological area overview. Other specialized climatological support is available from USAFETAC in accordance with current USAF regulations; see USAFETAC/TN–94/001, *Capabilities, Products, and Services of USAFETAC.*

In addition, data from weather satellites and radar provides detailed climatologies for the tropics. Weather satellite climatologies include total cloudiness (Sadler et al., 1984), highly reflective clouds (Garcia, 1985), and outgoing long-wave radiation (Janowiak et al., 1985). Climatologies based on weather-radar measurements determine seasonal and diurnal variations in convection over an area, and preferred areas for echoes to develop. These help local-area forecasting and were discussed in detail in 12.2.6.7.

12.2.4 Local Forecast Studies.

12.2.4.1 General. The most successful tropical forecast studies use all available observations (synoptic surface and upper-air charts, satellite data, weather-radar pictures or reports, surface observations, cross sections, checkerboards, thermodynamic diagrams, climatology, etc.). Units should try their hardest to obtain all of these. Each station should retain at least the last several years of weather observations, charts, and diagrams taken or prepared locally. More weather data, including satellite data for wider areas and longer periods, should be retained by centralized forecasting units to support forecast study and training. The almost complete destruction of unequalled tropical data collected during World War II severely handicapped tropical meteorology.

Forecast studies, regardless of results, should be well documented. Poor documentation has been a serious problem for military meteorology units in the tropics because of short tours and the resulting rapid staffing changes. In the past, useful techniques that were not documented or evaluated were discarded. When an unsuccessful effort to improve forecasting for a particular area was not recorded, it was doomed to be repeated.

Eliminating these deficiencies demands careful control of planning, preparation, evaluation, and documentation of *all* forecast studies.

It is beyond the scope of this report to survey the range of local and area forecast studies, whether successful or not. This would be more appropriate for regional manuals on tropical analysis and forecasting. Two examples must suffice. Brooks (1987) found that a significant increase in N-S pressure gradient over the western Gulf of Mexico gave about 48 hours warning of a winter outbreak of northerly winds over Honduras. Applying this relationship demands no experience. Palecek (1987) related occurrence of warm-season thunderstorms at Clark Air Base (15° N, 121° E) to prior wind speeds in three tropospheric layers, low-level moisture and depth, an inversion below 10,000 feet (3 km), and a complicated stability index. Eleven steps led to a forecast. Since synoptic changes had also to be considered, success must have depended on forecaster experience. In this section, examples of forecast studies are used in illustrating and evaluating efforts to solve tropical forecasting problems. These approaches are generally objective or synoptic and sometimes a combination of the two.

12.2.4.2 Objective Forecast Studies. Performance of a purely objective forecast technique is independent of the forecasting experience or the skill of the meteorologist using it.

• *Wind Forecasts.* One of the most widely applicable objective techniques was used by Lavoie and Wiederanders (1960) to forecast upper winds in the tropics by combining persistence and climatology in various ways, based on the equations:

$$U_f = (1 - R_{tu})U_c + R_{tu}U_p$$

$$V_f = (1 - R_{tv})V_c + R_{tv}V_p$$

where U and V are the zonal and meridional wind components and the subscripts f, c, and p denote the forecast, climatological, and persistence values, respectively. The lag-correlation coefficients for the zonal and meridional components are R_{tu} and R_{tv},

respectively. Accuracy is best when the lag-correlation coefficients are determined statistically for each level, season, and time period. It turned out that for 24-hour forecasts of 300-mb winds over the tropical Pacific, an average of the climatological wind and the persistence wind was almost as accurate as the optimum combination of the two. An evaluation made during May 1959 showed the statistical forecasts to be 11 percent better than persistence and 15 percent better than climatology (see Figure 12-4). The authors indicated that similar results could probably be obtained for other levels above the near-surface layer. This means that 24-hour wind forecasts for point locations can easily be prepared by taking the vector average of the current wind (for persistence) and the climatological wind. Reed (1967) used upper winds from Eniwetak (11° N, 162° E) in developing statistical wind forecast techniques, which he found applied throughout the tropical Pacific. His paper can be consulted for details on the theoretical basis of the method and for the lag-correlation coefficients for various levels and periods.

The method can be computerized to produce statistical wind forecasts or to combine persistence and climatology to use as first-guess fields in automated wind-analysis programs. For wind forecasts near and equatorward of the subtropical ridges, this statistical technique is seldom surpassed by either numerical or subjective techniques. Forecasters find this hard to believe, but in a test, they consistently forecast too much change in the winds and were regularly beaten by the statistics (Lavoie, 1963). This is a very good technique. Don't change persistence/climatology forecasts unless the evidence for doing so is very strong. Since this method always predicts a regression toward the mean, it should not be used where great value is attached to predicting large departures from normal. For example, in the region of the subtropical jet stream it could greatly reduce the strength of the jet core, possibly causing large route-wind forecast errors.

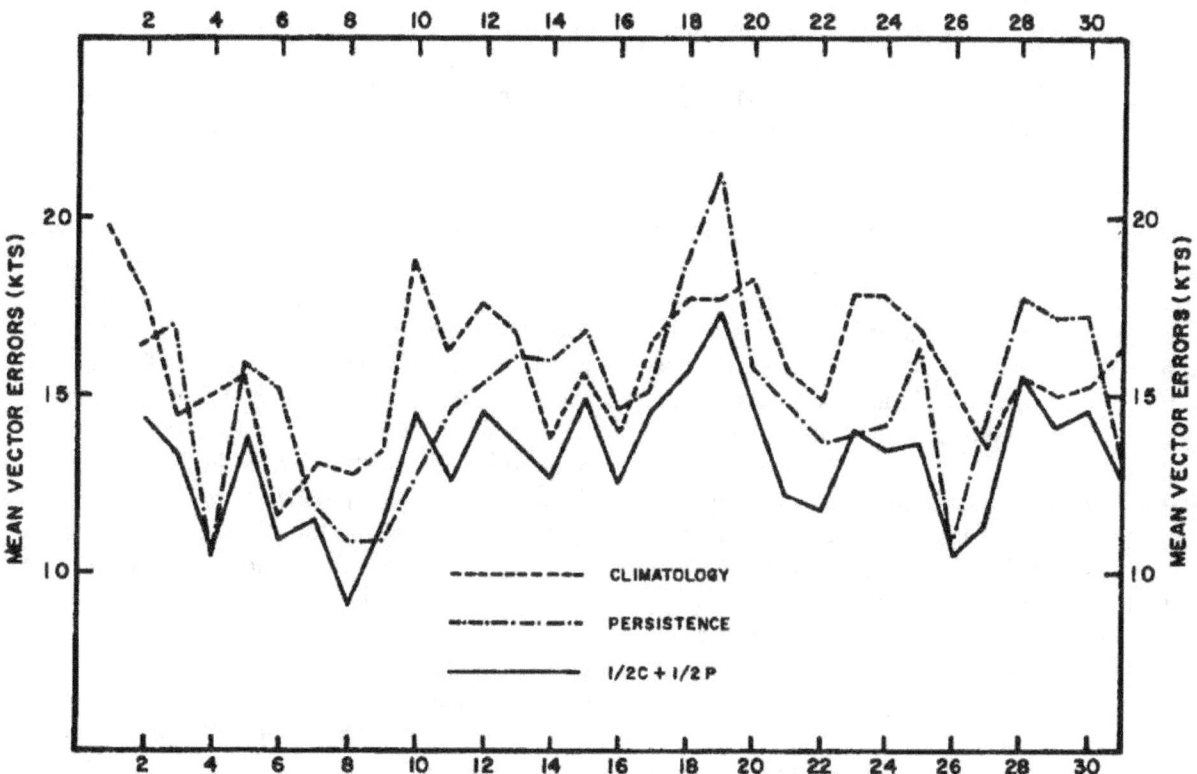

Figure 12-4. Mean May 1959 300-mb wind vector errors over the tropical Pacific for climatology, persistence, and forecasts, based on 50 percent climatology plus 50 percent persistence (after Lavoie and Wiederanders, 1960).

Combinations of persistence and climatology have been used successfully to forecast tropical cyclone movement in Miami, Guam, India, and Hong Kong, as well as in terminal weather forecasting (McCabe, 1961). Persistence-climatology combinations (even in higher latitudes) would test numerical predictions much more effectively than the separate persistence and climatology fields now used. In Section 12.3, 24-hour numerical predictions of upper winds in the tropics are compared to persistence/climatology forecasts as well as persistence and climatology forecasts.

Short-range forecasts of maximum surface wind gusts (excluding those produced by local convection) can be helped by an objective technique that relates observed surface winds to winds in the lowest 10,000 feet (3 km). It depends on increased mixing and momentum transfer during the time of maximum surface heating. Figure 12-5 illustrates the relationship.

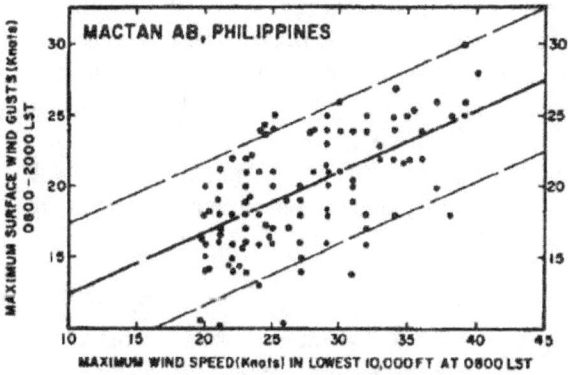

Figure 12-5. Relationship between maximum wind speed in the lowest 10,000 feet (3 km) at 0800L and maximum surface wind gust observed in subsequent 12 hours at Mactan AB (10° N, 124° E), Philippines (First Weather Wing, 1968)

This relationship allows maximum wind gusts to be estimated to within +/- 5 knots about 90 percent of the time. Similar relationships can be derived for other locations at which the surface wind undergoes a pronounced diurnal variation.

Statistically-based nomograms have been developed for U.S. bases in the western Pacific that are often affected by tropical cyclones. The nomograms relate expected sustained winds, average gusts, and peak gusts to tropical cyclone intensity, and bearing and distance from the base. Thus, local effects and distortions are allowed for in the nomograms.

• *Weather Forecasts.* Objectively-derived predictors based on sound physical principles have the best chance of success and for the relationships between predictors and forecasts to remain stable. For example, it has long been known that subsiding air ahead of tropical storms can produce a large area of unusually fair weather (Sadler et al., 1968).

The dynamic and subtly complex nature of tropical weather systems limits the value of objective methods to tropical weather forecasting. Even so, since there are few proven techniques for forecasting pressure, wind, temperature or moisture patterns, objective methods must generally rely on persistence of observed or derived elements, whereas in higher latitudes forecast variables are often included in objective forecasts. An objective technique usually works best when the element being forecast experiences a large diurnal variation; in general, the method works better over tropical continents than over tropical oceans.

During a showers regime, an areal forecast of scattered or isolated showers is usually satisfactory. Forecasting for an airfield is another matter. Over the southern United States (including Hawaii) in summer, it turned out that a local forecast of no rain was as good as the actual forecast, whether of rain or no rain. When probabilistic forecasts can be made, the climatological probability should be predicted, unless there is very strong contradictory synoptic evidence. Objective methods may be rather crude, often able to indicate only whether an event will or will not occur. Forecasting the onset time, duration, and/or intensity

must depend on other methods-the synoptic situation, climatology, etc. Thus, carefully developed objective aids should always be integrated with all the other aids the tropical forecaster has at his disposal.

12.2.4.3. Synoptic Studies. These can be divided into two categories: case studies and composite models. Synoptic case studies focus on the circulation and weather patterns over a specific area for a given period. Consecutive synoptic analyses at various levels are usually shown along with satellite images and auxiliary charts such as vertical time or space cross sections, thermodynamic diagrams, etc. Many of the synoptic models described in Chapter 8 were developed from synoptic case studies. Composite synoptic models average together individual synoptic analyses or observations so as to produce a "representative" model.

Case studies have preponderantly dealt with weather extremes, such as floods, tornadoes or hail—rare events. They never work as well as one might expect. The reason is that the investigator knew not only the initial situation, but the outcome as well. Thus, he almost unconsciously filtered and analyzed the data, ensuring that everything came out right in the end. Since filtering may distort perceptions of the synoptic sequence, an uncertainty is introduced that the forecaster cannot allow for. A case study benefits forecasting most if it represents typical situations recurring often.

In case studies, all available data should be integrated into the analyses. Some excellent examples are given in reports published by the Naval Weather Research Facility (NWRF, 1969; Ramage et al., 1969). These reports present the results of several workshops on the weather of southeast Asia during the northeast and southwest monsoons. For each day of the case studies, two adjacent large pages show large-scale surface analysis, local-area surface analysis, thermodynamic diagrams for key stations and temperature-advection charts (during northeast monsoon), upper-air analyses, weather satellite pictures, and pilot-report summaries. This representation clearly links circulation and weather patterns and shows topographical influences.

366

Each tropical forecast unit should maintain a notebook of case studies for reference and training. For each case study, the file should at least show the synoptic analyses at a few selected levels, satellite pictures, radar pictures, if available, and a record or summary of surface observations. The period of significant weather studied should start one or two days before the weather affected the area of interest and continue for at least one day after it ended. It would thus encompass the complete sequence, and show any relationships between weather and circulation that may be peculiar to the area. The frontispiece samples data collected and analyzed by JTWC immediately after the eye of Typhoon Russ passed 30 NM (55 km) south of Guam on 21 December 1990. Similar sequences might be prepared for fog episodes, squall lines, shear lines, troughs in the upper tropospheric westerlies, superpositions and duststorms.

Composite synoptic studies can be prepared in several ways, but all combine a limited number of similar synoptic (weather) situations and identify and quantify their most significant features. The days included in the composite may be determined by the circulation patterns or by occurrence of a certain type of weather over a particular area. Some examples of composite synoptic studies are given in the following paragraphs.

In studying the daily variation of convection around selected stations in southeast Asia, Conover (1967) developed a radar index (RI) that represented the percent echo coverage as determined from radar photographs or reports within 50 NM (90 km) of a station. Figure 12-6 shows RI values for Saigon/Tan Son Nhut AB (11° N, 107 °E) for June, July and August, 1967. Convection underwent large diurnal and day-to-day changes. After being averaged for each day, the RI values were separated into groups, with each group containing one-third of the cases. Figure 12-7 (opposite page) shows kinematic analyses of the composite mean resultant gradient-level winds for group 1 (above average convection) and group 3 (below average convection). On active days the monsoon trough is near its normal position and Saigon lies in westerly winds that are shearing cyclonically. Conversely, on inactive days, the monsoon trough is poorly defined and winds over Saigon are more southerly.

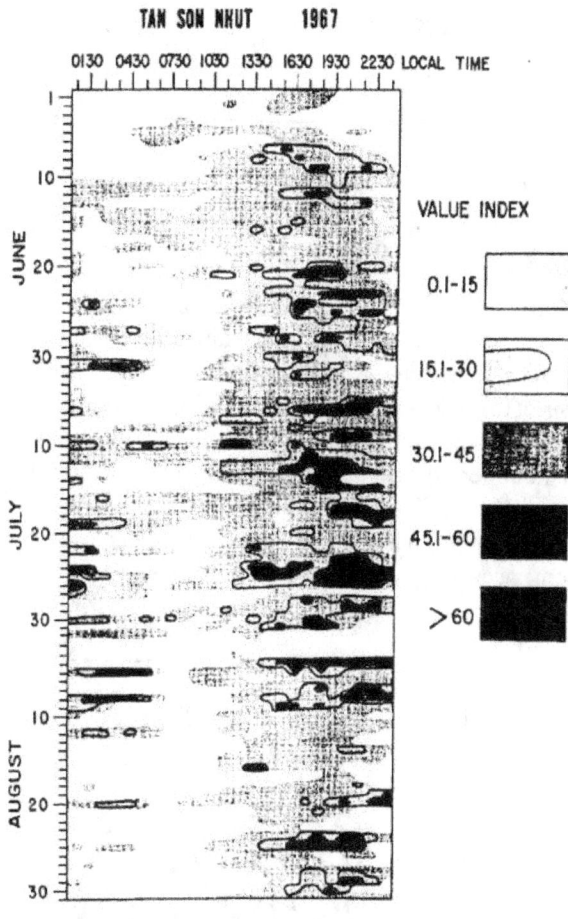

Figure 12-6. Hourly radar index percent coverage of radar echoes within 50 NM (90 km) of station for Saigon (11° N, 107° E) for June, July, and August 1967 (after Conover, 1967).

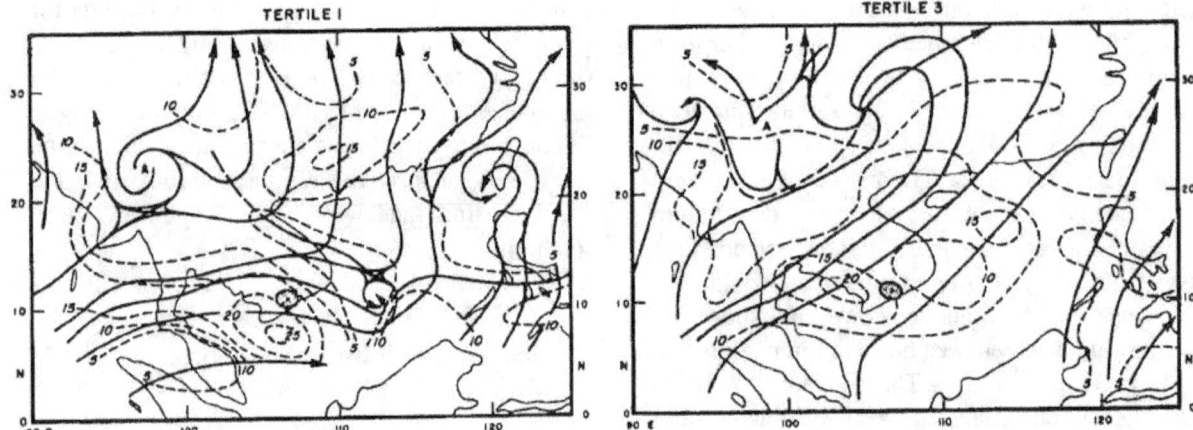

Figure 12-7. Kinematic analyses of the gradient-level resultant wind for: (A) days with above average (left) and (B) below average (right) radar-index values for the Saigon area (shown by shading). Isotachs are labeled in knots (after Conover, 1967).

Atkinson and Penland (1967) developed composite models of maximum wind gusts, rainfall, and flying weather at Kadena Air Base (26° N, 128° E) with respect to the location and intensity of tropical storms. For example, Figure 12-8 shows the probability during a 1-hour period with ceiling/visibility less than 1,500 feet and/or 3 miles (460 meters and/or 5 km) and less than 200 feet and/or a half mile (60 meters and/or 1 km) occurring at Kadena related to the position of a tropical storm with respect to Kadena. Similar

models can be prepared for other stations and for other weather categories.

Both objective and synoptic studies can help tropical forecasting. But there are limitations. Point forecasts of rainfall have little chance of success. Forecasts of above, near or below normal rainfall might be attempted, and these are more likely to be right for an area than for a point.

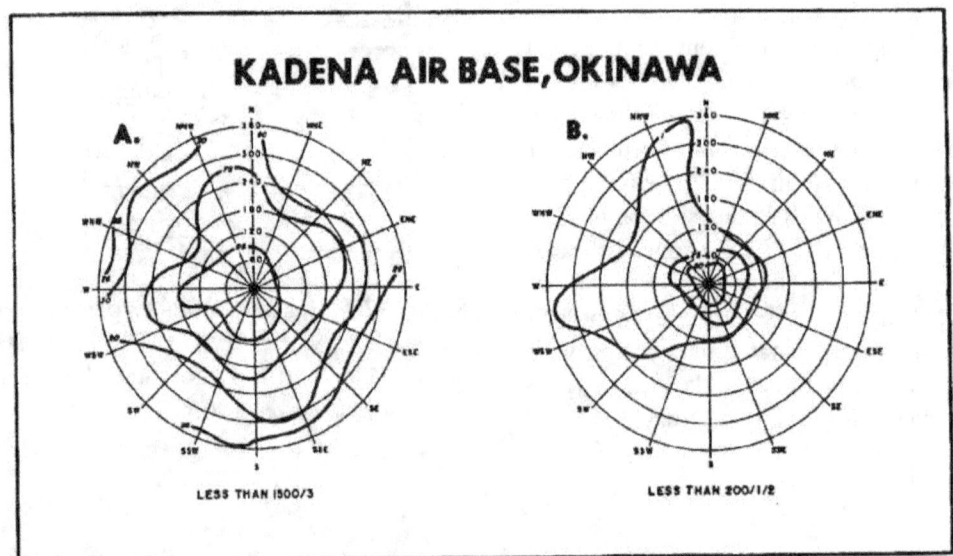

Figure 12-8. Percent probability of: (A) ceiling/visibility less than 1,500 feet (460 meters) and/or 3 miles (5 km), and (B) less than 200 feet (60 meters) and/or a half-mile (800 meters) during a 1-hour period at Kadena Air Base (26°N, 128°E), Okinawa, in relation to tropical cyclone-center location (adapted from Atkinson and Penland, 1967).

12.2.5 Use of Stability Indexes.

12.2.5.1. General. The discussion of temperature and water vapor in Chapter 5 pointed out that changes in vertical profiles of equivalent potential temperature (and hence changes in stability) are often the result rather than the cause of increased convection. This implies that conventional stability indexes (SIs) may not help convection forecasting much over tropical areas. Many tropical meteorologists who have had limited or no success with SI would agree. Since the largest day-to-day changes in vertical temperature and moisture distribution occur over the tropical continents during the transition months between the wet and dry seasons, SIs are likely to be most useful then. They are probably of least use over the tropical oceans where changes in mid-level moisture are often the result rather than the cause of convection, and over the tropical continents during the middle of the rainy season where stability changes relatively little from day-to-day.

12.2.5.2 Stability-Index Studies. Sansom (1963) reported that over East Africa stability indexes gave no better forecasts than persistence, while at Darwin (12° S, 131° E) they failed to aid summertime thunderstorm forecasting in an atmosphere that was always conditionally unstable (Hyson et al., 1964).

Subramanian and Jain (1966) compared the Showalter Index (Showalter, 1953) and Lifted Index (Galway, 1956) as aids to forecasting thunderstorms around New Delhi (29° N, 77° E) during the hot months (March to June) of 1963 to 1966. They compared the index values from the 0600L soundings to thunderstorm occurrence during the following 24 hours within distances of 50 and 100 miles (93 and 185 km) as determined from radar observations, and found (Figure 12-9) a relation between thunderstorm probabilities and SI values. Slightly better discrimination between thunderstorm and non-thunderstorm days is achieved by the Lifted Index, especially for the large negative cases. The authors failed to link SI values to the maximum heights

attained by thunderstorm cells. During the months studied, New Delhi is dry (March to June rainfall equals 4.41 inches (112 mm)) and a showers regime prevails. Vertical temperature and moisture profiles resemble those of the continental mid-latitudes for which the indexes were developed, but differ markedly from those of the tropical rainy season (see Figure 6-18).

12.2.5.3 Thunderstorm Wind Gusts. Not surprisingly, R.C. Miller (1967) found that southeast Asian thunderstorms resemble those of the southeastern United States, and that the strength of thunderstorm downrush wind gusts is a function of the lapse rate. He developed two indexes to quantify the relationship:

• *Dry-Instability Index (T_1).* This measures the low-level instability and is determined in two ways:

    (1) If the sounding has an inversion, the moist adiabat is followed from the warmest point on the inversion to the 600-mb level. The temperature difference between the intersection of the moist adiabat at 600 mb and the 600-mb free-air temperature is T_1. (The top of the inversion must be within 200 mb of the surface and is not expected to be eliminated by surface heating.)

    (2) If no inversion appears on the sounding (or if the inversion top is more than 200 mb above the surface), the forecast maximum temperature is lifted moist-adiabatically to 600 mb to determine T_1 as before.

• *Delta-T Index (T).* This measures instability in the middle and upper levels and is determined by lowering moist adiabatically the intersection of the wet-bulb curve and the 0° C isotherm to the surface, then finding the difference between the temperature at this point and the ambient surface temperature.

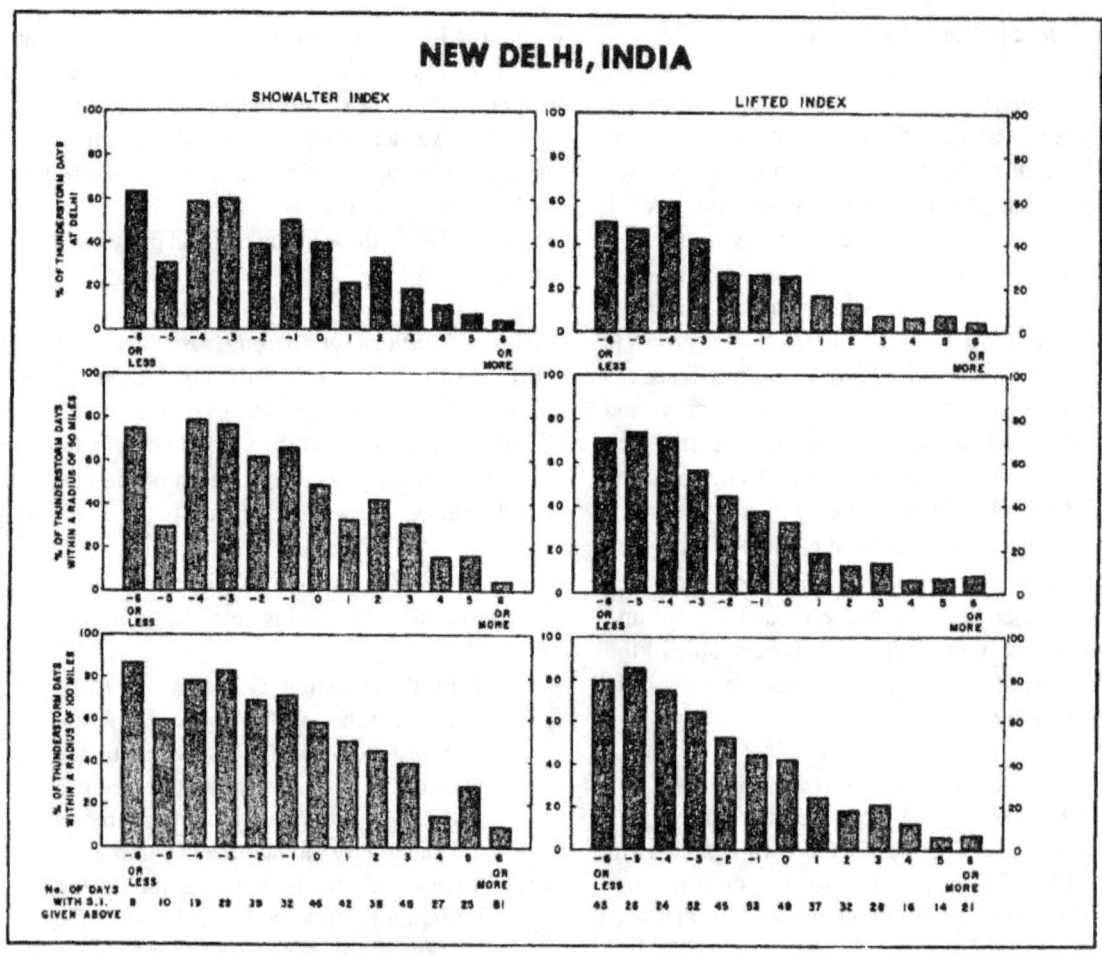

Figure 12-9. Relation between Showalter-Index and Lifted-Index values at 0600L and probability of thunderstorms during the following 24 hours at New Delhi (29° N, 77° E), and within radii of 50 and 100 NM (93 and 185 km) (adapted from Subramaniam and Jaim, 1966).

Miller compared wind-gust cases for southeast Asia and the southeastern United States to the simultaneous values of the T_1 and Delta-T indexes (Figure 12-10). The rough best fit is shown by the heavy dashed line, while the lighter dashed lines enclose the observed maximum and minimum values. These curves probably apply to other parts of the tropics. Each station can determine which curve best describes the air-mass variations that produce thunderstorms with gusty winds. If thunderstorms are preceded by intrusions of drier air between 700 and 500 mb, the Delta-T index will probably best estimate the wind gust. If thunderstorms are generated by low-level temperature increases (surface heating or low-level warm advection), then the T_1 index is recommended. But remember, a thunderstorm must first be forecast and SIs are not much good for that. Also, both indexes require the surface temperature at the time of the thunderstorm to be forecast. To realize the full potential of either index, moderate or heavy rain is needed.

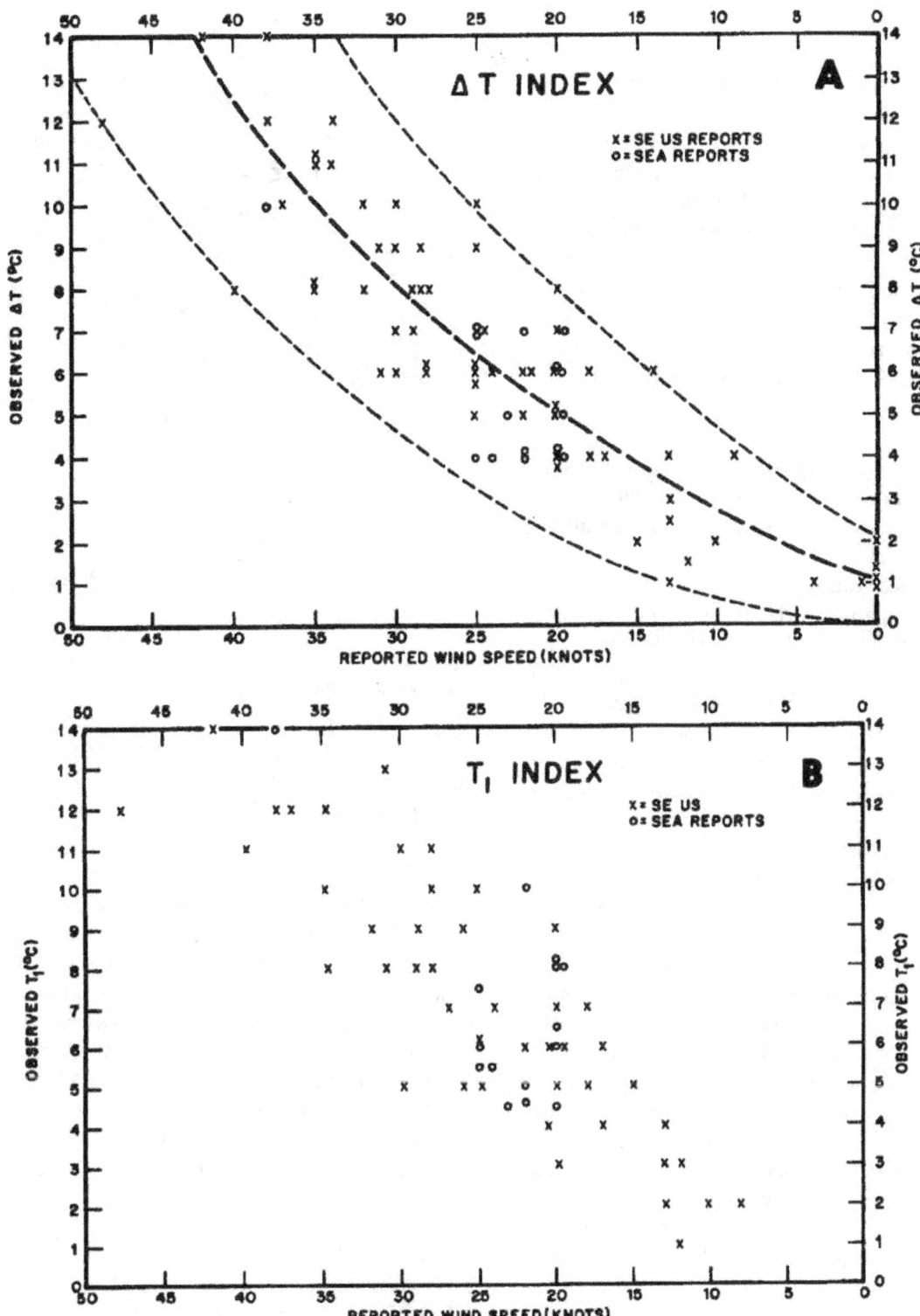

Figure 12-10. Reported maximum thunderstorm-produced surface wind gusts (knots) at stations in the southeastern United States and southeast Asia according to observed values of (A) the Delta-T Index, and (B) the T_i Index (adapted from R.C. Miller, 1967).

371

12.2.6 Radar.

12.2.6.1 General. Weather radar is the best aid to very short-range forecasts (up to 3 hours) in areas dominated by convection. It is used the same way in the tropics as in middle latitudes; the Federal Meteorological Handbook 1 and Whiton and Hamilton (1976) are applicable. This section supplements the guidance provided by the Handbook. As doppler radar becomes more common, indirect methods for obtaining wind velocities from echo movements, as described below, will give way to direct methods that measure motion of individual drops.

Forecasters should remember that rain not only *reflects* the radar signal but also *attenuates* it. Rain behind rain always appears lighter than it really is, and it often (especially in the farther eye-wall of a tropical cyclone) disappears altogether from the radar scope (see Figure 9-5). Three-centimeter radars were operated simultaneously at Udon (17° N, 103° E) and Ubon (15° E, 105° E) on the Korat Plateau, Thailand, during the wet seasons of 1967 and 1968 (Ing, 1971). Figure 12-11 shows echo frequencies. Each radar's zero isopleth passes through the other's maximum, while the line joining isopleths of equal frequency passes midway between the stations. Longer wavelength radars would do better, but attenuation would still be a problem. Weather satellite pictures could help evaluate location and intensity of attenuation.

12.2.6.2. Tropical Storm Tracking. Although national centers such as NHC and JTWC forecast tropical cyclone intensities and tracks, detailed forecasts of winds, rain and seas are still largely a local responsibility, and are based on continuous radar information. The literature on weather radar use in the tropics emphasizes tracking tropical storms (see Figure 9-5). For tropical cyclone-prone islands and coasts, the cost of weather radars is quickly amortized by the accurate and timely public warnings they provide.

The first radar indication of an approaching tropical storm is one of the outer rainbands, which appears like a squall line. The distance between the rainband

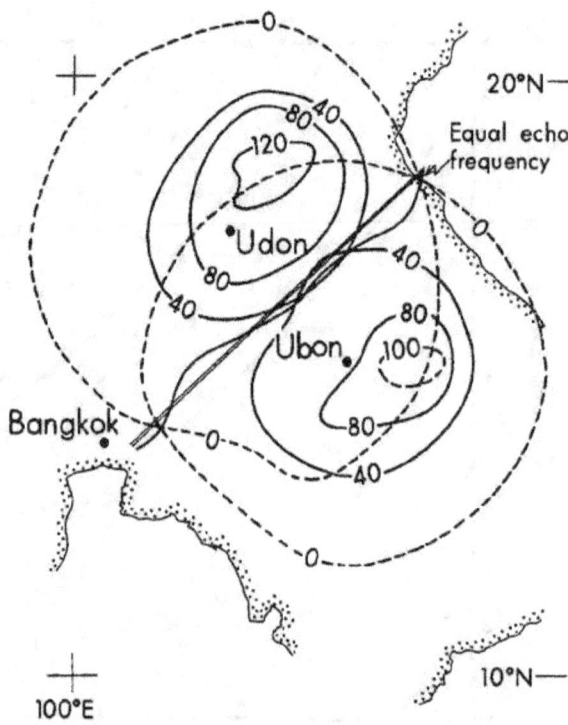

Figure 12-11. Radar-echo frequencies measures simultaneously by 3-cm radars located at Udon (17°N, 103°E) and Ubon (15°N, 105°E), June-August 1967 and 1968. The double line is equidistant from each station; the heavy line joins points of equal echo-frequency (adapted from Ing, 1971).

and the storm's eye can usually be determined from a recent satellite picture. This outer band may lie 50 NM (100 km) ahead of the first spiral band of the storm proper. In the outer portions of the storm, spiral bands are about 20 NM (40 km) apart; near the storm center, they tend to merge into the eye-wall cloud. The storm's eye appears as a nearly circular echo-free region that is usually from 10 to 50 NM (20 to 100 km) wide. Once 180° or more of the eye-wall cloud appears on the radarscope, the storm center can be located with an accuracy of about 5 NM (10 km) or better.

A well-defined eye may not always be visible on radar, even when the storm is within range, because heavy rain between the eye and the radar can severely attenuate the signal. Some intense storms show two concentric wall clouds.

Even though eye positions can be determined quite accurately by land-based radar once enough of the eye appears on the scope, these observations are of limited use to short-range forecasting. This is because frequent aircraft reconnaissance and radar fixes have revealed irregular small-scale fluctuations in the movement of eyes, making extrapolation very difficult. Jordan (1963) discussed these fluctuations, which range from periodic oscillations with a period of about 2 days down to nearly random fluctuations

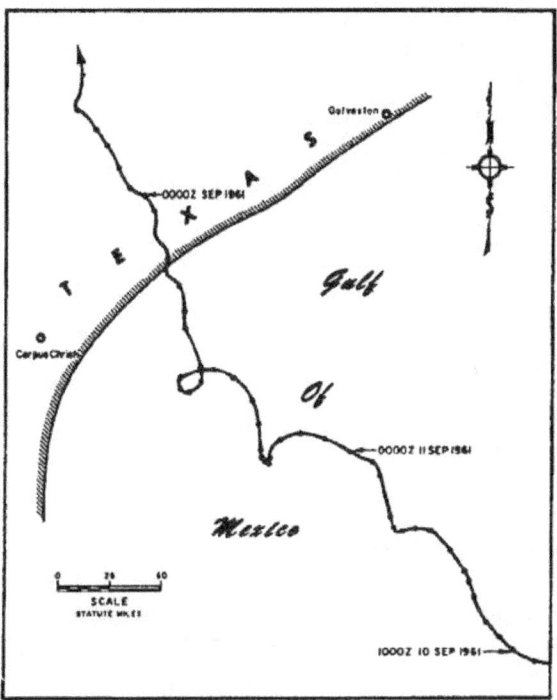

Figure 12-12. Hourly center positions of Hurricane Carla, 10-12 September 1961, based on radar fixes from Galveston (29 ° N, 95° W).

lasting less than an hour. An excellent example, from the track of Hurricane Carla on 10-12 September 1961, is shown in Figure 12-12. Because of such irregularities, Jordan concluded that observations separated by less than an hour cannot be relied on to indicate storm motion over the following 3 to 6 hours. Similarly, observations made a few hours apart are of limited use in indicating storm motion over the following 12 to 24 hours. Thus, to avoid being misled, forecasters must relate frequent storm-center fixes to other information such as the smoothed storm track during the past few days, steering currents, and

climatology. Six- to twelve-hour forecasts based on extrapolation of smoothed track positions of the previous 12 to 24 hours will usually better those extrapolated from fixes made over the previous several hours.

In tropical cyclone rainbands, individual cells move tangentially around the eye. New cells tend to develop on the upwind end of the spiral rainbands, become more stratiform farther downstream, and eventually dissipate near the downwind end of the band (Hardy et al., 1964). Figure 12-13 illustrates the movements of the rainbands and of the individual cells relative to the storm eye. The motion of a cell tracked for about 15 minutes, either at the end of a band or between bands, approximates the storm wind. Since in fully developed storms there is little wind shear between the near-surface layer and 20,000 feet (6 km) elevation, echo motion is close to lower-tropospheric wind speed. Wind speed varies considerably along a radius; thus, observations at differing distances from the center are required before the radial wind-speed profile can be determined. Also, as illustrated by Figure 9-2, wind speeds are asymmetrically distributed around the storm. In the lower troposphere, strongest winds usually occur to the right (left) of the storm's track in the NH (SH). Because of the earth's curvature, wind determination from echo movement should probably be limited to ranges of 100 NM (200 km) or less from the radar site. The Doppler radars now being deployed will allow direct determination of the winds from the motions of individual raindrops out to 125 NM (250 km).

12.2.6.3. Echo Movement. The same rules for forecasting echo movement in higher latitudes (Federal Meteorological Handbook I) apply in the tropics. The movement of single cells (or small clusters that may appear as one cell on the radar) is closely related to the wind velocity at 700 mb (or to the mean wind between 850 mb and 500 mb). Successive positions of small cells at 15- or 20-minute intervals determine cell motion. Averaging velocities of several cells gives a representative value and overcomes the handicap that individual cells generally last less than 30 to 60 minutes. For forecasting rainfall at a point, only cells at least 30 minutes upwind can be extrapolated with any confidence.

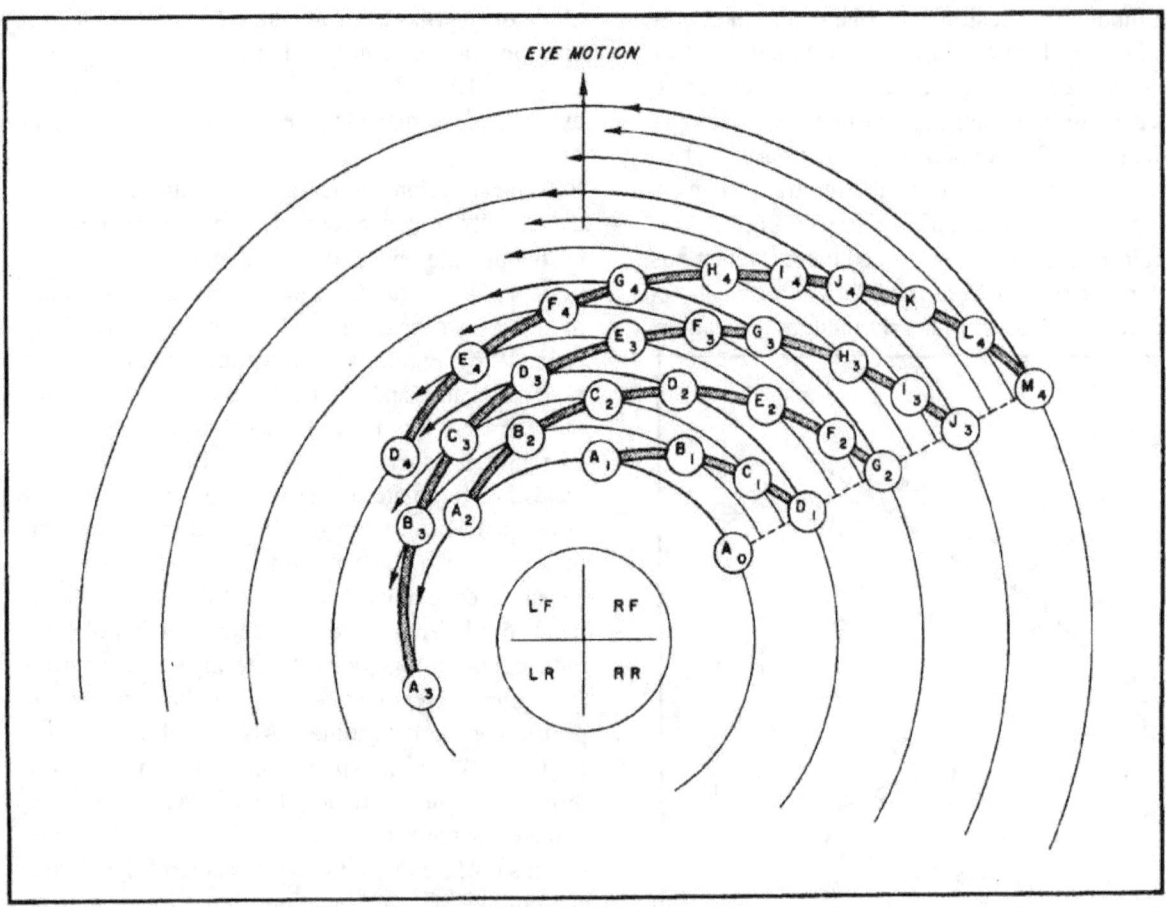

Figure 12-13. Kinematic model of a hurricane spiral rainband. A letter identifies an echo, while the subscript indicates units of time. Relative to the hurricane eye, new echoes develop radially outward (not necessarily at the same azimuth); once formed, they move in circular paths and eventually dissipate at the downward end of the rainband (after Hardy et al., 1964).

Tsonis and Austin (1981) analyzed 27 long-lived (>100 minutes) convective cells during GATE. They found that applying simple persistence of movement and area gave forecasts as good as more complex schemes, and that the best forecasts were made when the cell had already existed for 30 minutes and motion could be established. The best 2-hour forecasts erred 77 percent in areal rainfall. This represents optimal performance; forecasts of shorter-lived cells would do worse, and forecasts over land, worse still.

Tracing the movement of large groups of convective echoes 15 NM (30 km) or more across, and consisting of many thunderstorms or showers, rather than individual cells, helps prediction more because the groups usually maintain their identities for hours.

Unlike smaller cells, echo groups do not move with the wind at any one level or even with the mean wind. The results of Conover's (1967) study of echo groups wider than 50 NM (93 km) near Saigon (10° N, 107° E) from June through August, 1966, are presented in Figure 12-14, next page. Durations ranged from 1 to 20 hours, with most between 3 and 4 hours (Figure 12-14A). In about two-thirds of the cases, echo groups moved in the direction of the wind between 8,000 and 13,000 feet (2,400 and 4,000 meters) (Figure 12-14B). For the level of best agreement between the direction of echo movement and wind direction aloft, the ratio of echo speed to wind speed at that level was computed. Figure 12-14C shows that ratios of 0.3 to 1.0 predominate. The isolated cases of ratios greater than 1.3 probably reflect the effect of strong

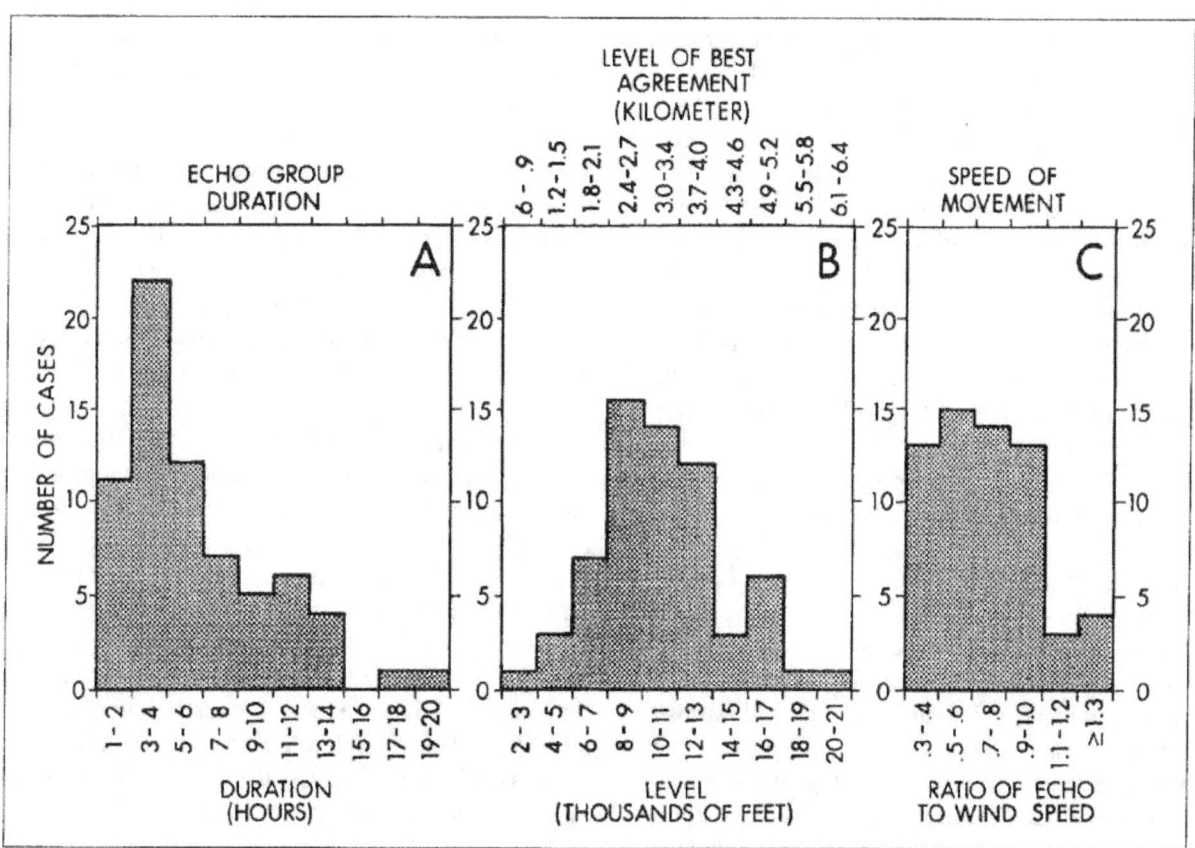

Figure 12-14. Statistics of large radar-echo groups (greater than 50 NM (93 km) wide)) near Saigon/Ho Chi Minh City (11° N, 107° E) during the southwest monsoon. (A) echo-group duration, (B) level of best agreement between echo-group motion and wind direction, and (C) ratio of echo-group speed to wind speed at level of best agreement (adapted from Conover, 1967).

wind shear and little change in wind direction in the vertical. Over West Africa, where this is common (Eldridge, 1957), new convection forms at the leading edge of the echo group and dissipates at the trailing edge. Overall, this causes the group to move faster than the environmental wind. The wide range in the relationship between echo group movements and winds aloft suggests ignoring the wind, and instead plotting successive positions of the centroids or boundaries of the group on the radar screen. Once a group has formed, its movement is relatively persistent; successive group positions (at half-hourly intervals) can be extrapolated for short-range forecasting.

As in higher latitudes, radar echoes in the tropics are often organized into squall lines (see 8.4.2) that tend to move to the right of the upper winds at an angle between 0 and 90 degrees. The line moves at approximately the speed of the component of the 700-mb wind normal to the line. At line speeds below 20 knots, the line may move slightly faster than the upper-wind component due to new cell growth ahead of the line.

Where wind shear is strong, as in the West African rainy season, squall lines may propagate westward above surface westerlies. As for large echo groups, extrapolation of past positions most accurately predicts line movement; however, when a line forms close upwind, upper winds need to be used in predicting its arrival.

12.2.6.4. Severe Weather Forecasting. Weather radar allows forecasters to monitor severe weather (e.g., heavy rain and strong downrush winds) caused by convection. Radars with an iso-echoing capability can pinpoint the most intense thunderstorm cells. The thunderstorm echo top can be determined from a range-height-indicator (RHI) scope or by tilting the radar antenna until the echo disappears. On any convective day, the potential maximum wind gusts as indicated by the Dry Instability (T_i) or Delta-T Indexes (see the previous section) will probably occur in the most intense thunderstorm cells. However, since intense cells cover only a small fraction of the area covered by a radar, the strongest gusts in that area are seldom measured by surface observers.

In a mature thunderstorm, strong downward-flowing currents spread out from beneath the cloud base as a pseudo-cold front. Therefore, thunderstorm-produced shifts in wind-direction may occur up to tens of miles from the actual thunderstorm. The strongest downrush winds occur in moderate to heavy rain beneath the cloud base. These microbursts (see Figure 10-18) can threaten airfield operations by producing strong crosswinds or by shifting headwinds into tailwinds during approach or takeoff. Forecasters must be alert to these hazards and establish reliable procedures to quickly advise pilots of rapid wind changes.

When a line of thunderstorms approaches a station, the radar scope may show a thin band or line ahead of the thunderstorm line. This precursor convection develops along density and convergence discontinuities at the leading edge of the cold air downrushing from the thunderstorms. Rakshit (1966) studied several thin lines at Calcutta (23° N, 88° E), and concluded:

• A single thin line occurred only with a thunderstorm (squall) line and preceded it by up to 15 NM (28 km).

• As the thin line passed over a station, the first gust was accompanied by a pressure jump and temperature fall, but there was no rain.

• The thin line had a fine-grained structure with no convective cells evident; it disappeared from the radar scope when the antenna was tilted by 2 or 3 degrees.

Since thin lines do not always precede thunderstorm lines, their absence does not signal absence of dangerous gusty winds.

A source of mid-tropospheric dry air is needed before evaporative cooling from rain can cause strong downdrafts and lead to squalls or microbursts. Thunderstorms embedded in a deep moist environment (rains regime) may persist for several hours (continuous thunderstorms); but lacking a dry air source, they cannot generate strong downdrafts.

12.2.6.5. Visibility Forecasting. Radars that are calibrated to make accurate reflectivity measurements can help in making rough estimates of visibility. Wilson (1968) used rainfall and visibility measurements collected from the Atlantic City meso-net during heavy rainstorms from August 1965 to 1966 to compute a correlation of 0.85 between visibility in km and rainfall rate R in mm hr^{-1}. The regression found by the least squares technique is $V = 12.5R^{-0.68}$. This resembles a regression determined by Atlas (1953) from drop-size distributions and liquid water content information - $V = 11.6R^{-0.63}$.

Rainfall rate and reflectivity values Z in mm^{6}m^{-3} have been related (Battan, 1959). $Z = 200R^{1.6}$ for steady rain and $Z = 486R^{1.37}$ for thunderstorm rain are based on average values for many storms (Federal Meteorological Handbook I); however, there is considerable variation from these relationships for individual cases. For example, Cheng and Kwong (1973) calculated $Z = 175R^{1.68}$ for summer at Hong Kong (22° N, 114° E), Woodley and Herndon's (1970) equation for Florida was $Z = 300R^{1.4}$, and in hurricanes $Z = 300R^{1.35}$ (Jorgensen and Willis, 1982).

From combining Wilson's *V-R* relationship and the *Z-R* relationship for thunderstorm or steady rain, the following formula relates visibility to radar reflectivity: $V = 266Z^{-0.5}$. This relationship, plotted in Figure 12-15 (next page), can be used to estimate the lowest surface visibility associated with the maximum radar reflectivity in a thunderstorm or steady rain. The most intense thunderstorm cells with reflectivity of 10^5 to 10^6 will lower visibilities to about 1,500 feet (460 meters) while in light rain (Z less than 10^4) visibility is usually 2 NM (3.7 km) or more.

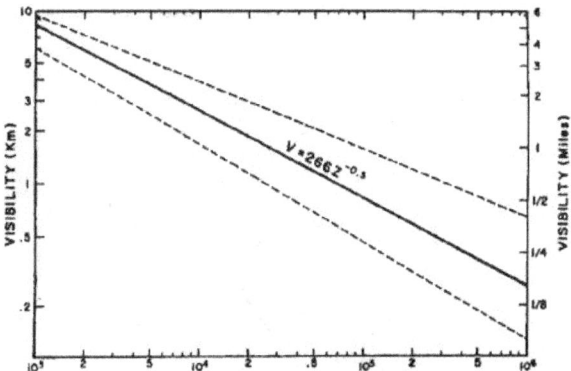

Figure 12-15. Surface visibility (*V*) in showers related to reflectivity (*Z*) measured by radar. Dashed lines define limits of *V-R* relationships based on observational studies (adapted from Wilson, 1968, Cataneo, 1969, and ESSA, 1967).

12.2.6.6. Rainfall Estimates. Weather-radar data has been used to estimate rainfall for river stage and flood forecasts, soil moisture and trafficability estimates, weather modification experiments, etc. The SSM/I, now aloft on DMSP, can estimate rainfall; resolutions are being improved. Even though radar estimates (at a point) leave a lot to be desired when compared to raingauge measurements, they often give a better idea of rainfall over an area than any practicable number of gauges distributed over the same area. This complex problem is discussed in some detail in the Federal Meteorological Handbook I, and in greater detail in many articles and technical reports, some of which are referenced in the Handbook. Proceedings of annual AMS Weather Radar Conferences and reports from the Illinois State Water Survey contain excellent papers.

12.2.6.7. Radar Climatology. According to the Federal Meteorological Handbook I, radar climatology provides:

• Frequency of precipitation echoes over inaccessible regions such as mountain and water areas.

• Estimates of seasonal variations of rainfall intensity.

• Frequency of echoes over specific regions as a function of hourly, monthly, or seasonal variations and differing synoptic situations.

• Locations of favored regions for echo development.

• Terrain effects on orographic rainfall during different synoptic situations.

The following paragraphs discuss some techniques for preparing radar climatologies and present examples of radar climatology studies. Doppler radar, such as the WSR-88D, can be programmed to record radar scans on video tape; these can then be played back and the climatology compiled. With other radars, the scope can be photographed by a video camera at specified time intervals. Time-lapse movies made from these tapes are excellent forecaster training aids. For example, they can depict the diurnal variation of convection around the radar station, or the distribution of rain around an approaching tropical storm. Great care must be taken to insure that the desired range settings, power adjustment, antenna elevations, etc., are properly recorded (along with the observation times) on the films. To increase the climatological usefulness of the data, the control settings should be standardized.

Despite hundreds of radars being deployed in the tropics, only a few dozen radar climatology studies have been made. The reason—at least until the advent of video—is that the work is unexciting and time-consuming.

Kulshrestha and Jain (1968) developed a radar climatology for the New Delhi (29° N, 77° E) area of India (Figure 12-16). Figure 12-16A shows that echo tops are highest during the summer monsoon (June through September) and lowest in winter. The diurnal variation of the area of maximum radar echo varies annually (Figure 12-16B). Possibly because of its prevailing fine weather, November is the only month that does not seem to fit the annual cycle. Figures 12-16C and D show the frequency distribution of individual echo widths during the hot season (March through May). The average echo was about 3.9 NM (7.2 km) wide and the average space between echoes was about 9.5 NM (17.7 km).

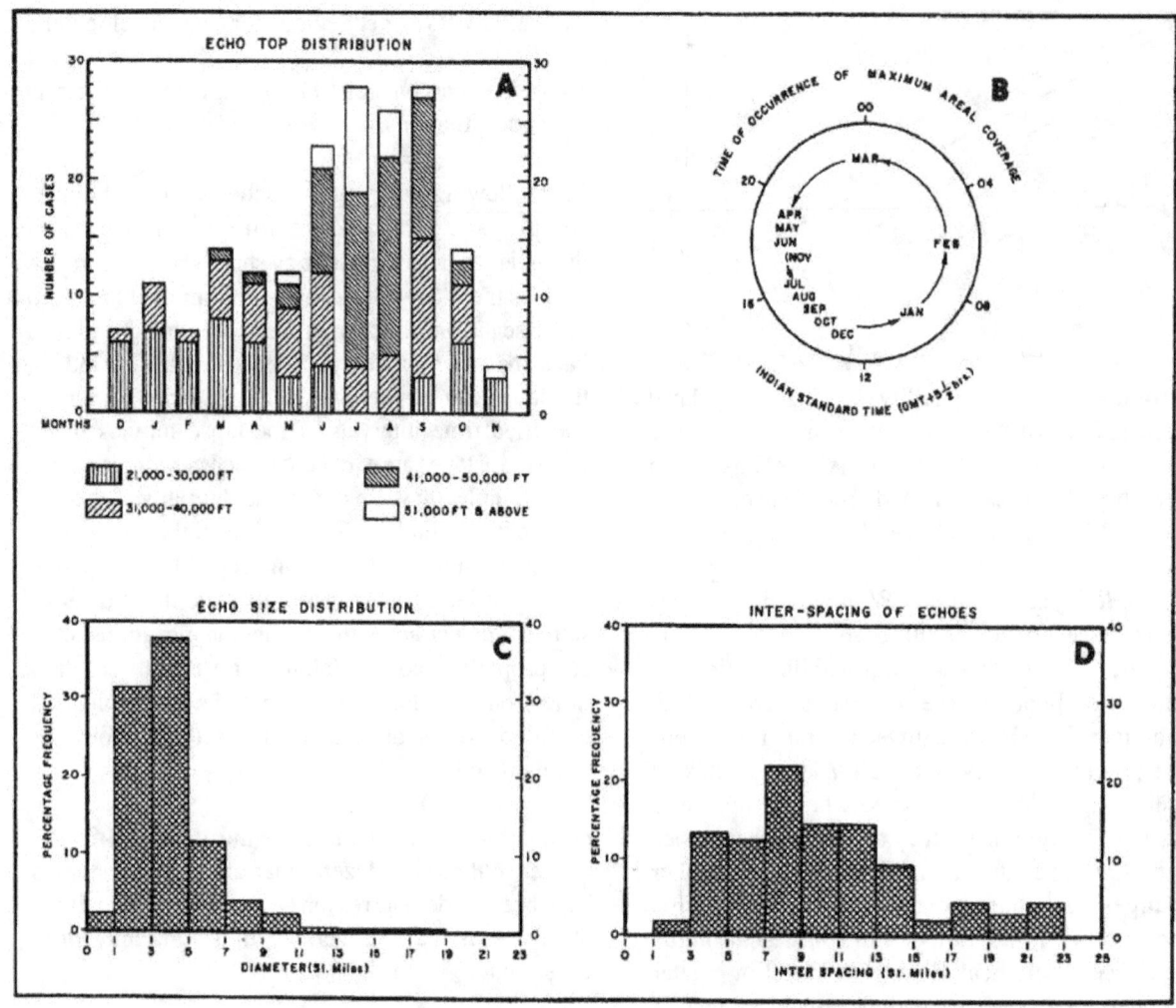

Figure 12-16. Radar climatology for New Delhi, India: (A) monthly distribution of echo tops, (B) time of occurrence of maximum areal coverage by month, (C) frequency distributions of echo sizes, and (D) interspacing of echoes during the hot season (March-May) (adapted from Kulshrestha and Jain (1968).

Gerrish (1971) used the University of Miami radar in developing a radar climatology of south Florida summer rain. Figure 12-17 (next page) shows some of the main results. The frequencies of echoes over land and water are shown in Figure 12-17B. During the day, echoes are more common over the land than over the sea, while the tallest echoes are confined to land. The areal/time patterns shown in Figure 12-17C provide a new model for summer convection in the Miami area. Over the ocean away from the coast the pattern is cellular, implying largely unorganized or random convection; however, near the coast and over land, patterns are organized; trends may persist for a day.

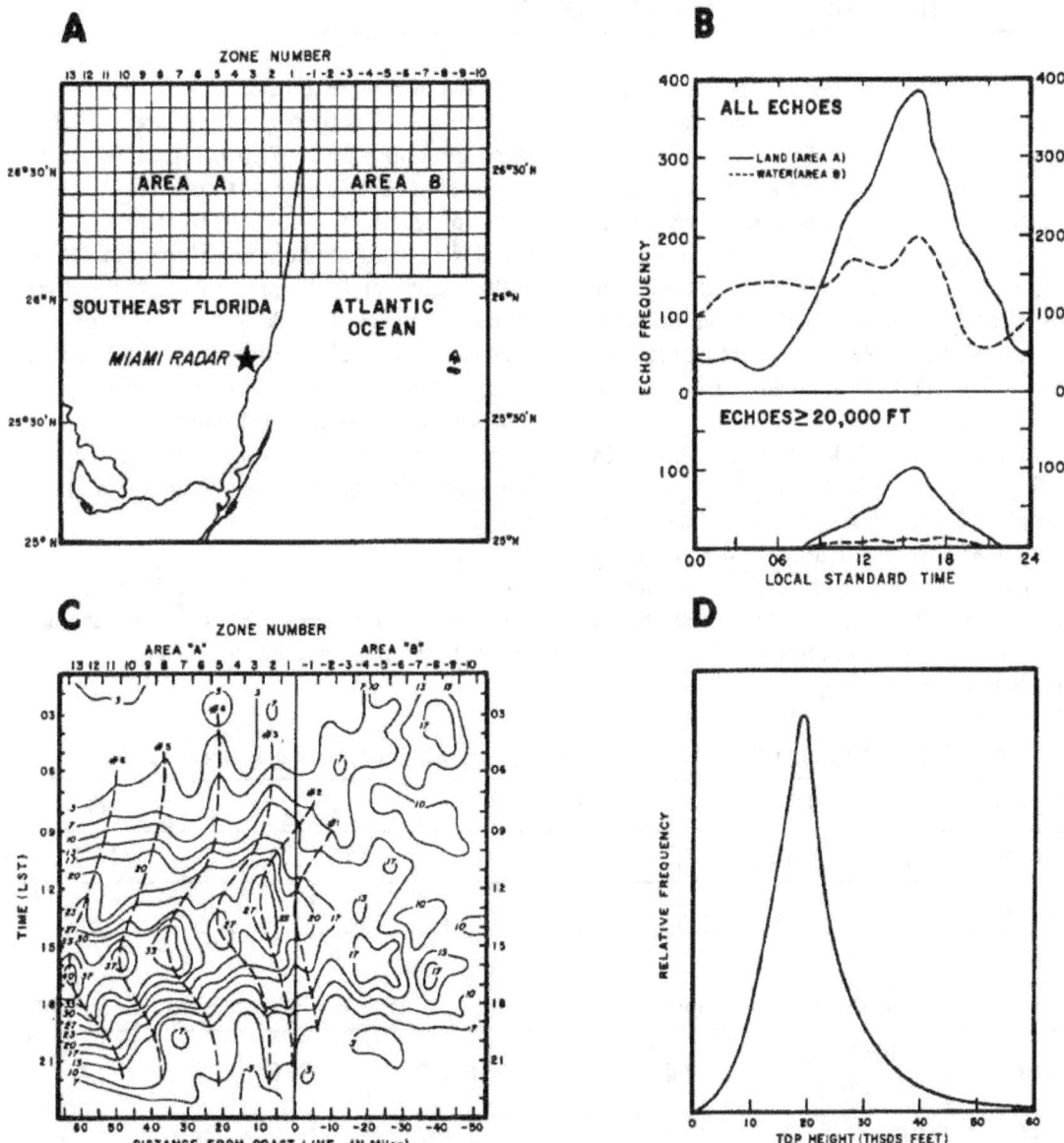

Figure 12-17. Radar climatology of the southeast Florida area: (A) areas and grid used to compile the climatology, (B) diurnal variation of echo frequency, (C) percent frequency variation by time of day and distance from the coast, and (D) relative frequencies of heights of echo tops for 1200 to 1800L (adapted from Gerrish (1971).

Separate maxima appear near sunrise in zones centered about 15 NM (28 km) apart; they seem to move inland in unison. Wind data suggests that line 2, at least during part of its history, is associated with the sea breeze (see Figure 7-12); however, the overall pattern with multiple lines of maximum frequency is much more complex than a model of a single sea-

breeze convergence line. Figure 12-17D shows the frequency distribution of echo heights for 1200 to 1800L. A pronounced mode is found between 18,000 and 20,000 feet (5.5 and 6 km); there are relatively few echo tops above 30,000 feet (9 km). There is little difference between land and sea.

12.3 TROPICAL NUMERICAL WEATHER PREDICTION

Numerical weather prediction (NWP) in mid-and high-latitudes has slowly improved. At NMC, atmospheric models have evolved from an initial, single-level barotropic model to the 18-layer global spectral model. A 28-layer model is being tested. Although this model includes the tropics, its skill is only marginally better than persistence at 24 hours. The ECMWF global spectral model also has problems in the tropics. Heckley (1985) reported that between 18° N and 18° S, the RMS 24-hour wind forecast error amounted to 12 knots at 200 mb and 6 knots at 850 mb, compared to 18 knots and 7 knots in persistence forecasts. But remember, this is "persistence" of the NWP analyzed fields that in turn depart as much or more from reality (11.4.3). Refining convective parameterization and increasing resolution between 1984 and 1985 (Tiedtke et al., 1988) failed to improve the 24-hour wind forecasts, but did give better results beyond 2 to 3 days.

In 12.2.4, the value of persistence/climatology forecasts was noted (Figure 12-4); it was suggested that they be compared to numerical predictions. USAFETAC did this for 32 tropical rawinsonde stations for March and April 1991. The NMC operational NWP made 24-hour wind forecasts at 0000 and 1200Z at 850 and 200 mb for each station location. Half (persistence + climatology) forecasts as well as persistence and climatology forecasts, were also made; all were compared to valid-time observed winds at each station—see Figures 12-18 and 12-19, pages 26 and 27). The averages of the average station errors (in knots) are shown in Tables 12-1 and 12-2.

From a sensitivity analysis, USAFETAC concluded that error differences of 0.2 knots at 850 mb and 0.4 knots at 200 mb would be significant.

TABLE 12-1. March and April 1991 average tropical wind forecasting errors in knots.

Level	NMC	1/2 (P + C)	P	C
850 mb	10.0	8.8	10.2	10.3
200 mb	19	20.1	22.4	26.1

TABLE 12-2. March and April 1991 average tropical wind forecasting errors in knots, stratified by latitude.

Latitude	Number of Stations	NMC	1/2 (P + C)	P	C
850 mb					
0-10°	7	10	7.8	9.3	8.6
10-20°	15	9.7	8.8	10	10.3
20-30°	20	10.4	9.6	11.1	11.5
200 mb					
0-10°	7	16.8	15.1	17.5	17.7
10-20°	13	19.3	21	21.3	25.3
20-30°	10	21.1	25.3	27.3	33

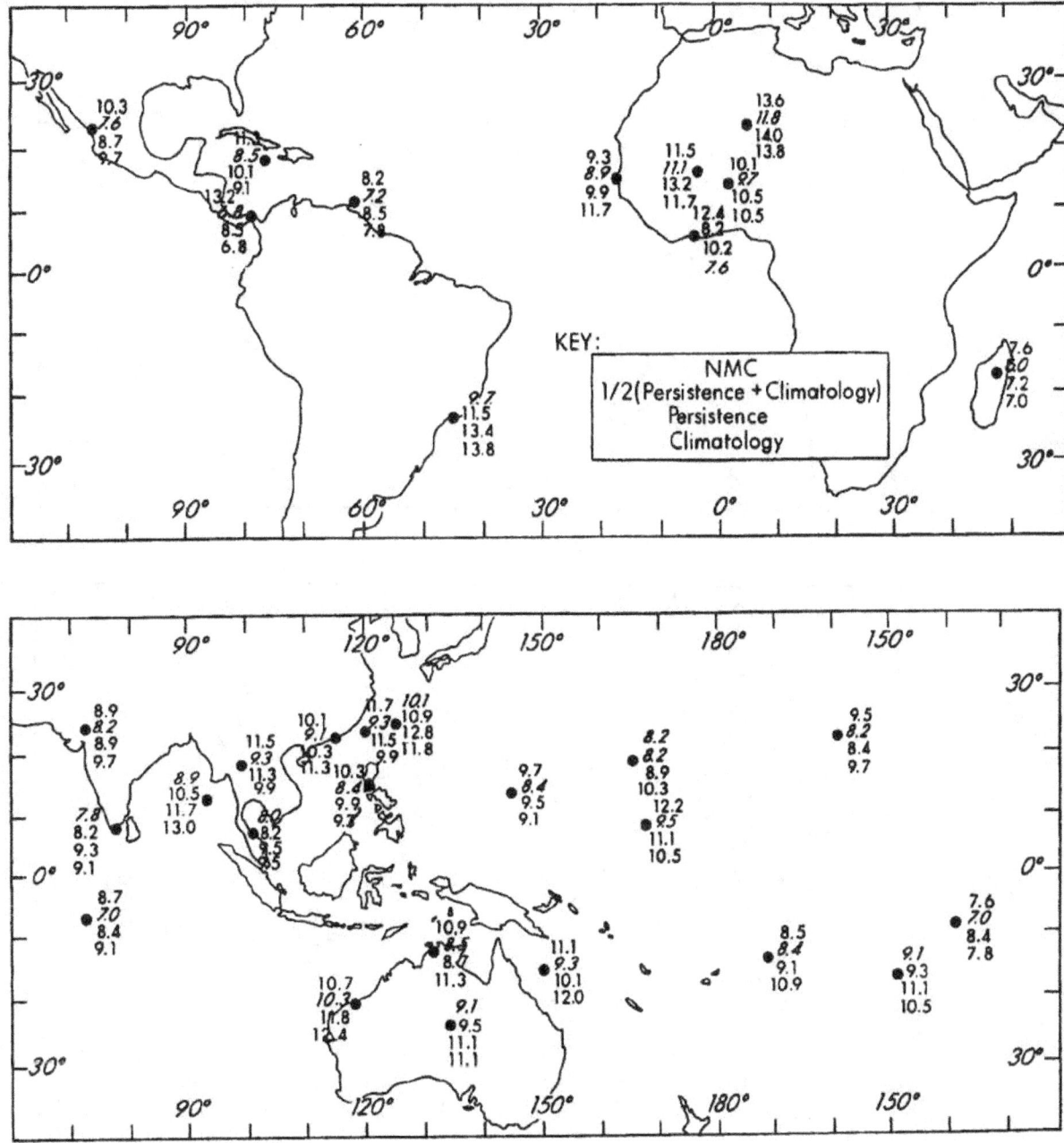

Figure 12-18. Wind forecasts for March and April 1991 at 32 tropical stations. Comparison of NMC numerical prediction of 850 mb winds 24 hours in advance to objective 1/2 (persistence + climatology); persistence ; and climatological predictions. Mean vector errors are in knots.

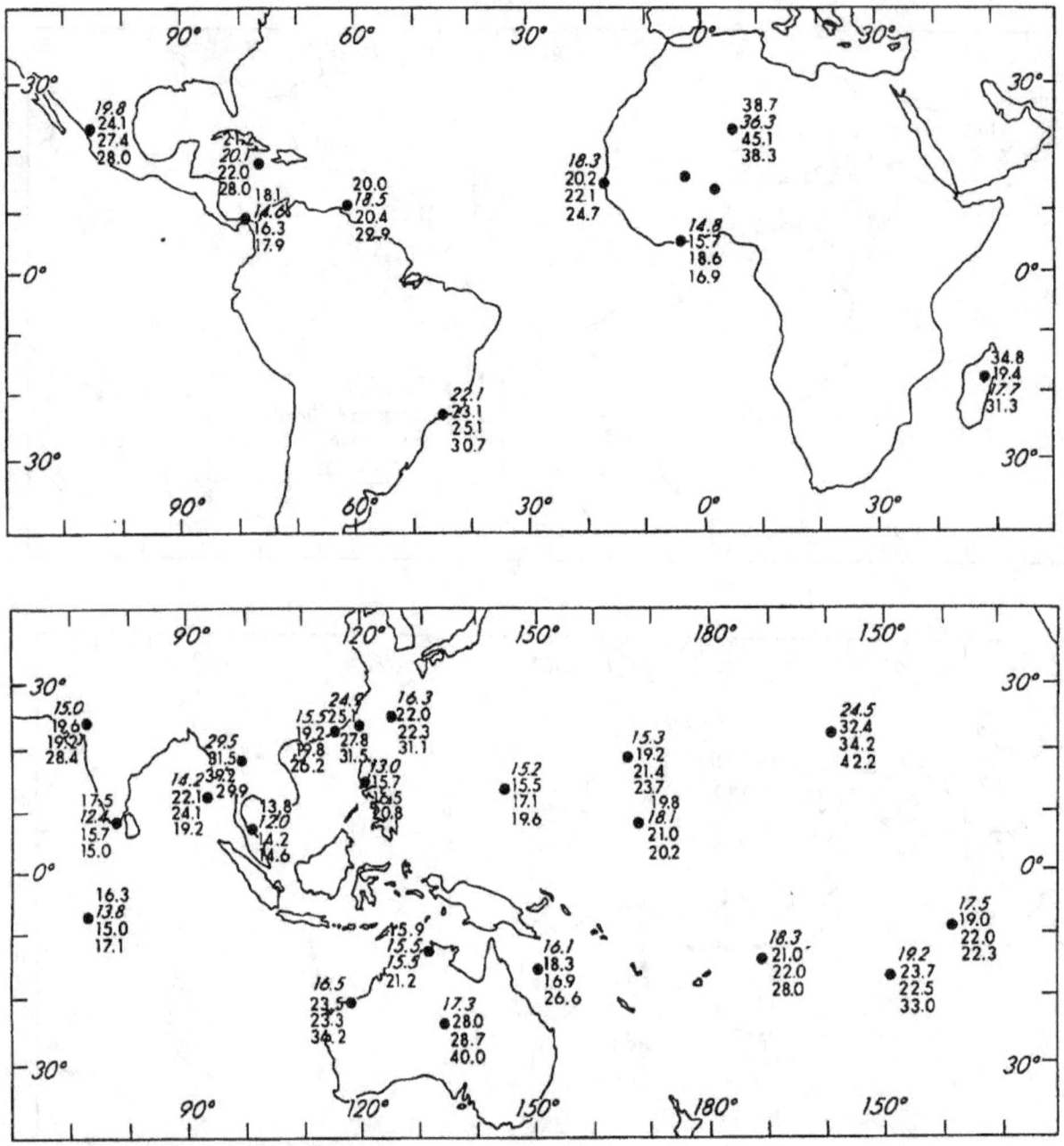

Figure 12-19. Wind forecasts for March and April 1991 at 32 tropical stations. Comparison of NMC numerical prediction of 200 mb winds 24 hours in advance to objective 1/2 (persistence + climotology); persistence; and climatological predictions. Mean vector errors are in knots.

At 850 mb, ½ (persistence + climatology) is superior, while simple persistence or climatology forecasts are almost as good as the numerical predictions. At 200 mb, numerical predictions are best, but even here, the improvement of 3 knots over persistence forecasts would hardly justify waiting several hours for a numerical product.

Table 12-2 supports the common assumption that numerical prediction skill is partly an inverse function of latitude. At 850 mb, ½ (P+C) is best at all latitudes, but its superiority to NMC diminishes with latitude. At 200 mb, ½ (P+C) is best between 0 and 10°, while NMC is 4 knots better between 20 and 30°.

These comparisons confirm the value of persistence/ climatology forecasts in the tropics, but ½ (P + C) may not be optimal for 24-hour wind forecasts; the best proportion of persistence and climatology could depend on altitude, location and season, and is worth determining. The best combinations for wind forecasts beyond 24 hours (presumably with the climatology contribution increasing with forecast time) should also be calculated. Besides being operationally valuable, such statistical forecast models should be used to test new numerical prediction techniques.

Numerical weather prediction models are using smaller and smaller grid spacing, more and more levels in the vertical, and shorter and shorter iterative time intervals. For forecasts to benefit, the number of observations must also increase. Satellites are being used (TIROS Operational Vertical Sounders, TOVS) to estimate the vertical distribution of mixing ratio. Tests have shown that when this information is included in the input of the ECMWF numerical model, rainfall forecasts in the tropics are slightly improved (Illari, 1989). Better resolutions attainable after 1993 with the new NOAA satellites may help more.

We still lack operational NWP models for the tropics. This means that in the near future, improvements in tropical analysis and forecasting will depend on subjective and statistical techniques.

12.4 MEDIUM AND LONG-RANGE FORECASTING

12.4.1 General. For this report, medium-range forecasts are those made for periods from 36 hours up to 1 week in advance; long-range forecasts are for periods longer than these. Medium-range forecasts may specify general cloudiness and rainfall, while long-range forecasts are less specific; e.g., rainfall above or below normal during a certain period, or a rainy season starting or ending earlier or later than normal. In the tropics, little progress has been made in either medium-range or long-range forecasting.

12.4.2 Medium-Range Forecasting (MRF). In parts of the NH subtropics during the cool season, some MRF skill may be achieved where weather can be linked to circulation features in mid-latitudes. For example, over southeast Asia, the occurrence and timing of northeast monsoon surges can be anticipated from numerical weather predictions of upper-trough passage across Siberia. Similarly, most of the heavy rain along the east coastal regions of Central America during winter stems from fronts or shear lines; forecasts can be made several days in advance on the basis of increasingly reliable numerical upper-air and surface prognoses over North America.

In 1971, Madden and Julian detected a 40- to 50-day oscillation in the zonal winds over the tropical Pacific. During FGGE, in 1979, the oscillation was well-marked and could also be found in cloud systems. Much research resulted that included as yet unconfirmed hypotheses on the origin and strength of the oscillation. Knutson and Weickmann (1987) list the following characteristics of the oscillation:

• It comprises global-scale tropical wind and convection anomalies.

• It is not strictly periodic but has a preferred time scale of about 30 to 60 days.

• It tends to propagate eastward at 6 to 12 knots over the tropical Indian and west Pacific Oceans.

• It has zonal wind anomalies in the lower troposphere that are out of phase with those in the upper troposphere.

• It occurs throughout the year.

The oscillation generally appears to start in the central equatorial Indian Ocean as a large convective mass that spawns smaller eastward-moving systems. Rather unexpectedly, the main cloud mass starts eastward across Indonesia and the western Pacific; its intensity fluctuates, but at least in hindsight, it can sometimes be followed around the globe. Figure 12-20 (next page) shows composite OLR anomalies accompanying passage of an oscillation in the equatorial region. Months may pass without an oscillation. At any time, convective clusters may move eastward along the equator at synoptic-scale intervals (Hayashi and Nakasawa, 1989) and so hide or distort the 30- to 60-day oscillation.

During the NH summer, other oscillations move northward from the equatorial Indian Ocean, taking about 20 days to reach 25° N. These can be linked to the monsoon cycle over India, where "active" and "break" conditions alternate. The ridge moving northward in Figure 8-58 is probably part of this sequence.

It is one thing to recognize passage of an oscillation after the fact, and another to anticipate it. Forecasters should wait until motion has been well established before extrapolating and predicting. Even then, subsequent variations in speed and intensity, as well as interaction with shorter-period oscillations, are likely, and the forecast should incorporate such a range.

The west-east oscillations are strongest between the central Indian Ocean and the date line; the south-north oscillations are strongest between the central Indian Ocean and south and southeast Asia.

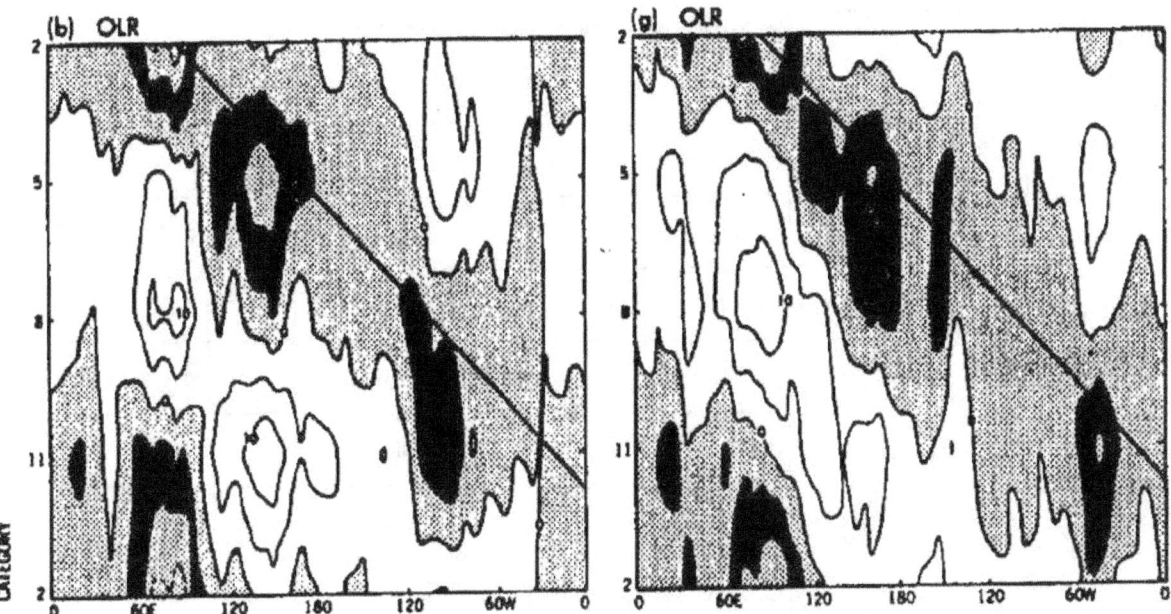

Figure 12-20. Composite 30-60 day anomalies of outgoing long-wave radiation between the equator and 10° N (isopleth interval 5 W m⁻²) for May to October (left side) and November to April (right side). The ordinate extends over approximately a 60-day interval (from Knutson and Weickmann, 1987).

Otherwise, during summer, about all that can be attempted for MRF is extrapolation of existing circulation and weather systems modified by climatology. To make accurate medium-range forecasts in some tropical areas, development and movement of tropical cyclones must be forecast days in advance, a feat well beyond the skills of today.

Meteorologists constantly seek patterns other than the astronomically-controlled annual and diurnal cycles. Since FGGE, the search has intensified in the tropics; most of the seekers agree that the 30- to 60-day cycle is real. However, rigorous analysis has questioned its statistical basis (Maliekal, 1990).

Forecasters should suspect any claim for a periodicity that is based on a relatively short time-series of data that has been averaged or filtered in some way. Remember, a periodicity that is not visually obvious in a long time-series plot of the raw data probably does not exist.

12.4.3. Long-Range Forecasting (LRF).

12.4.3.1 General. Flohn (1961) concluded that we lack:

• Useful, practical techniques for LRF.

• A fundamental and physically sound concept of the large-scale weather anomalies for LRF.

•An evaluation of the climatological data needed for any fundamental research directed toward LRF.

Flohn cautioned against assuring users that LRF is possible in the tropics or that successful techniques could be developed within a few years. Despite some research in the past several decades, his view still holds. Since dynamic techniques show little or no skill in making *short-range* forecasts in the tropics, it is easy to see why they have not been applied to LRF problems. Almost all attempts at LRF in the tropics have used statistical techniques, where the predictand (element to be forecast) is correlated to many predictors (usually observed values of a variety of weather elements) at various locations.

TABLE 12-3. Predictors for summer rainfall in India (Walker, 1924).

Area	Element	For Preceding
Peninsular India		
South American	Pressure	Apr-May
South Rhodesia	Rain	Oct-Apr
Dutch Harbor	Temperature	Dec-Apr
Java	Rain	Oct-Feb
Zanzibar	Rain	May
Cape of Good Hope	Pressure	Sep-Nov
Northwest India		
South American	Pressure	Apr-May
South Rhodesia	Rain	Oct-Apr
Dutch Harbor	Temperature	Mar-Apr
Equatorial	Pressure	Jan-May
Himalayan	Snow Depth	May

12.4.3.2. LRF of Summer Rainfall in India. The most famous attempts at LRF in the tropics were made by Sir Gilbert Walker and his disciples who tried to forecast seasonal rainfall in India using predictors from stations throughout the world. Walker (1924) computed several correlation coefficients between Indian seasonal rainfalls and possible predictors. He selected only a few pairs that appeared to be significantly correlated. The most important predictions were for the summer monsoon rainfall of Peninsular and Northwest India; the predictors selected for these are shown in Table 12-3, above.

These predictors gave a multiple correlation coefficient of 0.76 for both areas for the dependent data sample. Rao (1965) summarized Walker's work and later attempts at seasonal forecasting in India. He found that the correlation coefficients of most of the initial predictors chosen varied greatly from one decade to another. Figure 12-21 shows the variation with time of the best of the original predictors, South American pressure. This instability in Walker's initial formulae has caused them to be periodically revised since 1924. Showing the opposite trend to South American pressure, the correlation of March, April, May surface air temperature over western central India with Indian summer rainfall changed from slightly negative to significantly positive around 1940. Parthasarathy et al. (1990), who reported this, used

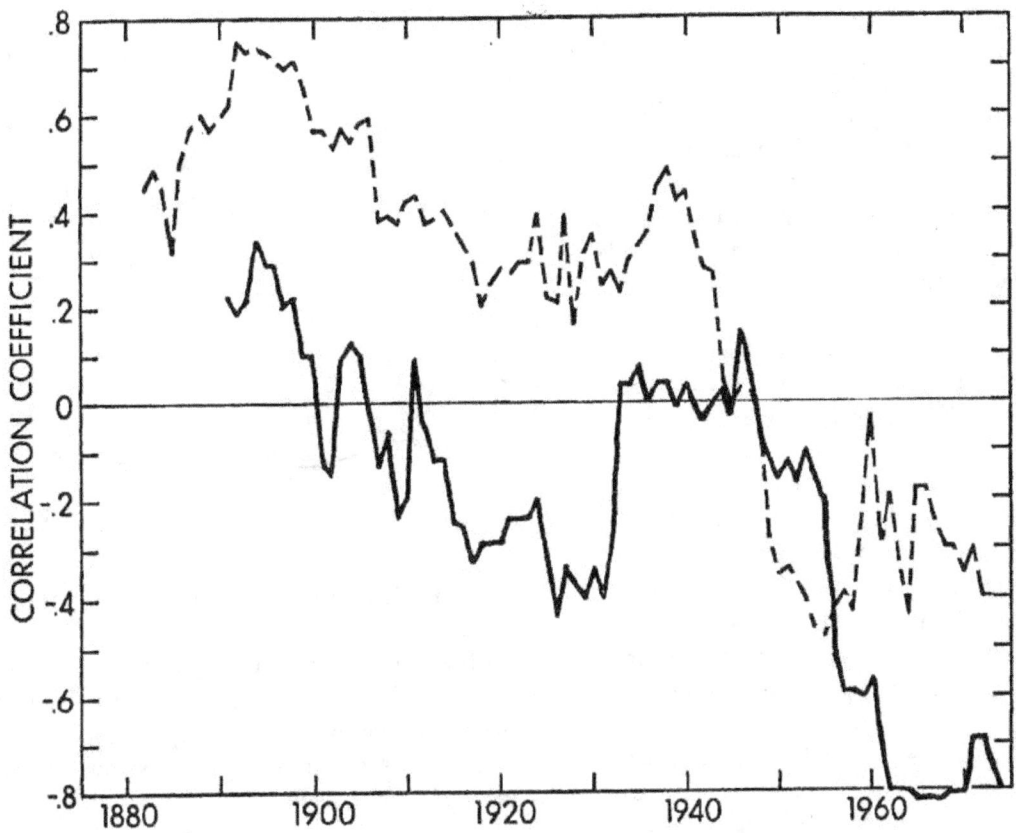

Figure 12-21. Fifteen-year running mean correlation coefficients between: (1) average of April-May surface pressure at three South American stations near 30° S and the subsequent June-September rainfall over Peninsular India (dashed line); (2) the difference between the January mean sea-level pressures at Irkutsk (52° N, 104° E) and Tokyo (36° N, 114° E) (solid line). Values are plotted at the mid-years of the 15-year periods.

dependent data for 1951 to 1980 to show that a relatively warm lower and middle troposphere over north and central India is followed by good monsoon rain. Whether such new relationships, summarized by Hastenrath (1990b), are statistically stable and help forecasts of seasonal rainfall remains to be seen.

Even though Walker was unsuccessful in LRF, in the process he discovered various global atmospheric oscillations or tendencies for pressure anomalies of opposite phase. The best known, the "Southern Oscillation," is a large negative correlation in surface-pressure anomalies between Djakarta (6° S, 107° E), Indonesia, and the eastern South Pacific. The associated El Niño and the possibilities it raises for

LRF are discussed in 6.3.5 (see Figure 6-25). Meteorological centers should receive monthly Climate Diagnostics Bulletins from the NMC Climate Analysis Center (1989). If a Bulletin indicates that El Niño or the opposing anti-Niño is well established, this information should be passed to the field, where the statistical rainfall relationships shown in Figure 6-25 can be applied to making LRFs. Even here, forecasters should be cautious. In the past, some apparently well-established Niños have suddenly and inexplicably quit. Although a vigorous Niño favors synoptic developments that will ensure its persistence, it does not guarantee them; should they be absent, then, as in 1975 and 1990, El Niño and its associated weather anomalies can quickly fade.

12.4.3.3 Atlantic Tropical Storms. Gray (1984a and b), using data from 1950 through 1982, discovered statistical relationships among the Atlantic tropical cyclone season and preceding (by the end of May) information on the quasi-biennial oscillation (QBO, see 4.3.5), El Niño, and average sea-level pressure over the Caribbean in April-May. For the latter two linkages, he gave the following reasonable physical explanations:

• During the QBO, Atlantic tropical cyclones are more frequent when 30-mb winds are westerly and increasing, rather than easterly and increasing. Gray gave no explanation.

• During El Niño, high SST over the eastern Pacific causes more deep convection there. The resultant outflow aloft enhances upper tropospheric westerlies over the Caribbean and western equatorial Atlantic. Consequently, the 200-mb anticyclonic flow necessary for tropical cyclones to develop is reduced.

• During the hurricane season in the Caribbean basin, below normal monthly mean sea-level pressure is associated with increased hurricane activity. Pressure anomalies tend to persist from spring through summer. Since the three predictors are not correlated with one another, they can be combined (using multiple regression analysis) to provide forecast equations that have more predictive skill than forecasts made using individual predictors. Gray developed prediction equations for

- Number of hurricanes

- Number of hurricanes and tropical storms

- Number of hurricane days

- Number of hurricane and tropical storm days

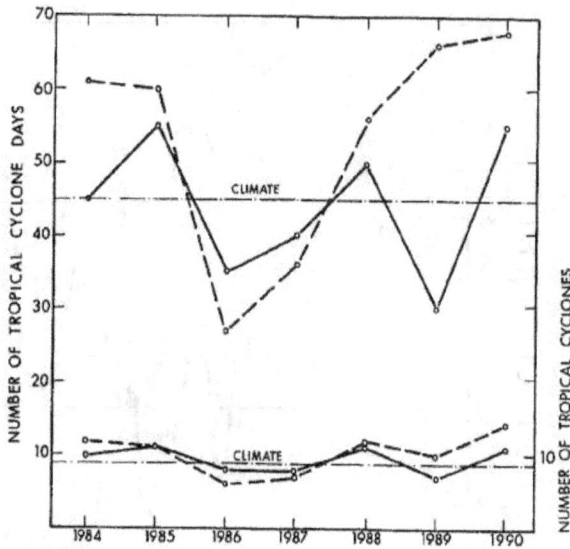

Figure 12-22. Predictions of North Atlantic hurricanes and tropical storms made each May for the following season by Gray (1990) (solid line), compared to the outcomes (dashed line) and to climatology (dot-dashed line). Upper curves show number of days with tropical cyclones; lower curves, number of tropical cyclones.

Gray made seven hurricane season predictions in late May 1984 through 1990 and updated them in late July. Performances of the May forecasts for number of hurricanes and tropical storms and number of days with hurricanes and tropical storms are shown in Figure 12-22. The July forecasts were somewhat better. The RMS errors of the forecasts differ little from those of climatological forecasts. In 1989, much too little activity was forecast. According to Gray (1989) the unusually heavy Sahel rainfall was responsible, causing the development of so many east and central Atlantic systems" Reminiscent of the Indian experience, Gray (1990) incorporated west Sahel rainfall in his predictions for 1990.

12.4.3.4 Other Forecasts. During the tropical cyclone season in the Australian region, Darwin (12° S, 131° E) surface-pressure anomalies are inversely related to the number of tropical cyclone days. Since Darwin winter pressure anomalies tend to persist, Nichols (1985) related them to subsequent tropical cyclone days by the regression equation:

Cyclone days =
224.5 - [11.6 (July to September pressure) - 1000 mb]

For northeastern Brazil, Hastenrath (1990c) found that half of the interannual variability of March-September rainfall can be predicted from the rainfall of the preceding October-January. The persistence predictors of Nicholls and Hastenrath are linked to the tendency of Southern Oscillation or El Niño anomalies to persist. This does not always happen.

For some years, May to October rainfall in Hong Kong (22° N, 114° E) was fairly successfully forecast in the previous February. This resulted from the January surface pressure gradient between Irkutsk (52° N, 104° E) and Tokyo (36° N, 140° E) being highly negatively correlated with the subsequent Hong Kong rainfall (Bell, 1976). There was no physical explanation. As Figure 12-21 showed, the correlation varied greatly since 1890. For the 15 years ending in 1988, it had fallen to -0.58 (Lau and Chan, 1990).

Many other simple and multiple regression techniques to forecast summer rainfall in Hong Kong have been developed and tested (Lau and Chan, 1990). All give forecast errors of about 12 inches (300 mm), or more than 15 percent of the normal summer rainfall of Hong Kong. The extremely wet summer of 1982 was underforecast, on average, by 30 inches (755 mm). Lau and Chan agreed with Reynolds (1978), who concluded that a regression equation with a correlation of less than 0.90 is unlikely to produce useful forecasts.

An even more dramatic example was reported by Parthasarthy et al. (1991), who correlated the surface pressure tendency at Bombay (19° N, 73° E) between the preceding winter and spring with subsequent summer monsoon rainfall over India. The correlation coefficient changed from negative between 1871 and 1901 to positive between 1902 and 1945; it has been negative since then. One of the major problems in deriving LRF techniques is that data is rarely, if ever, sufficient to allow a technique not only to be developed but to be tested on a large sample of independent data. For the present, tropical meteorologists should rely on climatology when they prepare long-range estimates of future weather conditions.

In summary, consistently skillful medium-range or long-range forecasts have not yet been achieved in the tropics. As a substitute, carefully tailored applied climatology often based on satellite data can usually provide sufficient information for operational planning. Thorough knowledge of diurnal variations and local influences on weather can also be incorporated.

12.5 PREPARING TO FORECAST IN A NEW AREA

The U.S. Air Force operates few tropical forecast offices, but future military contingencies in the tropics may suddenly demand forecasts for places that lack any but rudimentary published information on local and areal weather. What has been discussed in this report can be used to improve the chances of making successful forecasts. The work should first concentrate on the immediate season, and then move (ahead of the calendar) to successive seasons. Proceed to:

(1) Collect large-scale topographic maps of the area of interest (AOI). From the CD-ROM Defense Mapping Agency computer disks, plot 3-D representations of the AOI terrain (see Figure 7-38). On a larger scale, and for different aspects, plot views of the terrain surrounding key locations, such as airfields, staging zones, or ports. The 3-D images can greatly aid understanding of orographic effects and local circulations.

(2) From the latest Climate Diagnostics Bulletin published by the Climate Analysis Center, determine the stage of El Niño/Southern oscillation (ENSO) cycle and its prognosis. From Figure 6-25, decide if the AOI is likely to be affected. If it is, make sure that weather satellite and other data being analyzed come from earlier periods at the same stage of the ENSO cycle.

(3) Collect all available climatological data for the AOI and its environment from USAFETAC. Most comprise mean monthly and annual rainfalls. Data from analogous areas might be used; for example, eastern Madagascar, eastern north Luzon and northern Central America; coastal Angola and coastal Peru. Sea-surface temperature and tropical storm tracks and frequencies are useful.

(4) Collect satellite-based monthly climatologies of cloudiness (Sadler et al., 1984), frequency of highly reflective clouds (Garcia, 1985), and outgoing longwave radiation (Janowiak et al., 1985). Lightning climatologies may soon be available. From them,

deduce mean monthly cloudiness and precipitating clouds at key locations in the AOI. Compare these values with similarly determined values for stations with rainfall climatologies, and deduce the AOI rainfall climatology. From this, estimate frequency distributions of daily rainfalls using Tables 6-4 and 6-5, the coefficient of variation of mean monthly rainfall using Figure 6-37, and the chances of rain of different intensities using Table 6-7.

(5) From at least a month of geostationary IR images, determine the diurnal variation of deep clouds and their favored locations within the AOI. Construct distribution charts (see Figure 7-39) and link to the 3-D topographic representations. If time presses, use only near-dawn and mid-afternoon images to encompass the range of the variation. Combine IR and visible images to determine locations and diurnal variations of fog, stratus, dust, land-sea breezes, mountain-valley winds, and wind channeling.

(6) Using the same weather satellite images, including those from polar orbiting satellites, select a sequence of near-midnight and near-midday views in order to reduce or eliminate small-scale local distortions due to diurnal variations. This will help the meteorologist identify and follow synoptic and mesoscale disturbances and determine their climatologies.

(7) Deduce the likelihood of significant low-level jets from estimates of surface temperature gradients.

(8) As time permits, extend the analyses to subsequent seasons (ahead of the calendar).

In anticipation of urgent requirements, USAFETAC should maintain the climatologies mentioned in (4), on tapes as well as in Atlases. The Center should collect and extend at least 3 years' of all available geostationary satellite images in high-quality hardcopy and in video form. The excellent climatological files should be updated often and enhanced by unpublished data from remote places.

12.6 SUMMARY

This manual broadly surveys our knowledge of tropical meteorology. Lack of published research for South America and of recently-published research for Africa has unavoidably imposed a regional bias. The first edition covered the advances made since the tropical manual written by Palmer et al. (1955).

The second edition has taken advantage of tropical research conducted during GATE and FGGE, and including MONEX. Greatly improved climatology derived from satellite data and from growing interest in climate change has made this basis of tropical meteorology firmer than before. Despite a focused effort, progress in understanding and predicting tropical cyclones has been slow; progress elsewhere in the tropics, starting from much lower levels, has been more encouraging. To keep abreast of developments, tropical meteorologists must periodically review professional journal articles, technical reports, and other literature.

More progress has been made in diagnosing and understanding structure of the tropical atmosphere than in accurately predicting tropical circulation and weather patterns. In the tropics, and especially between 10° N and 10° S, following a weather system from one synoptic chart to the next is often possible only in retrospect. A surge is discovered after it has affected the weather. A squall line may be distinguished within patternless convection only after it starts to move, and then its chances of persisting can seldom be foretold. As a result, tropical forecasting is still handicapped. Thus, the best possible meteorological support must balance requirements and limitations. Tropical meteorologists should concentrate on techniques that are most likely to show forecasting skill. The following limitations should be clearly recognized:

• It is often possible to make an areal forecast of whether and where convection will be above or below normal; but it is rarely possible to pinpoint where and when rain will fall 30 minutes or an hour in advance.

• The convective, cellular nature of tropical rainstorms makes forecasts of point rainfall very unreliable.

• Rarely, when large-scale circulations (such as El Niño) become persistent, some medium- and long-range forecasts can be more skillful than chance, but for most of the time, over most of the tropics, medium- and long-range forecasts are useless.

• Tropical forecasting is handicapped by observation and forecast codes, which were designed for higher latitudes and have deficiencies when used in the tropics. For example, important information on cumuliform clouds such as shear-induced leaning, combining with layer clouds (as in shear lines), or evaporating tops (at an inversion) cannot be encoded. Holle (1975) proposed a new code to describe the state of the sky in the tropics but it was not accepted. Forecasts tend to be too stereotyped for many tropical stations or areas and there is excessive use of INTER or TEMPO groups. This needs major study.

Good analyses and diagnoses of the causes and precursors of present weather are prerequisite to accurate forecasts. Many times, however, the forecaster will be unable to account for some weather patterns. Hence, forecasting changes in these patterns will often fail. In general, weather systems that form and dissipate without moving are the most troublesome. Also, disturbances in the low-level wind-speed field, associated with horizontal shear or convergence in zonal currents, are trickier than disturbances associated with shifts in wind direction, such as vortexes or troughs.

Forecasting fine weather is easier than forecasting unsettled weather because of the former's larger time and space scales of divergence and vertical motion. The forecasts could reflect this by assigning higher probabilities when fine weather is expected. From another viewpoint, the rarer the meteorological event, the harder it is to forecast and the more important it is to the forecast user. Generally speaking, tropical forecasters must work harder than mid-latitude forecasters to achieve equal skill. Although the tropics lack good numerical weather predictions, meteorologists are making better wind forecasts because of more observations, better communications, and objective wind analysis and forecast programs. Short-range forecasts based on these analyses are usually accurate enough for aircraft operations since large-scale tropical circulation patterns change slowly.

Forecasting should be an integral part of meteorology, because it provides the acid test of hypotheses and is a prolific source of new ideas. Forecasters should also be researchers and researchers should test their findings by using them in forecasting. At the National Hurricane Center in Miami and to a lesser degree at the Joint Typhoon Warning Center in Guam, forecasters can undertake applied research in the "off-season." During summer 1990, when a field program to study tropical cyclone movement was based at JTWC (Elsberry, 1989), researchers regularly forecasted, and continually interacted with forecasters. In China, most forecasters work shift for 3 months and then spend the next 3 months on research. Research meteorologists instruct operational meteorologists and routinely work forecast shifts (Grice et al., 1986).

The Stormscale Operational and Research

Meteorology (STORM) program, planned for the 1990s over the United States (STORM Project Office, 1988), includes experimental forecast centers. In these, forecaster/researchers will use forecasting in applying the scientific method to solving meteorological problems.

Each of the tropical synoptic models described in Chapter 8 was developed from a few case studies and so tend to be unrepresentative. In a case study, the end as well as the beginning is known, and the range of the model and its exceptions—the statistics—are usually unknown. The forecaster using the model has neither the time to report on its value, nor to modify it on the basis of his experience. An Experimental Forecast Center would overcome these problems and continually correct and refine the models following the procedure illustrated in Figure 12-23.

In most tropical forecast offices, meteorologists are fully occupied on shift, with little time to tackle the host of local problems contributing so much to forecast errors. Great distances separate forecasters from researchers, who are thus deprived of evaluations and cut off from new ideas. Even so, on a very modest scale, turning tropical forecast offices into Experimental Forecast Centers would do more for forecasting skill than would scores of isolated investigators working on what they thought were important problems.

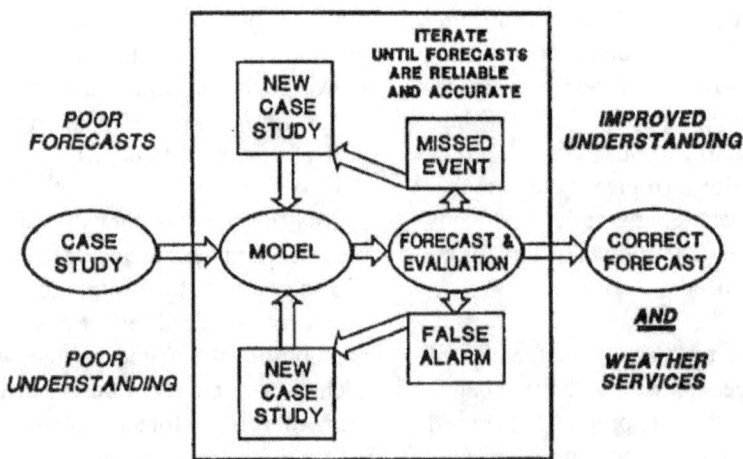

Figure 12-23. Iterative procedure for evaluating and modifying case studies through forecast performance.

REFERENCES

Adams, C.W., *Monthly Upper-air Means for Selected Stations of the Central Pacific 1956-1959,* Hawaii Inst. Geophys., HIG 64-24, 1964.

Air Weather Service, *Application of Meteorological Satellite Data in Analysis and Forecasting*, AWS TR-212, HQ AWS, Scott AFB, IL, 1969.

Air Weather Service, *Use of Asynoptic Data in Analysis and Forecasting*, AWS TR-225, HQ AWS, Scott AFB, IL, 1960.

Alaka, M.A., *Problems of Tropical Meteorology*, WMO Tech. Note 62, 1964.

Albright, M.D., E.E. Recker, and R.J. Reed, "The Diurnal Variation of Deep Convection in the Central Tropical Pacific During January-February 1979," *Mon. Wea. Rev.*, 113, pp 1663-1680, 1985.

Alexander, M.A., and S.D. Schubert, "Regional Earth-Atmosphere Energy Balance Estimates Based on Assimilations with a GCM," *J. Climate*, 3, pp 15-31, 1990.

Alvi, S.M.A., and K.G. Punjabi, "Diurnal and Seasonal Variations of Squalls in India," *Indian J. Meteor. Geophys.*, 17, pp 207-216, 1966.

Ananthakrishnan, R., V.R. Acharya, and A.R. Ramakrishna, "On the Criteria for Declaring the Onset of the Southwest Monsoon Over Kerala," *India Meteor. Dept. Forecasting Manual Part IV; 18: Monsoons of India*, 1967.

Ananthakrishnan, R., and J.M. Pathan, "North-South Oscillations of the Equatorial Trough and Seasonal Variations of Rainfall in the Tropics," *Proc. Symp. Tropical Meteor.*, Honolulu, Amer. Meteor. Soc., FV-1-6, 1970.

Anthes, R.A., *Tropical Cyclones, Their Evolution, Structure and Effects*, Meteor. Monogr., 1982.

Arakawa, H., "The Formation of Hurricanes in the South Pacific and the Outbreaks of Cold Air from the North Polar Regions," *J. Meteor. Soc. Japan*, Ser.2, 1, pp 1-6, 1940.

Ardanuy, P., "On the Observed Diurnal Oscillation of the Somali Jet," *Mon. Wea. Rev.*, 107, pp 1694-1700, 1979.

Argulis, R.P., *Tsunami*, WBTM-WR-48, ESSA Tech. Memo, 1970.

Aspliden, C.I., G.A. Dean, and H. Landers, *Satellite Study, Tropical North Atlantic, 1963*, Dept. Meteor. Florida State Univ, 1965-1967

Atkinson, G.D., *A Preliminary Estimate of Extreme Wind Speeds in Thailand*, 1st Weather Wing Tech. Study 3, 1966.

Atkinson, G.D., *Thunderstorms in Southeast Asia*, 1st Weather Wing Tech. Study 11, 1967a.

Atkinson, G.D., *Climatology of Significant Weather at Mactan AB, Philippines,* 1st Weather Wing Tech. Study 13, 1967b.

Atkinson, G.D., *Extreme Daily Rainfall Amounts in the Republic of Vietnam,* 1st Weather Wing Tech. Study 16, 1968.

Atkinson, G.D., and C.R. Holliday, "Tropical Cyclone Minimum Sea Level Pressure/Maximum Sustained Wind Relationship for the Western North Pacific," *Mon. Wea. Rev.,* 105, pp 421-427, 1977.

Atkinson, G.D., and H.E. Penland, *Typhoon Weather Models—Kadena AB, Okinawa,* 1st Weather Wing Tech. Study 12, 1967.

Atkinson, G.D., and J.C. Sadler, *Mean Cloudiness and Gradient-Level Wind Charts Over the Tropics, Vol. I (text), Vol. II (charts),* AWS TR 215, HQ Air Weather Service, Scott AFB, IL 1970.

Atlas, D., "Optical Extinction by Rainfall," *J. Meteor.,* 10, pp 486-488; 1953.

Atlas, R., S.C. Bloom, R.N. Hoffman, J.V. Ardizzone and G. Brin, "Space-based Surface Wind Vectors to Aid Understanding of Air-Sea Interactions," *Eos,* 72, 18, pp 201-208; 1991.

Augustine, J.A., "The Diurnal Variation of Large-Scale Inferred Rainfall over the Tropical Pacific Ocean during August 1979," *Mon. Wea. Rev.,* 112, pp 1745-1751; 1984.

Baer, F., "The Influence of Illumination on Cirrus Reports," *Bull. Amer. Meteor. Soc.,* 37, pp 366-367; 1956.

Barnes, G.M., "18th Conference on Hurricanes and Tropical Meteorology," *Bull. Amer. Meteor. Soc.,* 71, pp 558-570; 1990.

Barry, R.G., *Mountain Weather and Climate,* Methuen, London; 1981.

Bartlett, D., and J. Bartlett, "Africa's Skeleton Coast," *Nat. Geogr.,* 181, pp 54-85; 1992.

Battan, L.J., *Radar Meteorology,* U. Chicago Press; 1959.

Bechtold, P., J.-P. Pinty and P. Mascart, "A Numerical Investigation of the Influence of Large-Scale Winds on Sea-Breeze- and Inland-Breeze-Type Circulations," *J. Appl. Meteor.,* 30, pp 1268-1279; 1991.

Bedient H.A., and J. Vederman, "Computer Analysis and Forecasting in the Tropics," *Mon. Wea. Rev.,* 92, pp 565-567; 1964.

Bell, G.J., "Seasonal Forecasts of Hong Kong Summer Rainfall," *Weather,* 31, pp 208-212; 1976.

Bergeron, T., "Review of Tropical Hurricanes," *Quart. J. Roy. Meteor. Soc.,* 80, pp 131-164; 1954.

Bingham, H., *Across South America.* Houghton Mifflin, Boston; 1911.

Bjerknes, J, "Atmospheric Teleconnections from the Equatorial Pacific," *Mon. Wea. Rev.,* 97, pp 163-172, 1969.

Braak, C., "Het Climaat Van Nederlandsch Indie," *Magnet. Meteor. Observ., Batavia, Verhand.*, 8; 1921-29.

Brand, S., and C.P. Guard, "An Observational Study of Extratropical Storms Evolved from Tropical Cyclones in the Western North Pacific," *J. Meteor. Soc., Japan*, 57, pp 479-482; 1979.

Brier, G.W., and J. Simpson, "Tropical Cloudiness and Rainfall Related to Pressure and Tidal Variations," *Quart. J. Roy. Meteor. Soc.*, 95, pp 120-147; 1969.

Brill, K., and B. Albrecht, "Diurnal Variation of the Trade-Wind Boundary Layer," *Mon. Wea. Rev.*, 110, pp 601-613; 1982.

Brode, R.W., and M.K. Mak, "On the Mechanism of the Monsoonal Mid-Tropospheric Cyclone Formation," *J. Atmos. Sci.*, 35, pp 1473-1484; 1978.

Brook, R.R., "Koorin Nocturnal Low-Level Jet," *Bound. Layer Meteor.*, 32, pp 133-154; 1985.

Brooks. B., *Forecasting the "Atemporalado" in Honduras.* AWS/FM-87/001, HQ Air Weather Service, Scott AFB, IL; 1987.

Brooks, C.E.P., and N. Carruthers, *Handbook of Statistical Methods in Meteorology,* Her Majesty's Stationery Office, London; 1953.

Browner, S.P., W.L. Woodley and C.G. Griffith, "Diurnal Oscillation of the Area of Cloudiness Associated with Tropical Storms," *Mon. Wea. Rev.*, 105, pp 856-864; 1977.

Browning, K.A., *Foreword. Nowcasting: Mesoscale Observations and Short-Range Prediction*, ESA SP-165, pp xi-xii; 1981.

Bryson, R.A., and P.M. Kuhn, "Stress-Differential Induced Divergence with Application to Littoral Precipitation," *Erdkunde*, 15, pp 287-294; 1961.

Buchan, A., "Meteorology," *Encyl. Britannica, 9th Ed., Vol. 16,* pp 114-159; 1883.

Bunting, D.C., "Ship Anemometer Heights above Mean Water Level," *Naval Oceangr. Off. Informal Rept.*, pp 68-74, 1968.

Burpee, R.W., Some Features of Synoptic-Scale Waves Based on Compositing Analysis of GATE Data," *Mon. Wea. Rev.*, 103, pp 921-925; 1975.

Businger, S., "Arctic Hurricanes," *Amer. Scientist*, 79, pp 18-33; 1991.

Byers, H.R., *General Meteorology, Fourth Edition,* McGraw-Hill, New York; 1974.

Byers, H.R., and R.R. Braham, *The Thunderstorm,* U.S. Govt. Printing Off., Washington, D.C.; 1949.

Cadet, D., and M. Desbois, "A Case Study of a Fluctuation of the Somali Jet during the Indian Summer Monsoon," *Mon. Wea. Rev.*, 109, pp 182-187; 1981.

Caracena, F., R.L. Holle and C.A. Doswell III, *Microbursts: A Handbook for Visual Identification, Second Edition,* Dept. of Commerce, 1990.

Carlson, T.N., "Structure of a Steady-State Cold Low," *Mon. Wea. Rev.,* 95, pp. 763-777; 1967.

Carlson, T.N., "Synoptic Histories of Three African Disturbances that Developed into Atlantic Hurricanes," *Mon. Wea. Rev.,* 97, pp 256-276; 1969a.

Carlson, T.N., "Some Remarks on African Disturbances and Their Progress over the Tropical Atlantic," *Mon. Wea. Rev.,* 97, pp 716-726; 1969b.

Carr, F.H., "Mid-Tropospheric Cyclones of the Summer Monsoon," *Pure Appl. Geophys.,* 115, pp 1383-1412; 1977.

Carter, E.A., and L.A. Schuknecht, *Peak Wind Statistics Associated with Thunderstorms at Cape Kennedy, Florida,* NASA Contr. Rept. No. 61304; 1969.

Cataneo, R., "A Method for Estimating Rainfall Rate Radar Reflectivity Relationships," *J. Appl. Meteor.,* 8, pp 815-819; 1969.

Caviedes, C.N., "El Niño 1972: Its Climatic, Ecological, Human and Economic Implications," *Geogr. Rev.,* 65, pp 493-509; 1975.

Chaggar, T.S., "Réunion Sets New Rainfall Records," *Weather,* 39, pp 12-14; 1984.

Chan, J.C.L., and G.J. Holland, "Observing and Forecasting Tropical Cyclones — Where Next?" *Bull. Amer. Meteor. Soc.,* 70, pp 1560-1563; 1989.

Chang, C.-P., J.E. Erickson and K.M. Lau, "Northeasterly Cold Surges and Near-Equatorial Disturbances over the Winter MONEX Area during December 1974; Part I: Synoptic Aspects," *Mon. Wea. Rev.,* 107, pp 812-829; 1979.

Chang, C.-P., and K.M. Lau, "Northeasterly Cold Surges and Near-Equatorial Disturbances over the Winter MONEX Area during December 1974. Part II: Planetary-Scale Aspects," *Mon. Wea. Rev.,* 108, pp 298-312; 1980.

Chang, C.-P., and K.M. Lau, "Short-Term Planetary Scale Interactions over the Tropics and Midlatitudes during Northern Winter. Part I: Contrasts Between Active and Inactive Periods," *Mon. Wea. Rev.,* 110, pp 933-946; 1982.

Chang, C.-P., J.E. Millard and G.T.J. Chan, "Gravitational Character of Cold Surges during Winter MONEX," *Mon. Wea. Rev.,* 111, pp 293-307; 1983.

Chen, T.Y., "The Severe Rainstorms in Hong Kong during June 1966," *Roy. Obs. Hong Kong, Suppl. Meteor. Results;* 1969.

Chen, Y.-L., and T.A. Schroeder, "The Relationship among Local Winds, Synoptic-Scale Flow and Precipitation at Hilo during HAMEC," *Int. Conf. Monsoon and Mesoscale Meteor,* Taipei, Taiwan; 1986.

Cheng, P.T.T., and W.P. Kwang, *Radar Climatology of Hong Kong for the Years 1967-1969*, Roy. Obs., Hong Kong Tech. Note 34; 1973.

Cheng, S.S.M., *Dry Spells during Mid-Summer over South China*, Roy. Obs., Hong Kong Tech. Note 44; 1978.

Chin, P.C., and M.H. Lai, *Monthly Mean Upper Winds and Temperatures over Southeast Asia and the Western North Pacific*, Roy. Obs., Hong Kong Tech. Mem. 12; 1974.

Christian, H.J., R.J. Blakeslee and S.J. Goodman, "The Detection of Lightning from Geostationary Orbit," *J. Geophys. Res.*, 94, pp 13329-13337; 1989.

Chiu, P.-S., "A Contribution to the Upper-Air Climatology of Tropical South America," *J. Climatol.*, 5, pp 4-3-416; 1985.

Chu, P.-S., and J. Frederick, "Westerly Wind Bursts and Surface Heat Fluxes in the Equatorial Western Pacific in May 1982," *J. Meteor. Soc., Japan*, 68, pp 523-537; 1990.

Climate Analysis Center, *Climate Diagnostics Bulletin, August 1989*, NOAA/NWS/NMC; 1989.

Cobb, L.G., *Maximum Precipitation Values on the Lower Slopes of Mountainous Terrain, Research on Tropical Rainfall Patterns and Associated Mesoscale Systems*, Dept. Meteor., Texas A&M Univ.; 1966

Cobb, L.G., *The El Niño Phenomenon and Rainfall in Peru and Ecuador. Research on Tropical Rainfall Patterns and Associated Mesoscale Systems, Repts. 2 and 3.* Dept. Meteor., Texas A&M Univ.; 1967 and 1968.

Cobb, L.G., *The Annual and Daily Distribution of Rainfall in Southeast Asia: Research on Tropical Rainfall Patterns and Associated Mesoscale Systems*, Dept. Meteor., Texas A&M Univ., pp 53-75; 1968.

Cochemé, J., and P. Franquin, *An Agroclimatology Survey of a Semi-Arid Area in Africa South of the Sahara*, WMO Tech. Note 86; 1967.

Cole, A.E., and R.J. Donaldson, "Precipitation and Clouds: A Revision of Chapter 5, Handbook of Geophysics and Space Environments," *A.F. Surv. Geophys.*, 212, pp 1-108; 1968.

Coligardo, M., *An Investigation of Rainfall Variability and Distribution in Luzon and a Mesoscale Study of Rainfall of the Province of Laguna and Adjacent Areas, Philippines. Research on Tropical Rainfall Patterns and Associated Mesoscale Systems*, Dept. Meteor., Texas A&M Univ.; 1967.

Colòn, J.A., "A Study of Hurricane Tracks for Forecasting Purposes," *Mon. Wea. Rev.*, 81, pp 53-66; 1953.

Condray, P.M., and R.T. Edson, *The Caribbean Basin. An Electrooptical Climatology for the 8-12 Micron Band. Vol.1. Central America*, USAFETAC/TN-89/004, USAF Environmental Technical Applications Center, Scott AFB, IL; 1989.

Conner, W.C., R.H. Craft and D.L. Harris, "Empirical Methods for Forecasting the Maximum Storm Tide Due to Hurricanes and Other Tropical Storms," *Mon. Wea. Rev.*, 85, pp 113-116; 1957.

Conover, J.H., Studies of Clouds and Weather over Southeast Asia. *AFCRL Meteor. Branch* [not published]; 1967.

Conover, J.H., and J.C. Sadler, "Cloud Patterns as seen from Altitudes of 250 to 350 Miles - Preliminary Results," *Bull. Amer. Meteor. Soc.*, 41, pp 291-297; 1960.

Crutcher, H.L., and O.M. Davis, *U.S. Navy Marine Climatic Atlas of the World, Vol. VIII, the World,* Naval Weather Service Command, NAVAIR 50-1C-54; 1969.

Crutcher, H.L., and R.G. Quayle, *Mariners' Worldwide Climatic Guide to Tropical Storms at Sea,* NOAA/ EDS National Climatic Center; 1974.

Daniel, C.E.J., and A.H. Subramaniam, "Exceptionally Heavy Rain in South Kerala on 17/18 October, 1964," *Indian J. Meteor. Geophys.*, 17, pp 253-256; 1966.

Davidson, N.E., and G.J. Holland, "A Diagnostic Analysis of Two Intense Monsoon Depressions over Australia," *Mon. Wea. Rev.*, 115, pp 380-392; 1987.

Davidson, N.E., and B.J. McAvaney, "The ANMRC Tropical Analysis Scheme," *Aust. Meteor. Mag.*, 29, pp 155-168; 1981.

Defant, F., "Local winds," *Comp. Meteor.*, pp 655-672; 1951.

De Maria, M., M.B. Lawrence and J.T. Kroll, "An Error Analysis of Atlantic Tropical Cyclone Track Guidance Models," *Wea. Forecast.*, 5, pp 47-61; 1990.

Deshpande, D.V., "Heights of Cb Clouds over India during the Southwest Monsoon Season," *Indian J. Meteor. Geophys.*, 15, pp 47-54; 1964.

Deshpande, D.V., "Cirriform Clouds over India — Heights and Temperatures," *Indian J. Meteor. Geophys.*, 16, pp 635-644; 1965.

Detachment 8, 20th Weather Squadron, *Typhoon Information Package,* USAFETAC, 1988.

Dhar, O.N., "Cherrapunji Breaks the World Rainfall Record for 4-Day Duration and a List of Heavy Rainfall Stations in India," *Vayu Mandal,* July-Dec. 1977, pp 44-45; 1977.

Dhar, O.N., and P.R. Mhaisker, "Areal and Point Distribution of Rainfall Associated with Depressions/Storms on the Day of Crossing the East Coast of India," *Indian J. Meteor, Geophys.*, 24, pp 271-278; 1973.

Douglas, M.W., *The Structure and Dynamics of Monsoon Depressions,* Dept. Meteor., Florida State Univ., Rept. 86-15, 1987.

Drosdowsky, W., and G.J. Holland, "North Australian Cloud Lines," *Mon. Wea. Rev.*, 115, pp 2645-2659; 1987.

Dudhia, J., "Numerical Study of Convection Observed during the Winter Monsoon Experiment Using a Mesoscale Two-Dimensional Model," *J. Atmos. Sci.*, 46, pp 3077-3107; 1990.

Dunn, G.E., "Cyclogenesis in the Tropical Atlantic," *Bull. Amer. Meteor. Soc.*, 21, pp 215-229; 1940.

Duvel, J.P., "Convection over Tropical Africa and the Atlantic Ocean during Northern Summer. Part I: Interannual and Diurnal Variations," *Mon. Wea. Rev.*, 117, pp 2782-2799; 1989.

Dvorak, V.F., "Tropical Cyclone Intensity Analysis and Forecasting from Satellite Imagery," *Mon. Wea. Rev.*, 103, pp 420-430; 1975.

Dvorak, V.F., *Tropical Cyclone Intensity Analysis Using Satellite Data.* NOAA Tech. Rept. NESDIS 11; 1984.

Dvorak, V.F., and F. Smigielski, *A Workbook on Tropical Clouds and Cloud Systems Observed in Satellite Imagery.* NOAA/NESDIS; 1990.

Edson, R.T., and P.M. Condray, *The Caribbean basin. An Electrooptical Climatology for the 8-12 Micron Band. Vol. III. Northern South America,* USAFETAC/TN-89/006, USAF Environmental Technical Applications Center, Scott AFB, IL; 1989

Eigsti, S.L., *The Coastal Diurnal Wind Cycle at Port Aransas, Texas,* Atmos. Sci. Group, Univ. Texas Rept. 48; 1978.

Eldridge, R.H., "A Synoptic Study of West African Disturbances," *Q. J. Roy. Meteor. Soc.*, 83, pp 303-314; 1957.

Eliot, J., "*A Preliminary Discussion of Certain Oscillatory Changes of Pressure of Long Period and of Short Period in India*," *Mem. India Meteor. Dept.*, 16, pp 185-307; 1895.

Elrick, J.R., and A.C. Meade, *Weather Sensitivities of Electrooptical Weather Systems,* AWS/TN-87/003, HQ Air Weather Service, Scott AFB, IL; 1987.

Elsberry, R.L. (ed.), *A Global View of Tropical Cyclones,* Naval Postgraduate School; 1987.

Elsberry, R.L., *ONR Tropical Cyclone Motion Research Intiative; Field Experiment Planning Workshop,* Naval Postgraduate School, 6B-89-002; 1989.

Elsberry, R.L., G.J. Holland, H. Gerish, M. DeMaria, C.P. Guard and K. Emanuel, "Is There Any Hope for Tropical Cyclone Intensity Prediction? — A Panel Discussion," *Bull. Amer. Meteor. Soc.*, 73, pp 264-275; 1992.

Emmanuel. K.A., and R. Rotunno, "Polar Lows as Arctic Hurricanes," *Tellus*, 41A, pp 1-17; 1989.

Enfield, D.B., "El Niño, Past and Present," *Rev. Geophys.*, 27, pp 159-187; 1989.

ESSA, NASA and USAF, *U.S. Standard Atmosphere Supplement, 1966,* General Post Office, Washington, D.C.; 1967.

Estoque, M.A., "The Sea Breeze as a Function of the Prevailing Synoptic Situation," *J. Atmos. Sci.*, 19, pp 244-250; 1962.

Fernandez, W., "Environmental Conditions and Structure of the West African and Eastern Tropical Atlantic Squall Lines," *Arch. Meteor. Geophys. Bioclimatol., Ser. A*, 31, pp 71-89; 1982.

Fett, R.W., "Aspects of Hurricane Structure: New Model Considerations Suggested by TIROS and Project Mercury Observations," *Mon. Wea. Rev.*, 92, pp 43-60; 1964.

Fett, R.W., and W.A. Bohan, *Navy Tactical Applications Guide, Volume 6, Part 1 — Tropics, Weather Analysis and Forecast Applications, Meteorological Satellite Systems*, Bohan, Park Ridge, Illinois; 1986.

Findlater, J., "Observational Aspects of the Low-Level Cross-Equatorial Jet Stream," *Pure Appl. Geophys.*, 115, pp 1251-1262; 1977.

Fingerhut, W.A., "A Numerical Model of a Diurnally Varying Tropical Cloud Cluster Disturbance," *Mon. Wea. Rev.*, 106, pp 255-264; 1978.

First Weather Wing, *Objective Method to Forecast Gusty Surface Winds at Mactan AB, Philippines*, Tech. Study, 18; 1968.

Fletcher, R.D., "A Hydrometeorological Analysis of Venezuelan Rainfall," *Bull. Amer. Meteor. Soc.*, 30, pp 1-9; 1949.

Fletcher, R.D., "A Relationship Between Maximum Observed Point and Areal Rainfall Values," *Trans. Amer. Geophys. Union*, 31, pp 344-348; 1950.

Flohn, H., "Some Remarks on Long-Range Forecasting in Tropical and Subtropical Regions," *Proc. Symp. Tropical Meteorology in Africa*, Munitalp Foundation, Nairobi, pp 221-225; 1961.

Flohn, H., "Investigations on the Tropical Easterly Jet," *Bonner Meteor. Abhand.*, 4; 1964.

Flohn, H., "Studies on the Meteorology of Tropical Africa," *Bonner Meteor. Abhand.*, 5; 1965.

Flohn, H., "Tropical Circulation Patterns," *Bonner Meteor. Abhand.*, 15, pp 1-55; 1971.

Fox, T., "An Example of a Medium Level Westerly Wave over South and Central Africa," *Lusaka, Meteor. Notes, Ser. A*, 2; 1969.

Frank, N.L., "The Inverted V Cloud Pattern — an Easterly Wave?" *Mon. Wea. Rev.*, 97, pp 130-140; 1969.

Frank, N.L., "On the Energetics of Cold Lows," *Proc. Symp. Tropical Meteorology*, Honolulu, Amer. Meteor. Soc., EIV-1-6; 1970.

Frank, W.M., *The Structure and Energetics of the Tropical Cyclone*, Dept. Atmos. Sci., Colorado State Univ., Paper 258; 1976.

Frisby, E.M., *Hail Incidence in the Tropics,* US Army Electr. Comm. Tech. Rept., 2768; 1966.

Frisby, E.M., and H.W. Sansom, "Hail Incidence in the Tropics," *J. Appl. Meteor.,* 6, pp 339-354; 1967.

Frolow, S., "On Synchronous Variations of Pressure in Tropical Regions," *Bull. Amer. Meteor. Soc.,* 23, pp 239-254; 1942.

Fu, R., A.D. Del Genio and W.B. Rossow, "Behavior of Deep Convective Clouds in the Tropical Pacific Deduced from ISCCP Radiances," *J. Climate,* 3, pp 1129-1152; 1990.

Fujita, T.T., *The Downburst,* Univ. of Chicago; 1985.

Fujita, T.T., and F. Caracena, "An Analysis of Three Weather-Related Aircraft Accidents," *Bull. Amer. Meteor. Soc.,* 58, pp 1164-1181; 1977.

Fujita, T., T. Izawa, K. Watanabe and I. Imai, "A Model of Typhoon Accompanied by Inner and Outer Rainbands," *J. Appl. Meteor.,* 6, pp 3-19; 1967.

Fujita, T.T., K. Watanabe, K. Tsuchiya and M. Shimada, "Typhoon-Associated Tornadoes in Japan and New Evidence of Suction Vortices in a Tornado near Tokyo," *J. Meteor. Soc. Japan,* 50, pp 431-453; 1972.

Fukuda, H.T., *The End of a Rainy Season in the III Corps Tactical Zone of the Republic of Vietnam,* 1st Wea. Wing Sci. Serv. Spec. Tech. Rept.; 1968.

Fulks, J.R., "The Instability Line," *Comp. Meteor.,* pp 647-652; 1951.

Fullerton, C.M., and S.K. Wilson, *An Analysis of Four Showers with Rainfall Rates > 250 mm/hr,* Dept. Meteor, Univ. Hawaii, UHMET 75-04; 1975.

Gabites, J.F., *Low Latitude Analysis,* Proc. Interregional Seminar on Tropical Cyclones. Japan Meteor. Agency Tech. Rept. 21, pp 239-244; 1963.

Gage, K.S., B.B. Balsley, W.L. Ecklund, R.F. Woodman and S.K. Avery, "Wind-Profiling Doppler Radars for Tropical Atmospheric Research," *Eos,* 71, pp 1851-1854; 1990.

Galway, J.G., "The Lifted Index as a Predictor of Latent Instability," *Bull. Amer. Meteor. Soc.,* 37, pp 528-529; 1956.

Gan, T.L., "A Study of Some Heavy Rainfalls on the East Coast of Malaya during the Northeast Monsoon Season," *Mem. Malayan Meteor. Serv.,* 6; 1963.

Garcia, O., *Atlas of Highly Reflective Clouds for the Global Tropics: 1971-1983,* NOAA/ERL, Boulder, Colorado; 1985.

Garrett, A.J., "Orographic Cloud over the Eastern Slopes of Mauna Loa Volcano, Hawaii, Related to Insolation and Wind," *Mon. Wea. Rev.,* 108, pp 931-941; 1980.

Garstang, M., *A Study of the Rainfall Distribution of Trinidad, West Indies*, Woods Hole Oceanogr. Inst., 60-33; 1959.

Garstang, M., "Atmospheric Scales of Motion and Rainfall Distribution," *Proc. 1966 Army Conf. on Tropical Meteorology. Inst, Mar. Sci., Univ. Miami*, pp 24-35; 1966.

Garstang, M., N.E. LaSeur and C. Aspliden, *Equivalent Potential Temperature as a Measure of the Structure of the Tropical Atmosphere*, Dept. Meteor., Florida State Univ. Rept., 67-10; 1967.

Garstang, M., and 11 other authors, "Trace Gas Exchanges and Convective Transports over the Amazonian Rain Forest," *J. Geophys. Res.*, 93, pp 1528-1550; 1988.

Garstang, M., and 11 other authors, "The Amazon Boundary-Layer Experiment (ABLE 2B): A Meteorological Perspective," *Bull. Amer. Meteor. Soc.*, 71, pp 19-32; 1990.

Gentry, R.C., "Genesis of Tornadoes Associated with Hurricanes." *Mon. Wea. Rev.*, 111, pp 1793-1805; 1983.

Gerrish, H.P., *Mesoscale Studies of Instability Patterns and Winds in the Tropics*, Inst. Mar. Sci., Univ. Miami; 1967.

Gerrish, H.P., "A Model of Summer Convection in South Florida," *J. Appl. Meteor.*, 10, pp 949-957; 1971.

Ghosh, B.P., "A Radar Study on Thunderstorms and Convective Clouds around New Delhi during the South west Monsoon Season," *Indian J. Meteor. Geophys.*, 18, pp 391-396; 1967.

Giambelluca, T.W., The Structure of Cirrus Surges in the Tropical Central and Eastern Pacific during January and February 1979 [unpublished manuscript]; 1986.

Giambelluca, T.W., M.A. Nullet and T.A. Schroeder, *Rainfall Atlas of Hawaii*, State of Hawaii, Dept. Land, Natural Resources; 1986.

Gleeson, T.A., "Cyclogenesis in the Mediterranean Region," *Arch. Meteor. Geophys, Bioclimatol.,Ser. A*, 6, pp 154-171; 1954.

Godske, C.L., T. Bergeron, J. Bjerknes and R.C. Bundgaard, *Dynamic Meteorology and Weather Forecasting*, Amer. Meteor. Soc.; 1957.

Goldberg, R.A., M. Tisnado and R.A. Scofield, "Characteristics of Extreme Rainfall Events in Northwestern Peru during the 1982-83 Period," *J. Geophys. Res.*, 92, pp 14225-14241; 1987.

Goldbrunner, A.W., *Las Causas Meteorologicas de la Lluvias de Extraordinaria Magnitud en Venezuela, 2nd ed*, Minist. Defensa, Venezuela; 1963.

Golden, J.H., "Some Statistical Aspects of Waterspout Formation," *Weatherwise*, 26, pp 108-117; 1973.

Golden, J.H., "Life Cycle of Florida Keys' Waterspouts, I. Scale Interaction Implications for the Waterspout Life Cycle, II.," *J. Appl. Meteor.*, 13, pp 676-709; 1974.

Goodyear, H.V., *Frequency and Areal Distributions of Tropical Storm Rainfall in the United States Coastal Region on the Gulf of Mexico,* ESSA Tech. Rept. WB-7; 1968.

Gordon, A.H., and R.C. Taylor, "Numerical Steady-State Friction Layer Trajectories over the Oceanic Tropics as Related to Weather," *Int. Indian Ocean Exped. Meteor. Monogr.,* 7, East-West Center Press, Honolulu; 1970.

Gramzow, R.H., and W.K. Henry, "The Rainy Pentads of Central America," *J. Appl. Meteor.,* 11, pp 637-642; 1972.

Gray, W.M., "Global View of the Origin of Tropical Disturbances and Storms," *Mon. Wea. Rev.,* 96, pp 669-700; 1968.

Gray, W.M., *Fundamental Role of Cumulus Convection for Kinetic Energy Transformation in the Tropics and General Circulation,* Proc. Symp. on Tropical Meteorology, Honolulu, Amer. Meteor. Soc., DV, 1-8; 1970.

Gray, W.M., "Hurricanes: Their Formation, Structure and Likely Role in the Tropical Circulation," *Meteorology over the Tropical Oceans,* Roy. Meteor. Soc., pp 155-218; 1978.

Gray, W.M., "Atlantic Seasonal Hurricane Frequency: Part I: El Niño and 30 mb Quasi-biennial Oscillation Influences," *Mon. Wea. Rev.,* 112, pp 1649-1668; 1984a.

Gray, W.M., "Atlantic Seasonal Hurricane Frequency: Part II: Forecasting Its Variability," *Mon. Wea. Rev.,* 112, pp 1669-1683; 1984b.

Gray, W.M., "Tropical Cyclone Global Climatology," *WMO Tech. Doc. WMO/TD, 72, Vol.1,* pp 3-19; 1985.

Gray, W.M., *Summary of 1989 Atlantic Tropical Cyclone Activity and Seasonal Forecast Verification.* Dept. Atmos. Sci., Colorado State Univ.; 1989.

Gray, W.M., *Summary of 1990 Atlantic Tropical Cyclone Activity and Seasonal Forecast Verification.* Dept. Atmos. Sci., Colorado State Univ.; 1990.

Gray, W.M., and R.W. Jacobson, "Diurnal Variation of Deep Cumulus Convection," *Mon. Wea. Rev.,* 105, pp 1171-1188; 1977.

Greco, S., and eight other authors, "Rainfall and Surface Kinematic Conditions over Central Amazonia during ABLE 2B," *J. Geophys. Res.,* 95, pp 17,001-17,014; 1990.

Grice, G.K., J.D. Belville, C.F. Chappell, K.G. Crawford and R.A. Scofield, "Summary of the First Sino-American Academic Exchange Conference on Heavy Rain in Wuhan, China," *Bull. Amer. Meteor. Soc.,* 67, pp 851-855; 1986.

Griffiths, J.F., *Statistical rainfall analysis. Research on Tropical Rainfall Patterns and Associated Mesoscale Systems.* Dept. Meteor. Texas A&M Univ; 1964.

Griffiths, J.F., *The Analysis of Extreme Values: Extreme Value Analysis in Venezuela, Research on Tropical Rainfall Patterns and Associated Mesoscale Systems.* Dept. Meteor. Texas A&M Univ; 1967.

Gringorten, I.I., "Extreme-Value Statistics in Meteorology — A Method of Application," *AF Surv. Geophys.,* 125; 1960.

Grossman, R.L., and C.A. Friehe, "Vertical Structure of the Southwest Monsoon Low-Level Jet over the Central and Eastern Arabian Sea," *J. Atmos. Sci.,* 43, pp 3266-3272; 1986.

Guard, C.P., *A Study of Western North Pacific Tropical Storms and Typhoons That Intensify After Recurvature,* First Wea. Wing Tech. Note, 83/002; 1983.

Guard, C.P., "An Observational Assessment of the Southwest Monsoon East of East Asia," *Proc. Int. Conf. Climatology, Appl. Meteor.,* Manila; 1985.

Guard, C.P., Personal communication; 1990.

Gumbel, E.J., *Statistics of Extremes.* Columbia Univ. Press, New York; 1958.

Gupta, G.R., M.M. Nayasthani and R.K. Verma, "Clear Air Turbulence (CAT) in India and Neighbourhood," *Indian J. Meteor. Geophys.,* 23, pp 379-384; 1972.

Gutnick, M., "Climatology of the Tradewind Inversion in the Caribbean," *Bull. Amer. Meteor. Soc.,* 39, pp 410-420; 1958.

Hadley, G., "Concerning the Cause of the General Trade-Winds," *Phil. Trans.,* 29, pp 58-62; 1735.

Hamilton, R.A., and J.W. Archbold, "Meteorology of Nigeria and Adjacent Territory," *Quart. J. Roy. Meteor. Soc.,* 71, pp 231-264; 1945.

Hammer, R.M., "Cloud Development and Distribution around Khartoum," *Weather,* 25, pp 411-415; 1970.

Han, S.-C., *Rainfall over Oahu and Location of 300 mb Trough during Tradewind Regimes,* Hawaii Inst. Geophys, Rept. 70-28; 1970.

Hardin, V.S., *Accuracies of Radiosonde Data,* AWS TR-105-133, HQ Air Weather Service, Scott AFB, IL.,1955.

Hardy, K.R., D. Atlas and K.A. Browning, "The Structure of Hurricane Spiral Bands," *Proc. World Conf. on Radar Meteorology. Boulder,* pp 342-345; 1964.

Harris, B.E., and F.P. Ho, *Structure of the Troposphere over Southeast Asia during the Summer Monsoon Month of July,* Hawaii Inst. Geophys. Rept. 69-4; 1969.

Hastenrath, S., *Climate and Circulation of the Tropics,* D. Reidel, Dordrecht; 1985.

Hastenrath, S., "The Relationship of Highly Reflective Clouds to Tropical Climate Anomalies," *J. Climate,* 3, pp 353-365; 1990a.

Hastenrath, S., "Tropical Climate Prediction: A progress Report, 1985-1990," *Bull. Amer. Meteor. Soc.*, 71, pp 819-825; 1990b.

Hastenrath, S., "Prediction of Northeast Brazil Rainfall Anomalies," *J. Climate*, 3, pp 893-904; 1990c.

Hayashi, Y.-K., and T. Nakasawa, "Evidence of the Existence and Eastward Motion of Superclusters at the Equator," *Mon. Wea. Rev.*, 117, pp 236-243; 1989.

Hayden, C.M., "An Objective Analysis of Cloud Cluster Dimensions and Spacing in the Tropical North Pacific," *Mon. Wea. Rev.*, 98, pp 534-540; 1970.

Heckley, W.A., "Systematic Errors of the ECMWF Operational Forecasting Model in Tropical Regions," *Quart. J. Roy. Meteor. Soc.*, 111, pp 709-738; 1985.

Henry, W.K., "The Tropical Rainstorm," *Mon. Wea. Rev.*, 102, pp 717-725; 1974.

Hess, G.D., and K.T. Spillane, "Characteristics of Dust Devils in Australia," *J. Appl. Meteor.*, 29, pp 498-507; 1990.

Hill, E.L., W. Malkin and W.A. Schulz Jr., "Tornadoes Associated with Cyclones of Tropical Origin — Practical Features," *J. Appl. Meteor.*, 5, pp 745-763; 1966.

Hill, H.W., "The Weather of Lower Latitudes of the Southwest Pacific Associated with Passages of Disturbances in the Middle Latitude Westerlies," *Proc. Symp. on Tropical Meteorology. New Zealand Meteor. Serv.*, pp 352-365; 1964.

Hill, H.W., "The Precipitation in New Zealand Associated with the Cyclone of Early April 1968," *New Zealand J. Sci.*, 13, pp 641-662; 1970.

Hirst, A.C., and S. Hansterath, "Atmosphere-Ocean Mechanisms of Climate Anomalies in the Angola-tropical Atlantic Sector," *J. Phys. Ocean.*, 13, pp 1146-1157; 1983.

Holle, R.L., "Some Aspects of Tropical Oceanic Cloud Populations," *J. Appl. Meteor.*, 7, pp 173-183; 1968.

Holle, R.L., "Proposed New Code for State of Sky in the Tropics," *Bull. Amer. Meteor. Soc.*, 56, pp 55-58; 1975.

Holliday, C.R., and A.H. Thompson, "An Unusual Near-Equatorial Typhoon," *Mon. Wea. Rev.*, 114, pp 2674-2677; 1986.

Holliday, C.R., and K. Waters, "SSM/I Observation of Tropical Cyclone Gale Force Vicinity Winds," *Preprints, 4th Conf. Satellite Meteor. Oceanogr., Amer. Meteor. Soc.*, pp 267-270; 1989.

Hoover, R.A., "Empirical Relationships of the Central Pressures in Hurricanes to the Maximum Surge and Storm Tide," *Mon. Wea. Rev.*, 85, pp 167-174; 1957.

Hopwood, J.M., "A Note on the Quasi-Biennial Oscillation at Darwin," *Aust. Meteor. Mag.*, 16, pp 114-117; 1968.

Horel, J.D., A.N. Hahmann and J.E. Geisler, "An Investigation of the Annual Cycle of Convective Activity over the Tropical Americas," *J. Climate,* 2, pp 1388-1403; 1989.

Houze, R.A. Jr., "Structure and Dynamics of a Tropical Squall-Line System Observed during GATE," *Mon. Wea. Rev.,* 105, pp 1540-1567; 1977.

Houze, R.A. Jr., "Observed Structure of Mesoscale Convective Systems and Implications for Large-Scale Heating," *Quart. J. Roy. Meteor. Soc.,* 115, pp 425-461; 1989.

Hsu, S., "Coastal Air-Circulation System Observations and Empirical Model," *Mon. Wea. Rev.,* 98, pp 487-509; 1970.

Hsu, S.A., "Mesoscale Nocturnal Jet-Like Winds within the Planetary Boundary Layer over a Flat Open Coast," *Bound. Layer Meteor.,* 17, pp 485-494; 1979.

Hubert, L.F., and L.F. Whitney Jr., "Wind Estimates from Geostationary Satellite Pictures," *Mon. Wea. Rev.,* 99, pp 665-672; 1971.

Hurrell, J.W., and D.G. Vincent, "Relationships Between Tropical Heating and Subtropical Westerly Maxima in the Southern Hemisphere during SOP-1, FGGE," *J. Climate,* 3, pp 751-768; 1990.

Huschke, R.E. (ed.), *Glossary of Meteorology.* Amer. Meteor. Soc.; 1959.

Hyson, P., R.M. Leigh and R.C. Southern, "Observational and Forecasting Aspects of the Convection Cycle at Darwin," *Proc. Symp. on Tropical Meteorology. New Zealand Meteor. Serv.,* pp 306-315; 1964.

Illari, L., "The Quality of Satellite Precipitable Water Content Data and Their Impact on Analysed Moisture Fields," *Tellus,* 41A, pp 319-337; 1989.

India Meteorological Department, *ICAO Improved Short-Term High-Level Turbulence Reporting Programme, Final Report of Analysis.* Meteor. Off., Poona; 1968.

Ing, G., *Summer Circulation and Weather over the Korat Plateau,* Thailand. Dept. Meteor., Univ. Hawaii, UHMET 71-04; 1971.

Izawa, T., "On the Mean Wind Structure of Typhoon," *Meteor. Res. Inst. Japan Meteor. Agency,* Typhoon Res. Lab. Tech. Note 2; 1964.

Jalu, R., and J. Dettwiller, "Advection Froide vers les Basses Latitudes (Mars 1963)," *La Météorologie,* 7, pp 113-128; 1965.

Janowiak, J.E., A.F. Krueger and P.A. Arkin, *Atlas of Outgoing Longwave Radiation Derived from NOAA Satellite Data,* NOAA Atlas 6; 1985.

Jeandidier, G., and P. Rainteau, "Provision du Temps sur le Bassin du Congo," *Météor. Nat. Mongr., France,* 9.; 1957

Jelesnianski, C.P., "Numerical Computations of Storm Surges without Bottom Stress," *Mon. Wea. Rev.*, 94, pp 379-394; 1966.

Jelesnianski, C.P., "Numerical Computations of Storm Surge with Bottom Stress," *Mon. Wea. Rev.*, 95, pp 740-756; 1967.

Jelesnianski, C.P., *Special Program to List Amplitudes of Surges from Hurricanes: II General Track and Variant Storm Conditions,* Tech. Mem.II, NWS 7DL-46; 1974.

Jenkins, G.R., "Diurnal Variation of the Meteorological Elements," *Handb. Meteor.,* McGraw-Hill, New York, pp 746-753; 1945.

Johnson, A.H., "The Climate of Peru, Bolivia and Ecuador, In: *Climates of Central and South America,* Elsevier, Amsterdam, pp 147-218; 1976.

Johnson, D.H., "Rain in East Africa," *Quart. J. Roy. Meteor. Soc.,* 88, pp 1-19; 1962.

Johnson, D.H., *Forecasting Weather in West Africa,* WMO Tech. Note 64, pp 83-94; 1964.

Joint GARP Organizing Committee, *Appendix F, Report of the Study Group on Tropical Disturbances, Second Series*; 1969.

Joint Typhoon Warning Center, *1989 Annual Tropical Cyclone Report,* ; 1990.

Jordan, C.L., *Tracking of Tropical Cyclones,* Naval Wea. Res. Facil., NWRF 12-0763-075; 1963.

Jordan, C.L., and M. Shiroma, "A Record Rainfall at Okinawa," *Bull. Amer. Meteor. Soc.,* 40, pp 609-612; 1959.

Jorgensen, D.P., and P.T. Willis, "A Z-R Relationship for Hurricanes," *J. Appl. Meteor.,* 21, pp 356-366; 1982.

Joseph, P.V., D.K. Raipal and S.N. Deka, "'Andhi,' The Convective Duststorm of Northwest India," *Mausam,* 31, pp 431-442; 1980.

Kalu, A.E., "The African Dust Plume: Its Characteristics and Propagation across West Africa in Winter," In: *Saharan Dust (C. Morales, ed.),* Wiley, New York, pp 95-118; 1979.

Keen, R.A., "The Role of Tropical Cyclone Pairs in the Southern Oscillation," *Mon. Wea. Rev.,* 110, pp 1405-1416; 1982.

Keenan, T.D., and L.R. Brody, "Synoptic-Scale Modulation of Convection during the Australian Summer Monsoon," *Mon. Wea. Rev.,* 116, pp 71-85; 1988.

Keenan, T.D., J. McBride, G. Holland, N. Davidson and B. Gunn, "Diurnal Variations during the Australian Monsoon Experiment (AMEX) Phase II," *Mon. Wea. Rev.,* 117, pp 2535-2552; 1989.

Kelley, W.E. Jr., and D.R. Mock, "A Diagnostic Study of Upper Tropospheric Cold Lows over the Western North Pacific," *Mon. Wea. Rev.,* 110, pp 471-480; 1982.

Kiladis, G.N., and H.F. Diaz, "Global Climatic Anomalies Associated with Extremes in the Southern Oscillation," *J. Climate,* 2, pp 1069-1090; 1989.

Kloesel, K.A., and B.A. Albrecht, "Low-Level Inversions over the Tropical Pacific — Thermodynamic Structure of the Boundary Layer and the Above-Inversion Moisture Structure," *Mon. Wea. Rev.,* 117, pp 87-101; 1989.

Knutson, T.R., and K.M. Weickmann, "30-60 Day Atmospheric Oscillations: Composite Life Cycles of Convection and Circulation Anomalies," *Mon. Wea. Rev.,* 115, pp 1407-1436; 1987.

Kousky, V.E., "Frontal Influences on Northeast Brazil," *Mon. Wea. Rev.,* 107, pp 1140-1153; 1979.

Kousky, V.E., "Diurnal Variation of Rainfall in Northeast Brazil," *Mon. Wea. Rev.,* 108, pp 488-498; 1980.

Kousky, V.E., and M.A. Gan, "Upper Tropospheric Cyclonic Vortices in the Tropical South Atlantic," *Tellus,* 33, pp 538-550; 1981.

Kraus, E.B., "The Diurnal Precipitation Change over the Sea," *J. Atmos. Sci,,* 20, pp 551-556; 1963.

Krishnamurti, T.N., C.E. Levy, and H.-L. Pan, "On Simultaneous Surges in the Trades," J. Atmos. Sci., 32, pp 2367-2370; 1975.

Kulshrestha, S.M., and P.S. Jain, "Aspects of Radar Climatology of the Region around New Delhi in North India," *Proc. 13th Radar Meteorology Conf. Amer. Meteor. Soc.,* pp 294-297; 1968.

Kumar, S., "Interaction of Upper Westerly Waves with Intetropical Convergence Zone and Their Effect on the Weather over Zambia during the Rainy Season," *Mausam,* 30, pp 423-438; 1979.

Kuo, Y.-H., and G.T.-J. Chen, "The Taiwan Area Mesoscale Experiment (TAMEX): An Overview," *Bull. Amer. Meteor. Soc.,* 71, pp 488-503; 1990.

Kyle, A.C., *Longitudinal Variation of Large-Scale Vertical Motion in the Tropics,* Mass. Inst. Tech. MS thesis; 1970

Lajoie, F.A., "Cloud Top Equivalent Black-Body Temperatures and Tropical Cylone Forecasting in the Australian Region," *Aust. Meteor. Mag.,* 29, pp 73-88; 1981.

Lajoie, F.A., and I.J. Butterworth, "Oscillation of High-Level Cirrus and Heavy Precipitation around Australian Region Tropical Cyclones," *Mon. Wea. Rev.,* 112, pp 535-544; 1984.

Lambert, S.J., "A Comparison of Divergent Winds from the National Meteorological Center and the European Center for Medium Range Weather Forecasts Global Analyses for 1980-1986," *Mon. Wea. Rev.,* 117, pp 995-1005; 1989.

Lander, M., "On the Contribution of the Sea Surface Temperature Pattern to the Regional Differences in the Marine Boundary Layer Shear," *Proc. COADS Workshop, Boulder, Colorado,* pp 176-178; 1986.

Lander, M., "Evolution of the Cloud Pattern during the Formation of Tropical Cyclone Twins Symmetrical with Respect the Equator," *Mon. Wea. Rev.,* 118, pp 1194-1202; 1990.

Lander, M., Personal communication; 1991.

LaSeur, N.E., "Synoptic Analysis in the Tropics: The General Problem and Methods of Tropical Synoptic Analysis," *Proc. Symp. on Tropical Meteorology in Africa, Munitalp Foundation, Nairobi,* pp 7-13 and 24-34; 1960.

LaSeur, N.E., "Synoptic Models in the Tropics," *Proc. Symp. on Tropical Meteorology. New Zealand Meteor. Serv.,* pp 319-328; 1964a.

LaSeur, N.E., "Some Preliminary Results of the Barbados Field Program, a Statistical Measure of the Importance of Convection in Tropical Rainfall," *Proc. Army Conf. on Tropical Meteorology. Inst. Mar. Sci.,* Univ. Miami, pp 20-25; 1964b.

LaSeur, N.E., and H.F. Hawkins, "An Analysis of Hurricane Cleo (1958) Based on Data from Research Reconnaissance Aircraft," *Mon. Wea. Rev.,* 91, pp 694-709; 1963.

Lau, K.-H., and N.-G. Lau, "Observed Structure and Propagation Characteristics of Tropical Summertime Synoptic Scale Disturbances," *Mon. Wea. Rev.,* 118, pp 1888-1913; 1990.

Lau, R., and M.Y. Chan, "The Royal Observatory Long Range Forecast Methods, In: East Asian and Western Pacific Meteorology and Climate," (P. Sham and C.P Chang, eds.), *World Scientific, Singapore,* pp 485-493;1990.

Lauer, W., "Klimatische Grundzuge der Höhenstufung tropischer Gebirge," *Tagunsbericht und Wissenschlaftliche Abhandlungen.* F.Steiner, Innsbruck, pp 76-90; 1975.

Lavoie, R.L., *Research on the Meteorology of the Tropical Pacific and Its Applications,* Hawaii Inst. Geophys. Rept. 37; 1963.

Lavoie, R.L., and C.J. Wiederanders, *Objective Wind Forecasting over the Tropical Pacific,* Hawaii Inst. Geophys. Sci. Rept. 1; 1960.

Lee, T.F., "Dust Tracking Using Composite Visible/IR Images: A Case Study," *Weather and Forecasting,* 4, pp 258-263; 1989.

Leipper, D.F., "Observed Ocean Conditions and Hurricane Hilda, 1964," *J. Atmos. Sci.,* 24, pp 182-196; 1967.

Lenhardt, R.W., *Subsynoptic Atmospheric Variability up to 300 mb,* AFCRL Interim Notes Atmos. Properties, 78; 1967.

Leopold, L.B., "The Interaction of Tradewind and Sea Breeze," Hawaii. *J. Meteor.*, 6, pp 312-320; 1949.

Leroux, M., *Le Climat de l'Afrique Tropical*, Editions Champion, Paris; 1983.

Lettau, H., Dynamic and Energetic Factors Which Cause and Limit Aridity along South America's Pacific Coast. In: *Climates of Central and South America*, Elsevier, Amsterdam, pp 188-192; 1976.

Leverson, V.H., P.C. Sinclair and J.H. Golden, "Waterspout Wind, Temperature and Pressure Structure Deduced from Aircraft Measurements. *Mon. Wea. Rev.*, 105, pp 725-733; 1977.

Lighthill, J., and R.P. Pearce (eds.), *Monsoon Dynamics*, Cambridge Univ. Press; 1981.

Lim, J.T., *Characteristics of the Winter Monsoon over the Malaysian Region*, Dept. Meteor. Univ. Hawaii [PhD thesis]; 1979.

List, R.J., *Smithsonian Meteorological Tables (6th edition)*, Smithsonian Institution, Washington, D.C.; 1963.

Lockwood, J.G., "Probable Maximum 24-Hour Precipitation over Malaya by Statistical Methods," *Meteor. Mag.*, 96, pp 11-19; 1967.

Lord, S.J., and J.L. Franklin, "The Environment of Hurricane Debby (1982): Part I: Winds," *Mon. Wea. Rev.*, 115, pp 2760-2780; 1987.

Lorenz, E.N., *The Nature and Theory of the General Circulation of the Atmosphere*, WMO 218, TP 115; 1967.

Love, G., "Cross-Equatorial Influence of Winter Hemisphere Subtropical Cold Surges," *Mon. Wea. Rev.*, 113, pp 1487-1498; 1985a.

Love, G., "Cross-Equatorial Interactions during Tropical Cyclogenesis," *Mon. Wea. Rev.*, 113, pp 1499-1509; 1985b.

Lumb, F.E., "Topographic Influences on Thunderstorm Activity near Lake Victoria," *Weather*, 25, pp 404-410; 1970.

Lydolph, P.E., On the Causes of Aridity along a Selected Group of Coasts. In: *Coastal Deserts, Their Natural and Human Environments* (D.H.K. Amirau and A.W. Wilson, eds.), Univ. Arizona Press, pp 57-72; 1973.

Lyons, S.W., *Summer Weather over Haleakala, Maui*, Dept. Meteor. Univ. Hawaii, UHMET 79-09; 1979.

Lyons, S.W., "Origins of Convective Variability over Equatorial Southern Africa during Austral Summer," *J. Climate*, 4, pp 23-39; 1991.

Mackey, S., "How Typhoons Affect Tall Buildings — HK Study," *Far East Builder, January*, pp 37-40; 1969.

Madden, R.A., and P.R. Julian, "Detection of a 40-50 Day Oscillation in the Zonal Wind in the Tropical Pacific," *J. Atmos. Sci.*, 28, pp 702-708; 1971.

Madden, R.A., and E.J. Zipser, "Multi-Layered Structure of the Wind over the Equatorial Pacific during the Line Islands Experiment," *J. Atmos. Sci.*, 27, pp 336-342; 1970.

Maddox, R.A., "Large-Scale Meteorological Conditions Associated with Midlatitude, Mesoscale Convective Complexes," *Mon. Wea. Rev.*, 111, pp 1475-1493; 1983.

Mahrer, Y., and R.A. Pielke, "The Effects of Topography on Sea and Land Breezes in a Two-Dimensional Model," *Mon. Wea. Rev.*, 105, pp 1151-1162; 1977.

Mak, M.K., and J.E. Walsh, "On the Relative Intensities of Sea and Land Breezes," *J. Atmos. Sci.*, 33, pp 242-251; 1976.

Maliekal, J.A., *An Analysis of Temporal and Spatial Characteristics of Tropical Weather*, Dept. Meteor. Univ. Hawaii (PhD thesis); 1989.

Malkus, J.S., Large Scale Interactions. *The Sea, Vol. I.* Wiley, New York, pp 88-294; 1962.

Malkus, J.S., and H. Riehl, "Cloud Structure and Distributions over the Tropical Pacific," *Tellus,* 16, pp 275-287; 1964.

Maragos, J.E., G.B.K. Baines and P.J. Beveridge, "Tropical Cyclone Bebe Creates a New Land Formation on Funafuti Atoll," *Science,* 181, pp 1161-1164; 1973.

Martin, D., and O. Karst., "A Census of Cloud Systems over the Tropical Pacific: Studies in Atmos. Energetics Based on Aerospace Probings, Ann. Rept. —1968," *Space Sci. Engin. Cent.,* Univ. Wisconsin, pp 37-50; 1969.

Martin, D.W., B. Auvine and D. Suchman, "Synoptic Forcing and Control of Deep Convection on Day 261 of GATE," *Mon. Wea. Rev.,* 112, pp 1936-1959; 1984.

Martin D.W., and V.E. Suomi, "A Satellite Study of Cloud Clusters over the Tropical North Atlantic Ocean," *Bull. Amer. Meteor. Soc.,* 53, pp 135-156; 1972.

Martin, L.A., *An Investigation of the Rainfall Distribution for Stations in North and Central America; Research on Tropical Rainfall Patterns and Associated Mesoscale Systems.* Dept. Meteor., Texas A&M Univ.; 1964.

McBride, J.L., "The Australian Summer Monsoon," *Monsoon Meteorology,* Clarendon Press, Oxford, pp 60-92; 1987.

McBride, J.L., and T.D. Keenan, "Climatology of Tropical Cyclone Genesis in the Australian Region," *J. Climatol.,* 2, pp 13-33; 1982.

McCabe, J.T., (1) A Systematic Approach to the Use of Climatology and Persistence in Terminal Weather Forecasting. (2) An Application of the Climatology of Typhoon Movement to Operational Forecasting and Estimation of Typhoon Threat to a Particular Installation. *US Asian Wea. Symp.*, First Wea. Wing; 1961.

McCreary, F.E., "Stratospheric Winds over the Tropical Pacific Ocean (abstract)," *Bull. Amer. Meteor. Soc.*, 40, p 370; 1959.

McGarry, M.M., and R.J. Reed, "Diurnal Variations in Convective Activity and Precipitation during Phases II and III of GATE," *Mon. Wea. Rev.*, 106, pp 101-113; 1978.

McGurk, J.P., and D.J. Ulsh, "Evolution of Tropical Plumes in VAS Water Vapor Imagery," *Mon. Wea. Rev.*, 118, pp 1758-1766; 1990.

McPherson, R.D., "A Numerical Study of the Effect of a Coastal Irregularity on the Sea Breeze," *J. Appl. Meteor.*, 9, pp 767-777; 1970.

Meigs, P., "World Distribution of Coastal Deserts," In: *Coastal Deserts, Their Natural and Human Environment* (D.H.K. Amirau and A.W. Wilson, eds.), Univ. Arizona Press, pp 3-12; 1973.

Meisner, B.N., "Ridge-Regression — Time Extrapolation Applied to Hawaiian Rainfall Normals," *J. Appl. Meteor.*, 18, pp 904-912; 1979.

Menzel, W.P., and A. Chedin, "Summary of the Fifth International TIROS Operational Vertical Sounder (TOVS) Study Conference," *Bull. Amer. Meteor. Soc.*, 71, pp 691-693; 1990.

Meteorological Office, *Monthly Meteorological Charts of the Eastern Pacific Ocean*, H.M. Stationery Office, London; 1949.

Meteorological Office, *Monthly Meteorological Charts of the Eastern Pacific Ocean. 2nd Edition.*, H.M. Stationery Office, London; 1968.

Meteorology Working Group, *Meteorological Equipment Data Accuracies*, IRIG Document 110-64; 1965.

Meyencon, R., "Cyclone in the Mediterranean," *Mar. Wea. Log*, 27, pp 141-143; 1983.

Miller, B.I., "Rainfall Rates in Florida Hurricanes," *Mon. Wea. Rev.*, 86, pp 258-264; 1958.

Miller, B.I., "A Study of the Filling of Hurricane Donna (1960) over Land," *Mon. Wea. Rev.*, 92, pp 389-406; 1964.

Miller, B.I., "Characteristics of Hurricanes," *Science*, 157, pp 1389-1399; 1967.

Miller, D., and J.M. Fritsch, "Mesoscale Convective Complexes in the Western Pacific Region," *Mon. Wea. Rev.*, 119, pp 2978-2992; 1991.

Miller, F.R., and R.N. Keshavamurthy, "Structure of an Arabian Sea Summer Monsoon System," *Int. Indian Ocean Exped. Meteor. Monogr.*, 1, East-West Center Press, Honolulu; 1968.

Miller, R.C., "Forecasting Thunderstorm-Produced Surface Wind Gusts in Southeast Asia," *Milit. Wea. Warning Center Kansas City,* [unpub. rept.]; 1967.

Minnis, P., and E.F. Harrison, "Diurnal Variability of Regional Cloud and Clear Sky Parameters Derived from GOES Data: Part II November 1978 Cloud Distributions," *J. Climatol. Appl. Meteor.,* 23, pp 1012-1031; 1984.

Mishra, D.K., and M.S. Singh, "A Study of the Monsoon Depression Intensifying into Cyclonic Storm over Land," *Indian J. Meteor. Hydrol. Geophys.,* 28, pp 321-327; 1977.

Mooley, D.A., and H.L. Crutcher, *An Application of the Gamma Distribution Function to Indian Rainfall,* ESSA Tech. Rept. ESD 5; 1968.

Morgan, M.R., *Outbreaks of Antarctic Air in Relation to the Hurricane Season 1962-63,* Meteor. Branch Dept. Transportation, Canada, Tec. 571; 1965.

Morrrissey, M.L., "A Statistical Analysis of the Relationships Among Rainfall, Outgoing Longwave Radiation and the Moisture Budget during January-March 1979," *Mon. Wea. Rev.,* 114, pp 931-942; 1986.

Morrissey, M.L., M.A. Lander and J.A. Maliekal, "A Preliminary Evaluation of Ship Data in the Equatorial Western Pacific," *J. Atmos. Ocean. Technol.,* 5, pp 251-258; 1988.

Mukherjee, A.K., and K.P. Padmanabham, "Simultaneous Occurrence of Tropical Cyclones on Either Side of the Equator in the Indian Ocean Area," *Indian J. Meteor. Hydrol. Geophys.,* 28, pp 211-222; 1977.

Murakami, M., "Analysis of the Deep Convective Activity over the Western Pacific and Southeast Asia," *J. Meteor. Soc. Japan,* 61, pp 60-76; 1983.

Muramatsu, T., "Diurnal Variations of Satellite-Measured T_{BB} Areal Distribution and Eye-Diameter of Mature Typhoons," *J. Meteor. Soc. Japan,* 61, pp 77-90; 1983.

National Aeronautics and Space Administration, *Earth System Science, a Closer View,* NASA, Washingotn, D.C.; 1988.

National Research Council, "The New Year's Eve Flood on Oahu, Hawaii, Decmber 31, 1987 - January 1, 1988," *Natural Disasters Studies,* 1, National Academy Press, Washington, D.C. ; 1991.

Naujokat, B., "An Update of the Observed QBO of the Stratospheric Winds over the Tropics," *J. Atmos. Sci.,* 43, pp 1873-1877; 1986.

Naval Oceanography Command, *U.S. Navy Marine Climatic Atlas of the World. Vol.IX, World-Wide Means and Standard Deviations,* NAVAIR 50-1C-65; 1981.

Navy Weather Research Facility Staff, *The Diagnosis and Prediction of SE Asia Northeast Monsoon Weather,* Norfolk, Virginia; 1969.

Neal, A.B., *The Meteorology of the Australian Trades: Part 2: The Perturbing Effect of Willis Island on the Oceanic Trade Wind Flow*, Aust. Bur. Meteor., Tech. Rept., 5(2); 1973.

Neiburger, M., D.S. Johnson and C.-W. Chen, *Studies of the Structure of the Atmosphere over the Eastern Pacific Ocean in Summer: I. The Inversion over the Eastern North Pacific Ocean*. Univ. Calif. Pub. Meteor., 1, No. 1; 1961.

Neumann, C.J., "Mesoanalysis of a Severe South Florida Hailstorm," *J. Appl. Meteor.*, 4, pp 161-171; 1965.

Newell, R.E., "Climate and Ocean," *Amer. Sci.*, 67, pp 405-416; 1979.

Newell, R.E., J.W. Kidson, D.C. Vincent and G.J. Boer, *The General Circulation of the Tropical Atmosphere and Interaction with Extratropical Latitudes, Vol. 1.* MIT Press, Cambridge; 1972.

Nicholls, N., "Predictability of Interannual Variations of Australian Seasonal Tropical Cyclone Activity," *Mon. Wea. Rev.*, 113, pp 1144-1145; 1985.

Nicholson, S.E., "Rainfal and Atmospheric Circulation during Drought Periods and Wetter Years in West Africa," *Mon. Wea. Rev.*, 109, pp 2191-2208; 1981.

Nieuwolt, S., "Diurnal Variation of Rainfall in Malaya," *Ann. Assoc. Amer. Geogr.*, 58, pp 320-326; 1968.

NOAA/NESDIS, *Monthly Climatic Data for the World*; 1948.

NOAA/NESDIS, *The GOES User's Guide*; 1983.

Novlan, D.J., and W.M. Gray, "Hurricane-Spawned Tornadoes," *Mon. Wea. Rev.*, 102, pp 476-488; 1974.

Okulaja, F.O., "The Frequency Distribution of Lagos/Ikeja Wind Gusts," *J. Appl. Meteor.*, 7, pp 379-383; 1968.

Olascoaga, M.J., "Some Aspects of Argentine Rainfall," *Tellus*, 2, pp 312-318; 1950.

Orlanski, I., "A Rational Subdivision of Scales for Atmospheric Processes," *Bull. Amer. Meteor. Soc.*, 56, pp 527-530; 1975.

Orton, R., "Tornadoes Associated with Hurricane Beulah on September 19-23, 1967," *Mon. Wea. Rev.*, 98, pp 541-547; 1970.

Orville, R.E., and R.W. Henderson, "Global Distribution of Midnight Lightning," *Mon. Wea. Rev.*, 114, pp 2640-2653; 1986.

Osman, O.E., and S. Hastenrath, "On the Synoptic Climatology of Summer Rainfall over Central Sudan," *Arch. Meteor. Geophys. Bioclimatol., Ser. B*, 17, pp 297-324; 1969.

Palecek, D.A., Forecasting Thunderstorms at Clark AB, Philippines, March through October. [unpub. manuscript]; 1987.

Pallman, A.J., *The Synoptics, Dynamics and Energetics of the Temporal Using Satellite Radiation Data.* Dept. Geophys. St. Louis Univ.; 1968.

Palmer, C.E., "Tropical Meteorology," *Comp. Meteor.,* pp 859-880; 1951a.

Palmer, C.E., "On High-Level Cyclones Originating in the Tropics," *Trans. Amer. Geophys. Un.,* 33, pp 683-696; 1951b.

Palmer, C.E., "Tropical Meteorology," *Quart. J. Roy. Meteor. Soc.,* 78, pp 126-164; 1952.

Palmer, C.E., J.R. Nicholson and R.M. Shimaura, *An Indirect Aerology of the Tropical Pacific.* Inst. Geophys. Univ. Calif.; 1956.

Palmer, C.E., and W.D. Ohmstede, "The Simultaneous Oscillation of Barometers along and near the Equator," *Tellus,* 8, pp 495-507; 1956.

Palmer, C.E., C.W. Wise, L.J. Stempson and G.H. Duncan, *The Practical Aspect of Tropical Meteorology,* AWS Manual 105-48, Vol. 1, HQ Air Weather Service, Scott AFB, IL; 1955.

Parmenter, F.C., "A Southern Hemisphere Cold Front Passage at the Equator," *Bull. Amer. Meteor. Soc.,* 57, pp 1435-1440; 1976.

Parthasarathy, B., K. Rupa Kumar and A.A. Munot, "Evidence of Secular Variations in Indian Monsoon Rainfall-Circulation Relationship," *J. Climate,* 4, pp 927-938; 1991.

Parthasarathy, B., K. Rupa Kumar and N.A. Sontakke, "Surface and Upper Air Temperatures over India in Relation to Monsoon Rainfall," *Theor. Appl. Climatol.,* 42, pp 93-110; 1990.

Paulus, J.L.H., "Indian Ocean and Taiwan Rainfalls Set New Records," *Mon. Wea. Rev.,* 93, pp 331-335; 1965.

Pearson, A.D., and A.F. Sadowski, "Hurricane-Induced Tornadoes and Their Distribution," *Mon. Wea. Rev.,* 93, pp 461-464; 1965.

Pedgley, D.E., "Diurnal Variation of the Incidence of Monsoon Rainfall over the Sudan, Parts I and II." *Meteor. Mag.,* 98, pp 97-107, 129-134; 1969.

Petterssen, S., *Weather Analysis and Forecasting. Second Edition. Vol.1 Motion and Motion Systems.* McGraw-Hill, New York; 1956.

Pielke, R., "A Three-Dimensional Numerical Model of the Sea Breezes over South Florida," *Mon. Wea. Rev.,* 102, pp 115-139; 1974.

Pike, A.C., and C.J. Neumann, "The Variation of Track Forecast Difficulty Among Tropical Cyclone Basins," *Wea. Forecasting,* 2, pp 237-241; 1987.

Portig, W.H., "Thunderstorm Frequency and Amount of Precipitation in the Tropics," *Arch. Meteor. Geophys. Bioklimatol., Ser. B.,* 13, pp 21-35, 1963.

Portig, W.H., The Climate of Central America; *Climates of Central and South America.* Elsevier, Amsterdam, pp 405-478; 1976.

Powell, J., and D.E. Pedgley, "A Year's Weather at Termit, Republic of Niger," *Weather,* 24, pp 247-254; 1969.

Prasad, B., "Diurnal Variation of Rainfall in India," *Indian J. Meteor. Geophys.,* 21, pp 443-450; 1970.

Probert-Jones, J.R., "Meteorological Uses of Pulse Doppler Radar," *Nature,* 186, pp 271-273; 1960.

Prohaska, F.J., "New Evidence on the Climatic Controls along the Peruvian Coast," In: *Coastal Deserts Their Natural and Human Environments* (D.H.K. Amirau and A.W. Wilson, eds.), Univ. Arizona Press, pp 91-107; 1973.

Prümm, D., "Zeitliche Variationen Meteorologischen Grössen in der Wassernahen Luftschicht des Atlantischen Nordost-Passats," *Ber. Inst. Radiometeor. Mar. Meteor. Univ. Hamburg,* 23; 1974.

Puniah, K.B., "Heights of Tops of Cb Clouds over Southeast Asia," *Indian J. Meteor. Geophys.,* 24, pp 386-387; 1973.

Rainbird, A.F., *Weather Disturbances over Tropical Continents and Their Effects on Ground Conditions,* Dept. Atmos. Sci. Colorado State Univ.; 1968.

Rakshit, D.K., "A Radar Study of Thin Band or Precursor Line in Association with Thunderstorm Line," *Proc. 12th Conf. Radar Meteor. Amer. Meteor. Soc.,* pp 334-338; 1966.

Ram, N., "Frequency of Thunderstorms in India," *India Meteor. Dept. Sci. Notes,* 1, pp 49-55; 1929.

Ramage, C.S., "Diurnal Variation of Summer Rainfall over East China, Korea and Japan," *J. Meteor.,* 9, pp 83-86; 1952a.

Ramage, C.S., "Variation of Rainfall over South China through the Wet Season," *Bull. Amer. Meteor. Soc.,* 33, pp 308-311; 1952b.

Ramage, C.S., "Non-frontal Crachin and the Cool Season Clouds of the China Seas," *Bull. Amer. Meteor. Soc.,* 35, pp 404-411; 1954.

Ramage, C.S., *Surface Weather Chart Analysis in Low Latitudes,* Dept. Meteor. Univ. Hawaii; 1957.

Ramage, C.S., "The Subtropical Cyclone," *J. Geophys. Res.,* 67, pp 1401-1411; 1962.

Ramage, C.S., "Diurnal Variation of Summer Rainfall of Malaya," *J. Trop. Geog.,* 19, pp 62-68; 1964.

Ramage, C.S., "The Summer Atmospheric Circulation over the Arabian Sea," *J. Atmos. Sci.,* 23, pp 144-150; 1966.

Ramage, C.S., "Role of a Tropical "Maritime Continent" in the Atmospheric Circulation," *Mon. Wea. Rev.,* 96, pp 365-370; 1968.

Ramage, C.S., "Summer Drought over Western India," *Yearbook Assn. Pacific Coast Geographers,* 30, pp 41-54; 1969.

Ramage, C.S., *Monsoon Meteorology,* Academic Press, New York; 1971.

Ramage, C.S., "Structure of an Oceanic Near-Equatorial Trough Deduced from Research Aircraft Traverses," *Mon. Wea. Rev.,* 102, pp 754-759; 1974.

Ramage, C.S., "Effect of the Hawaiian Islands on the Tradewinds," *Prepr. Conf. Clim. and Energy: Clim. Aspects and Indust. Operations, Amer. Meteor. Soc.,* pp 62-67; 1978a.

Ramage, C.S., "Orographic, Mesoscale and Synoptic Influences Related to Mt. Waialeale Rainfall," *Bull. Amer. Meteor. Soc.,* 11, 1544 [abstr.]; 1978b.

Ramage, C.S., "Forecasting Tropical Cyclone Movement — Retrospect and Prospect," *Sel. Pap. 13th Tech. Conf. Hurricanes and Trop. Meteor. Amer. Meteor. Soc.,* pp 65-67; 1980.

Ramage, C.S., "Teleconnections and the Seige of Time," *J. Climatol.,* 3, pp 223-231; 1983.

Ramage, C.S., "Secular Change in Reported Surface Wind Speeds over the Ocean," *J. Appl. Meteor.,* 26, pp 525-528; 1987.

Ramage, C.S., L.R. Brody, R.F. Adler and S. Brand, *A Diagnosis of the Summer Monsoon of Southeast Asia,* NWRF Tech. Pap.,10-69; 1969.

Ramage, C.S., S.J.S. Khalsa and B.N. Meisner, "The Central Pacific Near-Equatorial Convergence Zone," *J. Geophys. Res.,* 86, pp 6580-6598; 1981.

Ramage, C.S., and C.R.V. Raman, *Meteorological Atlas of the International Indian Ocean Expedition, Vol. 2 Upper Air,* Nat. Sci. Found. (Charts); 1972.

Ramamurthi, K., *Some Aspects of the "Break" in the Southwest Monsoon in July and August,* India Meteor. Dept. FMU Rept. 4; 1969.

Ramanathan, K.R., and K.P. Ramakrishnan, "The Indian Southwest Monsoon and the Structure of Depressions Associated with It," *Mem. India Meteor. Dept.,* 26, pp 13-36; 1932.

Ranganathan, C., and K. Soundarajan, "A Study of a Typical Case of Interaction of an Easterly Wave with a Westerly Trough during the Postmonsoon Period," *Indian J. Meteor. Geophys.,* 16, pp 607-616; 1965.

Rao, K.N., "Seasonal Forecasting — India," *WMO-IUGG Symp. Res. Devel. Aspects of Long-Range Forecasting,* WMO, pp 17-30; 1965.

Rao, Y.P., "Southwest Monsoon," *India Meteor. Dept., Meteor. Monogr. Synop. Meteor.,* 1/76; 1976.

Reed, J.W., "Some Notes on Forecasting Winds Aloft by Statistical Methods," *J. Appl. Meteor.,* 6, pp 360-372; 1967.

Reed, R.J., *A Climatology of Wind and Temperatures in the Tropical Stratosphere Between 100 mb and 10 mb*, NWRF 26-0564-092; 1964.

Reed, R.J., W.J. Campbell, L.A. Rasmusson and D.G. Rogers, "Evidence of a Downward-Propagating Annual Wind Reversal in the Equatorial Stratosphere," *J. Geophys. Res.*, 66, pp 813-818; 1961.

Reed, R.J., and K.D. Jaffe, "Diurnal Variation of Summer Convection over West Africa and the Tropical Eastern Atlantic during 1974 and 1978," *Mon. Wea. Rev.*, 109, pp 2527-2534; 1981.

Reed, R.J., D.C. Norquist and E.E. Recker, "The Structure and Properties of African Wave Disturbances as Observed during Phase III of GATE," *Mon. Wea. Rev.*, 105, pp 317-333; 1977.

Reed, R.J., and E.E. Recker, "Structure and Properties of Synoptic-Scale Wave Disturbances in the Equatorial Western Pacific," *J. Atmos. Sci.*, 28, pp 1117-1133; 1971.

Refsdal, A., "Der Feuchtlabile Niederschlag," *Geophys. Publik. 5*, No. 12; 1930.

Rennick, M.A., "The Generation of African Waves," *J. Atmos. Sci.*, 33, pp 1955-1969; 1976.

Reynolds, G., "Two Statistical Heresies," *Weather*, 33, pp 74-76; 1978.

Riehl, H., *Waves in the Easterlies*, Misc. Rept. 17, Dept. Meteor. Univ. Chicago; 1948.

Riehl, H., "Some Aspects of Hawaiian Rainfall," *Bull. Amer. Meteor. Soc.*, 30, pp 176-187; 1949.

Riehl, H., "On the Role of the Tropics in the General Circulation of the Atmosphere," *Tellus*, 2, pp 1-17; 1950.

Riehl, H., *Tropical Meteorology*, McGraw-Hill, New York; 1954.

Riehl, H., "Upper Air Observations in the Tropics," *World Wea. Watch Planning Rept.*, 1, WMO; 1966.

Riehl, H., *Southeast Asia Monsoon Study*, Dept. Atmos. Sci. Colorado State Univ.; 1967.

Riehl, H., *Climate and Weather in the Tropics*, Academic Press, New York; 1979a.

Riehl, H., "Occurrence and Structure of the Equatorial Trough Zone in Venezuela," *Quart. J. Roy. Meteor. Soc.*, 105, pp 217-229; 1979b.

Riehl, H., L. Cruz, M. Mata and C. Muster, "Precipitation Characteristics of the Venezuela Rainy Season," *Quart. J. Roy. Meteor. Soc.*, 99, pp 746-757; 1973.

Riehl, H., and J.S. Malkus, "On the Heat Balance and Maintenance of Circulation in the Trades," *Quart. J. Roy. Meteor. Soc.*, 83, pp 21-29; 1957.

Riehl, H., and J.S. Malkus, "On the Heat Balance in the Equatorial Trough Zone. *Geophysica*, 6, pp 503-537; 1958.

Rodgers, E.B., S.W. Chang, J. Stout, J. Steranka and J.-J. Shi, "Satellite Observations of Variations in Tropical Cyclone Convection Caused by Upper-Tropospheric Troughs," *J. Appl. Meteor.*, 30, pp 1163-1184; 1991.

Rodgers, E.B., J. Stout, J. Steranka and S. Chang, "Tropical Cyclone-Upper Atmospheric Interaction as Inferred from Satellite Total Ozone Observations," *J. Appl. Meteor.*, 29, pp 934-954; 1990.

Ropelewski, C.F., and M.S. Halpert, "Global and Regional Scale Precipitation Patterns Associated with the El Niño/Southern Oscillation," *Mon. Wea. Rev.*, 115, pp 1606-1626; 1987.

Ropelewski, C.F., and M.S. Halpert, "Precipitation Patterns Associated with the High Index Phase of the Southern Oscillation," *J. Climate*, 2, pp 268-284;1989.

Ruprecht, E., and W.M. Gray, "Analysis of Satellite-Observed Cloud Clusters, Papers I and II," *Tellus*, 28, pp 391-425; 1976.

Sadler, J.C., "Tropical Cyclones of the Eastern North Pacific as Revealed by TIROS Observations," *J. Appl. Meteor.*, 3, pp 347-366; 1964.

Sadler, J.C., "The Feasibility of Global Tropical Analyses," *Bull. Amer. Meteor. Soc.*, 46, pp 118-130; 1965.

Sadler, J.C., "On the Origin of Tropical Vortices," *Proc. Working Panel Dyn. Meteor.*, NWRF 12-1167-132, pp 39-76; 1967.

Sadler, J.C., "Average Cloudiness in the Tropics from Satellite Observations," *Int. Indian Ocean Exped. Meteor. Monogr. 2*, East-West Center Press, Honolulu, 1969.

Sadler, J.C., "The Monsoon Circulation and Cloudiness over the GATE Area," *Mon. Wea. Rev.*, 103, pp 369-387; 1975a.

Sadler, J.C., *The Upper Tropospheric Circulation over the Global Tropics*, Dept. Meteor. Univ. Hawaii, UHMET 75-05; 1975b.

Sadler, J.C., *Tropical Cyclone Initiation by the Tropical Upper Tropospheric Trough*, Dept. Meteor. Univ. Hawaii, UHMET 75-02; 1976a.

Sadler, J.C., *Typhoon Intensity Changes in the Pacific Storm Fury Area*, Dept. Meteor. Univ. Hawaii, UHMET 76-02; 1976b.

Sadler, J.C., W.R. Brett, B.E. Harris and F.P. Ho, *Forecasting Minimum Cloudiness over the Red River during the Summer Monsoon*, Hawaii Inst. Geophys. Rept. 68-16; 1968.

Sadler, J.C., and B.J. Kilonsky, "Meteorological Events in the Central Pacific during 1983 Associated with the 1982-83 El Niño," *Trop. Ocean Atmos. Newsl.*, 21, pp 3-5; 1983.

Sadler, J.C., and B.J. Kilonsky, "Deriving Surface Winds from Satellite Observations of Low-Level Cloud Motions," *J. Climate Appl. Meteor.*, 24, pp 758-769; 1985.

Sadler, J.C., B. Kilonsky, L. Oda and A. Hori, *Mean Cloudiness over the Global Tropics from Satellite Observations*, NAVENVPREDRSCHFAC Contractor Rept. CR 84-09; 1984.

Sadler, J.C., and M. Lander, "Regional Scale Circulations Produced by the Orography of Central America," *Proc. COADS Workshop, Boulder, Colorado*, pp 169-171; 1986.

Sadler, J.C., M.A. Lander, A.M. Hori and L.K. Oda, *Tropical Marine Climatic Atlas, Vol I: Indian Ocean and Atlantic Ocean*, Dept. Meteor. Univ. Hawaii, UHMET 87-01 (charts); 1987.

Sadler, J.C., and J.T. Lim, "Monitoring Monsoon Outflow from Geosynchronous Data." In: *Monsoon Dynamics*, Cambridge Univ. Press, London, pp 80-96; 1979.

Sadler, J.C., and L.K. Oda, *The Synoptic (A) Scale Circulations during the Second Phase of GATE 17 July - 19 August 1974*, Dept. Meteor. Univ. Hawaii, UHMET 78-02; 1979.

Sadler, J.C., and T.C. Wann, *Mean Upper Tropospheric Flow over the Global Tropics*, AWS Tech. Rept. 83/002, Vol. II (charts); 1984.

Sanders, F., "Quasi-Geostrophic Diagnosis of the Monsoon Depression of 5-8 July 1979," *J. Atmos. Sci.*, 41, pp 538-552; 1984.

Sands, W., *A Note on Severe Thunderstorms in SEA*, 1st Wea. Wing Tech. Topics, 1-4; 1969.

Sansom, H.W., *A Moisture/Stability Index as an Aid to Local Forecasting in East Africa*, East African Meteor. Dept. Tech. Memo, 9; 1963.

Saucier, W.J., *Principles of Meteorological Analysis*, Univ. Chicago Press; 1955.

Saxton, D.W., "The Nature and Causes of Clear-Air Turbulence," *AIAA Paper 66-966*; 1966.

Schroeder, T.A., "Hawaiian Waterspout-Tornado of 26 November 1975," *Weatherwise*, 29, pp 172-177; 1976.

Schroeder, T.A., "Meteorological Analysis of an Oahu Flood," *Mon. Wea. Rev.*, 105, pp 458-468; 1977.

Schroeder, T.A., *Mesoscale Structure of Hawaiian Rainstorms*, Dept. Meteor. Univ. Hawaii, UHMET 78-03; 1978.

Schroeder, T.A., B.J. Kilonsky and W. Kreisel, "Der Tagesgang des Niederschlages auf den Hawaii-Inseln," *Erdkunde*, 32, pp 89-101; 1978.

Seagraves, M.A., "Weather and Environmental Effects on Electro-optical Devices," *U.S. Army Laboratory Command, Atmos. Sci. Lab., TR-0254*; 1989.

Seck, A., "The Heug or Dry Rainy Season in Senegal," *Ann. Geogr.*, 385, pp 225-246; 1962.

Sellers, W.D., *Physical Climatology*, Univ. Chicago Press; 1965.

Sharma, A.K., A.T.C. Chang and T.W. Wilheit, "Estimation of the Diurnal Cycle of Oceanic Precipitation from SSM/I Data," *Mon. Wea. Rev.*, 119, pp 2168-2175; 1991.

Sharma, K.K., "Squalls at Nagpur," *Indian J. Meteor. Geophys.*, 17, pp 77-82; 1966.

Shepard, P.A., and M.H. Omar, "The Wind Stress over the Ocean from Observations in the Trades," *Quart. J. Roy. Meteor. Soc.*, 78, pp 583-589; 1952.

Sherretz, L.A., J.G. Pitko and R.A. Dorr, *Hawaiian Weather Forecast Studies,* Dept. Meteor. Univ. Hawaii, UHMET 71-03; 1971.

Shin, K.-P., G.R. North, Y.-S. Ahn and P.A. Arkin, "Time Scales and Variability of Area-Averaged Tropical Oceanic Rainfall, *Mon. Wea. Rev.*, 118, pp 1507-1516; 1990.

Shoemaker, D.N., W.M. Gray and J.D. Sheafer, "Influence of Synoptic Track Aircraft Reconnaissance on JTWC Tropical Cyclone Forecast Errors," *Weather and Forecasting,* 5, pp 503-507; 1990.

Short, D.A., and J.M. Wallace, "Satellite-inferred Morning-to-evening Cloudiness Changes," *Mon. Wea. Rev.*, 108, pp 1160-1169; 1980.

Showalter, A.K., "A Stability Index for Thunderstorm Forecasting," *Bull. Amer. Meteor. Soc.*, 34, pp 250-252; 1953.

Simpson, J.S., and A.S. Dennis, *Cumulus Clouds and Their Modification,* NOAA Tech. Mem. ERL OD-14; 1972.

Simpson, R.H., "Evolution of the Kona Storm, a Subtropical Cyclone," *J. Meteor.*, 9, pp 24-35; 1952.

Simpson, R.H., "The Search for More Adequate Tools to Describe the Circulation in Tropical and Equatorial Latitudes," *Proc. Working Panel on Tropical Dynamic Meteorology, Monterey, CA,* NWRF Rept. No. 12-1167-132, pp 5-15; 1967.

Simpson, R.H., *Synoptic Analysis Models for the Tropics,* AWS Tech. Rept. 217, pp 189-200; 1969.

Simpson, R.H., "Mean Layer Analysis for Tracking and Prediction of Disturbances in the Tropics," *Proc. Symp. on Tropical Meteor., Honolulu, Amer. Meteor. Soc.,* G IV, pp 1-7; 1970a.

Simpson, R.H., *A Reassessmment of the Hurricane Prediction Problem,* ESSA Tech. Memo WBTM SR-50; 1970b.

Simpson, R.H., N. Frank, D. Shideler and H.M. Johnson, "Atlantic Tropical Disturbances, 1967," *Mon. Wea. Rev.*, 96, pp 251-259; 1968.

Simpson, R.H., and H. Riehl, *The Hurricane and Its Impact,* Louisiana State Univ. Press, Baton Rouge; 1981.

Singh, R., "On the Occurrence of Tornadoes and Their Distribution in India," *Mausam,* 32, pp 307-314; 1981.

Smith, J.S., "The Hurricane-Tornado," *Mon. Wea. Rev.*, 93, pp 453-459; 1965.

Smith, W.S., "High-altitude Conk Out," *Nat. Hist.*, 92, No. 11, pp 26-34; 1983.

Solot, S.B., *The Meteorology of Central Africa*, AWS Tech. Rept., 105-50; 1943.

Sourbeer, R.H., and R.C. Gentry, "Rainstorm in Southern Florida, January 21, 1957," *Mon. Wea. Rev.*, 89, pp 9-16; 1961.

Southern, R.L., W.R. Kininmonth and N.R. Prescod, "Derivation of Convective Forecasting Models for Northern Australia from a Climatology of Lightning Discharges," *Proc. Conf. Summer Monsoon, Southeast Asia, NWRF,* pp 239-254; 1970.

Spillane, K.T., "The Winter Jet Stream of Australia and Its Turbulence," *Aust. Meteor. Mag.*, 16, pp 64-71; 1968.

Sreenivasaiah, B.N., and K.N. Sur, "A Study of the Duststorms of Agra," *Mem. India Meteor. Dept.*, 27, pp 1-30; 1939.

Stechnovsky, D.I., and T.S. Krouskova, "GARP Report of the Fifth Session of the Joint Organizing Committee, Bombay, 1-5 February 1971, Annex E of Appendix;" 1970

Stephens, G.L., "On the Relationship Between Water Vapor over the Oceans and the Sea Surface Temperature," *J. Climate*, 3, pp 634-645; 1990.

Steranka, J., E.B. Rogers and R.C. Gentry, "The Diurnal Variation of Atlantic Tropical Cyclone Cloud Distribution Inferred from Geostationary Satellite Infrared Measurements," *Mon. Wea. Rev.*, 112, pp 2338-2344; 1984.

Stone, R.G., *A Compendium on Cirrus and Cirrus Forecasting,* AWS Tech. Rept., pp 105-130; 1957.

Stone, R.G., G.E. Emmons, J.E. Miller and R.N. Culman, *Caribbean Weather Characteristics and Methods of Analysis,* Air Corps Tactical School, Maxwell Field, Alabama; 1942.

STORM Project Office, *Storm Watch,* NCAR, Boulder; 1988.

Stowe, L.L., Y.Y.M. Yeh, T.F. Eck, C.G. Wellemeyer and H.L. Kyle, "Nimbus-7 Global Cloud Climatology: Part II: First Year Results," *J. Climate*, 2, pp 671-709; 1989.

Subba Reddy, E.V., "A Study of Squalls at Madras Airport," *Mausam*, 36, pp 91-96; 1985.

Subramaniam, D.V., and P.S. Jain, "Stability Index and Area Forecasting of Thunderstorms," *Proc. 12th Conf. on Radar Meteor. Amer. Meteor. Soc.*, pp 156-159; 1966.

Sugg, A.L., *A Mean Storm-Surge Profile,* ESSA Tech. Memo WBTM-SR-49; 1969.

Sutton, L.J., "Haboobs," *Quart. J. Roy. Meteor. Soc.*, 51, pp 25-30; 1925.

Takahashi, T., "Hawaiian Hailstones — 30 January 1985," *Bull. Amer. Meteor. Soc.*, 68, pp 1530-1534; 1987.

Taljaard, J.J., "Synoptic Meteorology of the Southern Hemisphere," *Meteorology of the Southern Hemisphere. Amer. Meteor. Soc.,* pp 139-213; 1972.

Taniguchi, G., *The Effects of Typhoons on Weather at Clark Air Base,* 1st Wea. Wing Tech. Study, 10; 1967.

Tao, S.-Y., and L.-X. Chen, "A Review of Recent Research on the East Asian Summer Monsoon in China," *Monsoon Meteorology,* Clarendon Press, Oxford, pp 60-92; 1987.

Tao, W.-K., J. Simpson and S.-T. Soong, "A Numerical Study of a Squall Line over the Taiwan Strait during TAMEX IOP 2," *Mon. Wea. Rev.,* 119, pp 2677-2698; 1991.

Thomas, S.I.T., and V.K. Raghavendra, "Heights of Cumulonimbus Cloud Tops over the Deccan Plateau and Adjoining Plains of Andhra Pradesh and East Maharashtra — A Radar Study," *Indian J. Meteor. Hydrol. Geophys.,* 28, pp 479-482; 1977.

Tiedtke, M., W.A. Heckley and J. Slingo, "Tropical Forecasting at ECMWF: the Influence of Physical Parameterization on the Mean Structure of Forecasts and Analyses," *Quart. J. Roy. Meteor. Soc.,* 114, pp 639-664; 1988.

Trenberth, K.E., "Atmospheric Quasi-Biennial Oscillations," *Mon. Wea. Rev.,* 108, pp 1370-1377; 1980.

Trenberth, K.E., and J.G. Olsen, "An Evaluation and Intercomparison of Global Analyses from the National Meteorological Center and the European Center for Medium Range Weather Forecasts," *Bull. Amer. Soc.,* 69, pp 1047-1057; 1988.

Trewartha, G.T., *The Earth's Problem Climates, 2nd Edition,* Univ. Wisconsin Press; 1981.

Tsonis, A.A., and G.L. Austin, "An Evaluation of Extrapolation Techniques for the Short-term Prediction of Rainfall Amounts," *Atmos. Ocean,* 19, pp 54-65; 1981.

Tuller, S.E., "World Distribution of Mean Monthly and Annual Precipitable Water," *Mon. Wea. Rev.,* 96, pp 785-797; 1968.

Turton, J., and S. Nicholls, "A Study of the Diurnal Variation of Stratocumulus Using a Multiple Mixed Layer Model," *Quart. J. Roy. Meteor. Soc.,* 113, pp 969-1010; 1987.

U.S. Navy Weather Research Facility, *An Examination of the Distribution of Hurricane Forecast Errors Using Probability Ellipses,* NWRF 12-1063-81; 1963.

U.S. Weather Bureau, *Generalized Estimates of Probable Maximum Precipitation and Rainfall Frequency Data for Puerto Rico and Virgin Islands,* USWB Tech. Paper 42; 1961.

U.S. Weather Bureau, *Rainfall-frequency Atlas of the Hawaiian Islands,* USWB Tech. Paper 43; 1962.

U.S. Weather Bureau, *Synoptic Meteorology as Practiced by the National Meteorological Center,* NAWAC Manual; 1963.

U.S. Weather Bureau Hydrometeorological Section and U.S. Army Corps of Engineers, Engineering Department, *Maximum Possible Precipitation over the Panama Canal Basin,* Hydrometeor. Rept.,4; 1943.

Van Bemmelen, W., "Land- und Seebrise in Batavia," *Beitr. z. Phys., freien Atmos., 10*, pp 169-177; 1922.

Velden, C.S., "Observational Analyses of North Atlantic Tropical Cyclones from NOAA Polar-Orbiting Satellite Microwave Data," *J. Appl. Meteor., 28*, pp 59-70; 1989.

Velden, C.S., B.M. Goodman and R.T. Merrill, "Western Pacific Tropical Cyclone Intensity Estimation from NOAA Polar-orbiting Satellite Microwave Data," *Mon. Wea. Rev., 119*, pp 159-168; 1991. Verploegh, G., "Observation and Analysis of the Surface Wind over the Ocean," *Mededel. Verhand., 89*; 1967.

Vincent, D.G., "Circulation Features over the South Pacific during 10-18 January 1979," *Mon. Wea. Rev., 110*, pp 981-993; 1982.

Vincent, D.G., "Cyclone Development in the South Pacific Convergence Zone during FGGE, 10-17 January 1979," *Quart. J. Roy. Meteor. Soc., 111*, pp 155-172; 1985.

Vojtesak, M.J., K.P. Martin and G. Myles, *SWANEA (Southwest Asia and Northeast Africa). A Climatological Study. Vol.1 — The Horn of Africa,* USAFETAC/TN-90/004; 1990.

Vojtesak, M.J., K.P. Martin, G. Myles and M.T. Gilford, *SWANEA (Southwest Asia and Northeast Africa). A Climatological Study. Vol. II — The Middle East Peninsula,* USAFETAC/TN-91/002/; 1991.

Vonder Haar, T., and A.H. Oort, "New Estimate of Annual Poleward Energy Transport by Northern Hemisphere Oceans," *J. Phys. Oceanogr., 2*, pp 169-172; 1973.

von Ficker, H., "Die Passatinversion," *Veröff. Meteor. Inst. Berl.,* (1) 4, pp 1-33; 1936.

Walker, G.T., "Correlations in Seasonal Variations of Weather. X. Applications to Seasonal Forecasting," *Mem. India Meteor. Dept., 24*, pp 333-345; 1924.

Wallace, J.M., "General Circulation of the Tropical Lower Stratosphere," *Rev Geophys. Space Phys., 11*, pp 191-222; 1973.

Wallace, J.M., and F.R. Hartranft, "Diurnal Wind Variations, Surface to 30 Kilometers," *Mon. Wea. Rev., 97*, pp 446-455; 1969.

Wallace, J.M., T.P. Mitchell and C. Deser, "The Influence of Sea-Surface Temperature on Surface Wind in the Eastern Equatorial Pacific: Seasonal and Interannual Variability," *J. Climate, 2*, pp 1492-1499; 1989.

Walters, K.R., Sr., *A Descriptive Climatology for Baledogle, Somalia,* USAFETAC/TN-88/001; 1988.

Walters, K.R., Sr., A.G. Korik and M.J. Vojtesak, *The Caribbean Basin — A Climatological Study,* USAFETAC/TN-89/003; 1989.

Walters, K.R., Sr., and W.F. Sjoberg, *The Persian Gulf Region — A Climatological Study,* USAFETAC/TN-88/002; 1988.

Wang, S.-T., H. Cheng, C.-H. Hsu and T.-K. Chiou, "Environmental Conditions for Heavy Precipitation during May-June over Taiwan Area," *Conf. Wea. Anal. Forecasting, Taipei. Meteor. Soc. Taipei,* pp 55-72; 1985.

Warren, S.G., C.J. Hahn, J. London, R.M. Chervin and R.L. Jenne, *Global Distribution of Total Cloud Cover and Cloud Type Amounts over the Ocean,* NCAR Tech. Note TN-317-STR; 1989.

Weldon, R., *Cloud Patterns and the Upper Air Wind Field,* AWS/TR-79/003, 1979.

West, R.C., *The Pacific Lowlands of Colombia, a Negroid Area of the American Tropics,* Louisiana State Univ. Press, Baton Rouge; 1957.

Wexler, R.L., *Annual Versus Daily Rainfall: Southeast Asia,* AWS Tech. Rept., 196, pp 226-242; 1967.

Whitfield, M.B., and S.W. Lyons, "An Upper-tropospheric Cold Low over Texas during Summer," *Wea. Forecasting,* 7, pp 89-106; 1992.

Whiton, R.C., and R.E. Hamilton, *Radarscope Interpretation: Severe Thunderstorms and Tornadoes,* AWS-TR-76-266; 1976.

Wiederanders, C.M., *Analysis of Monthly Mean Resultant Winds for Standard Pressure Levels over the Pacific,* Hawaii Inst. Geophys. Rept., 13; 1961.

Williams, K.T., and W.M. Gray, "Statistical Analysis of Satellite-observed Tradewind Cloud Clusters in the Western North Pacific," *Tellus,* 25, pp 313-336; 1973.

Williams, M., and R.A. Houze,Jr., "Satellite-observed Characteristics of Winter Monsoon Cloud Clusters," *Mon. Wea. Rev.,* 115, pp 505-519; 1987.

Wilson, J.W., "Use of Radar in Short-period Terminal Weather Forecasting," *Proc. 13th Radar Meteor. Conf., Amer. Meteor. Soc.,* pp 400-407; 1968.

Wilson, J.W., R.D. Roberts, C. Kessinger and J. McCarthy, "Microburst Wind Structure and Evaluation of Doppler Radar for Airport Wind Shear Detection," *J. Climate Appl. Meteor.,* 23, pp 898-915; 1984.

Winner, D.C., *Climatological Estimates of Clock-hour Rainfall Rates,* AWS Tech. Rept. 202; 1968.

Woodley, W.L., and A. Herndon, "A Raingauge Evaluation of the Miami Reflectivity-rainfall Rate Relationship," *J. Appl. Meteor.,* 9, pp 258-263; 1970.

Working Group on Numerical Experimentation, Report of the Third Session. WMO/TD 223; 1987.

World Meteorological Organization, *International Cloud Atlas (abridged),* WMO; 1956a.

World Meteorological Organization, *World Distribution of Thunderstorm Days,* WMO No. 21 TP 6 Part 2; 1956b.

World Meteorological Organization, *International Meteorological Tables,* WMO No. 186 TP 94; 1966.

World Meteorological Organization, *Operational Techniques for Forecasting Tropical Cyclone Intensity and Movement*, WMO 528; 1979.

World Meteorological Organization, *Weather Reporting; Observing Stations*, WMOP/OMM 9; 1986.

World Meteorological Organization, *World Weather Watch: Fourteenth Status Report on Implementation*, WMO 714; 1989.

Wylie, D., B.B. Hinton and K. Kloesel, "The Relationship of Marine Stratus Clouds to Wind and Temperature Advections," *Mon. Wea. Rev.*, 117, pp 2620-2625; 1989.

Xian, Z., and R.A. Pielke, "The Effects of Width of Landmasses an the Development of Sea Breezes," *J. Appl. Meteor.*, 30, pp 1280-1304; 1991.

Xu, J.M., Chinese Polar Orbiting Meteorological Satellite. In: *East Asia and Western Pacific Meteorology and Climate* (P. Sham and C.P. Chang, eds.), World Scientific, Singapore, pp 131-138; 1990.

Yang, S., and P.J. Webster, "The Effect of Summer Tropical Heating on the Location and Intensity of the Extratropical Westerly Jet Streams," *J. Geophys. Res.*, 95, pp 18,705-18,721; 1990.

Zehr, R.M., "The Diurnal Variation of Deep Convective Clouds and Cirrus with Tropical Cyclones," *Pre-prints, 17th Conf. on Hurricanes and Tropical Meteorology, April 7-10, Miami, Florida. Amer. Meteor. Soc.*, pp 276-279; 1987.

Zehr, R.M., "Improving Objective Satellite Estimates of Tropical Cyclone Intensity," *Preprints, 4th Conf. Satellite Meteorology and Oceanography, May 16-19, 1989, San Diego, CA. Amer. Meteor. Soc.*, J25-J28; 1989.

Zhao, S., and G.A. Mills, "A Study of a Monsoon Depression Bringing Record Rainfalls over Australia. Part II: Synoptic-diagnostic Description," *Mon. Wea. Rev.*, 119, pp 2074-2094; 1991.

Zipser, E.J., "The Role of Organized Unsaturated Convective Downdrafts in the Structure and Rapid Decay of an Equatorial Disturbance," *J. Appl. Meteor.*, 8, pp 799-814; 1969.

Zipser, E.J., "Mesoscale and Convective-scale Downdrafts as Distinct Components of Squall-line Structure," *Mon. Wea. Rev.*, 105, pp 1568-1589; 1977.

ABBREVIATIONS and ACRONYMS

AB	Air Base
AFCRL	Air Force Cambridge Research Laboratories (now Geophysics Directorate of Philips Laboratory; was also formerly Air Force Geophysics Laboratory (AFGL), then Geophysics Laboratory (GL).
AFGWC	Air Force Global Weather Central (USAF)
AIAA	American Institute of Aeronautics and Astronautics
AIREP	Aircraft report
AMEX	Australian Monsoon Experiment
AMS	American Meteorological Society
AMSU	Advanced microwave sounding unit
ANMRC	Australian Numerical Meteorology Research Center (now part of the Bureau of Meteorology Research Center, BMRC)
AOI	Area of interest
ASDAR	Aircraft to satellite data relay
ATEX	Atlantic Tradewind Experiment
ATS	Applications Technology Satellite (NASA)
AWN	Automated Weather Network
AWDS	Automated Weather Distribution System (AWS)
AWS	Air Weather Service
AWSTR	Air Weather Service Technical Report, now AWS/TR
BOMEX	Barbados Oceanographic and Meteorological Experiment
CAT	Clear air turbulence
CC	Conditional climatology
CDROM	Compact disc read only memory
CI	Contingency index
CIRES	Cooperative Institute for Research in Environmental Sciences
cm	Centimeter (0.39 inches)
CNES	Centre National D'Etudes Spatiales
DMSP	Defense Meteorological Satellite Program
DoD	Department of Defense
D-value	Difference between the "standard" height of a pressure surface and its actual height
ECMWF	European Center for Medium Range Weather Forecasts
EDIS	Environmental Data and Information Service (NOAA)
EDS	Environmental Data Service (NOAA)
EMEX	Equatorial Mesoscale Experiment
ENSO	El Niño Southern Oscillation Cycle
EOS	Earth Observing Satellite (NASA)
EOSAT	Earth Observation Satellite Company
ERS	Earth Remote Sensing
ESA	European Space Agency
ESSA	Environmental Sciences Services Administration (absorbed by NOAA in 1970)
EUMETSAT	European Meteorological Satellite

FAA	Federal Aviation Administration
FGGE	First GARP Global Experiment
FLR	Fractional low radiance
FNOC	Fleet Numerical Oceanography Center
FY-1B	Fen Yung meteorological satellite (China)
GARP	Global Atmospheric Research Program
GATE	GARP Atlantic Tropical Experiment, 1974
GMS	Geostationary Meteorological Satellite (Japan)
GOES	Geostationary Operational Environmental Satellite (NASA/NOAA)
GTS	Global Telecommunications System
HARP	Hawaiian Rainband Project
HK	Hong Kong
ICAO	International Civil Aviation Organization
IIOE	International Indian Ocean Expedition
INSAT	Geostationary Indian National Satellite
INTER	Intermittent (in WMO Terminal Airfield Meteorological Code)
IRIG	Inter-Range Instrumentation Group
ITCZ	Intertropical convergence zone
IWRS	Improved Weather Reconnaissance System
JTWC	Joint Typhoon Warning Center (USAF and US Navy)
K	Kelvin = °C + 272.16
km	kilometer (= 3281 feet; 0.54 nautical miles)
L	Local (time)
LIE	Line Islands Experiment, 1967
LLJ	Low-level jet
LRF	Long-range forecasting
m	meter (3.28 feet)
mb	millibar
MCC	Mesoscale convective complex
METAR	Meteorological Aviation Report
METEOR	Meteorological satellite (USSR)
METEOSAT	Geostationary Meteorological Satellite (ESA)
MIT	Massachusetts Institute of Technology
mm	millimeter (0.039 inches)
MONEX	Monsoon Experiments of FGGE
MOS	Marine Observations Satellite (Japan)
MRF	Medium-range forecasting
NASA	National Aeronautics and Space Administration
NASDA	National Space Development Agency (Japan)
NAWAC	National Weather Analysis Center

NCAR	National Center for Atmospheric Research
NETWC	Near-equatorial tradewind convergence
NESS	National Environmental Satellite Service (now part of NESDIS)
NESDIS	National Environmental Satellite, Data, and Information Service
NEXRAD	Next Generation Weather Radar (WSR-88D)
NH	Northern Hemisphere
NHC	National Hurricane Center (NOAA)
NM	Nautical mile (1.85 kilometers)
NMC	National Meteorological Center (NOAA)
NOAA	National Oceanic and Atmospheric Administration
NOAA-Next	Polar-orbiting meteorological satellites (1993-)
NOARL	Naval Oceanographic and Atmospheric Research Laboratory
NOMSS	National Oceanographic and Meteorological Satellite System
NPOES	National Polar-Orbiting Environmental Satellite
NWP	Numerical weather prediction
NWRF	Naval Weather Research Facility; became Naval Environmental Prediction and Research Facility (NEPRF), now Naval Oceanographic and Atmospheric Research Laboratory-WEST (NOARL-WEST)
NWS	National Weather Service
OLR	Outgoing longwave radiation
PIBAL	Pilot-balloon observation
POES	Polar-Orbiting Environmental Satellite
QBO	Quasi-biennial oscillation
RAWIN	Radio wind observation
RECCO	Reconnaissance Code
RH	Relative humidity
RHI	Range height indicator (radar)
RI	Radar index
RMS	Root mean square
RUSSWO	Revised Uniform Summary of Surface Weather Observations (AWS), now called Surface Observation Climatic Summaries (SOCS)
s	second
SATCOM	Satellite telecommunications
SEA	Southeast Asia
SH	Southern Hemisphere
SI	Stability index
SMONEX	Summer Monsoon Experiment of FGGE
SMOS	Surface Meteorological Observation Summaries (USN)
SOCS	Surface Observation Climatic Summaries
SPCZ	South Pacific convergence zone
SSM/I	Special sensor microwave/imager
SST	Sea surface temperature
SSWWS	Seismic Sea Wave Warning System

STJ	Subtropical jet stream
STORM	Stormscale Operational and Research Meteorology Program
TAMEX	Taiwan Area Mesoscale Experiment, 1987
TEJ	Tropical easterly jet
TEMPO	Temporary (in WMO Terminal Airfield Meteorological Code)
TIROS	Television Infrared Observational Satellite
TOPCAT	Project to study clear air turbulence (CAT) near the subtropical jet over Australia
TOPEX/ POSEIDON	Ocean Topography Experiment (U.S./France)
TOVS	TIROS - Operational Vertical Sounder
TRMM	Tropical Rainfall Measuring Mission
TUTT	Tropical upper-tropospheric trough
USAF	United States Air Force
USAFETAC	USAF Environmental Technical Applications Center
USN	United States Navy
USSA	United States standard atmosphere
USWB	United States Weather Bureau
UTC	Universal Time Cordinated ("Z" or Greenwich Mean Time)
V	Visibility
W	watts
WBAN	Weather Bureau, Air Force, Navy
WBFC	Weather Bureau Forecast Center
WMO	World Meteorological Organization
WMONEX	Winter Monsoon Experiment of FGGE
WPC	Weather plotting chart
WSR-88D	Weather Surveillance Radar - 88 Doppler
WWW	World Weather Watch
Z	See UTC
Z-R	Reflectivity-rainfall relationship (radar)

Geographical Index

A

Addis Ababa 127
Africa 1, 4, 5, 7, 20, 21, 25, 26, 30, 35, 51, 53, 55, 57, 59, 61, 62, 72-76, 84, 87, 89, 95, 97, 101, 102, 108, 109, 111, 114, 117, 119, 120, 121, 122, 123, 124, 125, 127, 129, 130, 140, 144, 160, 169, 189, 192, 193, 201, 206, 211-213, 218-223, 227, 228, 230, 238, 253, 254, 257, 262, 263, 274, 275, 291, 293, 295-297, 308, 309, 319, 345, 346, 357, 369, 375, 391
    East Africa 97, 116, 117, 120, 222, 291, 369
    Equatorial Africa 5, 7, 160, 206
    Horn of Africa 293
    North Africa 4, 21, 53, 55, 57, 75, 76, 89, 108, 119, 122, 130, 218, 219, 225, 227, 230, 296, 297
    Southern Africa 4, 51, 53, 57, 73, 74, 76, 102, 120, 121, 218, 227, 228, 295, 345
    West Africa 62, 87, 109, 123, 124, 144, 189, 192, 193, 201, 211-213, 221, 222, 225, 238, 254, 292, 295, 375
Agra 299
Ahmedabad 110, 111
Altiplano 21, 166
Amazon 1, 13, 21, 97, 101, 138, 140, 169, 172, 238
Andersen Air Base 37
Andes 48, 51, 76, 114, 166, 213, 237, 238
Angola 102, 117, 292, 295, 390
Annam Mountain 293, 339
Arabia 21, 55, 63, 102-104, 120, 297-300
Arabian Sea 1, 6, 53, 55, 57, 62, 97, 191, 193, 196, 198, 200, 201, 206, 255, 259, 260
Arakan Mountain 87
Argentina 142
Asia 1, 4, 20, 21, 25, 26, 30, 35, 45, 49, 55, 57, 60, 61, 67, 72, 74, 75, 76, 82, 89, 92, 93, 97, 114, 119-122, 124, 125, 127, 128, 142, 144, 167, 177, 214, 220, 222, 224, 225, 237, 239, 240, 260, 276, 282, 284, 293, 297, 336, 339-342, 345, 357, 366, 367, 369, 371, 384
    Central Asia 220, 293
    East Asia 45, 49, 72, 75, 76, 121, 125, 220, 222
    Asia Minor 167
    South Asia 89, 92, 121, 124, 125, 345, 384
    Southeast Asia 1, 21, 55, 82, 93, 97, 142, 144, 177, 214, 237, 239, 240, 274, 276, 282, 284, 293, 309, 339, 357, 366, 367, 369-371, 384
    Southwest Asia 4, 114, 122, 222, 297, 357
Atlantic 1-3, 5, 6, 34, 39, 48, 51, 58-61, 81, 84, 85, 92, 95, 97, 105, 106, 114-117, 127, 154, 160, 171, 188, 193, 194, 201, 202, 206, 220, 232-234, 237, 238, 242, 248, 250, 253-255, 258-260, 262-267, 269-271, 274, 301, 305, 311, 312, 325, 337, 355, 358, 379, 388
    Eastern Atlantic 6, 253
    Equatorial Atlantic 117, 388
    South Atlantic 48, 51, 97, 127, 206, 234, 248, 260, 309
    Tropical Atlantic 3, 81, 95, 97, 105, 106, 171, 254, 263
Atlas Mountain 53
Australia 7, 20, 25, 30, 35, 51, 53, 57, 59-63, 72, 74, 76, 83, 114, 120, 125, 127, 128, 189, 201, 213, 218, 220, 222, 223, 227, 250, 255, 259, 268, 270, 283, 291, 301, 305, 309, 349, 350, 389
    Northern Australia 51, 53, 57, 120, 125

B

Baledogle 168
Bangkok 80, 282
Bangladesh 76, 283
Barbados 82, 142
Bay of Bengal 53, 55, 57, 76, 134, 189-191, 220, 255, 258-260, 283
Biak 37
Bight of Benin 160
Bolivia 21, 140, 237
Bombay 62, 109-111, 121, 127, 133, 134, 169, 230, 232, 389
Borneo 185, 214, 215, 220
    North Borneo 214, 215, 220
Brazil 30, 180, 206, 218, 238, 389
    Northeast Brazil 180, 238
Burao 168
Burma 87, 120, 190, 240, 292

C

Calcutta 282, 376
Cambodia 146, 147
Canton Island 64, 117, 318, 319
Cape Kennedy 295
Caracas 151
Caribbean 45, 57, 79, 114, 136, 202, 218, 253, 263, 266, 292, 309, 388
Caroline Islands 5, 260
Caspian Sea 120
Cebu City 281, 282
Central America 25, 26, 30, 35, 57, 86, 89, 114, 136-138, 140, 142, 145, 146, 151, 207, 218, 292, 293, 309, 318, 357, 384, 390
Chad 297
Cherrapunji 149
Chiangmai 226, 282
China 1, 22, 30, 53, 73, 86, 89, 92, 102, 109, 112, 113, 121, 122, 125, 126, 131, 172, 190, 201, 206, 214-217, 225, 240, 339, 343, 392
    Central China 122, 126
    South China 62, 86, 89, 102, 112, 113, 125, 131, 172, 189, 190, 201, 206, 214-217, 225, 240
    Southeast China 1, 92
    West China 121
Christmas Island 37, 117
Clark Air Base 46, 282, 303, 359, 364
Colombia 87, 237
Congo Basin 101, 221
Costa Rica 136

D

Daly Waters 63
Danang 359, 361
Darwin 37, 63, 65, 116, 369, 389
Djakarta 105, 159, 387
Djibouti 301
Doha 296
Don Muang 353, 354

E

Easter Island 116
Ecuador 115, 117, 292
Egypt 222
El Chichon 315
El Salvador 127, 136
Eniwetak 46, 154, 155, 364
Entebbe 178
Ethiopia 127
Europe 26, 30, 34, 35, 120, 322

F

Fanning Island 230
Fiji 223, 312
Florida 27, 88, 101, 112, 142, 150, 151, 162-164, 266, 285-287, 290, 295, 312, 376, 378, 379
French Polynesia 117
Funafuti 312

G

Galveston Bay 162, 373
Gan 64, 206, 218
Ganges Valley 240, 255
Great Indian Desert 111
Great Rift 4
Guadelupe 85
Guam 213, 245, 256, 259, 266, 268-270, 284, 293, 365, 367, 392
Guantanamo Bay 305
Guatemala 167
Guayaquil 102
Gulf of Aden 167
Gulf of Guinea 297
Gulf of Mexico 22, 27, 266, 311, 364, 373

Gulf of Panama 237
Gulf of Thailand 276, 295
Gulf of Tomini 183
Gulf Stream 67, 102
Guyanas 157

H

Haifeng 113
Hawaii 21, 27, 30, 86, 88, 112, 114, 117, 127, 144, 146, 149, 151, 158, 159, 165, 179, 193, 201, 202, 213, 214, 225, 227, 229, 230, 237, 253, 276, 284, 285, 290, 292, 318, 357, 366
Hilo 127, 158, 159, 314
Himalayas 21, 76, 119, 121, 123, 126, 140, 220, 225, 240
Hispaniola 305
Honduras 136, 207, 218, 364
Hong Kong 22, 35, 85, 102, 125, 131, 156, 214, 215, 217, 218, 259, 284, 301, 303, 318, 365, 376, 389
Honolulu 127, 167, 168, 234, 276
Houston 218

I

India 5, 21, 45, 53, 60, 62, 87, 101, 105, 109-112, 114, 117, 120, 121, 124, 126, 127, 130, 133, 189, 193, 196-198, 200, 220, 230, 232, 240, 259, 276, 277, 282, 283, 290-295, 307-309, 318, 346, 355, 365, 370, 377, 378, 384, 386-389
Indian Ocean 3, 5, 40, 48, 53, 55, 59, 60, 62, 74, 83, 87, 97, 114, 119, 120, 124, 125, 203, 206, 220, 222, 234, 249, 250, 255, 257, 259, 266, 305, 384
Indochina 112, 114, 120, 125, 126, 225
Indonesia 5, 7, 20, 35, 51, 53, 114, 116, 120, 222, 234, 275, 384, 387
Isthmus of Tehuantapec 218
Ivory Coast 102

J

Jamaica 127
Japan 58, 62, 67, 131, 217, 242, 288, 339
Jodhpur 110, 111
Johnston Island 253, 351, 352

K

Kadena 37, 149, 152, 246, 304, 368
Kano 296
Karachi 121
Kauai 21, 37, 113, 149, 284
Kenya 166, 291, 292
Kerala 133
Key West 287
Khartoum 222, 296, 299
Khasi Hills 87
Kingston 127
Kisumu 178
Konkan 133
Koolau Range 213
Korat 177, 282, 372
Kuroshio Current 67, 102
Kwajalein 46, 234

L

La Asuncion 151
Lagos 305
Lake Maracaibo 178
Lake Nicaragua 178
Lake Okeechobee 164
Lake Victoria 164
Lake Volta 165
Libya 222
Lihue 84, 229
Lima 102, 172
Line Island 209-211, 230, 309
Luang Prabang 154, 155
Luzon 303, 390

M

Macassar Strait 183
Mactan Air Base 281, 365
Madagascar 62, 89, 120, 390
Madras 295
Majuro Atoll 176
Malacca Strait 182
Malaysia 5, 112, 113, 151, 153, 180, 181, 218, 220
Manila 105
Maritime Continent 5, 7, 8, 9, 220, 274
Marshall Island 1, 253
Maui 166, 202, 284
Mauna Loa 159, 165, 166
Mauritius 259
Mediterranean Sea 167, 260
Mekong River 146
Merida 218
Mexico 22, 55, 120, 218, 292, 309, 311
Micronesia 260
Middle East 62, 122, 309
Midway Island 205
Mindanao 5
Molokai 37, 284
Mosquito Gulf 183
Mount Galung 315
Mount Pinatubo 315
Mozambique Channel 220, 292
Mt. Cameroon 144
Mt. Haleakala 166
Mt. Kenya 166
Mt. Waialeale 21, 113, 144, 214, 229, 236

N

Nagpur 293, 294
Namibia 62, 102
New Caledonia 264
New Delhi 110, 111, 369, 370, 377, 378
New Guinea 5, 7, 89, 223, 224
New Zealand 264
Nha Trang 151
Nicaragua 136, 218
North America 25, 26, 30, 35, 60-62, 67, 73, 75, 142, 384
Northwest Cape 74
Nubian Desert 301, 302

O

Oahu 127, 142, 213, 227, 229, 284
Obbia 168
Okinawa 67, 149, 304, 339, 368

P

Pacific 3, 28, 39, 45, 46, 48, 55, 57, 58, 60, 85, 95, 97, 100, 114-117, 119, 127, 129, 136, 151, 203, 207, 230, 234, 255, 257, 305, 337, 349, 364, 365, 384
    Central Pacific 6, 7, 85, 95, 117, 240, 253, 274
    Eastern Pacific 6, 61, 85, 89, 92, 95, 114, 117, 171, 263, 266, 274, 343, 388
    Equatorial Pacific 20, 22, 115, 116
    North Pacific 4, 5, 34, 48, 57, 60, 84, 95, 96, 114, 193, 194, 196, 201, 206, 249-252, 255, 258-260, 262-267, 269-271, 301, 305, 325, 336, 337, 339-343
    South Pacific 7, 48, 60, 83, 95, 119, 206, 227, 248, 249, 305, 387
    Southeast Pacific 28, 116, 260
    Southwest Pacific 25, 26, 30, 35, 230, 259, 266, 270
    West Pacific 2, 35, 45, 57, 83, 95, 151, 256, 259, 384
    Western Pacific 20, 25, 27, 34, 116, 117, 119, 257, 270, 301, 303, 307, 320, 366, 384
Pago Pago 296
Pakistan 104, 124, 196, 255

Panama 136, 147, 151, 218, 237
Papeete 80
Papua 5
Pass Christian 312
Persian Gulf 170, 297, 298, 299, 301
Peru 22, 102, 115, 117, 292, 390
Philippines 55, 87, 89, 92, 102, 112, 114, 130, 221, 240, 245, 260, 281, 282, 292, 303, 318, 365
Piarco 147
Pinatubo 315
Piura 37
Pleiku 282
Pohnpei 37
Puerto Chicama 116
Puerto Rico 112, 151, 356

Q

Queensland 291, 292
Quito 166

R

Red Sea 301, 302
Reunion Island 149
Ryukyu Islands 92

S

Sahara 57, 193, 296, 297
Sahel 130, 297, 388
Saigon/Ho Chi Minh City/Tan Son Nhut 81, 82, 141, 216, 282, 359-362, 367, 368, 374, 375
San Juan 37, 142, 305
San Salvador 127, 135, 136, 151
Senegal 227
Shanghai 72
Singapore 64, 72, 112, 113, 215, 216
Soctrang 282
Solomon Island 5
Somalia 62, 73, 102-104, 168, 170, 220, 301
Songkla 282
South America 4, 20, 21, 25, 26, 30, 35, 48, 53, 55, 57, 58, 61, 62, 73, 74, 89, 94, 114, 116, 117, 121, 145, 146, 151, 157, 166, 172, 213, 218, 220, 222, 237, 238, 386, 387, 391
South China Sea 4-7, 53, 86, 102, 190, 201, 206, 214-217, 225, 240, 260, 274, 275
Sri Lanka 35
St. Helena Island 127
Sudan 211, 222
Sulawesi 183-185
Sumatra 295

T

Tahiti 80, 234
Taipei 37
Taiwan 1, 76, 86, 125, 126, 149, 225, 293, 303
Taiwan Strait 293
Tarawa Island
Texas 88, 288, 290, 373
Thailand 37, 80, 120, 144, 146, 147, 177, 276, 278-280, 282, 292, 295, 354, 372
Tibesti massif 297
Tibet 89, 109, 114, 119, 121, 123, 126, 140
Tucson 321
Turkey 121, 222

U

Ubon 372
Udon 372
United States 8, 30, 35, 37, 120, 125, 151, 248, 283, 286-288, 290, 295, 296, 301, 305, 311-313, 366, 369-371, 392

C-5

V

Venezuela 1, 142, 151, 214, 238
Vietnam 151, 214, 282, 284, 339
Virgin Islands 151
Viti Levu 312

W

Waikoloa Beach 158
Wake Island 171
Washington Island 234
West Bengal 283
Western Ghats 87, 200, 318
Willis Island 154, 319

Y

Yangjiang 113
Yemen 301
Yucatan 183

Z

Zambia 227, 295

Subject Index

A

Advection 102, 162, 230, 366, 370
Aircraft 10, 30-34, 105, 168, 188, 198, 200, 201, 206, 210, 234-238, 242, 245, 249, 262, 265-268, 271, 282, 286, 287, 295, 296, 305, 307-310, 314-317, 322, 324-325, 328, 334-345, 347, 357, 373, 392
    Aircraft reports (AIREP) 30, 57, 316, 317, 324, 332, 334, 335, 345, 347, 357
Air mass 187, 213, 316, 319, 330
Air-sea interaction 10, 343
Air temperature 8, 21, 67-71, 75, 77, 84, 101, 103, 154, 172, 356, 369, 386
Altimeter 362
Altocumulus 105, 106, 108, 172, 310,
Altostratus 105, 108, 172, 207, 223, 229, 230, 310
Anabatic wind (*See* Upslope) 165, 180, 181
Analysis 276, 295, 303, 316-318, 320, 322, 324, 326-332, 334-344, 346, 348-355
    automated 316, 349
    composite 302
    computer-aided 350
    kinematic 48, 103, 113, 198, 199, 243, 319, 320, 326, 329-332, 339-342, 345-347, 349, 350, 351, 355, 358, 367, 368
    numerical weather prediction 5, 6
    objective 328
    post- 188, 268, 316, 339, 351
    rainfall 355
Andhi 299
Anemometer 154, 301, 319-321
Anomalies 45, 46, 117, 118, 245, 262, 337, 384-389
Anticyclones 4, 15, 21, 48, 51, 53, 55, 60, 61, 67, 74, 84, 86, 89, 122, 125, 189, 221, 326, 328, 339
    oceanic 55, 67, 86
    subtropical 74, 84, 221
Aridity (*See* Desert) 21
Asymptote 5, 106, 108, 213, 326, 327, 328, 330, 332, 358
Atlases 363, 390
Atolls 321
Aziab 297

B

Baroclinicity 191, 197, 201
Barotrophic 189, 201, 316, 380
Beaufort Number 320
Brightness 95, 96
Buffer zone 3
Buoys 27, 28, 317, 343, 345

C

Checkerboard diagram 353
Cirrocumulus 105, 106, 108
Cirrostratus 105, 106, 108, 203, 206, 247, 308
Cirrus 34, 97, 101, 105, 106, 108, 174, 175, 186, 201, 202, 203, 206, 210, 224, 234, 243, 244, 246, 256, 282, 308, 321, 324, 338, 347
Climatology 35, 39, 48, 63, 79, 127, 139, 186, 189, 218, 219, 222, 241, 249, 252, 259, 267, 270, 285, 287, 294, 317, 328, 335-337, 346-349, 358, 359, 361-366, 373, 377-383, 385, 388-391
    conditional 359, 361
    data 218, 219
    of tropical cyclones 249
    radar 377-379
    rainfall 390
Clouds 3, 4, 12, 21, 34, 84-86, 89, 92, 95, 97-100, 102, 105, 107, 108, 113-117, 121, 125, 126, 146, 160, 164, 166, 171, 173, 176, 182, 186, 187, 193, 195, 200, 202-205, 207-209, 212, 213, 217, 221, 223, 224, 227, 228, 234, 235 237, 246, 284-288, 290, 296, 298, 301, 307, 310, 318, 321, 323, 324, 339, 362, 363, 372, 390, 391
    cloud clusters 6, 92, 93, 95-97, 125, 174, 175, 238, 339, 358
    cloud droplets 98, 99, 113, 213
    cloud motion 34, 262, 267, 316, 325, 334, 343-345
    deep 92, 95, 114, 125, 390
Coasts 2, 21, 66, 67, 74, 84, 101, 102, 104, 115, 117, 120, 121, 136, 154, 157, 161, 169, 172, 178, 182, 183, 186, 295, 301, 311-313, 319, 372

Coefficient of variation 130, 390

Cold lows 1-3, 188, 189, 201-203, 205-207, 262, 263

Condensation 11, 19, 20, 21, 62, 84, 98, 99, 101, 102, 112, 121, 160, 173, 180, 193, 194, 202, 204, 220, 245, 260, 265

Conduction 11, 22

Confluence 326-328, 332

Continuity 3, 7, 18, 19, 121, 187, 188, 208, 213, 214, 297, 328, 330, 338, 339, 346-348, 351

Convection 5, 7, 11, 21, 23, 24, 82, 84, 85 89-91, 98, 100, 104, 105, 108, 112, 114-117, 138, 140, 157, 162, 175, 176, 182, 183, 186, 192, 193, 198, 202-206, 210, 211, 213, 214, 217, 218, 220-222, 224, 225, 227, 232, 237, 238, 240, 242, 248, 262-264, 268, 288, 296, 299, 308, 316-319, 327, 339, 343, 347, 353, 358, 359, 363, 365, 367, 369, 372, 375, 376, 377, 378, 384, 388, 391

Convergence 1-7, 13, 48, 51, 53, 85, 89, 95, 97, 104, 106, 108, 109, 112-114, 125, 150, 151, 157, 160, 162, 166, 167, 169, 175, 176, 178, 180, 181, 195, 208, 213, 214, 220-225, 227, 229, 230, 234, 236-238, 252, 253, 262-264, 274, 281, 326, 332, 336, 339, 358, 376, 379, 391

Convergence zone 5-7, 51, 95, 114, 175, 222, 223, 236

    Intertropical Convergence Zone (ITCZ) 5, 6

    Near-equatorial Tradewind Convergence (NETWC) 3, 5-7, 13, 15, 20, 48, 49, 53, 55, 85, 92, 95, 114, 115, 125, 178, 213, 221, 230, 234-239, 252, 274, 336, 339, 358

    South Pacific Convergence Zones (SPCZ) 7, 55, 175, 222-225, 227

Cooling 2, 11, 19, 20, 63, 67, 73, 84, 85, 100-102, 119, 124, 156, 165-167, 169-178, 180, 186, 213, 245, 317, 318, 322, 376

Coriolis force 48, 73, 119, 157, 180, 260

Correlation 47, 148, 269, 311, 364, 376, 386, 387, 389

Cross section 195, 197, 205, 216, 236, 242, 244, 296, 330, 336, 346, 350, 351, 353, 363, 366

    space 366

    time 350, 353

Cumulonimbus 3, 5-7, 12, 15, 23, 82, 92, 94, 97, 101, 105, 106, 108, 109, 111, 126, 162, 166, 184, 198, 202, 206-209, 246, 247, 282, 283, 290, 299, 310, 338

Cumulus 7, 23, 24, 82, 84, 86, 89, 92, 94, 97-101, 105, 106, 108, 109, 113, 115, 160, 162, 165, 171-173, 186, 193, 210, 230, 238, 287, 288, 290, 307, 310, 316, 338

Currents (See Ocean) 13, 14, 67, 213, 214, 329

Curvature 3, 97, 255, 264, 265, 270, 328, 336

Cusp (See Singular point) 327

Cycle 20, 24, 39, 63, 65, 92, 116, 119, 126, 153-155, 157, 162, 165, 167, 170, 173, 174, 176, 178, 180, 183, 186, 220, 244, 318, 319, 339, 353

    annual 20, 39, 63, 65, 92, 116

Cyclogenesis 53, 201, 214, 220, 221, 260, 345

Cyclones 15, 20-22, 33-36, 40, 48, 51, 53, 55, 57, 60, 61, 67, 74, 82, 84, 86, 89, 95, 98, 101, 105, 116, 117, 119, 122, 123, 125, 126, 131, 149, 151, 153, 174, 175, 187, 188, 189, 190-198, 200-202, 206, 207, 220, 221, 225, 241-246, 248-255, 257-267, 269-273, 282, 287, 288, 293, 295, 301, 303, 305, 307, 310-312, 316, 319, 322, 325-328, 335, 338, 339, 343-347, 355

    extra-tropical (See Typhoons) 33

    hybrid 126, 189, 201, 206, 207

    mid-tropospheric 48, 123, 189, 191-193, 198-200, 206, 316, 343, 345

    subtropical 22, 193, 195, 196, 198, 206

    tropical (See Tropical cyclones)

    upper tropospheric 105

    wave 332

    West African 123, 189, 192, 225

D

Data 1, 2, 10, 14, 15, 20, 25-28, 30, 33-38, 48, 77, 89, 96, 105, 107, 111, 114, 117, 127, 133, 138, 140, 141, 143, 146, 147, 150, 152, 156, 159, 170, 171, 175, 178, 183, 186-189, 193-195, 200, 205-208, 214, 218, 219, 242, 247, 249, 251, 252, 258, 263, 265-268, 271, 272, 276, 281, 282, 284, 290-293, 295, 304, 308, 311, 312, 316, 317, 321-323, 325-330, 334-339, 343-348, 350, 351, 353, 355-359, 362, 363, 366, 367, 377, 379, 385-391

    collection 308, 316, 317

Decoupling 223, 230

Depressions 1, 4, 6, 7, 98, 119, 123, 125, 189-191, 197, 200, 201, 204, 205, 207, 218, 220, 224, 230, 234, 238, 248, 252, 254, 255, 258, 262, 266, 328, 351, 355

    monsoon 1, 6, 7, 98, 123, 189, 190, 191, 196, 200, 201, 220, 230, 234, 238, 255, 258, 355

    tropical 119, 207, 218, 224, 248, 252, 254, 328

Desert 74, 89, 111, 114, 115, 117, 119, 121, 122, 124, 126, 172, 218, 296, 297, 299, 301-302

Difluence 326, 327, 332

Diurnal variation 79, 94, 97, 98, 154-159, 161, 163-165, 167, 169, 171-186, 213, 238, 275, 280, 285, 288, 295, 310

    of cloud 171, 172

    of dewpoint 209

    of pressure 157

    of rainfall 176-178, 180, 183, 186, 213

    of wind 157

Divergence 74, 85, 87, 96, 97, 104, 107, 108, 112, 122, 124, 125, 150, 160, 166, 167, 171, 176, 180, 187, 195, 203, 204, 208, 218, 222, 225, 240, 243, 262, 264, 296

Downdraft 100, 210, 211, 230, 295, 301

Dwnslope flow 165, 166, 178, 182

Downwelling 73
Drizzle 104, 172
Dropwindsondes 30
Dust 102, 301
Dust devils 296, 301
D-value 30, 281

E

Easterly wave 97
Electrooptical device 101, 104, 299
El Niño 2, 7, 10, 20, 22, 48, 74, 92, 115-118, 253, 256, 387-391
Entrainment 99, 113
Error 28, 34, 148, 276, 280, 282
    Root-mean-square (RMS) 322-325, 334, 349, 350, 380, 388
Evaporation 11, 12, 20, 22, 98, 100, 135, 209, 210, 245
Extreme wind 273, 293, 295, 303, 305, 306
Expedition 1, 62, 84
    Hawaiian Rainband Project (HARP) 165
    International Indian Ocean Expedition (IIOE) 62, 357
Experiment 1, 92, 125, 171, 175, 187, 209, 230, 309, 350, 377, 392
    Atlantic Tradewind Experiment (ATEX) 171, 357
    Australian Monsoon Experiment (AMEX) 125, 357
    Barbados Oceanographic Meterological Experiment (BOMEX) 357
    First GARP Global Experiment (FGGE) 384, 385, 391
    GARP Atlantic Tropical Experiment (GATE) 1, 92, 98, 127, 171, 175, 193, 209, 254, 263, 357, 374, 391
    Line Islands Experiment (LIE) 175, 209, 230, 309
    Summer Monsoon Experiment (SMONEX) 1
    Taiwan Area Mesoscale Experiment (TAMEX) 1, 125, 357
    Winter MONEX (WMONEX) 1, 216, 349

F

Flood 1, 22, 112, 149, 182, 186, 208, 220, 311, 366, 377
Fog 2, 74, 84, 101, 102, 104, 105, 157, 160, 172, 176, 186, 296, 301, 322, 338, 353, 359, 362, 367, 390
    advection 102
    radiation 101, 172
Freezing level 75, 290, 310
Friction 21, 48, 62, 63, 74, 100, 159, 160, 167, 169, 195, 242, 288, 293, 296, 303, 314, 343
Front 5, 23, 33, 37, 62, 66, 100, 113, 123, 125, 126, 157, 160, 167, 187, 188, 194, 208-210, 213, 214, 216-221, 223-227, 239, 241, 248, 256, 286-288,
    290, 293, 296, 297, 299, 328, 330-332, 338, 339, 346, 347, 356, 367, 376, 384
    cold 23, 66, 187, 188, 208, 210, 213-218, 220, 223, 224, 239, 290, 296, 297, 330, 332, 376
    frontal analysis 1, 347, 356
Frontogenesis 194
Funnel cloud 283-287

G

General circulation 10, 15, 62
Geopotential height 78
Global Atmospheric Research Program (GARP) 1, 92

H

Haboob 222, 296, 299
Hadley cell 82, 113, 214, 218, 220, 296
Hail 227, 273, 290-292, 366
Harmattan 297
Heat 4-7, 10-14, 17, 19-24, 40, 51, 53, 55, 57, 62, 67, 82, 84, 87, 89, 98, 111, 112, 119-125, 130, 154, 155, 180, 186, 189, 192-194, 200,
    201, 206, 213, 220-222, 225, 237, 245, 246, 255, 260, 264, 265, 299, 303, 336, 349
    heat energy 14, 82, 245
    heat low 2, 4, 6, 24, 40, 48, 51, 53, 55, 57, 87, 123, 189, 200, 221, 237, 299, 366
    heat trough 4, 6, 13, 19-20, 40, 48, 51, 62, 87, 89, 111, 120, 121, 124-125, 130, 189, 192, 193, 201, 213, 222, 299
    latent 62, 82, 98, 192, 194, 245, 265
    sensible 11, 12, 13, 98, 245, 264

High. *(See* Anticyclone)

Hodograph 3, 158, 167

Humidity *(See* Moisture) 30, 67, 72, 75, 76, 78, 85, 99, 103, 154, 157, 205, 209, 275, 322, 323, 325

Hurricane-generated tornado 287

Hurricane 1, 2, 21, 22, 30, 36, 97, 151, 152, 188, 189, 191, 206, 207, 224, 241-245, 248, 253-255, 257, 260, 262-266, 272, 287-289, 301, 303, 305, 311-313, 328, 355, 373, 374, 376, 388, 392

    Bebe (1972) 312

    Beulah (1967) 287, 288

    Camille (1969) 301, 311, 312

    Carla (1961) 373

    Cleo (1958) 245

    Debby (1982) 271

    Donna (1960) 151, 243

    Isbell (1964) 288

I

Ice 99, 100, 105, 290

Icing 273, 309, 310, 324

Indian summer 89, 117, 134, 189, 386

Indraft 327

Inertial navigation 324, 334

Infrared (IR) 4, 33-35, 93, 94, 101, 104, 126, 164, 174-175, 182, 184-186, 206, 212-213, 217, 224, 240, 246, 247, 268, 297, 298, 300-302, 315, 321, 390

Insolation *(See* Radiation) 4, 11, 104, 111, 154, 165, 180, 292, 297, 307

Inversion 4, 62, 63, 89, 100, 101, 103-105, 108, 112-115, 144, 155, 158, 160, 161, 165, 166, 169, 171, 172, 195, 216, 217, 222, 227, 229, 230, 234, 236, 238, 284, 287, 296, 323, 364, 369, 391

    subsidence 4, 62, 104, 105, 144, 158, 172, 195, 217, 296

    surface 62, 63, 169

Isallobar 353, 355

Isogon 232

Isohyet 112-114, 145

Isotach 48, 159, 213, 215, 223, 226, 232, 358, 368

J

Jet stream 17, 19-21, 58, 60, 62, 73, 105, 122, 206, 214, 217, 222-224, 264, 286, 298, 307, 309, 324, 338, 348, 364

    easterly *(See* Upper tropospheric) 21, 60

    low-level 2, 62, 125, 168-170, 296, 297

    Somali 168, 170

    subtropical *(See* Upper tropospheric) 17, 19, 20, 58, 62, 105, 309, 364

    upper tropospheric 264

K

Katabatic flow *(See* Downslope) 180

Kelvin wave 116, 117, 220

Kinetic energy 23, 96

L

Land breeze 154, 157, 159, 160-162, 164, 166, 167, 170, 172, 176, 178, 182, 183, 281

Lapse rate 2, 109, 111, 112, 167, 295, 301, 369

Lightning detection *(See* Thunderstorms) 276

M

Mass-flow 19

Mei-Yu front 113, 125, 126

Mesoscale 1, 96, 98, 101, 113, 125, 145, 146, 296, 299, 308

Mesoscale convective complexes 113, 125

Microbursts 37, 295, 296, 307, 376

Mixing 4, 8, 22, 63, 101, 104, 167, 169, 172, 186, 260, 265, 319, 383

    vertical 63, 104, 167-169, 186

Model 1, 2, 4, 5, 36, 38, 39, 113, 117, 158-164, 171, 172, 175, 176, 187-191, 193-195, 197, 199-201, 203-207, 209, 211, 213, 215, 217, 219, 221, 223-225, 227, 229, 231, 233, 235, 237, 239, 242, 244, 254, 262, 267, 268, 270, 272, 279, 280, 296, 303, 304, 308, 311, 312, 316, 325, 327, 330, 332, 339, 343, 345-347, 350, 356, 357, 366, 368, 374, 378-380, 383, 392

    numerical 1, 117, 161-163, 172, 175, 187, 308, 311, 383

synoptic 2, 187-189, 191, 193, 195, 197, 199, 201, 203, 205, 207, 209, 211, 213, 215, 217, 219, 221, 223, 225, 227, 229, 231, 233, 235, 237, 239,316, 366, 392

Moisture 5, 76, 81, 82, 84, 92, 94, 96, 101,102, 144, 151, 156, 172, 198, 202, 206, 208, 222, 225, 227, 229, 230, 236, 237, 283, 287, 295, 308, 325, 339, 351, 353, 364, 366, 369, 377

Monsoon 1, 3-7, 10, 15, 20, 40, 51, 53, 55, 57, 60, 62, 73, 74, 86, 87, 89, 92, 97, 98, 101, 102, 105, 108, 111-114, 117, 119-126, 131, 133, 134, 144, 146, 176-178, 183, 189-191, 193, 196, 197, 200, 201, 206, 208, 211, 213, 214, 216, 220-223, 230, 234, 237, 238, 240, 252, 253, 255, 258, 260, 262, 263, 282, 293, 295, 299, 318, 328, 332, 336, 337, 339, 345, 358, 366, 367, 375, 377, 384, 386, 387, 389

   monsoon trough 87, 108, 112, 125, 189, 190, 191, 196, 197, 200, 201, 206, 220, 221, 238, 240, 252, 253, 255, 258, 262, 263, 328, 336, 339, 358, 367

   southwest 53, 97, 101, 105, 134, 144, 146, 220, 318, 328, 339, 345

   summer 89, 112, 117, 120, 121, 123-125, 133, 134, 144, 189, 193, 213, 214, 221, 222, 240, 282, 299, 377, 386, 389

   winter 97, 102, 113, 114, 119, 120, 123, 222, 230, 332

Mountain 2, 48, 53,62, 63, 76, 85-87, 97, 101, 113,114, 121, 125, 136, 138, 142, 144, 149, 155, 157,158, 161, 162, 165-167, 176, 178, 180-183, 211-214, 218, 227, 238, 290, 293, 295, 303, 305, 307,308, 318,319, 322, 339,377, 390

N

Nephanalysis 340, 341, 342, 347

Neutral point (See Singular point) 327, 328, 330, 336

Nimbostratus 310

"Northers" 281

Numerical models 1, 38, 117, 161-163, 172, 175, 187, 262, 267, 270, 308, 311, 383

Numerical Weather Prediction (NWP) 1, 34, 36, 269, 325, 350, 357, 380, 383, 384, 392

O

Observation 1, 4, 5, 7, 10, 15, 25-33, 35, 37-40, 48, 81, 89, 104, 105, 117, 144, 148, 157, 159, 162, 168, 170, 174, 180, 187, 188, 201, 204-206, 210, 213, 218, 230, 241, 242, 249, 251, 253, 267, 268, 270, 274, 275, 281, 282, 284, 285, 299, 308, 309, 316-318, 320-326, 334, 343, 345, 346, 350, 351, 353-358, 363, 366, 367, 369, 373, 377, 383, 391, 392

   aircraft (ASDAR) 30, 206, 210, 268, 308, 324

   aircraft reconnaissance 249, 317, 325

   lightning 274

   radar 369

   satellite 4, 10, 241,249, 251, 267

   ship 27,35, 89,201

   surface 25,104, 144, 205, 281,318, 321, 324, 343, 346, 354, 363, 367

   upper-air 267, 316,322

Orographic effect 208, 282, 390

Oscillation 2, 6, 10, 15, 24, 45, 63, 116, 117,240, 303, 373, 384, 387-390

   north-south 45

   quasi-biennial (QBO) 2, 6,17, 18, 24, 63, 388

   southern 116, 117, 387, 389, 390

Outdraft 211, 327

P

Pentad 39, 131-133, 135-140, 239

Persistence 7, 264, 267, 270, 357, 364, 365, 366, 369, 374, 380-383, 387, 389

Pilot Balloon (PIBAL) 29, 159, 168, 180, 323, 324, 345

Polar outbreak 22, 67, 121, 222, 296

Precipitable water 76, 77

Precipitation 12, 20, 121, 144, 290, 377

Pressure 28, 30, 34, 39-49, 51, 53,55,57,59,60-63,65, 73, 75, 78, 82, 86, 153, 154, 156, 157, 169-171, 174, 176, 187, 189, 192, 194, 195, 198, 202, 209, 215, 218-221, 225, 230, 237, 240, 275, 293, 294, 301, 303, 311-313

   minimum 267

   pressure change 45, 46, 221, 262, 293, 294, 318, 355

   pressure tendency 389

Profiler 37

Profiles 72, 78, 81, 168

   of temperature 78

   of wind 168

R

Radar 4, 35, 37, 82, 101, 213, 246, 282, 287, 290, 296, 299, 307, 310, 316, 317, 322, 324, 334, 346, 355, 357, 358, 363, 367-369, 372,373, 375-379

   Doppler radar 324, 334, 372, 373, 377

   index 82, 367

   reflectivity 376

Radiation 4, 11, 13, 20, 21, 101, 105, 108, 112, 114, 115, 138, 154, 156, 157, 171-174, 176, 183, 186, 225, 363, 385, 390
    direct 176
    ongoing longwave (OLR) 114, 115, 138, 139, 223, 224, 230, 234, 384
Radiosondes 105, 112, 324, 325
Radiowind (RAWIN) 29, 48, 57, 334, 345, 349
Rain forest 61
Rainband 151, 152, 165, 244, 245, 246, 287, 372, 373, 374
Rainfall 4, 5, 8, 10, 21, 26, 35, 36, 88-90, 92, 94, 96, 98, 100, 102, 104-106, 108-118, 120-136, 138, 140-152, 154, 155, 166, 167, 171, 172, 175-183, 186, 187, 194, 199, 206, 207, 213-215, 218, 220, 222, 227, 229, 232, 234, 238-240, 246, 275, 304, 319, 338, 355, 358, 359, 362, 368, 369, 373, 374, 376, 377, 383, 384, 386-391
    distribution 133, 143, 144, 150, 152, 199, 206
    diurnal variation of 176-178, 180, 183, 186, 213
    extremes 149
    frequency 142, 177, 178
Rainy season 57, 72, 80, 127, 130, 131, 133, 135, 136, 138, 141, 142, 154, 178, 227, 281, 351, 353, 356, 369, 375, 384
Rawinsondes 1, 334
Reconnaissance 1, 30, 34, 105, 201, 242, 245, 249, 262, 263, 265, 267, 271, 303, 317, 325, 328, 335, 346, 373
Research flight 89, 234
Ridge 3, 17, 34, 40, 48, 51, 53, 57-62, 75, 86, 111, 121, 122, 125, 131, 197, 203-205, 211, 216, 218, 239, 240, 253, 255, 265, 272, 286, 297, 323, 326, 328, 329, 332, 335, 336, 338, 339, 345, 357, 364, 375, 384
    subequatorial 60, 61
    subtropical 17, 34, 40, 51, 53, 58, 59, 60, 61, 75, 121, 197, 255, 265, 271, 297, 326, 328, 332, 336, 364
Root-mean-square (RMS) 322-325, 334, 349, 350, 380, 388
Runoff 12, 135

S

Satellite (meterological) 1, 2, 4, 5, 28, 30, 33-36, 38, 89, 92, 94-98, 101, 104, 105, 108, 112-114, 122, 125, 166, 169, 174-176, 183, 186-189, 191, 194, 201, 202, 206 208, 210, 213, 214, 217, 218, 225, 230, 232, 234-237, 240 244, 246, 247, 249, 251, 254, 256, 259, 262, 264, 265, 267, 268, 272, 274, 299, 301, 309, 315-317, 321, 322, 324-326, 328, 330, 335, 338, 339, 343, 344, 350, 355-358, 363, 366, 367, 372, 383, 389-391
    cloud motion 344
    data 2, 10, 30, 33, 35, 38, 89, 189, 207, 268, 316, 325, 326, 338, 339, 347, 348, 355-357, 363, 389, 391
    Defense Meteorological Satellite Program (DMSP) 33, 35, 104, 191, 200, 212, 256, 274, 275, 298, 300, 302, 321, 325, 345, 377
    Geostationary Meteorological Satellite (GMS) 93, 126, 183, 184, 185, 217, 224, 240, 247, 321
    Geostationary Operational Meteorological Satellite (GOES) 33, 34, 36, 85, 94, 171, 182, 196, 201, 202, 214, 220, 235-237, 252, 276, 321
    microwave image 33, 246, 325, 345
    visible image 85, 104, 105, 191, 196, 200, 201, 214, 237, 256, 301, 302, 390
Sea breeze 21, 74, 86, 157-163, 166-170, 172, 176, 178, 180, 181, 183, 295, 301, 319, 338, 379, 390
Sferic (See Lightening detection) 276
Shamal 297-299
Shear 2, 6, 7, 45, 94, 96, 97, 99, 100, 109-111, 114, 125, 168, 189, 190, 202, 208, 210, 211, 213, 214, 221, 222, 234, 237, 242, 244, 253, 260-262, 264, 283, 288, 301, 307-309, 321, 330, 334, 335, 339, 343, 355, 358, 367, 373, 375, 384, 391
    cyclonic 96, 97, 214, 221, 234, 260
    horizontal 391
    shear line 2, 34, 45, 114, 208, 213, 214, 237, 244, 321, 330, 339, 367, 384, 391
    vertical 211, 288
Showers 6, 7, 94, 100, 101, 109, 111-113, 123, 146, 147, 160, 162, 176, 178, 183, 188, 205, 211, 221, 227, 282, 293, 295, 297, 318, 319, 322, 362, 366, 369, 374, 377
Singular point 327, 329, 330, 351
Skew - T (aerological) diagram 209
Southern Oscillation 116, 117, 387, 389, 390
Specific heat 20, 67
Squall lines 2, 6, 7, 23, 95, 97, 112, 123, 173, 176-178, 186, 193, 208-211, 213, 222, 238, 239, 286, 295, 339, 367, 375
Stability 82, 84, 92, 96, 112, 144, 160, 162, 167, 175, 183, 192, 193, 200, 202, 211, 220, 222, 287, 292, 316, 353, 357, 358, 364, 369, 376, 386
    indexes 82, 369
Standard deviation 7, 39, 40-44, 48, 67-71, 77, 80, 130, 142, 153, 269, 276, 278, 318
Storm tide 241
Stratocumulus 84, 89, 92, 101, 105, 106, 108, 115, 171, 172, 186, 193
Stratosphere 6, 10, 24, 63, 65, 307
Stratus 82, 84, 85, 89, 101, 102, 104-106, 108, 109, 112, 113, 157, 172, 186, 203, 206, 207, 223, 227, 229, 247, 301, 308, 310, 322, 338, 353, 390
Stream function 326, 333
Streamline (See Kinematic analysis) 19, 48, 167, 191, 192, 210, 222, 316, 324, 326-330, 332, 334, 336, 343, 346-348, 351, 358
Sumatra 181, 182, 295
Superposition 208, 230, 234, 367
Surface wind 7, 20, 21, 34, 35, 45, 48, 49, 62, 73, 87, 102, 103, 105, 107, 108, 112, 116, 123, 133, 157, 158, 165, 167-171, 194, 195, 201, 214, 216, 220, 222, 223, 242, 243, 245, 247-249, 256, 273, 288, 293, 297, 301, 303, 307, 319, 320, 324, 325, 331, 338, 343, 345, 347, 349, 365, 366, 371
Surges 2, 7, 97, 126, 208, 211, 214, 216-218, 220-222, 236, 238, 239, 296, 297, 299, 310, 313, 339, 345, 355, 358, 384
    cold 172, 217, 297
    storm (ocean) 310, 311, 313

T

Temperature 4, 5, 8, 19-22, 28, 30, 34, 35, 89, 96, 99-105, 107, 109, 110, 116, 119, 121, 122, 124, 125, 150, 153-157, 159-162, 165, 168, 171, 172, 176, 186, 187, 198, 201, 202, 206, 208-210, 213, 214, 216, 218, 219, 229, 242, 245, 247, 260 262, 264, 267, 275, 287, 293, 294, 297, 301, 304, 308, 310, 316, 318, 319, 322-325, 329, 333, 343, 345, 351, 356, 362, 363, 366, 369, 370 376, 386, 390
    equivalent potential 81, 83, 84, 369
    gradient 5, 20, 62, 66, 67, 74, 75, 100, 104, 121, 122, 124, 201, 214, 245, 264, 297, 316, 319, 390
    maximum 267, 369
    minimum 154
    potential 8, 80-84, 209, 210, 369
    sea surface (SST) 105, 154
    wet bulb potential 209
Terminology 3, 189, 248
Thermal Wind 2, 62, 100, 168, 323, 343
Thermocline 73, 74, 116
Thermodynamic diagram (See Skew - T diagram) 353, 363, 366
Thunderstorm (See Lightening) 2, 5, 7, 20, 97, 100, 101, 109-113, 125, 150, 151, 162, 173, 175, 176, 178, 182, 183, 203, 207, 208, 213, 222, 223, 227, 236-238, 273-283, 287, 290, 293-296, 299, 305, 307, 310, 318, 319, 322, 353, 359, 362, 364, 369-371, 374, 376
    continuous 112, 113, 125, 150-151, 222, 281, 293, 322, 376
    probability of 280, 370
    thunderstorm duration 281, 282
    thunderstorm wind 293, 369
Tide 176, 241, 311
    atmospheric 176
    storm 241
Tornadoes 37, 101, 273, 283-288, 296, 366
    hurricane-generated 287
Tradewind 3, 5-7, 12, 13, 19-21, 89, 92, 95, 96, 100, 108, 113-116, 125, 142, 144, 154, 155, 158, 161, 166-168, 171, 172, 175, 176, 178, 182, 183, 188, 205, 206, 208, 213, 214, 220-224, 227, 229, 230, 234-238, 252, 253, 260, 262-264, 274, 284, 286, 287, 318, 319, 323, 326, 328, 332, 336, 338, 339, 347, 358
    boundary layer 171
    cloud 85, 86, 171, 172
    rainfall 113, 214
Trajectories 236, 326
Tropical cyclone (See Hurricanes and Typhoons) 1, 5-7, 33-36, 89, 95, 98, 101, 105, 116, 117, 119, 123, 125, 131, 149, 151, 153, 174, 175, 187-190, 193-195, 198, 201, 206, 220, 221, 241-253, 255, 257-267, 269, 271-273, 282, 287, 288, 293, 295, 301, 303, 305, 307, 311, 312, 316, 319, 322, 325-328, 335, 338, 339, 346, 347, 366, 385, 388, 391
    dissipation 2, 264
    forecasting 242, 266, 267, 355, 357
    formation 260, 355
    frequency 252
    intensity 249, 268, 366
    movement 48, 259, 265-267, 357, 365, 392
    rain 373
    rainband 373
Tropical depression 119, 207, 218, 224, 248, 252, 254, 328
Tropical storms (See Tropical cyclones) 151, 189, 224, 252, 253, 255, 258, 260, 262, 263, 265, 328, 339, 356, 366, 368, 372, 388
    dot 339, 343
Tropopause 63, 79, 195, 242, 309
Trough 3, 4, 7, 34, 105, 108, 117, 120, 122-126, 189, 195, 201, 203, 205, 213, 223, 225, 227, 230, 234, 237, 252, 260, 262, 283, 292, 308, 309, 318, 326, 328, 329, 335-339, 358, 367, 391
    heat 4, 6, 13, 19, 20, 40, 48, 51, 62, 87, 89, 111, 120, 121, 124, 125, 130, 189, 192, 193, 201, 213, 222, 299
    low-latitude 20, 51, 87, 89, 111, 120, 121, 124, 125, 130, 189, 192, 193, 201, 213, 222, 299
    mid-latitude 188, 335
    monsoon 87, 108, 112, 125, 189, 190, 191, 196, 197, 200, 201, 206, 220, 221, 238, 240, 252, 253, 255, 258, 262, 263, 328, 336, 339, 358, 367
    tropical upper tropospheric (TUTT) 7, 39, 57, 58, 60, 80, 203-206, 262, 263
    upper tropospheric westerlies 7, 20, 203, 213, 367, 388
Tsunami 312-314
Turbulence 30, 155, 168, 237, 273, 296, 305, 307-309, 324
    clear air (CAT) 307-309
    convective 307
    mechanical 307
Typhoon 2, 117, 125, 149, 151, 152, 174, 188, 190, 201, 203, 205, 240-242, 244-248, 252, 253, 256, 259, 260, 262-264, 266, 271, 288, 293, 301, 303, 307, 328, 339, 367, 392
    Flo (1990) 243, 247, 262
    Gordon (1989) 263
    Jack (1989) 264
    Kate (1970) 260

Kit (1957) 303
Nelson (1988) 246
Tip (1979) 303

U

Upslope flow 166, 181
Upwelling 66, 67, 73, 74, 84, 85, 92, 102, 104, 108, 117, 121, 245, 264

V

Vapor pressure 99
Vertical motion 85, 92, 124, 144, 187, 189, 195, 199, 229, 350, 358, 391
Volcanic ash 273, 315
Vortex (*See* Cyclone, Singular point, Tropical cyclone) 86, 95, 97, 108, 187-189, 191-193, 201, 204, 205, 230, 234, 238, 252-254, 262, 263, 327, 328, 338, 391
Vorticity 96, 97, 189, 200, 203, 206, 208, 220

W

Water vapor (*See* Moisture and Wind: zonal) 66-68, 70, 72, 74-76, 78, 80, 82, 84, 86
Waterspout 101
Wave 2, 4, 187, 188, 206, 208, 216, 223, 232, 237, 262, 273, 307, 308, 312, 314, 326, 345, 350, 356
    gravity 216, 308, 311, 312, 350
    mountain 307
    ocean 273, 311-312
Wind (*See* Tradewinds) 1-7, 153-155, 157-162, 164-176, 178, 180-183, 186-190, 192, 194-198, 201, 202, 204-211, 213-218, 220-227, 229, 230, 233-238, 241-249, 252, 253, 256, 260-265, 267, 268, 272-274, 281, 283, 284, 286-288, 293-299, 301, 303-309, 311-313, 316-332, 334-351, 355, 356, 358, 359, 362, 364-366, 368-376, 379-382, 390-392
    downslope 21, 157, 165, 166, 181, 182
    gradient 49, 50, 52, 54, 56
    lake 164
    local 7, 48, 104, 167, 168, 176, 178, 180, 186, 317
    maximum 30, 160, 189, 194, 242, 264, 301, 303, 311, 312, 320, 334, 338, 365, 366, 368, 376
    meridional 364
    mountain-valley 165-167, 176, 178, 319, 390
    offshore 157, 159, 281
    onshore 152, 313
    resultant 7, 15, 39, 48, 62, 63, 120, 122, 334, 336-338, 368
    upslope (anabatic) 165, 180
    zonal 63, 64, 65, 197, 261, 384
    wind gusts 248, 295, 301, 305, 306, 365, 366, 368, 369, 371, 376
    wind shear 6, 7, 37, 94, 97, 100, 109, 110, 168, 189, 202, 210, 221, 242, 253, 260, 262, 264, 283, 301, 307-309, 316, 334, 335, 344, 355, 358, 373, 375
    wind stress 73, 312
Windward Slope 213, 214

DISTRIBUTION

SECRETARY OF THE AIR FORCE/ST, 1670 AIR FORCE PENTAGON, WASHINGTON DC 20330-1670 1
USAF/XOW, 1490 PENTAGON RM BD927, WASHINGTON DC 20330-1490 1
HQ USAF/XOWX, 1490 AF PENTAGON, WASHINGTON DC 20330-1490 1
HQ USAF XOWP, 1490 AIR FORCE PENTAGON, WASHINGTON DC 20330-1490 2
HQ USAF XORR, 1480 AIR FORCE PENTAGON, WASHINGTON DC 20330-1480 1
USAF XOWR, 1490 PENTAGON RM BD866, WASHINGTON DC 20330-1490 1
USAF/XOOOW, 1490 PENTAGON RM BD927, WASHINGTON DC 20330-1480 1
AFOSR/NM, 110 DUNCAN AVE, STE B115, BOLLING AFB DC 20332-0001 1
AFOSR/NL, BOLLING AFB DC 20332-5000 1
OL A, AFCOS, FT RICHIE MD 21719-5010 1
USAF/CADS, 232 AVE A WEST, LITTLE ROCK AFB AR 72099-5153 1

AWS/CC, 102 WEST LOSEY ST ROOM 105, SCOTT AFB IL 62225-5206 2
AWS/XO, 102 LOSEY ST BLDG 105, SCOTT AFB IL 62225-5206 3
AWS/XOO, 102 WEST LOSEY ST, SCOTT AFB IL 62225-5000 1
AWS/XOS, 102 WEST LOSEY ST BLDG 1521 RM 105, SCOTT AFB IL 62225-5206 1
AWS/XOT, 102 W LOSEY ST BLDG 1521 RM 105, SCOTT AFB IL 62225-5206 1
AWS/XOTT, 102 WEST LOSEY ST, SCOTT AFB IL 62225-5205 1
AWS/XOXT, 102 LOSEY ST BLDG 1521, SCOTT AFB IL 62225-5206 1
AWS/SYD, 102 WEST LOSEY ST RM 105, SCOTT AFB IL 62225-5206 1
AWS/XT, 102 WEST LOSEY ST, SCOTT AFB IL 62225-5206 1
AWSTL, FL4415 859 BUCHANAN ST, SCOTT AFB IL 62225-5118 150
OL K, AWS NEXRAD OSF, 3200 MARSHALL DR STE 100, NORMAN OK 73072-8028 1
OL N, AWS, C/O ARL (AMSRL-BE-W) BLDG 1646 RM 24, WHITE SANDS MSSL RNGE NM 88002-5501 1
OL S, AWS , 720 IRWIN AVE (STOP 8202), FALCON AFB CO 80912-7210 1
OL-F, AWS (SMC/CIA), 2420 VELA WAY SUITE 1467 A 8, LOS ANGELES AFB CA 90245-4659 1
COMBAT WEATHER FACILITY, 595 INDEPENDENCE RD BLDG 91027, HURLBURT FLD FL 32544-5618 1
USAFETAC/DOS, 859 BUCHANAN ST, SCOTT AFB IL 62225-5116 1
OL A, USAFETAC, 151 PATTON AVENUE RM 120, ASHEVILLE NC 28801-5002 3
AFGWC/DO, 106 PEACEKEEPER DR STE 2N3 MBB 39, OFFUTT AFB NE 68113-4039 10
AFGWC/SY, 106 PEACEKEEPER DR STE 2 N3 MBB 39, OFFUTT AFB NE 68113-4039 10
DET 7, AFGWC, 7940 5TH ST 2ND FLOOR BLDG 10 NORTH SIDE, TINKER AFB OK 73145-9195 1

ACC DOW, 205 DODD AVE, STE 101, LANGLEY AFB VA 23665-2789 1
ACC DOWRX, 205 DODD AVE, STE 101, LANGLEY AFB VA 23665-2789 1
ACC AOS/AOW (WSU), 205 DODD AVE, STE 203A, LANGLEY AFB VA 23665-8495 1
AFFOR/WX, ATTN AFDIS POC, APO AA 34042-5000 1
SOUTHEAST AIR DEFENSE SECTOR, ATTN: DON, 164 ALABAMA AVE, TYNDALL AFB FL 32403-5015 1
BASE WEATHER STATION (BLDG 50), PO BOX 450011 29080 WILBUR WRIGHT BLVD, SELFRIDGE ANGB MI 48045-0011 ... 1
BUCKLEY BASE WEATHER STATION, BLDG 909 RM 101, BUCKLEY ANGB CO 80011-9599 1
WEATHER READINESS TRAINING CENTER (WRTC), PO BOX 465 RTE 1, CAMP BLANDING, STARKE FL 32091-9703 1
CSC SDFW 3580 D. AVE BLDG T-650, TINKER AFB OK 73145-9155 1
94 OG/OGW, 1477 MIMOSA DR, DOBBINS ARB GA 30069-4821 1
JSOC WEATHER, 32744 ALDISH DR, FT BRAGG NC 28307-50000 1
USCENTAF/A3-DOOW, 524 SHAW DRIVE, SHAW AFB SC 29152-5029 1
USSOUTHAF/A3-DOOSM, 5340 EAST GAFFORD WAY, DAVIS MONTHAN AFB AZ 85707-4224 1
WESTERN AIR DEFENSE SECTOR/DOCN, 852 LINCOLN BLDG, MCCHORD AFB WA 98438-1317 1
WESTOVER BASE WEATHER STATION, BLDG 7091 RM 123, WESTOVER ARB MA 01022-5000 1
1 WEATHER SQ/CC, AIRPORT DR BLDG 3082, FT LEWIS WA 98433-5000 1
1 OSS OSW, 190 FLIGHTLINE RD STE 100, LANGLEY AFB VA 23665-2299 1
2 WEATHER FLIGHT, BLDG 130, ANDERSON WAY, FT MCPHERSON GA 30330-5000 1
2 OSS/OSW, 105 LINDBERGH ROAD EAST, BARKSDALE AFB LA 71110-2163 1
3 WEATHER SQ, GRAY AAF BLDG 90049 RM 118 CLARKE RD, FT HOOD TX 76544-5076 1
3 ASOG, BLDG 1001 STE W312 761ST TANK BATTALION RD, FT HOOD TX 76544-5056 1
OL B, 3 WEATHER SQ, BLDG 5003 RM 101 FORNEY AAF, FT LEONARD WOOD MO 65473-5862 1
OL C, 3 WS, BLDG 91251 RM 16 LIBBY AAF, FT HUACHUCA AZ 85613-6660 1
OL-A, 3 WEATHER SQ, POST RD BLDG 4907 RM 104, FT SILL OK 73503-5100 1
4 OSS/OSW, 1980 CURTIS AVE BLDG 4507 STE 140, SEYMOUR JOHNSON AFB NC 27531-2524 1
5 OSS/OSW, 221 FLIGHT LINE DR UNIT 2, MINOT AFB ND 58705-5021 1
6 OSS/OSW, 7709 HANGAR LOOP DR STE 2, MACDILL AFB FL 33621-5205 1
6 WEATHER FLT, BASOPS BLDG CAIRNS AAF, FT RUCKER AL 36362-5162 1
OL-A, 6 WEATHER FLT, MAP RTE 3 BOX 303, TROY AL 38081-9384 1
7 OSS/OSW, 674 ALERT AVE, DYESS AFB TX 79607-1774 1
9 OSS/OSW, 7900 ARNOLD AVE STE 100, BEALE AFB CA 95903-1217 1
10 ASOS/ASW, 743 RAY PL BLDG 743 MARSHALL AAF, FT RILEY KS 66442-5317 1
12 ASOS/ASW/CC/IM, BLDG 2405 CHAFFEE ROAD, FT BLISS TX 79916-6700 1

12 ASOS/OSW (B FLT), SLEWITZKE ST BLDG 11210 RM 103 BIGGS AAF, FT BLISS TX 79918-5000 1
13 ASOS/ASW, BUTTS AAF BLDG 9601 RM 113, FT CARSON CO 80913-6403 1
15 ASOS/ASW, BLDG 7755 WRIGHT AAF, FT STEWART GA 31314-5067 1
OL-A, 15 ASOS/ASW, STRANCH ST BLDG 1252 RM 113, HUNTER AAF GA 31409-5193 1
16 ASOS/ASW, BLDG 5214 PILOT ST, FT KNOX KY 40121-5540 1
17 ASOS ASW (C FLT), LAWSON AAF BLDG 2485 RM 110, FT BENNING GA 31905-6034 1
18 ASOG/WSO, 259 MAYNARD STREET STE K, POPE AFB NC 28308-2787 1
18 WEATHER SQ/CC, PRAGER ST BLDG AT-3551, FT BRAGG AI NC 28307-5000 1
OL A, 18 WEATHER SQ, 6970 BRITTEN DRIVE STE 101, FT BELVOIR VA 22060-5132 1
OL B, 18 WEATHER SQ, CONDON RD BLDG 2408, FT EUSTIS VA 23604-5252 1
19 ASOS/ASW, 7163 HOTEL AVENUE, FT CAMPBELL AI KY 42223-6114 1
20 ASOS, 2065 HANGAR ACCESS RD, FT DRUM NY 13602-5042 1
20 OSS/OSW, 408 KILLIAN AVENUE, SHAW AFB SC 29152-5047 1
21 ASOS/ASW, POLK AAF BLDG 4226, FT POLK LA 71459-6250 1
23 OSS/OSW, 3393 SURVEYOR ST STE A, POPE AFB NC 28308-2797 1
24 WS, UNIT 0640, APO AA 34001-5000 1
27 OSS OSW, 110 E SEXTANT AV STE 1040, CANNON AFB NM 88103-5322 1
28 OSS OSW, 1291 RYAN ST STE 105, ELLSWORTH AFB SD 57706-4801 1
49 OSS OSW, 1801 8TH ST BLDG 571, HOLLOMAN AFB NM 88330-8023 1
51 CBCS/CTFW, 575 10TH STREEET, ROBINS AFB GA 31098-6345 1
53 CBCS/CTFW (AFDIS), 575 10TH STREET, ROBINS AFB GA 31098-2236 1
55 OSS/OSW, 513 SAC BLVD STE 101, OFFUTT AFB NE 68113-2094 1
57 OSS/OSW, 6278 DEPOT RD STE 102, NELLIS AFB NV 89191-7256 1
65 OSS/WX, UNIT 8025, APO AE 09720-8025 1
75TH RGR (ATTN: SWO), FT BENNING GA 31905-5000 1
96 CCSG SCTXD, 201 WEST EGLIN STE 236, EGLIN AFB FL 32542-6829 1
314 OSS/OSW, 2740 1ST ST BLDG 120, LITTLE ROCK AFB AR 72099-5060 1
319 OSS/OSW, 695 STEEN AVE STE 106, GRAND FORKS AFB ND 58205-6244 1
347 OSS/OSW, 8227 KNIGHTS WAY STE 1062, MOODY AFB GA 31699-1899 1
355 OSS/OSWF, 4360 S PHOENIX ST BLDG 4820, DAVIS MONTHAN AFB AZ 85707-4638 1
366 OSS/OSW, 665 THUNDERBOLT ST BLDG 262 RM 11, MT HOME AFB ID 83648-5401 1
416 OSS/OSW, 592 HANGAR RD BLDG 1000 STE 121, GRIFFISS AFB NY 13441-4520 1
509 OSS OSW, 745 ARNOLD AVE STE 1A, WHITEMAN AFB MO 65035-5026 1
DET 1, 549 CTS/WX, 661 7TH ST BICYCLE LAKE AAF BLDG 6212, FORT IRWIN CA 92310-5000 1
608 COS/DOOW, 245 DAVIS AVE EAST BLDG 5546 STE 245, BARKSDALE AFB LA 71110-2279 1
615 AMOG/DOMW, 575 WALDRON STREET, TRAVIS AFB CA 94535-2150 1

AETC/XOSW, 1F ST STE 2, RANDOLPH AFB TX 78150-4325 1
AFIT/ENP (CAPT GOLDIZEN), 2950 P ST BLDG 640, WRIGHT-PATTERSON AFB OH 45433-7765 1
AFIT/ENP (MSGT RAHE), 2950 P ST BLDG 240, WRIGHT-PATTERSON AFB OH 45433-7765 1
AFIT CIR, WRIGHT-PATTERSON AFB OH 45433-6583 1
AFIT LDEE, 2950 P ST BLDG 640, WRIGHT-PATTERSON AFB OH 45433-7765 1
AU/ACSC (MAJOR MUOLO/DEA), 225 CHENNAULT CIRCLE, MAXWELL AFB AL 36112-6426 1
AUL/LSE, BLDG 1405 600 CHENNAULT CIRCLE, MAXWELL AFB AL 36112-6424 1
12 OSS DOW, H 08, 1350 5TH ST EAST, RANDOLPH AFB TX 78150-4410 1
14 OSS DOW, 595 1ST ST STE # 3, COLUMBUS AFB MS 39710-4201 1
42 OS/OSWF, 220 WEST ASH BLDG 844, MAXWELL AFB AL 36112-6608 1
45 AS/OSFWX, 817 H ST STE 102, KEESLER AFB MS 39534-2452 1
47 OSS/DOW, 541 1ST ST STE 2, LAUGHLIN AFB TX 78843-5210 1
56 OSS/OSW, 14185 WEST FALCON, LUKE AFB AZ 85309-1629 1
64 OSS/DOW, 145 N DAVIS DR, REESE AFB TX 79489-5029 1
71 OSS/OSW, 301 GRITZ ST STE 52, VANCE AFB OK 73705-5412 1
80 OSS/DOW, 620 J AVENUE STE 3, SHEPPARD AFB TX 76311-2553 1
97 OSS/WXF, 603 E AVE STE 1, ALTUS AFB OK 73523-5023 1
325 OSS/OSW, STOP 22 408 FLIGHTINE RD, TYNDALL AFB FL 32403-5048 1
333 TCHTS TTCJB, 600 FIRST STREET STE 101, KEESLER AFB MS 39534-2494 1
334 TRS TTMV, 700 H ST BLDG 4332, KEESLER AFB MS 39534-2499 150
558 FTS, 2065 1ST DRIVE WEST, RANDOLPH AFB TX 78150-4351 1

AFMC/DOW, 4225 LOGISTICS AVE STE 2, WRIGHT-PATTERSON AFB OH 45433-5714 1
NAIC DXLA, 4115 HEBBLE CREEK ROAD STE 9, WRIGHT-PATTERSON AFB OH 45433-5613 3
NAIC TATW, 4115 HEBBLE CREEK ROAD STE 33, WRIGHT-PATTERSON AFB OH 45433-5637 1
DET 3, AFFTC/DOSW, PO BOX 19070, LAS VEGAS NV 89132-0070 1
AFTAC/TMKS, 1030 SOUTH HIGHWAY A1A, PATRICK AFB FL 32925-3002 1
AFTAC/TNRE, 1030 SOUTH HIGHWAY A1A, PATRICK AFB FL 32925-3002 1

AFTAC/TNLW, 1030 SOUTH HIGHWAY A1A, PATRICK AFB FL 32925-3002 1
AFTAC/TMR, STE 1450 1300 N 17TH STREET, ARLINGTON VA 22209 1
AFOTEC/WE, 8500 GIBSON BLVD SE, KIRTLAND AFB NM 87117-5558 1
AL/EQS, 139 BARNES DRIVE STE 2, TYNDALL AFB FL 32403-5323 1
AL/OEBE ARMSTRONG LABORATORY, 2402 EAST DRIVE, BROOKS AFB TX 78235-5114 1
ASC/WE, BLDG 91 3RD ST, WRIGHT-PATTERSON AFB OH 45433-6503 1
PHILLIPS LABORATORY, ATTN BRAIN NEWTON, 29 RANDOLPH RD, HANSCOM AFB MA 01731-3010 1
ESC/ENS, BLDG 1704 ROOM 106, HANSCOM AFB MA 01731-5000 1
ESC/WE, 11 EGLIN ST, HANSCOM AFB MA 01731-2122 1
HQ AFCESA/CEOM, 139 BARNES DRIVE STE 1, TYNDALL AFB FL 32403-5319 1
HQ AFOTEC/TKC (AFDIS), 8500 GIBSON BLVD SE, KIRTLAND AFB NM 87117-5206 1
PHILLIPS LAB/LIMI, 3550 ABERDEEN AVE SE, KIRTLAND AFB NM 87117-5776 1
PL/LIAF, 3550 ABERDEEN AVE SE, KIRTLAND AFB NM 87117-5776 1
PL/GPAA, 29 RANDOLPH RD, HANSCOM AFB MA 01731-3010 1
PL/TSML, 5 WRIGHT ST, HANSCOM AFB MA 01731-3004 1
PL/GPOA, 29 RANDOLPH RD, HANSCOM AFB MA 01731-3010 1
PL/GP, 29 RANDOLPH ROAD, HANSCOM AFB MA 01731-3010 1
PL/GPAA (HSTX), 29 RANDOLPH RD, HANSCOM AFB MA 01731-3010 1
PL/WE, 3350 ABERDEEN AVENUE, KIRTLAND AFB NM 87117-5776 1
ROME LAB TECH LIB, 26 ELECTRONICS PKY BLDG 106 CORRIDOR W STE 262, GRIFFISS AFB NY 13441-4514 1
SMC/SDEW, 160 SKYNET ST STE 2315, LOS ANGELES CA 90245-4683 1
WL/AARI-3, 2690 C ST STE 1, WRIGHT-PATTERSON OH 45433-7408 1
46 TW/TSWG, 211 W EGLIN BLVD STE 128, EGLIN AFB FL 32542-5429 1
46 TEST GROUP WE, 871 DEZONIA DRIVE BLDG 1183, HOLLOMAN AFB NM 88330-7715 1
46 OSS/OSWA, 601 W CHOCTAWHATCHEE AVE STE 60, EGLIN AFB FL 32542-5719 1
72 OSS/OSW, 3800 A AVE BLDG 240, TINKER AFB OK 73145-9108 1
75 OSS/OSW, 5970 SOUTHGATE DR, HILL AFB UT 84056-5232 1
76 OSS OSW, 303 LUKE DR, STE 1, KELLY AFB TX 78241-5638 1
77 OSS/OSW, 3028 PEACEKEEPER STE 4, MCCLELLAN AFB CA 95652-1020 1
78 OSS/OSW, 250 EAGLE ST STE 202, ROBINS AFB GA 31098-2602 1
88 WF/OSWL, 2130 8TH ST STE 11, WRIGHT-PATTERSON AFB OH 45433-7552 1
88 WF/OSWB, 5291 SKEEL AVENUE STE 1, WRIGHT-PATTERSON AFB OH 45433-5231 1
88 WF/OSWA, 2049 MONAHAN WAY BLDG 91, WRIGHT-PATTERSON AFB OH 45433-7204 1
412 OSS/OSW, 85 S FLIGHTLINE RD BLDG 1200, EDWARDS AFB CA 93524-6460 1
647 SVS/SVBMA (JASON SMITH), 98 BARKSDALE ST, HANSCOM AFB MA 01731-1807 1

AFRES/DOTSC 155 2ND ST, ROBINS AFB GA 31098-1635 1
301 OD/DOBW, CARSWELL ARB TX 76127-5000 1
304 ARRS/DOOR, PORTLAND IAP OR 97218-2797 1
434 OSS/ATWX, BLDG S 28 HOOSIER BLVD, GRISSOM ARB IN 46971-5000 1
482 OG/OSAW, 360 CORAL SEA BLVD, HOMESTEAD AFS FL 33039-1299 1
910 AG/OSA USAFR (KEN GOULD), 3976 KING GRAVES RD YOUNGSTOWN/WARREN MAP, VIENNA OH 44473-0910 1
924 OSS/OSA, 9707 AIRSIDE ROAD, AUSTIN TX 79719-2557 1

AFSOC DOOWO, 100 BARTLEY ST, HURLBURT FLD FL 32544-5273 1
16 OSS/OGSW, 150 BENNETT AVE BLDG 90730, HURLBURT FIELD FL 32544-5727 1

HQ AFSPACECOM/DOOW, 150 VANDENBERG ST STE 1105, PETERSON AFB CO 80914-4200 1
CAPE CANAVERAL FORECAST FACILITY/ROCC, BLDG 81900, CAPE CANAVERAL AFS FL 32925-6537 1
ADF WE, STOP 77, 18201 E DEVILS THUMB AVE, AURORA CO 80011-9536 1
DET 2, SMC/TDOR (WEATHER) ONIZUKA ASN, 1080 LOCKHEED WAY BOX 044, SUNNYVALE CA 94089-1235 1
1ST CACS/CC, 1 NORAD ROAD STE 3105, CHEYENNE MTN AS CO 80914-6009 1
14 AF/DOW, 747 NEBRASKA AVE STE 22, VANDENBERG AFB CA 93437-6268 1
21 OSS/OSW, 125 W HAMILTON AVE, PETERSON AFB CO 80914-1220 1
30 WS, 900 CORRAL RD BLDG 21150, VANDENBERG AFB CA 93437-5002 1
30 WS/DOV, 900 CORAL RD BLDG 21150, VANDENBERG AFB CA 93437-5001 1
45 WS /DOO, 1201 MINUTEMAN ST, PATRICK AFB FL 32925-3238 1
50 OSS/OSW, 300 O'MALLEY AVE STE 146, FALCON AFB CO 80912-3026 1
50 WS/DOWO, 300 O'MALLEY STE 146, FALCON AFB CO 80912-7160 1
90 OSS/DOW, 7505 SABER RD BLDG 1250 RM 1AF, F E WARREN AFB WY 82005-2684 1
341 OSS/DOW, 7224 FLIGHTLINE DR ROOM 209, MALMSTROM AFB MT 59402-7526 1
SWC/DOB (STOP 8202), 720 IRWIN AVE, FALCON AFB CO 80912-7210 1

AMC/DOWR, 402 SCOTT DR UNIT 3A1, SCOTT AFB IL 62225-5302 1
AMC/DOWO, 402 SCOTT DR UNIT 3A1, SCOTT AFB IL 62225-5302 1
AMWC/WCOXI, 5656 TEXAS AVENUE, FT DIX NJ 08640-5000 1
TACC/WXF, 402 SCOTT DRIVE RM 132, SCOTT AFB IL 62225-5029 1
22 OSS/OSW, 53435 KANSAS CT STE 110, MCCONNELL AFB KS 67221-3720 1

60 OSS/WXF, 611 E STREET, TRAVIS AFB CA 94535-5024 1
62 OSS/OSW, 1172 E STREET RM 127, MCCHORD AFB WA 98438-1008 1
89 OSS/OSW, 1240 MENOHER DR BLDG 1220, ANDREWS AFB MD 20762-6511 1
92 OSS/OSW, 901 WEST BOSTON STE 115, FAIRCHILD AFB WA 99011-8529 1
305 OSS/OSW, 1730 VANDENBERG AVENUE, MCGUIRE AFB NJ 08641-5509 1
375 OSS/OSW, 433 HANGAR RD, SCOTT AFB IL 62225-5029 1
377 ABW OTW, 3400 CLARK AVE SE, KIRTLAND AFB NM 87117-5776 1
436 OSS/OSW, 501 EAGLE WAY ST, DOVER AFB DE 19902-7504 1
437 OSS/OSW, 221 S BATES ST ROOM 130, CHARLESTON AFB SC 29404-5426 1
615 AMOG/DOMW, 575 WALDRON ST, TRAVIS AFB CA 94535-2150 1
621 AMO/WXC, 1907 EAST ARNOLD AVENUE RM 415, MCGUIRE AFB NJ 08641-5613 1
722 OSS/OSW, 2645 GRAEBER ST STE 3, MARCH AFB CA 92518-2264 1

NGB XOOSW, MAIL STOP 18, ANDREWS AFB MD 20762-6008 1
ANGRC/DOSW, 3500 FETCHET AVENUE, ANDREWS AFB MD 20762-5157 1
104 WEATHER FLIGHT, 2701 EASTERN BLVD, BALTIMORE MD 21220-2899 1
105 WEATHER FLIGHT TNANG, 240 KNAPP BOULEVARD, NASHVILLE TN 37217-2538 1
107 WEATHER FLIGHT, 26000 SOUTH ST BLDG 1516, SELFRIDGE ANGB MI 48045-5024 1
110 WEATHER FLIGHT, HQ 131 TFW, 10800 NATURAL BRIDGE RD, BRIDGETON MO 63044-2371 1
111 WEATHER FLIGHT, 14657 SNEIDER STREET, ELLINGTON ANGB TX 77034-5586 1
113 WEATHER FLIGHT, 824 E VANATTI COURT, TERRE HAUTE IN 47803-5012 1
116 WEATHER FLIGHT, 307 6TH STREET, MCCHORD AFB WA 98439-1201 1
120 WEATHER FLIGHT, 19089 BRECKENBRIDGE AVE, AURORA CO 80011-9527 1
121 COS OTW, BLDG 500, RICKENBACKER ANGB OH 43217-5005 1
121 SG/EM, RICKENBACKER IAP 7556 SOUTH PERIMETER ROAD, COLUMBUS OH 43217 1
121 WEATHER FLIGHT, 3252 E PERIMETER ROAD, ANDREWS AFB MD 20762-5011 1
122 WEATHER FLIGHT, 400 RUSSELL AVENUE, NEW ORLEANS NAS LA 70143-5200 1
123 WEATHER FLIGHT, 6801 CORNFOOT ROAD, PORTLAND OR 97218-2797 1
125 WEATHER FLIGHT, 4200 N 93RD EAST AVENUE, TUILSA OK 74115-1699 1
126 WEATHER FLIGHT, 1919 EAST GRANGE AVE, MILWAUKEE WI 53207-6298 1
127 WEATHER FLIGHT, P.O. BOX 19061 FORBES ANGB, TOPEKA KS 66619-5000 1
131 WEATHER FLIGHT, 1 TANK DESTROYER BLVD BOX 35, BARNES ANGB MA 01085-1385 1
134 ARG/XP, TN ANG, MCGHEE-TYSON AB TN 37901-5000 1
140 WEATHER FLIGHT (PAANG), 201 FAIRCHILD STREET, WILLOW GROVE ARS PA 19090-5320 1
146 WEATHER FLIGHT, 300 TANKER ROAD #4254 PITTSBURG IAP, CORAOPOLIS PA 15108-4254 1
154 WEATHER FLIGHT, CAMP ROBINSON, NORTH LITTLE ROCK AR 72118-2200 1
156 WEATHER FLIGHT, 5225 MORRIS FIELD DRIVE, CHARLOTTE NC 28208-5797 1
159 WEATHER FLIGHT, RT 1, BOX 465 CAMP BLANDING, STARKE FL 32091-9703 1
164 WEATHER FLIGHT, RICKENBACKER IAP 7556 SOUTH PERIMETER ROAD, COLUMBUS OH 43217-5910 1
165 WEATHER FLIGHT, 1019 OLD GRADE LANE, LOUISVILLE KY 40213-2678 1
169 FG SW WEATHER STOP 6, MCENTIRE ANGB, 1325 SOUTH CAROLINA RD, EASTOVER SC 29044-5006 1
181 WEATHER FLIGHT, 8150 W JEFFERSON BLVD, DALLAS TX 75211-9570 1
195 WEATHER FLIGHT, 106 MULCAHEY DRIVE BLDG 106, PORT HUENEME CA 93041-4003 1
200 WEATHER FLIGHT, 291 THUNDERBOLT STREET ROOM 8, SANDSTON VA 23150-2513 1
202 WEATHER FLIGHT, BLDG 3138, OTIS ANGB MA 02542-5001 1
203 WEATHER FLIGHT, 125 PINEGROVE ST FT INDIANTOWN GAP, ANNVILLE PA 17003-5154 1
204 WEATHER FLIGHT, 3306 FEIEBELKORN ROAD, MCGUIRE AFB NJ 08641-6004 1
207 WEATHER FLIGHT, 3912 W MINNESOTA ST, INDIANAPOLIS IN 46241-4064 1
208 WEATHER FLIGHT, 206 AIRPORT DR, ST PAUL MN 55107-4098 1
209 WEATHER FLIGHT, 2210 W 35 STREET, BLDG 9 RM 119, AUSTIN TX 78703-1222 1
210 WEATHER FLIGHT, 1280 SOUTH TOWER DRIVE, ONTARIO ANGB CA 91761-7627 1

NATIONAL SEVERE STORM FORECAST CENTER, 601 E 12TH ST, RM 1728 KANSAS CITY MO 64106 1
NCDC LIBRARY, 151 PATTON AVENUE LIBRARY, ASHEVILLE NC 28801-2733 1
NGDC/NOAA (ATTN: AF LIAISON OFFICER) , MAIL CODE E/GC2 325 BROADWAY, BOULDER CO 80333-3328 1
NIST PUBS PRODUCTION, RM A635 ADMIN BLDG, GAITHERSBURG MD 20899 1
NOAA CENTRAL LIBRARY, 1315 EAST WEST HIGHWAY, SSMC 3 2ND FLOOR, SILVER SPRING MD 20910 1
NOAA/MASC LIBRARY MC5, 325 BROADWAY, BOULDER CO 80303-3328 1
NOAA/NWS W/OM21, SSMC-2 ROOM 13148 1325 EAST-WEST HIGHWAY, SILVER SPRING MD 20910-3283 1
NOAA SHIP OREGON II R332, 3209 FREDERIC STREET, PASCAGOULA MS 39567 1
NOAA/NWS W/OSD, SSMC- 2 ROOM 12220 1325 EAST-WEST HIGHWAY, SILVER SPRING MD 20910-3283 1
NOAA/NWS NATIONAL HURRICANE CENTER, 1320 SOUTH DIXIE HWY RM 531, CORAL GABLES FL 33146 1
NWS W/OM21, 1325 EAST-WEST HWY RM 13208, SILVER SPRING MD 20910 1
NWS W/OSD, BLDG SSM C-2 EAST-WEST HWY, SILVER SPRING MD 20910 1
NWS TRAINING CENTER, 617 HARDESTY, KANSAS CITY MO 64124 1

US COAST GUARD RES & DEV CTR (AFDIS POC), 1082 SHENNCOSSETT RD, GROTON CT 06340-6096 1
G3 SUPPORT METEOROLOGY, AIR COMMAND HEADQUARTERS, WESTWIN MANITOBA CANADA R3J0T0 1
BUREAU OF METEOROLOGY TRAINING CENTER, GPO BOX 1289K MELBOURNE AUSTRAILIA 3001 1

CFFC, CANADIAN FORCES BASE EDMONTON, P.O. BOX 10500, EDMONTON ALBERTA CANADA T5J4J5 1
METEOROLOGY & OCEANOGRAPHY, NATIONAL DEFENCE HEADQUARTERS, 222 NEPEAN STREET VANIER BUILDING, OTTAWA, ONTARIO CANADA K1A0K2 1
SCIENCE NORTH, 100 CHEMIN LAKE ROAD, SUDBURY, ONTARIO CANADA P3E5S9 1
CANADIAN FORCES FORECAST CENTRE, TRENTON, 8TH WING, ASTRA ONTARIO CANADA K0K 1B0 1

PACAF/DOW, 25 E ST STE I232, HICKAM AFB HI 96853-5426 1
PACAF/DOWO, 25 E ST STE I232, HICKAM AFB HI 96853-5426 1
3 OSS/WE, 7TH ST BLDG 32235, ELMENDORF AFB AK 99506-3097 1
3 ASOS/WEATHER, 3112 BROADWAY STE 7, EIELSON AFB AK 99702-1850 1
DET 1, 3 ASOS/GEW, BLDG 1558, FT WAINWRIGHT AK 99703-5200 1
8 OSS/OSW, UNIT 2139 BLDG 2858, APO AP 96264-2139 1
15 OSS/OSW, 800 HANGAR AVE, HICKAM AFB HI 96853-5244 1
18 OSS/OSTL, UNIT 5177 BOX 10, APO AP 96363-5177 1
18 OSS/OSW, UNIT 5177 BOX 40, APO AP 96368-5177 1
25 ASOS/DOW, 1102 WRIGHT AVE, WHEELER AAF HI 96854-5200 1
OL A, 25 ASOS, BRADSHAW AFB HI, APO AP 96556-5000 1
35 OSS/OSCL, UNIT 5011, APO AP 96319-5011 1
35 OSS/OSW, UNIT 5011, APO AP 96319-5011 1
36 OSS/OSW, UNIT 14035 BOX AF, APO AP 96543-4035 1
DET 1, 36 OSS/OSJ, PSC 489 BOX 20, FPO AP 96536-0051 1
51 OSS/OSW, UNIT 2163, APO AP 96278-2163 1
199 WEATHER FLIGHT, 1102 WRIGHT AVENUE, WHEELER AAF HI 96854-5200 1
354 OSS/OSW, 1215 FLIGHT LINE AVE STE 2, EIELSON AFB AK 99702-1520 1
374 OSS/OSW, UNIT 5222, APO AP 96328-5222 1
OL A, 374 OSS, UNIT 45007, APO AP 96343-0085 1
607 WEATHER SQUADON/DOOF, UNIT 15173, APO AP 96205-0108 1
607 WEATHER SQUADRON/DOO, UNIT 15173 BLDG 1506, APO AP 96205-0108 2
607 COS/DOW, UNIT 2072, APO AP 96278-2072 1
DET 1, 607 WEATHER SQUADRON, UNIT 15674, APO AP 96258-0674 1
DET 2, 607 WEATHER SQUADRON, UNIT 15200 BLDG S 819, APO AP 96271-0136 1
OL A DET 1, 607 WEATHER SQUADRON, UNIT 15675, APO AP 96257-0675 1
OL A, 607 WEATHER SQUADRON, UNIT 15630 BLDG 1610, APO AP 96208-0195 1
OL A DET 2, 607 WEATHER SQUADRON, UNIT 15673, APO AP 96218-0673 1
OL B DET 1, 607 WEATHER SQUADRON, UNIT 15118, APO AP 06224-0420 1
OL B, 607 WEATHER SQUADRON, BLDG S 252 UNIT 15242, APO AP 96205-0015 1
OL C, 607 WEATHER SQUADRON, BLDG S 3101 RM 4, APO AP 96297-0626 1
611 OSS/OSW, 6900 9TH ST STE 205, ELMENDORF AFB AK 99506-2250 1
C/O FT RICHARDSON NCOIC, 611 OSS/WE, 6900 9TH STREET STE 205, ELMENDORF AFB AK 99506-2250 1

USSPACECOM J3W, 250 S PETERSON BLVD STE 116, PETERSON AFB CO 80914-3220 1
USCENTCOM CCJ3-W, 7115 S BOUNDARY BLVD BLDG 540 , MACDILLAFB FL 33609-7001 1
USSPACECOM J3W, 250 S PETERSON BLVD STE 116, PETERSON AFB CO 80914-3220 1
USTRANSCOM TCJ3/ODM, 508 SCOTT DR BLDG 1900, SCOTT AFB IL 62225-5357 1
USCINCPAC (J37), BOX 13, CAMP H.M. SMITH HI 96861-5025 1
USCINCPAC J64, ATTN : LT COL MOTTIS, BOX 64029, CAMP SMITH HI 96861-4029 1
HQ USEUCOM ECJ33-OD-WE, UNIT 30400 BOX 1000, APO AE 09128-4209 1
USSOCCENT SOCJ2- SWO, 7115 S BOUNDARY DRIVE MACDILL AFB FL, 33621-5101 1
USSOCOM SOJ3 OW, 7701 TAMPA POINT BLVD, MACDILL AFB FL 33621-5323 1
USSOUTHCOM SWO, UNIT 0640, APO AA 34001-5000 1
USSTRATCOM J 315, 901 SAC BLVD STE 1B29, OFFUTT AFB NE 68113-6700 1
OL A, SOCOS/WX, BLDG AT 3275 BAY 50, FT BRAGG NC 28307-5203 1

ARMED FORCES MEDICAL INTELLIGENCE CTR, INFO SVCS DIV BLDG 1607 FT DETRICK, FREDERICK MD 21702-5004 1
ARMY RES LAB, BATTLEFIELD ENVIRONMENT DIR AMSRL-BE, WHITE SANDS MISSILE RANGE NM 88002-5501 1
ARMY TRAINING AND DOCTRINE COMMAND, ATDO-IW (ATTN: SWO), FT MONROE VA 23651-5000 1
USA CBT SYS TEST ACTIVITY, MET BR, BLDG 1134 ATTN: STECS-PO-OM, ABERDEEN PROVING GRND MD 21005-5059 ... 1
COMMANDER US ARMY PACIFIC (APIN-OPW), FT SHAFTER HI 96858-5100 1
COMMANDER, FORCES COMMAND, AFIN-ICW, FT MCPHERSON GA 30330-6000 1
CDR USASOC, ATTN: AOIN-ST, FT BRAGG NC 28307-5200 1
DIRECTOR USA-CETEC, ATTN: GL-AE, FT BELVOIR VA 22060-5546 1
DIRECTOR, USA REDSTONE TECHNICAL TEST CENTER, ATTN: STERT-TE-F-MT, REDSTONE ARSENAL AL 35898-8052 1
FIRST US ARMY, ATTN STAFF WEATHER OFFICER, FT MEADE MD 20755-7300 1
SECOND US ARMY AFKD-OPI-W (AFDIS POC), FT GILLEM GA 30050-5000 1
HQ 5TH U.S. ARMY, AFKB-OP (SWO), FT SAM HOUSTON TX 78234-7001 1
SIXTH US ARMY AFKC-OP-IS-SWO (AFDIS POC), PRESIDIO, SAN FRANCISCO CA 94129-5000 1
HQ DA DCS OPS AND PLANS, ATTN: DAMO-ZD RM 3A538, 400 ARMY PENTAGON, WASHINGTON DC 20330-5000 1
HQ ARCENT, AFRD-DSO-SWO, FT MCPHERSON GA 30330-7000 1
LOS ALAMITOS AAF (MR ADAMS), BLDG 1 AFRC 11200 LEXINGTON DR, LOS ALAMITOS CA 90720-5001 1

NATIONAL RANGE DIRECTORATE, MET BR ATTN: STEWS-NR-DA-F, WHITE SANDS MISSILE RANGE NM 88002-5504 1
TECHNICAL LIBRARY, DUGWAY PROVING GROUND, DUGWAY UT 84022-5000 1
TEXCOM FSTD, ATTN: CSTE-TFS-SP, FT SILL OK 73503-6100 1
USA TECOM, ATTN: AMSEL-TC-AM(BE) C O NVESD, FT BELVOIR VA 22060-5677 1
USA DUGWAY PROVING GROUND, TROPICAL TEST SITE UNIT 7140 ATTN: STEDP-MT-TM-TP, APO AA 34004-5000 1
USA TECOM, ATTN: AMSEL-RD-NV-VMD (MET), FT BELVOIR VA 22060-5677 1
USA ARMY ENGINEER TOPOGRAPHIC LAB, ATTN: CEETL-TD, FT BELVOIR VA 22060-5677 1
USA INTELLIGENCE CTR (WEATHER SUPPORT TEAM), ATTN ATZS CDI-W, FT HUACHUCA AI AZ 85613-6000 1
USARSPACE (MOSC-OO), 1670 N NEWPORT RD STE 121, COLORADO SPRINGS CO 80916-2749 1
1CC AZSB-GTFD, AH-64 CSM ATTACK, FT CAMPBELL AI KY 42223-5000 1
160TH SOAR(A) , ATTN: ADAV-ST-FS(MR LEVARN) 6950 38TH STREET, FT CAMPBELL KY 42223-1291 1
CAC/SWO, ATTN; ATZL-CAW-E 415 SHERMAN, FT LEAVENWORTH KS 66207-1344 1
DA, COLD REGIONS RESEARCH AND ENGINEERING LABORATORY, ATTN TIM BALDWIN, HANOVER NH 03755-1290 1
US ARMY ENGINEER SCHOOL, ATSE-CD-TV (ATTN: MR WESTERMEIER), FT LEONARD WOOD MO 65473 1
COMMANDANT, USA JFKSWCS, NCOA, FT BRAGG NC 28307-5200 1
COMMANDER, USA INTELLIGENCE CTR, ATTN ATZS-EHE, FT HUACHUCA AI AZ 85613-6000 1
USAARIEM/BIOPHYSICS MEDICAL MODELING DIVISION, KANSAS STREET, NATICK MA 01760-5007 1
HQ 629TH MI BN (CEWI), 29TH ID (LIGHT), 7100 GREENBELT ROAD, GREENBELT MD 20770-3398 1

USAFA DFP, 2354 FAIRCHILD DR STE 2A6, USAF ACADEMY CO 80840-5701 1
USAFA/DFEG, 2354 FAIRCHILD HALL STE 6K12, USAF ACADEMY CO 80840-6238 1
USAFWS/WSC, 4455 DEVLIN DR, NELLIS AFB NV 89141-6545 1
34 OSS/OSW, 9265 AIRFIELD DRIVE, STE 1, USAF ACADEMY CO 80840-2060 1
54 OSS/OSW, AIR FIELD DR. BLDG 9206, USAF ACADEMY CO 80840-5000 1

AFSOUTH (CMFWC CAPT STRAYER), PSC 813 BOX 136, FPO AE 09620-5000 1
DET 1, AFBS (AFN WEATHER), UNIT 25708, APO AE 09242-5000 1
HQ V CORPS, UNIT 29355, APO AE 09014-5000 1
HQ, USAREUR/SWO, UNIT 29351 BLDG 12, APO AE 09014 1
USAFE/DOW, UNIT 3050 BOX 15 BLDG 546 ROOM 306, APO AE 09094-5015 1
USAFE/DOWR, UNIT 3050 BOX 15 BLDG 201 ROOM 305, APO AE 09094-5015 1
USAFE AOS/AOXR, UNIT 1010 BOX 570, APO AE 09094-5015 1
3 AF/DOW, UNIT 4840, APO AE 09459-4840 1
5 ATAF WEA OFFICE (LTC CERASUOLO), 5 ATAF WEATHER CENTRE, 36100 VINCENZA ITALY 1
16 AF WE, UNIT 6170, APO AE 09601-6170 1
17 AF/WE, UNIT 8495, APO AE 09094-8495 1
31 OSS/OSW, UNIT 6170 BLDG 904, APO AE 09601-6170 1
39 OSS/OSW, UNIT 1075 BOX 275, APO AE 09824-0275 1
48 OSS/DOM, UNIT 5245 BOX 390, APO AE 09464-5390 1
52 OSS/WEF, UNIT 8870 BOX 270, APO AE 09126-0270 1
86 OSS/OSW, UNIT 8495, APO AE 09094-8495 1
86 OSS DOWB/WEATHER SUPPORT UNIT, UNIT 8230, APO AE 09094-8230 1
86 OSS/OSWA/BASE WEATHER STATION, UNIT 3230, APO AE 09094-3230 1
86 OSS/OSWC, UNIT 4070, APO AE 09136-4070 1
100 OSS/DOW, UNIT 4965 BLDG 500, APO AE 09459-4965 1
617 WS, UNIT 29351 BLDG 12, APO AE 09014-5000 1
A FLT, 617 WS, UNIT 29231, APO AE 09102-3737 1
DET 2, 617 WS, UNIT 20200 BLDG 1310, APO AE 09165-9616 1
DET 3, 617 WS, CMR 416 BOX S, APO AE 09140-9998 1
DET 4, 617 WS, UNIT 7890 EUROPEAN FORECAST CENTER, APO AE 09126-7890 1
DET 5, 617 WS, CMR 454 UNIT 31020, APO AE 09250-0047 1
DET 6, 617 WS, UNIT 29632, APO AE 09096-5000 1
DET 7, 617 WS, UNIT 28130, APO AE 09114-5000 1
DET 8, 617 WS, UNIT 29719, APO AE 09028 1
DET 9, 617 WS, UNIT 28216, APO AE 09173-5000 1
DET 10, 617 WS, UNIT 26410 BLDG 543 RM 111, APO AE 09182-0006 1
OL A DET 2, 617 WS, UNIT 24304, APO AE 09252 1
OL A, 617 WS, C/O HHD BDE, APO AE 09157 1
OL B, 617 WS, CMR 423, APO AE 09107-5000 1
OL C, 617 WS, CMR 445 BOX 260, APO AE 09046 2
OL D, 617 WS, 21 TAACOM UNIT 23203, APO AE 09263 1
OL E, 617 WS, UNIT 31401 BOX 6, APO AE 09630-0006 1
4404 OSS(P)/OSW, ATTN AFDIS POC, APO AE 09894-0408 1
4409 OG/WE, UNIT 66200 BOX 100, APO AE 09852-6200 1

HQ USMC (CODE ASL-44), 2 NAVY ANNEX, WASHINGTON DC 20380-1775 1
COMMANDING GENERAL, (ATTN: WSO), OPERATION MCAS PSC BOX 8011, CHERRY POINT NC 28533-0011 1
COMMANDING GENERAL, (ATTN: WSO) MCAS, PO BOX 9500 MCAS EL TORO, SANTA ANA CA 92706-4000 1
COMMANDING OFFICER, (ATTN: WSO), MWSS 374 MCAS TUSTIN PSC BOX 109033, SANTA ANA CA 92710-9033 1

COMMANDING OFFICER, (ATTN: WEATHER SERVICE OFFICER), MWSG 17 UNIT 37190, FPO AP 96603-7190 1
COMMANDING OFFICER, (ATTN: WEATHER SERVICE OFFICER), UNIT 35202, FPO AP 96372-5202 1
COMMANDING OFFICER, (ATTN: WSO), MWSS 271 PSC BOX 8078, MCAS CHERRY POINT NC 28533-6033 1
COMMANDING OFFICER, (ATTN: WSO) MWSG 27 PSC BOX 8082, MCAS CHERRY POINT NC 28533-0082 1
COMMANDING OFFICER, (ATTN: WSO), MWSS 171 UNIT 37201, FPO AP 96603-7201 1
COMMANDING OFFICER, (ATTN: WSO), MWSS 371 MCAS YUMA PO BOX 99210, YUMA AZ 85369-9210 1
COMMANDING OFFICER, (ATTN: WSO), MWSS 272 PSC BOX 21032, JACKSONVILLE NC 28545-1031 1
COMMANDING OFFICER, (ATTN: WX SERVICE OFFICER), MWSS 273 PSC BOX 66038, BEAUFORT SC 29904-6038 1
COMMANDING OFFICER, (ATTN: WX SERVICE OFFICER), MCAS TUSTIN PO BOX 105000, SANTA ANA CA 92710-5000 1
COMMANDING OFFICER, (ATTN: WX SERVICE OFFICER), MWSS 172 UNIT 37206, FPO AP 96603-7206 1
COMMANDING OFFICER, (ATTN: WX SVC OFF), MCAS NEW RIVER PSC BOX 21001, JACKSONVILLE NC 28545-1001 1
COMMANDING OFFICER, (ATTN: WSO) AVIATION GROUND SPT ELEMENT, MCAGCC 29 PALMS BOX 788285, TWENTY NINE PALMS CA 92278-8285 1
COMMANDING OFFICER, (ATTN: WX SERVICE OFFICER) MCAS, PO BOX 55010, BEAUFORT SC 29904-5010 1
COMMANDING OFFICER, (ATTN: WX SERVICE OFFICER) MCAF, BOX 63061 MCBH, KANEOHE BAY HI 96863-3601 1
COMMANDING OFFICER, (ATTN: WX SERVICE OFFICER) MCAS, BOX 555151, CAMP PENDELTON CA 92055-5151 1
COMMANDING OFFICER, (ATTN: WSO) MCAS, MWSS 373 MCB PO BOX 555861, CAMP PENDELTON CA 92055-5861 1
COMMANDING OFFICER, (ATTN: WEATHER SERVICE CHIEF) MCAF, 2100 ROWELL ROAD, QUANTICO VA 22134-5063 1
COMMANDING OFFICER, (ATTN: WSO), MWSG 37 MCAS EL TORO PSC BOX 99009, SANTA ANA CA 92709-9009 1
COMMANDING OFFICER, (ATTN: WO), MWSS 373 MCAS EL TORO PSC BOX 99011, SANTA ANA CA 92709-9011 1
COMMANDING OFFICER, (ATTN: WEATHER SERVICE OFFICER), PSC 561 BOX 1876, FPO AP 96310-1876 1
COMMANDING OFFICER, (ATTN: WSO), MWSS 274 PSC BOX 8079, MCAS CHERRY POINT NC 28533-0079 1
COMMANDING OFFICER, (ATTN: WSO), MWSG 37 MCAS EL TORO PSC BOX 99009, SANTA ANA CA 92709-9009 1

NAVLANTMETOC DET BRUNSWICK COMPONENT, BLDG 429 RM 306, USN SUB BASE, GROTON CT 06349-5100 1
COMNAVMETOCCOM/512 (LCDR FORD), 1020 BALCH BLVD, STENNIS SPACE CENTER MS 39529-5005 1
COMNAVMETOCCOM (CODE N433), 1020 BALCH BLVD, STENNIS SPACE CENTER MS 39529-5005 1
COMNAVMETOCCOM, CODE N332, STENNIS SPACE CTR MS 39529-5001 1
COMNAVMETOCCOM, CODE N312, STENNIS SPACE CTR MS 39529-5000 1
COMNAVSPECWARCOM (CODE N27 FORCE OCEANOGRAPHER), 2000 TRIDENT WAY , SAN DIEGO CA 92155-5599 1
COMSECONDFLEET, (CODE J335), FPO AE 09506-6000 1
COMSIXFLEET, (CODE N312) CDR MCGEE, FPO AE 09501-6002 1
OFFICER IN CHARGE, FLENUMMETOC DET ASHEVILLE, 151 PATTON AVENUE, ASHEVILLE NC 28801-5014 1
FLENUMMETOCCEN, ATTN: DAVE HUFF, MONTEREY CA 93943-5005 1
LIBRARIAN, FLENUMMETOCEN, MONTEREY CA 93943-5005 1
NAVAL AIR WARFARE CENTER-WEAPONS DIV, GEOPHYSICAL SCIENCES BRA CODE 32AF, POINT MUGU CA 93042-5001 1
NAVAL POSTGRADUATE SCHOOL CODE MR/HY (MR HANEY), 589 DYER RD BLDG 235, MONTEREY CA 93943-5114 1
NAVAL RESEARCH LABORATORY, MONTEREY CA 93943-5006 1
NAVAL RESEARCH LABORATORY, CODE 4323, WASHINGTON DC 20375 1
NAVAL RESEARCH LABORATORY, CODE 4180, WASHINGTON DC 20375 1
NAVAL POSTGRADUATE SCHOOL, CHMN DEPT OF METEOROLOGY CODE 63, MONTEREY CA 93943-5000 1
LIBRARY, NAVAL REASERCH LABORATORY, 7 GRACE HOPPER AVENUE, MONTEREY CA 93943-5052 1
COMMANDING OFFICER, NAVEURMETOCCEN, PSC 819 BOX 31, FPO AE 09645-3200 1
OFFICER IN CHARGE, NAVEURMETOCDET, PSC 814 BOX 22, FPO AE 09865 1
OFFICER IN CHARCE, NAVEURMETOCDET, PSC 817 BOX 13, FPO AE 09622-0800 1
OFFICER IN CHARGE, NAVEURMETOCDET, PSC 812 BOX 3380, FPO AE 09627-3380 1
NAVICECEN (LT KLEIN), 4251 SUITLAND RD FOB#4, WASHINGTON DC 20395 1
COMMANDING OFFICER, NAVICECEN, 4301 SUITLAND ROAD FOB #4, WASHINGTON DC 20395-5108 1
NAVLANTMETOCDET, PSC 1001 BOX 35-W, FPO AE 09508-0014 1
NAVLANTMETOCDET, PATUXENT RIVER NAS MD 20670-5103 1
NAVLANTMETOCEN, CODE 50, ATTN: MET TEAM 931 3RD AVE MCCADY BLDG, NORFOLK NAS VA 23511-2394 1
NAVLANTMETOCEN (LT OSTWALD), 931 3 RD AVE MCCADY BLDG, NORFOLK NAS VA 23511-2394 1
NAVOCEANO/CODE 541, 1002 BALCH BLVD, STENNIS SPACE CENTER MS 39533-5001 1
NAVOCEANO, CODE 9220, STENNIS SPACE CTR MS 39529-5001 1
NAVOCEANO, CODE N2513 1002 BALCH BLVD, STENNIS SPACE CTR MS 39522-5001 1
NAVOCEANO, CODE N2513 DORIS LEWIS 8100 2ND FLOOR 1002 BALCH BLVD, STENNIS SPACE CTR MS 39522-5001 500
NAVOCEANO, CODE N25131 8100 RM 203D, STENNIS SPACE CTR MS 39522-5001 2
NAVPACMETOCCEN WEST GUAM/JTWC (AFDIS POC), PSC 489 BOX 2, FPO AP 96563-0051 1
NAVPACMETOCCEN, ATTN: TECH LIBRARY BOX 113, PEARL HARBOR HI 96860-7000 1
NAVPACMETOCFAC, NAS NORTH ISLAND, SAN DIEGO CA 92135-5130 1
MET, NAVPACMETOCFAC SAN DIEGO (AFDIS POC), PO BOX 357076, NAS NORTH ISLAND CA 92135-7076 1
NRAD (CODE 423/JIM BROYLES), 1045 MONTEREY VISTA WAY, ENCINITAS CA 92024 2
OCEANOGRAPHER OF THE NAVY, US NAVAL OBSERVATORY BLDG 1 3450 MASS AVE, WASHINGTON DC 20392-5421 1
US PACIFIC FLEET (N3WX) CSC/WILLIAM LITTLE, 250 MAKALAPA DRIVE, PEARL HARBOR HI 96860-7000 1

AL/HR-DOKL FL2870, 7909 LINDBERG DR RM 239 BLDG 578 BROOKS AFB TX 78235-5352 1
AFFTC TECH LIBRARY FL2806 412 TW/TSTL 307 E POPSON AVE BLDG 1400 RM 106 EDWARDS AFB CA 93524-6630 1
TECHNICAL LIBRARY FL2825 203 W EGLIN BLVD STE 300 EGLIN AFB FL 32542-6843 1
ROME LAB TECH LIBL CORRIDOR W STE 262 26 ELECTRONIC PKWY BLDG 106, GRIFFISS AFB NY 13441-4514 1

TECHNICAL LIBRARY FL2051 SA-ALC/CNL 485 QUENTIN ROOSEVELT RD BLDG 171 KELLY AFB TX 782416425 1
PHILIPS LAB TECHNICAL LIBRARY FL2809 PL/DOSUL 3400 ABERDEEN AVE SE BLDG 419 KIRTLAND AFB NM 87117-5776 1
AIR UNIVERSITY LIBRARY FL3386 AUL/LD 600 CHENNAULT CIRCLE BLDG 1405 MAXWELL AFB AL 36112-6424 1
AUL/LSE BLDG 1405 600 CHENNAULT CIRCLE MAXWELL AFB AL 36112-6424 1
HQ SSC/RMMI FL3100 201 E MOORE DR BLDG 856 RM 1701 MAXWELL AFB GUNTER ANNEX AL 36114-3005 1
TECH LIBRARY FL2513 45 SW CSR 5123 1030 S HWY A1A BLDG 989 RM A1-S3 BOX 4127 PATRICK AFB FL 32925-0127 1
TECHNICAL INFO CTR FL7050 AL/EQ-TIC 139 BARNES DR STE 2 BLDG 1120 TYNDALL AFB FL 32403-5323................... 1
TECHNICAL LIBRARY FL2827 30 SPW/XPOT 806 13TH ST STE A BLDG 7015 VANDENBERG AFB CA 93437-6111 1
WRIGHT LAB LIBRARY FL2802 WL/DOC 2690 C ST STE 4 BLDG 22 WRIGHT-PATTERSON AFB OH 45433-7411 1
USAFA LIBRARY FL7000 HQ USAFA/DFSEL 2354 FAIRCHILD DR STE 3A10 USAF ACADEMY CO 80840-6214 1

DEFENSE INTELLIGENCE AGENCY DIA D1W 1B DIAC RM A4-130 WASHINGTON DC 20340-6612..................................... 1
DTIC-FDAC CAMERON STATION ALEXANDRIA VA 22304-6145 3
NASA-MSFC-ELAF HUNTSVILLE AL 35812-5000 1
NASA-MSFC-ES44 HUNTSVILLE AL 35812-5000 1
NASA GODDARD MAIL CODE 916 GREENBELT MD 20771 1
OCFM, 8455 COLESVILLE ROAD STE 1500 SILVER SPRING MD 20910-5000 1
BUREAU OF METEOROLOGY TRAINING CENTER GPO BOX 1289K MELBOURNE AUSTRALIA 3001 1

DR COLIN S RAMAGE 1420 ACADIA ST DURHAM NC 27701 .... 15*
MR GARY ATKINSON 5513 SW 18TH ST TOPEKA KS 66604 1
MR MIKE GENTRY PO BOX 58425 HOUSTON TX 77258 1
DR WILLIAM M GRAY DEPT OF ATMOSPHERIC SCIENCES COLORADO STATE UNIVERSITY FORT COLLINS CO 80523.... 1
DR CAROLE J HAHN CIRES UNIVERSITY OF COLORADO CAMPUS BOX 216 BOULDER CO 80309 1
DR BARRY HINTON SPACE SCIENCE AND ENGINEERING CENTER UNIVERSITY OF WISCONSIN-MADISON 1225 WEST
DAYTON STREET MADISON WI 53706 1
DR DENNIS W MOORE DIRECTOR, JIMAR UNIVERSITY OF HAWAII 1000 POPE ROAD HONOLULU HI 96822 1
MR LOUIS ODA 1616 LIHOLIHO STREET HONOLULU HI 96822 1
DR RICHARD E ORVILLE DEPT OF ATMOSPHERIC SCIENCES SUNYA 1400 WASHINGTON AVE ALBANY NY 12222 1
DR THOMAS A SCHROEDER DPT OF METEOROLOGY UNIV OF HAWAII 2525 CORREA ROAD HONOLULU HI 96822.......... 1
MR PATRICK SHAM DIRECTOR ROYAL OBSERVATORY HONG KONG 1 1
DR NORMAN K WAGNER DEPT OF CIVIL ENGINEERING UNIVERSITY OF TEXAS AUSTIN TX 78712-1076 1
MR SCOTT WOODRUFF NOAA/ERL 325 BROADWAY BOULDER CO 80303-3328.. 1
DR X ZIANG DATA PROCESSING SYSTEM DIVISION WORLD WEATHER WATCH DEPT WORLD METEOROLOGICAL
ORGANIZATION 41 GIUSEPPE-MOTTA CASE POSTALE NO 5 GENEVA 20 SWITZERLAND 1
INSTITUTE OF ATMOSPHERIC PHYSICS UNIVERSITY OF ARIZONA ROOM 542 PAS BUILDING TUCSON AZ 85721 1
DEPT OF ATMOSPHERIC SCIENCES UCLA 405 HILGARD AVE LOS ANGELES CA 90024 1
DEPT OF METEOROLOGY UNIVERSITY OF CHICAGO 5734 ELLIS AVE CHICAGO IL 60637 1
DEPT OF ATMOSPHERIC SCIENCES COLORADO STATE UNIVERSITY FORT COLLINS CO 80523 1
DEPT OF METEOROLOGY FLORIDA STATE UNIVERSITY TALLAHASSEE FL 32306 1
DEPT OF METEOROLOGY UNIVERSITY OF HAWAII 2525 CORREA ROAD HONOLULU HI 96822 1
DEPT OF METEOROLOGY UNIVERSITY OF MARYLAND COLLEGE PARK MD 20742-2425 1
CENTER FOR METEOROLOGY AND PHYSICAL OCEANOGRAPHY DEPT OF EARTH, ATMOSPHERIC AND PLANETARY
SCIENCES MASSACHUSETTS INSTITUTE OF TECHNOLOGY CAMBRIDGE MA 02139 1
DEPRTMENT OF METEOROLOGY AND PHYSICAL OCEANOGRAPHY UNIVERSITY OF MIAMI 4600 RICKENBACKER CAUSEWAY
KEY BISCAYNE FL 33149 1
DEPT OF METEOROLOGY NAVAL POSTGRADUATE SCHOOL MONTEREY CA 93940 1
DEPT OF ATMOSPHERIC SCIENCES SUNYA 1400 WASHINGTON AVE ALBANY NY 12222 1
DEPT OF MARINE, EARTH AND ATMOSPHERIC SCIENCES NORTH CAROLINA STATE UNIVERSITY RALEIGH NC 27695 .. 1
DEPT OF METEOROLOGY PENNSYLVANIA STATE UNIVERSITY 503 WALKER BUILDING UNIVERSITY PARK PA 16802 1
DEPT OF EARTH AND ATMOSPHERIC SCIENCES GEOSCIENCES BUILDING PURDUE UNIVERSITY WEST LAFAYETTE IN
47907 1
DEPT OF EARTH AND ATMOSPHERIC SCIENCES ST LOUIS UNIVERSITY 221 N GRAND BLVD ST LOUIS MO 63103 1
METEOROLOGY DEPT TEXAS A & M UNIVERSITY COLLEGE STATION TX 77843... 1
DEPT OF ENVIRONMENTAL SCIENCES CLARK HALL UNIVERSITY OF VIRGINIA CHARLOTTESVILLE VA 22903................. 1
DEPT OF ATMOSPHERIC SCIENCES UNIVERSITY OF WASHINGTON AK-40 LIBRARY SEATTLE WA 98195 1
DEPT OF METEOROLOGY UNIVERSITY OF WISCONSIN-MADISON 1225 WEST DAYTON STREET MADISON WI 53706 1

This Page Intentionally Left Blank

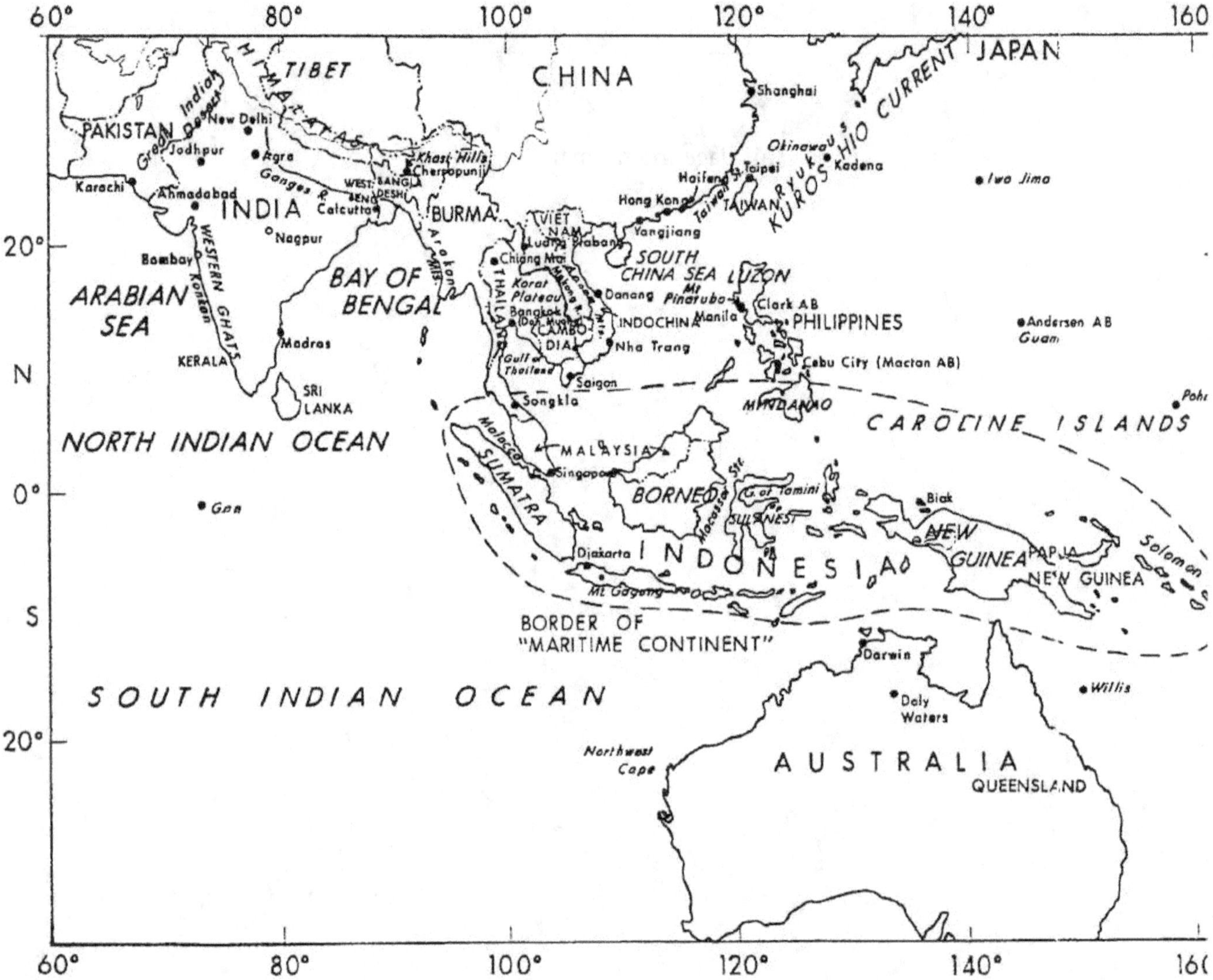

Figure 1-5a. Locator chart, Eastern Hemisphere. The dashed line encloses the "Maritime Continent."

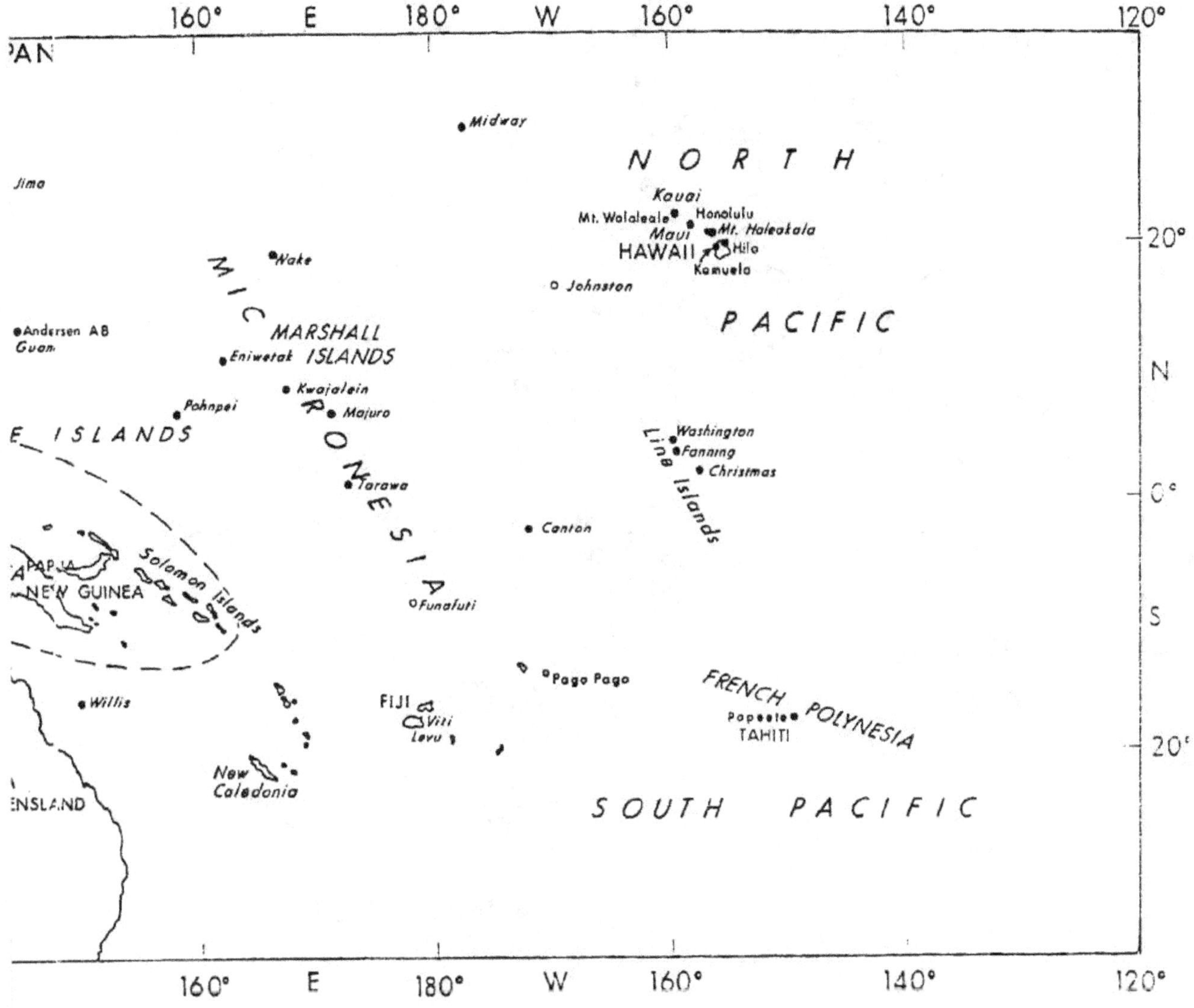

160° E 180° W 160° 140° 120°

PAN

• Midway

N O R T H

Jima

Kauai
Mt. Walaleale• •Honolulu
Maui •Mt. Haleakala — 20°
HAWAII ⌂•Hilo
Kamuela

•Wake

MIC

• Johnston P A C I F I C

•Andersen AB
Guam.

MARSHALL
•Eniwetak ISLANDS N

R O

•Pohnpei • Kwajalein
• Majuro

Washington
Line Islands •Fanning
• Christmas
C°

E ISLANDS

N E

•Tarawa

S I A

• Canton

PAPUA
NEW GUINEA Solomon Islands

•Funafuti
S

• Pago Pago FRENCH POLYNESIA
Papeete•
TAHITI

•Willis FIJI
•Viti
Levu — 20°

New
Caledonia SOUTH PACIFIC

ENSLAND

160° E 180° W 160° 140° 120°

(3)

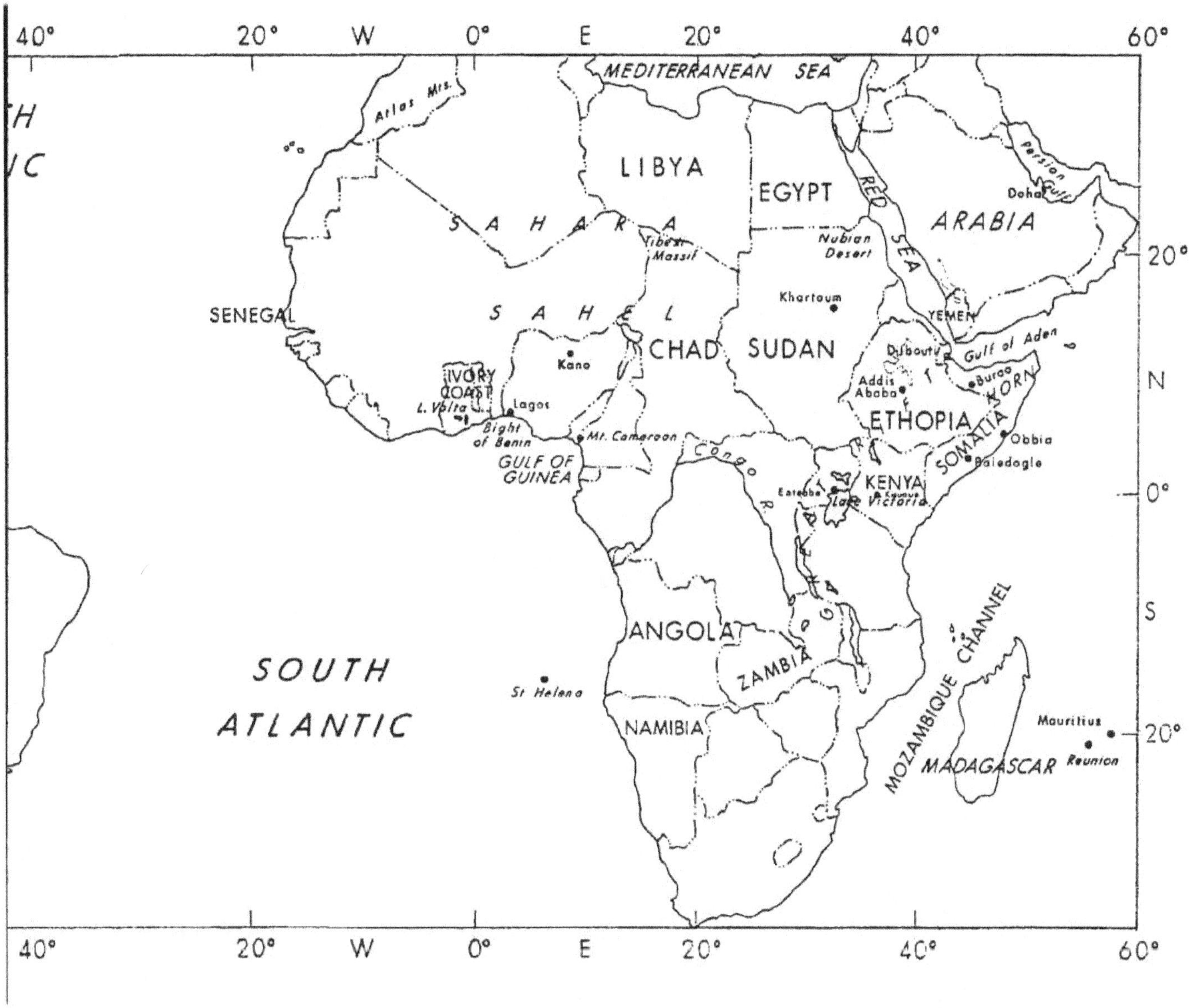

Figure 1-5b. Locator chart, Western Hemisphere.